BIOLOGICAL REACTIVE INTERMEDIATES V

Basic Mechanistic Research in Toxicology and Human Risk Assessment

ADVANCES IN EXPERIMENTAL MEDICINE AND BIOLOGY

Recent Volumes in this Series

Volume 379
SUBTILISIN ENZYMES: Practical Protein Engineering
Edited by Richard Bott and Christian Betzel

Volume 380
CORONA- AND RELATED VIRUSES: Current Concepts in Molecular Biology and Pathogenesis
Edited by Pierre J. Talbot and Gary A. Levy

Volume 381
CONTROL OF THE CARDIOVASCULAR AND RESPIRATORY SYSTEMS IN HEALTH AND DISEASE
Edited by C. Tissa Kappagoda and Marc P. Kaufman

Volume 382
MOLECULAR AND SUBCELLULAR CARDIOLOGY: Effects of Structure and Function
Edited by Samuel Sideman and Rafael Beyar

Volume 383
IMMUNOBIOLOGY OF PROTEINS AND PEPTIDES VIII: Manipulation or Modulation of the Immune Response
Edited by M. Zouhair Atassi and Garvin S. Bixler, Jr.

Volume 384
FATIGUE: Neural and Muscular Mechanisms
Edited by Simon C. Gandevia, Roger M. Enoka, Alan J. McComas, Douglas G. Stuart, and Christine K. Thomas

Volume 385
MUSCLE, MATRIX, AND BLADDER FUNCTION
Edited by Stephen A. Zderic

Volume 386
INHIBITORS TO COAGULATION FACTORS
Edited by Louis M. Aledort, Leon W. Hoyer, Jeanne M. Lusher, Howard M. Reisner, and Gilbert C. White II

Volume 387
BIOLOGICAL REACTIVE INTERMEDIATES V: Basic Mechanistic Research in Toxicology and Human Risk Assessment
Edited by Robert Snyder, James J. Kocsis, I. Glenn Sipes, George F. Kalf, David J. Jollow, Helmut Greim, Terrence J. Monks, and Charlotte M. Witmer

---

A Continuation Order Plan is available for this series. A continuation order will bring delivery of each new volume immediately upon publication. Volumes are billed only upon actual shipment. For further information please contact the publisher.

BIOLOGICAL REACTIVE INTERMEDIATES V

Basic Mechanistic Research in Toxicology and Human Risk Assessment

Edited by

Robert Snyder
Rutgers University
Environmental and Occupational Health Sciences Institute
Piscataway, New Jersey

James J. Kocsis
Thomas Jefferson University
Philadelphia, Pennsylvania

I. Glenn Sipes
University of Arizona
Tuscon, Arizona

George F. Kalf
Thomas Jefferson University
Philadelphia, Pennsylvania

David J. Jollow
Medical University of South Carolina
Charleston, South Carolina

Helmut Greim
GSF Institute of Toxicology
Technical University
Munich, Germany

Terrence J. Monks
University of Texas
Austin, Texas

Charlotte M. Witmer
Rutgers University
Environmental and Occupational Health Sciences Institute
Piscataway, New Jersey

Springer Science+Business Media, LLC

Library of Congress Cataloging-in-Publication Data

Biological reactive intermediates V : basic mechanistic research in toxicology and human risk assessment / edited by Robert Snyder ... [et al.].
p. cm. -- (Advances in experimental medicine and biology ; v. 387)
"Proceedings of the Fifth International Symposium on Biological Reactive Intermediates, held January 4-8, 1995, in Munich, Germany"--T.p. verso.
Includes bibliographical references and index.

1. Biotransformation (Metabolism)--Congresses. 2. Biochemical toxicology--Congresses. 3. Poisons--Metabolism--Congresses. I. Snyder, Robert, 1935- . II. International Symposium on Biological Reactive Intermediates (5th : 1995 : Munich, Germany) III. Series.
[DNLM: 1. Toxicology--congresses. 2. Biotransformation--congresses. W1 AD559 v.387 1996 / WM 203 W324 1996]
RA1220.B54 1996
615.9--dc20
DNLM/DLC
for Library of Congress 95-46530
CIP

Proceedings of the Fifth International Symposium on Biological Reactive Intermediates, held January 4–8, 1995, in Munich, Germany

ISBN 978-1-4757-9482-3 **ISBN 978-1-4757-9480-9 (eBook)**
DOI 10.1007/978-1-4757-9480-9

Originally published by Plenum Press, New York in 1996
Softcover reprint of the hardcover 1st edition 1996

10 9 8 7 6 5 4 3 2 1

PREFACE

Much of organic chemistry is based on the ability of suitably structured chemicals to bind together through the formation of covalent bonds. Biochemistry is replete with examples of enzymatically catalyzed reactions in which normal body constituents can be linked through covalent bonds during the process of intermediary metabolism. The finding that xenobiotic chemicals that enter the body from the environment, are metabolized to highly reactive species, and then covalently react with cellular macromolecules to induce toxic and carcinogenic effects was an observation that spawned the research featured in the Fifth International Symposium on Biological Reactive Intermediates (BRI V).

The group of investigators that became fascinated with this process and its significance in terms of human health began their discussions in Turku, Finland (1975), and continued them at Guildford, England (1980), College Park, Maryland (1985), Tucson, Arizona (1990), and Munich, Germany (1995). Among the results were a series of reports listed below, as well as the book for which this serves as the Preface.

- Jollow, D.J., Kocsis, J.J., Snyder, R. and Vainio, H. (eds), Biological Reactive Intermediates: Formation, Toxicity and Inactivation, Plenum Press, NY, 1975.
- Snyder, R., Park, D.V., Kocsis, J.J., Jollow, D.V., Gibson, G.G. and Witmer, C.M. (eds), Biological Reactive Intermediates II: Chemical Mechanisms and Biological Effects, Plenum Press, N.Y., 1982.
- Kocsis, J.J., Jollow, D., Witmer, C.M., Nelson, J.O., Snyder, R. (eds), Biological Reactive Intermediates III: Mechanisms of Action in Animal Models and Human Disease Plenum Press, NY, 1986.
- Witmer, C.M., Snyder, R., Jollow, D.J., Kalf, G.F., Kocsis, J.J. and Sipes, I.G. (eds) Biological Reactive Intermediates IV: Molecular and Cellular Effects and Their Impact on Human Health. Plenum Press, NY, 1991.

The Fifth International Symposium on Biological Reactive Intermediates was organized by two committees, one U.S. and one European, which consulted with 60 colleagues from around the world to develop the program. The U.S. Committee consisted of Professors R. Snyder and C.M. Witmer, Rutgers University; D.J. Jollow, Medical University of South Carolina,; I. G. Sipes, University of Arizona; G.F. Kalf and J.J. Kocsis, Thomas Jefferson University; and T.J. Monks, University of Texas. The European Committee consisted of Professors H. Greim (Local Chairman), GSF-Forschungszentrum für Umwelt und Gesundheit and Technical University of Munich, H. Kappus, Free University of Berlin, H. Bolt, University of Dortmund, F. Oesch, University of Mainz, and F. Wiebel, GSF-Forschungszentrum für Umwelt und Gesundheit.

BRI V was a five-day symposium featuring 45 invited speakers plus a total of 97 poster presentations. Subjects covered included the chemistry and formation of biological reactive intermediates, cellular damage and control of gene expression by biological reactive intermediates, impact on cellular redox state, nitric oxide, organ specific effects of biological reactive intermediates, organ-organ interactions, modeling of bioactivation reactions, and linking mechanistic studies to risk assessment. The presentations were followed by intensive discussions.

The organizers are delighted at the support provided by the Deutsche Forschungsgemainschaft, the Deutsche Gesellschaft für experimentelle und klinische Pharmakologie und Toxikologie, BASF AG, Basotherm, Bayer AG, Böhringer (Ingelheim), Böhringer (Mannheim), Hoescht AG, E. Merck AG, Schering Forschungsgesellschaft mbH, the Environmental and Occupational Health Science Institute (EOHSI) of Rutgers–The State University of New Jersey, The University of Medicine and Dentistry of New Jersey/Robert Wood Johnson Medical School, and the Joint Graduate Program in Toxicology.

Chairpeople had a great responsibility at BRI V. Because of the intense interest in the scientific presentations it was essential that the timing of papers and discussions were closely controlled. Of greater importance was their role in leading and contributing to the discussion. We thank our chairpeople for their dedicated efforts: Professors S. Orrenius, Karolinska Institute, Stockholm, F. Wiebel, C.M. Witmer, P. Dansette, University René Decartes, K.W. Bock, University of Tübingen, and D. Henschler, University of Würzburg. Leaders of the poster discussion sessions were Professors G.G. Gibson, University of Surrey, P. Dansette, D.J. Reed, Oregon State University, H. Sies, University of Düsseldorf, H. Kappus, and I.G. Sipes.

Dr. James R. Gillette of the Laboratory of Chemical Pharmacology of the National Institutes of Health, Bethesda, MD was the guest of honor at the symposium. For years he has been a leader in biochemical pharmacology and toxicology, was among the first to recognize and publish on the covalent binding phenomenon, has made seminal contributions to research in this area, and has trained many of the scientists currently working on biological reactive intermediates. It is with pleasure and respect that we dedicate this book to him.

The administrators and secretarial staff of the GSF and of the Toxicology Division of EOHSI were heavily involved in planning, correspondence, preparation of all of the written material for the symposium, and for the handling of financial matters. They performed their tasks with a high degree of professionalism. It would not have been possible to mount this symposium without their excellent assistance. The leaders of this effort were Ms. Bernadine Chmielowicz and Mr. Kevin Kimberlin at EOHSI, and Fr. Erika Kreppel at the GSF.

The International Symposium on Biological Reactive Intermediates has moved from venue to venue, and from country to country. Never has the setting been as stimulating as BRI V in Munich in January. The weather was cold, the city sparkled with snow and lights, and the science was exciting. We are indebted to Professor Greim, the faculty of the Toxicology Institute of the GSF, and to the staff of the Hospital Rechts der Isar, for providing a hospitable atmosphere in which we could meet, discuss our research, and learn from each other.

Modern toxicology continues to grow at an amazing pace. New insights into mechanisms of toxicity and pathology are made possible by an ever-widening circle of talented scientists, representing different but related disciplines. A new vocabulary based on recent advances in toxicology such as the application of molecular biology to help solve toxicological problems, and the role of nitric oxide in biological responses to chemicals, mixed with the vocabulary related to the solution of long standing toxicological problems. The common thread among them is the intense desire to prevent or cure human disease caused by chemicals in the various environments where exposure is likely. BRI V brought together a group of colleagues from many countries to discuss their work in a spirit of scientific comradeship that enhanced communication, led to the development of new research collaborations, and proved once again to be the correct forum for the interchange of ideas among the active researchers in this field.

Robert Snyder
David J. Jollow
I. Glenn Sipes
George F. Kalf
James J. Kocsis
Terrence J. Monks
Charlotte M. Witmer
Helmut Greim
Hermann Bolt
Hermann Kappus
Franz Oesch
Fredrich Wiebel

CONTENTS

Session I

Session II

Session III

Session IV

Session V

Session VI

Session VII

Session VIII

Session IX

NEW BIOLOGICAL REACTIVE INTERMEDIATES

Metabolic Activation of Thiophene Derivatives

Daniel Mansuy and Patrick M. Dansette

Laboratoire de Chimie et Biochimie Pharmacologiques et Toxicologiques (URA 400)
Université Paris V
45 Rue des Saints-Pères, 75270 Paris Cedex 06, France

INTRODUCTION

Metabolic activation of benzene rings as well as that of many aromatic heterocycles such as furans generally involves the intermediate formation of arene oxides as electrophilic reactive metabolites. Much less was known until very recently about the metabolic oxidative activation of thiophene derivatives (Rance, 1989). A first report on the metabolism of thiophene itself in rats has shown the formation of thiophene-derived mercapturates in urine and proposed the intermediate formation of a thiophene epoxide (Bray et al., 1971). More recently, it has been shown that the major oxidized metabolites of several 2-aroylthiophenes, such as tienilic acid (Mansuy et al., 1984) were derived from 5-hydroxylation of their thiophene ring (Neau et al., 1990).

Our interest in the detailed mechanism of the metabolic activation of thiophene derivatives came from a study of the origin of the hepatotoxic effects of a thiophene-containing drug, tienilic acid (TA) (Fig. 1) and of its isomer TAI (Fig. 2).

RESULTS AND DISCUSSION

Metabolic Activation of Tienilic Acid and Its Isomer

Tienilic acid (TA) is an uricosuric diuretic drug which was found devoid of any direct hepatotoxic effects in laboratory animals but which has led in humans to rare cases (1 patient for 10,000) of hepatitis of the immunoallergic type (Poupon et al., 1980 ; Zimmermann et al., 1984). A possible mechanism for metabolic activation of TA and for the appearance of such toxic effects has been proposed (Mansuy, 1990 ; Beaune et al., 1994) (Fig. 1). It is based on data showing that the main human liver cytochrome P450, P450 2C9, that is responsible

Biological Reactive Intermediates V
Edited by R. Snyder et al., Plenum Press, New York, 1996

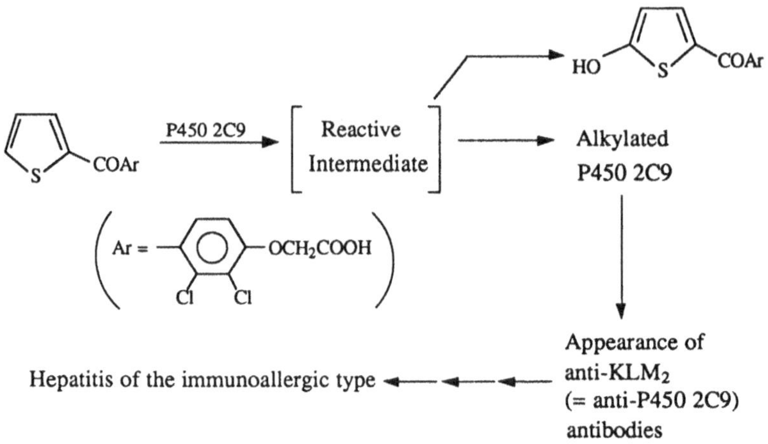

Figure 1. Possible mechanism for the first steps of tienilic acid-induced hepatitis.

for the oxidation of TA into 5-hydroxy-TA and reactive metabolites able to covalently bind to P450 2C9, is the antigen responsible for the appearance of autoantibodies (called anti-LKM_2 antibodies) in the serum of patients suffering from TA-induced hepatitis. The mechanism of presentation of alkylated P450 2C9 to the immune system remains to be determined. It could occur after alkylation of P450 2C9 in the endoplasmic reticulum, and either transfer of alkylated P450 2C9 to the hepatocyte membrane or lysis of the hepatocyte. Alternatively, it could occur after alkylation of P450 2C9 present in low amounts in the hepatocyte membrane (Beaune et al., 1994). Subsequent steps could be the formation of P450 2C9-anti -LKM_2 (anti P450 2C9) antibody complexes at the hepatocyte membrane, and destruction by the immune system of the hepatocytes activated by these complexes. Many questions about this mechanism remain to be answered ; one of them concerns the nature of the reactive metabolite formed during the P450 2C9-dependent oxidation of TA.

The isomer of tienilic acid, TAI, is also mainly metabolized by P450 2C9 in human liver microsomes with an extensive formation of reactive metabolites that covalently bind to microsomal proteins (Dansette et al., 1991). These reactive metabolites have been trapped by thiol-containing nucleophiles such as mercaptoethanol or N-acetylcysteine (Mansuy et al., 1991) (Fig. 2). The structure of the primary metabolite obtained upon oxidation of TAI by rat liver microsomes in the presence of mercaptoethanol clearly shows that the reactive intermediate involved in TAI oxidation is an unstable, highly electrophilic thiophene sulfoxide. This intermediate reacts with thiols by a Michael-type addition at position 2 of its thiophene ring leading to the dihydrothiophene sulfoxide isolated and shown in Fig. 2. In fact, this dihydrothiophene sulfoxide is not stable in the presence of an excess of thiol nucleophile. It undergoes a series of reactions leading eventually to the final dihydrothiophene metabolite that bears the thiol residue at position 4 of the thiophene ring and is very stable in the medium (Lopez-Garcia et al., 1993). *In vivo*, the nucleophile that is present in large amounts in the hepatocyte is glutathione. Thus, one should expect that the final stable metabolite of TAI *in vivo* should be the dihydrothiophene derived from TAI with a glutathione or N-acetyl-cysteine residue bound at position 4 of its thiophene ring. Accordingly, we have detected the appearance of the latter metabolite (mercapturate derived from TAI) in the urine of rats treated with TAI (P. Valadon, P. Dansette et D. Mansuy, to be published).

These results provided the first evidence for the formation of thiophene sulfoxides as reactive intermediates of thiophene compounds *in vitro* and *in vivo*.

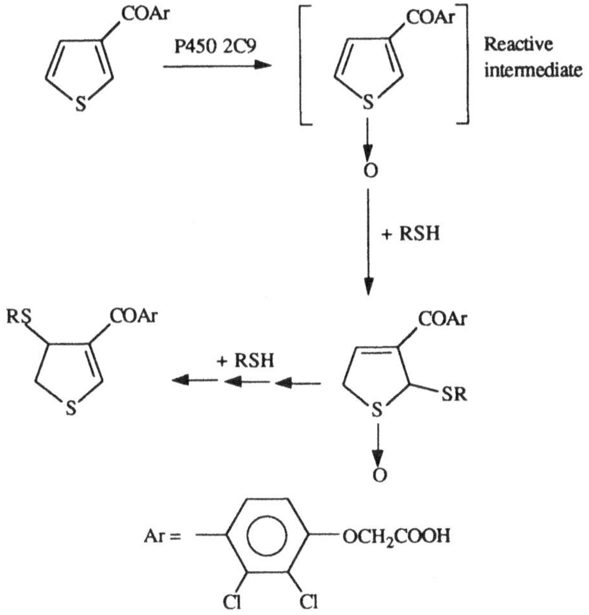

Figure 2. Metabolic oxidative activation of the isomer of tienilic acid (TAI) and trapping of its reactive intermediate by thiols.

Metabolic Oxidation of Thiophene Itself in Vivo and in Vitro

The major urinary metabolite of thiophene in rats, which corresponds to 20% of the administered dose, has been isolated and completely characterized (Dansette et al., 1992). It is also a dihydrothiophene bearing a N-acetylcysteine residue bound at position 2 of the thiophene ring (Fig. 3). It should derive from thiophene sulfoxide, a primary reactive intermediate formed by metabolic oxidation of thiophene, which reacts with glutathione by a Michael-type addition at position 2 of the thiophene ring; this reaction is possibly catalyzed by a glutathione transferase. Further usual transformation of the glutathione adduct to the corresponding mercapturate would lead to the isolated metabolite.

Quite recently, metabolic oxidation of thiophene by rat liver microsomes was found to lead to compounds of different structure. They are stereoisomers coming from the Diels-Alder dimerization of thiophene sulfoxide, and their structure has been determined by IR, ^{1}H and ^{13}C NMR and mass spectrometry (P. Dansette, D. Mansuy, to be published) (Fig. 3). Altogether, these results show that, even in the case of thiophene itself, the corresponding thiophene sulfoxide plays a central role in the oxidative metabolism of thiophene both *in vivo* and *in vitro*. *In vivo*, because of the presence of high concentrations of glutathione and of glutathione transferase, it is likely that thiophene sulfoxide is rapidly trapped by this nucleophile leading to the mercapturate metabolite isolated in urine. However a small amount of the Diels Alder dimer of thiophene was also found in urine. On the contrary, when generated *in vitro* by rat liver microsomes and in the absence of an efficient nucleophilic trapping system, thiophene sulfoxide undergoes a Diels-Alder-type dimerization.

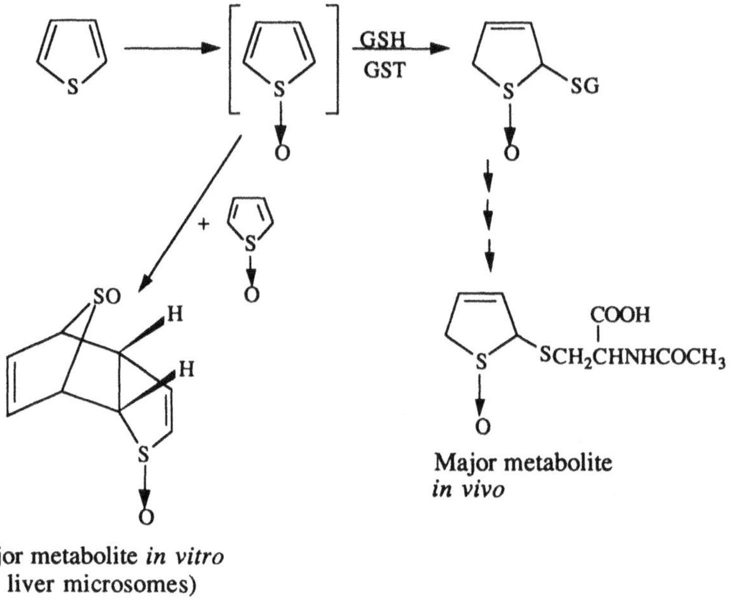

Figure 3. Metabolic oxidation of thiophene either *in vivo* (rats) or *in vitro* (rat liver microsomes).

Inactivation and Alkylation of P450 2C9 upon Metabolic Activation of Tienilic Acid

A detailed study of the oxidation of TA and TAI by yeast-expressed P450 2C9 showed that TA, but not TAI, acts as a mechanism-based inhibitor of P450 2C9 (Lopez-Garcia et al., 1994). Covalent binding of TA metabolites to microsomal proteins occurs in parralel with enzyme inactivation and was only partially inhibited by the presence of glutathione in the reaction medium. However, it is noteworthy that glutathione does not protect P450 2C9 from inactivation. TA exhibits all the characteristics of a mechanism-based inactivator of P450 2C9 with a partition ratio of 12. A specific covalent binding of about 1 mol TA metabolite per mol P450 2C9 occurs before complete loss of enzyme activity (in incubations performed in the presence of glutathione).

A plausible mechanism for the suicide inactivation of P450 2C9 upon TA oxidation is shown in Fig. 4. Considering the results observed for the oxidative metabolism of TAI and thiophene, the first step should be the S-oxidation of TA by P450 2C9. The thiophene sulfoxide of TA should be particularly electrophilic because of the presence of an electron-withdrawing keto substituent at position 2 of the thiophene ring. Position 5 of the thiophene ring of TA sulfoxide is especially electrophilic and able to react with any nucleophile present in P450 2C9 active site. Its reaction with H_2O as a nucleophile would lead to 5-hydroxy-TA eventually, whereas its reaction with a nucleophilic group of an aminoacid residue of the P450 2C9 active site would result in the alkylation and inactivation of P450 2C9. The immunoblot detection of alkylated P450 2C9 in yeast microsomes incubated with TA and NADPH as the only protein recognized by an antibody raised in rabbits against TA covalently bound to bovine serum albumin is in good agreement with this mechanistic proposition (Lecoeur et al., 1994).

Thiophene Sulfoxides: A New Class of Reactive Metabolites

The aforementioned results indicate that thiophene sulfoxides could play a central role in the oxidative metabolism of thiophene compounds. They constitute a new class of

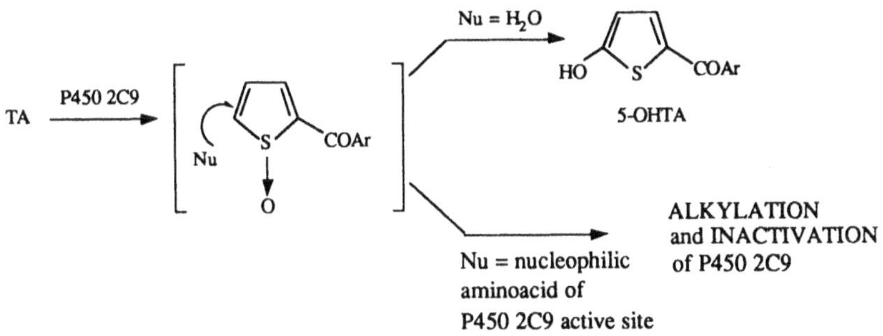

Figure 4. Proposed mechanism of the alkylation and inactivation of P450 2C9 during oxidation of tienilic acid (from Lopez-Garcia et al., 1994).

reactive metabolites whose toxicological consequences remain to be established. In fact, very few data are available on the preparation and properties of thiophene sulfoxides in the chemical literature (Raasch, 1985). So far, only thiophene sulfoxides bearing two bulky substituents at position 2 and 5 have been prepared and characterized. Quite recently, the first complete X-ray structure of such a sulfoxide, 2,5-thiophene-1-sulfoxide, has been obtained (Pouzet et al., 1995). It shows a pyramidal structure of the thiophene ring but with a relatively small deviation from planarity. Only few data are presently available about the chemical reactivity of thiophene sulfoxides, the reaction that has been studied most often being their condensation with dienophiles according to Diels-Alder reactions. The above results suggest that, besides Diels-Alder dimerizations that were observed in the metabolism of thiophene itself *in vitro* (Fig. 3), reactions of thiophene sulfoxides with nucleophiles play a key role in the fate of these reactive intermediates under *in vivo* and *in vitro* conditions. This is especially true in the case of thiophene sulfoxides bearing electron-withdrawing substituents, as with TAI or TA, which are highly electrophilic. However, more data are necessary to really understand the intrinsic properties of thiophene sulfoxides and the consequences of these properties in toxicology.

REFERENCES

Beaune, P., Pessayre, D., Dansette, P., Mansuy, D. and Manns, M., 1994, Autoantibodies against cytochrome P450: Role in human diseases, *Advances in Pharmacology*. 30:199-245.

Bray, H.G., Carpanini F.M.B.and Waters, B.D., 1971, The metabolism of thiophene in the rabbit and in the rat, *Xenobiotica* 1:157-168.

Dansette, P.M., Amar, C., Valadon, P., Pons, C., Beaune, P.H. and Mansuy, D., 1991, Hydroxylation and formation of electropholic metabolites of tienilic acid and its isomer by human liver microsomes: Catalysis by a P450 IIC different from that responsible for mephenytoin hydroxylation, *Biochem. Pharmacol.* 41:553-560.

Dansette, P.M., Thang, D.C., El Amri, H. and Mansuy, D., 1992, Evidence for thiophene-S-oxide as a primary reactive metabolite of thiophene *in vivo*: Formation of a dihydrothiophene sulfoxide mercapturic acid, *Biochem. Biophys. Res. Commun.* 186:1624-1630.

Lecoeur, S., Bonierbale, E., Challine, D., Gautier, J.C., Valadon, P., Dansette, P.M., Catinot, R., Ballet, F., Mansuy D. and Beaune, P.H., 1994, Specificity of in vitro covalent binding of tienilic acid metabolites to human liver microsomes in relationship to the type of hepatotoxicity: Comparison with two directly hepatotoxic drugs, *Chem. Res. Toxicol.* 7:434-442.

Lopez-Garcia, M.P., Dansette, P.M., Valadon, P., Amar, C., Beaune, P.H., Guengerich, F.P. and Mansuy, D., 1993, Human liver P450s expressed in yeast as tools for reactive metabolite formation studies: Oxidative activation of tienilic acid by P450 2C9 and P450 2C10, *Eur. J. Biochem.* 213:223-232.

Lopez-Garcia, M.P., Dansette, P.M. and Mansuy, D., 1994, Thiophene derivatives as new mechanism-based inhibitors of cytochromes P450: Inactivation of yeast expressed human liver cytochrome P450 2C9 by tienilic acid, *Biochemistry* 33:166-175.

Mansuy, D., Dansette, P., Foures, C., Jaouen, M., Moinet, G. and Bayer, N., 1984, Metabolic hydroxylation of the thiophene ring: Isolation of 5-hydroxytienilic acid as the major urinary metabolite in rat and man, *Biochem. Pharmacol.* 33:1429-1435.

Mansuy, D., 1990, Formation of reactive metabolites and appearance of anti-organelle antibodies in man, *Advances in Exp. Med. and Biol.* 283:133-137.

Mansuy, D., Valadon, P., Erdelmeier, I., Lopez-Garcia, M.P., Amar, C., Girault, J.P. and Dansette, P.M., 1991, Thiophene-S-oxides as new reactive metabolites: Formation by cytochrome P450-dependent oxidation and reaction with nucleophiles, *J. Am. Chem. Soc.* 113:7825-7826.

Neau, E., Dansette, P.M., Andronik, V. and Mansuy, D., 1990, Evidence for 5-hydroxylation of 2-aroyl-thiophenes as a general metabolic pathway using a simple UV-visible assay, *Biochem. Pharmacol.* 39, 1101-1107.

Poupon, R., Homberg, J.C., Abuaf, N., Petit, J., Bodin, F. and Darnis, F., 1980, Atteintes hépatiques dues à l'acide ténilique. Six observations avec présence d'anticorps antireticulum endoplasmique, *Nouv. Presse Med.* 9:1881-1884.

Pouzet, P., Erdelmeier, I., Ginderow, D., Mornon, J.P., Dansette, P. and Mansuy, D., 1995, Thiophene-S-oxides: Convenient preparation, first complete structural characterization and unexpected dimerization of one of them, 2,5-diphenyl-1-oxide, *JCS Chem. Comm.* in press.

Raasch, M.S., 1985, Chemistry of heterocyclic compounds, thiophene and its derivatives, (Gronowitz S., edit.) Wiley, New-York, 44:871.

Rance, D.J., 1989, Sulfur-containing drugs and related organic compounds, chemistry, biochemistry and toxicology: Metabolism of sulfur functional groups (Damani L.A. Edit.) 1 part B, Ellis Horwood Ldt, Chichester, 217-268.

Zimmerman, H.J., Lewis, J.H., Ishak, K.G. and Maddrey, W., 1984, Ticrynafen-associated hepatic injury: Analysis of 340 cases, *Hepatology* 4:315-323.

ACTIVATION OF TOXIC CHEMICALS BY CYTOCHROME P450 ENZYMES

Regio- and Stereoselective Oxidation of Aflatoxin B_1

F. Peter Guengerich,[1] Yune-Fang Ueng,[1] Bok-Ryang Kim,[1] Sophie Langouet,[4] Brian Coles,[2] Rajkumar S. Iyer,[3] Ricarda Thier,[1] Thomas M. Harris,[3] Tsutomu Shimada,[1] Hiroshi Yamazaki,[1] Brian Ketterer,[2] and Andre Guillouzo[4]

[1] Department of Biochemistry and Center in Molecular Toxicology
[2] Cancer Research Campaign Molecular Toxicology Group
University College and Middlesex School of Medicine
London W1P 6DB, United Kingdom
[3] Department of Chemistry and Center in Molecular Toxicology
Vanderbilt University School of Medicine, Nashville, Tennessee 37232
[4] Unité de Recherches Hépatologiques, INSERM U49
Hôpital de Pontchaillon, 35011 Rennes Cédex, France

INTRODUCTION

Cytochrome P450 (P450) enzymes are involved in the oxidations of numerous steroids, eicosanoids, alkaloids, and other endogenous substrates. These enzymes are also the major ones involved in the oxidation of potential toxicants and carcinogens such as those encountered among pollutants, solvents, and pesticides, as well as many natural products. A proper understanding of the basic mechanisms by which the P450 enzymes oxidize such compounds is important in developing rational strategies for the evaluation of the risks of these compounds.

Transformation of chemicals by P450 enzymes can lead to activation. Some of the major routes are shown in Fig. 1 (Guengerich and Shimada, 1991). For instance, amines (especially arylamines) are oxidized to hydroxylamines (reaction **1**), which can be esterified with a good leaving group to lead to alkylation of macromolecules. Oxidation of alkenes and aromatic rings forms epoxides, which are prone to reaction with nucleophiles due to ring strain (reaction **2**). However, as we will see later, epoxides vary enormously in their activity. Sometimes alkanes are oxidized to alkenes prior to epoxidation, as demonstrated in the case of urethan (Guengerich and Kim, 1991). Carbon hydroxylation is generally a detoxication step, rendering materials more polar and readily excreted. However, in the case of nitrosamines, α-hydroxylation is followed by non-enzymatic rearrangement to alkyl diazohydroxides

Biological Reactive Intermediates V
Edited by R. Snyder et al., Plenum Press, New York, 1996

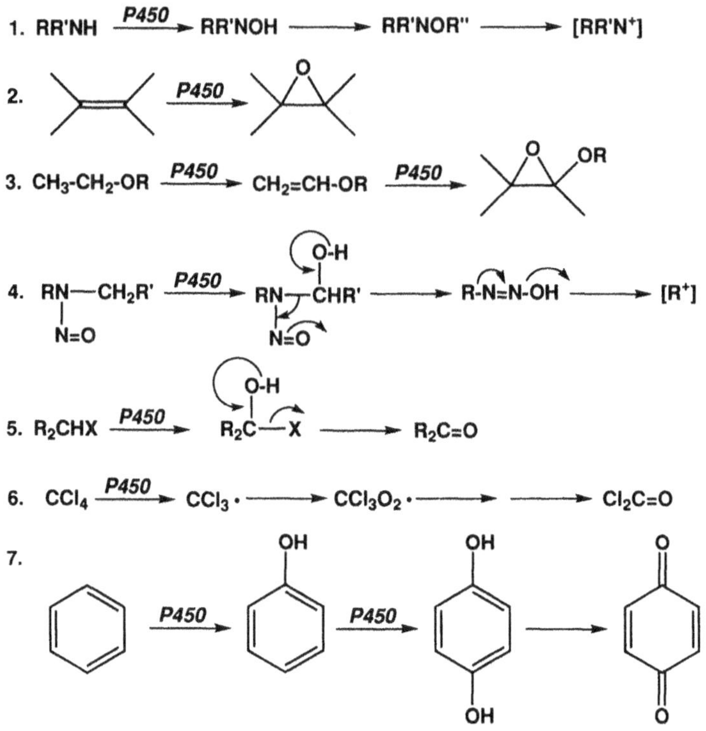

Figure 1. Examples of activation reactions catalyzed by P450 enzymes (Guengerich and Shimada, 1991).

(reaction **4**). (In reactions **1** and **4**, cations are shown in brackets as reactive intermediates but should not be construed as indicating S_N1 chemistry.) Also, carbon hydroxylation of alkyl halides can lead to carbonyl formation (reaction **5**); e.g., phosgene production from $CHCl_3$ (Pohl et al., 1981). Sometimes reduction reactions are catalyzed by P450s and generate reactive products, such as the conversion of CCl_4 to phosgene (reaction **6**). Finally, catechols and hydroquinones formed by hydroxylations can be readily oxidized to quinones (reaction **7**). The chemistry involved in P450 catalysis is relatively invariant and may be understood in terms of oxygen activation, odd-electron abstraction (of a non-bonded or p electron or a hydrogen atom), and oxygen rebound (radical recombination) (Ortiz de Montellano, 1986; Guengerich, 1991; Guengerich and Macdonald, 1993). The rate-limiting steps in the basic P450 catalytic cycle appear to vary depending upon the particular enzyme and substrate (Guengerich et al., 1995), but the fundamental chemical paradigms can be used to rationalize the different reactions that have been observed (Ortiz de Montellano, 1986; Guengerich and Macdonald, 1990).

It should be emphasized that P450s catalyze detoxication reactions as well as activations, and whether a particular P450 is detrimental or beneficial depends upon the situation the host is presented with (Guengerich, 1992). In recent years many of the human P450 enzymes have been extensively characterized and it is now possible to define which individual human P450 enzymes contribute most to a particular reaction through the use of correlation, chemical and antibody inhibition, purification, and expression studies (Guengerich and Shimada, 1991; Guengerich, 1995). Of course, results of such *in vitro*

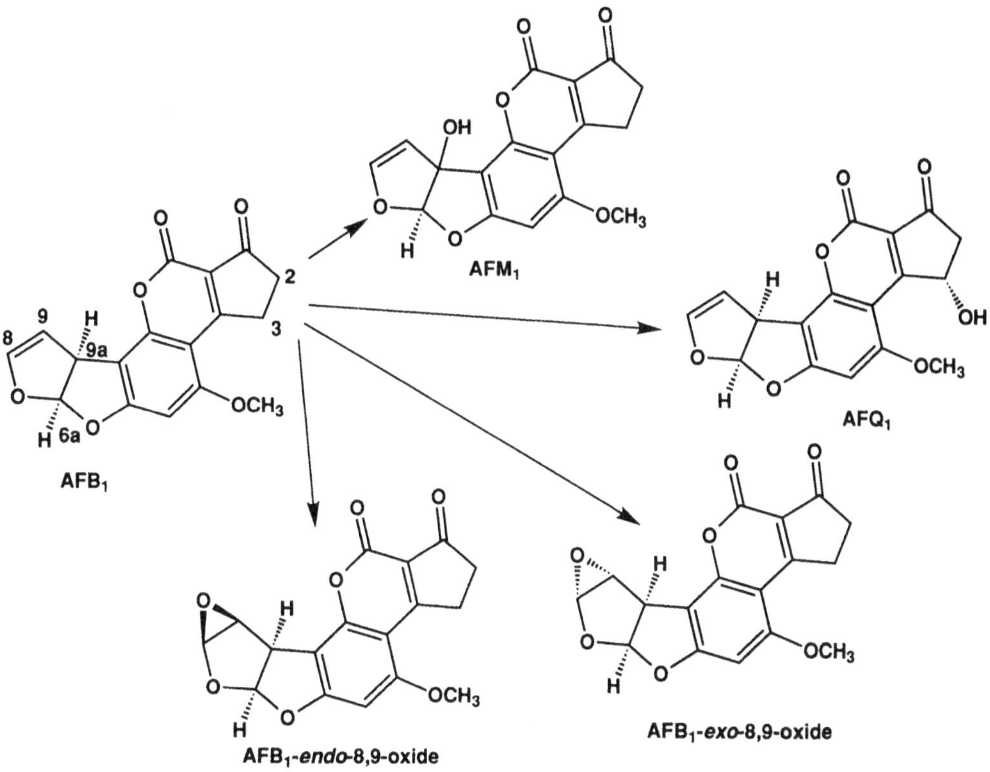

Figure 2. Major routes of AFB_1 oxidation by human liver P450s.

studies must be considered in the context of *in vivo* considerations. However, these approaches appear to be very powerful in evaluating the relevance of species comparisons.

During the past five years we have been studying aspects of the metabolism and reactivity of the products of aflatoxin B_1 (AFB_1) a fungal natural product that is one of the most potent hepatocarcinogens known (Fig. 2). The current paradigm provides insight into the significance of roles of different P450s, the importance of stereochemistry, interactions with conjugating enzymes, and the posing of questions relative to tissue localization of enzymes and relevance to humans.

RESULTS AND DISCUSSION

Reactivity and Genotoxicity of AFB_1 and its Epoxide Isomers

AFB_1 is the most potent of the aflatoxins and related benzofurans in most genotoxicity assays; it is also the most hepatocarcinogenic (Busby and Wogan, 1984). All evidence presented to date indicates that AFB_1 itself is not particularly genotoxic and the 8,9-epoxide is the only genotoxic product, as judged by either base-pair mutations (e.g., *Salmonella typhimurium* TA1535 revertants), frameshift mutations (*S. typhimurium* TA1538 revertants), or induction of the bacteria SOS response (*umu* assay in *S. typhimurium* TA1535/pSK1002) (Busby and Wogan, 1984; Garner et al., 1972; Baertschi et al., 1989).

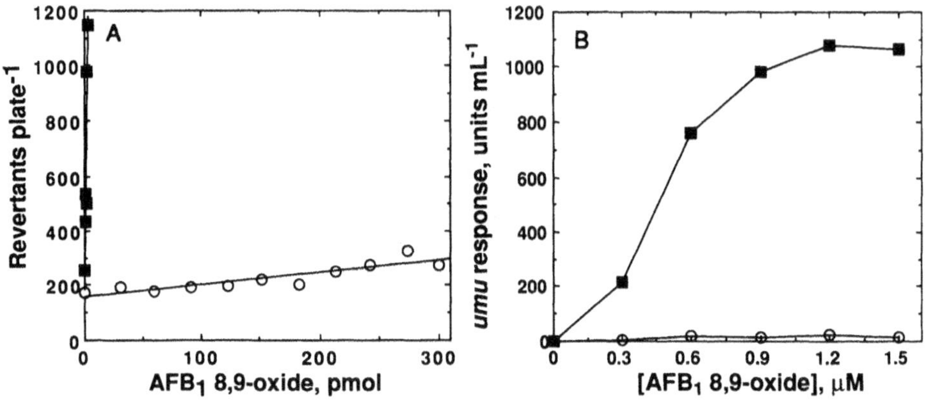

Figure 3. Genotoxicity of stereoisomers of AFB_1-8,9-oxide. (A) Revertants in *Salmonella typhimurium* TA100 (per plate) (Iyer et al., 1994). (B) *umu* gene response in *S. typhimurium* TA1535/pSK1002 (Ueng et al., 1994). *exo* (■); *endo* (○).

Synthetic methods have been used to prepare the unstable AFB_1-8,9-epoxide (Baertschi et al., 1988; Iyer and Harris, 1993) and resolve the *exo* and *endo* stereoisomers (Raney et al., 1992a; Iyer et al., 1994). The *exo* isomer is at least 10^3 times more genotoxic than the *endo* (Fig. 3), in spite of the longer half-life of the *endo* isomer in buffer solution (Iyer et al., 1994; Ueng et al., 1995). These results are rationalized by results showing the obligatory intercalation of the planar rings of AFB_1 (oxide) between DNA bases and the S_N2 nature of the reaction with the guanine N7 atom (Iyer et al., 1994).

The different AFB_1-8,9-epoxide stereoisomers also show selective reaction with glutathione (GSH) with individual forms of GSH S-transferase (GST) (Raney et al., 1992b).

Roles of Individual P450s in AFB_1 Genotoxicity

Although peroxidases and lipoxygenases have been reported to oxidize AFB_1 (Battista and Marnett, 1985), the P450s are much more active in catalyzing the reaction. Among rat P450s, the male-specific P450 2C11 has been reported to be the most active (Shimada et al., 1987).

In a series of studies with human liver microsomes, a major role for P450 3A4 in AFB_1 activation was implicated in studies involving correlation, immunoinhibition, and selective inhibition with chemical inhibitors (Shimada and Guengerich, 1989; Shimada et al., 1989). Purified P450 3A4 was also active, and the same pattern was seen with aflatoxin G_1 and sterigmatocystin (Shimada and Guengerich, 1989). In the results presented by others, P450 3A4 appeared to be the enzyme involved to the greatest extent in AFB_1 activation (Forrester et al., 1990; Aoyama et al., 1990).

The conclusion about the extent of involvement of P450 3A4 in AFB_1 activation is due, in part, to the high level of expression of P450 3A4 in human liver, relative to other P450s (Shimada et al., 1994; Guengerich, 1995). Attempts to compare the catalytic activities of different recombinant P450s are often limited by the lack of information about levels of actual holoenzyme. Also, in many cases it is unknown whether the microenvironment of the enzyme is most optimal. We expressed human P450s 1A2 and 3A4 in *Escherichia coli*, purified the enzymes, quantified them using difference spectroscopy, and reconstituted them in systems that appear to be most optimal for several catalytic activities (Gillam et al., 1993; Sandhu et al., 1994). P450 3A4 was more active than P450 1A2 at an AFB_1 concentration

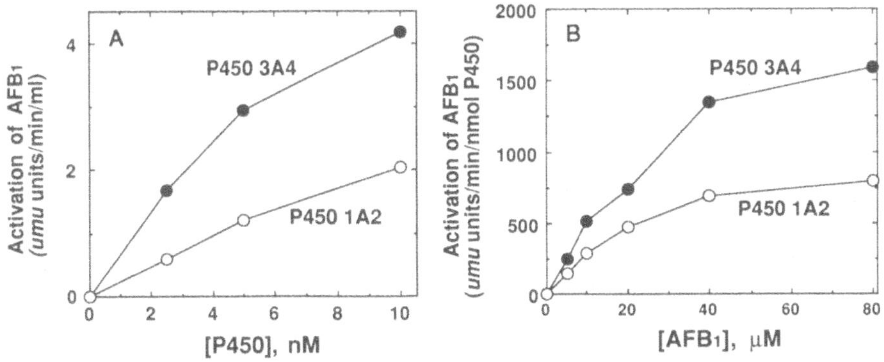

Figure 4. *umu* genotoxic response to AFB_1 activation by recombinant P450s 1A2 and 3A4 (Ueng et al., 1995). (A) Response as a function of P450 concentration (with [AFB_1] at 10 mM). (B) Response as a function of AFB_1 concentration (with [P450] at 5 nm).

of 10 mM (Fig. 4A). Over a range of concentrations extending from 5 to 80 mM, P450 3A4 was always more active than P450 1A2 (Fig. 4B).

In repeated studies in this and other laboratories, α-naphthoflavone (αNF) has been shown to stimulate AFB_1 genotoxicity when added to human liver microsomes (Buening et al., 1978; Buening et al., 1981; Shimada and Guengerich, 1989; Shimada et al., 1989).

Formation of AFB_1 Products by Different P450s

P450s can catalyze AFB_1 detoxication as well as activation, and an appropriate understanding of the system requires the identification of specific products and their contributions. Previous correlation studies with human liver microsomes suggested that

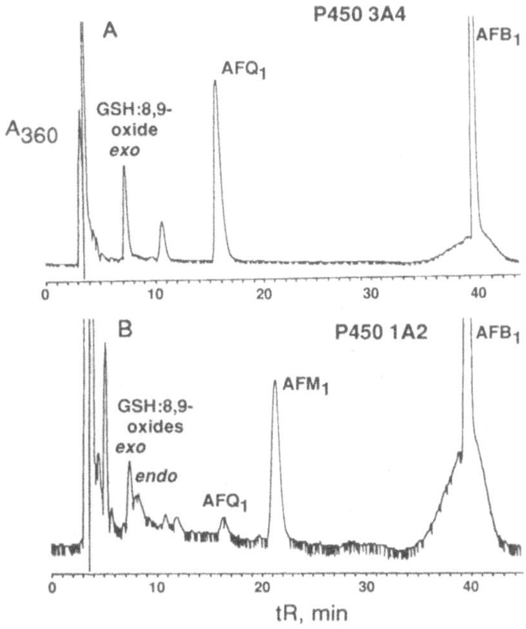

Figure 5. AFB_1 oxidation products formed by recombinant P450 3A4 (A) and P450 1A2 (B). In both cases the AFB_1 concentration was 60 mM and the incubation time was 20 min. Sufficient mouse GST were present to trap epoxides in both cases. Identified peaks are indicated on the chromatogram.

P450 3A4 forms aflatoxin Q_1 (AFQ_1) and P450 1A2 forms aflatoxin M_1 (AFM_1), both of which are considered detoxication products because of their decreased genotoxicity relevant to AFB_1 (Raney et al., 1992c). AFM_1 and AFQ_1 are stable and easily measured by HPLC. Reaction of the AFB_1-8,9-epoxides with GSH with excess GST yields the diastereomeric GSH-conjugates, which can be separated by HPLC (Raney et al., 1992a; Raney et al., 1992b).

Recombinant P450 3A4 formed only AFQ_1 and the *exo* form of AFB_1-8,9-epoxide (Fig. 5A; a small amount of *exo* epoxide was not trapped in this particular experiment). In contrast, P450 1A2 formed mainly AFM_1, a small amount of AFQ_1, and roughly equal amounts of *both* isomers of AFB_1-8,9-epoxide (Fig. 5B). These patterns were similar at all concentrations of AFB_1 used, down to 2 mM (Ueng et al., 1995). In another experiment, recombinant P450 1A1 (Guo et al., 1994) was found to form *only* AFM_1 at a rate ~1/10 that of P450 1A2. We have also found that recombinant P450 3A5 (85% identical to P450 3A4) also catalyzes AFQ_1 formation and epoxidation, but the ratio is approximately opposite that of P450 3A4 (Gillam et al., 1995).

The effects of αNF on reactions were investigated (Fig. 6). αNF stimulates AFB_1 (*exo*) epoxidation and inhibits 3α-hydroxylation in human liver microsomes (Raney et al., 1992c; Ueng et al., 1995). αNF was a very potent inhibitor of *all* reactions catalyzed by recombinant human P450 1A2 (Fig. 6B). However, the same pattern of stimulation of epoxidation and inhibition of 3α-hydroxylation was seen with recombinant P450 3A4s in the microsomes (Fig. 6C).

Plots of rates of AFB_1 epoxidation and 3α-hydroxylation *vs* AFB_1 concentration are sigmoidal with both human liver microsomes and with P450 3A4 (Ueng et al., 1995; Guengerich et al., 1994). These become more hyperbolic in the presence of αNF (Guengerich et al., 1994). The physical interpretation of these phenomena is not clear yet, but our working hypothesis is that P450 3A4 is behaving as an allosteric enzyme and αNF is an effector (Guengerich et al., 1994).

AFB_1 Metabolism in Human Hepatocyte Cultures

Hepatocytes were obtained from a GSH μ positive liver sample and cultured in the absence of other cells (Guillouzo et al., 1985). These were treated with 3-methylcholanthrene (3MC, 5 mM) or Oltipraz (50 mM). At time points of 24 and 48 h (starting time), AFB_1 (5 mM) was added to the cultures for 8 h and then products in the medium were analyzed by HPLC (Raney et al., 1992b; Raney et al., 1992c; Ueng et al., 1995). In the untreated culture the GSH-AFB_1 oxide level was 0.04 mM and the AFM_1 level was 0.06 mM (in the "24 h"

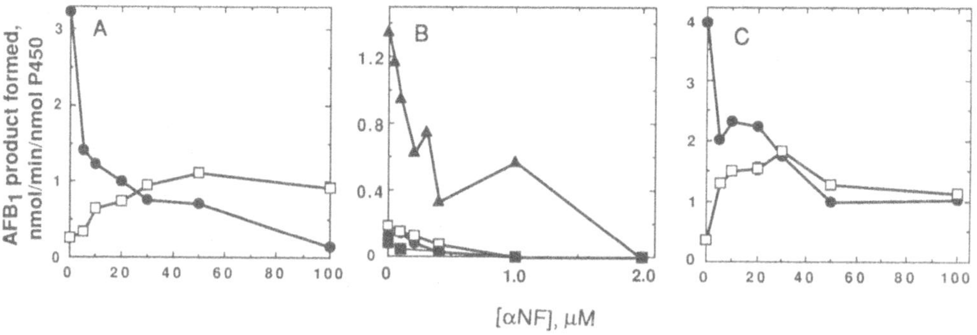

Figure 6. Effects of αNF on rates of formation of AFB_1 oxidation products with human liver microsomes (A) or a reconstituted enzyme system containing recombinant P450 1A2 (B) or P450 3A4 (C). AFQ_1 (●), AFM_1 (▲), AFB_1-*exo*-8,9-oxide (□), AFB_1-*endo*-8,9-oxide (■).

incubation). In the "48 h" incubation of these cells, the GSH-AFB_1 oxide level was 0.43 mM and the AFM_1 level was 0.36 mM. The GSH-AFB_1 oxide conjugates appeared to be *exo* and *endo* in the ratio of ~1:1.

The cells treated with 3MC showed a pattern similar to the untreated after 24 h, apparently before induction occurred. After 48 h, though, the AFB_1 incubation yielded 1.0 mM GSH:AFB_1 oxide conjugate (mostly *endo*) and 0.75 mM AFM_1. When cells were treated with Oltipraz, the levels of GSH-AFB_1 oxide conjugates and AFM_1 was <0.01 mM after 24 h. After 48 h they were respectively, 0.07 and <0.01 mM and most of the GSH-AFB_1 oxide conjugate was *exo*.

These results may appear complex but can be interpreted in terms of results of studies with the P450s and GSTs. In untreated cells the level of P450 1A2 is low but rises with continued culture, yielding AFM_1 and the *endo* AFB_1 oxide. This pattern is exacerbated by treatment with 3MC. However, other P450s (3A4) also rise in level in the untreated cultures, yielding *exo* AFB_1 oxides. Oltipraz appears to either down-regulate P450 genes or inhibit the enzymes. It appears to be quite dramatic with P450 1A2.

Conclusions

The stereochemistry of epoxidation of AFB_1 is of considerable significance in determining the course of AFB_1 toxicity. As in the case of many of the polycyclic hydrocarbons (Wood et al., 1984), it is critical to know which isomer is formed. Different P450s can oxidize AFB_1 regioselectively. P450 3A4 can attack AFB_1 at either of the end rings to cause activation (8,9-epoxidation) or detoxication (3α-hydroxylation) (Figs. 2, 5). Although P450 3A4 is often considered a promiscuous enzyme, we found that P450 1A2 was not as distinct, forming both AFM_1 and AFQ_1. The oxygen appears to be inserted from the "plane" of the olefin to give both epoxide isomers (Figs. 2, 5). The weak genotoxicity of the *endo* AFB_1 epoxide argues that this transformation might even be considered a detoxication process. Thus all epoxides are not equal, even of AFB_1.

The finding that GSH-AFB_1 epoxide conjugates can be identified in human hepatocytes is of special significance, in that low levels of GST activity towards AFB_1 oxides have been reported but the significance in human cells has been unclear (Raney et al., 1992b; Eaton and Gallagher, 1994). Oltipraz appears to have two important properties, that of inducing human GSTs (Raney et al., 1992b; Meyer et al., 1993) and also inhibiting human P450s. The cell studies suggest that the latter may be of even greater significance than the former.

Placing these findings into the overall significance of AFB_1-induced cancer is difficult, even in the absence of considerations about hepatitis B virus infection. AFB_1 is first encountered in the digestive tract, and the small intestine contains considerable P450 3A4 in most individuals. Epoxidation and binding to macromolecules in the small intestine might be considered a protective effect, since these cells are sloughed rapidly. Since levels of AFB_1 exposure are so low, it may seem remarkable that any reaches the liver. Once there, different enzymes compete and the result can be *exo* 8,9-oxide formation or any of a host of detoxication reactions. Apparently GST action does have a role. Molecular epidemiology studies on the significance of these different pathways are challenging but provide an opportunity to study the significance of an interesting series of P450 oxidations and GST conjugations.

ACKNOWLEDGMENTS

This work was supported in part by United States Public Health Service grants CA44353, ES03755, ES00267, and ES07028 and grants from the Campaign for Cancer Research, the Ministry of Education, Science, and Culture of Japan, and the Deutsches Forschungsgemeinschaft.

REFERENCES

Aoyama, T., Yamano, S., Guzelian, P.S., Gelboin, H.V., and Gonzalez, F.J.,1990, 5 of 12 forms of vaccinia virus-expressed human hepatic cytochrome-P450 metabolically activate aflatoxin B_1, *Proc. Natl. Acad. Sci. USA* 87:4790-4793.

Baertschi, S.W., Raney, K.D., Shimada, T., Harris, T.M., and Guengerich, F.P.,1989, Comparison of rates of enzymatic oxidation of aflatoxin B_1, aflatoxin G_1, and sterigmatocystin and activities of the epoxides in forming guanyl-N^7 adducts and inducing different genetic responses, *Chem. Res. Toxicol.* 2:114-122.

Baertschi, S.W., Raney, K.D., Stone, M.P., and Harris, T.M.,1988, Preparation of the 8,9-epoxide of the mycotoxin aflatoxin B_1: The ultimate carcinogenic species, *J. Am. Chem. Soc.* 110:7929-7931.

Battista, J.R. and Marnett, L.J.,1985, Prostaglandin H synthase-dependent epoxidation of aflatoxin B_1, *Carcinogenesis* 6:1227-1229.

Buening, M.K., Chang, R.L., Huang, M.R., Fortner, J.G., Wood, A.W., and Conney, A.H.,1981, Activation and inhibition of benzo(*a*)pyrene and aflatoxin B_1 metabolism in human liver microsomes by naturally occurring flavonoids, *Cancer Res.* 41:67-72.

Buening, M.K., Fortner, J.G., Kappas, A., and Conney, A.H.,1978, 7,8-Benzoflavone stimulates the metabolic activation of aflatoxin B_1 to mutagens by human liver, *Biochem. Biophys. Res. Commun.* 82:348-355.

Busby, W.F. and Wogan, G.N., 1984, Aflatoxins, in: *Chemical Carcinogens* (Searle, C.E. , ed.) Amer. Chem. Soc., Washington, D.C., pp. 945-1136.

Eaton, D.L. and Gallagher, E.P.,1994, Mechanisms of aflatoxin carcinogenesis, *Annu. Rev. Pharmacol. Toxicol.* 34:135-172.

Forrester, L.M., Neal, G.E., Judah, D.J., Glancey, M.J., and Wolf, C.R.,1990, Evidence for involvement of multiple forms of cytochrome P-450 in aflatoxin B_1 metabolism in human liver, *Proc. Natl. Acad. Sci. USA* 87:8306-8310.

Garner, R.C., Miller, E.C., and Miller, J.A.,1972, Liver microsomal metabolism of aflatoxin B_1 to a reactive derivative toxic to *Salmonella typhimurium* TA 1530, *Cancer Res.* 32:2058-2066.

Gillam, E.M.J., Baba, T., Kim, B-R., Ohmori, S., and Guengerich, F.P.,1993, Expression of modified human cytochrome P450 3A4 in *Escherichia coli* and purification and reconstitution of the enzyme, *Arch. Biochem. Biophys.* 305:123-131.

Gillam, E.M.J., Guo, Z., Ueng, Y-F., Yamazaki, H., Cock, I., Reilly, P.E.B., Hooper, W.D., and Guengerich, F.P.,1994, Expression of cytochrome P450 3A5 in *Escherichia coli*: Effects of 5' modifications, purification, spectral characterization, reconstitution conditions, and catalytic activities, *Arch. Biochem. Biophys.* , in press.

Guengerich, F.P. and Kim, D-H.,1991, Enzymatic oxidation of ethyl carbamate to vinyl carbamate and its role as an intermediate in the formation of 1,N^6-ethenoadenosine, *Chem. Res. Toxicol.* 4:413-421.

Guengerich, F.P. and Macdonald, T.L., 1993, Sequential electron transfer oxidation reactions catalyzed by cytochrome P-450 enzymes, in: *Advances in Electron Transfer Chemistry, Vol. 3* (Mariano, P.S. , ed.) JAI Press, Greenwich,CT, pp. 191-241.

Guengerich, F.P. and Macdonald, T.L.,1990, Mechanisms of cytochrome P-450 catalysis, *FASEB J.* 4:2453-2459.

Guengerich, F.P. and Shimada, T.,1991, Oxidation of toxic and carcinogenic chemicals by human cytochrome P-450 enzymes, *Chem. Res. Toxicol.* 4:391-407.

Guengerich, F.P., 1995, Human cytochrome P450 enzymes, in: *Cytochrome P450* (Ortiz de Montellano, P.R., ed.) Plenum Press, New York, in press.

Guengerich, F.P., Bell, L.C., and Okazaki, O.,1995, Interpretations of cytochrome P450 mechanisms from kinetic studies, *Biochimie*, in press.

Guengerich, F.P., Kim, B-R., Gillam, E.M.J. and Shimada, T., 1994, Mechanisms of enhancement and inhibition of cytochrome P450 catalytic activity, in: *Proceedings, 8th Int. Conf. on Cytochrome P450: Biochemistry, Biophysics, and Molecular Biology* (Lechner, M.C., ed.) John Libbey Eurotext, Chichester, U.K., pp. 97-101.

Guengerich, F.P.,1991, Reactions and significance of cytochrome P-450 enzymes, *J. Biol. Chem.* 266:10019-10022.

Guengerich, F.P.,1992, Metabolic activation of carcinogens, *Pharmacol. Ther.* 54:17-61.

Guillouzo, A., Beaune, P., Gascoin, M. N., Begue, J. M., Campion, J. P., Guengercih, F. P., and Guguen-Guillouzo, C., 1985, Maintenance of cytochrome P-450 in cultured adult human hepatocytes, *Biochem. Pharmacol.* 34:2991-2995.

Guo, Z., Gillam, E.M.J., Ohmori, S., Tukey, R.H., and Guengerich, F.P.,1994, Expression of modified human cytochrome P450 1A1 in *Escherichia coli*: Effects of 5' substitution, stabilization, purification, spectral characterization, and catalytic properties, *Arch. Biochem. Biophys.* 312:436-446.

Iyer, R., Coles, B., Raney, K.D., Thier, R., Guengerich, F.P., and Harris, T.M.,1994, DNA adduction by the potent carcinogen aflatoxin B_1 mechanistic studies, *J. Am. Chem. Soc.* 116:1603-1609.

Iyer, R.S. and Harris, T.M.,1993, Preparation of aflatoxin B_1 8,9-epoxide using *m*-chloroperbenzoic acid, *Chem. Res. Toxicol.* 6:313-316.

Meyer, D.J., Coles, B., Harris, J., Gilmore, K.S., Raney, K.D., Harris, T.M., Guengerich, F.P., Kensler, T.W. and Ketterer, B., 1993, Induction of rat liver GSH transferases by 1,2-dithiole-3-thione illustrates both anticarcinogenic and tumor-promoting properties, in: *Antimutagenesis and Anticarcinogenesis Mechanisms III* (Bonzetti, G., Hayatsu, H., DeFlora, S., Waters, M.D. and Shenkel, D.M. , eds.) Plenum Press, New York, pp. 171-179.

Ortiz de Montellano, P.R., 1986, Oxygen activation and transfer, in: *Cytochrome P-450* (Ortiz de Montellano, P.R. , ed.) Plenum Press, New York, pp. 217-271.

Pohl, L.R., Branchflower, R.V., Highet, R.J., Martin, J.L., Nunn, D.S., Monks, T.J., George, J.W., and Hinson, J.A.,1981, The formation of diglutathionyl dithiocarbonate as a metabolite of chloroform, bromotrichloromethane, and carbon tetrachloride, *Drug Metab. Dispos.* 9:334-339.

Raney, K.D., Coles, B., Guengerich, F.P., and Harris, T.M.,1992a, The *endo* 8,9-epoxide of aflatoxin B_1: A new metabolite, *Chem. Res. Toxicol.* 5:333-335.

Raney, K.D., Meyer, D.J., Ketterer, B., Harris, T.M., and Guengerich, F.P.,1992b, Glutathione conjugation of aflatoxin B_1 *exo* and *endo* epoxides by rat and human glutathione S-transferases, *Chem. Res. Toxicol.* 5:470-478.

Raney, K.D., Shimada, T., Kim, D-H., Groopman, J.D., Harris, T.M., and Guengerich, F.P.,1992c, Oxidation of aflatoxins and sterigmatocystin by human liver microsomes: Significance of aflatoxin Q_1 as a detoxication product of aflatoxin B_1, *Chem. Res. Toxicol.* 5:202-210.

Sandhu, P., Guo, Z., Baba, T., Martin, M.V., Tukey, R.H., and Guengerich, F.P.,1994, Expression of modified human cytochrome P450 1A2 in *Escherichia coli*: Stabilization, purification, spectral characterization, and catalytic activities of the enzyme, *Arch. Biochem. Biophys.* 309:168-177.

Shimada, T. and Guengerich, F.P.,1989, Evidence for cytochrome $P\text{-}450_{NF}$, the nifedipine oxidase, being the principal enzyme involved in the bioactivation of aflatoxins in human liver, *Proc. Natl. Acad. Sci. USA* 86:462-465.

Shimada, T., Iwasaki, M., Martin, M.V., and Guengerich, F.P.,1989, Human liver microsomal cytochrome P-450 enzymes involved in the bioactivation of procarcinogens detected by *umu* gene response in *Salmonella typhimurium* TA1535/pSK1002, *Cancer Res.* 49:3218-3228.

Shimada, T., Nakamura, S., Imaoka, S., and Funae, Y.,1987, Genotoxic and mutagenic activation of aflatoxin B_1 by constitutive forms of cytochrome P-450 in rat liver microsomes, *Toxicol. Appl. Pharmacol.* 91:13-21.

Shimada, T., Yamazaki, H., Mimura, M., Inui, Y., and Guengerich, F.P.,1994, Interindividual variations in human liver cytochrome P450 enzymes involved in the oxidation of drugs, carcinogens, and toxic chemicals: Studies with liver microsomes of 30 Japanese and 30 Caucasians, *J. Pharmacol. Exp. Ther.* 270:414-423.

Ueng, Y-F., Shimada, T., Yamazaki, H., and Guengerich, F.P.,1994, Oxidation of aflatoxin B_1 by bacterial recombinant human cytochrome P450 enzymes, *Chem. Res. Toxicol.* , in press.

Wood, A.W., Chang, R.L., Levin, W., Thakker, D.R., Yagi, H., Sayer, J.M., Jerina, D.M., and Conney, A.H.,1984, Mutagenicity of the enantiomers of the diastereomeric bay-region benzo(c)phenanthrene 3,4-diol-1,2-epoxides in bacterial and mammalian cells, *Cancer Res.* 44:2320-2324.

INVESTIGATING THE ROLE OF THE MICROSOMAL EPOXIDE HYDROLASE MEMBRANE TOPOLOGY AND ITS IMPLICATION FOR DRUG METABOLISM PATHWAYS*†

Thomas Friedberg,‡ Bettina Löllmann, Roger Becker, Romy Holler, Michael Arand, and Franz Oesch

Institute of Toxicology, University of Mainz
Obere Zahlbacher Straße 67, D-55131 Mainz, Germany

SUMMARY

The microsomal epoxide hydrolase (mEH) catalyzes the hydrolysis of reactive epoxides which are formed by the action of cytochromes P450 from xenobiotics. In addition the mEH has been found to mediate the transport of bile acids. For the mEH it has been shown that it is cotranslationally inserted into the endoplasmic reticulum. Here we demonstrate that the amino-terminal twenty amino acid residues of this protein serve as its single membrane anchor signal sequence and that the function of this sequence can be also supplied by a cytochrome P450 (CYP2B1) anchor signal sequence.

In addition we present data showing that the membrane anchor signal sequence of the mEH is dispensable for the catalytic activity of this protein. Our results indicate that it might be feasable to invert the topology of the mEH in the membrane of the endoplasmic reticulum without affecting the catalytic activity of this protein. With this strategy it will be possible to investigate whether the membrane topology of xenobiotic metabolizing enzymes is important for their role in chemical carcinogenesis.

* This work was supported by the Deutsche Forschungsgemeinschaft (SFB 302).

† Abbreviations: ER, endoplasmic reticulum; mEH, microsomal epoxide hydrolase; δmEH, truncated epoxide hydrolase; CYP/EH; cytochrome P450 / mEH fusionprotein; PAGE, polyacrylamide gel electrophoresis; PMSF, phenylmehylsulfonylfluoride.

‡ To whom correspondence should be addressed (Present address): Biomedical Research Centre, University of Dundee, Ninewells Hospital, Dundee DD1 9SY, UK; Tel: + 44 382 632621; FAX: + 44 382 69993.

Biological Reactive Intermediates V
Edited by R. Snyder et al., Plenum Press, New York, 1996

INTRODUCTION

The microsomal epoxide hydrolase (EC 3.3.2.3) plays a central role in the metabolism of several carcinogenic polycyclic aromatic hydrocarbons (1-4). Cytochromes P450 and the microsomal epoxide hydrolase (mEH)[1] catalyze in this metabolism a cascade of events culminating in the formation of dihydrodiol epoxides which are thought to be the ultimate carcinogenic metabolites of certain polycyclic aromatic hydrocarbons. It has been shown unequivocally for several microsomal cytochromes P450, that they are cotranslationally inserted into the endoplasmic reticulum (ER) (5) and are mainly exposed to its cytosolic face (6,7). A cotranslational insertion into the ER was also shown for the mEH and the amino-terminus of this protein is not proteolytically processed during its insertion into membranes (8,9) . There is evidence that this enzyme has a similar orientation in the membrane to that of microsomal cytochromes P450 (10,11). However, a recent report (12) demonstrated that the mEH was identical with a bile acid transport protein which could assume two opposite orientations within the membrane.

Theoretical calculations predict that the first twenty amino acid residues of the mEH might serve as a membrane insertion signal sequence, however, other transmembrane domains have also been suggested for this enzyme (13) . In this report we have tried to find out whether or not the amino-terminus of the mEH serves as its single membrane anchor signal sequence in order to start to address the question which structural features of the mEH determine its orientation in the membrane of the ER.

For a cytochrome P450, namely the bovine 17 α steroid-hydroxylase, it has been shown by its heterologous expression in COS 1 cells that an anchor signal sequence is important for the catalytic activity of this protein (10) . However, for several cytochromes P450 it has been demonstrated in a yeast system (14) and in bacterial expression systems (15,16), that their catalytic activities are retained in the absence of a membrane insertion signal. For cytochromes P450 the structural constraints for catalytic activity might be rather strict as this activity is not only dependent on the proper formation of the catalytic centers and on the incorporation of heme, but also on their interaction with the cytochrome P450 reductase and it appears that cytochromes P450 do not fold correctly in a mammalian expression system upon deletion of their membrane anchor sequence. In the case of the microsomal epoxide hydrolase one might expect less structural constraints as this enzyme requires neither a prosthetic group nor a second protein for catalytic activity. In this work we demonstrate, that the natural membrane insertion signal of the mEH is *not* necessary for the catalytic activity of this protein in mammalian cells.

MATERIALS AND METHODS

Construction of the Templates Coding for the Modified mEH Proteins

In this work we prepared templates coding for the mEH, for an mEH in which the first twenty amino acids of the N-terminus of the mEH were replaced by the membrane signal anchor sequence of CYP2B1 (the resulting fusion protein is defined as CYP2B1/mEH) and for an mEH with a deletion of the first twenty N-terminal amino acids (this protein is defined as δmEH). The construction of these templates has been described (17).

Cell Free Transcription and Translation

The SP6 promoter plasmids containing the various cDNA constructs were directly used for the cell free coupled transcription/translation which uses reticulocyte lysate

(Promega) as the translation system. Conditions for the coupled transcription / translation reaction in the presence of ^{35}S methionine were as described by the manufacturer (Promega). The protein biosynthesis was performed either in the presence or in the absence of dog pancreas microsomes (Promega) or alternatively, these membranes were added posttranslationally after EDTA (5 mM final) had been added to the translation assay. The translation products were analysed by SDS-PAGE[1] analysis (18).

Heterologous Expression of mEH Constructs in BHK21 Fibroblasts

The BHK21 Syrian hamster kidney fibroblast cell line was used for the heterologous expression of the various mEH cDNA constructs, as it is known that these cells are almost devoid of an endogenous mEH enzyme activity (19) . In addition we were not able to detect the endogenous expression of the mEH in these cells by immunoblotting. The cDNAs in the expression vector pMPSV were cotransfected together with the plasmid LK444 which confers G418-resistance into the fibroblasts as described. G418-resistant clones were analysed by immunoblotting using a polyclonal anti-rat mEH antiserum. Three clones having the highest level of expression for either the mEH or the CYP2B1/mEH fusion protein or the mEH with a deleted N-terminus were chosen for further analysis.

Membrane Integration Assay

In order to assay for the integration of the mEH and the modified mEH constructs, dog pancreas microsomes present during the in vitro translation of these proteins and membranes from fibroblasts expressing these proteins were alkaline extracted as described (17).

Determination of Cellular mEH Enzyme Activity

The mEH enzyme activity of the various BHK21 fibroblast cell lines was determined as follows: The fibroblasts were grown in cell culture, isolated by mild trypsinolysis and washed with DMEM-medium containing 10% FCS followed by phosphate buffered saline. The cells were resuspended in KBS (10 mM NaPi, pH 7.4, 110 mM KCl, 1mM EDTA, 60 U /ml Trasylol, 0.5 mM PMSF) and were lysed by sonication on ice. The enzyme activity of the lysed cells was determined with 0.6 to 1 mg of cellular protein using benzo[a]pyrene 4, 5-oxide as substrate. Under these conditions the enzyme activity was linear with time and protein.

RESULTS

Construction of Templates Coding for Modified mEH Proteins

Modifications of the N-terminus of the mEH were performed. In one expression construct the first 20 amino acids of the mEH were removed and in the other construct these sequences were replaced by the membrane anchor of CYP2B1. The resulting proteins were defined as δmEH and CYP2B1/mEH respectively. The N-terminal sequence of the mEH contains a negatively charged amino acid residue at position 4 followed by a hydrophobic core of 16 amino acid residues. This core precedes a stretch of fifteen amino acid residues containing positively and negatively charged amino acids followed by the remaining peptide chain. In the mEH with a shortened N-terminus, the negatively charged amino acid residue and the hydrophobic core is removed. In the CYP2B1/mEH fusion protein a negatively charged amino acid residue at position 2 is followed by a hydrophobic core of 19 residues. This core is followed by a stretch of fifteen amino acid residues containing five positive charges followed by the remaining peptide chain.

Cell Free Translation of mEH and of Modified mEH Proteins

In order to study whether the N-terminal sequence of the mEH is necessary for the cotranslational insertion and anchoring of this protein into microsomal membranes, we translated the mEH, the truncated mEH and the CYP2B1/mEH fusion protein in the absence and in the presence of dog pancreas microsomes. The translation mixtures were subsequently subjected to alkaline treatment and the membrane fractions were recovered by centrifugation. To assay for non-specific binding of the translated proteins to membranes, the membranes were also added posttranslationally before centrifugation. The mEH and the CYP2B1/mEH fusionprotein which were translated in the presence of the microsomes were found almost exclusively in the microsomal pellet whereas the shortened mEH, translated under identical conditions, was alkaline extracted into the supernatant after centrifugation. Alkaline extraction of translational mixtures which did not contain membranes or which had been supplied with membranes post-translationally, followed by SDS-PAGE analysis, showed that approximately equal amounts of the in vitro translated mEH and of the CYP2B1/mEH fusion protein were found in the supernatant and in the pellet after centrifugation. Under these conditions the shortened mEH was mainly found in the supernatant. In another experiment the mEH and the modified mEH proteins were translated in the presence of dog pancreas microsomes and the membranes were treated either with an alkaline solution as above, or with 20 mM Tris buffer pH 8.0. Again as for the experiment described above the mEH and the CYP2B1/mEH fusion protein were not alkaline-extracted from the membranes whereas the truncated mEH was. The treatment of the membranes at pH 8.0, however, extracted much less shortened mEH from the membranes than did the alkaline treatment, indicating that under near neutral conditions the shortened mEH remained associated with the membranes.

Membrane Integration of the mEH and the Modified mEH Proteins in Fibroblasts

In order to verify whether the mode of membrane integration found for the mEH and the modified mEH proteins in a cell free translation system faithfully reflected this process in an intact cell, we expressed these proteins in BHK21 hamster kidney fibroblasts and analysed for the integration of the mEH, the CYP2B1/mEH fusion protein and the shortened mEH into cellular membrane fractions. The membranes were either treated with an alkaline solution or with a near neutral solution (see Materials and Methods) and were isolated by centrifugation. The resulting supernatants and pellets were analysed by SDS-PAGE followed by immunoblotting using an anti-mEH antiserum as the first antibody. After alkaline and after neutral treatment of the membrane fraction, the mEH and the CYP2B1/mEH fusion protein remained mainly integrated in the membrane fractions. The shortened mEH was, however, extracted from the membrane fractions mainly at alkaline pH and not so much at near neutral pH conditions. Thus it can be said that in the intact mammalian cell as well as in a cell free system, the mEH as well as the CYP2B1/mEH fusion protein are integrated into the membranes whereas the truncated mEH is only peripherally associated with membranes.

Catalytic Function of the mEH and of the Modified mEH Proteins

The catalytic function of the mEH, the CYP2B1/mEH fusion protein and of the truncated mEH was evaluated in cells which stably express these proteins heterologously.

As a control this enzyme activity was determined in the parental cell line. The mEH enzyme activity was determined using benzo[*a*]pyrene 4,5-oxide. This enzyme assay was shown to be highly sensitive. Enzyme activity was determined after the cells had been broken by means of a short ultrasonication at 4°C. The mEH enzyme activity of the cellular homogenates is given in Table 1. Cells expressing the mEH or the modified mEH proteins had an mEH enzyme activity which was clearly higher than that of BHK21 cells. Cells expressing the mEH had a four fold and a 28 fold higher mEH enzyme activity as compared to the CYP/EH or the δmEH expressing cells.

In order to correlate the mEH enzyme activity of the various cell lines with the level of the heterologously expressed proteins, we determined their level of expression by immunoblotting using a serial dilution of the cellular homogenates and of a homogeneous epoxide hydrolase preparation. The first antibody in this assay was an antiserum directed against the mEH. For quantification, the fluorography of the immunoblot was scanned and the signal intensities obtained for the mEH and the modified mEH proteins were integrated. The signal intensities were linear with protein. By comparison of the signal intensities obtained for the purified mEH with those obtained for the heterologously expressed proteins, we estimated the content of these proteins in the cellular homogenates. These values are given in Tab. 1. In this experiment we found that the cellular expression level of the CYP/EH fusionprotein and of the δmEH was lower than the heterologous expression of the mEH by a factor of two and 20 fold respectively. Thus the catalytic turnover number of the δμEH is only by about 30% less than that of the mEH, whereas the CYP/EH displayed 40% of the catalytic activity of the mEH (see Tab. 1).

Table 1. mEH enzyme activity and concentration of mEH related proteins in the parental BHK21 cell line and in BHK 21 cells which express the mEH and modified mEH proteins heterologously[&]

Cell line	Specific mEH enzyme activity pmole/min/mg	Conc. of mEH related protein mg/mg cellular protein	Relative catalytic turnover number #
BHK21	1.2 ± 0.4	n.d	n.d
BHK-mEH	266.0 ± 5.6	1.0	1.0
BHK-CYP/mEH	65.4 ± 3.4	0.6	0.4
BHK-dmEH	9.3 ± 0.9	0.05	0.7

[&] The table is taken from ref. 17

defined as the ratio of the catalytic turnover number of the mEH in BHK-mEH cells to the catalytic turnover number of the modified cellular mEH proteins and calculated from the specific mEH activities found in the cellular protein extracts and the cellular concentration of the heterologous expressed proteins.

The cellular mEH enzyme activity towards benzo[a]pyrene 4,5-oxide was measured. The enzyme activity of each cell line was determined in three separate pools of cells. The enzyme assays were done in triplicate for each pool.

The cellular concentration of the heterologously expressed proteins was determined by immunoblotting of the cellular homogenates from a single cell-harvest of each cell line. Serial dilutions (four concentrations) of the cellular protein of each cell line were analysed in two separate immunoblots. As standard 10.0; 5.0; 2.5 and 1.25 ng of a homogeneous mEH preparation were mixed with 50 μg of cellular protein of BHK 21 cells and analysed by immunoblotting. The optical density of the immunosignals was integrated. The intensity of the immunblot signals was linear with the amount of protein analysed. The first antibody was an anti-rat mEH antiserum raised in goat. nd= non detectable.

DISCUSSION

Based on theoretical calculations, the N-terminal twenty amino acid residues of the mEH are the most hydrophobic domain of the human, rat and rabbit mEH (10) . Hydropathy analysis using two different methods indicated that the mEH was either anchored only via its N-terminal region into the membrane (10) or contained several transmembrane loops (13). Some experimental evidence supports the former model. Craft et al.(10) employed a membrane impermeant fluorescent probe which binds covalently to proteins, in order to evaluate whether the mEH is mainly exposed to the lumen of the ER or to its cytosolic side. They found that this probe labelled the mEH but not UDP-glucuronosyltransferases, which are thought to be entirely exposed to the luminal face of the ER. Thus these authors concluded on the basis of this experiment and the hydropathy profile of the mEH that this protein is exclusively exposed to the cytosolic face of the ER and is anchored into the membrane most likely via its amino-terminus. However, very recently (12) a bile acid transport protein, which was identified to be identical with the mEH, was found to assume two topological orientations in the membrane as evidenced by the accessibility of an epitope to its cognate antibody and by protease protection experiments in the presence of an intact membrane.

In the present work, we were able to show in a cell free system, and in fibroblasts, that the 20 N-terminal amino acid residues of the mEH function as its sole membrane anchor signal sequence. In addition we demonstrated that the function of this signal can also be mediated by a cytochrome P450 (CYP2B1) membrane anchor sequence.

Our results do not establish whether the mEH and the CYP2B1/mEH fusion protein were localized in the endoplasmic reticulum of the BHK21 hamster kidney fibroblasts, or in other subcellular membrane compartments, as we did not attempt to perform a subcellular fractionation of the fibroblast membranes. However studies employing immuno-fluorescence microscopy of BHK21 fibroblasts which express the mEH heterologously, indicate that this protein was at least largely associated with membrane structures resembling endoplasmic reticulum and / or Golgi membranes.

In order to study whether the twenty N-terminal amino acid residues of the mEH are not only necessary for its cotranslational insertion into cellular membranes but also for the catalytic activity of the mEH, we determined the mEH enzyme activity of the rat mEH and the modified mEH proteins in the above mentioned fibroblasts, which stably express these proteins heterologously. We used benzo[*a*]pyrene 4,5-oxide as the substrate for this assay. The stable expression system was preferred to a transient expression system because it is known that in the latter system the heterologous expression level of a protein might vary from one transfection to the next (20) , whereas in the stable expression system these difficulties are not encountered, once a stable expressing cell clone has been selected.

We found that the expression level of the δmEH was rather low. Screening of further clones which had been transfected with the template coding for the δmEH did not yield clones expressing higher levels of this protein. Also others have found that the expression of a truncated membrane protein (cytochrome P450) which was devoid of a membrane anchor signal sequence lead to a five fold reduction in the expression level of this protein as compared to the full length membrane protein (21) . This might be due to toxic effects of the truncated protein for the cell expressing it or due to degradation of the truncated protein.

The δmEH expressing cells had a 28 fold lower mEH enzyme activity as compared to the mEH expressing cells. However, the cellular expression level of the δmEH was also found to be 20 fold lower than that of the mEH (Tab. 1). We conclude from these results that the catalytic activity of the δmEH is approximately as high as that of the mEH. Therefore the membrane anchor signal sequence of the mEH, unlike the analogous sequences in cytochromes P450, appears to be dispensable for the catalytic activity of the mEH in a mammalian system.

In contrast to the low expression of the δmEH in BHK21 cells, a rather high level of expression was achieved for the CYP/EH fusion protein. However, comparison of the mEH enzyme activity found for the CYP/EH expressing cells with the expression level of this protein (Tab. 1) indicates that the catalytic activity of the CYP/EH is by a factor of two lower than that of the mEH. However, we found some variability of the mEH enzyme activity of the CYP/EH expressing cells. Therefore we cannot conclude whether the CYP/EH fusion protein has exactly the same catalytic activity as the mEH. Nevertheless our data indicate that the membrane anchor of the mEH can be functionally replaced by another membrane anchor.

The results presented in this paper open up the possibility of altering the membrane orientation of the mEH by site directed mutagenesis of charged amino acid residues flanking the mEH membrane anchor signal sequence and of determining the consequences of this alteration for the catalytic activity of this protein. More importantly the impact of the cellular orientation of the mEH on its toxicological function can thus be studied.

ACKNOWLEDGMENT

We thank especially Dr. C. B. Kasper for the generous gift of the plasmid pEH52. We thank Mrs. I. Ihrig-Biedert for her valuable technical assistance.

This work was generously supported by the German Research Council (DFG): Grant Nr. SFB 302 / A70.

REFERENCES

1. Davies, R.L., Crespi C.L., Rudo K., Turner T.R. and Langenbach R. (1989) Development of a human cell line by selection and drug-metabolizing gene transfection with increased capacity to activate promutagens. *Carcinogenesis*, **10**, 885-891.
2. Oesch, F. (1973) Mammalian epoxide hydrases. Inducible enzymes catalysing the inactivation of carcinogenic and cytotoxic metabolites derived from aromatic and olefinic compounds. *Xenobiotica*, **3**, 305-340.
3. Grover, P.L. (1986) Pathways involved in the metabolism and activation of polycyclic aromatic hydrocarbons. *Xenobiotica*, **16**, 915-931.
4. Wood, A.W., Levin W., Lu A.Y.H., Yagi H., Hernandez O., Jerina D.M. and Conney A.H. (1976) Metabolism of benzo (a) pyrene and benzo (a) pyrene derivatives to mutagenic products by highly purified hepatic microsomal enzymes. *J. Biol. Chem.*, **251**, 4882-4890.
5. Bar-Nun, S., Kreibich G., Adesnik M., Alterman L., Negishi M. and Sabatini D.D. (1980) Synthesis and insertion of cytochrome P-450 into endoplasmic reticulum membranes. *Proc. Natl. Acad. Sci. USA.*, **77**, 965-969.
6. De Lemos- Chiarandini, C., Frey A.B., Sabatini D.D. and Kreibich G. (1987) Determination of the membrane topology of the phenobarbital-inducible rat liver cytochrome P-450 isoenzyme PB-4 using site specific antibodies. *J. Cell. Biol.*, **104**, 209-219.
7. Szczesna-Skorupa, E., Browne N., Mead D. and Kemper B. (1988) Positive charges at the NH2 terminus convert the membrane anchor signal peptide of cytochrome P-450 to a secretory signal peptide. *Proc. Natl. Acad. Sci. USA.*, **85**, 738-742.
8. Okada, Y., Frey A.B., Guenthner T.M., Oesch F., Sabatini D.D. and Kreibich G. (1982) Studies on the biosynthesis of microsomal membrane proteins. *Eur. J. Biochem.*, **122**, 393-402.
9. Gonzales, F.J. and Kasper C.B. (1980) *Biochem. Biophys. Res.Commun.*, **93**, 1254-1258.
10. Craft, J.A., Baird S., Lamont M. and Burchell B. (1990) Membrane topology of epoxide hydrolase. *Biochem. Biophys. Acta*, **1046**, 32-39.

11. Waechter, F., Bentley P., German M., Oesch F. and Stäubli W. (1982) Immuno-electron-microscopic studies on the subcellular distribution of rat liver epoxide hydrolase and the effect of phenobarbitone and 2-acetamidofluorene treatment. *Biochem. J.*, **202**, 677-686.
12. Alves, C., Vondippe P., Amoui M. and Levy D. (1993) Bile acid transport into hepatocyte smooth endoplasmic reticulum vesicles is mediated by microsomal epoxide hydrolase, a membrane protein exhibiting two distinct topological orientations. *J Biol Chem*, **268**, 20148-20155.
13. Porter, T.D., Beck T.W. and Kasper C.B. (1986) Complementary DNA and amino acid sequence of rat liver microsomal, xenobiotic epoxide hydrolase. *Arch. Biochem. Biophys.*, **248**, 121-129.
14. Yabusaki, Y., Murakami H., Sakaki T., Shibata M. and Ohkawa H. (1988) Genetically engineered modification of P450 monooxygenases: Functional analysis of the amino-terminal hydrophobic region and hinge region of the P450/reductase fused enzyme. *DNA*, **7**, 701-711.
15. Sagara, Y., Barnes H.J. and Waterman M.R. (1993) Expression in Escherichia coli of functional P450c17 lacking its hydrophobic amino-terminal signal anchor. *Arch. Biochem. Biophys.*, **304**, 272-278.
16. Larson, J.R., Coon M.J. and Porter T.D. (1993) Alcohol inducible cytochrome P450IIE1 lacking the hydrophobic NH2-terminal segment retains catalytic activity and is membrane bound when expressed in E. coli. *J. Biol. Chem.*, **266**, 7321-7324.
17. Friedberg, T., Lollmann B., Becker R., Holler R. and Oesch F. (1994) The microsomal epoxide hydrolase has a single membrane signal anchor sequence which is dispensable for the catalytic activity of this protein. *Biochem J*, **303**, 967-972.
18. Laemmli, U.K. (1970) Cleavage of structural proteins during the assembly of the head of bacteriophage T4. *Nature*, **227**, 680-685.
19. Glatt, H.R., Gemperlein I., Setiabudi F., Platt K.L. and Oesch F. (1990) Expression of xenobiotic metabolising enzymes in propagatable cell cultures and induction of micronuclei by 13 compounds. *Mutagenesis*, **5**, 241-249.
20. Clark, B.J. and Waterman M.R. (1991) Heterologous expression of mammalian P450 in COS cells. In Waterman, M.R. and Johnson, E.F. (eds), *Methods in Enzymology*. Vol. 206. Academic Press, San Diego. pp. 100-108.
21. Clark, B.J. and Waterman M.R. (1991) The hydrophobic amino-terminal sequence of bovine 17a hydroxylase is required for the expression of a functional hemoprotein in COS 1 cells. *J. Biol. Chem.*, **266**, 5898-5904.

CYTOTOXICITY OF NITRIC OXIDE AND HYDROGEN PEROXIDE

Is There a Cooperative Action?

Iosif Ioannidis,[1] Thomas Volk,[2] and Herbert de Groot[1]

[1] Institut für Physiologische Chemie
Universitätsklinikum
Hufelandstr. 55, 45122 Essen, Germany
[2] Abteilung für Anesthäsiologie und Intensivmedizin
Universitätsklinikum Charité
Schumannstr. 20/21, 10117 Berlin, Germany

INTRODUCTION

Nitric oxide (•NO) is well defined as an important effector molecule in biological systems. Secreted from various cell types, •NO contributes to a variety of physiological and pathophysiological processes [22]. Despite of its function in the control of blood pressure, platelet aggregation and neurotransmission, which are mediated via the activation of the soluble guanylate cyclase, •NO has been shown to be a potent modulator of the cytotoxic activity of macrophages as well [12; 21]. The mechanisms by which the bactericidal and the tumoricidal potential of •NO is defined are still poorly understood. Diverse molecular targets have been postulated to exist for an attack of •NO on the cell surface as well as inside the cells such as thiol groups of proteins yielding S-nitrosothiols [14]. In addition to the membrane-linked targets, •NO is suggested to inhibit the mitochondrial respiratory chain, DNA and protein synthesis and iron metabolism [11]. Besides these effects, which could be defined as direct effects of •NO, an increasing number of studies postulates that the cytotoxicity of •NO is enhanced by chemical interactions with oxygen and reactive oxygen species to form other potentially toxic radicals. For example, •NO reacts with the superoxide anion ($O_2^{\bullet-}$) forming the peroxynitrite anion ($ONOO^-$) which decays, once protonated, to the very reactive hydroxyl radical (•OH) and to nitrogen dioxide (•NO_2) [1]. Increased reactivity due to an interplay of •NO with $O_2^{\bullet-}$ via the $ONOO^-$ pathway is supported by studies in biological as well as in chemical systems [28; 25; 4; 5].

Here, we present results on the damaging effect of •NO-releasing compounds in cooperation with reactive oxygen species generated both in the extracellular space by $O_2^{\bullet-}$- and H_2O_2-generating systems, and intracellularly by inhibition of the mitochondrial respiratory chain. Since the half-life of •NO strongly decreases with increasing concen-

Biological Reactive Intermediates V
Edited by R. Snyder et al., Plenum Press, New York, 1996

trations of molecular oxygen we further studied whether hypoxic conditions might affect ·NO reactivity.

METHODS

Fu5 hepatoma cells and sinusoidal liver endothelial cells from rat as well as L929 fibroblast cells from mouse were treated and harvested with small modifications as described elsewhere [13; 26; 6]. For the experiments cells in confluency were used. On the day of experimentation, medium was removed and the cells were washed with and kept in Krebs-Henseleit buffer (pH 7,4, supplemented with 20 mM Hepes and 10 mM glucose) at 37°C (air/CO_2, 19:1). Hypoxic conditions were initiated by addition of nitrogen-saturated (CO_2/N_2, 1:19) Krebs-Henseleit buffer [2]. The experiments were started by adding the substances as indicated. S-Nitroso-N-acetyl-DL-penicillamine (SNAP) and sodium nitroprusside (SNP) were used as ·NO-donating compounds, and 3-morpholinosydnonimine-*N*-ethylcarbamide (SIN-1) as a compound which releases both ·NO and $O_2^{\bullet-}$. Glucose oxidase/glucose and xanthine oxidase/hypoxanthine were applied to produce H_2O_2 and $O_2^{\bullet-}$/H_2O_2, respectively. KCN, antimycin A and rotenone were given to inhibit cellular respiration. LDH Leakage was used to indicate cell viability [16]. Comparative experiments with trypan blue exclusion and phase contrast cell counts gave the same results.

RESULTS AND DISCUSSION

Cytotoxicity of ·NO-Donating Substances

All ·NO donors induced a concentration dependent cytotoxicity against sinusoidal liver endothelial cells. A significant loss in the viability of these cells of about 30% occurred between 2 and 4 h of incubation when the cells were exposed to 5 mM SIN-1, 5 mM SNAP and 20 mM SNP (Fig.1). Similar results were obtained in incubations with Fu5 hepatoma cells. The toxicity of SIN-1, SNP and SNAP is suggested to be mediated by the reactivity of ·NO since this molecule is known to be released during the decomposition of these compounds [14]. The denitrosylated substances had no influence on cell viability in agreement with data reported previously [19]. In contrast to endothelial cells and to Fu5 hepatoma cells, SNP and SNAP exhibited only slight effects against L929 fibroblast cells even at high

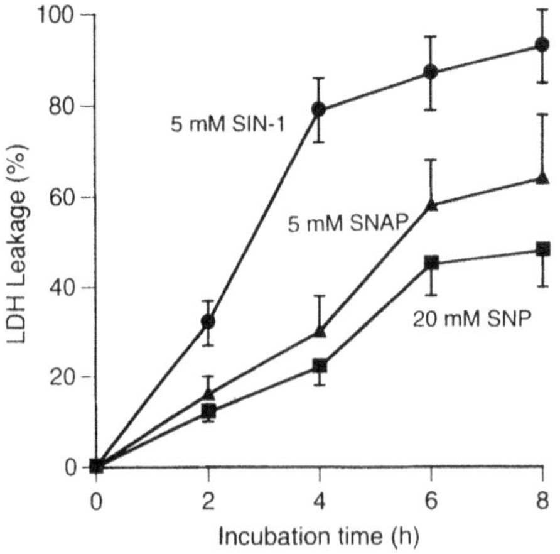

Figure 1. Toxicity of ·NO-donating compounds against sinusoidal liver endothelial cells. Cells (10^5/ml) were exposed to 5 mM SIN-1, 5 mM SNAP and 20 mM SNP. Cell injury was estimated by LDH Leakage. Data represent means ± S.E.M. from 3-6 experiments.

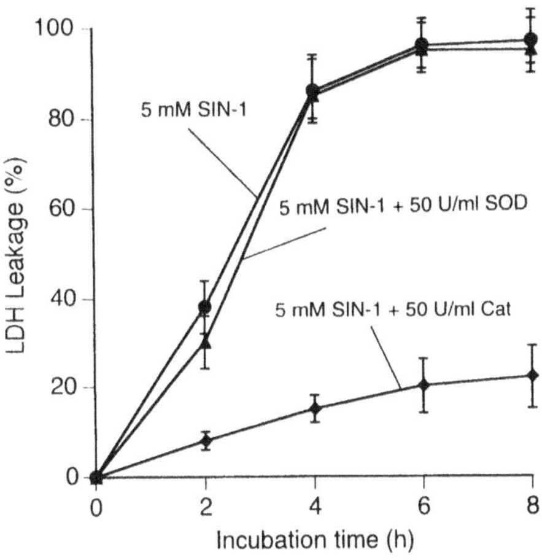

Figure 2. Effects of superoxide dismutase and catalase on SIN-1-induced toxicity against sinusoidal liver endothelial cells. Cells (10^5/ml) were exposed to 5 mM SIN-1 together with 50 U/ml superoxide dismutase (SOD) or 50 U/ml catalase (Cat). Cell injury was estimated by LDH Leakage. Data represent means ± S.E.M. from 3 experiments.

concentrations (20 mM and 5 mM, respectively), suggesting that this cell type might be resistant to the sole reactivity of ·NO. Only SIN-1, at 20 mM, induced toxic effects decreasing the viability of L929 fibroblast cells to about 50% after 6 h (data not shown).

Taking into account that concentrations of ·NO donors were used which produced similar amounts of ·NO as measured by the hemoglobin method [7], SIN-1 was the most effective of them regardless of the cell line studied. Since SIN-1 not only liberates ·NO but also $O_2^{\bar{\cdot}}$ the results give rise to the hypothesis that ·NO interacts with reactive oxygen species to produce a higher toxic potential. Besides the direct effects of ·NO, which could affect cell metabolism by the sole reaction of the molecule with critical targets [23], it has been suggested that $O_2^{\bar{\cdot}}$ reacts with ·NO to produce the reactive peroxynitirite ($ONOO^-$) anion [1]. In our experiments, the destructive potential of 5 mM SIN-1 against endothelial cells and Fu5 hepatoma cells was reduced by catalase by almost 70% after 8 h of incubation, while superoxide dismutase, which accelerates the spontaneous dismutation of $O_2^{\bar{\cdot}}$ to H_2O_2, did not influence cytotoxicity (Fig.2). Hence, our results provide no evidence for a significant role of $ONOO^-$ formation in ·NO cytotoxicity. Probably H_2O_2 rather than $O_2^{\bar{\cdot}}$ promotes the ·NO toxicity by SIN-1 [15].

Superoxide dismutase and catalase at activities up to 200 U/ml had no influence on the toxicity of both, 20 mM SNP and 5 mM SNAP (data not shown), against these cells, suggesting that $O_2^{\bar{\cdot}}$ as well as H_2O_2 were not produced.

Cytotoxicity of Enzymatically Produced Reactive Oxygen Species

Reactive oxygen species like $O_2^{\bar{\cdot}}$ and H_2O_2 are well known to react with cell compartments leading to tissue injury [10]. Liver sinusoidal endothelial cells, Fu5 hepatoma cells as well as L929 fibroblast cells showed high sensitivity when exposed to systems producing $O_2^{\bar{\cdot}}$ and H_2O_2. A strong decrease in the viability of all three cell types was induced by 20 mU/ml glucose oxidase/10 mM glucose and 20 mU/ml xanthine oxidase/1 mM hypoxanthine (60-80% after 8 h of incubation, each; data not shown). Similar to the SIN-1 experiments presented, superoxide dismutase activities up to 200 U/ml had no influence on cell damage induced by xanthine oxidase/hypoxanthine, which generates both $O_2^{\bar{\cdot}}$ and H_2O_2. Complete protection was observed at catalase activities down to 1.3 U/ml in incubations with xanthine oxidase/hypoxanthine as well as with glucose oxidase/glucose, which produces only H_2O_2. These results are in line with data reported for other cell lines [20; 27] and support the hypothesis that H_2O_2 rather than $O_2^{\bar{\cdot}}$ initiates the toxic effects against the cells.

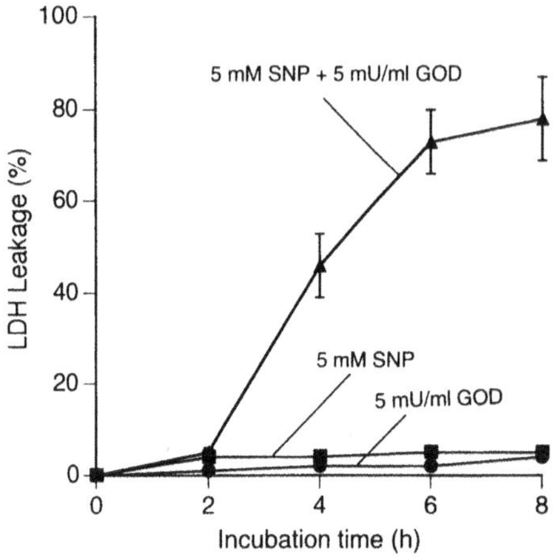

Figure 3. Cooperative action of ·NO and H_2O_2 in the killing of sinusoidal liver endothelial cells. Cells (10^5/ml) were exposed to 5 mU/ml glucose oxidase (GOD)/10 mM glucose and/or 5 mM SNP. Cell injury was estimated by LDH Leakage. Data represent means ± S.E.M. from 3 experiments.

Cytotoxicity of both ·NO and H_2O_2

The combined action of subtoxic concentrations of both H_2O_2 and ·NO led to a markedly enhanced toxicity in all three cell types. For example, in the presence of 5 mU/ml glucose oxidase/10 mM glucose and 5 mM SNP, which alone did not influence the viability of endothelial cells, cytotoxicity was about 70% after 6 h of incubation (Fig.3). Similar results were observed in experiments with subtoxic concentrations of xanthine oxidase/hypoxanthine and SNAP or glucose oxidase/glucose and SNAP with the three cell types. The damaging combinations were not inhibited when superoxide dismutase was added but were reduced clearly in the presence of catalase, suggesting again a cooperative action between ·NO and H_2O_2. However, the mechanism of this interaction remains unclear. Enhanced cytotoxicity of ·NO and H_2O_2 might occur due to the production of other, highly reactive oxygen species like hydroxyl radicals or singlet oxygen [17; 24]. In addition to such a chemical interaction, the molecules ·NO and H_2O_2 could act independently on different compartments of the target cells thus leading to stronger cytotoxicity.

Influence of Endogeneously Released Reactive Oxygen on ·NO Reactivity

Since ·NO due to its lipophilic character could easily pass through membranes into a cell and affect cell metabolism, we were interested to clarify the role of intracellularly produced reactive oxygen in the ·NO toxicity. Therefore we exposed endothelial cells to potent inhibitors of the respiratory chain in order to accelerate the production of $O_2^{\bullet-}/H_2O_2$ [3]. Regardless of the ·NO donor used, KCN, antimycin A, and rotenone, applied in non-toxic doses, increased the •NO-mediated cell death by 30-40% (Table 1), suggesting that also endogenously produced reactive oxygen species potentiate the effects of ·NO on cell metabolism. However, under these conditions the loss of ATP has to be considered. The reduced energy status might facilitate the reactivity of ·NO inside the cells.

·NO Under Hypoxic Conditions

It is well known that the half-life of ·NO is limited to a few seconds because of the spontaneous reaction of ·NO with molecular oxygen [18]. The product of the reaction of these two molecules in aqueous solutions is nearly exclusively NO_2^- [9]. Therefore, oxygen might also modulate the effectiveness of ·NO against cells, suggesting that ·NO loses its radical character by this reaction. We therefore studied the toxicity of ·NO against endothelial

Table 1. Effects of respiratory chain inhibitors on the ·NO-mediated toxicity against sinusoidal liver endothelial cells

·NO Donor	Inhibitor	LDH Leakage (%)
SNAP	none	39±6
	KCN	79±8
	Antimycin A	72±9
	Rotenone	70±12
SNP	none	35±9
	KCN	71±8
	Antimycin A	74±11
	Rotenone	68±7

Cells (10^5/ml) were exposed for 6 hours to 5 mM SNAP or 20 mM SNP together with 1 mM KCN, 1 M antimycin A or 0.5 M rotenone. Cell injury was estimated by LDH Leakage. Data represent means S.E.M. from 4-5 experiments.

Table 2. Toxicity of ·NO under authentic hypoxic conditions against sinusoidal liver endothelial cells

·NO Donor	O_2 content	LDH Leakage (%)
SNAP	Normoxia	36±7
	Hypoxia	85±12
SNP	Normoxia	37±6
	Hypoxia	62±9
SIN-1	Normoxia	85±11
	Hypoxia	17±5

Cells (10^5/ml) were exposed for 6 hours to 5 mM SNAP, 20 mM SNP or 5 mM SIN-1. The hypoxic incubations were performed in nitrogen-saturated Krebs-Henseleit buffer. Cell injury was estimated by LDH Leakage. Data represent means S.E.M. from 3-4 experiments.

cells under authentic hypoxic conditions. Under these conditions, SNP as well as SNAP, induced a stronger loss of viability of the cells compared to normoxic conditions (Table 2). In the absence of the ·NO donors the cells were not sensitive to hypoxia. In contrast to SNP and SNAP, SIN-1 was almost ineffective against the cells under hypoxic conditions, demonstrating that O_2 is required for the decomposition of the molsidomine to ·NO and $O_2^{\cdot -}$ as reported previously [8].

The role of O_2 in ·NO toxicity remains unclear. Regarding the high affinity of ·NO to O_2 and its derivatives, yielding reactive products such as $^{\cdot}NO_2$, $^{\cdot}OH$ and/or $ONOO^-$, hypoxic conditions might change the mechanism of the cytotoxicity of ·NO.

REFERENCES

1. Beckman, J. S., T. W. Beckman, et al. (1990). "Apparent hydroxyl radical production by peroxynitrite: Implications for endothelial injury from nitric oxide and superoxide." *Proc. Natl. Acad. Sci. U.S.A.* **87**: 1620-1624.

2. Brecht, M., C. Brecht, et al. (1993). "Improvement of the energetic situation of hypoxic hepatocytes by calcium channel blockers." *Res. Commun. Chem. Pathol. Pharmacol.* **82**: 185-198.
3. Cadenas, F., A. Boveris, et al. (1977). "Production of superoxide radicals and hydrogen peroxide by NADH-ubiquinone reductase and ubiquinol-cytochrome c reductase from beef heart mitochondria." *Arch. Biochem. Biophys.* **180**: 248-257.
4. Darley-Usmar, V. M., N. Hogg, et al. (1992). "The simultaneous generation of superoxide and nitric oxide can initiate lipid peroxidation in human low density lipoprotein." *Free Rad. Res. Commun.* **17**: 9-20.
5. de Groot, H., U. Hegi, et al. (1993). "Loss of α-tocopherol upon exposure to nitric oxide or the sydnonimine SIN-1." *FEBS Lett.* **315**: 139-142.
6. Fast, D. J., R. C. Lynch, et al. (1993). "Interferon-γ, but not interferon-αβ, synergizes with tumor necrosis factor-α and lipid A in the induction of nitric oxide production by murine L929 cells." *J. Interferon Res.* **13**: 271-277.
7. Feelisch, M. and E. Noack (1987). "Correlation between nitric oxide formation during degradation of organic nitrates and activation of guanylate cyclase." *Eur. J. Pharmacol.* **139**: 19-30.
8. Feelisch, M., J. Ostrowski, et al. (1989). "On the mechanism of NO release from sydnonimines." *J. Cardiovasc. Pharmacol.* **14 (Suppl. 11)**: S13-S22.
9. Feelisch, M. (1991). "The biochemical pathways of nitric oxide formation from nitrovasodilators: Appropriate choice of exogenous NO donors and aspects of preparation and handling of aqueous NO solutions." *J. Cardiovasc. Pharmacol.* **14 (11)**: 25-33.
10. Halliwell, B. and J. M. C. Gutteridge (1989). *Free radicals in biology and medicine*. Oxford, Clarendon Press.
11. Henry, Y., C. Ducrocq, et al. (1991). "Nitric oxide, a biological effector." *Eur. Biophys. J.* **20**: 1-15.
12. Hibbs, J. B. J., R. R. Taintor, et al. (1988). "Nitric oxide: A cytotoxic activated macrophage effector molecule." *Biochem. Biophys. Res. Commun.* **157**: 87-94.
13. Hugo-Wissemann, D., I. Anundi, et al. (1991). "Differences in glycolytic capacity and hypoxia tolerance between hepatoma cells and hepatocytes." *Hepatology* **13**: 297-303.
14. Ignarro, L. J., H. Lippton, et al. (1981). "Mechanism of vascular smooth musce relaxation by organic nitrates, nitrites, nitroprusside and nitric oxide: Evidence for the involvement of S-nitrosothiols as active intermediates." *J. Pharmacol. Exp. Ther.* **218**: 739-749.
15. Ioannidis, I. and H. de Groot (1993). "Cytotoxicity of nitric oxide in Fu5 rat hepatoma cells: Evidence for co-operative action with hydrogen peroxide." *Biochem. J.* **296**: 341-345.
16. Jauregui, H. O., N. T. Hayner, et al. (1981). "Trypan blue dye uptake and lactate dehydrogenase in adult rat hepatocytes: Freshly isolated cells, cell suspensions and primary monolayer cultures." *In Vitro* **17**: 1100-1110.
17. Kanner, J., S. Harel, et al. (1991). "Nitric oxide as an antioxidant." *Arch. Biochem. Biophys.* **289**: 130-136.
18. Kelm, M. and J. Schrader (1990). "Control of coronary vascular tone by nitric oxide." *Circ. Res.* **66**: 1561-1575.
19. Kröncke, K.-D., H.-H. Brenner, et al. (1993). "Pancreatic islet cells are highly suspectible towards the cytotoxic effects of chemically generated nitric oxide." *Biochem. Biophys. Acta* **412**: 1-9.
20. Link, E. M. and P. A. Riley (1988). "Role of hydrogen peroxide in the cytotoxicity of the xanthine/xanthine oxidase system." *Biochem. J.* **249**: 391-399.
21. Marletta, M. A., P. S. Yoon, et al. (1988). "Macrophage oxidation of L-arginine to nitrite and nitrate: Nitric oxide is an intermediate." *Biochemistry* **27**: 8706-8711.
22. Moncada, S., R. M. J. Palmer, et al. (1991). "Nitric oxide: Physiology, pathophysiology, and pharmacology." *Pharmacol. Rev.* **43**: 109-142.
23. Nathan, C. (1992). "Nitric oxide as a secretory product of mammalian cells." *FASEB J.* **6**: 3051-3064.
24. Noronha-Dutra, A. A., M. M. Epperlein, et al. (1993). "Reaction of nitric oxide with hydrogen peroxide to produce potentially cytotoxic singlet oxygen as a model for nitric oxide-mediated killing." *FEBS Lett.* **321**: 59-62.
25. Radi, R., J. S. Beckman, et al. (1991). "Peroxynitrite-induced lipid peroxidation: The cytotoxic potential of superoxide and nitric oxide." *Arch. Biochem. Biophys.* **288**: 481-487.
26. Rauen, U., M. Hanssen, et al. (1993). "Energy-dependent injury to cultured sinusoidal endothelial cells of the rat liver in UW solution." *Transplantation* **55**: 469-473.
27. Rubin, R. and J. L. Farber (1984). "Mechanisms of the killing of cultured hepatocytes by hydrogen peroxide." *Arch. Biochem. Biophys.* **228**: 450-459.
28. Wang, J. F., P. Komarov, et al. (1991). "Contribution of nitric oxide synthase to luminol-dependent chemiluminescence generated by phorbol-ester-activated Kupffer cells." *Biochem. J.* **279**: 311-314.

DEVELOPMENT OF A MASS SPECTROMETRIC ASSAY FOR 5,6,7,9-TETRAHYDRO-7-HYDROXY-9-OXIMIDAZO [1,2-*a*]PURINE IN DNA MODIFIED BY 2-CHLORO-OXIRANE

Michael Müller,[1] Frank Belas,[2] Hitoshi Ueno,[1] and F. Peter Guengerich[1]

[1] Department of Biochemistry and Center in Molecular Toxicology
Vanderbilt University School of Medicine
Nashville, Tennessee 37232
[2] Department of Pharmacology
Vanderbilt University School of Medicine
Nashville, Tennessee 37232

INTRODUCTION

Vinyl chloride is recognized as a human carcinogen. 2-Chlorooxirane, its biological reactive intermediate formed by the P450 enzyme system in the liver, and the rearrangement product chloroacetaldehyde give rise to the etheno (ε) DNA adducts. This type of modified DNA bases appears to be the molecular basis for the genotoxic effects of the parent compound, and site specific mutagenesis studies (Basu et al., 1993) show its mutagenicity. There is evidence that ε adducts might not only be formed by vinyl chloride metabolites but also be present at a certain level due to endogeneous sources (Misra et al., 1994).

Our previous studies (Guengerich et al., 1993) on the mechanisms of adduct formation by 2-chlorooxirane with guanosine and calf thymus DNA revealed a new DNA adduct. Initial attack of the exocyclic amino group at the unsubstituted methylene moiety of the oxirane yields an aminal, which subsequently cyclizes to form 5,6,7,9-tetrahydro-7-hydroxy-9-oximidazo[1,2-*a*] purine (I) (Fig. 1). This product was fully characterized and due to its unique stability and high levels of formation in the nucleoside model reactions has to be considered as a DNA adduct in its own right.

In order to further quantitate the *in vitro* and *in vivo* formation, we developed an LC/ESI-MS (liquid chromatography/electrospray ionization-mass spectrometry) method. This involved the synthesis of an internal trilabeled adduct standard, establishing the conditions of selective reaction monitoring (SRM), and the measurement of adduct formation in calf thymus DNA after *in vitro* treatment with 2-chlorooxirane.

Biological Reactive Intermediates V
Edited by R. Snyder et al., Plenum Press, New York, 1996

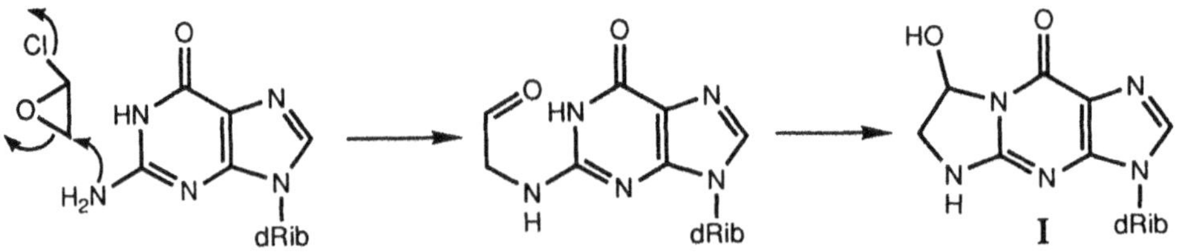

Figure 1. Formation of 5,6,7,9-tetrahydro-7-hydroxy-9-oximidazo[1,2-*a*] purine (I) from 2-chlorooxirane.

RESULTS AND DISCUSSION

Synthesis of Trilabeled ($^{13}C_3$) Adduct Standard (I)

Heavy isotope labeled guanine ($^{13}C_3$) was synthesized according to known procedures (Lazarus et al., 1982; Cain et al., 1946; Sharma et al., 1983; Gmeiner et al., 1988).

The adduct synthesis involved direct fluorination of the guanine to fluorohypoxanthine, its adduction with aminoacetaldehde dimethyl acetal, subsequent hydrolysis of the acetal and ring closure to yield the desired 5,6,7,9-tetrahydro-7-hydroxy-9-oximidazo[1,2-*a*] purine (I) (Fig. 2). This product was characterized by mass spectrometry (*m/z* 197 (MH^+, 100 %)) and ^{1}H-NMR ([^2H_2O] δ 3.67 (dd, 1 H, H-6); 4.04 (dd, 1 H, H-6); 6.32 (dd, 1 H, H-7), 8.66 (d, 1 H, H-2)). Correct insertion of labels was checked by ^{13}C-NMR ([^2H_2O] δ 109.3 (d, C-9a); 137.4 (s, C-2); 153.9 (d, C-9)).

The yields for the initial fluorination were lower compared to the same reaction with unlabeled material (11 versus 53 %), which was attributed to the high amount of inorganic salts still present in the starting material (isotopically labeled guanine). Nevertheless, the devised method is of general use for the synthesis of heavy isotope labeled N^2-guanine base adducts, which can be prepared just by reacting fluorohypoxanthine with the appropriate amine.

LC/ESI-MS (Liquid Chromatography/Electrospray Ionization-Mass Spectrometry)

LC/ESI-MS has developed to a mature analytical technique from its origins some 20 years ago and promises to set new standards in the characterization and quantification of

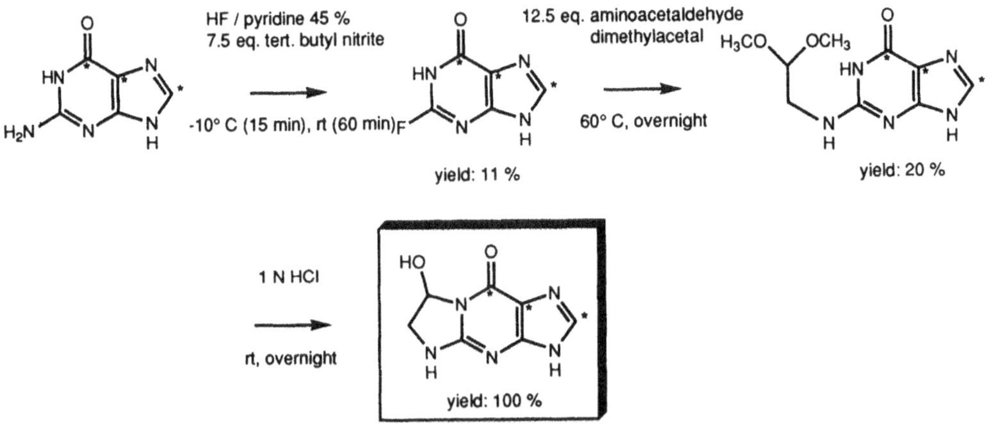

Figure 2. Synthesis of the internal adduct standard (I).

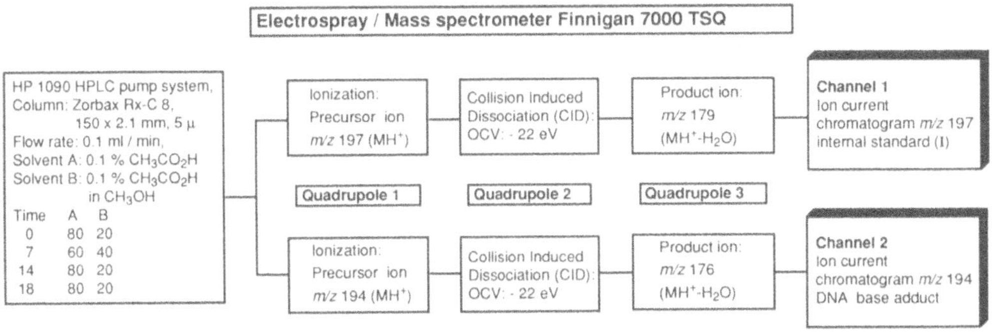

Figure 3. LC MS/MS selective reaction monitoring (SRM) experiment. See text.

biomolecules (Fenn et al., 1990). Most recently investigators have used this method to determine the structures of DNA oligomers with and without adducts as well as for rapid DNA sequencing (Reddy et al., 1994; Potier et al., 1994; Little et al., 1994). In combination with heavy isotope labeled internal standards using SRM, new limits of detection of a specific adduct in modified DNA seem to be achievable.

The SRM experiment involved HPLC separation of the DNA adduct and its labeled standard, generation of precursor ions (usually the MH^+ ions) and introduction of these ions into a collision cell yielding defined product ions (in this case MH^+-H_2O). Mass spectrometric data was collected in quadrupole 1 and 3 and combined to a signal on channel 1 (internal standard) and channel 2 (adduct) (Fig. 3). Only an adduct eluting at the same time with the heavy isotope marker and giving the same pattern of ionization in both mass spectrometers could cause the appropriate coincidental signal (Fig. 4). Therefore, SRM established three levels of selectivity for the given DNA adduct: HPLC retention time, specific precursor ion, and unique product ion formation, enabling us to rule out any other adduct also present in the sample.

As an on-line analytical technique LC/ESI-MS in the SRM mode offers the advantage of a direct sample introduction without extensive sample workup or derivatization compared

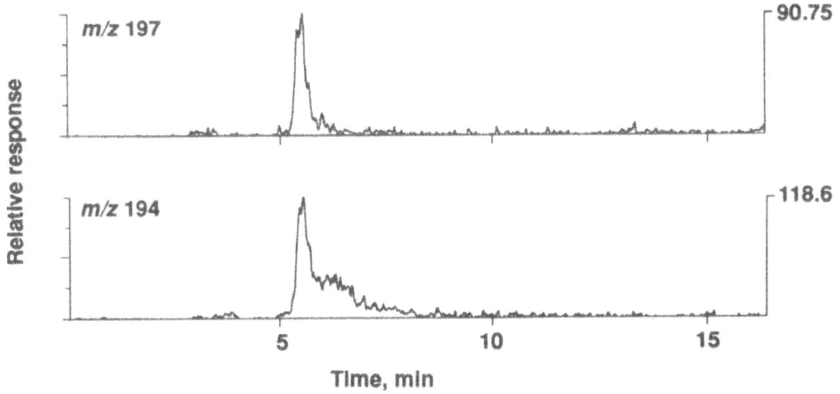

Figure 4. Ion current chromatograms resulting from analysis of a DNA sample treated with 75 mM 2-chlorooxirane. Internal standard (I) (5 ng, $^{13}C_3$) was added to the sample and the *m/z* 194/197 ratio was calculated from relative responses.

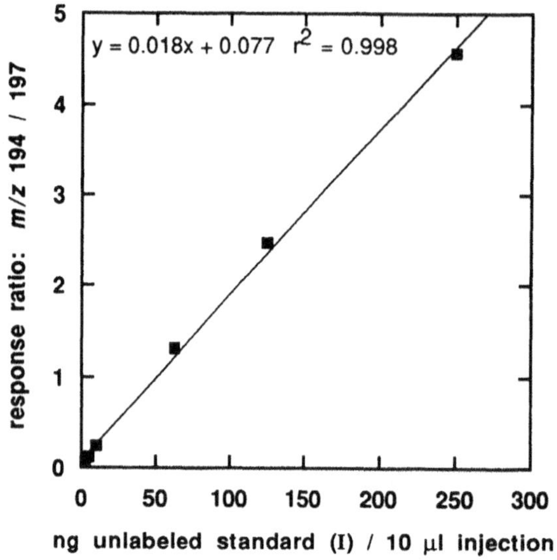

Figure 5. Standard curve for HPLC/MS analysis of (I).

to gas chromatography/electron capture negative chemical ionization-mass spectrometry and a very short analysis time compared to ^{32}P-postlabeling, methods commonly used for the detection and quantification of DNA adducts.

Formation of 5,6,7,9-Tetrahydro-7-Hydroxy-9-Oximidazo[1,2-*a*] Purine (I) in Calf Thymus DNA by 2-Chlorooxirane

To quantitate the adduct formation of compound (I) in calf thymus DNA a standard curve was established by injecting known amounts of the unlabeled base adduct spiked with a fixed amount of internal standard and determining the ratios of the signals at *m/z* 194 and 197 (Fig. 5). The system was able to pick up as little as 5 ng of the adduct on column without any difficulty.

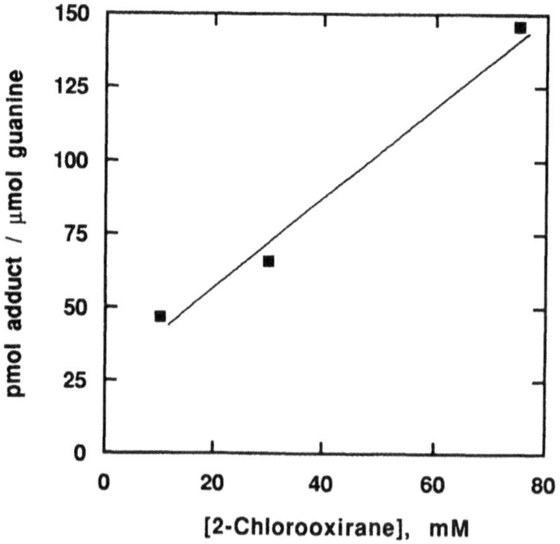

Figure 6. DNA adduct formation as a function of 2-chlorooxirane added.

Calf thymus DNA was treated with 2-chlorooxirane as described previously (Guengerich et al., 1993). Prior to the acid depurination step the internal standard was added to the samples. A fraction containing the released DNA bases was recovered after Amicon Centricon 30 filtration and an aliquot was directly subjected to LC/ESI-MS analysis.

The adduct was selectively identified to be present in 2-chlorooxirane treated calf thymus DNA, confirming our previous results. Moreover the pre-liminary quantitative results (Fig. 6) suggest that it is formed at least at the same level as the 1-N^2-ε-guanine adduct under similar conditions.

Conclusions

We were able to show the formation of the new 2-chlorooxirane derived adduct 5,6,7,9-tetrahydro-7-hydroxy-9-oximidazo[1,2-*a*] purine (I) in calf thymus DNA using LC/ESI-MS in the SRM-mode. Thorough quantification and comparison to the 1-N^2-ε-guanine adduct in other *in vitro* systems and *in vivo* remain future goals. In addition careful evaluation whether compound (I) could be due to endogeneous sources or might be an unique biomarker for vinyl chloride exposure needs to be done. The LC/ESI-MS method has high selectivity, sensitivity and analytical throughput and is considered to offer a powerful tool to address these questions.

ACKNOWLEDGEMENTS

This work was supported by United States Public Health Service grants CA4453 and ES00267.

REFERENCES

Basu, A.K., Wood, M.L., Niedernhofer, L.J., Ramos, L.A., and Essigmann, J.M. (1993) Mutagenic and genotoxic effects of three vinyl chloride-induced DNA lesions: 1,N^6-ethenoadenine, 3,N^4-ethenocytosine and 4-amino-5-(imidazol-2-yl)imidazole *Biochemistry* 32, 12793 - 12801

Cain, C.K., Mallette, M.F., and Taylor, E.C., Jr. (1946) Pyrimido[4,5-*b*]pyrazines. I. Synthesis of 6,7-symmetrically Substituted Derivatives *J. Am. Chem. Soc.* 68, 1996 - 1999

Fenn, J.B., Mann, M., Meng, C.K., and Wong, S.F. (1990) Electrospray ionization - principles and practice *Mass Spec. Rev.* 9, 37 - 70

Gmeiner, H.W., and Poulter, C.D. (1988) An efficient synthesis of [8-^{13}C] adenine *J. Org. Chem.* 53, 1322 - 1323

Guengerich, F.P., Persmark, M., and Humphreys, W.G. (1993) Formation of 1,N^2- and N^2,3-ethenoguanine from 2-halooxiranes: Isotopic labeling studies and isolation of a hemiaminal derivative of N^2-(2-oxoethyl)guanine *Chem. Res. Toxicol.* 6, 635 - 648

Lazarus, R.A., Sulewski, M.A., and Benkovic, S.J. (1982) Synthesis of [4*a*-^{13}C]-6-methyltetrahydropterin *J. Label. Compds. Radiopharm.* 19, 1189 - 1195

Little, D.P., Chorush, R.A., Speir, J.P., Senko, M.W., Kelleher, N.L., and McLafferty, F.W. (1994) Rapid sequencing of oligonucleotides by high-resolution mass spectrometry *J. Am. Chem. Soc.* 116, 4893 - 4897

Misra, R.R., Chiang, S.-Y., and Swenberg, J.A. (1994) A comparison of two ultrasensitive methods for measuring 1,N^6-etheno-2'-deoxyadenosine and 3,N^4-etheno-2'-deoxycytidine in cellular DNA *Carcinogenesis* 15, 1647 - 1652

Potier, N., van Dorsselaer, A., Cordier, Y., Roch, O., and Bischoff, R. (1994) Negative electrospray ionization mass spectrometry of synthetic and chemically modified oligonucleotides *Nucleic Acids Res.* 22, 3895 - 3903

Reddy, D.M., Rieger, R.A., Torres, M.C.. and Iden, C.R. (1994) Analysis of synthetic oligodeoxynucleotides containing modified components by electrospray ionization mass spectrometry *Analyt. Biochemistry* 220, 200 - 207

Sharma, M., Alderfer, J.L., and Box, H.C. (1983) Synthesis of morpholinium [^{13}C] formate and its application in the synthesis of [8-^{13}C] purine base *J. Label. Compds. Radiopharm.* 20, 1219 - 1225

6

REACTIONS OF REACTIVE METABOLITES WITH HEMOPROTEINS—TOXICOLOGICAL IMPLICATIONS

Covalent Alteration of Hemoproteins

Yoichi Osawa,*[1] Kashime Nakatsuka,[1] Mark S. Williams,[2]
James T. Kindt,[1] and Mikiya Nakatsuka[1]

[1] Molecular and Cellular Toxicology Section
Laboratory of Molecular Immunology, NHLBI
[2] Experimental Immunology Branch, NCI
National Institutes of Health, Bethesda, Maryland 20892-1760

INTRODUCTION

It was suggested in 1971 based on studies with the hepatotoxic agent, bromobenzene, that metabolism of chemicals to reactive intermediates could play a central role in toxicological processes involved with xenobiotics (Brodie et al., 1971). Over the subsequent 20 years, studies on a variety of chemicals have supported such a notion due to the correlation of the level of protein bound metabolites with toxicity (Hinson & Roberts, 1992). Only in a few cases, however, have the proteins involved or the nature of the covalent modifications been determined and in no case has a direct toxicological role for the altered proteins been shown. One of the best characterized cellular targets of reactive intermediates are the liver microsomal P450 cytochromes (Ortiz de Montellano & Correia, 1983; Halpert & Stevens, 1991), which in many instances are the enzymes responsible for the generation of the reactive metabolites. It is known that at least three pathways exist for the covalent alteration of the cytochrome P450, one that involves alteration of the heme, a second that involves alteration of the protein, and a third that involves the crosslinking of heme to the protein (Scheme I) (Osawa & Pohl, 1989). Although the alterations of the heme or the protein have been well documented (Ortiz de Montellano, 1990; Halpert, 1982; Halpert, Miller, & Gorsky, 1985; Roberts et al., 1994; Roberts, Hopkins, Alworth, & Hollenberg, 1993; Bryant, Skipper, Tannenbaum, & Maclure, 1987; Gorelick, Hutchins, Tannenbaum, & Wogan, 1989), the protein bound heme adducts have not been structurally defined, due to the complexity of the cytochrome P450 system and the potential reactions that can occur after initial heme

* To whom correspondence should be addressed

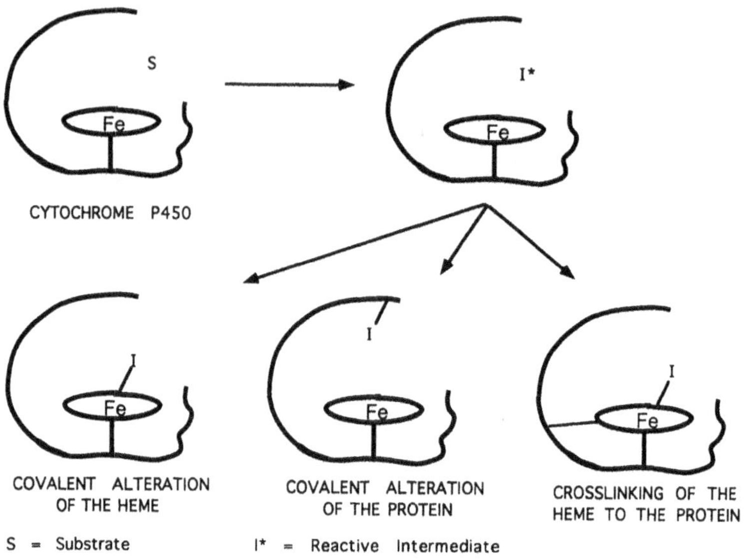

Scheme 1. Pathways for alteration of cytochrome P450.

alteration. At present, only an active site peptide bound to a one-carbon fragment of the initial heme prosthetic group has been identified in a reaction of cumene hydroperoxide with P450 2B1 (Yao, Falick, Patel, & Correia, 1993). On the other hand, studies on myoglobin or hemoglobin and their interaction with chemically simple molecules such as H_2O_2 (Catalano, Choe, & Ortiz de Montellano, 1989) or $CBrCl_3$ (Osawa, Highet, Bax, & Pohl, 1991; Kindt, Woods, Martin, Cotter, & Osawa, 1992; Osawa, Darbyshire, Steinbach, & Brooks, 1993) have proved to be more fruitful in elucidating the mechanism of formation of the protein bound heme adducts. These studies on myoglobin and hemoglobin as well as more recent ones on nitric oxide synthase will be described in this report, along with a discussion of the implications of these reactions in pathological conditions such as myocardial ischemia reperfusion injury and red cell hemolysis.

EXPERIMENTAL PROCEDURES

Treatment of human red cells with $CBrCl_3$. Samples of human blood were collected in vacucontainers containing EDTA and subsequently diluted 1:25 (v/v) with phosphate buffered saline (PBS). Four mL of this cell suspension was placed in a glass 12 mL test tube and bubbled with argon for 10 minutes. The tube was capped with a rubber stopper and the head space was flushed with argon. After addition of 2.0 μL of $CBrCl_3$, the tube was placed in a shaking incubator set at 37°C for 20 min. The cells were subsequently washed with PBS and resuspended into 2.5 mL of PBS and placed in an incubator at 37°C. An aliquot (20 μL) was taken at various times and analyzed by HPLC for altered heme products as previously described (Kindt, Woods, Martin, Cotter, & Osawa, 1992). An aliquot (100 μL) was also taken and spun at 800 x g for 5 min and a portion (20 μL) of the supernatant was analyzed by HPLC to determine the extent of hemolysis (Kindt, Woods, Martin, Cotter, & Osawa, 1992). An aliquot (200 μL) was also taken and diluted 2-fold with 0.4% methyl violet in PBS and analyzed for inclusion bodies.

RESULTS AND DISCUSSION

Myoglobin

Myoglobin catalyzes the reductive debromination of $CBrCl_3$ to trichloromethyl radical in a reaction analogous to that for cytochrome P450 catalyzed reductive dehalogenation reactions (Scheme II). By limiting the amount of reducing equivalents to stoichiometric levels, one equivalent of the reactive intermediate, the trichloromethyl radical, was formed and subsequently reacted with the heme of myoglobin. Altered heme products were isolated by the use of HPLC and characterized by the use of mass spectrometry, peptide mapping, and 1- and 2-D proton- and ^{13}C- NMR. Three dissociable products, an acrylic acid **3**, a trichloromethyl alcohol **4**, and bis-trichloromethyl **2** adducts of the ring I vinyl group (Scheme III), were formed in approximately equal amounts and together accounted for 40% of the heme that was altered (Osawa, Fellows & Highet, 1992). A protein bound product **1**, which accounted for 60% of the heme that was altered (Osawa, Fellows & Highet, 1992),

$$\text{Myoglobin } (Fe^{+2}) + CBrCl_3 \longrightarrow \bullet CCl_3 + Br^-$$

$$\text{P450 } (Fe^{+2}) + CCl_4 \longrightarrow \bullet CCl_3 + Cl^-$$

Scheme 2. Reductive dehalogenation reactions catalyzed by myoglobin and cytochrome P450.

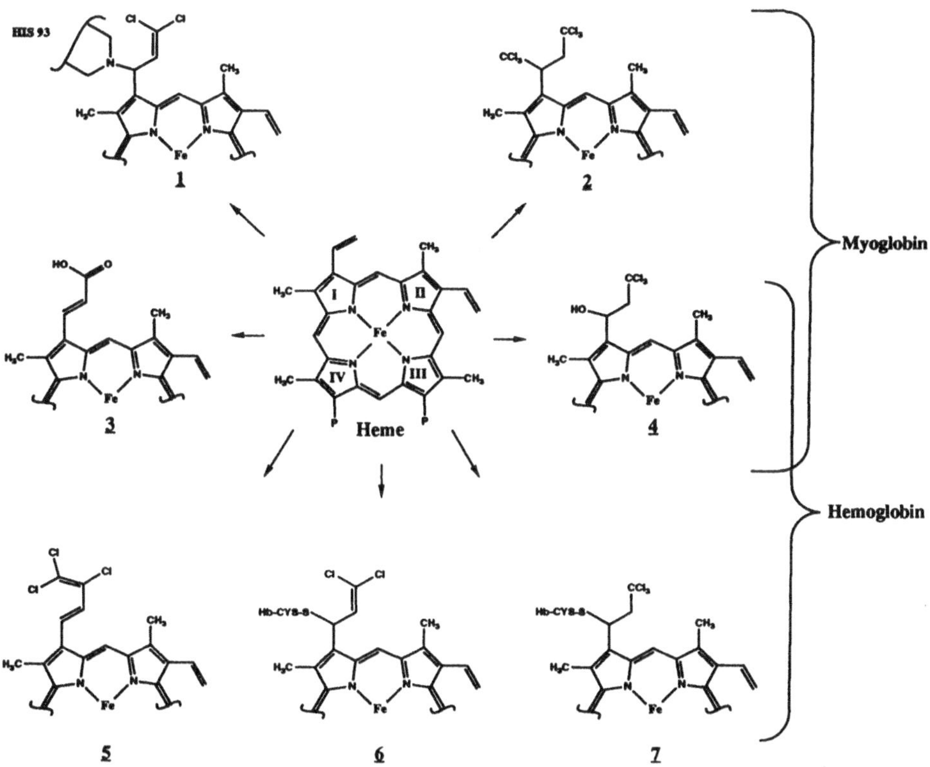

Scheme 3. Structures of the major altered heme products formed from the reaction of $CBrCl_3$ with myoglobin and hemoglobin.

was due to the crosslinking of histidine residue 93, the proximal residue of the heme iron, to the ring I vinyl group, which was also altered by the addition of a carbon and two chlorines from $CBrCl_3$ (Osawa, Highet, Bax, & Pohl, 1991). The formation of the protein bound heme adduct caused a translocation of the heme in the active site and loss of the normal proximal ligation to histidine 93 and religation to other histidine residues (Osawa, Highet, Bax, & Pohl, 1991; Osawa, Darbyshire, Steinbach, & Brooks, 1993). This "histidine shuffle" led to alteration of the three dimensional structure of the protein as evidenced by enhanced proteolysis of the altered protein relative to that of the native protein (Osawa & Pohl, 1989). This change also led to transformation of myoglobin from an oxygen storage protein to an oxidase, capable of generating H_2O_2 in the presence of reducing equivalents from ascorbate, glutathione, or a metmyoglobin reducing system (Osawa, Darbyshire, Steinbach, & Brooks, 1993). Molecular modeling studies indicated that the access of water to the active site may have been responsible for the oxidase activity (Osawa, Darbyshire, Steinbach, & Brooks, 1993). Although the protein bound heme adduct was formed in large amounts after a single catalytic cycle, multiple catalytic cycles led to the degradation of the protein bound adduct (Osawa, Darbyshire, Steinbach, & Brooks, 1993). If similar reactions occur with P450 cytochromes, these results could explain in part the difficulties encountered with identification and characterization of protein bound heme adducts from the cytochrome P450 enzymes. Furthermore, the redox capability of the altered protein may be involved in the hepatotoxicity caused by some liver toxins such as CCl_4 or $CBrCl_3$.

H_2O_2 treatment of ferric myoglobin leads to the formation of an Fe^{IV} complex and a radical species that is found localized on the protein (Tew & Ortiz de Montellano, 1988; Catalano, Choe, & Ortiz de Montellano, 1989). Although the nature of the reactive intermediate generated after H_2O_2 treatment is different from that of the $CBrCl_3$ treatment, both reactions lead to the crosslinking of the heme to protein and transformation to an oxidase (Osawa & Korzekwa, 1991). In the case of H_2O_2, tyrosine residue 103 is involved in the crosslinking reaction to heme (Catalano, Choe, & Ortiz de Montellano, 1989). In analogy to that found for the protein bound adduct formed from the $CBrCl_3$ reaction, the protein bound heme adduct formed from the H_2O_2 was eventually destroyed after redox cycling or addition of excess H_2O_2 (Osawa & Korzekwa, 1991).

The reaction of myoglobin with H_2O_2 may be involved in pathological conditions such as myocardial ischemia reperfusion injury where H_2O_2 and other oxidants are known to be produced in excess and myoglobin is found in large amounts. Isolated heart preparations, which have undergone ischemia reperfusion injury, have been shown to contain myoglobin in the Fe^{IV} state (Arduini, Eddy, & Hochstein, 1990). This higher oxidation state of myoglobin has been shown to be reduced back innocuously by cellular reductants to the ferric state (Galaris, Cadenas, & Hochstein, 1989; Arduini et al., 1992). However, during excess oxidative stress a protein bound heme adduct of myoglobin may be formed, which may contribute to ischemia reperfusion injury in myocytes by acting as an oxidase. In this regard it has recently been found that introduction of the oxidatively altered myoglobin into human fibroblasts by osmotic lysis of pinosomes caused a loss of cell viability (Osawa, Y. and Williams, M.S. manuscript in preparation).

Hemoglobin

Hemoglobin, a tetrameric complex of two α and two β chains with each of the monomers having a high degree of similarity to myoglobin, also metabolizes $CBrCl_3$ to the trichloromethyl radical and leads to the regiospecific alteration of the heme prosthetic group, similar to that found for myoglobin (Scheme III compounds **3-7**) (Kindt, Woods, Martin, Cotter, & Osawa, 1992; Osawa et al., 1994; Osawa, Fellows & Highet, 1992). Two of the three dissociable products, an acrylic acid **3** and trichloromethyl alcohol **4** adducts are

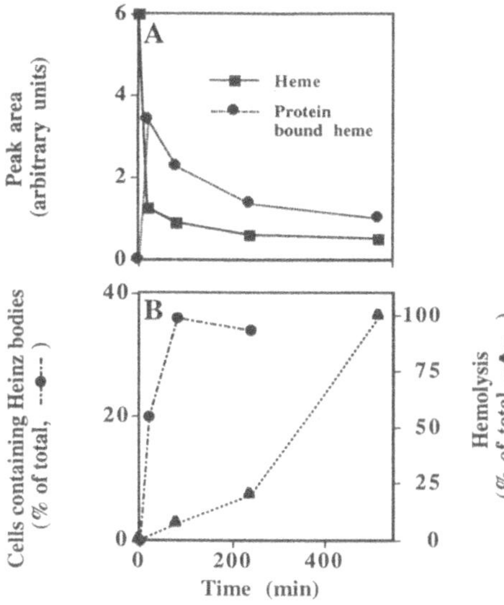

Figure 1. Effect of $CBrCl_3$ on human red cells in vitro. Panel A: Amount of heme and protein bound heme; Panel B: Extent of formation of Heinz bodies and hemolysis. The cells were treated and subsequently analyzed as described under "Experimental Procedures".

identical to that formed from the myoglobin reaction, whereas the third dissociable product, compound **5,** is a novel adduct that is formed by attack of two trichloromethyl radicals on the heme moiety. The bistrichloromethyl adduct **2**, which is formed in the myoglobin reaction is not formed in the hemoglobin reaction. Two major protein bound products **6** and **7** involving a crosslink between cysteine residue 93 of the β chain and the heme moiety have also been characterized (Kindt, Woods, Martin, Cotter, & Osawa, 1992). Cysteine residue 93, which is not present in myoglobin, has previously been shown to be a target of electrophilic reactive intermediates of carcinogens (Haugen, 1989), and appears to react more avidly with the heme cationic intermediate than the proximal histidine residue. The protein bound adduct **6** is highly similar to that found for myoglobin (adduct **1**), whereas adduct **7** has an additional chlorine substituent. The profile of products formed indicate the highly similar yet non-identical nature of the active sites of myoglobin and hemoglobin.

In analogy to that found for myoglobin multiple redox events caused the degradation of the protein bound adducts of hemoglobin, which may be important in the lysis of red blood cells (Kindt, Woods, Martin, Cotter, & Osawa, 1992). Red blood cells are known to form hemichromes that in turn apparently catalyze the formation of Heinz inclusion bodies, and eventually result in hemolysis (Winterbourn, 1990). Recently, it was noted that the absorption spectra of the partially purified protein bound heme adducts of hemoglobin were highly similar to that reported for hemichromes (Kindt, Woods, Martin, Cotter, & Osawa, 1992). To investigate this further, human red cells were treated with $CBrCl_3$ and subsequently monitored for heme alteration, protein bound heme formation, Heinz body formation, and hemolysis (Figure 1). The formation of the protein bound heme adducts were early events followed by Heinz body formation and subsequent hemolysis. These findings suggest that the alteration of the hemoglobin, in part to protein bound products, may be involved in the formation of hemichromes and Heinz bodies. Similar reactions may explain how abnormal hemoglobins with a propensity to autooxidize lead to Heinz body formation and hemolysis (Rachmilewitz, 1974; Winterbourn, 1990).

The oxidative reactions of hemoglobins may also be of pharmacological importance as chemically modified human hemoglobins are currently being developed as red cell substitutes for use in trauma patients (Winslow, 1992). This is of special concern as these modified hemoglobins would lack the protection of the cellular antioxidant defenses that

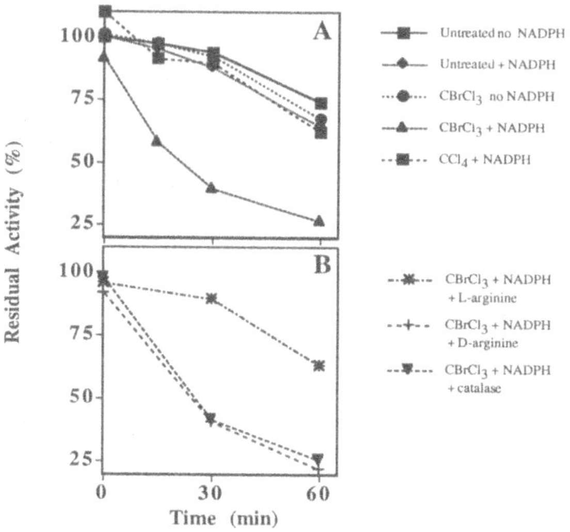

Figure 2. Effect of $CBrCl_3$ on purified recombinant rat neuronal nitric oxide synthase. A cDNA, which contained the full reading frame for rat neuronal NOS was kindly provided by Drs. C. Lowenstein and S. Snyder (Johns Hopkins University). Rat neuronal NOS was recombinantly expressed with the use of a baculovirus system and subsequently purified (M. Nakastuka D. R. Rotrosen, C. L. Yeung, and Y. Osawa, manuscript in preparation) and stored at -80°C at a concentration of 0.13 mg/mL in 50 mM Tris-HCl, pH 7.4, containing 10 % (v/v) glycerol, 0.1 M NaCl, 0.1 mM EDTA, 10 μM tetrahydrobiopterin (from B. Schirks Laboratories, Jona, Switzerland), 0.1 mM dithiothreitol, 0.1 mM PMSF, 0.5 μM leupeptin, 0.5μM pepstatin, 10 μM FMN, 10 μM FAD, 3.6 μM hemin chloride, and 5 μM L-arginine. The reaction mixture contained recombinant NOS (13 μg of protein/mL), 2 mM $CaCl_2$, 10 μg/mL calmodulin, 100 μM NADPH, 10 units/mL superoxide dismutase, and where indicated $CBrCl_3$ (1 mM) or CCl_4 (1 mM) in a total volume of 100 μL of 40 mM potassium phosphate, pH 7.4. The reaction mixtures were incubated on ice and aliquots (20 μL) were taken at indicated times and added to an assay mixture containing 2 mM $CaCl_2$, 10 μM tetrahydrobiopterin, 10 μg/ml calmodulin, 1 mM NADPH, 10 units/ml superoxide dismutase, 100 units/ml catalase, 100 μM oxyhemoglobin, 30 μM L-[^{14}C(U)]-arginine (NEN, 220.8 mCi/mmol) in a total volume of 100 μL of 40 mM potassium phosphate, pH 7.4. The assay mixture was incubated for 3 min at room temperature at which time 120 μl of cold 0.1 M citric acid, pH 2.5, was added. The amount of radiolabelled citrulline was quantified by an HPLC method as previously described (Osawa, Davila, Meyer & Nakatsuka, 1994). The specific activity of the isolated NOS was 402 nmol/min/mg of protein. Panel A: As indicated some of the mixtures were without NADPH. Panel B: Where indicated arginine was added at a final concentration of 50 μM and catalase was at 100 units/ml.

presumably limit oxidative damage to hemoglobin within native red blood cells. One of these potential blood substitutes, a bis(3,5-dibromosalicyl)fumarate modified human hemoglobin was found to undergo oxidative modification in part to form protein bound products, whereas another agent, mono-(3,5-dibromosalicyl)fumarate modified human hemoglobin was found to be highly resistant to oxidative damage (Osawa, Darbyshire, Meyer, & Alayash, 1993). In light of the redox activity of the altered protein and the degradation of the altered heme and subsequent release of iron, these reactions appear to be important factors for consideration in evaluating the safety of these agents, especially due to the large amounts of compound to be used and the compromised status of the patients.

Nitric Oxide Synthase

Nitric oxide synthase (NOS), which catalyzes the metabolism of arginine to citrulline and nitric oxide was recently shown to be a cytochrome P450 like hemoprotein (White &

Marletta, 1992; Stuehr & Ikeda-Saito, 1992; McMillan et al., 1992). Xenobiotics such as phencyclidine and $CBrCl_3$, which are known to suicide inactivate liver microsomal P450 cytochromes, have been shown to suicide inactivate NOS activity in rat brain cytosolic preparations (Osawa, Davila, Meyer & Nakatsuka, 1994; Osawa & Davila, 1993). With the use of a reconstituted system containing purified recombinant NOS, we showed that $CBrCl_3$ inactivates neuronal NOS in an NADPH-dependent manner, whereas in the same system CCl_4 had no effect on enzyme activity presumably due to the inabiltiy of NOS to metabolize CCl_4 (Figure 2 Panel A). L-arginine, but not D-arginine, could protect from the inactivation due to $CBrCl_3$ indicating an active site involvement in the inactivation (Figure 2 Panel B). Catalase had no affect.on the inactivation due to $CBrCl_3$, indicating that accumulated H_2O_2 was not a likely mechanism for inactivation (Figure 2 Panel B). The recent finding that N^G-methyl-L-arginine, a suicide inactivator of macrophage NOS, acts in part to alter the heme prosthetic group of the enzyme (Olken, Osawa, & Marletta, 1994) indicates that the heme is an important target in the modulation of NOS by xenobiotics, in analogy to other hemoproteins. Studies with the use of the reconstituted NOS systems should greatly aid in identifying compounds that can inhibit this family of enzymes as well as elucidate the mechanism of this inhibition. Furthermore, these studies should help in evaluating the potential role of NOS in chemically induced toxicities.

CONCLUSION

The covalent alteration of hemoproteins, such as cytochrome P450, myoglobin, hemoglobin, and nitric oxide synthase, by biological reactive intermediates may play a role in disruption of a variety of physiological processes regulated by these proteins and lead to pathological conditions. Knowledge of the structural basis of these processes should ultimately lead to a better understanding of the mechanisms of action of these hemoproteins and to the design of safer drugs and environmental chemicals.

ACKNOWLEDGMENTS

We are grateful to Dr. Lance R. Pohl for critically reviewing this manuscript.

REFERENCES

Arduini, A., Mancinelli, G., Radatti, G.L., Damonti, W., Hochstein, P., & Cadenas, E. (1992). Reduction of sperm whale ferrylmyoglobin by endogenous reducing agents: Potential reducible loci of ferrylmyoglobin. *Free Radical Biology and Medicine*, **13**, 449-454.

Arduini, A., Eddy, L., & Hochstein, P. (1990). Detection of ferryl myoglobin in the isolated ischemic rat heart. *Free Radical Biology and Medicine*, **9**, 511-513.

Brodie, B.B., Reid, W.D., Cho, A.K., Sipes, I.G., Krishna, G., & Gillette, J.R. (1971). Possible mechanism of liver necrosis caused by aromatic organic compounds. *Proc Natl Acad Sci U S A*, **68**, 160-164.

Bryant, M.S., Skipper, P.L., Tannenbaum, S.R., & Maclure, M. (1987). Hemoglobin adducts of 4-aminobiphenyl in smokers and nonsmokers. *Cancer Research*, **47**, 602-608.

Catalano, C.E., Choe, Y.S., & Ortiz de Montellano, P.R. (1989). Reactions of the protein radical in peroxide-treated myoglobin: Formation of a heme-protein cross-link. *Journal of Biological Chemistry*, **264**, 10534-10541.

Galaris, D., Cadenas, E., & Hochstein, P. (1989). Redox cycling of myoglobin and ascorbate: A potential protective mechanism against oxidative reperfusion injury in muscle. *Arch Biochem Biophys*, **273**, 497-504.

Gorelick, N.J., Hutchins, D.A., Tannenbaum, S.R., & Wogan, G.N. (1989). Formation of DNA and hemoglobin adducts of fluoranthene after single and multiple exposures. *Carcinogenesis*, **10**, 1579-1587.

Halpert, J. (1982). Further studies of the suicide inactivation of purified rat liver cytochrome p-450 by chloramphenicol. *Mol Pharmacol*, **21**, 166-172.

Halpert, J., Miller, N.E., & Gorsky, L.D. (1985). On the mechanism of the inactivation of the major phenobarbital-inducible isozyme of rat liver cytochrome P-450 by chloramphenicol. *Journal of Biological Chemistry*, **260**, 8397-8403.

Halpert, J.R. & Stevens, J.C. (1991). Cytochrome P-450 as a target of biological reactive intermediates. *Adv Exp Med Biol*, **283**, 105-109.

Haugen, D.A. (1989). Charge-shift strategy for isolation of hemoglobin-carcinogen adducts formed at the β93 cysteine sulfhydryl groups. *Chemical Research in Toxicology*, **2**, 379-385.

Hinson, J.A. & Roberts, D.W. (1992). Role of covalent and noncovalent interactions in cell toxicity: effects on proteins. *Annu Rev Pharmacol Toxicol*, **32**, 471-510.

Kindt, J.T., Woods, A., Martin, B.M., Cotter, R.J., & Osawa, Y. (1992). Covalent alteration of the prosthetic heme of human hemoglobin by $BrCCl_3$: Cross-linking of heme to cysteine residue 93. *Journal of Biological Chemistry*, **267**, 8739-8743.

McMillan, K., Bredt, D.S., Hirsch, D.J., Snyder, S.H., Clark, J.E., & Masters, B.S.S. (1992). Cloned, expressed rat cerebellar nitric oxide synthase contains stoichiometric amounts of heme, which binds carbon monoxide. *Proc Natl Acad Sci U S A*, **89**, 11141-11145.

Olken, N.M., Osawa, Y., & Marletta, M.A. (1994). Characterization of the inactivation of nitric oxide synthase by N^G-methyl-L-arginine: evidence for protein modification and heme loss. *Biochemistry*, In press.

Ortiz de Montellano, P.R. (1990). Free radical modification of prosthetic heme groups. *Pharmacol.Ther.*, **48**, 95-120.

Ortiz de Montellano, P.R. & Correia, M.A. (1983). Suicidal destruction of cytochrome P-450 during oxidative drug metabolism. *Annu Rev Pharmacol Toxicol*, **23**, 481-503.

Osawa, Y., Fellows, C., Meyer, C.M., Woods, A., Castoro, J.A., Cotter, R.J., Wilkins, C., & Highet, R.J. (1994). Structure of the novel heme adduct formed during the reaction of human hemoglobin with $BrCCl_3$ in red cell lysates. *Journal of Biological Chemistry*, **269**, 15481-15487.

Osawa, Y., Darbyshire, J.F., Meyer, C.M., & Alayash, A.I. (1993). Differential susceptibilities of the prosthetic heme of hemoglobin-based red cell substitutes. *Biochem Pharmacol*, **12**, 2299-2305.

Osawa, Y., Darbyshire, J.F., Steinbach, P.J., & Brooks, B.R. (1993). Metabolism-based transformation of myoglobin to an oxidase by $BrCCl_3$ and molecular modeling of the oxidase form. *Journal of Biological Chemistry*, **268**, 2953-2959.

Osawa, Y. & Davila, J.C. (1993). Phencyclidine, a psychotomimetic agent and drug of abuse, is a suicide inhibitor of brain nitric oxide synthase. *Biochem Biophys Res Commun*, **194**, 1435-1439.

Osawa, Y., Davila, J.C., Meyer, C.M. & Nakatsuka, M. (1994). Mechanism based inactivation of nitric oxide synthase, a P450-like enzyme, by xenobiotics. In M.C. Lechner (Ed.). *Cytochrome P-450: biochemistry, biophysics, and molecular biology* pp. 459-462. John Libbey Eurotext, Paris.

Osawa, Y., Fellows, C. & Highet, R.J. (1992). Use of stable and radioactive isotopes to identify reactive metabolites and target macromolecules associated with toxicities of halogenated hydrocarbons. In E. Buncel & G.W. Kabalka (Eds.). *Synthesis and application of isotopically labelled compounds 1991, proceedings of the fourth international symposium, Toronto, Canada* pp. 415-420. Elsevier Science, Amsterdam.

Osawa, Y., Highet, R.J., Bax, A., & Pohl, L.R. (1991). Characterization by NMR of the heme-myoglobin adduct formed during the reductive metabolism of $BrCCl_3$: Covalent bonding of the proximal histidine to the ring I vinyl group. *Journal of Biological Chemistry*, **266**, 3208-3214.

Osawa, Y. & Korzekwa, K. (1991). Oxidative modification by low levels of HOOH can transform myoglobin into an oxidase. *Proc Natl Acad Sci U S A*, **88**, 7081-7085.

Osawa, Y. & Pohl, L.R. (1989). Covalent bonding of the prosthetic heme to protein: A potential mechanism for the suicide inactivation or activation of hemoproteins. *Chemical Research in Toxicology*, **2**, 131-141.

Rachmilewitz, E.A. (1974). Denaturation of the normal and abnormal hemoglobin molecule. *Seminars in Hematology*, **11**, 441-462.

Roberts, E.S., Hopkins, N.E., Zaluzec, E.J., Gage, D.A., Alworth, W.L., & Hollenberg, P.F. (1994). Identification of active-site peptides from 3H-labeled 2-ethynylnaphthalene-inactivated P450 2B1 and 2B4 using amino acid sequencing and mass spectrometry. *Biochemistry*, **33**, 3766-3771.

Roberts, E.S., Hopkins, N.E., Alworth, W.L., & Hollenberg, P.F. (1993). Mechanism-based inactivation of cytochrome P450 2B1 by 2-ethynylnaphthalene: Identification of an active-site peptide. *Chemical Research in Toxicology*, **6**, 470-479.

Stuehr, D.J. & Ikeda-Saito, M. (1992). Spectral characterization of brain and macrophage nitric oxide synthases. *Journal of Biological Chemistry*, **267**, 20547-20550.

Tew, D. & Ortiz de Montellano, P.R. (1988). The myoglobin protein radical: Coupling of Tyr-103 to Tyr-151 in the H_2O_2-mediated cross-linking of sperm whale myoglobin. *Journal of Biological Chemistry*, **263**, 17880-17886.

White, K.A. & Marletta, M.A. (1992). Nitric oxide synthase is a cytochrome P-450 type hemoprotein. *Biochemistry*, **31**, 6627-6631.

Winslow, R.M. (1992). *Hemoglobin-based red cell substitutes*. The Johns Hopkins University Press, Baltimore.

Winterbourn, C.C. (1990). Oxidative denaturation in congenital hemolytic anemias: The unstable hemoglobins. *Seminars in Hematology*, **27**, 41-50.

Yao, K., Falick, A.M., Patel, N., & Correia, M.A. (1993). Cumene hydroperoxide-mediated inactivation of cytochrome P450 2B1. Identification of an active site heme-modified peptide. *Journal of Biological Chemistry*, **268**, 59-65.

IMMUNOCHEMICAL DETECTION OF DRUG-PROTEIN ADDUCTS IN ACETAMINOPHEN HEPATOTOXICITY*

Jack A. Hinson,[†1] Dean W. Roberts,[1,2] N. Christine Halmes,[1] Jennifer D. Gibson,[1] and Neil R. Pumford[1]

[1] Division of Toxicology
University of Arkansas for Medical Sciences
Little Rock, Arkansas 72205
[2] Division of Biochemical Toxicology
National Center for Toxicological Research
Jefferson, Arkansas 72079

INTRODUCTION

In overdose the over-the-counter analgesic acetaminophen may produce a centrilobular hepatic necrosis (1,2). In addition this toxicity may be accompanied by nephrotoxicity (1,3). The mechanism of hepatotoxicity has been studied extensively in experimental animals. It was shown that inhibition of the drug metabolizing enzymes resulted in a diminution of the toxicity, whereas induction of the drug metabolizing enzymes resulted in an increase in the toxicity (4). These data indicated that metabolism of acetaminophen was critical in the development of the toxicity. Moreover, administration of radiolabeled acetaminophen resulted in covalent binding of radiolabel to protein in the necrotic hepatocytes (5). *In vitro* experiments revealed that the metabolism was by a cytochrome P-450 dependent mechanism (6). It was postulated that covalent binding of this metabolite to critical proteins was the mechanism of the hepatotoxicity.

In addition, it was shown hepatic glutathione was a detoxification mechanism *in vivo*. Administration of a toxic dose of radiolabeled acetaminophen resulted in depletion of hepatic glutathione before covalent binding occurred, whereas administration of a nontoxic dose of acetaminophen did not deplete hepatic glutathione and covalent binding did not occur (7). Thus, therapeutic doses of acetaminophen are safe because the reactive metabolite is

---

* Supported by NIH Grant 1R01 GM48749

† To whom reprints should be addressed at Division of Toxicology, Slot 638, University of Arkansas for Medical Sciences, 4301 W. Markham Street, Little Rock, AR 72205

Biological Reactive Intermediates V
Edited by R. Snyder et al., Plenum Press, New York, 1996

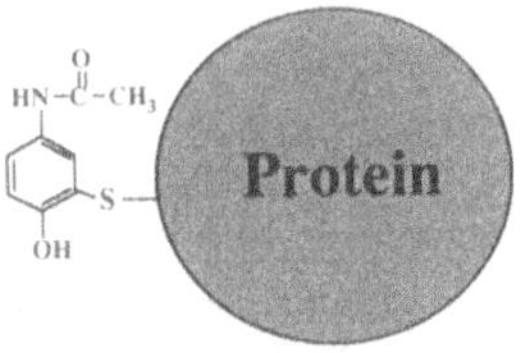

Figure 1. Acetaminophen-Protein Adduct.

efficiently detoxified by glutathione. These findings have led to the utilization of N-acetylcysteine, a compound which repletes glutathione, as an effective antidote (8).

Subsequent to these initial findings it has been determined that the reactive metabolite is N-acetyl-p-benzoquinone imine. The compound has been synthesized and shown to react with glutathione to form the same product that is formed *in vivo* (9-11). Additionally, the reaction of glutathione with the metabolite has been shown to be catalyzed by a glutathione transferase (12). Also, the metabolite has been shown to be formed by cytochrome P-450s IA2, 3A4, and 2E1 (13-15), and covalently binds to protein as the 3-(cystein-S-yl)acetaminophen protein adduct (Figure 1) (16).

Even though early work suggested that covalent binding to critical proteins and alteration of critical functions resulted in death of the cell, the mechanism has not been delineated; however, the hypothesis has generated much research. Various functions and enzymatic activities have been investigated to determine which alterations occurred before toxicity and which alterations were a result of the toxicity. These investigations include the effects of acetaminophen hepatotoxicity on energy production by mitochondria (17), on calcium metabolism (18), and on plasma membrane enzyme activities (19). Thus far this approach has not defined the critical target(s) of the reactive metabolite.

In the late 1980's our laboratory (20-22) as well as the laboratory of Drs. Cohen and Khairallah at the University of Connecticut (23-25) initiated a new approach to studying the importance of covalent binding in the hepatotoxicity of acetaminophen. This approach utilized immunochemical methods to define the critical targets in the toxicity. We raised a polyclonal antibody against the mercapturic acid of acetaminophen (3-(N-acetylcystein-S-yl)acetaminophen) and developed immunochemical assays which were specific for acetaminophen covalently bound to cysteine groups on protein. Cohen and coworkers developed similar assays using an antibody which recognized acetaminophen protein adducts.

In this manuscript we describe the utilization of this polyclonal antibody to study covalent binding of acetaminophen to cysteine groups on protein and hepatotoxicity. Immunohistochemical staining of sections of livers from acetaminophen treated mice have allowed us to study covalent binding of acetaminophen in individual hepatocytes as well as development of the hepatotoxicity (26). Also, utilizing Western immunoblot analyses to determine which proteins contain acetaminophen adducts (27), the cytosolic 56 kDa selenium binding protein has been isolated and identified as a major protein to which acetaminophen binds (28).

MATERIALS AND METHODS

Immunoassays

To develop a polyclonal antibody which was specific for acetaminophen covalently bound to cysteine groups on protein, the immunogen 3-(N-acetylcystein-S-yl)acetaminophen-Keyhole Limpet hemocyanin was synthesized. Initially, N-acetyl-p-benzoquinone imine was synthesized by oxidation of acetaminophen with silver oxide and then reacted to N-acetylcysteine. The resulting acetaminophen-N-acetylcysteine conjugate (acetaminophen-mercapturate) was isolated and covalently attached to the antigenic protein Keyhole Limpet hemocyanin (KLH) using a soluble carbodiimide coupling agent. Rabbits were subsequently immunized with the N-acetylcysteine-acetaminophen-KLH conjugate. From these rabbits one antiserum was obtained which was highly specific for acetaminophen-cysteine adducts (20).

The polyclonal antiserum was used to study the formation of acetaminophen-cysteine adducts in acetaminophen hepatotoxicity. Three immunochemical procedures were utilized to study acetaminophen covalently bound to protein. These procedures were the competitive ELISA (21,22), Western immunoblot (27), and immunohistochemical analyses (26). Western immunoblot assays utilized electrophoretic separation of protein with SDS PAGE as initially described by Laemmli followed by electrophoretic transfer to polyvinylidene difluoride membranes. Immunohistochemical analyses were performed using a microwave fixation technique and localization of acetaminophen protein adducts was by modification of the unlabeled antibody enzyme method described by Sternberger (26). The staining procedure utilized the anti-acetaminophen-N-acetylcysteine antibody. Liver sections were also stained with Mayer's hematoxylin for conventional histological analyses.

Effect of Selenium on Acetaminophen Toxicity

Toxicity studies were performed using male B6C3F1 mice. Mice were fasted overnight (5:00 PM) and the experiment begun at 9:00 AM. Na_2SeO_3 was administered to some mice 24 hours prior to acetaminophen administration. Acetaminophen was administered to mice in saline whereas control animals received only saline. Animals were anesthetized under a carbon dioxide atmosphere and blood removed from the retro-orbital sinus. Subsequently, the animals were killed by cervical dislocation and the livers surgically removed. Serum was separated by centrifugation and analyzed for ALT using a kit from Sigma Chemical Co.

Protein Isolation and Analyses

Proteins were isolated using the following columns: DEAE-Sepharose fast flow anion exchange column (20 X 1), hydroxyapatite HPLC column (7.6 X 100 mm, Rainin), and TSK DEAE-5-PW column (75 X 7.5mm, Beckman). Sequence analyses were performed by automated Edman degradation (28).

RESULTS

To determine the relationship between covalent binding of acetaminophen to hepatic protein and the development of hepatotoxicity, mice were treated with toxic doses of the drug. Subsequently mice were sacrificed, the livers were removed, and sections of the livers were stained immunohistochemically for visualization of acetaminophen covalently bound to cysteine groups on protein (26). Acetaminophen-protein adducts were observed in the innermost layer of cells surrounding the central vein as early as 15 minutes following a hepatotoxic dose of acetaminophen (400 mg/kg). By 30 minutes there was a 90% depletion of hepatic glutathione and acetaminophen-protein adducts were evident in the centrilobular region. By one hour following this dose acetaminophen-protein adducts reached their maximum extent and were found exclusively in the centrilobular region of the liver

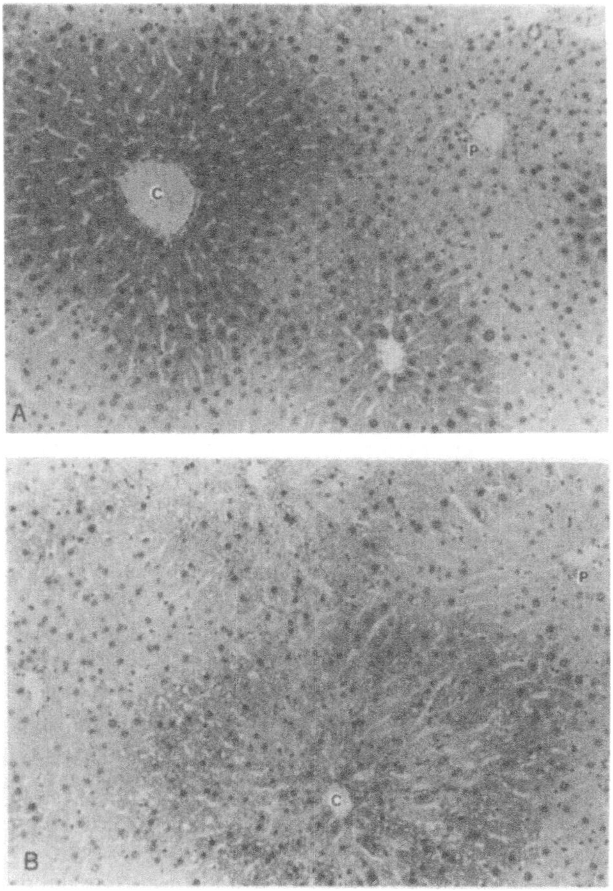

Figure 2. A. Mouse liver sections after toxic doses of Acetaminophen: Acetaminophen (400 mg/kg) at one hour. The immunohistochemical stain (dark areas) reveals adetaminophen-protein adducts in hepatocytes throughout centrilobular and midzonal areas. B.Mouse liver sections after toxic doses of Acetaminophen: Acetaminophen (400 mg/kg) at six hours. The immunohistochemical stain is decreased as protein adducts are lost as vacuoles develop and necrotic cells disintegrate.

(Figure 2a). By two hours vacuolization and shrinking of hepatocytes were prominent (Figure 2b). These events correlated with increases in ALT levels and acetaminophen-protein adducts in serum which occurred as a result of hepatocyte lysis. By six hours after treatment substantial loss of adducts from the necrotic cells was prominent.

In an effort to better understand the molecular basis of the toxicity, livers from acetaminophen-treated mice were analyzed by Western immunoblots for the presence of acetaminophen covalently bound to individual proteins (27). These analyses indicated that there were a large number of proteins to which acetaminophen covalently bound; however, adducts appeared to be in greater abundance in certain protein fractions. The major protein to which acetaminophen appeared to bind was in liver cytosol and had a molecular weight of approximately 55 kDa (Figure 3). Other proteins which contained acetaminophen adducts were in: cytosol (78 kDa, 81 kDa, 87 kDa, 98 kDa), mitochondria (39 kDa, 50 kDa, 66 kDa, 79 kDa), and plasma membrane (72 kDa, 82 kDa, 115 kDa, 118 kDa). Microsomes contained only low levels of adducts.

The major protein to which acetaminophen covalently bound was a cytosolic protein with a molecular weight of approximately 55 kDa (Figure 3) (27). To better understand the importance of this protein in acetaminophen hepatotoxicity, it was isolated from mice treated with hepatotoxic doses of acetaminophen (28). Sequence analysis was performed on 7 internal peptides of the trypsin digest of the major peak described above. A total of 85 amino

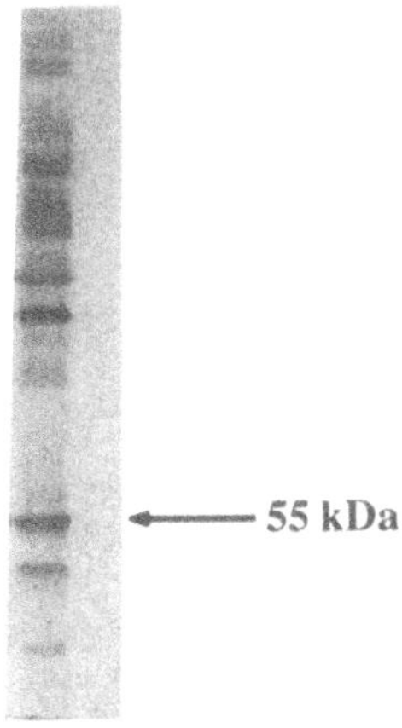

Figure 3. Western immunoblot of liver cytosol from acetaminophen-treated mice. Mice were treated with a toxic dose of acetaminophen (400 mg/kg) and sacrificed at 2 hours (A). Control mice (C) received only saline.

acids were sequenced and these data indicated a 97% sequence homology with the previously deduced sequence of a mouse liver cDNA encoding for a selenium binding protein of 56 kDa (29). This protein has been previously shown to become associated with radiolabeled selenium when radiolabeled sodium selenite was administered to animals (30). The codon for the amino acid selenocysteine is not present, and the mechanism of selenium association is not understood. Also, its function is unknown, but it has been postulated to be a regulatory protein important in the anticarcinogenic properties of selenium.

Studies were performed in an effort to determine if the 56 kDa selenium binding protein was a determinant in the hepatotoxicity. Previous work by Schnell et al. showed that prior administration of sodium selenite to animals decreased the hepatotoxicity of acetaminophen. Even though these investigators showed that selenium administration decreased the amount of drug metabolized to N-acetyl-p-benzoquinone imine, it seemed plausible that prior interaction of selenium with the 56 kDa selenium binding protein might result in a selective decrease in sites available for binding of the reactive metabolite of acetaminophen. If covalent binding of acetaminophen to the 56 kD selenium binding protein was important in the hepatotoxicity, then prior selenium administration may alter the toxicity by this mechanism. To determine if prior selenium administration selectively altered covalent binding to the 56 kDa selenium binding protein, sodium selenite was administered to mice prior to administration of hepatotoxic doses of acetaminophen. Sodium selenite was administered to mice at a dose of 20.7 umol/kg, and after 24 hours the mice were administered marginally hepatotoxic doses of acetaminophen (200 mg/kg and 300 mg/kg). Mice were sacrificed at 2 hours and 4 hours, times before significant lysis of hepatocytes. As shown in Figure 4, prior selenium administration decreased covalent binding of acetaminophen to all hepatic cytosolic proteins at 2 and 4 hours at doses of 200 mg/kg and 300 mg/kg. A decrease was observed in covalent binding to the 56 kDa selenium binding protein; however, the percent decrease did not appear to be significantly different from the per cent decrease observed in other proteins. These data suggest that binding of the reactive metabolite of acetaminophen to the 56 kDa selenium binding protein is not significantly altered by prior selenium administration. Furthermore, the data support the conclusion of Schnell et al. (31) that selenium decreases the proportion of the dose of acetaminophen converted to the reactive metabolite *in vivo* and explains why he observed that selenium decreases the hepatotoxicity of acetaminophen.

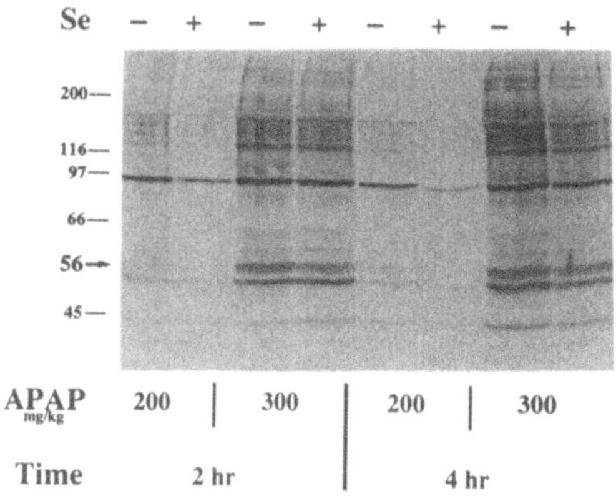

Figure 4. Western immunoblot of hepatic cytosol showing effect of selenium on acetaminophen covalent binding. Selenium (Na_2SeO_3, 20.7 umol/kg) was administered to mice 24 hours prior to actetaminophen administration. Acetaminophen covalent binding in the hepatic cytosol was determined at 2 and 4 hours following acetaminophen administration (200 mg/kg and 300 mg/kg).

DISCUSSION

In this manuscript we summarize some of our data utilizing immunochemical methods to understand mechanisms of acetaminophen hepatotoxicity. Immunohistochemical assays have allowed us to study the individual hepatocytes which contain acetaminophen-protein adducts. These data coupled with traditional methods of histological analyses have given us a better understanding of the time progression of adduct formation as it relates to the histopathological progression of the toxicity (26). The conclusions from these studies indicated: 1) acetaminophen covalent binding to protein occurred significantly before development of the hepatotoxicity, 2) covalent binding occurred in the innermost layer of hepatocytes and before total hepatic glutathione was depleted, 3) covalent binding occurred in the centrilobular zones and these zones were the sites of the ensuing toxicity, 4) lysis of hepatocytes occurred with the release of acetaminophen-protein adducts and the appearance of acetaminophen-adducts in serum, and 5) even though it appeared that the majority of the hepatocytes which contained acetaminophen-protein adducts became necrotic, adducts were detected in a small number of actively dividing hepatocytes and in macrophage-like cells in the regenerating liver. Taken as a whole, these data support the concept that covalent binding of the reactive metabolite of acetaminophen to critical protein is the mechanism of the hepatotoxicity; however, it appears that a small amount of binding may occur without cell death.

We have also focused our attention on understanding which proteins become derivitized following administration of toxic doses of acetaminophen to mice. In initial work we have isolated the major protein to which acetaminophen covalently binds (28). Sequence analysis of 85 amino acids gave a 97% homology with the previously deduced sequence of the hepatic 56 kDa selenium binding protein (29). Similar data have been obtained by Bartolone et al (32). This protein has been previously studied extensively in the laboratory of Medina at Baylor University. Their interest relates to understanding the anticarcinogenic properties of selenium and they showed that this protein becomes radiolabeled when $[^{75}Se]Na_2SeO_3$ is administered to animals or certain cell cultures. The mechanism of selenium binding to the protein is not understood; a codon specific for selenocysteine was not found. The function of the protein is unknown.

Peptides from 55 kDa APAP Adduct	% Homology 56 kDa SBP	% Homology 56 kDa ABP
L H K	100	100
G T W E K P G Q A S P M G Y	87	100
H N V M V S T E W A A P N V F K D G F N P A H V E	100	100
I F V W D W Q R H E I I Q T L	100	100
V I Q V P S K	100	100
Q Y D I S N P Q K P	100	100
L Y A T T S L Y S D	90	100

Figure 5. Comparison of deduced sequences of the 56 kDa selenium- and acetaminophen-binding proteins to the sequence of the isolated protein. Sequence data for the protein are from Pumford et al. (28). The deduced sequence data are from Lanfear et al. (33).

In further work we examined the possibility that prior selenium administration may selectively affect the covalent binding of acetaminophen to this protein *in vivo*. Administration of selenium 24 hours before acute doses of acetaminophen resulted in a decrease in total covalent binding to protein. It has been previously reported by Schnell et al. that acetaminophen hepatotoxicity and metabolic activation are decreased in the rat by prior selenium administration (31). Examination of specific binding to the 56 kDa selenium binding protein indicated that binding to this protein was decreased; however, the decrease appeared to be at the same level of decrease as observed for the other proteins to which acetaminophen covalently binds. These data suggest that the binding of selenium to the 56 kDa selenium binding protein is independent of the mechanism of hepatotoxicity of acetaminophen.

At present we do not know the significance of covalent binding of acetaminophen to the 56 kDa selenium binding protein. We have suggested three possible scenarios (28). The first postulation is that the protein is important in the functioning of the cell and covalent binding to the protein may be the mechanism of the toxicity. Understanding this postulation will require a determination of the function of the protein. The second postulation is that the 56 kDa selenium binding protein may serve as a secondary detoxification mechanism after glutathione depletion. This protein may contain significant SH groups which react with various reactive metabolites. The third postulation is that acetaminophen binding to this protein is not related to toxification or detoxification.

Of interest is the recent report by Lanfear et al. (33) that the 56 kDa selenium binding protein and the 56 kDa acetaminophen binding protein are different (Figure 5). They cloned a full-length cDNA encoding of the 56 kDa acetaminophen binding protein and determined that there was a 100% homology with the sequence which we reported for the protein which binds acetaminophen (28). Moreover, the deduced sequence of the 56 kDa acetaminophen binding protein differed at only 14 residues from the 56 kDa selenium binding protein. Using reverse transcription/PCR with oligonucleotide primers, they showed that the acetaminophen binding protein is mainly expressed in liver, whereas the 56 kDa selenium binding protein is expressed in liver, kidney, and to a lesser extent, lung. They conclude that the two proteins are regulated independently. The importance of this protein in acetaminophen toxicity will require further investigations.

LITERATURE CITED

1. Hinson, J.A. 1980. Biochemical toxicology of acetaminophen. Revs. Biochem. Toxicol. 2:103-129.

2. Boyd, E.M., and Bereczky, G.M. 1966. Liver necrosis from paracetamol. Brit. J. Pharmacol. 26:606-614.
3. Davidson, D.G.D., and Eastham, W.N. 1966. Acute liver necrosis following overdose of paracetamol. Br. Med. J. 5512:497-499.
4. Mitchell, J.R., Jollow, D.J., Potter, W.Z., Davis, D.C., Gillette, J.R., and Brodie, B.B. 1973. Acetaminophen-induced hepatic necrosis. I. Role of drug metabolism. J. Pharmacol. Exp. Ther. 187:185-194.
5. Jollow, D.J., Mitchell, J.R., Potter, W.Z., Davis, D.C., Gillette, J.R., and Brodie, B.B. 1973. Acetaminophen-induced hepatic necrosis. II. Role of covalent binding *in vivo*. J. Pharmacol. Exp. Ther. 187:195-202.
6. Potter, W.Z., Davis, D.C., Mitchell, J.R., Jollow, D.J., Gillette, J.R., and Brodie, B.B. 1973. Acetaminophen-induced hepatic necrosis. III. Cytochrome P-450-mediated covalent binding *in vitro*. J. Pharmacol. Exp. Ther. 187:203-210.
7. Mitchell, J.R., Jollow, D.J., Potter, W.Z., Gillette, J.R., Brodie, B.B. 1973. Acetaminophen-induced hepatic necrosis. IV. Protective role of glutathione. J. Pharmacol. Exp. Ther. 187:211-217.
8. Prescott, L.F. 1983. Paracetamol overdosage. Pharmacological considerations and clinical management. Drugs 25:290-314.
9. Blair, I.A., Boobis, A.R., Davies, D.S., and Cresp, T.M. 1980. Paracetamol oxidation: Synthesis and reactivity of N-acetyl-p-benzoquinone imine. Tetrahedron Lett. 21:4947-4950.
10. Potter, D.W. and Hinson, J.A. 1986. Reactions of N-acetyl-p-benzoquinone imine with reduced glutathione, acetaminophen, and NADPH. Mol. Pharmacol. 30:33-41.
11. Hinson, J.A. Monks, T.J., Hong, M. Highet, R.J., and Pohl, L.R. 1982. 3-(Glutathion-S-yl)acetaminophen: A biliary metabolite of acetaminophen. Drug Metab. Dispos. 10:47-50
12. Coles, B. Wilson, I., Wardman, P., Hinson, J.A., Nelson, S.D., and Ketterer, B. 1988. The spontaneous and enzymatic reaction of N-acetyl-p-benzoquinoneimine with glutathione: A stopped-flow kinetic study. Arch. Biochem. Biophys. 264:253-260.
13. Harvison, P.J., Guengerich, F.P., Rashed, M.S., and Nelson, S.D. 1988. Cytochrome P-450 isozyme selectivity in the oxidation of acetaminophen. Chem. Res. Toxicol. 1:49-52.
14. Raucy, J.L., Lasker, J.M., Lieber, C.S., and Black, M. 1989. Acetaminophen activation by human liver cytochromes P450IIE1 and P450IA2. Arch. Biochem. Biophys. 271:270-283.
15. Patten C.J., Thomas, P.E., Guy, R.L., Lee, M., Gonzalez, F.J., Guengerich, F.P., and Yang, C.S. 1993. Cytochrome P450 enzymes involved in acetaminophen activation by rat and human liver microsomes and their kinetics. Chem. Res. Toxicol. 6:511-518.
16. Hoffman, K.J., Streeter, A.J., Axworthy, D.B., and Baille, T.A. 1985. Structural characterization of the major covalent adduct formed *in vitro* and *in vivo* between acetaminophen and mouse liver proteins. Mol. Pharmacol. 27:566-573.
17. Donnelly, P.J., Walker, R.M., and Racz, W.J. 1994. Inhibition of mitochondrial respiration *in vivo* is an early event in acetaminophen-induced hepatotoxicity. Arch. Toxicol. 68:110-118.
18. Moore, M. Tjhor, H., Moore, G., Nelson, S., Orrenius, S. 1985. The toxicity of acetaminophen and N-acetyl-p-benzoquinone imine in isolated hepatocytes is associated with thiol depletion and increased cytosolic Ca_{2+}. J. Biol. Chem. 260:13035-13040.
19. Tsokos-Kuhn, J.O., Hughes, H., Smith, C.V., and Mitchell, J.R. 1988. Alkylation of the liver plasma membrane and inhibition of the Ca_{2+} ATPase by acetaminophen. Biochem. Pharmacol. 37:2125-2131.
20. Roberts, D.W., Pumford, N.R., Potter, D.W., Benson, R.W., and Hinson, J.A. 1987. A sensitive immunochemical assay for acetaminophen-protein adducts. J. Pharmacol. Exp. Ther. 241:182-189.
21. Pumford, N.R., Hinson, J.A., Potter, D.W., Rowland, K.L., Benson, R.W., and Roberts, D.W. 1989. Immunochemical quantitation of 3-(cystein-S-yl)acetaminophen adducts in serum and liver proteins of acetaminophen-treated mice. J. Pharmacol. Exp. Ther. 248:182-189.
22. Pumford, N.R., Roberts, D.W., Benson, R.W., and Hinson, J.A. 1990. Immunochemical quantitation of 3-(cystein-S-yl)acetaminophen protein adducts in subcellular liver fractions following a hepatotoxic dose of acetaminophen. Biochem. Pharmacol. 40:573-579.
23. Bartolone, J.B., Sparks, K., Cohen, S.D., and Khairallah, E.A. 1987. Immunochemical detection of acetaminophen-bound liver proteins. Biochem. Pharmacol. 36:1193-1196.
24. Bartolone, J.B., Birge, R.B., Sparks, K., Cohen, S.D., and Khairallah, E.A. 1988. Immunochemical detection of acetaminophen covalent binding to proteins. Biochem. Pharmacol. 37:4763-4774.
25. Bartolone, J.B., Cohen, S.D., and Khirallah, E.A. 1989. Immunohistochemical localization of acetaminophen-bound liver proteins. Fundam. Appl. Toxicol. 13:859-862.
26. Roberts, D.W., Bucci, T.J., Benson, R.W., Warbritton, A.R., McRae, T.A., Pumford, N.R., and Hinson, J.A. 1991. Immunohistochemical localization and quantification of the 3-(cystein-S-yl)acetaminophen-protein adduct in acetaminophen hepatotoxicity. Am. J. Pathol. 138:359-371.

27. Pumford, N.R., Hinson, J.A., Benson, R.W., and Roberts, D.W. 1990. Immunoblot analysis of protein containing 3-(cystein-S-yl)acetaminophen adducts in serum and subcellular liver fractions from acetaminophen-treated mice. Toxicol. Appl. Pharmacol. 104:521-532.
28. Pumford, N.R., Martin, B.M., and Hinson, J.A. 1992. A metabolite of acetaminophen covalently binds to the 56 kDa selenium binding protein. Biochem. Biophys. Res. Comm. 182:1348-1355.
29. Bansal, M., Mukhopadhyay, T., Scoitt, J., Cook, R.G., Mukhopadhyay, R., and Medina, D. 1990. DNA sequencing of a mouse liver protein that binds selenium: Implications for selenium's mechanism of action in cancer prevention. Carcinogenesis 11:2071-2073.
30. Bansal, M.P. Oborn, C.J., Danielson, K.G., and Medina, D. 1989. Evidence for two selenium-binding proteins distinct from glutathione peroxidase in mouse liver. Carcinogenesis 10:541-546.
31. Schnell, R.C., Park, K.S., Davies, M.H., Merrick, B.A., and Weir, S.W. 1988. Protective effects of selenium on acetaminophen hepatotoxicity of the rat. Toxicol. Appl. Pharmacol. 95:1-11.
32. Bartolone, J.B., Birge, R.B., Bulera, S.J., Bruno, M.K., Nishanian, E.V., Cohen, S.D., and Khairallah, E.A. 1992. Purification, antibody production, and partial amino acid sequence of the 58 kDa acetaminophen-binding liver proteins. Toxicol. Appl. Pharmacol. 113:19-29.
33. Lanfear, J., Fleming, J., Walker, M., and Harrison, P. 1993. Different patterns of regulation of the genes encoding the closely related 56 kDa selenium- and acetaminophen-binding proteins in normal tissues and during carcinogenesis. Carcinogenesis 14:335-340.

8

AN OXIDANT SENSOR AT THE PLASMA MEMBRANE

Axel Knebel, Mihail Iordanov, Hans J. Rahmsdorf, and Peter Herrlich

Forschungszentrum Karlsruhe
Institut für Genetik
Postfach 3640, D-76021 Karlsruhe, Germany

INTRODUCTION

The expression of genes is predominantly determined by conditions of the microenvironment of cells. Prime examples of such regulation are found in embryonic development of all multicellular organisms and also in the adult when various cytokines and hormones exert highly inducer-specific influences on genes. The naturally occurring regulating agents interact with specific receptors: e.g., the retinoids, vitamin D3, thyroid hormones and the steroid hormones with appropriate nuclear receptors (RARs, RXRs, VDR and the specific steroid hormone receptors), or the members of the large TGFβ and FGF families with their respective cell surface receptors. While nuclear receptors act as transcription factors themselves and select their genes by receptor-specific recognition elements, the growth factors induce, through their cell surface receptors, a complex process of signal transduction to the nucleus (for reviews see Beato, 1989; Karin, 1994; Gilbert, 1994; Angel and Herrlich, 1994; McCormick, 1995; Howe and Weiss, 1995; Ullrich and Simon, 1995).

It has been known for a long time that adverse agents cause dramatic changes of gene expression. Cells in culture and in the multicellular organism react to heat shock, heavy metals, radiation, toxic plant metabolites, carcinogenic chemicals and various man-made agents that have not occurred during evolution. To exert a positive influence on an elaborate physiological process such as gene expression, the physical or chemical agents must be able to interact with a cellular macromolecule that translates the interaction into a "language" understood by the cell's "machinery". Various laboratories have put in efforts to find the primary interacting macromolecules. Successful searches led to the discovery of a heavy metal regulated transcription factor (Heuchel et al., 1994), of the aromatic hydrocarbon receptor which is also a transcription factor (reviewed by Okey et al., 1994; Swanson and Bradfield, 1993), of the heat shock factors (Lis and Wu, 1993; Sistonen et al., 1994), and of the UV induced autophosphorylation of growth factor receptors (Sachsenmaier et al., 1994; Schieven et al., 1994).

Biological Reactive Intermediates V
Edited by R. Snyder et al., Plenum Press, New York, 1996

OXIDANTS AND ANTIOXIDANTS

A wealth of data has drawn attention to the significance of maintaining reducing conditions in cells and to the fight against reactive oxygen intermediates. Cells maintain reducing conditions by an excess of -SH group molecules and by the synthesis of enzymes that deal with oxygen such as superoxide dismutase and catalase (for review see Cadenas, 1989; Joenje, 1989). The cellular compartments contain an excess of free -SH groups although there may be differences between compartments. It is thought that the reduced glutathione level is kept at around 5 mM which depends on glutathione synthesis and a net of reducing steps, e.g., by glutathione reductase. Severe reduction of glutathione level is lethal (Meister and Anderson, 1983) suggesting that spontaneously occurring oxidants can then cause lethal damage. Enforced expression of metallothionein, another -SH reagent, protects cells from the effects of a number of toxic agents (Kelley et al., 1988; Kaina et al., 1990). Since intracellular -SH-rich proteins such as metallothioneins may also act as metal scavengers and stores for zinc, it is not clear at what level the protection occurs: neutralization of a reactive toxic intermediate or improved repair of damage which may involve zinc-requiring enzymes.

Recent evidence has linked oxidants to human disease. Oxidants can cause carcinogenic mutations and arguments have been raised that endogenous metabolic oxygen intermediates may represent a predominant factor in human cancer (Cerutti, 1985; Ames, 1989). Various links seem to exist between oxidant state and cellular survival. For example, neural and glial cell survival after injury can be improved by treatment with the antioxidant N-acetyl-cysteine which by its acetyl group readily penetrates into cells and increases reducing power (Staal et al., 1990). Also overexpression of Cu/Zn superoxide dismutase delays neuronal apoptosis (Greenlund et al., 1995). Mutations in Cu/Zn superoxide dismutase, on the other hand, have been found to be associated with familial amyotrophic lateral sclerosis (Deng et al., 1993; Rosen et al., 1993), and other neurological disorders go along with oxidative stress, Alzheimer's disease being a recent addition.

Interestingly, not only can oxidants induce damage of decisive structures such as genes, oxidant treatments of cells in culture also trigger changes of gene expression. Several observations focus on the activities of the transcription factors NFκB and AP-1. In vitro, AP-1 (Fos and Jun) requires a reduced cysteine in its DNA binding domain for activity (Abate et al., 1990; Xanthoudakis et al., 1992; Oehler et al., 1993). Also NFκB DNA binding activity has been reported to depend on a reduced cysteine (Toledano and·Leonard, 1991; Molitor et al., 1991). In vivo, treatment of cells with the antioxidant N-acetylcysteine induces AP-1 activity. The same type of antioxidant treatment, however, abolishes NFκB activation in response to many agents while oxidants induce NFκB activity (Meyer et al., 1993; Schenk et al., 1994). Interestingly, also transient expression of thioredoxin or exogenous addition to the culture medium exert such antioxidant effects (Schenk et al., 1994). These experiments do not prove redox regulation of specific proteins in vivo since the redox state cannot be monitored in vivo at the transcription factors to be regulated. Activity enhancement or inhibition could be the result of signal transduction from a redox-regulated "sensor" to NFκB or AP-1. The best evidence for a role of intracellular reactive oxygen intermediates stems from experiments manipulating superoxide dismutase and catalase expression in cells (Schmidt et al., 1995). Overexpression of Cu/Zn superoxide dismutase supposed to increase H_2O_2 levels, enhances TNF induced NFκB activity, while catalase which destroys H_2O_2, inhibits NFκB activation.

MEMBRANE SENSORS FOR OXIDANTS

The aforementioned experiments have not defined the nature of the "oxidant sensor". Since reactive oxygen intermediates including H_2O_2 diffuse rapidly, both into and out of the cell, as well as through cellular compartments, such "sensors" could be located anywhere in the cell. Stimulated by our own and other laboratories' analyses of the UV response (Sachsenmaier et al., 1994; Schieven et al., 1994), we speculated that cells would carry oxidant-sensitive structures on their surface. We would like to report here the detection of such H_2O_2 sensitive receptor molecules in the plasma membrane.

Irradiating cells in culture with short or long-wave length ultraviolet light causes the activation of several growth factor receptors within fractions of a minute. Direct evidence for the activation has been presented for the EGF receptor (Sachsenmaier et al., 1994). The receptor is triggered to autophosphorylate at tyrosines as if it were bound by its ligand EGF (Chen et al., 1987). UV induces, however, in an apparent absence of ligand. EGF receptor activation leads to subsequent binding of Shc protein to the receptor (Knebel et al., unpublished), indicating that autophosphorylation occurred at the appropriate sites, and to the activations of Ras, Raf, ERK-1 and ERK-2, and to transcription of genes such as c-fos (Devary et al., 1992; Radler-Pohl et al., 1993; Sachsenmaier et al., 1994). Also for the PDGF receptor, UV induced phosphorylation has been detected (Sachsenmaier et al., unpublished). Indirect so-called cross-downmodulation evidence has indicated that many more growth factor receptors are activated in response to UV irradiation (see examples in Sachsenmaier et al., 1994).

In order to assay for possible sensitivity of growth factor receptors to oxidant treatment, we used cultured rat and mouse cell lines that overexpress an individual receptor. These cell lines were kindly provided by Dr. Gordon Gill (San Diego), Dr. Michael J. Weber (Charlottesville) and Dr. Axel Ullrich (München). Immediately and at various times after treatment with oxidants such as H_2O_2 or potassium permanganate, cells were lysed and either total proteins analysed by SDS-PAGE followed by Western blotting with phosphotyrosine specific antibodies, or the receptor was precipitated with a specific antibody. The precipitate was then solubilized in SDS sample buffer, resolved by SDS-PAGE and the phosphorylation at tyrosines was determined by Western blotting with a phosphotyrosine specific antibody. Such analyses revealed efficient oxidant-induced phosphorylation of the receptors for insulin, PDGF and EGF. The induced autophosphorylation of the EGF receptor will be documented below.

H_2O_2 INDUCED INCREASED PHOSPHORYLATION OF THE EGF RECEPTOR

Within 2 min. after H_2O_2 addition (150 μM) to cultured rat-1 cells that had been stably transfected with an expression clone for the human EGF receptor (Wasilenko et al., 1990), the EGF receptor is nearly maximally phosphorylated (Fig. 1). The response is similar in magnitude and kinetics to the induced activation by UVC. The magnitude is comparable to that achieved by the addition of 0.5 ng/ml EGF. The EGF response is, however, slower, with maximum level reached at 5 min. (shown for 2 ng/ml in Fig. 1). Also potassium permanganate (at 150 μM) causes elevated EGF receptor phosphorylation (Fig. 2).

The increased phosphorylation could result from ligand independent conformational activation of the tyrosine kinase function, from H_2O_2 induced dimerization of EGF receptor subunits, or from inhibition of a phosphatase, or from some other downmodulating principle. The mechanism has yet to be explored. Other laboratories have found inhibition of phosphotyrosine phosphatase activity by H_2O_2 in vivo, leading to enhanced phosphorylation of

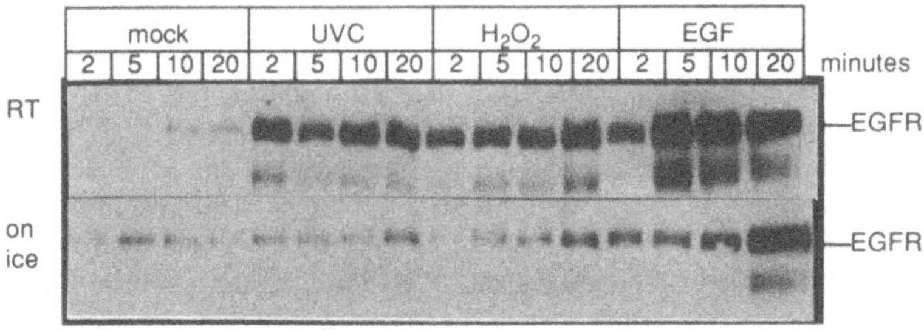

Figure 1. Temperature-dependence of UVC, H_2O_2 and EGF induced tyrosine phosphorylation of the EGF-receptor. Rat-1-HER cells (Wasilenko et al., 1990) overexpressing the human wild type EGF-receptor were starved in DMEM without phenol-red and without FCS for 4 hours. The cells were preincubated with 1mM $NaVO_4$ for 1 hour and then either irradiated without removing the culture medium with 1000 J/m^2 UV (254 nm), treated with 150 μM H_2O_2 or 2 ng/ml human EGF. The experiment was performed in duplicate at 25°C (at room temperature) and at 4°C (on ice). At the indicated time points the cells were washed with ice cold PBS and immediately lysed in 2x sample buffer (Laemmli). The proteins were resolved by SDS-PAGE and transfered on an Immobilon membrane. The relative amount of tyrosine phosphorylation was detected in Western Blot Analysis using PY20 antibodies (Transduction Laboratories, Lexington, Kentucky). The position of the EGF-receptor was determined by immunoprecipitation of the receptor, followed by Western Blot Analysis (not shown). Equal loading of the gel was ascertained by staining with Coomassie-blue (not shown).

the insulin receptor at tyrosine residues and to the phosphorylation of target proteins of this receptor (Heffetz et al., 1992; Sullivan et al., 1994). Fig. 1 shows that at 4° the basal level of tyrosine phosphate at the receptor is slightly increased, perhaps due to inhibition of dephosphorylation, and that the UVC and H_2O_2 induced phosphorylations are severely reduced and retarded. Since the catalytic activity induced by EGF does not seem inhibited (although delayed), a step other than the actual kinase reaction seems inhibited. As a speculative argument, the reduced membrane fluidity may affect the removal of an inhibitory protein (e.g., of a phosphatase) or the kinetics of dimerization of the EGF receptor subunits.

N-ACETYL CYSTEINE BLOCKS THE UV AND OXIDANT INDUCED EGF RECEPTOR AUTOPHOSPHORYLATION

The autophosphorylation of receptors can explain how oxidants induce genes. Is this a major pathway of gene induction? If it were, -SH reagents known to block oxidant

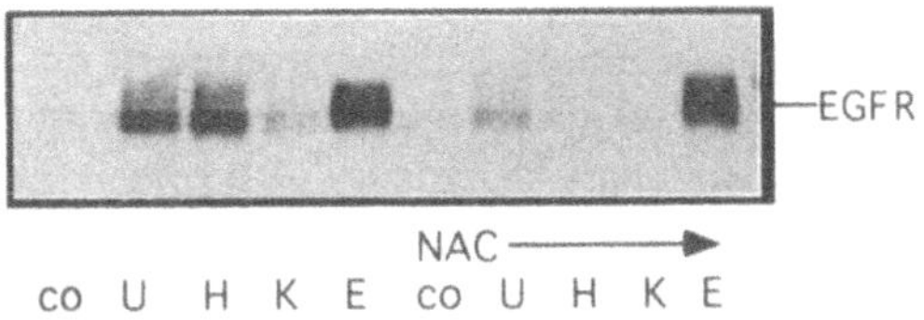

Figure 2. N-acetyl-cysteine inhibits EGF-receptor autophosphorylation. Confluent dishes of B82-L-cells (Chen et al., 1987), stably transfected with the human EGF-receptor, were starved in DMEM without fetal calf serum and phenol red for 4 hours. The cells were pretreated for 60 minutes with 1mM $NaVO_4$, then mock-treated or treated with 30mM N-acetyl-cysteine for 30 seconds. The cells were then induced with 1500 J/m^2 UVC, 150μM H_2O_2, 150 μM $KMnO_4$ or 2ng/ml human EGF for 5 minutes and harvested in 2 x sample buffer (Laemmli). Phosphorylation of the EGF-receptor at tyrosine residues was determined as described in Figure 1. Equal loading was ascertained by Coomassie-blue staining (not shown). co = non treated cells, U = UVC, H = H_2O_2, K = potassium permanganate, E = EGF, NAC = N-Acetylcysteine.

dependent gene expression should also obliterate instantaneously the oxidant induced EGF receptor phosphorylation. This is indeed the case. N-Acetyl cysteine at 30 mM almost totally prevents the activation of the EGF receptor by either UVC, H_2O_2 or potassium permanganate (Fig. 2). Also, EGF induced activation is sensitive to N-acetylcysteine, in the experiment shown only marginally. Since also agents such as glutathione that are not effectively and instantly taken up by cells block both gene induction and EGF receptor phosphorylation (not shown), we propose that the oxidant pathway through the cell surface receptors is a major mechanism of oxidant induced gene regulation.

ACKNOWLEDGEMENTS

We are grateful to Drs. Gordon Gill, Michael Weber and Axel Ullrich for providing cell lines overexpressing growth factor receptors.

REFERENCES

Abate, C., Patel, L., Rauscher III, F. J., and Curran, T. (1990). Redox regulation of Fos and Jun DNA-binding activity in vitro. Science 249, 1157-1161.

Ames, B. N. (1989). Endogenous oxidative DNA damage, aging and cancer. Free Radic. Res. Commun. 7, 121-128.

Angel, P., and Herrlich, P. (1994). The Fos and Jun families of transcription factors. In: "CRC-press", Boca-Raton, FL, U.S.A.

Beato, M. (1989). Gene regulation by steroid hormones. Cell 56, 335-344.

Cadenas, E. (1989). Biochemistry of oxygen toxicity. Annu. Rev. Biochem. 58, 79-110.

Cerutti, P. A. (1985). Prooxidant states and tumor promotion. Science 227, 375-381.

Chen, W. S., Lazar, C. S., Poenie, M., Tsien, R. Y., Gill, G. N., and Rosenfeld, M. G. (1987). Requirement for intrinsic protein tyrosine kinase in the immediate and late actions of the EGF receptor. Nature 328, 820-823.

Deng, H.-X., Hentati, A., Tainer, J. A., Iqbal, Z., Cayabyab, A., Hung, W.-Y., Getzoff, E. D., Hu, P., Herzfeldt, B., Roos, R. P., Warner, C., Deng, G., Soriano, E., Smyth, C., Parge, H. E., Ahmed, A., Roses, A. D., Hallewell, R. A., Pericak-Vance, M. A., and Siddique, T. (1993). Amyotrophic lateral sclerosis and structural defects in Cu,Zn superoxide dismutase. Science 261, 1047-1051.

Devary, Y., Gottlieb, R. A., Smeal, T., and Karin, M. (1992). The mammalian ultraviolet response is triggered by activation of Src tyrosine kinases. Cell 71, 1081-1091.

Gilbert, S. F. (1994). Developmental Biology. (Sunderland, MA, U.S.A.: Sinauer Associates), pp. 531-622.

Greenlund, L. J. S., Deckwerth, T. L., and Johnson Jr., E. M. (1995). Superoxide dismutase delays neuronal apoptosis: A role for reactive oxygen species in programmed neuronal death. Neuron 14, 303-315.

Heffetz, D., Rutter, W. J., and Zick, Y. (1992). The insulinomimetic agents H_2O_2 and vanadate stimulate tyrosine phosphorylation of potential target proteins for the insulin receptor kinase in intact cells. Biochem. J. 288, 631-635.

Heuchel, R., Radtke, F., Georgiev, O., Stark, G., Aguet, M., and Schaffner, W. (1994). The transcription factor MTF-1 is essential for basal and heavy metal-induced metallothionein gene expression. EMBO J. 13, 2870-2875.

Howe, L. R., and Weiss, A. (1995). Multiple kinases mediate T-cell-receptor signaling. TIBS 20, 59-64.

Joenje, H. (1989). Genetic toxicology of oxygen. Mutation Res. 219, 193-208.

Kaina, B., Lohrer, H., Karin, M., and Herrlich, P. (1990). Overexpressed human metallothionein IIA gene protects Chinese hamster ovary cells from killing by alkylating agents. Proc. Natl. Acad. Sci. U.S.A. 87, 2710-2714.

Karin, M. (1994). Signal transduction from the cell surface to the nucleus through the phosphorylation of transcription factors. Curr. Opinion in Cell Biol. 6, 415-424.

Kelley, S. L., Basu, A., Teicher, B. A., Hacker, M. P., Hamer, D. H., and Lazo, J. S. (1988). Overexpression of metallothionein confers resistance to anticancer drugs. Science 241, 1813-1815.

Lis, J., and Wu, C. (1993). Protein traffic on the heat shock promoter: Parking, stalling, and trucking along. Cell 74, 1-4.

McCormick, F. (1995). Ras signaling and NF1. Curr. Opinion in Gen. and Dev. 5, 51-55.

Meister, A., and Anderson, M. E. (1983). Glutathione. Ann. Rev. Biochem. 52, 711-760

Meyer, M., Schreck, R., and Baeuerle, P. A. (1993). H_2O_2 and antioxidants have opposite effects on activation of NF-κB and AP-1 in intact cells: AP-1 as secondary antioxidant-responsive factor. EMBO J. 12, 2005-2015.

Molitor, J. A., Ballard, D. W., and Greene, W. C. (1991). κB-specific DNA binding proteins are differentially inhibited by enhancer mutations and biological oxidation. The New Biol. 3, 987-996.

Oehler, T., Pintzas, A., Stumm, S., Darling, A., Gillespie, D., and Angel, P. (1993). Mutation of a phosphorylation site in the DNA-binding domain is required for redox-independent transactivation of AP1-dependent genes by v-Jun. Oncogene 8, 1141-1147.

Okey, A. B., Riddick, D. S., and Harper, P. A. (1994). Molecular biology of the aromatic hydrocarbon (dioxin) receptor. TiPS 15, 226-232.

Radler-Pohl, A., Sachsenmaier, C., Gebel, S., Auer, H.-P., Bruder, J. T., Rapp, U., Angel, P., Rahmsdorf, H. J., and Herrlich, P. (1993). UV-induced activation of AP-1 involves obligatory extranuclear steps including Raf-1 kinase. EMBO J. 12, 1005-1012.

Rosen, D. R., Siddique, T., Patterson, D., Figlewicz, D. A., Sapp, P., Hentati, A., Donaldson, D., Goto, J., O'Regan, J. P., Deng, H.-X., Rahmani, Z., Krizus, A., McKenna-Yasek, D., Cayabyab, A., Gaston, S. M., Berger, R., Tanzi, R. E., Halperin, J. J., Herzfeldt, B., Van den Bergh, R., Hung, W.-Y., Bird, T., Deng, G., Mulder, D. W., Smyth, C., Laing, N. G., Soriano, E., Pericak-Vance, M. A., Haines, J., Rouleau, G. A., Gusella, J. S., Horvitz, H. R., and Brown Jr, R. H. (1993). Mutations in Cu/Zn superoxide dismutase gene are associated with familial amyotrophic lateral sclerosis. Nature 362, 59-62.

Sachsenmaier, C., Radler-Pohl, A., Zinck, R., Nordheim, A., Herrlich, P., and Rahmsdorf, H. J. (1994). Involvement of growth factor receptors in the mammalian UVC response. Cell 78, 963-972.

Schenk, H., Klein, M., Erdbrügger, W., Dröge, W., and Schulze-Osthoff, K. (1994). Distinct effects of thioredoxin and antioxidants on the activation of transcription factor NF-kB and AP-1. Proc. Natl. Acad. Sci. U.S.A. 91, 1672-1676.

Schieven, G. L., Mittler, R. S., Nadler, S. G., Kirihara, J. M., Bolen, J. B., Kanner, S. B., and Ledbetter, J. A. (1994). ZAP-70 tyrosine kinase, CD45, and T cell receptor involvement in UV- and H_2O_2-induced T cell signal transduction. J. Biol. Chem. 269, 20718-20726.

Schmidt, K. N., Amstad, P., Cerutti, P. and Baeuerle, P.A. (1995). The roles of hydrogen peroxide and superoxide as messengers in the activation of transcription factor NF-κB. Chemistry & Biology 2, 13-22.

Sistonen, L., Sarge, K. D., and Morimoto, R. I. (1994). Human heat shock factors 1 and 2 are differentially activated and can synergistically induce hsp70 gene transcription. Mol. Cell. Biol. 14, 2087-2099.

Staal, F. J. T., Roederer, M., Herzenberg, L. A., and Herzenberg, L. A. (1990). Intracellular thiols regulate activation of nuclear factor κB and transcription of human immunodeficiency virus. Proc. Natl. Acad. Sci. U.S.A. 87, 9943-9947.

Sullivan, S. G., Chiu, D. T., Errasfa, M., Wang, J. M., Qi, J. S., and Stern, A. (1994). Effects of H_2O_2 on protein tyrosine phosphatase activity in HER14 cells. Free Radic. Biol. Med. 16, 399-403.

Swanson, H. I., and Bradfield, C. A. (1993). The AH-receptor: Genetics, structure and function. Pharmacogenetics 3, 213-230.

Toledano, M. B., and Leonhard, W. J. (1991). Modulation of transcription factor NF-κB binding activity by oxidation-reduction in vitro. Proc. Natl. Acad. Sci. U.S.A. 88, 4328-4332.

Ullrich, A. and Simon (1995). Cell regulation. Curr. Opinion in Cell Biol. 7, 145-196.

Wasilenko, W. J., Nori, M., Testerman, N., and Weber, M. J. (1990). Inhibition of epidermal growth factor receptor biosynthesis caused by the *src* oncogene product, $pp60^{v\text{-}src}$. Mol. Cell. Biol. 10, 1254-1258.

Xanthoudakis, S., Miao, G., Wang, F., Pan., Y.-C. E., and Curran, T. (1992). Redox-activation of Fos-Jun DNA binding activity is mediated by a DNA repair enzyme. EMBO J. 11, 3323-3335.

IDENTIFICATION OF HYDROGEN PEROXIDE AS THE RELEVANT MESSENGER IN THE ACTIVATION PATHWAY OF TRANSCRIPTION FACTOR NF-κB

Kerstin N. Schmidt,[1] Paul Amstad,[2] Peter Cerutti,[2] and Patrick A. Baeuerle[1]

[1] Institute of Biochemistry
University of Freiburg
D-79104 Freiburg, Germany
[2] Department of Carcinogenesis
Swiss Institute for Experimental Cancer Research
CH-1066 Épalinges s. Lausanne, Switzerland

INTRODUCTION

The inducible higher eukaryotic transcription factor NF-κB is activated by a large variety of distinct simuli [1-3]. In unstimulated cells, this factor resides in a latent form in the cytoplasm [4]. Latency is achieved by association of the DNA-binding NF-κB dimer with an inhibitory subunit, called IκB [5]. IκB suppresses DNA-binding and nuclear transport of NF-κB. Upon stimulation of cells, IκB is phosphorylated and proteolytically degraded [6-9]. Both reactions are required for activation [10]. The released NF-κB is then translocated to the nucleus where it initiates transcription of target genes. Among the numerous proteins which are induced by a concerted action of NF-kB with other transcription factors are cytokines, chemokines, cell adhesion molecules, hematopoetic growth factors and receptors, histocompatibility antigens and acute phase proteins [1-3]. While NF-κB may be indispensable as inducer of many immediate-early inflammatory and immune reactions, the transcription factor is likely to play a fatal role in certain diseases and syndromes that involve an abberrant expression of inflammatory cytokines [22-24].

A large number of different viruses and bacteria have been reported to activate NF-κB [1-3]. Among the endogenous signals activating NF-κB are inflammatory cytokines that are newly produced in response to various pathogenic stimulations, such as tumor necrosis factors α and β and interleukin-1 [3]. In view of this variety of pathogen and pathogen-related stimuli, a common intracellular messenger is hard to envision. Several lines of evidence suggested that reactive oxygen intermediates (ROIs) serve as messengers for most if not all stimuli: (i) the inhibition of NF-κB activation by several structurally unrelated antioxidans, (ii) the induction of NF-κB by H_2O_2 treatment in some cell lines and (iii) the observation that many conditions

Biological Reactive Intermediates V
Edited by R. Snyder et al., Plenum Press, New York, 1996

activating NF-κB are also known to induce oxidative stress [11-20]. In order to identify the relevant ROI species and to gain more direct evidence for an involvement of ROIs as messengers, we investigated whether changes in the level of enzymes that control intracellular ROI levels affect the activation of transcription factor NF-κB.

MATERIALS AND METHODS

Cell Culture and Treatments

The mouse epidermal cell line JB6 clone 41, the catalase overexpressing subclone JB6-Cat4 and the Cu/Zn-SOD-overexpressing subclone JB6-SOD15 were characterized in

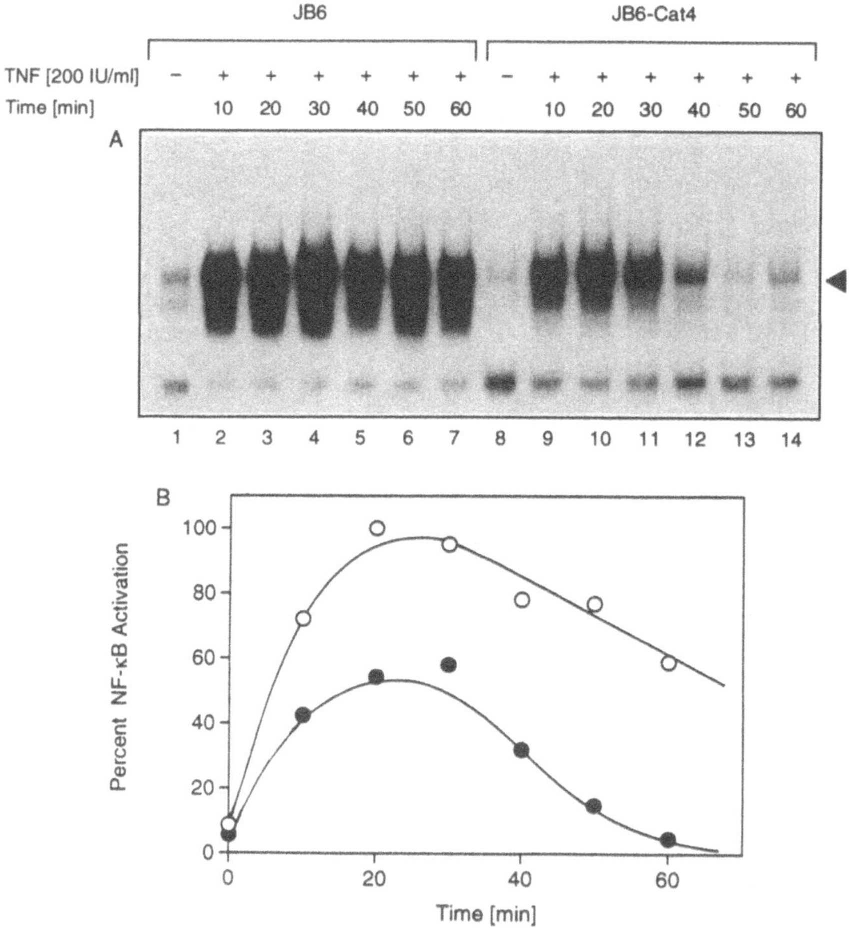

Figure 1. The activation of NF-κB by TNF in cells stably overexpressing catalase. (**A**) Cell cultures of the parental line JB6 clone 41 (lanes 1-7) and of the JB6-Cat4 clone (lanes 8-14) were left untreated (lanes 1 and 8) or stimulated with 200 IU/ml TNF for the indicated periods of time (lanes 2-7 and 9-14). Total cell extracts were prepared and analyzed for NF-κB-specific DNA binding activity using EMSA. A filled arrowhead indicates the position of the TNF-inducible protein-DNA complex. (**B**) Quantitation of the radioactivity in the NF-κB—DNA complexes by β imaging. Open circles, JB6 clone 41 cells; filled circles, JB6-Cat4 cells. The maximal level of NF-κB activity obtained with JB6 clone 41 cells was set to 100%.

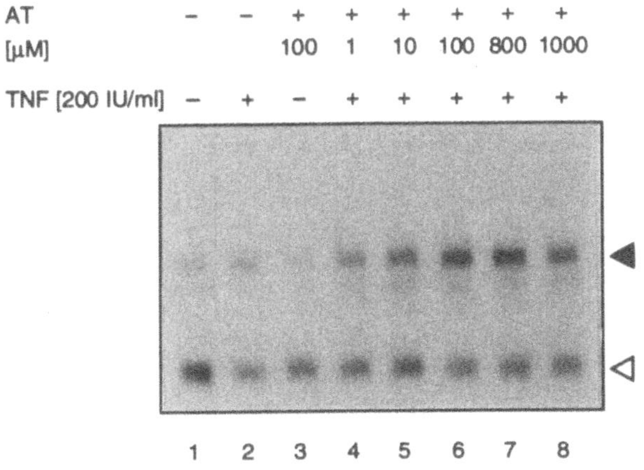

Figure 2. The effect of aminotriazole on the TNF-inducibility of NF-κB in JB6-Cat4 cells. The indicated amounts of aminotriazole (AT) were added to the cell cultures 1 h before the 1 h-treatment with 200 IU/ml TNF (lanes 4-8). The filled arrowhead indicates the position of the NF-κB—DNA complex, the open arrowhead the position of a non-specific complex.

detail previously [20,21]. Cells were grown as described [20,21]. For activation of NF-κB, cells were treated for the indicated periods of time with 200 IU/ml of human, recombinant TNF-a (Boehringer, Mannheim) or 400 nM of the sodium salt of okadaic acid (Calbiochem). For inhibition of catalase, cells were incubated with the indicated concentrations of 3-amino-1,2,4-triazole (AT) 1 h before stimulation with 200 IU/ml TNF. The treatment of JB6 clone 41 and JB6-SOD15 cells with purified catalase from bovine liver (Boehringer, Mannheim) was performed for 30 h at final concentrations of the enzyme between 3 and 333 IU/ml followed by stimulation with 200 IU/ml TNF for 20 min.

Cell Extracts

In order to monitor the activity state of all the cellular NF-κB, total cell extracts were prepared and used for electrophoretic mobility shift assays as described previously [26]. Protein concentrations in supernatants of cell lysates, i.e., cell extracts, were determined by an assay based on the Coomassie Brilliant Blue reaction (Bio-Rad).

Electrophoretic Mobility Shift Assays

Cell extracts with equal amounts of protein were used in EMSAs. The DNA binding conditions for NF-κB were described in detail elsewhere [27]. The radioactivity in NF-κB–DNA complexes was quantitated by β imaging (Molecular Dynamics imager).

RESULTS AND CONCLUSIONS

In this study, we provide evidence that H_2O_2 is essential for the activation of NF-κB by tumor necrosis factor α (TNF) and okadaic acid. Cell lines stably overexpressing the H_2O_2-degrading enzyme catalase were deficient in activating NF-κB. Addition of the catalase inhibitor aminotriazole restored the inducibility of the transcription factor. Stable overexpression of cytoplasmic Cu/Zn-dependent superoxide dismutase (SOD), an enzyme

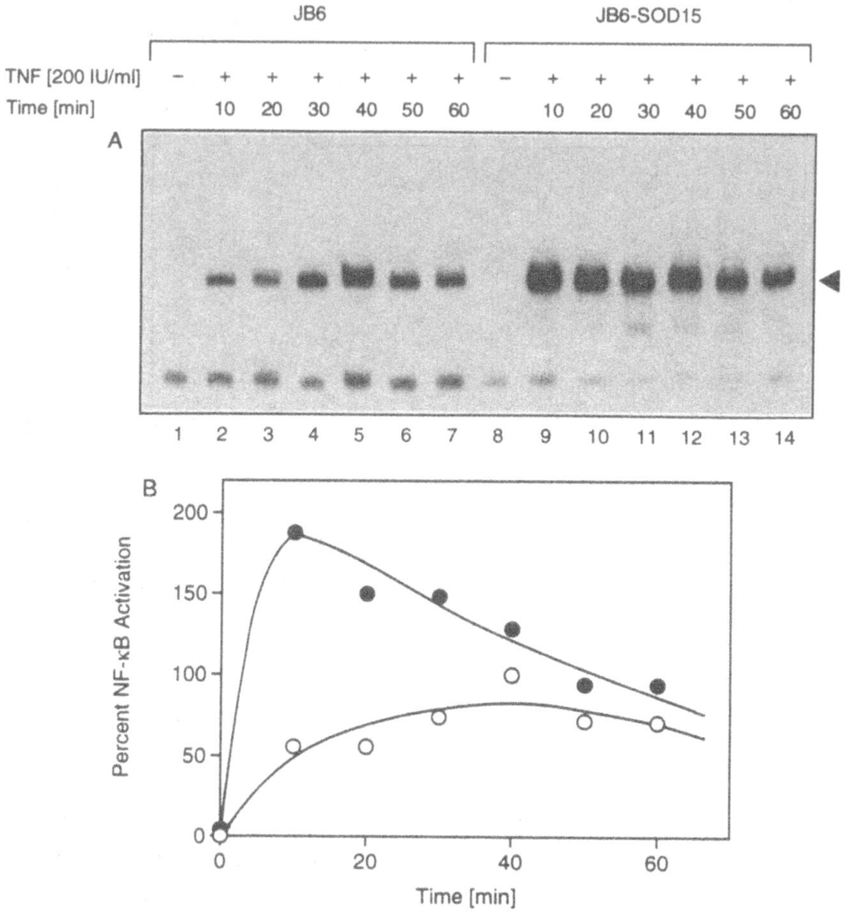

Figure 3. The activation of NF-κB by TNF in cells stably transfected with Cu/Zn superoxide dismutase. (A) Cell cultures of the parental line JB6 clone 41 (lanes 1-7) and of the SOD-overexpressing line JB6-SOD15 (lanes 8-14) were left untreated (lanes 1 and 8) or stimulated for 1 h with 200 IU/ml TNF for the indicated periods of time. Total cell extracts were prepared and analysed for NF-κB DNA binding activity using EMSA. (B) Quantitation of the radioactivity in the NF-κB—DNA complexes by β imaging. The maximal NF-κB activity seen with JB6 clone 41 cells (40 min-value) was set to 100%.

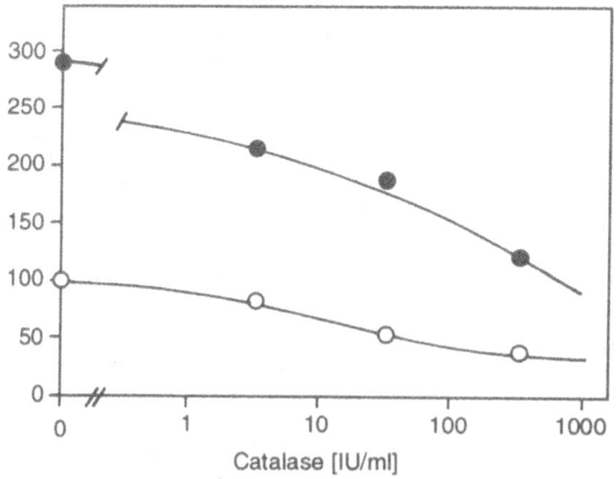

Figure 4. The effect of exogenously added catalase on NF-κB activation by TNF in JB6 clone 41 and JB6-SOD15 cells. Cell cultures were incubated for 30 h with the indicated IU/ ml of purified bovine liver catalase and then induced for 20 min with 200 UI/ml TNF. Total cell extracts were prepared and analysed for NF-κB activity by EMSA. Open circles, JB6 clone 41 cells; filled circles, JB6-SOD15 cells. The amount of ^{32}P-radioactivity in the NF-κB—DNA complexes was quantitated by β imaging and is shown plotted in a dose/response curve. The value found for JB6 clone 41 cells under control conditions was set to 100%.

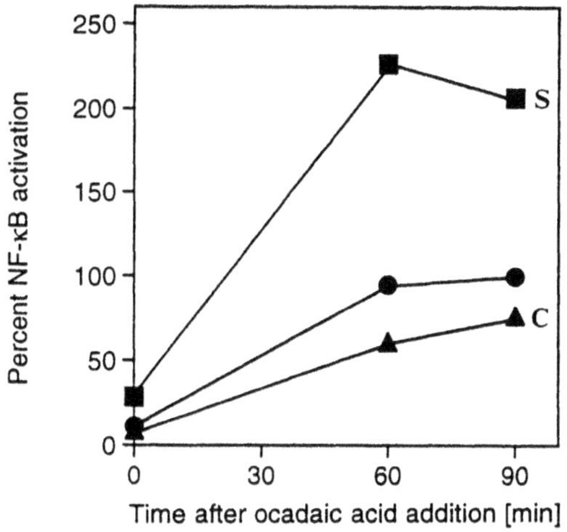

Figure 5. The activation of NF-κB by okadaic acid in the three cell lines used. Cell cultures from JB6 clone 41, JB6-Cat4 and JB6-SOD15 cells were treated for the indicated periods of time with 400 nM okadaic acid. Total cell extracts were prepared and equal amounts of protein analysed by EMSA for NF-κB DNA binding activity. The radioactivity in NF-κB—DNA complexes was quantitated by β imaging and is shown plotted as kinetic. The value found for JB6 clone 41 cells after 60 min was set to 100%. Filled circles, JB6 clone 41 cells; filled triangles, JB6-Cat4 cells (C); filled squares, JB6-SOD15 cells (S).

enhancing the production of H_2O_2 from superoxide, had the opposite effect: NF-κB activation by TNF and okadaic acid was potentiated. In parental and transfected cell lines, the level of inducible cytoplasmic NF-κB—IκB complex was the same, indicating that stable overexpression of catalase and SOD affected the posttranslational activation of NF-κB but not its synthesis (data not shown).

Our data show that a particular ROI species, H_2O_2, is involved as a messenger in the TNF- and okadaic acid-induced activation of the pathogen-inducible transcription factor NF-κB. Superoxide is only indirectly involved as a source for H_2O_2. These data explain the inhibitory effects of many antioxidative compounds on the activation of NF-κB and its target genes. Normal levels of catalase appear insufficient to completely remove H_2O_2 which is overproduced in response to various stimuli. This is a prerequisite for the accumulation of the ROI and its use as an intracellular messenger molecule of the pathogen response.

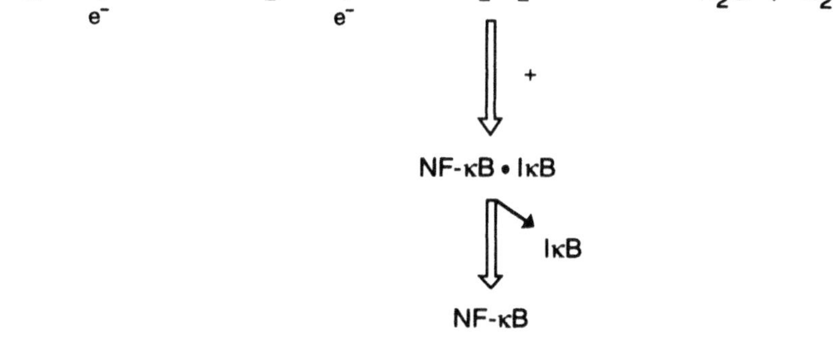

Figure 6. A model summarizing the results of this study.

REFERENCES

1. Baeuerle, P. A. (1991). *Biochem. Biophys. Acta* **1072**, 63-8.
2. Grilli, M., Jason, J.-S. & Lenardo, M.J. (1993). *Int. Rev. Cytol.* **143**, 1-62.
3. Baeuerle, P. A. & Henkel, T. (1994). *Annu. Rev. Immunol.* **12**, 141-179.
4. Baeuerle, P. A. & Baltimore, D. (1988a). *Cell* **53**, 211-217.
5. Baeuerle, P. A. & Baltimore, D. (1988b). *Science* **242**, 540-546.
6. Sun, S. C., Ganchi, P. A., Ballard, D. W., & Greene, W. C. (1993).*Science* **259**, 1912- 1915.
7. Brown, K., Park, S., Kanno, T., Franzoso, T. & Siebenlist, U. (1993). *Proc. Natl. Acad. Sci. U.S.A.* **90**, 2532-2536.
8. Beg, A. A., Ruben, S. M. Scheinman, R. I., Haskill, S., Rosen, C. A. & Baldwin, A. J. (1992).*Genes Dev.* **6**, 1899-1913.
9. Henkel, T., Machleidt, T., Alkalay, I. Krönke, M. Ben-Neriah, Y. & Baeuerle, P. A. (1993). Nature **365**, 182-185.
10. Traenckner, E. B.-M., Wilk, S. & Baeuerle, P. A. (1994). *EMBO J.* **13**, 101-109.
11. Schreck, R. & Baeuerle, P. A. (1991). *Trends in Cell Biol.* **1**, 39-42.
12. Schreck, R., Albermann, K. & Baeuerle, P.A. (1992). *Free Rad. Res. Comms.* **17**, 221-237.
13. Meyer, M., Schreck, R., Müller, J. M. & Baeuerle, P. A. (1993). *Oxidative Stress on Cell Activation and Viral Infection.* (C. Pasquier et al., ed), pp. 217-235, Birkhäuser Verlag AG, Basel.
14. Schreck, R., Meier, B., Männel, D., Dröge, W. & Baeuerle, P.A. (1992). *J. Exp. Med.* 175, 1181-1194.
15. Meyer, M., Schreck, R. & Baeuerle, P.A. (1993).*EMBO J.* **12**, 2005-2015.
16. Schreck, R., Rieber, P. & Baeuerle, P.A. (1991). *EMBO J.* **10**, 2247-2258.
17. Staal, F. J. T., Roederer, M. & Herzenberg, L. A. (1991). *Proc. Natl. Acad. Sci. USA* **87**, 9943-9947.
18. Mihm, S., Ennen, J., Pessara, U., Kurth, R. & Droege, W. (1991). *AIDS* **5**, 497-503.
19. Menon, S.D., Qin, S., Guy, G.R. & Tan, Y.H. (1993). *J. Biol. Chem.* **268**, 26805-26812.
20. Amstad, P., Peskin, a. Shah, G., Mirault, M.-E., Moret, R., Zbinden, I. & Cerutti, P. (1991). *Biochem. J.* **30**, 9305-9113.
21. Amstad, P., Moret, R. & Cerutti, P. (1993). *J. Biol. Chem.* **269**, 1-4.
22. Sendtner, M. & Thoenen, H. (1994). *Curr. Biol.* **4**, 1036-1039.
23. Avraham, K. B., Schickler, M., Sapoznikov, D. Yarom, R. & Groner,Y. (1988). *Cell* **53**, 211-217.
24. Rosen, D.R. et al. (1993). *Nature* **362**, 59-62.
25. Halliwell, B. & Gutteridge, J.M.C. (1989). *Free Radical in Biology and Medicine.* (2nd edn) Clarendon Press, Oxford.
26. Schreiber, E., Matthias, P., Müller, M.M. & Schaffner, W. (1989). *Nucl. Acids Res.* **17**, 6419.
27. Zabel, U., Schreck, R. & Baeuerle, P.A. (1991). *J. Biol. Chem.* **266**, 252-260.

10

REDOX REGULATION OF AP-1

A Link between Transcription Factor Signaling and DNA Repair

Steven Xanthoudakis[1*] and Tom Curran[2]

[1] Neurogenetics Program, Department of CNS
[2] The Roche Institute of Molecular Biology
Hoffmann-La Roche Inc. Nutley, New Jersey 07110

ABSTRACT

The inducible transcription factor, AP-1, is a heterodimeric leucine zipper complex containing the protein products of the *fos* and *jun* protooncogenes. The DNA binding activity of Fos and Jun is regulated *in vitro* by a posttranslational mechanism involving reduction-oxidation. Redox regulation is mediated through a conserved cysteine residue located in the DNA binding domain of both proteins. Oxidation or chemical modification of the cysteine has an inhibitory effect on AP-1 DNA binding activity. Conversely, reduction of this residue by chemical reducing agents or by a ubiquitous nuclear redox factor (Ref-1), purified and cloned from human cells, stimulates AP-1 DNA binding activity. In addition, recombinant Ref-1 stimulates the DNA binding activity of several other classes of redox regulated transcription factors. Immunodepletion studies indicate that Ref-1 is the major AP-1 redox activity in Hela cells. Interestingly, Ref-1 is a bifunctional protein; it also possesses an apurinic/apyrimidinic (A/P) endonuclease DNA repair activity. However, the redox and DNA repair activities of Ref-1 are physically and biochemically distinguishable. Ref-1 may represent a novel component of the signal transduction processes that regulate eukaryotic gene expression in response to cellular stress.

POSTTRANSLATIONAL REGULATION OF AP-1 BINDING ACTIVITY BY A REDOX MECHANISM

It has become apparent in recent years that proto-oncogenes function in multiple aspects of signal transduction. Their protein products include peptide growth factors, cell surface receptors, G proteins, and protein kinases (1-3). In addition, a subset of proto-onco-

* Corresponding author: (201) 235-7387, FAX (201) 235-7617.

genes, exemplified by c-*fos* and c-*jun*, encode nuclear transcription factors that have been implicated in a number of signal transduction cascades associated with growth, differentiation, neuronal excitation and cellular stress (4, 5). These genes provide a useful paradigm for the investigation of stimulus-evoked alterations in gene expression. As members of a complex immediate-early multi-gene family, c-*fos* and c-*jun* are thought to function in coupling a variety of short-term extracellular stimuli originating at the cell surface to long-term changes in cellular phenotype by modulating the expression of specific target genes (5). Fos and Jun family members form an array of homodimeric and heterodimeric protein complexes, collectively referred to as "AP-1", that bind to DNA regulatory elements containing the activator protein-1 (AP-1) site. Dimerization of the proteins via the leucine zipper domain (a coiled-coil α-helix) brings into juxtaposition regions in each protein which are rich in basic amino acids and that form a bipartite DNA binding domain (6).

Several mechanisms are thought to regulate the assembly, targeting and functional specificity of the various AP-1 complexes. These include: differential expression of individual family members, protein-protein interactions with other classes of transcription factors (7-12), conformational alterations (13-15) and altered DNA-binding specificities of the different heterodimers (16-18). Superimposed upon these mechanisms is the important influence of posttranslational modification. The levels of AP-1 DNA-binding activity and the transcriptional responses from AP-1 elements can be augmented in the absence of *de novo* protein synthesis, presumably through modification of pre-existing AP-1 proteins (19-21). Several years ago we fortuitously identified an unusual posttranslational modification involving reduction/oxidation (redox) that regulates the DNA binding activity of Fos and Jun *in vitro* (22). We found that recombinant Fos and Jun proteins, purified from *E. coli* under non-reducing conditions, failed to bind to AP-1 oligonucleotides in a sequence-specific manner, while those produced in reticulocyte lysates bound efficiently to DNA. Interestingly, supplementation of the purified bacterial AP-1 preparations with crude reticulocyte lysates restored the DNA binding activity of recombinant Fos and Jun (22). Analysis of these two protein expression systems revealed that efficient binding of Fos-Jun heterodimers and Jun homodimers to the AP-1 site was dependent on the presence, in the binding assay, of high concentrations (>5mM) of DTT or β-mercaptoethanol (22). Thus, we reasoned that the redox effect on AP-1 DNA binding activity was perhaps being mediated through specific cysteine residues in Fos and Jun (23). Examination of the primary amino acid sequence of the proteins revealed the presence of two cysteine residues located at similar positions in each of the proteins; the first, referred to as C1, was embedded in the highly basic DNA binding domain, and the second, referred to as C2, was located immediatedly C-terminal to the leucine zipper domain. Independent mutagenesis of both these cysteines demonstrated that redox regulation of AP-1 binding activity mapped to the C1 cysteine and that the sulfhydryl form of the cysteine in both Fos and Jun was necessary to achieve optimal DNA binding activity (23). The critical cysteine is flanked by basic amino acids (KCR) and is conserved in all of the Fos- and Jun-related proteins (6) (Figure 1), including those identified in *Drosophila* (24), and in several of the ATF/CREB proteins (6). Substitution of this residue with a serine resulted in increased DNA binding activity and a loss of redox regulation *in vitro* (23). Alkylation of the cysteine with N-ethylmaleimide inhibited DNA binding, while the DNA-bound form of the protein complex was protected from modification, indicating that the C1 cysteine lies in close proximity to the DNA (15, 23). In the absence of an appropriate reducing agent, the C1 cysteine is presumably oxidized to a state that does not permit the protein complex to bind DNA, although we have shown that this inactivation does not result from the formation of intra- or intermolecular disulphide bridges (23).

C1 *CYS*

Fos	LSPEEEEKRRIRRERNKMAAA<u>KCR</u>NRRRELTDTLQAETDQLEDEKSALQTEIANLLKEKEKLEFILAAH
Fra 1	LSPEEEERRRVRRERNKLAAA<u>KCR</u>NRRKELTDFLQAETDKLEDEKSGLQREIEELQKQKERLELVLEAH
Fra 2	LSPEEEEKRRIRRERNKLAAA<u>KCR</u>NRRRELTDRLQAETDQLEEEKSGLQKEIAELQKEKEKLEFMLVAH
Fos B	LTPEEEEKRRVRRERNKLAAA<u>KCR</u>NRRRELTDRLQAETDQLEEEKAELESEIAELQKEKERLEFVLVAH
Jun	PPLSPIDMESQERIKAERKRMRNRIAAS<u>KCR</u>KRKLERIARLEEKVKTLKAQNSELASTANMLREQVAQLKQKVMNH
Jun B	PPVSPINMEDQERIKVERKRLRNRLAAT<u>KCR</u>KRKLERIARLEDKVKTLKAENAGLSSAAGLLREQVAQLKQKVMTH
JUN D	PPLSPIDMDTQERIKAERKRLRNRIAAS<u>KCR</u>KRKLERISRLEEKVKTLKSQNETLASTASLLREQVAQLKQKVLSH

Figure 1. Conservation of the KCR Motif Among the Fos and Jun Families of Transcription Factors. Amino acid sequences corresponding to the DNA binding domain and leucine zipper motif of different Fos- and Jun-related proteins are aligned. The position of the conserved C1 cysteine residue and flanking basic amino acids are indicated.

A CELLULAR PROTEIN STIMULATES REDOX-DEPENDENT AP-1 DNA BINDING

Several lines of evidence suggest that redox regulation plays an important physiological role *in vivo*. First, the transforming v-*jun* oncogene contains a naturally-occurring mutation of *cys* to *ser* at the C1 position (25). Introduction of this mutation into c-*jun* enhances its transforming potential (26). A similar amino acid substitution in Fos augments both its DNA binding activity as well as its transforming activity (27). Hence, the oncogenic potential of *fos* and *jun* may be increased through deregulation of redox control. Second, preparations of AP-1 proteins, purified from mammalian cells by oligonucleotide affinity chromatography, exhibit the same redox-mediated enhancement in DNA binding activity as the *E. coli*-derived recombinant Fos and Jun proteins (22). Thus, redox control of AP-1 DNA binding appears to be a general property of AP-1 proteins and it is not an artifact associated with polypeptides expressed in *E. coli*. Third, and perhaps most important, we discovered that in the absence of chemical reducing agents, AP-1 DNA binding could be dramatically stimulated by pretreating Fos and Jun with nuclear extracts prepared from a variety of mammalian cells and tissues (22, 28). The nuclear redox activity, designated Ref-1 (Redox factor-1), was initially shown to be sensitive to both proteolysis and heat treatment (28). Inhibition of this activity by alkylating agents suggested that Ref-1 contained free sulfhydryl moeities and that these were necessary for stimulating AP-1 DNA binding activity (22). Additional studies aimed at characterizing Ref-1 in crude nuclear extracts established that Ref-1 augmented AP-1 DNA binding without altering the dimerization properties of Fos and Jun or the physical interaction of the protein complex with the AP-1 site (28). The redox activity of Ref-1 could be enhanced in the presence of thioredoxin, thioredoxin reductase and NADPH, implying that Ref-1 may participate in a redox cycle (23).

To further investigate the role of Ref-1 in AP-1 redox control and signal transduction it was necessary to purify and characterize the activity. A seven hundred fold purification of Ref-1 from Hela cell nuclear extracts was achieved by conventional column chromatography (29). Throughout the course of purification Ref-1 activity was monitored using electrophoretic mobility shift assays. Truncated Fos and Jun polypeptides were used as substrates for Ref-1 in order to distinguish stimulation of AP-1 DNA binding activity from any enrichment in cellular AP-1 activity that was endogenous to the nuclear extract. Immunoblotting analysis of Ref-1 activity isolated from different stages of the purification process indicated that Ref-1, a 37 kD polypeptide, was antigenically distinct from Fos and Jun, although it did co-purify with AP-1-related proteins through most of the fractionation (29).

A reverse genetics approach was used to isolate the gene encoding the human Ref-1 protein (30). Briefly, analysis of the amino acid sequence of purified Ref-1 enabled us to deduce a partial DNA sequence and to use mixed-primer PCR amplification to obtain a short fragment corresponding to the N-terminal domain. This fragment was used as a probe to screen a phage library from which several overlapping cDNA clones were isolated. The largest cDNA cloned contained a single 318 amino acid open reading frame that, when expressed in reticulocyte lysates, encoded a polypeptide with a molecular weight identical to that of the purified Ref-1 protein. Subsequent studies revealed that *ref*-1 is ubiquitously expressed as a 1.6 kb mRNA from a single gene located on chromosome 14 (30, 31).

To confirm the identity of the cloned *ref*-1 cDNA it was necessary to demonstrate that the product of the cloned *ref*-1 gene could stimulate AP-1 DNA binding activity in a reconstituted system comprised of purified components. To this end, the cloned *ref*-1 cDNA was expressed in *E.coli* and purified to near homogeneity as a fusion protein (30). Biochemical analysis showed that recombinant Ref-1 (rRef-1) was functionally indistinguishable from the cell-derived activity(30). The recombinant protein displayed all of the characterisitc features associated with the cellular redox activity including redox-dependent stimulation of AP-1 DNA binding and a thioredoxin-mediated enhancement of its redox activity. In addition, rRef-1 was capable of stimulating the DNA binding activity of several other known redox-regulated transcription factors (eg. Myb, ATF-1/-2, p65NF-κb), suggesting that it may play a more general role in the signal transduction processes that regulate gene expression. Immunofluorescence studies using affinity purified antibodies raised against the recombinant Ref-1 protein showed that Ref-1 localizes to the nuclear compartment of cells. Furthermore, experiments in which Ref-1 was immunodepleted from crude nuclear extracts indicated that Ref-1 constitutes the major AP-1 redox component in Hela cells.

A ROLE FOR Ref-1 IN TRANSCRIPTION-REPAIR COUPLING

Since Ref-1 was purified and cloned on the basis of its AP-1 redox activity, we did not anticipate that a database search for homologous sequences would identify Ref-1 as a DNA repair enzyme (30). This additional function was discovered based on its sequence identity to a group of DNA repair proteins designated as class II hydrolytic apurinic/apyrimidinic (A/P) endonucleases (32-36) (figure 2). Until recently, Ref-1, like other enzymes in its class, was thought to function mainly in the repair of DNA lesions generated by exposure to reactive oxygen radicals (37). These enzymes act by hydrolyzing DNA at the phosphate bond 5' to abasic sites. The identification of a DNA repair protein that also had the ability to regulate the DNA-binding activity of different transcription factors suggested a novel link between the processes of oxidative stress DNA repair and transcription regulation.

To examine the relationship between the redox and DNA repair functions of Ref-1, we generated a series of N- and C-terminal deletion mutants and assayed them for each of the two activities *in vitro* (38). Our results indicated that the redox and DNA repair activities of Ref-1 are encoded by distinct domains. Sequences in the N-terminal region of Ref-1 are required for the redox activity, while the DNA repair activity requires conserved C-terminal sequences. These domains are functionally independent and can be genetically dissociated without significantly disrupting the individual activities. The high degree of sequence similarity shared by the C-terminal half of Ref-1 and other functionally related repair proteins in bacteria (Exo III, Exo A) (39, 40) and flies (Rrp1) (36) is consistent with the assignment of the A/P endonuclease activity to the C-terminal region. On the other hand, the redox

```
       1                                               50                                                    100
Ref-1  .......... .......... .......... .......... ..........    .......... .......... .......... .......... ..........
HAP-1  .......... .......... .......... .......... ..........    .......... .......... .......... .......... ..........
mAPEX  .......... .......... .......... .......... ..........    .......... .......... .......... .......... ..........
BAP-1  .......... .......... .......... .......... ..........    .......... .......... .......... .......... ..........
 RRP1  MPRVKAVKKQ AEALASEPTD PTPNANGNGV DENADSAAEE LKVPAKGKPR    ARKATKTAVS AENSEEVEPQ KAPTAVARGK KKQPKDTDEN GQMEVVAKPK
EXO A  .......... .......... .......... .......... ..........    .......... .......... .......... .......... ..........
EXOIII .......... .......... .......... .......... ..........    .......... .......... .......... .......... ..........

       101                                             150                                                   200
Ref-1  .......... .......... .......... .......... ..........    .......... .......... .......... .......... ..........
HAP-1  .......... .......... .......... .......... ..........    .......... .......... .......... .......... ..........
mAPEX  .......... .......... .......... .......... ..........    .......... .......... .......... .......... ..........
BAP-1  .......... .......... .......... .......... ..........    .......... .......... .......... .......... ..........
 RRP1  GRAKKATAEA EPEPKVDLPA GKATKPRAKK EPTPAPDEVT SSPPKGRAKA    EKPTNAQAKG RKRKELPAEA NGGAEEAAEP PKQRARKEAV PTLKEQAEPG
EXO A  .......... .......... .......... .......... ..........    .......... .......... .......... .......... ..........
EXOIII .......... .......... .......... .......... ..........    .......... .......... .......... .......... ..........

       201                                             250                                                   300
Ref-1  .......... .......... .......... .......... ..........    .......... .......... .......... .......... ..........
HAP-1  .......... .......... .......... .......... ..........    .......... .......... .......... .......... ..........
mAPEX  .......... .......... .......... .......... ..........    .......... .......... .......... .......... ..........
BAP-1  .......... .......... .......... .......... ..........    .......... .......... .......... .......... ..........
 RRP1  TISKEKVQKA ETAAKRARGT KRLADSEIAA ALDEPEVDEV PPKAASKRAK    KGKMVEPSPE TVGDFQSVQE EVESPPKTAA APKKRAKKTT NGETAVELEP
EXO A  .......... .......... .......... .......... ..........    .......... .......... .......... .......... ..........
EXOIII .......... .......... .......... .......... ..........    .......... .......... .......... .......... ..........

       301                                             350                                                   400
Ref-1  .......... .......... .......... .......... .....MPKRG    KKGAVAEDGD ELRTEPEAKK SKTAAKKNDK EAAGEGPALY EDPPDQKTSP
HAP-1  .......... .......... .......... .......... .....MPKRG    KKGAVAEDGD ELRTEPEAKK SKTAAKKNDK EAAGEGPALY EDPPDHKTSP
mAPEX  .......... .......... .......... .......... .....MPKRG    KKAA.ADDGE EPKSEPETKK SKGAAKKTEK EAAGEGPVLY EDPPDQKTSP
BAP-1  .......... .......... .......... .......... .....MPKRG    KKGAVVEDAE EPKTEPEAKK SKAGAKKNEK EAVGEGAVLY EDPPDQKTSP
 RRP1  KTKAKPTKQR AKKEGKEPAP GKKQKKSADK ENGVVEEEAK PSTETKPAKG    RKKAPVKAED VEDIEEAAEE SKPARGRKKA AAKAEEPDVD EESGSKTTKK
EXO A  .......... .......... .......... .......... ..........    .......... .......... .......... .......... ..........
EXOIII .......... .......... .......... .......... ..........    .......... .......... .......... .......... ..........

       401                                             450                                                   500
Ref-1  SGKPAT.... .......... ......LKIC SWNVDGL... ......RAWI    KKKGLDWVKE EAPDILCLQE TKCSEN...K LPAELQE..L PGLSHQYWSA
HAP-1  SGKPAT.... .......... ......LKIC SWNVDGL... ......RAWI    KKKGLDWVKE EAPDILCLQE TKCSEN...K LPAELQE..L PGLSHQYWSA
mAPEX  SGKSAT.... .......... ......LKIC SWNVDGL... ......RAWI    KKKGLDWVKE EAPDILCLQE TKCSEN...K LPAELQE..L PGLTHQYWSA
BAP-1  SGKSAT.... .......... ......LKIC SWNVDGL... ......RAWI    KKKGLDWVKE EAPDILCLQE TKCSEN...K LPVELQE..L SGLSHQYWSA
 RRP1  AKKAETKTTV TLDKDAFALP ADKEFNLKIC SWNVAGL... ......RAWL    KKDGLQLIDL EEPDIFCLQE TKCAND...Q LPEEVTR..L PGYHPYWLCM
EXO A  .......... .......... ......MKLI SWNIDSLNAA LTSDSARAKL    SQEVLQTLVA ENADIIAIQE TKLSAKGPTK KHVEILEELF PGYENTWRSS
EXOIII .......... .......... ......MKFV SFNINGL... ......RARP    HQLE.AIVEK HQPDVIGLQE TKVHDDMFPL ..EEVAKLGY NVFYH.....

       501                                             550                                                   600
Ref-1  P.SDKEGYSG VGLL.SRQCP LKVSYGI..G EEEHDQEGRV IVAEFDSFV.    ...LVTAYVP NAG..RGLVR LEYRQRWDEA FRKFLK.GLA SRKPLVLCGD
HAP-1  P.SDKEGYSG VGLL.SRQCP LKVSYGI..G DEEHDQEGRV IVAEFDSFV.    ...LVTAYVP NAG..RGLVR LEYRQRWDEA FRKFLK.GLA SRKPLVLCGD
mAPEX  P.SDKEGYSG VGLL.SRQCP LKVSYGI..G EEEHDQEGRV IVAEFESFV.    ...LVTAYVP NAG..RGLVR LEYRQRWDEA FRKFLK.DLA SRKPLVLCGD
BAP-1  P.SDKEGYSG VGLL.SRQCP LKVSYGI..G EEEHDQEGRV IVAEYDAFV.    ...LVTAYVP NAG..RGLVR LEYRQRWDEA FRKFLK.GLA SRKPLVLCGD
 RRP1  P.G...GYAG VAIY.SKIMP IHVEYGI..G NEEFDDVGRM ITAEYEKFY.    ...LINVYVP NSG..RKLVN LEPRMRWEKL FQAYVK.KLD ALKPVVICGD
EXO A  QEPARKGYAG TMFLYKKELT PTISFPEIGA PSTMDLEGRI ITLEFDAFF.    ...VTQVYTP NAG..DGLKR LEERQVWDAK YAEYLA.ELD KEKPVLATGD
EXOIII ...GQKGHYG VALL.TKETP IAVRRGFPGD DEEAQR..RI IMAEIPSLLG    NVTVINGYFP QGESRDHPIK FPAKAQFYQN LQNYLETELK RDNPVLIMGD

       601                                             650                                                   700
Ref-1  LNVAHEEIDL RNPKGNKK.. ....NAGFTP QERQGFGELL QAVALADSFR    HLYPNTPYAY TFWTYM.MNA RSKNVGWRLD YFLLSHSL.. ..LPALCDSK
HAP-1  LNVAHEEIDL RNPKGNKK.. ....NAGFTP QERQGFGELL QAVPLADSFR    HLYPNTPYAY TFWTYM.MNA RSKNVGWRLD YFLLSHSL.. ..LPALCDSK
mAPEX  LNVAHEEIDL RNPKGNKK.. ....NAGFTP QERQGFGELL QAVPLADSFR    HLYPNTAYAY TFWTYM.MNA RSKNVGWRLD YFLLSHSL.. ..LPALCDSK
BAP-1  LNVAHEEIDL RNPKGNKK.. ....NAGFTP QERQGFGELL QAVPLTDSFR    HLYPNTAYAY TFWTYM.MNA RSKNVGWRLD YFLLSQSV.. ..LPALCDSK
 RRP1  MNVSHMPIDL ENPKNNTK.. ....NAGFTQ EERDKMTELL .GLGFVDTFR    HLYPDRKGAY TFWTYM.ANA RARNVGWRLD YCLVSERF.. ..VPKVVEHE
EXO A  YNVAHNEIDL ANPASNRR.. ....SPGFTD EERAGFTNLL .ATGFTDTFR    HVHGDVPERY TWWAQRSKTS KINNTGWRID YWLTSNRI.. ..ADKVTKSD
EXOIII MNISPTDLDI GIGEENRKRW LRTGKCSFLP EEREWMDRLM .SWGLVDTFR    HANPQTADRF SWFDYRSK.G FDDNRGLRID LLLASQPLAE CCVETGIDYE

       701        720
Ref-1  IRSKALGSDH CPITLYLAL*
HAP-1  IRSKALGSDH CPITLYLAL*
mAPEX  IRSKALGSDH CPITLYLAL*
BAP-1  IRSKALGSDH CPITLYLAL*
 RRP1  IRSQCLGSDH CPITIFFNI*
EXO A  MIDSGARQDH TPIVLEIDL*
EXOIII IRSMEKPSDH APVWATFRR*
```

Figure 2. Sequence Alignment Between Ref-1 and Related A/P Endonucleases From Other Organisms. The amino acid sequence alignment between human Ref-1, bovine BAP-1, murine APEX (mAPEX), drosophila Rrp1, *E. coli* exonuclease III (EXOIII), and *S. pneumoniae* exonuclease A (EXO A) is shown. Note that Ref-1 and HAP-1 represent identical proteins. Sequence alignments were generated using the Genetics Computer Group (GCG) Bestfit and Pileup protein analysis software packages.

activity of Ref-1 localizes to a region that is unique to the mammalian enzyme. The N-terminal 62 amino acids of Ref-1 are not present in ExoA (40), ExoIII (39) or Rrp1 (36). Our studies indicate that this short N-terminal region and sequences immediately adjacent to it are critical for Ref-1 redox activity. At least one of the bacterial repair proteins, ExoIII, was found to lack any detectable redox activity *in vitro* (41).

The redox and DNA repair activities of Ref-1 may be subject to different regulatory influences. We have found that chemical alkylation or oxidation of cysteine sulfhydryls in Ref-1 inhibits the redox activity without affecting its DNA repair activity (38). This is consistent with the observation that formation of a stable cross-linked complex between Ref-1 and Jun is dependent on the presence of reduced cysteines in both proteins (38). In the case of Jun this interaction can only occur through the redox active cysteine located in the DNA binding domain. At the moment, the site of interaction within Ref-1 has yet to be mapped, although other studies have shown that cys^{65}, which is located in the redox domain of Ref-1, is critical for redox-dependent stimulation of Jun DNA binding activity (41). Taken together with the observation that the redox activity of Ref-1 can be directly regenerated by thioredoxin (23, 30), these findings suggest that Ref-1 may be subject to posttranslational regulation. Clearly, this mode of control would act to ensure a rapid response to the changing redox state of the cell. Changes in the redox state of specific cysteines could serve as a mechanism to selectively modulate the redox activity of Ref-1 without altering its endonuclease activity. The endonuclease activity would be required even under basal growth conditions to enable the cells to cope with the high rate of spontaneous base loss that occurs in the genome on a continuous basis (42).

It is not clear why Ref-1 has evolved two apparently distinct activities that function in DNA repair and transcription. One possible reason is that the two activities may act coordinately to initiate DNA repair and also stimulate the DNA binding activity of latent AP-1 proteins during oxidative stress. Under these conditions thioredoxin or a functionally related activity would maintain Ref-1 in a reduced and thereby redox-active state. The net effect would be to augment the cellular response to adverse oxidative conditions. Rapid activation of AP-1 and other transcription factors by Ref-1, followed by the expression of specific target genes, could act to minimize cellular damage caused by oxidative stress (4). Indeed, transcriptional and posttranslational induction of AP-1 DNA binding activity has been shown to accompany treatment of cells with agents that induce oxidative stress (43-45). Given the ability of Ref-1 to associate with AP-1 (29), another possibility is that Ref-1 may be preferentially directed to AP-1 target genes. This would confer an advantage to the cell by increasing the rate of repair to AP-1 responsive genes following oxidative damage.

Finally, further elucidation of the precise physiological role of Ref-1 awaits additional studies using *in vivo* systems. Unfortunately, transfection studies attempting to address this issue have been hampered by difficulties encountered in trying to establish cell lines that over-express Ref-1 in a constitutive or inducible manner. Furthermore, since Ref-1 is ubiquitously expressed at high basal levels, additional expression encoded by an exogenous allele may not yield any new information. Therefore, as an alternative approach to study the function of Ref-1 *in vivo*, we have initiated experiments to disrupt the *ref*-1 locus using a gene targeting strategy. Analysis of both cultured cells and whole animals lacking a functional *ref*-1 gene will help further our understanding of the biological role of Ref-1.

REFERENCES

1. Reddy, E. P., Skalka, A. M. &Curran, T. (1988) *The Oncogene Handbook* (Elsevier Science Publishers, Amsterdam).
2. Hunter, T. (1991) *Cell.* 64, 249-270.

3. Cantley, L. C., Auger, K. R., Carpenter, C., Duckworth, B., Graziani, A., Kapeller, R. &Soltoff, S. (1991) *Cell*. 64, 281-302.
4. Holbrook, N. J. & Fornace Jr., A. J. (1991) *The New Biologist*. 3, 825-833.
5. Morgan, J. I. & Curran, T. (1991) *Ann. Rev. Neurosci.* 14, 421-451.
6. Kerppola, T. & Curran, T. (1991). In *Current Opinion in Structural Biology*, eds. (Current Biology, London), pp. 71-79.
7. Diamond, M. I., Miner, J. N., Yoshinago, S. K. &Yamamoto, K. R. (1990) *Science*. 249, 1266-1272.
8. Gaub, M.-P., Bellard, M., Scheuer, I., Chambon, P. &Sassone-Corsi, P. (1990) *Cell*. 43, 1267-1276.
9. Jonat, C., Rahmsdorf, H. J., Park, K.-K., Cato, A. C. B., Gebel, S., Ponta, H. &Herrlich, P. (1990) *Cell*. 62, 1189-1204.
10. Owen, T. A., Bortell, R., Yocum, S. A., Smock, S. L., Zhang, M., Abate, C., Shalhoub, V., Aronin, N., Wright, K. L., van Wijnen, A. J., Stein, J. L., Curran, T., Lian, J. B. &Stein, G. S. (1990) *Proc. Natl. Acad. Sci. USA*. 87, 9990-9994.
11. Schule, R., Rangarajan, P., Kliewer, S., Ransone, L. J., Bolado, J., Yang, N., Verma, I. M. &Evans, R. M. (1990) *Cell*. 62, 1217-1276.
12. Yang-Yen, H. F., Chambard, J.-C., Sun, Y. L., Smeal, T., Schmidt, T. J., Drouin, J. &Karin, M. (1990) *Cell*. 62, 1205-1215.
13. Kerppola, T. K. & Curran, T. (1991) *Science*. 254, 1210-4.
14. Kerppola, T. K. & Curran, T. (1991) *Cell*. 66, 317-26.
15. Patel, L., Abate, C. &Curran, T. (1990) *Nature*. 347, 572-575.
16. Kerppola, T. K. & Curran, T. (1994) *Oncogene*. 9, 675-84.
17. Ryseck, R.-P. & Bravo, R. (1991) *Oncogene*. 6, 533-542.
18. Hai, T. & Curran, T. (1991) *Proc. Natl. Acad. Sci. USA*. 4143, 3720-3724.
19. Welham, M. J., Wyke, J. A., Lang, A. &Wyke, A. W. (1990) *Oncogene*. 5,
20. Chiu, R., Imagawa, M., Imbra, R. J., R., B. J. &Karin, M. (1987) *Nature*. 329, 648-651.
21. Angel, P., Imagawa, M., Chiu, R., Stein, B., Imbra, R. J., Rahmsdorf, H. J., Jonat, C., Herrlich, P. &Karin, M. (1987) *Cell*. 49, 729-739.
22. Abate, C., Rauscher III, F. J., Gentz, R. &Curran, T. (1990) *Proc. Natl. Acad. Sci. USA*. 87, 1032-1036.
23. Abate, C., Patel, L., Rauscher III, F. J. &Curran, T. (1990) *Science*. 249, 1157-1161.
24. Perkins, K. K., Admon, A., Patel, N. &Tjian, R. (1990) *Genes and Dev*. 4, 822-834.
25. Maki, Y., Bos, T. J., Davis, C., Starbuck, M. &Vogt, P. K. (1987) *Proc. Natl. Acad. Sci. USA*. 84, 2848-2852.
26. Morgan, I. M., Asano, M., Havarstein, L. S., Ishikawa, H., Hiiragi, T., Ito, Y. &Vogt, P. K. (1993) *Oncogene*. 8, 1135-40.
27. Okuno, H., Akahori, A., Sato, H., Xanthoudakis, S., Curran, T. &Iba, H. (1993) *Oncogene*. 8, 695-701.
28. Abate, C., Luk, D. &Curran, T. (1990) *Cell. Grow. & Diff.* 1, 455-462.
29. Xanthoudakis, S. & Curran, T. (1992) *EMBO J.* 11, 653-656.
30. Xanthoudakis, S., Miao, G., Wang, F., Pan, Y.-C. E. &Curran, T. (1992) *EMBO J.* 11, 3323-3335.
31. Robson, C. N., Hochhauser, D., Craig, R., Rack, K., Buckle, V. J. &Hickson, I. D. (1992) *Nucleic Acids Res*. 20, 4417-21.
32. Robson, C. N., Milne, A. M., Pappin, D. J. C. &Hickson, I. D. (1991) *Nucl. Acids Res.* 19, 1087-1092.
33. Robson, C. N. & Hickson, I. D. (1991) *Nucl. Acids Res.* 19, 5519-5523.
34. Seki, S., Hatsushika, M., Watanabe, S., Akiyama, K., Nagao, K. &Tsutsui, K. (1992) *Biochem. Biophys. Acta*. 1131, 287-299.
35. Demple, B., Herman, T. &Chen, D. S. (1991) *Proc. Natl. Acad. Sci. USA*. 88, 11450-11454.
36. Sander, M., Lowenhaupt, K. &Rich, A. (1991) *Proc. Natl. Acad. Sci. USA*. 88, 6780-6784.
37. Doetsch, P. W. & P., C. R. (1990) *Mutation Res*. 236, 173-201.
38. Xanthoudakis, S., Miao, G. &Curran, T. (1994) *PNAS*. 91, 23-27.
39. Saporito, S. M., Smith-White, B. J. &Cunningham, R. P. (1988) *J. Bacteriol.* 170, 4542-4547.
40. Puyet, A., Greenberg, B. &Lacks, S. A. (1989) *J. Bacteriol.* 171, 2278-2286.
41. Walker, L. J., Robson, C. N., Black, E., Gillespie, D. &Hickson, I. D. (1993) *Mol. Cell. Biol.* 13, 5370-5376.
42. Ames, B. N. (1987) *Ann. Intern. Med.* 107, 526-545.
43. Crawford, D., Zbinden, I., Amstad, P. &Cerutti, P. (1988) *Oncogene*. 3, 27-32.
44. Stein, B., Rahmsdorf, H. J., Steffen, A., Litfin, M. &Herrlich, P. (1989) *Mol. Cell. Biol.* 9, 5169-5181.
45. Devary, Y., Gottlieb, R. A., Lau, L. F. &Karin, M. (1991) *Mol. Cell. Biol.* 11, 2804-2811.

THE TRANSCRIPTION FACTOR TCF/Elk-1*

A Nuclear Sensor of Changes in the Cellular Redox Status

Judith M. Müller,[1] Michael A. Cahill,[2] Alfred Nordheim,[2] and Patrick A. Baeuerle[1]

[1] Institute of Biochemistry
Albert-Ludwig-University
D- 79104 Freiburg, Germany
[2] Institute for Molecular Biology
Hannover Medical School
D- 30623 Hannover, Germany

ABSTRACT

The ternary complex factor (TCF) and the serum response factor (SRF) are nuclear transcription factors which are essential for efficient signal transduction via the serum response element SRE in the c-*fos* promoter. Their activation leads to a rapid induction of c-*fos* gene expression. Activation of mitogen-activated protein kinases (MAPK) by signalling cascades and subsequent TCF phosphorylation are known to be essential steps in this transcriptional activation.

In transient transfections we could show activation of a SRE dependent promoter following treatment of HeLa cells with either the oxidant H_2O_2 or various antioxidants. This activation is dependent on the presence of both an intact SRE as well as a TCF binding site. In gel shifts we observed changes in the migration of ternary complexes which were due to hyperphosphorylation of TCF/Elk-1. In-gel kinase assays showed an activation of MAPK Ser/Thr kinase activity by both oxidant and antioxidant stimuli, which temporally correlates with the appearance of hyperphosphorylated TCF/Elk-1. Thus antagonistic intracellular redox changes can lead to the same functional modulation of the nuclear transcription factor TCF/Elk-1 that was previously described following mitogenic stimuli.

* Abreviations: **BHA** butylated hydroxyanisole, **FCS** fetal calf serum, **kDa** kilo dalton, **MAPkinase** mitogen activated proteine kinase, **MBP** myelin basic protein, **NAC** N-acetyl L-cysteine, **PPase** phosphatase, **PDTC** pyrrolidine dithiocarbamate, **ROI** reactive oxygen intermediates, **SRE** serum response element, **SRF** serum responsive factor, **TCF** ternary complex factor, **wt** wildtype.

Biological Reactive Intermediates V
Edited by R. Snyder et al., Plenum Press, New York, 1996

INTRODUCTION

Serum induced activation of the c-*fos* gene is mediated by factors binding to the SRE (serum response element). The SRE represents an important c-*fos* promoter element that serves as a nuclear entry point for extracellular signals such as mitogens, protein kinase C activation, UV and ionizing radiation (reviewed in 41). These stimuli activate signalling cascades leading to the rapid induction of c-*fos* gene expression via the activation of kinases like PKC, Raf, MEK, MAPK. Simultaneous binding to the c-*fos* SRE and adjacent sequences by the proteins SRF (serum response factor) and TCF (ternary complex factor) was shown to be essential for signal uptake by the SRE. TCF requires SRF to be bound specifically to the SRE [10,11,13,14,39]. TCFs have been identified as members of the Ets family of transcription factors and include Elk-1 and SAP-1 (reviewed in 21). Activation by signalling cascades leads to TCF phosphorylation and subsequent promoter stimulation[10,14,16,42].

The level of reactive oxygen intermediates is finely regulated in living cells. Changes which increase or reduce levels of oxygen and its intermediates are associated with numerous pathological and physiological conditions[12,36]. Changes in the levels of reactive oxygen intermediates induced by either oxidants or antioxidants, also activate the c-*fos* and other SRE-containing promoters[5,23].

In order to identify signalling pathways and transcription factors activated upon changes in intracellular ROI/redox level we examined the c-*fos* gene induction by oxidants or antioxidants in detail.

RESULTS AND DISCUSSION

Transient transfection studies showed that both the oxidant H_2O_2 and the antioxidant pyrrolidine dithiocarbamate (PDTC) stimulate transcription of a SRE-dependent reporter gene in HeLa cells (Fig. 1). 150 µM H_2O_2, 60 µM PDTC or 10% FCS caused an approximately 4-fold stimulation of luciferase generated light units following transfection of a SRE dependent reporter plasmid. PMA at a concentration of 50ng/ml was the strongest inducer

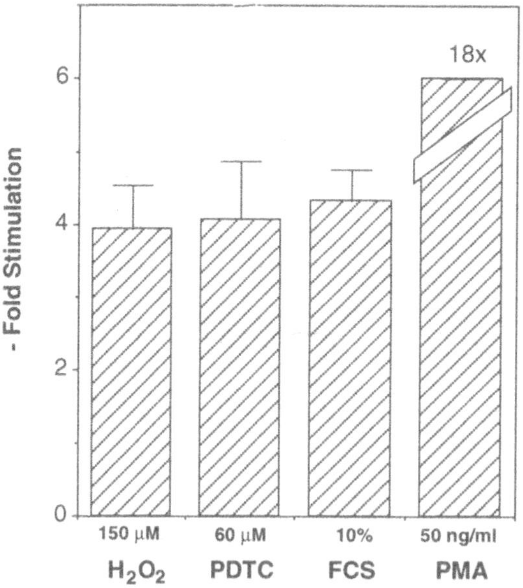

Figure 1. H_2O_2, PDTC, FCS and PMA activate c-*fos* SRE-dependent transcription in HeLa-cells. HeLa cells transfected with the pSRE$_2$-tk80-luc plasmid were serum starved (H_2O_2, FCS, and PMA) or grown in 5% FCS (PDTC) followed by stimulation with the indicated substances. The average and standard deviation derived from 4 independent transfections is shown. Activation is expressed as fold-induction above uninduced levels.

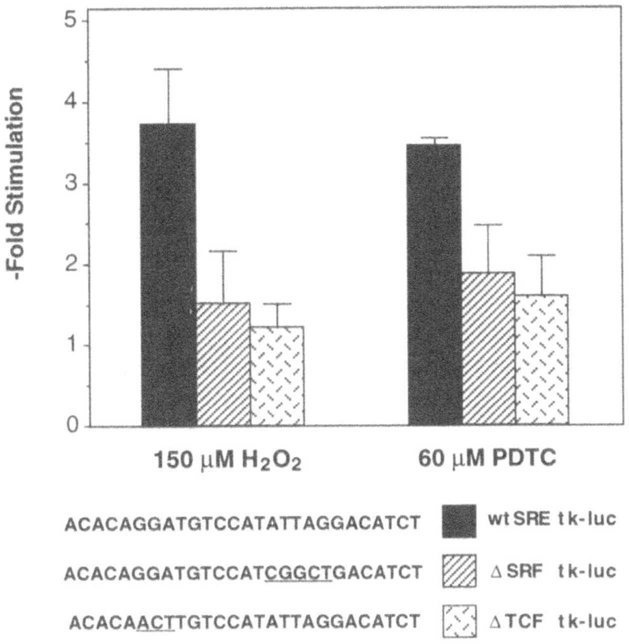

Figure 2. Intact SRF- and TCF-binding sites are required for oxidant- and antioxidant-mediated transactivation in transfected HeLa-cells. Very little activation by H_2O_2 or PDTC was seen when the transfected reporter plasmid contained either a mutated SRF-binding site (ΔSRF) or a non-functional TCF-binding site (ΔTCF).

(18-fold) under these conditions. Using reporter constructs containing mutated binding-sites (Fig. 2) we demonstrated that both the binding site for TCF and the binding site for SRF were required for the stimulation by H_2O_2 or PDTC. Mutation of the SRF binding site led

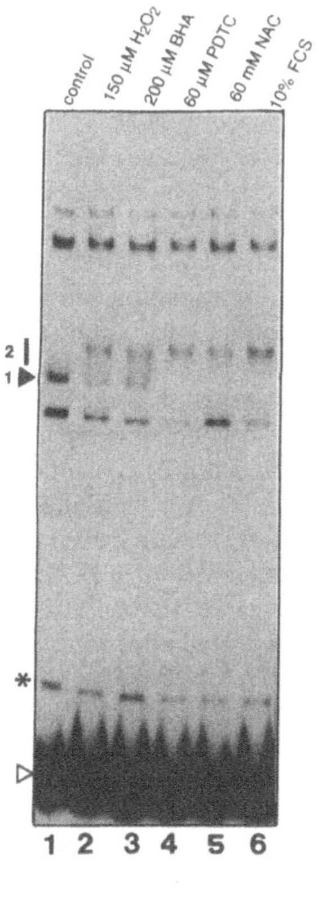

Figure 3. Induced mobility changes of TCF-DNA complexes after HeLa cell stimulation with oxidants and antioxidants. Serum-starved HeLa cells were stimulated with H_2O_2, various antioxidants, or FCS as detailed below. Binding reactions contained 5 μg of total cell extract with coreSRF$_{132-222}$ (SRF) and ^{32}P-labelled c-*fos* SRE probe. 1: control, 2: 150 μM H_2O_2 (40 min.), 3: 200 μM BHA (15 min.), 4: 60 μM PDTC (30 min.), 5: 60 mM NAC (30 min.), 6: 10% FCS (15 min.). The positions of SRF complexes and the SRF/TCF uninduced (1) and induced (2) ternary complexes are indicated. The asterisks depicts the SRE/coreSRF complex. The open arrowhead points to free SRE probe.

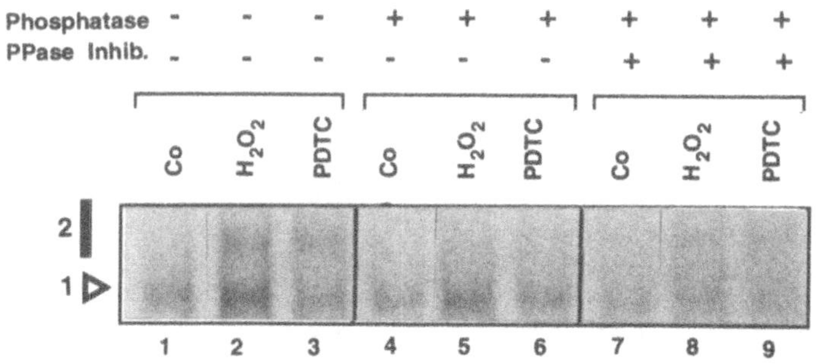

Figure 4. The altered mobility of the ternary TCF-DNA complex is reversed by treatment with potato acid phosphatase. Dialysed whole cells extracts were treated with (lane 4-9) or without (lane 1-3) potato acid phosphatase in the presence (lane 7-9) or absence (lane 1-6) of a phosphatase inhibitor cocktail. 4 μg of stimulated (150μM H_2O_2, 60μM PDTC) or unstimulated control extracts were then used in a bandshift (see Fig. 3). Only Elk-1/SRF complexes are shown.

to a 2.4-fold reduction of H_2O_2 and 1.8-fold reduction of PDTC stimulation compared to the wild type reporter plasmid. Similarily the mutated TCF site reduced stimulation to 3.0-fold or 2.1-fold, respectively. This indicates that a SRE-dependent ternary complex is critical for receiving the redox-dependent signals.

In a bandshift assay, TCF-DNA complexes can be assayed using a ^{32}P labelled SRE DNA fragment and a SRFcore *in-vitro*-translated protein in a native polyacrylamide gel. Upon stimulation with either H_2O_2 or various structural unrelated antioxidants such as PDTC, BHA, or NAC, a ternary complex with reduced migration ability could be detected. Stimulation with FCS led to the appearance of complexes of similarly reduced mobility. Hyperphosphorylation of TCF, which correlates precisely with c-*fos* promoter activation, has previously been observed after stimulation with FCS or EGF[16]. The complexes induced by H_2O_2, PDTC, and NAC in HeLa cells are formed predominantly by Elk-1 or antigenically closely related TCFs since they are abolished[24] by an antiserum specific for Elk-1 which does not react with the related TCFs[40] SAP-1a and SAP-1b.

The retardation of the Elk-1/SRFcore band in bandshifts could be reversed by treatment of the H_2O_2 or PDTC stimulated HeLa cell extracts with potato acid phosphatase, but not in the presence of phosphatase inhibitors (Fig. 4). We conclude that Elk-1 is hyperphosphorylated following stimulation with the oxidant H_2O_2 or with antioxidants. To verify whether c-*fos* transcription occurred following redox imbalancing we looked at c-*fos* mRNA levels in Northern blots. PDTC, BHA, FCS and H_2O_2 treatment led to the accumulation of c-*fos* mRNA (data not shown). This increase temporally followed the appearence of the hyperphoshorylated ternary complex, reflecting redox-induced c-*fos* mRNA accumulation after SRE activation. The induction of c-*fos* mRNA has been reported after H_2O_2 or UV stimulation[25,29].

We then analysed Elk-1 hyperphoshorylation upon treatment of HeLa cells with PDTC, FCS or H_2O_2 as a function of time. Kinetics showed as was expected from previous reports[22,42] that FCS treatment caused rapid and transient Elk-1 hyperphosphorylation (Fig. 5, lanes 12-16). Using H_2O_2 or PDTC (Fig. 5, lanes 2-11) the reduced mobility complex appeared after 15 to 30 min and disapeared shortly after 60min (data not shown). Treatment with BHA showed a slower migrating Elk-1 band after only 5 min, which was no longer detctable after 30min[24]. Superoxide producing agents such as paraquat or pyrogallol[4] induced no change in Elk-1 mobility at various concentrations for up to 60 minutes (data not shown). We concluded that the oxidative response is more specific for H_2O_2 rather than for general oxidative stress. These obsevations are streghtened by data from transient transfections, where 100μM paraquat had no effect[24] on a SRE dependent luciferase reporter.

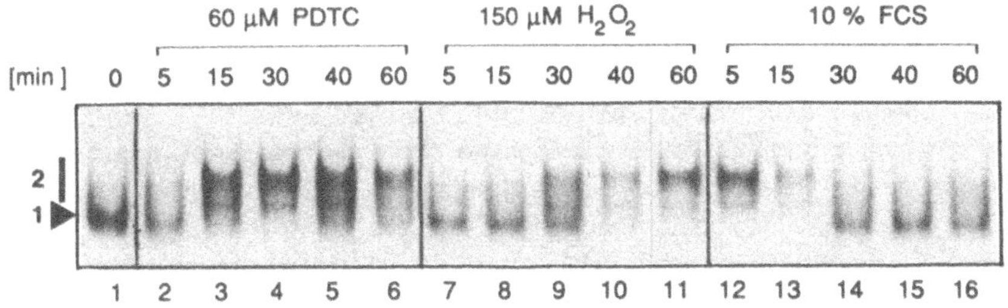

Figure 5. Kinetics of modified Elk-1 appearance following PDTC, H_2O_2 and FCS stimulation. Serum-starved HeLa cells were treated with 60 μM PDTC, 150 μM H_2O_2 or 10% FCS for the indicated times. 5 μg of whole cell extract were used in a binding reactions as in Fig. 3. Only Elk-1/SRF complexes are shown.

Protein synthesis is not required for these redox responses, because in the presence of cycloheximide H_2O_2 or PDTC induced similar shifts (data not shown). Both oxidants and antioxidants therefore rapidly activate SRE-dependent gene expression via induction of kinase activity and following hyperphoshorylation of TCF/Elk-1.

We then asked the question whether changes in the concentraion of ROI levels can activate MAPkinases. Activated ERK1 (p44 MAPK) and ERK2 (p42 MAPK) kinase can phosphorylate TCF/Elk-1 after mitogenic stimuli such as FCS[16]. Figure 6 shows an in-gel kinase asssays using copolymerized myelin basic protein (MBP) as known substrate for MAPkinases. Oxidant as well as antioxidant stimuli lead to a strong increase of MAPKinase activity as can be seen by enhanced ^{32}P incooperation. Temporally this activation can be seen directly before[24] Elk-1 hyperphosphorylation appears in band shifts. This and the finding that Elk-1 expression vectors with mutated MAPkinase sites lead to drastically reduced SRE dependent luciferase stimulation[24] let us to conclude that signals in response to oxidants or antioxidants converge at or above the MAPkinase level in the signal cascade and lead to c-*fos* promoter activation via hyperphosphorylation of TCF/Elk-1.

SUMMARY

We could demonstrate that in HeLa cells TCF (ternary complex factor) is modulated by oxidants as well as by antioxidants (Fig. 3). Using specific antibodies we could reveal its identity as Elk-1. Upon stimulation with H_2O_2 or PDTC the SRFcore/Elk-1 complex migrates

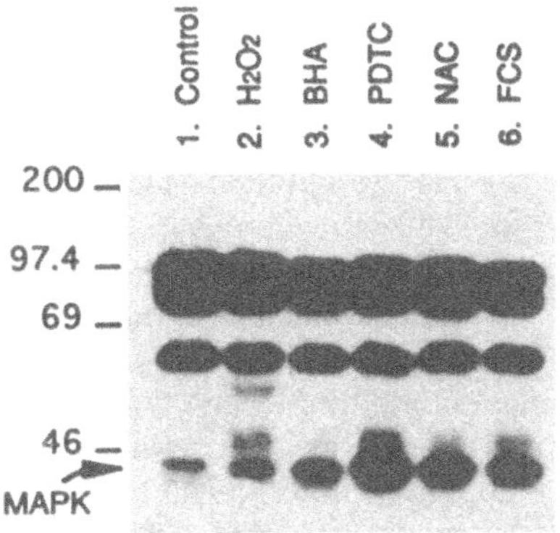

Figure 6. MAPkinases are activated by redox disturbances in HeLa cells. Whole cell lysates (6μg) were analysed by in-gel kinase assay using MBP (myelin basic protein) as the substrate. The positions of ^{14}C molecular weight marker proteins and induced Elk-kinases are indicated. The same extracts as in Fig. 3 were used in this experiment.

slower in a bandshift, this pattern could be reversed to the unstimulated migration pattern by potato acid phosphatase treatment (Fig. 4). This shows that the transcription factor TCF/Elk-1 is rapidly hyperphoshorylated after treatment with agents that imbalance ROI levels, a phenomenon previously described for FCS stimulation.

In transient transfections we could show that the transcription factor TCF/Elk-1 responds to changes of the cellular redox status induced by treatment with the oxidant H_2O_2 or the antioxidants PDTC and activates a SRE dependent reporter plasmid (Fig. 1). Activation via the SRE of the *c-fos* promoter has been descibed as essential for mRNA induction by signals such as serum or EGF and is dependent on intact SRE and TCF binding site[34]. The activation of transcription by oxidants or antioxidants is equally dependent on the intact SRF and TCF binding site in the reporter plasmid, because no stimulation could be observed, when the SRE or the TCF-binding sites were mutated (Fig. 2).

The hyperphosphoylation of TCF/Elk-1 correlates temporally with the activation of MAPK Ser/Thr kinase activity as could be demonstrated by activation of p42 and p44 MAPK in in-gel kinase assays, using myelin basic protein (MBP) as the substrate (Fig. 6). The Elk-1 MAPkinase phosphorylation sites relevant for transcriptional activation have been identified[22] as Ser383 and Ser389. Mutations in these serines repress activation of a SRE driven reporter plasmid by redox stimuli in cotransfections[24]. This strongly suggests that the *c-fos* regulation via enhanced or reduced levels of ROIs is transmitted to Elk-1 via MAPkinases. Future studies will determine which parts of this signalling cascade are shared by mitogenic and redox- dependent stimuli, and whether there are differences in the reponse to oxidants or antioxidants.

Antagonistic intracellular redox changes may thus use the same MAPkinases to obtain an identical functional modulation and modification of this nuclear transcription factor. Other cellular transcription factors, such as NF-κB[32,33] or AP-1[23], are also activated in specific redox situations. The cytoplasmic NF-κB complex is activated by release of an inhibitory subunit IκB only in response to hydroxy radicals but not to antioxidants[31]. Conversly AP-1 that can be activated both transcriptionally and posttranscriptionally is activated more strongly by antioxidants than by oxidants (reviewed in 26). In concert with these inducible cytoplasmic transcription factors TCF/Elk-1 as a nuclear sensor could lead to activation of a specific set of genes depending on the specific "ROI/redox situation".

MATERIAL AND METHODS

For transfections 9 x 10^5 HeLa cells per 60 mm dish were transfected as described[14] followed by serum deprivation for 14h in low iron MEM medium, or incubation with 5% FCS for PDTC stimulation. After stimulation with 150 μM H_2O_2, 60 μM PDTC, 10% FCS or 50 ng/ml PMA for 6 hours the cells were washed twice with ice cold PBS, harvested, lysed and measured for luciferase-generated light units[16] and protein content. Luciferase generated light units were normalised to protein content and are shown as fold stimulation above levels measured from uninduced simultaneously transfected cells. The plasmids ptk80-luc, pSRE$_2$-tk80-luc[16], pEL$_2$-tk80 (ΔTCF) and pM$_2$-tk80 (ΔSRE) were gifts of W. Ernst. pEL$_2$-tk80 and pM$_2$-tk80 were identical to pSRE$_2$-tk80-luc except mutations in the SRE (see Fig 2), which drastically reduce the DNA binding of TCF/Elk-1 or SRF [34].

For whole cell extracts[3] HeLa cells were starved in 0% FCS/MEM for 12 to 14 hours before stimulation. Following stimulation cells were washed twice with ice cold PBS containing 2 mM Na_3VO_4, 10 mM NaF, 500 μM PMSF, and 2 μg/ml Aprotinin. Extracts were done exactly as described[42], frozen in liquid nitrogen and stored at -80°C. EMSA binding reactions[42] contained 0.3 μl *in vitro* translated coreSRF$_{132\text{-}222}$, and 10,000 cpm of ^{32}P labelled *c-fos* SRE probe in 20 μl.

Prior to the use of potato acid phosphatase[27] (PAP) whole cell extracts were dialysed against a 100 times volume of a PAP buffer for 2 hours. 5µg of dialysed extract were then incubated for 3 h at 30°C with 0.05 units of dialysed PAP (Boehringer) in the presence or absence of a phosphatase inhibitor cocktail (10 mM NaF, 0.4 mM Navanadate, 2 mM Na_2pNPP, 1 mM Na_2betaglycerophosphate, 1.5 mM Na_2MoO_4) and then used in bandshift reactions.

In-gel kinase assays were performed as described[6], with the exception that 10 mM DTT was used instead of 5 mM β-mercaptoethanol during all gel incubations. Myelin basic protein (Sigma) was copolymerised in a 8% acrylamide separating gel at 0.5 mg/ml. Frozen whole cell extracts (6 µg) were diluted into Laemmli loading buffer immediately followed by incubation at 85°C for 5 min and electrophoresis.

ACKNOWLEDGEMENTS

We thank H. L. Pahl for critical reading the manuscript, M. Baccarini and R. Hipskind for helpful discussion and instructions on in-gel kinase assays and phosphatase treatments. R. Janknecht and W. Ernst provided Elk-1 mutants, recombinant Elk-1 protein, and luciferase reporter plasmids, and R. Zinck provided antibodies and GST plasmids for expression of in-gel kinase substrates. Part of this study has been submitted elsewere. This study is in partial fulfillment of the doctoral theses of J.M.M. and M.A.C.

REFERENCES

1. Abate, C. Patel, L. Rauscher, F. J. III, & Curran, T. *Science* **249**, 1157-1161 (1990).
2. Amstad, P. A. Krupitza, G. & Cerutti, P. A. *Cancer Res.* **52**, 3952-3960 (1992).
3. Baccarini, M. Sabatini, D. M. App, H. & Stanley, E. R. *EMBO J.* **9**, 3649-3657 (1990).
4. Bergelson, S. Pinkus, R. & Daniel, V. *Oncogene* **9**, 565-571 (1994).
5. Cerutti, P.A. *Science* **227**, 375-381 (1985).
6. Chao, T. -S. O. Byron, K. L. Lee, K. -M. Villereal, M. & Rosner, M. R. *J. Biol. Chem.* **267**, 19876-19883 (1992).
7. Chen, R.-H. Sarnecki, C. & Blenis, J. *Mol. Cell. Biol.* **12**, 915-927 (1992).
8. Dröge, W. et al., *FASEB J.* **8**, 1131-1138 (1994).
9. Gamou, S. and Shimizu, N. *FEBS Letters* **357**, 161-164 (1995).
10. Gille, H. G. Sharrocks, A. D. & Shaw, P. E. *Nature* **358**, 414-417 (1992).
11. Graham, R. & Gilman, M. *Science* **151**, 189- 192 (1991).
12. Halliwell, B., & Gutteridge, J.M.C. *Free Radicals in Biology and Medicine* (Clarendon Press, Oxford, 1989).
13. Herrera, R.E., Shaw, P.E., & Nordheim, A. *Nature* **340**, 68-70 (1989).
14. Hill, C. S. Marais, R. John, S. Wynne, J. Dalton, S. & Treisman, R. *Cell* **73**, 395-406 (1993).
15. Hug, H. and Sarre, T.F. *Biochem. J.* **291**, 329-343 (1993).
16. Janknecht, R. Ernst, W. H. Pingoud, V. & Nordheim, A. *EMBO J.* 12, 5097-5104 (1993).
17. Karin, M. *Curr. Opin.. Cell Biol.* **6**, 415-424 (1994).
18. Kuge, S. & Jones, N. *EMBO J.* **13**, 655-664 (1994).
19. Kyriakis, J. M. et al., *Nature* **369**, 156-160 (1994).
20. Li, Y. & Jaiswal, A. K. *Biochem. Biophys. Res. Comm.* **188**, 992-996 (1992).
21. Macleod, K. Leprince D. & Strehlin, D. *Trends Biochem.Sci.* **17**, 251-256 (1992).
22. Marais, R. Wynne, J. & Treisman, R. *Cell* **73**, 381-393 (1993).
23. Meyer, M. Schreck, R. & Baeuerle, P. A. *EMBO J.* **12**, 2005-2015 (1993).
24. Müller, J. M. Cahill, M. A. Baeuerle, P. A. & Nordheim, A. *submitted* .
25. Nose, K. et al. *Eur. J. Biochem.* **201**, 99-106 (1991).
26. Pahl, H. L. & Baeuerle, P. A. *Bioessays* **16**, 497-502 (1994).
27. Papavassiliou, A. G. & Bohmann, D. *Meth. Mol. & Cell. Bio.* **3**, 149- 152 (1990).

28. Perez-Albuerne, E. D. Schatteman, G. Sanders, L. K. & Nathans, D. *Proc. Natl. Acad. Sci. USA* **90**, 11960-11964 (1993).
29. Rao, G. N. Lasségue, B. Griendling, K. K. Wayne, R. & Berk, B. C. *Nucl. Acids Res.* **21**, 1259-1263 (1993).
30. Sachsenmaier, C. et al., *Cell* **78**, 963-972 (1994).
31. Schmidt, K. N., Amstad, P., Cerutti, P. A. & Baeuerle, P. A. *Chem .& Biol.* **2**, 13-22 (1995).
32. Schreck, R. Meier, B. Männel, D. N. Dröge, W. & Baeuerle, P, A. *J. Exp. Med.* **175**, 1181-1194 (1992).
33. Schreck, R. Rieber, P. & Baeuerle, P. A. *EMBO J.* **10**, 2247-2258 (1991).
34. Shaw , P. E. Schröter, H. & Nordheim, A. *Cell* **56**, 563-572 (1989).
35. Shibanuma, M. Kuroki, T. & Nose, K. *Oncogene* **3**, 17-21 (1988).
36. Sies, H. *Eur. J. Biochem.* **215**, 213-219 (1993).
37. Stein, B. Rahmsdorf, H. J. Steffen, A. Litfin, M. & Herrlich, P. *Mol. Cell. Biol.* **9**, 5169-5181 (1989).
38. Stevenson, M. A. Pollock, S. S. Coleman, C. N. & Calderwood, S. K. *Cancer Res.* **54**, 12-15 (1994).
39. Treisman, R. *Cell* **46**, 567-574 (1986).
40. Treisman, R. *Curr. Opin. Genet. & Dev.* **4**, 96-101 (1994).
41. Treisman, R. *Trends Biochem.Sci.* **17**, 423-426 (1992).
42. Zinck, R. Hipskind, R. A. Pingoud, V. & Nordheim, A. *EMBO J.* **12, 2377-2387 (1993).**

SOME ACTIVITIES OF THE OXIDANT CHROMATE

Hyoung-Sook Park,[1] Sean O'Connell,[2] Saul Shupack,[1*] Edward Yurkow,[1] and Charlotte M. Witmer[1]

[1] Graduate Program in Toxicology
Environmental and Occupational Health Sciences Institute
Rutgers University
Piscataway, New Jersey 08855-1179
[2] Heartland Biotechnologies
Davenport, Iowa 52806-1341

INTRODUCTION

Hexavalent chromium [Cr(VI)] compounds are recognized mutagens and carcinogens (IARC, 1980), although the mechanism of action is unknown. Chromium compounds commonly exist in the Cr(VI) and trivalent ([Cr(III)] states, with the Cr(III) state as the most stable form of chromium. Cr(III) compounds differ from Cr(VI) not only in that they are more stable but they exist as cations in octahedral form (with six ligands) and cannot cross membranes readily. They are considered to be non-bioavailable and non-carcinogenic. Cr(VI) compounds, which exist as the oxides or the oxyanions and thus readily cross membranes, are rapidly reduced in the body (both intra- and extracellularly) to the stable Cr(III) via Cr(V) and Cr(IV). Some synthetic Cr(III) complexes, with lipid soluble organic anions as ligands such as the picolinate, cross membranes fairly readily. Cr(III) complexes formed *in situ* gain entry to the nucleus, as only Cr(III) is detected bound to DNA. Thus it may be that only the Cr(III) formed *in situ* is toxic. Cr(VI) compounds are not mutagenic without metabolic reduction, (Petrilli and De Flora, 1978). The intermediates, Cr(V) and Cr(IV), are being investigated as putative toxins/carcinogens of Cr(VI). Cr(V) has been detected by electron paramagnetic resonance (EPR) on reduction of Cr(VI) in cells in culture and *in vitro* (O'Brien *et al.*,1981; Goodgame and Joy, 1986; Shi and Dalal, 1988; Aiyar *et al.*, 1988; Sugiyama, 1991; Witmer *et al*, 1994). Both Cr(V) and Cr(IV) can disproportionate so that the reduction of Cr(VI) is not a straightforward process, A two-electron reduction of Cr(VI) is catalyzed by DT diaphorase, which also decreases the mutagenicity of Cr(VI) *in vitro*.

---

* Saul Shupack (Villanova University, Villanova, PA) contributed a great deal to this work before his untimely death in 1994)

Biological Reactive Intermediates V
Edited by R. Snyder et al., Plenum Press, New York, 1996

We have demonstrated the production of reactive oxygen species (ROS) after adding $K_2Cr_2O_7$ to lung A549 cells. We also reported the production of three unidentified free radical signals, concomitant with Cr(V), using EPR (Witmer *et al.*, 1994). ROS include the ·OH and $O_2^{\cdot-}$ radicals and H_2O_2, all of which are toxic, and contribute to oxidative stress. Oxidative stress is a complex condition following production of ROS and the resultant damage differs with affected tissues. ROS cause various types of damage such as inactivation of enzymes, mutations, reprogramming of genes (O'Halloran, 1993), apoptosis and are the putative toxins in reperfusion injury to ischemic tissue (Granger *et al.*,1982). Whether oxidative stress is a primary factor in Cr(VI) toxicity is being investigated. Recently. Shi and Dalal (1994) have demonstrated that Cr(V) complexes, produced on reduction of Cr(VI), react with H_2O_2 to produce ·OH radicals in a Fenton-like reaction and that Cr(IV) complexes can also redox cycle (Shi *et al.*,1994). Thus it is difficult to separate the effects of any one of these species *in vivo*.

The objective of the present studies was to study the effects of antioxidants, including vitamins, on Cr-induced ROS in two cell types, H4 Hepatoma cells and A549 lung cells, as well as the effects of Cr(VI) on the cell cycle of A549 cells. A549 cells are human lung alveolar epithelial tumor cells and H4 cells are rat hepatoma cells. The lung is the target of Cr-induced carcinogenesis and the liver is a major site of xenobiotic metabolism.

METHODS AND MATERIALS

Chemicals and Cells

Both A549 and H4 cells were obtained from the American Type Culture Collection (Rockville, MD) and were maintained in a 95 % air, 5% CO_2 atmosphere, in RMPI medium (Gibco, Grand Island, NY) for A549 cells or Swim's medium (Sigma Chemical Co., St. Louis, MO) for H4 cells, at 37^0 C, pH7.2, with 10% fetal calf serum and penicillin (90 µg/ml) and streptomycin (90 U/ml). Other treatment of the A549 cells was as described in Witmer *et al.* (1994). The 2'7'dichlorofluorescein (reduced) (DCFH) and the diacetate (DCFH-DA) were purchased from Molecular Probes (Eugene, OR). Chromium picolinate (98% purity) was generously donated by Nutrition 21 (San Diego, CA). The inorganic Cr(II) and Cr(III) chlorides were purchased from Aldrich Chemical Co. (Milwaukee, WI). The synthetic Cr compounds were synthesized by Saul Shupack, using the method of Krumpole *et al.* (1978) for the Cr(V)butyrato compound and standard methods for the Cr(III) tetraphenylporphyrin compound. Vitamins were purchased from Sigma Chemical Company (St. Louis, MO). All other compounds were purchased from Fisher Chemical Company (Pittsburgh, PA). All chemicals were of the best grade available.

Detection of Reactive Oxygen Species (ROS) and Propidium Iodide-Treated DNA

ROS, such as H_2O_2 and other peroxides detected by oxidized DCFH were detected in the cultured cells using the EPICS Profile flow cytometer (Coulter Electronics, Hieleach, FL) equipped with an Omnichrome 25 mW argon laser. The oxidized dye is stimulated at 488 nm and fluorescence was detected using a 525 nm low band pass filter. Propidium iodide fluorescence was detected with the flow cytometer with a filter at 630 nm. Analysis of histograms and calculations of ROS values and DNA content were made using Coulter software, EPICS CYTOLOGIC. Positive control ROS studies were carried out with H_2O_2 and phorbol myristic acetate (data not shown).

Cell Cycle Measurements

The effects of $K_2Cr_2O_7$ on specific phases of the cell cycle were determined using the DNA intercalating agent, propidium iodide which fluoresces at 630 nm. After exposure of the cells to the dichromate for 3, 24 and 48 hr, the cells were fixed in ethanol (EtOH), the propidium iodide added and the fluorescence determined in the flow cytometer. Fluorescence is proportional to DNA content thus the resulting histograms (using the Coulter software) represent the relative numbers of cells in the G_0/G_1, S and G_2 + M phases of the cell cycle.

EXPERIMENTAL

Effects of Chromium Species on ROS Production (H4 Cells)

Chromium compounds of several oxidation states were added to the H4 cells and the ROS determined, as described. This procedure was based on Bass *et al.* (1983). Briefly, in these experiments, the serum-deprived cells collected with EDTA, washed and suspended in Hank's bufered saline solution (HBSS), pH 7.4. The H4 cells (2 x 10^6) were then loaded with the dye (final concentration, 5 μM DCFH-DA; final volume in the tube, 2.2 ml) and the mixture equilibrated by shaking at 37^0 for 20 min to allow uptake of the dye and the removal of the acetate groups by cellular esterases. The chromium compound was added (zero time) and the mixture incubated at 37°C for 20 min, or other time periods, as indicated. Samples were then immediately analyzed for ROS in the flow cytometer and the results recorded. The following compounds at 10, 30, 100 and 300 μM (final concentration as Cr content) were added to the H4 cells; (1) $CrCl_2$; (2) trivalent compounds as (a) inorganic $CrCl_3$ [$Cr(III)_i$] (b) Cr(III) picolinate [pic], (c) the synthetic compound, Cr(III) tetraphenylporphyrin chloride complex [TPP]; (3) synthetic sodium bis(2-hydroxy-2 methyl butyrato)oxochromate V [Cr(V)], and (4) potassium dichromate, $K_2Cr_2O_7$, [Cr(VI)]. Pic is a complex with picolinate anions occupying three of the six Cr(III) ligand positions; both pic and TPP have increased membrane permeability compared with the inorganic chromium chlorides.

Effects of Antioxidant Vitamins on Chromate-Induced ROS

Non-toxic concentrations of vitamins C (ascorbate) and E (α tocopherol) (as the succinate) were used for these ROS experiments. Toxicity of the vitamins was tested by the incubation of vitamin C (10-500 μM) and vitamin E (5-100 μM) for 20 hr with the cultured cells. The cell number was determined by counting viable cells in the Coulter counter (Coulter Electronics, Hieleah, FL). Viability was determined by Trypan Blue exclusion. Concentrations of vitamin C (ascorbate) up to 300 μM were not toxic to the A549 cells while vitamin E was toxic above 20 μM (84% survival at 20 μM). (See Table 1). Thus 20 μM vitamin E and up to 200 μM concentrations of vitamin C were used in these studies.

In the first group of experiments, the antioxidant vitamins C and\or E were incubated with the A549 cells for 20 hr prior to loading the cells with the dye, DCFH-DA. Vitamin C was added at a 200 μM final concentration and vitamin E at 10 and 20 μM. After incubation with either vitamin (or both) the cells were trypsinized, harvested and suspended in phosphate buffer prior to addition of the dichromate (zero time). ROS measurements were made in samples incubated with the dichromate for 20 and 30 minutes. Dichromate concentrations were 100 μM and 200 μM Cr). Control samples contained phosphate buffer or the appropriate vitamin.

In another group of experiments, vitamin C (200 μM) and/or vitamin E (20 μM) were incubated with serum deprived A549 cells 60 min prior to the addition of the dye. The cells were not washed to remove the vitamins prior to measuring the ROS, although they were recovered

Table 1. Effects of Chromium Species on H4 Cells (ROS Production)

Sample conc; μM	Mean fluorescence	Log conversion	Percent positive
Pic			
10	114.01	6.19	—
30	106.33	4.64	—
100	92.71	2.91	—
300	86.60	2.34	—
CrVI			
10	116.97	6.90	1.65
30	125.39	9.19	16.95
100	129.77	11.00	25.90
300	142.84	17.57	47.63
$CrCl_3$			
10	119.78	7.69	9.79
30	112.74	5.99	—
100	113.52	6.19	—
300	112.84	5.99	—
$CrCl_2$			
10	112.90	5.99	—
30	110.90	5.56	—
100	104.20	4.32	—
300	97.40	3.36	—
TPP			
10	178.73	64.15	83.82
30	193.42	106.20	91.62
100	172.35	49.87	79.86
300	156.12	28.04	64.23
CrV			
10	148.95	21.80	58.82
30	154.59	27.05	58.94
100	174.68	55.55	78.46
300	190.78	98.78	85.19
Control			
1	116.27	6.65	—
2	137.65	14.67	—

Values were obtained using EPICS CYTOLOGIC. Readings were taken 20 min after addition of chromium compounds. See Text for details.
Pic - Cr picolinate (CrIII).
TPP - tetraphenylporphyrin, Cr(III).
CrV - sodium bis(2-hydroxy-2-methyl butyrato)oxychromate V.
CrVI- potassium dichromate.
Concentrations are for Cr content.

with EDTA and growth medium was removed before resuspending the cells in HBSS. The dye was incubated with the cells for 20 min at 37°C prior to addition of the dichromate, as above. In these "limited incubation" studies, the dichromate was added at 20, 100 and 200 μM dichromate. ROS readings were taken after 10, 25 and 39 min of incubation with chromate.

Effects of Catalase and Superoxide Dismutase (SOD) on Chromate-induced ROS

The effects of the antioxidants, catalase (1000 units) and/or superoxide dismutase (SOD)(100 units), on the $K_2Cr_2O_7$-induced ROS were studied using a short preincubation time of 60 min to prevent loss of enzymatic activity. Catalase and/or SOD were added to serum-deprived cells (at 37°C) 60 min prior to the dye addition (80 min prior to dichromate treatment). The cells were washed and resuspended in phosphate buffer and dichromate added (100 μM; 200 μM Cr), followed by incubation at 37°C. Readings were taken 15, 30 and 50 min after dichromate addition. Controls were cells treated with PBS or the appropriate enzyme. As in previous experiments, addition of either Cr(VI) or antioxidants alone to the dye had no effect on the oxidation state of the dye, as determined in the Perkin-Elmer fluorimeter (data not shown).

Chromate Effects on the Cell Cycle of A549 Cells

The effects of $K_2Cr_2O_7$ on specific phases of the cell cycle of A549 cells were determined using the DNA intercalating agent, propidium iodide, which fluoresces at 630 nm. The A549 cells (approximately $2x10^6$ cells) in logarithmic growth phase were exposed to $K_2Cr_2O_7$ (1,5,10 and 20 μM) for 3, 24 and 48 hr. At the end of the chromium exposure, the cells were washed with PBS (Ca^{2+} and Mg^{2+} free) twice and then trypsinized and resuspended in 1.5 ml of Ca^{2+}- Mg^{2+}-free PBS. The cell suspension was then added to 3 ml volumes of ice-cold EtOH, with increasing concentrations to a final EtOH concentration of 70%, to fix the cells. The cells were then centrifuged (1,000 rpm) and the supernatants discarded. The fixed cells were washed twice with the same PBS and resuspended in 1.0 ml of propidium iodide/sodium citrate solution (50 μg PI/ml and 0.1% sodium citrate solution). RNAse (10 μl,180 units) was added and the mixture was incubated for 18 hr at 4°C. The cells were then washed, resuspended in 1.0 ml PBS and analyzed for fluorescence at 630 nm in the flow cytometer.

RESULTS

Effects of Chromium Species on ROS (H4 Cells)

Table 1 shows the mean, log transformation and % positive values of the ROS produced by the several species of Cr on addition to H4 cells. These results are from a representative experiment; the study was carried out three times. The ROS-producing potency of the Cr compounds, with the H4 cell line, was as follows (at 30 and 100 μM concentrations): Cr (III) as TPP > Cr(V) as the bis α-hydroxy carboxylic acid complex > Cr(VI) > $Cr(III)_i$ = Cr(II) = Pic = control, as seen in Table 1. At 300 μM concentration, Cr(V) > TPP, Cr(III). (At 10 μM only CrV and TPP were above background). Despite this reversal of potency of Cr(V) and TPP at the highest concentration, the results show that the inorganic chlorides of Cr(II) and Cr(III) cause only background levels of ROS but the TPP complex of Cr(III) causes ROS while the Pic causes no ROS formation. Pic and $CrCl_3$ were used as negative controls in many experiments as the ROS values were equal to or less than those from buffer only. The ROS produced with the synthetic compounds were dose-dependent (data not shown). The effects in this experiment were maximal at the 20 min time period.

Effects of Antioxidants on chromate-induced ROS (A549 cells)

The results of these experiments are shown in Figures 1 and 2 and Tables 2 and 3.

Vitamin C reduced the chromate-induced ROS in A549 cells in the experiments in which it was preincubated for 20 hr with the cells prior to addition of the dichromate. (Figure 1). The same results were obtained with serum-deprived cells with only 60 minutes of preincuation with vitamin C. In other experiments in which the cells were not serum deprived and the cells were

Table 2. Effect of vitamin C and vitamin E on the growth of A549 cells

Treatment	Concentration µM	% surviving cells 20 hr
Control	—	100
Vitamin C	10	132.0
	20	107.7
	100	95.8
	200	92.2
	300	104.3
	500	13.5
Vitamin E	5	98.4
	10	105.0
	20	84.0
	50	14.1
	100	0

Experiments were carried out in duplicate.
The vitamin was incubated with the A549 cells for 20 hours, cells were then washed and treated as described in the text.

trypsinized, there was an increase in the ROS on addition of Cr (data not shown). Vitamin E had no significant effect on the chromate-induced ROS with 20 hour preincubation (Figure 2), while with only 60 min preincubation, it significantly reduced the Cr-induced ROS. The maximal reduction of the ROS was with vitamins C and E both preincubated with the cells prior to addition of dichromate. Vitamin E studies have been inconsistent with some results showing increases in ROS. However, in each case with different results there were slightly different conditions in the assays, and in each case the results were reproducible. Table 3 shows results from a representative

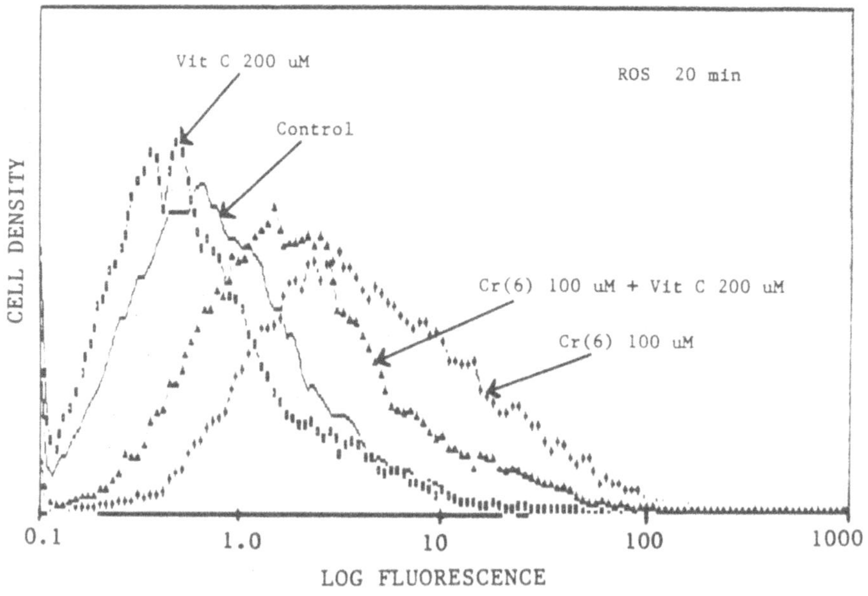

Figure 1. Effect of Vitamin C on chromate-induced ROS in A549 cells. Reproduced from Witmer *et al.*, 1994. See text for details.

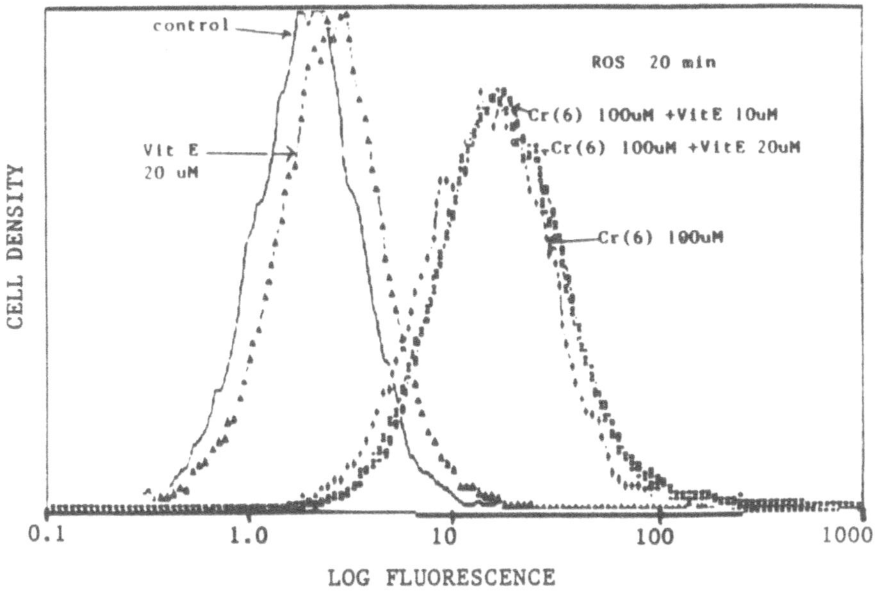

Figure 2. Effect of Vitamin E succinate on chromate-induced ROS in A549 cells. Reproduced from Witmer *et al.*,1994. See text for details.

experiment. Vitamin C effects varied with conditions also, but was not as sensitive as vitamin E to changes such as the addition of EDTA vs trypsin, etc.

Effects of Catalase and SOD on Chromate-Induced ROS

These effects are shown in Figure 3. Catalase and SOD together are seen to have no effect on the ROS in this experiment. SOD had no effect at any time period or under any

Table 3. Effects of vitamins C and E on Cr-induced ROS in lung A549 cells. 60 min preincubation.

	ROS (Mean)		
Sample	10 min	25 min	39 min
Control	1.86	2.18	2.32
Chromate (200 μM)	6.19	8.95	10.52
Vitamin C (200 μM)	1.98	2.27	2.46
Vitamin C + Chromate	3.24	4.47	4.96
Vitamin E (20μM)	1.42	1.74	1.93
Vitamin E + Chromate	4.63	6.72	7.86
Vitamin C + Vitamin E	1.39	1.49	1.48
Vitamin C + Vitamin E + Chromate	2.57	3.49	3.67

Chromate - potassium dichromate, $K_2Cr_2O_7$.
Concentrations are for Cr content (200 μM in all samples).
Values are mean fluorescence, using CYTOLOGIC software.
Vitamins were added 60 min prior to addition of the dye.
Cells were equilibrated with the dye for 20 min, chromate was then added (zero time) and readings were made at the times indicated.
Vitamin concentrations were constant.

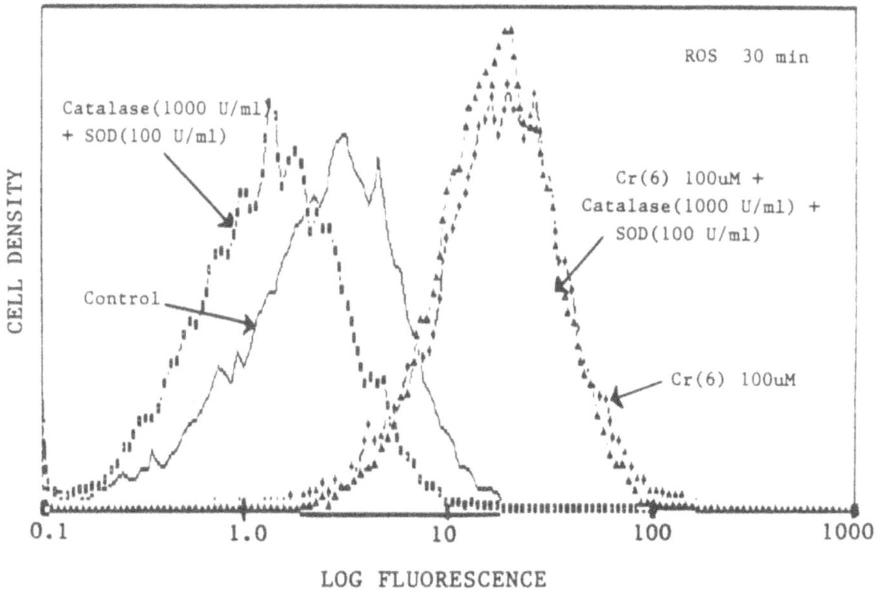

Figure 3. Effects of catalase/SOD on ROS in A549 cells. See text for details.

conditions when added prior to the dichromate. Catalase under the conditions of this experiment gave identical results as the catalase and SOD together. Catalase, however, when added to serum-deprived cells reduced the number of Cr(VI)-induced ROS substantially (data not shown), but SOD had little or no effect on these species. The difference in conditions had a definite effect on the results of ROS production but they were never completely obliterated. The dye is specific for H_2O_2 (or any peroxide) production, although catalase is specific for H_2O_2.

Effects of Chromate on the Cell Cycle

The results of the cell cycle studies are found in Table 4 and Figure 4. Under the conditions of these experiments, the S phase of the cell cycle in A549 cells was increased by the Cr(VI) addition (Figure 4). This build-up was caused by Cr(VI) in concentrations up

Table 4. Effects of Potassium Dichromate on Cell Cycle Progression in A549 Cells.

Concentration	Percentage of cells in cell cycle+		
$K_2Cr_2O_7$	G_0/G_1	S	$G_1 + M$
untreated	61.9	27.0	10.7
1 μM	54.7	31.0	14.4
5 μM	48.9	30.5	20.4
10 μM	45.6	43.5	11.1
20 μM	75.3	22.5	negligible

Logarithmically growing cells were treated with 1, 5, 10 and 20 μM potassium dichromate for 24 hr.
Values were obtained using CYTOLOGIC software. See text for details.

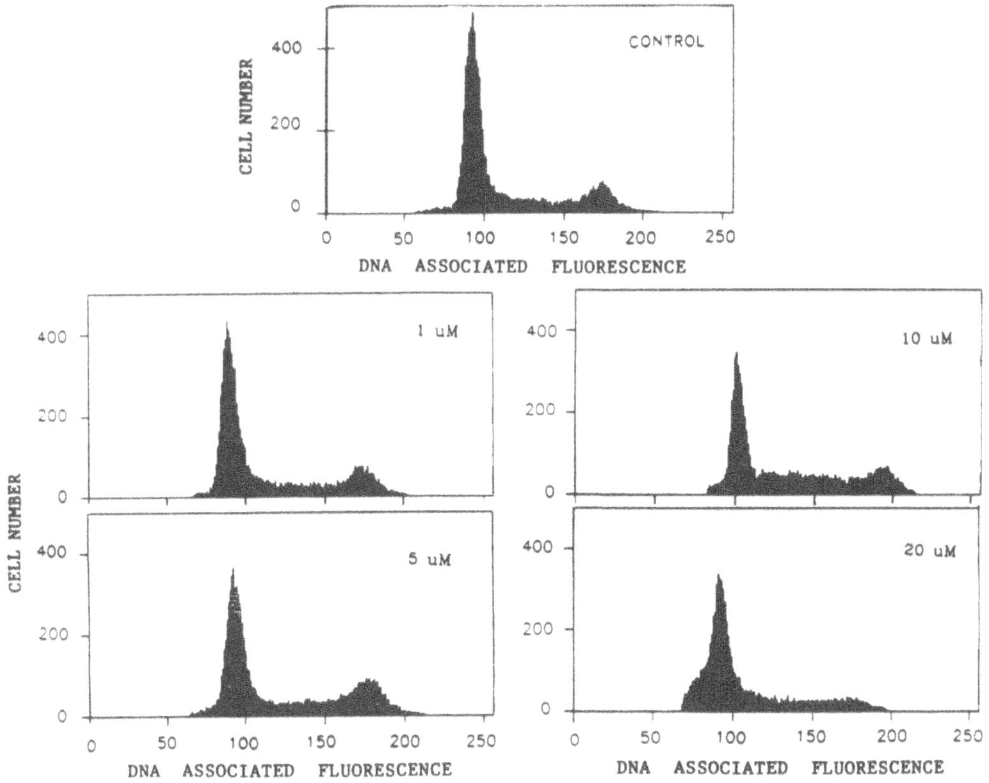

Figure 4. DNA patterns after dichromate exposure. Exposure of the cells to dichromate was for 24 hr. Values were obtained using CYTOLOGIC software. See Text for details.

to and including 20 μM. There is an increase in G_2+M at 10 μM concentration indicating toxicity. Previous studies showed that the dichromate is toxic at 10 uM. There is a prominent shoulder on the left of G_o/G_1 phase at 20 μM Cr(VI) (Figure 4) with 24 hr treatment which is indicative of apoptosis. The results with 48 hour exposure are similar in that the S phase is increased from 11.5% in untreated cells to 31% with 48 hour exposure to 20 μM dichromate (data not shown). Figure 4 shows the same data in graphic form.

DISCUSSION

This work was carried out to determine the conditions affecting ROS production after treatment of cells with various species of chromium and to investigate the effects of Cr(VI) on the cell cycle. Our finding that TPP, a synthetic Cr (III) porphyrin compound, caused ROS when added to H4 cells was unexpected. Although it is relatively membrane permeable, its redox potential is not known. It may have a potential similar to the Fe(III) porphyrins and thus act as an oxidizing agent. An example of a toxic form of Cr(III) is that of $Cr(NO_3)_3$ which on inhalation causes functional changes in macrophages. However, macrophages produce $O_2^{\cdot -}$ as a normal function after ingestion of xenobiotics and tissues which are not part of the immune system lack the NADPH oxidase, the enzyme involved. Other Cr(III) compounds, such as pic (Cr picolinate) caused no ROS production and in fact had lower levels than buffer controls in this system. Cr picolinate is freely available as a dietary supplement and its apparent lack of toxicity is supported by its lack of resultant ROS production. The action of Cr(V) as an oxidant was expected and supports the findings of Shi and Dalal (1994) that Cr(V) complexes, formed from Cr(VI) compounds, can redox cycle and form free radicals. Thus Cr(V) complexes can contribute in the Cr(VI) toxicity. The various disproportionations which are possible from Cr(V) and Cr(IV) species complicate our interpretation of present data. It may be that the ROS detected with these studies is less than that actually formed, as $O_2^{-\cdot}$ although a potential oxidizing agent, is not detected by the DCDF dye. Also, if Cr(V) reacts rapidly with the H_2O_2, some H_2O_2 will be removed before it can oxidize the dye. The DCDF dye used to detect ROS is considered to be specific for peroxides.

The finding of decrease in ROS with catalase is a general reaction reported by others, identifying the presence of H_2O_2 in the ROS, although the results were not consistent in our studies. In cells which were serum-deprived the ROS were reduced by catalase although the ROS (and thus the H_2O_2) were not completely removed. Catalase cannot enter the cells, but H_2O_2 is freely diffusible and the equilibrium of the peroxide is apparently altered by the extracellular catalase, with more H_2O_2 exiting into the medium. It is surprising that the effect of catalase was not greater, and not consistent; apparently serum interferes with some of the reactions. Other peroxides appear to be formed. The lack of effect of vitamin E is not surprising as vitamin is membrane-bound and works more efficiently when vitamin C is present. The increased amount of ROS in some experiments with vitamin E present may be a result a formation of alpha tocopherol radical without its reduction by either GSH of ascorbate.

A complicating factor in interpretation of the data from these experiments is that the glutathione (GSH) content of cells in culture has an important effect on the amount of ROS produced from chromium (Yurkow *et al.*,unpublished); depleting the GSH content increases the ROS detected. GSH is known to regenerate ascorbate from the oxidized form and by complexing any trace elements such as Cr(III) and to react directly with the dihydroxyascorbate under certain conditions. The GSH content of the cells was not determined in the present work and may have been very varied. Further modifying effects of GSH on the cellular response to vitamins C and E are not known. A high residue of ascorbate in cells preincubated for 60 min prior to chromate addition appeared to favor the formation of the ascorbate radical in the Cr(VI) reduction. These findings require further studies.

The effects of chromate on the S phase of the cell cycle are somewhat different from those of Bakke *et al.* (1984) who demonstrated that at the lowest concentration of Cr(VI) that caused arrest of cell cycle progression (1-2 μM $K_2Cr_2O_7$) in NHIK 3025 cells, there was a prolongation of G2 phase of the cell cycle. This is a cell line from hyperplastic human cervical tissue. When $K_2Cr_2O_7$ concentration was increased to >2 μM, the cells were arrested in S phase. The indication of apoptosis in these A549 cells with chromate treatment also differs markedly from the findings of Manning *et al.* (1994) who reported apoptosis as a response to Na_2CrO_4 in CHO cells only at 150 μM concentrations. Costa's group (Costa *et al.*,1982) concluded that several divalent toxic metals blocked the cycle in S phase; Cr(III) was not studied. Cr(III) is the species of Cr which binds to DNA and it appears to interfere with DNA or RNA synthesis.

The effects of Cr(VI) administration on DNA are more complex than appears from the studies with cells in culture. Our laboratory has demonstrated that ip treatment of rats with potassium dichromate results in changes (both increases and decreases) in levels of hepatic and lung isozymes of CYP450 (Faria *et al.*,1993; Witmer *et al.*,1994; Faria *et al.*,1995). These changes are sex specific and were shown by Western blots to be changes in protein content. This may be an indirect effect as other work (Kim and Yurkow, unpublished) has recently shown that Cr(VI) causes persistent activation of mitogen activated kinase

(MAP kinase) in the H4 cells. Differences in response would be expected in different tissues and this difference should be explored. It is also not known whether the mechanism of cell death from Cr(VI) may be primarily from apoptosis.

The mechanism of H_2O_2 formation following Cr(VI) to cells is not known and many reactions which follow reduction of Cr(VI) actually remove H_2O_2 from cells (e.g., redox cycling between Cr(V) and Cr(VI). Whether the attack of Cr(VI) on mitochondria causes sufficient leakage of H_2O_2 to account for the ROS measured is something which requires further study. Recent findings that ROS activate SOD in mammalian cells are relevant to these findings. How much H_2O_2 is produced by this activation in these cells is not known. However, the fact that Cr(VI) causes oxidative stress which has its own cascade of toxicity remains an important fact, which suggests that epigenetic effects of Cr(VI) may also be very important in the carcinogenic effects of hexavalent chromium.

REFERENCES

Aiyar, J., Borges, K.M., Floyd, R.A. and Wetterhahn, K.E. (1988).Role of chromium(V), glutathione thiyl radical and hydroxyl radical intermedaites in chromium(VI)-induced DNA damage. Toxicol Environ Chem. 22:135-148.

Bakke, O., Jakobsen, K. and Eik-Nes, K.B. (1984). Concentration-dependent effects of potassium dichromate on the cell cycle. Cytometry 5:482-486.

Bass, D.A., Parce, J.W., DeChatelet, L.R., Szfjda, P., Seeds, M.C. and Thomas, M. (1983). Flow cytotometric studies of oxidative product formation by neutrophils graded response to membrane stimulation. J. Immunol. 130: 1910-1917.

Costa, M., Cantoni, O., de Mars, M. and Swartzendruber, D.E. (1982). Toxic metals produce an S-phase specific cell cycle block. Res. Commun. Chem. Pathol. Pharmacol. 38: 405-414.

Faria, E.C., Sadrieh, N. and Witmer, C.M. (1993) Effects of hexavalent chromium [Cr(VI)] on cytochrome P-450 isozymes. The Toxicologist 13:8.

Faria, E., Sadrieh, N. Thomas, P.E. and Witmer, C.M. (1995). Sex and organ specific effects of hexavalent chromium [Cr(VI)] on cytochrome P450 (P450) isozymes. The Toxicologist 15:312.

Goodgame, D.M.L. and Joy, A.M. (1986). Relatively long-lived chromium (V) species are produced by action of glutathione on carcinogenic chromium(VI). J. Inorganic Biochem. 26: 219-224.

Granger, D.N., Hollwarth, M. A.,and Parks, D.A. (1982). Ischemia reperfusion injury: Role of oxygen-derived free radicals. Acta Physiol. Scand. 126, Suppl 548: 47-63.

IARC (1980). Chromium and chromium compounds. In: IARC monographs for carcinogenic risk of chemicals to humans; Some metals and metallic compounds. Lyon: International Agency for Research on Cancer; 23:205-323.

Krumpole, M., DeBoer, B.G. and Rocek, J. (1978). A stable chromium (V) compound. Synthesis, properties, and crystal structure of potassium bis(2-hydroxy-2-methylbutyrato)-oxochromate(V)monohydrate. J. Am. Chem. Soc. 100:145-153.

Manning, F.C.R., Blankenship, L.J., Wise, J.P., Xu, J., Bridgewater, L.C., and Patierno, S.P. (1994). Environ. Hlth. Perspect. 102 (Suppl. 3):159-167.

O'Brien, P., Barrett, J. and Swanson, F. (1985). Chromium (V) can be generated in the reduction of chromium (VI) by glutathione. Inorg. Chim. Acta 108: L19-L20.

O'Halloran, T.V. (1993). Transition metals in control of gene expression. Science 261: 715-725.

Petrilli, F.L. and DeFlora, S. (1978). Metabolic deactivation of hexavalent chromium mutagenicity. Mutat. Res. 54:139-147.

Shi, X. and Dalal, N.L. (1988). The mechanism of chromate reduction by glutathione: ESR evidence for the glutathionyl radical and an isolable Cr(V) intermediate. Biochem. Biophys, Res. Commun. 156 137-142.

Shi, X. and Dalal, N.S. (1994). Generation of hydroxyl radical by chromate in biologically relevant systems: Role of Cr(V) complexes versus tetraperoxochromate(V). Environ. Health Perspect. 102:231-236.

Shi, X.,Mao,Y, Knapton, A.D., Ding,M., Rojanasakul, Y., Gannett, p.M., Dalal, N. and Liu, K. (1994). Reaction of Cr(VI) with ascorbate and hydrogen peroxide generates hydroxyl radicals and causes DNA damage: Role of a Cr(IV)-mediated Fenton-like reaction. Carcinogenesis 15:2475-2478.

Sugiyama, M.(1991). Effects of vitamins on Cr(VI)-induced damage. Environ. Health perspect. 92:63-71.

Witmer, C., Faria, E., Park, H.S., Sadrieh, N., Yurkow, E., O'Connell, S., Sirak, A. and Schleyer, H. (1994). *In Vivo* Effects of Chromium. Environ. Health Persepct.102: 169-176.

REACTIVE DOPAMINE METABOLITES AND NEUROTOXICITY

Implications for Parkinson's Disease

Teresa G. Hastings, David A. Lewis, and Michael J. Zigmond

University of Pittsburgh
Departments of Neurology, Psychiatry, and Neuroscience
568 Crawford Hall, Pittsburgh, Pennsylvania 15260

INTRODUCTION

Parkinson's disease occurs in approximately one percent of individuals over the age of 55. It is characterized by the presence of tremor, rigidity, and bradykinesia. Pathologically, the hallmark of Parkinson's disease is the progressive degeneration of the dopamine (DA)-containing neurons of the nigrostriatal projection. Neurological deficits associated with the loss of DA neurons do not appear until the degeneration is extensive, presumably due to compensatory properties of the remaining DA neurons and their targets (Bernheimer et al., 1973; Zigmond et al., 1984). The mechanism responsible for the degenerative process is not known, although factors such as genetic predisposition and environmental toxins have been proposed to play a role (Agid et al., 1993). There is also growing evidence that endogenous factors such as oxidative stress and DA itself may contribute to the neurodegenerative process (Cohen, 1983; Agid et al., 1993; Zigmond et al., 1992).

A role for DA-induced toxicity to DA neurons is supported by several lines of evidence. First, like the neurotoxin 6-hydroxydopamine, DA is an unstable molecule that will oxidize to form reactive oxygen species and quinones (Graham, 1978). Second, the accumulation of the dark pigment neuromelanin in DA-containing cells of the substantia nigra is positively correlated with the susceptibility of these neurons to degradation in Parkinson's disease (Hirsch et al., 1988; Kastner et al., 1992). Neuromelanin is the polymerization product of oxidized DA, suggesting that oxidation is occurring within DA-containing neurons. Likewise, free cysteinyl-DA, the cysteine conjugated form of oxidized DA, has been found in DA-rich regions of the brain in a variety of species and used as an index of the rate of DA autoxidation (Fornstedt et al., 1990a). Levels of free cysteinyl-catechols are increased under conditions known to promote oxidative stress, such as aging (Fornstedt et al., 1990b) or ascorbate deficiency (Fornstedt and Carlsson, 1991). Third, the exposure of cells to DA, either *in vivo* or *in vitro*, has neurotoxic consequences (Graham et al., 1978; Michel and Hefti, 1990; Filloux and Townsend, 1993). Finally, pharmacological depletion

Biological Reactive Intermediates V
Edited by R. Snyder et al., Plenum Press, New York, 1996

of DA prior to exposure to methamphetamine (Schmidt et al., 1985) or an ischemic event (Weinberger et al., 1985) attenuates the neurotoxic effects in striatum.

DA may act as a toxin when the catechol ring of the DA molecule is oxidized in the presence of oxygen and transition metal ions to form reactive oxygen species and DA quinones (Graham, 1978).Depending upon whether it is a one- or two- electron transfer, either superoxide anion or hydrogen peroxide may be formed. The interaction of these reactive oxygen species with transition metal ions such as iron forms the most reactive free radical species, the hydroxyl radical. The hydroxyl radical will react immediately with a variety of macromolecules including proteins, lipids, and DNA, with potentially cytotoxic consequences (Halliwell, 1992).

The cytotoxicity of DA and related catechol analogues also has been attributed to the reactive quinone form of the oxidized molecule (Graham et al., 1978). The electron-deficient DA quinone will react readily with nucleophilic compounds such as sulfhydryl groups. Sulfhydryl groups are the strongest nucleophiles within the cell and exist in the form of free cysteine, glutathione (GSH), or cysteinyl residues of proteins. Since cysteinyl residues are often at the active site of a protein, the covalent binding of DA to the associated sulfhydryl group is likely to inactivate critical protein functions, resulting in deleterious effects upon the cell (Graham et al., 1978).

A variety of mechanisms, involving both small molecules and enzymes, exist to protect against the damaging effects of reactive species. There is evidence to suggest that some of these defense mechanisms may be defective in Parkinson's disease. When compared to age-matched controls, post-mortem brain tissue from parkinsonian patients has increased iron (Sofic et al., 1991) without a parallel increase in ferritin (Mann et al., 1994) in substantia nigra, suggesting the potential for an increased availability of free iron to promote oxidation reactions. Other studies showed decreased levels of GSH (Perry et al., 1982) and GSH-peroxidase activity (Kish et al., 1985). Superoxide dismutase levels are increased in parkinsonian substantia nigra (Martilla et al., 1988; Saggu et al., 1989), which may have deleterious effects given the decreased ability to eliminate H_2O_2. Increased lipid peroxidation also was characteristic of parkinsonian substantia nigra (Dexter et al., 1989; 1994). These findings suggest that conditions may be favorable to promote the oxidation of DA. Indeed, free cysteinyl-DA has been found in human substantia nigra, caudate, and putamen, with the ratio of cysteinyl-DA to DA being significantly higher in patients with evidence of depigmentation such as occurs in Parkinson's disease (Fornstedt et al., 1989).

Therefore, we hypothesize that conditions exist under which DA is likely to form reactive DA metabolites whenever there is an increased availability of DA, or a decrease in the antioxidant buffering capacities that normally maintain DA in the reduced state. To explore this hypothesis we have examined the ability of DA to oxidize and covalently bind to striatal proteins both *in vitro* and *in vivo*. We have chosen to focus on the DA quinones rather than oxygen free radical formation because the quinone is more specific to the oxidation of DA. And, we are focussing on the quinone binding to protein as opposed to forming free cysteinyl conjugates because we believe that quinone binding to protein is not only a more sensitive index of DA oxidation as compared to free cysteinyl-DA conjugates, but also a direct measure of a potentially cytotoxic event.

METHODS

Isolation of Protein-Bound Radioactivity

All studies were performed using male Sprague-Dawley rats (275-350 g; Zivic-Miller, Inc., Allison Park, PA). Rats were killed by decapitation and the brain was rapidly dissected. Neostriatal slices (350 μm) were prepared and incubated at 37°C for up to 120 min in 1 ml of standard Krebs buffer (117 mM NaCl, 4.7 mM KCl, 1.2 mM $MgCl_2$, 1.25 mM $CaCl_2$, 1.2 mM NaH_2PO_4, 25 mM $NaHCO_3$, 11.5 mM dextrose) containing ^{3}H-DA (10 nM - 100 μM). In some experiments, the incubation medium also contained GSH (0.01, 0.1, or 1.0 mM). The tubes were gassed continuously with 95% O_2 and 5% CO_2 during the incubation. Immediately after incubation with ^{3}H-DA, the slices were rinsed in Krebs buffer, homogenized in 0.1 N perchloric acid containing 0.2 mM sodium bisulfite, and centrifuged (38,000 x g, 15 min, 4°C). The resulting protein pellet was washed three times by resuspension in 10% trichloroacetic acid and recentrifuged to remove any noncovalently-bound tritium. The final pellet was solubilized in 0.1 N NaOH and then neutralized aliquots were analyzed by liquid scintillation spectroscopy. Identical aliquots were used to determine protein content (Bradford, 1976).

Isolation of Cysteinyl-Catechols *in Vitro*

In a separate experiment, protein isolated from eight neostriatal slices incubated for 60 min in Krebs buffer containing 100 μM ^{3}H-DA was subjected to acid hydrolysis (1 ml 6 N HCl, 105°C, 24 h, under vacuum), alumina extraction, and then analysis on HPLC with electrochemical detection (Hastings and Zigmond, 1994). Cysteinyl-catechols were identified by comparison with cysteinyl-DA and cysteinyl-dihydroxyphenylacetic acid (DOPAC) standards that had been enzymatically synthesized and isolated (Rosengren et al., 1985). Fractions eluted from the HPLC system were collected and assayed for tritium content.

Cysteinyl-Catechol Formation *in Vivo*

Rats (300-350 g) were stereotaxically injected into the striatum (coordinates: + 0.8 mm AP; ± 2.7 mm ML from bregma; - 5.0 mm DV from dura) with DA (0.05-1.0 μmole/ 2 μl over 20 min). Additional animals received DA combined just before injection with an equimolar concentration of ascorbate or GSH. Non-surgically treated rats served as controls in the biochemical analyses.

At 24 h or 7 d following the intrastriatal injection of DA, animals were killed by decapitation and approximately 30 mg of striatal tissue surrounding the injection site was dissected out and used for analyses of free and protein-bound cysteinyl-catechols. Striatal tissue was homogenized in 1 ml of 0.1 N perchloric acid containing 0.2 mM sodium bisulfite, 1 mM ascorbate, and 1 mM dithiothreitol. Homogenates were centrifuged (34,000 x g, 15 min) to separate the acid-precipitated protein pellet from the acid-soluble components of the tissue. Following centrifugation, the acid-soluble supernatant was used for the analysis of free cysteinyl-catechols. The protein pellet was washed by resuspension and recentrifugation in the homogenizing solution, and then subjected to acid-hydrolysis as previously described (Hastings and Zigmond, 1994). Following an extraction of the catechol moieties with alumina, the cysteinyl-catechol derivatives were analyzed on HPLC with electrochemical detection.

Histochemical Analyses

Histochemical analyses on striatal tissue were performed 7 days following intrastriatal DA injections. Animals were deeply anesthetized and perfused transcardially with cold 4% paraformaldehyde. Coronal sections (40 μm) were labeled with a variety of markers. Antibodies against tyrosine hydroxylase (TH) and serotonin were used to identify dopaminergic terminals and serotonergic terminals, respectively, and an antibody against synaptophysin was employed as a general marker for nerve terminals in the striatum. The specificity

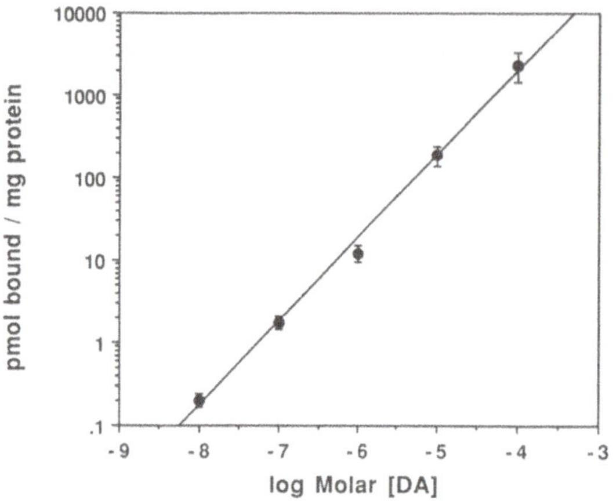

Figure 1. Concentration-dependent binding of DA to neostriatal protein. Neostriatal slices were incubated with ^{3}H-DA (0.01 - 100 μM) for 60 min at 37°C in a standard Krebs buffer. The amount of radioactivity incorporated into the protein was determined for each concentration according to procedures described in Methods (mean ± SEM; n = 3). Data collected in three separate experiments have been combined and presented on a semilog plot fitted to an exponential curve ($R^2 >$ 0.99). (From Hastings and Zigmond, 1994. Reprinted by permission.)

of each antibody in immunocytochemical studies has been previously demonstrated (Lewis et al., 1994; Glantz and Lewis, 1994; Barr et al., 1987; McLean and Shipley, 1987) Adjacent sections were stained for Nissl substance to examine general cellular architecture and gliosis in the region of the injection. Lesion size was defined as the largest cross-sectional area of diminished TH immunoreactivity in the striatum for each injection. Similar measures of the extent of tissue damage and gliosis were made on the adjacent Nissl-stained sections.

RESULTS AND DISCUSSION

Binding of ^{3}H-DA to Protein *in Vitro*

The incubation of striatal slices with ^{3}H-DA (0.01 - 100 μM) for 60 min resulted in tritium bound to protein. The amount of radiolabeled DA incorporated into protein was directly proportional to the amount of DA in the incubation medium (Figure 1). This finding is consistent with previous reports of the binding of catechols to proteins (Saner and Thoenen, 1971; Maguire et al., 1974; Scheulen et al., 1975; Rotman et al., 1976; Kato et al., 1986). Binding of 3H-DA to protein was also shown to be dependent upon the incubation period and the presence of antioxidants (Figure 2). The presence of either GSH or ascorbate in the

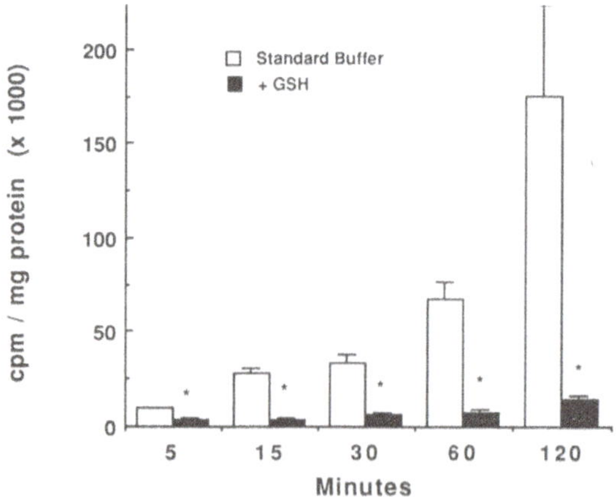

Figure 2. Effect of GSH on the binding of tritium to protein. Striatal slices were incubated with 60 nM ^{3}H-DA for the indicated times at 37°C in a standard Krebs buffer or buffer containing GSH (0.01, 0.10, or 1.00 mM). The amount of radioactivity incorporated into protein was determined at each time point according to procedures described in Methods. The results of all three GSH concentrations were similar and therefore combined (mean ± SEM; n=3). Student *t*-test * $p < 0.05$, significantly different from standard buffer. (From Hastings and Zigmond, 1994. Reprinted by permission.)

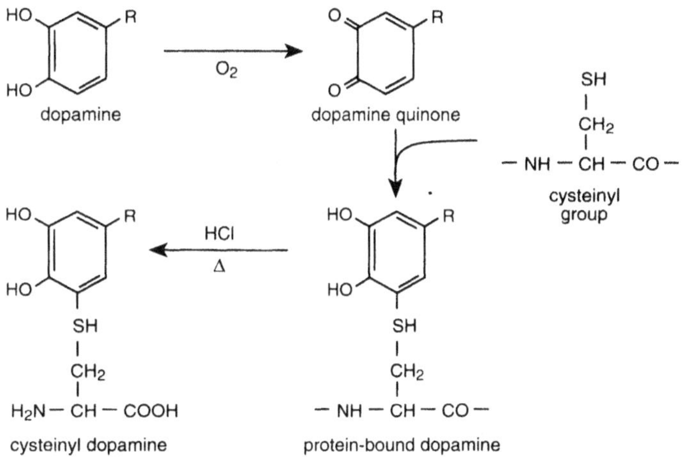

Figure 3. Reaction scheme depicting the stages of DA oxidation and binding to cysteinyl residues on protein followed by acid hydrolysis to isolate the modified amino acid residue.

incubation medium decreased tritium binding to protein during incubation of slices with 60 nM ^{3}H-DA. Binding was reduced by approximately 90%, with GSH concentrations as low as 10 μM , when compared to the standard buffer alone (Figure 2). Antioxidants may act to reduce the quinone derivative of DA back to the parent catechol, thereby preventing nucleophilic addition into the oxidized catechol ring. GSH, on the other hand, may also act as a nucleophile, competing with protein sulfhydryls to bind the oxidized quinone. By either mechanism, our results demonstrate that DA must oxidize to the catechol quinone form prior to its binding to protein.

Characterization of the DA-Protein Conjugate after Incubation with a High Concentration of DA

DA quinones are proposed to bind to cysteinyl residues on protein (Figure 3). To demonstrate that this is the case in our reaction system and to determine whether the protein adduct may be detected, we sought to isolate the catechol-modified amino acid residue ($R=CH_2\text{-}CH_2\text{-}NH_3^+$) from DA-exposed tissue. Striatal slices were exposed to 100 μM ^{3}H-DA for 60 min to incorporate a high level of covalently bound tritium. Protein was isolated from these slices by acid precipitation, then hydrolyzed, alumina-extracted, and analyzed by HPLC with electrochemical detection. Our results showed two major electroactive peaks that eluted in positions corresponding to cysteinyl-DOPAC and cysteinyl-DA standards (Figure 4A). Fractions were collected continuously as they eluted off the HPLC and were analyzed for tritium content by liquid scintillation spectroscopy. The radioactive profile demonstrated that the majority (72%) of the tritium was contained within these two major electroactive peaks (Figure 4B). It was not surprising that most of the tritium was found to be associated with cysteinyl-catechols, since under physiological conditions the sulfhydryl group on cysteine represents the major nucleophilic site on proteins. These observations are similar to those seen *in vitro* in studies of enzyme-catalyzed oxidation of DA (Ito et al., 1988; Hastings, 1995).

Our results show that in addition to isolating cysteinyl-DA, we also were isolating cysteinyl-DOPAC. This suggests that whereas some DA is oxidizing to a quinone, other DA molecules are converted to DOPAC via monoamine oxidase. This DOPAC then oxidizes to

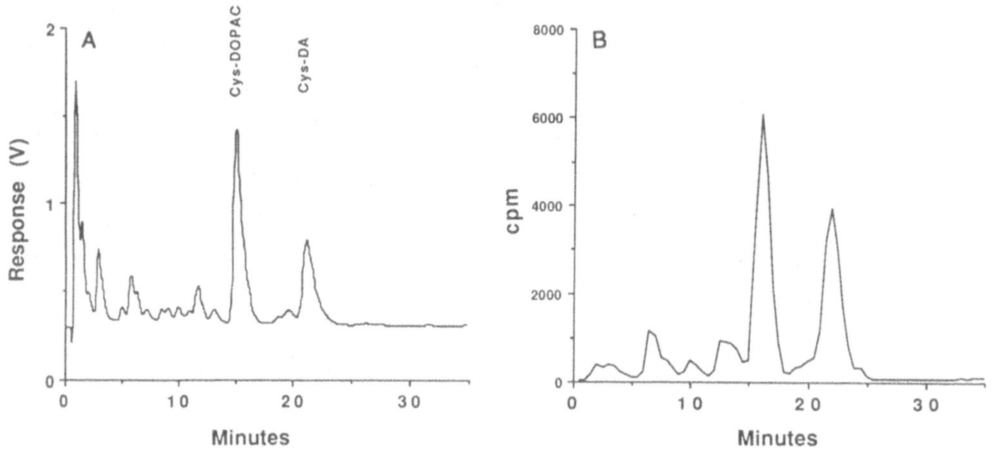

Figure 4. HPLC analysis of cysteinyl-catechols isolated from protein exposed to a high concentration of DA in vitro. Striatal slices were incubated with 100 μM ^{3}H-DA for 60 min. The acid-precipitated protein was hydrolyzed, alumina-extracted, and analyzed on HPLC. Fractions eluting off the HPLC were analyzed for tritium content. A. Electrochemical profile of the hydrolyzed protein indicating peaks which eluted in positions corresponding to the elution positions of cysteinyl-catechol standards. B. The corresponding radioactive profile of the hydrolyzed protein. (From Hastings and Zigmond, 1994. Reprinted by permission.)

form a DOPAC quinone which binds to protein. These assumptions are supported by the fact that cysteinyl-DA is not a substrate for monoamine oxidase (Carstam et al., 1990). The potential toxicity of DA metabolites has been proposed previously based on their ability to oxidize to reactive species similar to that of the parent catechol (Tse et al., 1976). Our results

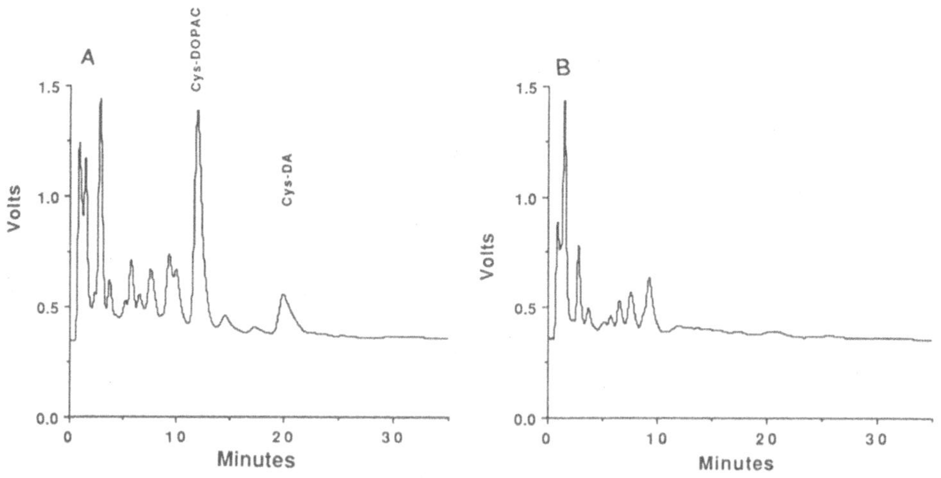

Figure 5. HPLC analysis of cysteinyl-catechols isolated from protein exposed to a high concentration of DA in vivo. Rats were injected intrastriatally with 1.0 μmole DA (A) or NaCl (B). Striatal tissue isolated 24 h later was analyzed for protein-bound cysteinyl-catechols. The representative HPLC chromatograms show (A) the presence of protein-bound cysteinyl-DA and cysteinyl-DOPAC isolated from striatum exposed to exogenous DA and (B) the absence of cysteinyl-catechols in the contralateral striatum exposed to NaCl.

suggest the potential for DOPAC as well as DA to act as a neurotoxin by binding to protein under oxidative conditions.

Selective Neurotoxicity was Associated with DA Oxidation *in Vivo*

We sought to determine whether the oxidation of DA and binding to protein would also occur *in vivo* (Hastings et al., 1994). High concentrations of DA (1.0 μmol in 2 μl) were injected into striatum and the tissue surrounding the injection site were analyzed for protein-bound cysteinyl-catechols 24 h later. Cysteinyl-DA and cysteinyl-DOPAC were isolated from DA-injected striatum but not from the contralateral striatum that was injected with an equimolar concentration of NaCl (Figure 5). When varying concentrations of DA (0.05-1.0 μmol in 2 μl) were injected into striatum, tissue analysis showed that protein-bound cysteinyl-DA and cysteinyl-DOPAC rose in proportion to the concentration of DA injected ($R<0.84$), increasing by least 100-fold above baseline at the highest dose, indicating that a large amount of DA and DOPAC had oxidized to the reactive quinone intermediates. Protein-bound cysteinyl-catechols were also 10-100-fold higher than the free cysteinyl-catechol levels 24 h following DA exposure, suggesting that the bound forms are a more sensitive measure of the level of catechol oxidation within tissue.

Histological analyses at 7 d showed cellular damage and gliosis at the injection site, and a surrounding region of selective loss of TH immunoreactivity. There was no detectable loss of other terminal markers, such as synaptophysin or serotonin immunoreactivity, within this region and the cellular architecture did not appear to be disrupted, suggesting that the loss of TH immunoreactivity is specific to DA terminals in striatum. As in the case of cysteinyl catechol levels, the size of the region with decreased TH immunoreactivity was dependent upon the concentration DA that the tissue was exposed to ($R<0.87$). This suggests a causal relationship between the amount of quinone binding to protein and the resulting neurotoxicity. Our results showing selective loss of DA terminals parallel those of Filloux and Townsend (1993) who observed that an intrastriatal injection of DA resulted in the selective loss of radioligand binding to the high affinity DA transporter, another protein indicative of DA terminals, as well as a centralized region of nonselective toxicity. These observations are also consistent with the loss of TH immunoreactivity resulting from an actual degeneration of DA terminals, an interpretation supported by our observation that tissue DA levels are decreased by 31% 7 days after a DA injection.

To assess the role of DA oxidation in DA-induced toxicity, we supplied the tissue with antioxidants during the exposure to DA. We observed that the co-administration of DA (1.0 μmol) with an equimolar concentration of ascorbate or glutathione reduced protein cysteinyl-catechol formation by 50-60% (Figure 6). Nissl stain revealed a central region of nonspecific damage similar to that seen in striatal tissue injected with DA alone. However, the surrounding zone of decreased TH immunoreactivity, representing the selective loss of DA terminals, was no longer observed when DA was administered in the presence of antioxidants. The fact that antioxidants reduced both the lesion size and the amount of cysteinyl-catechols provides additional support for the causal connection between DA oxidation and toxicity to DA terminals.

SUMMARY AND CONCLUSIONS

Our results demonstrate that (1) an oxidation product of DA, presumably the quinone, binds to protein in a time- and concentration-dependent manner, (2) both DA and its metabolite, DOPAC, oxidize and bind to cysteinyl residues on protein following exposure of striatal tissue to exogenous DA, either *in vitro* and *in vivo*, (3) exposure of striatal tissue

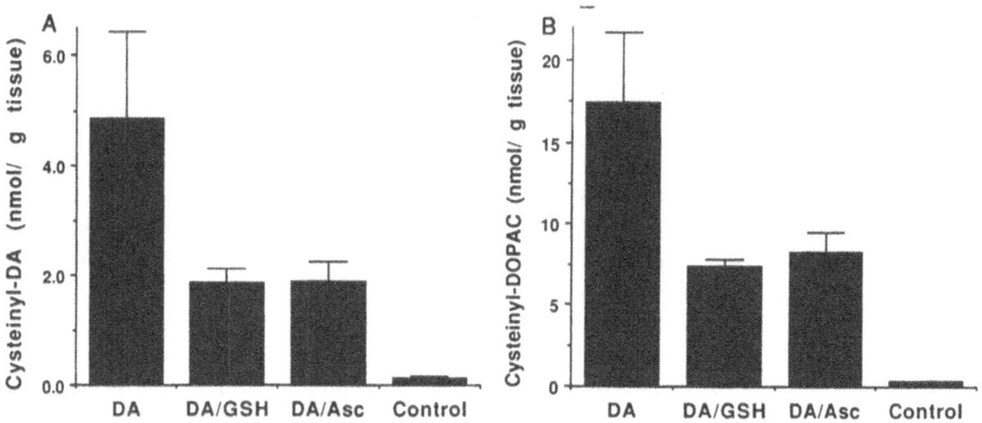

Figure 6. Effect of co-administration of DA and antioxidants on protein-bound cysteinyl-catechol levels. Protein-bound cysteinyl-DA (A) and cysteinyl-DOPAC (B) were measured in striatal tissue 24 h following intrastriatal injections of 1.0 μmole DA alone (DA), DA combined with equimolar GSH (DA/GSH) or ascorbate (DA/Asc), and compared to control tissue. All values are mean ± S.E.M. (n=4).

to exogenous DA results in both centralized nonspecific toxicity and a surrounding region of selective toxicity to DA terminals, and (4) the presence of antioxidants reduce protein-bound cysteinyl-catechol formation and block the selective toxicity to DA terminals resulting from DA exposure.

We hypothesize that DA-induced toxicity to DA neurons may be a contributing factor to the neurodegenerative process that occurs in Parkinson's disease. We believe that the toxicity is caused by an imbalance between the rate of DA oxidation and the inactivation of the resulting reactive metabolites. There is evidence to suggest that conditions characterized by a decreased antioxidant capacity exist in brains of patients with Parkinson's disease (see Introduction). Likewise, data indicate that after partial nigrostriatal lesions compensatory mechanisms induced in the remaining DA terminals, as well as therapeutic interventions with L-DOPA, may increase the availability of DA and thus create additional oxidant stresses that may contribute to the progression of the disease (Bernheimer et al., 1973; Zigmond et al., 1992). Further research will be necessary to test these hypotheses and if correct, to determine whether antioxidant treatment will be useful in ameliorating the neurodegenerative process.

ACKNOWLEDGMENTS

The authors thank Ariel Rabinovic and Richard Whitehead for excellent technical assistance. This research was supported in part by U.S.P.H.S. grants NS19608, MH00058, and MH00519.

REFERENCES

Agid Y., Ruberg M., Javoy-Agid R., Hirsch E., Raisman-Vozari R., Vyas S., Baucheux B., Michel P., Kastner A., Blanchard B., Damier P., Villares J. and Zhang, P. (1993) Are dopaminergic neurons selectively vulnerable to Parkinson's disease? In: Narabayashi H., Ngatsu T., Yanagisawa N. and Mizuno, Y. (eds.), *Advances in Neurology*, Vol. 60, pp 148-164.

Barr G.A., Eckenrode T.C. and Murray M. (1987) Normal development and effects of early deafferentation on choline acetyltransferase, substance P and serotonin-like immunoreactivity in the interpeduncular nucleus. *Brain Res.* 418, 301-313.

Bernheimer H., Birkmayer W., Hornykiewicz O., Jellinger K. and Seitelberger, F. (1973) Brain dopamine and the syndromes of Parkinson and Huntington: Clinical, morphological and neurochemical correlations. *J. Neurol. Sci.* 20, 415-455

Bradford M.M. (1976) A rapid and sensitive method for the quantitation of microgram quantities of protein utilizing the principle of dye-binding. *Anal. Biochem.* 72, 248-254.

Carstam R., Brinck C., Fornstedt B., Rorsman H. and Rosengren E. (1990) 5-S-Cysteinyldopac in human urine. *Acta Derm. Venereol.* (Stockh). 70, 373-377.

Cohen G. (1983) The pathobiology of Parkinson's disease: Biochemical aspects of dopamine neuron senescence. *J. Neural Transm.* 19 (Suppl), 89-103.

Dexter D.T., Carter C.J., Wells F.R., Javoy-Agid F., Lees A., Jenner P. and Marsden C.D. (1989) Basal lipid peroxidation in substantia nigra is increased in Parkinson's disease. *J. Neurochem.* 52, 381-389.

Dexter D.T., Holley A.E., Flitter W.D., Slater T.F., Wells F.R., Daniel S.E., Lees A.J., Jenner P. and Marsden C.D. (1994) Increased levels of lipid hydroperoxides in the parkinsonian substantia nigra: an HPLC and ESR study. *Movement Disorders* 9, 92-97.

Filloux F. and Townsend J.J. (1993) Pre- and post-synaptic neurotoxic effects of dopamine demonstrated by intrastriatal injection. *Exp. Neurol.* 119, 79-88.

Fornstedt B., Bergh I., Rosengren E. and Carlsson, A. (1990a) An improved HPLC-electrochemical detection method for measuring brain levels of 5-S-cysteinyldopamine, 5-S-cysteinyl-3,4-dihydroxyphenylalanine, and 5-S-cysteinyl-3,4-dihydroxyphenylacetic acid. *J. Neurochem.* 54, 578-586.

Fornstedt B., Brum A., Rosengren E. and Carlsson A. (1989) The apparent autoxidation rate of catechols in dopamine-rich regions of human brains increases with the degree of depigmentation of substantia nigra. *J. Neural Transm.* 1, 279-295.

Fornstedt B., Pileblad E. and Carlsson A. (1990b) *In vivo* autoxidation of dopamine in guinea pig striatum increases with age. J. Neurochem. 55, 655-659.

Fornstedt B. and Carlsson A. (1991) Vitamin C deficiency facilitates 5-S-cysteinyldopamine formation in guinea pig striatum. *J. Neurochem.* 56, 407-414.

Glantz L.A. and Lewis D.A. (1994) Synaptophysin and not Rab3A immunoreactivity is specifically reduced in the prefrontal cortex of schizophrenic subjects. *Soc. Neurosci. Abstr.* 20, 622.

Graham D.G., Tiffany S.M., Bell W.R. Jr. and Gutknecht W.F. (1978) Autoxidation versus covalent binding of quinones as the mechanism of toxicity of dopamine, 6-hydroxydopamine, and related compounds toward C1300 neuroblastoma cells *in vitro*. *Mol. Pharmacol.* 14, 644-653.

Graham D.G. (1978) Oxidative pathways for catecholamines in the genesis of neuromelanin and cytotoxic quinones. *Mol. Pharmacol.* 14, 633-643.

Halliwell B. (1992) Reactive oxygen species and the central nervous system. *J. Neurochem.* 59, 1609-1623.

Hastings T.G., Lewis D.A. and Zigmond M.J. (1994) Intrastriatally administered dopamine: Evidence of selective neurotoxicity associated with dopamine oxidation. *Soc. Neurosci. Abstr.* 20, 413.

Hastings T.G. and Zigmond M.J. (1994) Identification of catechol-protein conjugates in neostriatal slices incubated with ^{3}H-dopamine: Impact of ascorbic acid and glutathione. *J. Neurochem.* 63, 1126-1132.

Hastings T.G. (1995) Enzymatic oxidation of dopamine: The role of prostaglandin H synthase. *J. Neurochem.* 64, 919-924.

Hirsch E., Graybiel A.M. and Agid Y.A. (1988) Melanized dopaminergic neurons are differentially susceptible to degeneration in Parkinson's disease. *Nature* 334, 345-348.

Ito S., Kato T. and Fujita K. (1988) Covalent binding of catechols to proteins through the sulphydryl group. *Biochem. Pharmacol.* 37, 1707-1710.

Kastner A., Hirsch E.C., Lejeune O., Javoy-Agid F., Rascol O. and Agid Y. (1992) Is the vulnerability of neurons in the substantia nigra of patients with Parkinson's disease related to their neuromelanin content? *J. Neurochem.* 59, 1080-1089.

Kato T., Ito S. and Fujita K. (1986) Tyrosinase-catalyzed binding of 3,4-dihydroxyphenylalanine with proteins through the sulfhydryl group. *Biochim. Biophys. Acta* 881, 415-421.

Kish S.J., Morito C. and Hornykiewicz O. (1985) Glutathione peroxidase activity in Parkinson's disease brain. *Neurosci. Lett.* 58, 343-346.

Lewis D.A., Melchitzky D.S., Haycock J.W. (1994) Expression and distribution of two isoforms of tyrosine hydroxylase in macaque monkey brain. *Brain Res.* 656, 1-13.

Maguire M.E., Goldmann P.H. and Gilman A.G. (1974) The reaction of [^{3}H]norepinephrine with particulate fractions of cells responsive to catecholamines. *Mol. Pharmacol.* 10, 563-581.

Mann V.M., Cooper J.M., Daniel S.E., Srai K., Jenner P., Marsden C.D. and Schapira A.H.V. (1994) Complex I, iron, and ferritin in Parkinson's disease substantia nigra. *Ann. Neurol.* 36, 876-881.

Martilla R.J., Lorentz H., Rinne U.K. (1988) Oxygen toxicity protecting enzymes in Parkinson's disease. *J. Neurol. Sci.* 86, 321-31.

McLean J.H. and M.T. Shipley (1987) Serotonergic afferents to the rat olfactory bulb: I. Origins and laminar specificity of serotonergic inputs in the adult rat. *J. Neurosci.* 7, 3016-3028.

Michel P.P. and Hefti F. (1990) Toxicity of 6-hydroxydopamine and dopamine for dopaminergic neurons in culture. *J. Neurosci. Res.* 26, 428-435.

Perry T.L., Godin D.V. and Hansen S. (1982) Parkinson's disease: A disorder due to nigral glutathione deficiency? *Neurosci. Lett.* 33, 305-310.

Rosengren E., Linder-Eliasson E. and Carlsson A. (1985) Detection of 5-S-cysteinyldopamine in human brain. *J. Neural Transm.* 63, 247-253.

Rotman A., Daly J.W. and Creveling C.R. (1976) Oxygen-dependent reaction of 6-hydroxydopamine, 5,6-dihydroxytryptamine, and related compounds with proteins *in vitro*: A model for cytotoxicity. *Mol. Pharmacol.* 12, 887-899.

Saggu H., Cooksey J., Dexter D., Wells F.R., Lees A., Penner P. and Marsden C.D. (1989) A selective increase in particulate superoxide dismutase activity in parkinsonian substantia nigra. *J. Neurochem.* 53, 692-697.

Saner A. and Thoenen H. (1971) Model experiments on the molecular mechanism of action of 6-hydroxydopamine. *Mol. Pharmacol.* 7, 147-154.

Scheulen M., Wollenberg P., Bolt H.M., Kappus H. and Remmer H. (1975) Irreversible binding of dopa and dopamine metabolites to protein by rat liver microsomes. *Biochem. Biophys. Res. Commun.* 66, 1396-1400.

Schmidt D.J., Ritter J.K., Sonsalla P.K., Hanson R. and Gibb J.W. (1985) Role of dopamine in the neurotoxic effects of methamphetamine. *J. Pharmacol. Exp. Ther.* 233, 539-544.

Sofic E., Paulus W., Jellinger K., Reiderer P. and Youdim M.B.H. (1991) Selective increase of iron in substantia nigra zona compacta of Parkinsonian brains. *J. Neurochem.* 56, 978-982.

Tse D.C.S., McCreery R.L. and Adams R.N. (1976) Potential oxidative pathways of brain catecholamines. *J. Med. Chem.* 19, 37-40.

Weinberger J., Nieves-Rosa J. and Cohen G. (1985) Nerve terminal damage in cerebral ischemia: Protective effect of alpha-methyl-para-tyrosine. *Stroke* 16, 864-870.

Zigmond M.J., Acheson A.L., Stachowiak M.K. and Stricker E.M. (1984) Neurochemical compensation after nigrostriatal injury in an animal model of preclinical Parkinsonism. *Arch. Neurol.* 41, 856-890.

Zigmond M.J., Hastings T.G. and Abercrombie, E.D. (1992) Neurochemical responses to 6-hydroxydopamine and L-DOPA therapy: Implications for Parkinson's disease. *Ann. NY Acad. Sci.* 648, 71-86.

SELECTIVE DEPLETION OF MITOCHONDRIAL GLUTATHIONE CONTENT BY PIVALIC ACID AND VALPROIC ACID IN RAT LIVER

Possibility of a Common Mechanism

U. Zanelli,[1] P. Puccini,[2] D. Acerbi,[2] P. Ventura,[2] and P. G. Gervasi[1]

[1] Istituto di Mutagenesi e Differenziamento
CNR Pisa via Svezia 10 Pisa 56100 Italy
[2] Chiesi Farmaceutici
via Palermo 26 Parma 36036 Italy

INTRODUCTION

Pivalic acid (PIV) or trimethylacetic acid is a compound thought to be harmless and widely used for pro-drug production, such as pivampicillin and pivaloyloxymethyl dopa ester (Binderup et al. 1971, Wickers et al. 1985). These pro-drugs administrated to rat and monkey undergo an hydrolysis releasing free PIV, which is excreted in urine mostly unchanged or as carnitine or glucuronic acid conjugates.

In man the conjugation with carnitine is dominant and a carnitine deficiency, especially in long-term therapy with these drugs, has been reported (Melegh et al. 1987). However, an oxidative route of PIV leading to reactive intermediates has not been described.

In the present study, we have examined the time-course and the dose-dependent effects of PIV and valproic acid (VPA) administration on glutathione (GSH) content in liver homogenate and mitochondria of adult rats. The investigation on VPA was performed for comparative reasons, as it has been reported that an electrophilic intermediate of VPA metabolism formed in mitochondria, the (E)-2,4-diene VPA, is responsible for the formation of a GSH-conjugated metabolite found in human urine (Kassahun et al. 1991). Thus, as the cytosolic and mitochondrial pools of GSH are distinct (Meredith and Reed 1982), a VPA specific mitochondrial GSH depletion was expected.

Biological Reactive Intermediates V
Edited by R. Snyder et al., Plenum Press, New York, 1996

MATERIALS AND METHODS

Chemicals

PIV and VPA sodium salts were purchased from Fluka (Buchs, Switzerland); all other chemicals were obtained from commercial sources.

Animals and Drug-Treatment. Male Sprague Dawley rats weighting 240-300g were administered a single dose of 200, 400 or 600 mg/Kg sodium pivalate suspended in corn oil or sodium valproate in 0.1M phosphate buffer either by oral gavage or ip injection. Animals were killed by CO_2 asphyxia always at the same hour (2 p.m.) to avoid influence of circadian cycle on GSH content (Bélanger et al. 1991), livers were collected and homogenates prepared. Mitochondrial fractions were isolated as described by Yohana and Tampo (1987).

Enzymatic Assays. GSH content was determined with Ellman's reagent following deproteinization with 5% trichloroacetic acid (Habig et al. 1974). Data are averages ± S.D. (bars) of three or more rats. Protein content was assayed according to Lowry et al. (1951).

RESULTS

GSH content in both mitochondria and homogenates of liver was determined after administration of a single dose of PIV and VPA. Control rats showed a GSH content of 6.0 ± 0.3 µmol/gr tissue in liver homogenate and 18.2 ± 0.9 µgr/mgr protein in the mitochondria fraction. Treatments with 200mg/g of PIV or VPA did not significantly modify the GSH content in either homogenate or mitochondria (data not shown).

Administration(ip) of 600mg/Kg PIV resulted in severe depletion of the total GSH content (to about 65% of control value) between 3h and 5h (fig. 1A). A more acute depletion (to about 50% of control value) was shown in mitochondria 5h after the PIV treatment (Fig. 1B). VPA, at the same dose, provoked, both in homogenate and mitochondria, a GSH depletion similar to that of PIV. After either treatment, the complete cytosolic and mitochandrial GSH recovery was reached within 24h. But in mitochondria reduced GSH levels were still present 15h after the administration of either drug.

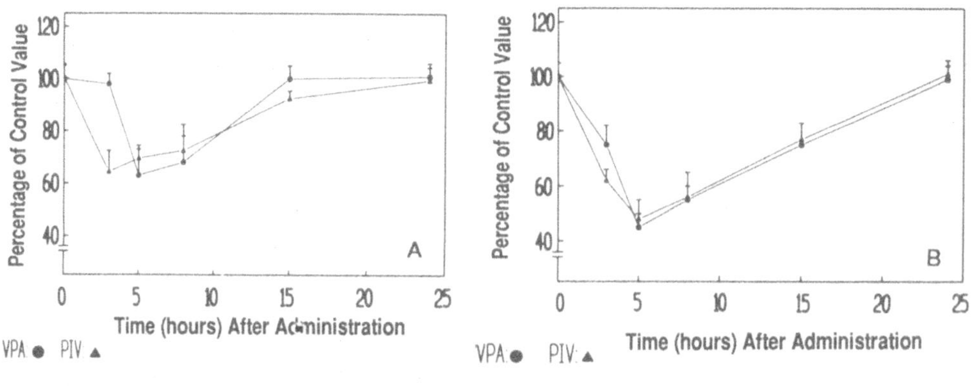

Figure 1. A. Depletion of GSH in whole homogenate after single ip administration of 600mg/kg of drug. B. Depletion of GSH in mitochondrial fraction after single ip administration of 600mg/kg of drug.

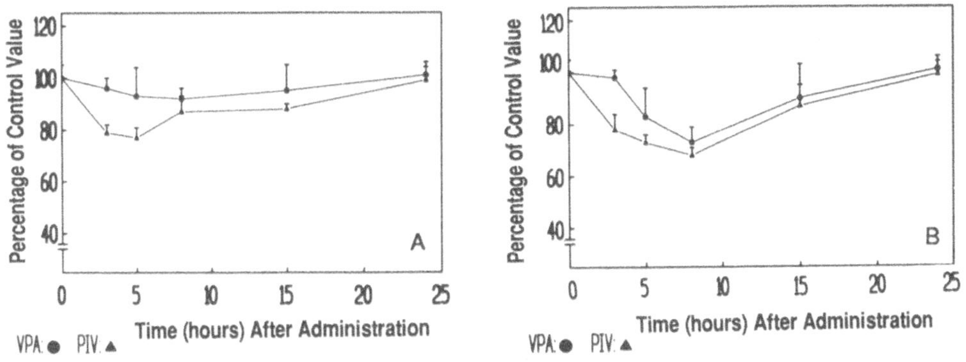

Figure 2. A. Depletion of GSH in whole homogenate after single oral dose of 600mg/kg of drug. B. Depletion of GSH in mitochondrial fraction after single oral dose of 600mg/kg of drug.

When 600mg/g PIV and VPA were administered by gavage, only PIV caused a small depletion of GSH in the homogenate (Fig. 2A). In mitochondria (Fig. 2B), however, both PIV and VPA induced a similar and persistent GSH depletion: the greatest amounting to 70% of control value, was found at 8h after both drug treatments.

400mg/g ip doses of PIV and VPA produced, in whole homogenate, only a slight depletion of GSH (Fig. 3A). In mitochondria, the GSH depletion was greater and more prolonged than in homogenates; the GSH content was reduced to about 70% of control value (Fig. 3B) and the GSH depletion was still present 14h after drug injection. A dose of 400mg/Kg PIV or VPA, given by gavage, altered GSH content neither in liver homogenate nor in mitochondria (data not shown).

DISCUSSION

Collectively taken these data have demonstrated that VPA is able to induce GSH depletion and that this effect is more pronounced in mitochondria due to the formation, in this organelle, of an electrophilic metabolite, the (E)-2,4-diene VPA (Kassahun et al. 1991). Indeed, it should be kept in mind that the GSH content in mitochondria is about 15% of the total GSH in the hepatocytes and that the mitochondrial and cytosolic GSH pools are distinct.

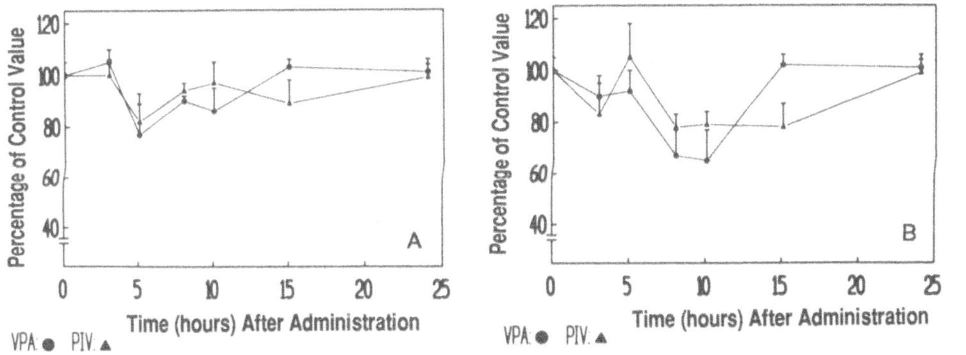

Figure 3. A. Depletion of GSH in whole homogenate after single ip administration of 400mg/kg of drug. B. Depletion of GSH in mitochondrial fraction after single ip administration of 400mg/kg of drug.

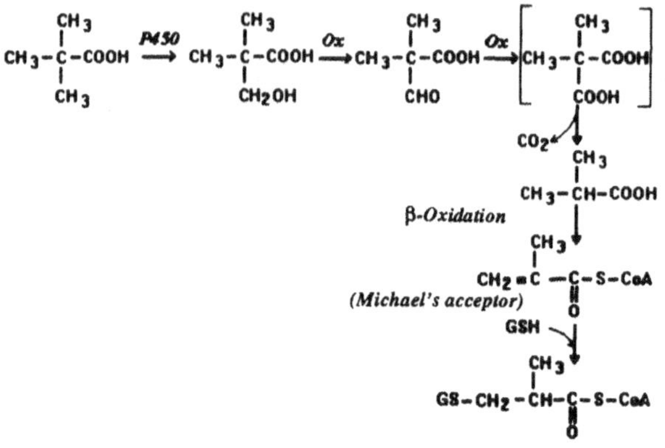

Scheme 1. Postulated PIV metabolism forming Michael's Acceptor.

Only depletions of mitochondria GSH content greater than 50% have resulted in a net flow of GSH from cytosol to mitochondria (Meredith and Reed 1982).

PIV reduced GSH content in a similar manner to VPA both in mitochondria and whole homogenate and, in particular, greater GSH depletions were observed in mitochondria. The ability of PIV to reduce GSH content was unexpected since PIV has no functional groups able to conjugate GSH and, so far, oxidative metabolism of PIV to form GSH reactive intermediates has not been described.

In order to explain this finding, we have postulated that PIV could undergo oxidative metabolism in microsomes and mitochondria: the isobutyl acid could be formed by microsomes and thereafter undergo mitochondrial β-oxidation to yield an α-β-unsaturated methacrylic-CoA ester able to give, as in the case of VPA, the Michael reaction with GSH (Scheme 1). In keeping with this hypothesis, the ip administration of 400mg/g sodium isobutyrate decreased the GSH content in mitochondria after 5h to about 70% but no depletion was observed in whole homogenates. Although the presence and the toxicological importance of PIV-dependent GSH depletion in man remains to be determined, particularly at therapeutic doses, PIV might not be as inert as it was supposed and thereafter its use in prodrugs should be reevaluated.

REFERENCES

Bélanger, P.M., Desgagné, M. and Bruguerolle, B. (1991). Temporal variations in microsomal lipid peroxidation and glutathione concentration of rat liver. *Drug Metabol. Dipos.* **19:** 241-244.

Binderup, E, Gotfrendsen, W.O. and Roholt, J., (1971). Orally active cefaloglycin esters. *Journal of Antibiotics* **24:** 767-773.

Habig, W.H., Pabst, M.J. and Jacoby W.B. (1974). The first enzymatic step in mercapturic acid formation. *J. Biol. Chem.* **249:** 7130-7139.

Kassahun, K., Farrel, K. and Abbot, F. (1991). Identification and characterization of the glutathione and N-acetylcysteine conjugates of (E)-2-propyl-2,4-pentadienoic acid, a toxic metabolite of valproic acid, a toxic, in rat and humans. *Drug Metabol. Dispos.* **19:** 525-535.

Lowry, O.H., Rosebrough, N.J., Farr, A.L. and Randall, R.J. (1951). Protein measurement with Folin phenol reagent. *J.Biol. Chem.* **193:** 265-275.

Melegh, B., Kerner, J. and Bieber, L.L. (1987). Pivampicillin-promoted excretion of pivaloyl carnitine in humans. *Biochem. Pharmacol.* **36:** 3405-3409.

Meredith, M.J. and Reed, D.J. (1982). Status of the mitochondrial pool of glutathione in the isolated hepatocytes. *J. Biol: Chem.* **257:** 3747-3753.

Ruff, L.J. and Brass, E.P. (1991). Metabolic effects of pivalate in isolated rat hepatocytes. *Toxicol. App. Pharmacol.* **110:** 295-302.

Wickers, S., Duncan, C.A.H., Withe, S.D., Ramjit, H.G., Smith, J.L., Walker, S.W., Flynn, H.Q. and Arison, B.H. (1985). Carnitine and glucuronic acid conjugates of pivalic acid. *Xenobiotica* **15:** 453-258.

Yohana, M. and Tampo, Y. (1987). Bromosulfophthalein abolishes glutathione-dependent protection against lipid peroxidation in rat liver mitochondria. *Biochem. Pharmacol.***36:** 2831-2837.

REGULATION OF THE SUPEROXIDE RELEASING SYSTEM IN HUMAN FIBROBLASTS

Beate Meier

Chemisches Institut
Tierärztliche Hochschule Hannover
Bischofsholer Damm 15, 30173 Hannover, Germany

SUMMARY

Fibroblasts release low amounts of reactive oxygen intermediates (ROIs) upon contact with appropriate stimuli, which have regulatory functions in inter- and intracellular signal transduction. Ca^{2+}/calmodulin are involved in the activation process of the ROI-generating system. A rise of intracellular Ca^{2+} is necessary, but not sufficient to trigger the release of ROIs. As a second signal, protein phosphorylation seems to be involved in the activation process.

INTRODUCTION

Fibroblasts release low amounts of ROIs in response to appropriate stimuli continuously for more than four hours in nmolar amounts (1-3), which act as intra- and intercellular messengers. The primary species released is O_2^-, synthesized by an NAD(P)H-oxidase which is structurally and genetically distinct from the NADPH-oxidase in polymorphonuclear granulocytes, macrophages or B-lymphocytes (4-7). The regulation of the ROI generating system in fibroblasts or other cells, not belonging to the phagocyte system, is yet unknown.

In the present study, we give evidence that O_2^- release is regulated by the rise of intracellular Ca^{2+} and calmodulin as well as protein phosphorylation.

MATERIAL AND METHODS

Cell lines derived from primary cultures of human fibroblasts were established by the method of explantate culture as described (1) and propagated in DMEM medium, supplemented with 5 % heat-inactivated fetal calf serum (FCS) in 5 % CO_2 in air at 37°C. To determine the O_2^- and H_2O_2 formation, fibroblasts were cultured as monolayer on all four

Biological Reactive Intermediates V
Edited by R. Snyder et al., Plenum Press, New York, 1996

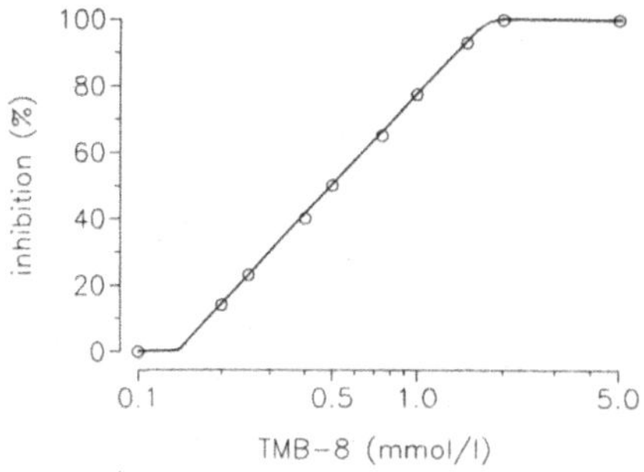

Figure 1. Inhibition of O_2^- production in fibroblasts by TMB-8. Platelet activating factor (5 µmol/l) was added as stimulus to the fibroblasts and O_2^- release was determined in the presence of different amount of TMB-8.

sides of glass cuvettes to confluence. Formation of O_2^- was determined photometrically at 550 nm by the reduction of cytochrome c (8) and H_2O_2 fluorimetrically by the peroxidase-mediated oxidation of scopoletin (9) as described (1). Rise of intracellular Ca^{2+}-ion concentration was determined with adherent fibroblasts using Quin-2AM as fluorescent Ca^{2+}-chelator.

RESULTS

All substances known so far to trigger O_2^- release in human fibroblasts cause an increase of the intracellular Ca^{2+} concentration (not shown). This stimulated O_2^- release was prevented upon addition of TMB-8 (fig.1) or Quin-2/AM, Ca^{2+}-chelators able to penetrate cell membranes, whereas non-penetrative Ca^{2+}-chelators like EDTA, EGTA or DETAPAC up to 10 mmol/l were ineffective as were the Ca^{2+}-channel blockers verapamil (5 µmmol/l) or tetrodotoxin (5µ mmol/l), which block the voltage dependent "early" Ca^{2+}- and Na^+-channels. Also trace metal ions, blocking the voltage dependent insensitive "late" Ca^{2+}-/K^+- and proton-channels ($ZnCl_2$, which $NiCl_2$, $MnCl_2$, $CoCl_2$, $CdCl_2$) or $La(NO_3)_3$) up to 1 mmol/l, showed no effect on the stimulated O_2^- release.

Calmodulin inhibitors or antagonists like trifluoperazine or W-7 inhibited the O_2^- and H_2O_2 release in fibroblasts (fig.2).

A rise of intracellular Ca^{2+}-ion concentration alone is not sufficient to trigger the activation of the NAD(P)H-oxidase as several substances which induce the liberation of intracellular Ca^{2+} are unable to induce the formation of O_2^- radicals, like thapsigargin, ruthenium red, caffein, carbachol or the chemotactic peptide V-G-V-A-P-G, a repetition fragment of elastin.

Thus a second signal, likely the activation of a protein kinase, is necessary to activate the fibroblast ROI-releasing system, because the phosphatase inhibitors okadaic acid (fig.3)

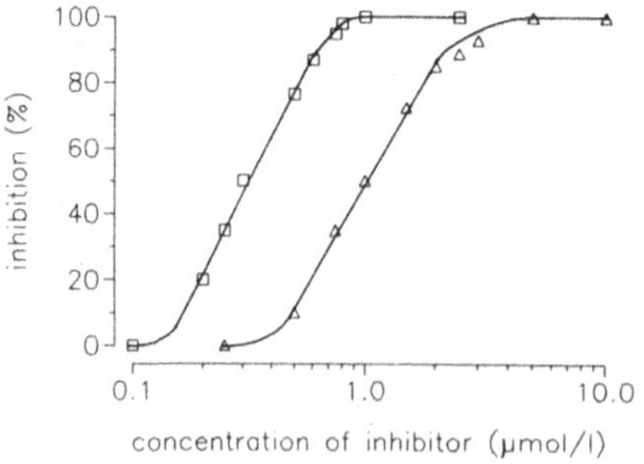

Figure 2. Inhibition of O_2^- production in fibroblasts by trifluoperazine or W-7. Platelet activating factor (5 µmol/l) was added as stumulus to the fibroblasts and the O_2^- release was determined. 2 min after O_2^- release was observable W-7 (Δ) or trifluoperazine (□) were added and inhibitions were determined.

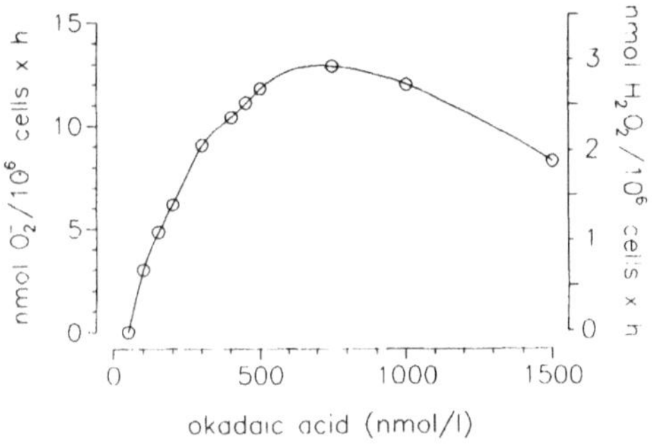

Figure 3. Concentration dependence of O_2^- and H_2O_2 formation by fibroblasts stimulated with okadaic acid. Primary cultures of human fibroblasts were incubated with okadaic acid, and O_2^- and H_2O_2 release were determined.

or calyculin A (fig.4) trigger the release of ROIs or amplify their amount upon contact with "activated" cells.

In contrast to the phagocyte-NADPH-oxidase system the fibroblast-NAD(P)H-oxidase is not affected by protein kinase C activators or inhibitors. The diacylglycerols sn-1-oleoyl-2-acetyl-glycerol or sn-1,2-di-octanoyl-glycerol, both potent activators of the Ca^{2+}-ion and phospholipid-dependent protein kinase C (100 µmol/l), did not cause a stimulated O_2^- and H_2O_2 release in human fibroblasts. Moreover, staurosporine or calphostin c both potent inhibitors of this enzyme, did not prevent stimulation of the fibroblast system. PMA at optimal concentrations of about 2 µmol/l is a weak stimulant for fibroblasts and caused less than 1 % O_2^- release in comparison to Ca-ionophore A 23187 or ionomycin (unpublished). Nanomolar concentrations of PMA sufficient for complete activation of protein kinase C were ineffective in triggering O_2^- release in fibroblasts. After depletion of the fibroblasts of protein kinase C by a sustained incubation with PMA, the fibroblasts still had the capacity to release ROIs upon contact with appropriate stimuli.

Also tyrosine kinase inhibitors (genistein, herbimycin A) or inhibitors of g-proteins (pertussis toxin, cholera toxin) or activators like F^- did not show any effect on O_2^- release in fibroblasts, making their participation in the regulation process unlikely.

DISCUSSION

As in neutrophils an increased cytosolic Ca^{2+} concentration might be a triggering signal for O_2^- release in fibroblasts. However in contrast to the NADPH-oxidase of phagocytes extracellular Ca^{2+} is not involved in the activation process. Ca^{2+} is suggested to stimulate O_2^- release in PMNs independently in two synergistic ways, either directly by activating NADPH-oxidase possibly through a calmodulin - protein kinase regulated mechanism or by activating protein kinase C, a Ca^{2+}-activated, phospholipid-dependent protein

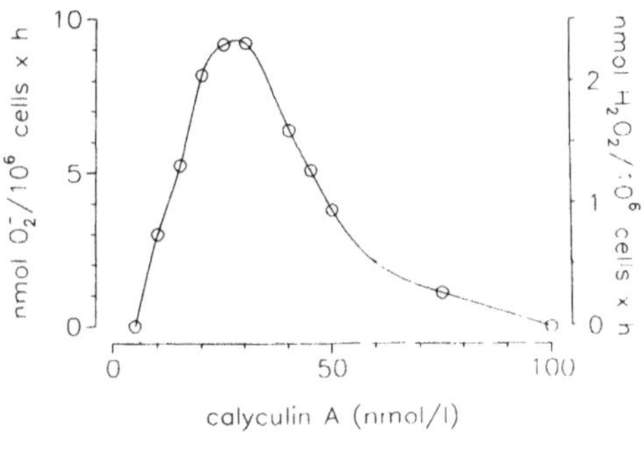

Figure 4. Concentration dependence of O_2^- and H_2O_2 formation in fibroblasts stimulated with calyculin A. Primary cultures of fibroblasts were incubated with calyculin A, and O_2^- release and H_2O_2 release were determined.

kinase. PMA and other phorbol esters are believed to act as structural analogues of diacylglycerol which, in association with phospholipid, bind and activate protein kinase C directly without raising intracellular Ca^{2+} (10). As shown in fig.1, the dual calmodulin and protein kinase C inhibitors, trifluoperazine and W-7, inhibited O_2^- and H_2O_2 release by fibroblasts stimulated with Ca-ionophores, PAF or cytokines (IL-1, TNF), suggesting that calmodulin is also involved in the activation of the NAD(P)H-oxidase in fibroblasts. However, there are indications that these substances also might directly inhibit neutrophil NADPH-oxidase (10). In contrast to neutrophils, diacylglycerols known as potent activators of protein kinase C, did not induce O_2^- formation in fibroblasts. All these observations suggest that in contrast to leukocytes an activation of protein kinase C is not involved in the regulation of the ROI-releasing system in fibroblasts. However, a yet unidentified protein kinase seems to trigger the release of ROIs in fibroblasts possibly by phosphorylation of serine/threonine residues, because okadaic acid as well as calyculin A, both inhibitors of protein phosphatase 2A, induce or amplify the release of ROIs. Both systems, the regulation by Ca^{2+}/calmodulin and protein phosphorylation, make a precise regulation of the ROI-releasing systems in fibroblasts a likely possibility. The distinct regulation mechanisms in comparison to the phagocyte NADPH-oxidase is physiologically important due to the different functions of both systems.

REFERENCES

1. Meier, B., H. H. Radeke, S. Selle, M. Younes, H. Sies, K. Resch and G. G. Habermehl (1989) Biochem. J. 263, 539-545
2. Meier, B., H.H. Radeke, S. Selle, H.-H. Raspe, H. Sies, K. Resch and G.G. Habermehl (1990) Free. Rad. Res. Commun. 8, 149-160
3. Meier, B., S. Selle, H.H. Radeke, G.G. Habermehl, K. Resch and H. Sies (1990) Biol. Chem. Hoppe-Seyler. 371, 1021-1025, 1990
4. Meier, B., A.R. Cross, J.T. Hanock, F.J. Kaup and O.T.G. Jones (1991) Biochem J. 275, 241-245
5. Meier, B., A.J. Jesaitis, A. Emmendörfer, J. Roesler and M.T. Quinn (1993) Biochem. J. 289, 481-486
6. Emmendörfer, A., J. Roesler, J. Elsner, E. Raeder, M.-L. Lohmann-Mathes and B. Meier (1993) Eur. J. Haematol. 51, 223-227
7. Jones,S.A., J.D. Wood, M.J. Coffey and O.T.G. Jones (1994) FEBS Lett. 355, 178-182
8. McCord, J. M. and I. Fridovich (1969) J. Biol. Chem. **244**, 6049-6055
9. Root, R.K., J. Metcalf, N. Oshino and B. Chance (1975) J. Clin. Invest. **55**, 945-955
10. Cross, A.R. (1990) Free Rad. Biol. Med. 8, 71-93

STUDIES ON THE MECHANISM OF PHOTOTOXICITY OF BAY y 3118 AND OTHER QUINOLONES

U. Schmidt and G. Schlüter

FB Toxicology, BAYER AG
D 42096 Wuppertal Germany

1. INTRODUCTION

The quinolone antibiotic BAY y 3118 is a photochemically labile compound, and irradiation with UV-A light leads to its rapid decomposition. In vivo, following oral drug administration to guinea pigs, irradiation of the animals not only results in a severe local phototoxicity, but also leads to systemic phototoxic reactions [1].

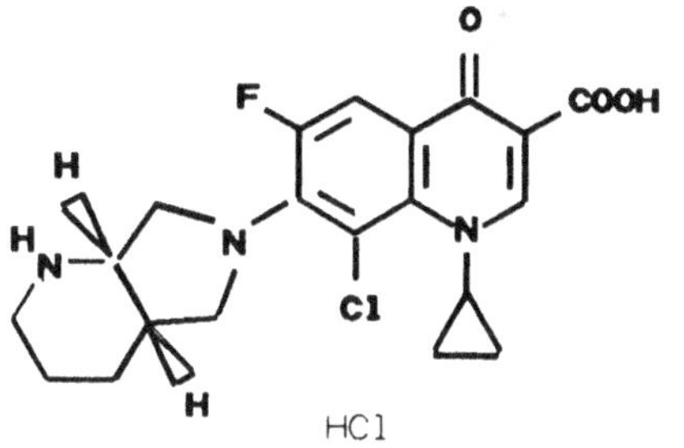

BAY y 3118 (MG: 442.3)
S,S-ENANTIOMER

Scheme 1

- **cutaneous phototoxicity**
  (guinea pig, 20 - 100 mg/kg p.o. + UV-A 20 J/cm[2])
- **systemic phototoxicity**
  (guinea pig, 50 - 100 mg/kg p.o. + UV-A 20 J/cm^2 morbid after 4-5 days)
- **systemic toxicity**
  (guinea pig, 10 mg/kg i.v. BAY y 3118 photo degradation products, death after 30 sec.)
- **species differences in phototoxicity**
  (rat guinea pig, human ?)

It is discussed in the literature [2-5], that quinolones are efficient producer of singlet oxygen in the presence of light and oxygen. This active oxygen specie has been postulated to be responsible for the observed phototoxic reaction.

Other mechanisms, such as production of toxic photo degradation products, and toxic reactive intermediates of energy transfer to a biological entity are possible.

It was shown that α-tocopherol (α-t) is a very efficient radical scavenging compound, so we used the depletion of α-t as a marker reaction for radical formation.

2. METHODS AND RESULTS

2.1. Photolability of Quinolones (in Vitro)

I. We irradiated quinolones with UV-A (15-16 mW/cm^2) in concentration of 1, 2, 5, 10 and 20 μM (in methyl alcohol-solution) for 30 min in the presence of α-tocopherol (10 μM). These concentrations are comparable to the in vivo situation. After the irradiation we measured the concentration of the unreacted α-tocopherol by HPLC with fluorescence detection (λ 294 nm/328 nm).

Fig. 1 shows the efficient reaction of BAY y 3118-degradation products with α-t in an almost equimolar way and the stability or no reaction of ciprofloxacin. The stability of ofloxacin and a stronger radical formation for nalidixic acid was also shown, which is in agreement with literature. Under this condition no reaction for sparfloxacin and lomefloxacin occured.

II. In a further experiment we irradiated a fixed concentration (10 μM) of the same quinolones but for different time periods (0, 5, 15, 30 and 60 min.). The ranking of photolability was the same as in the first experiment, with the exception, that lomefloxacin after 60 min. also lowered the α-t-concentration.

III. In the next experiment we investigated the question if glutathione could influence the reaction with α-T. We used 500 μM-GSH, which is also comparable to the in vivo situation, in the presence of again 10 μM α-T. The reaction with BAY y 3118 and UV-A took place in methyl alcohol-solution. But there was no difference in the reaction, with and without GSH.

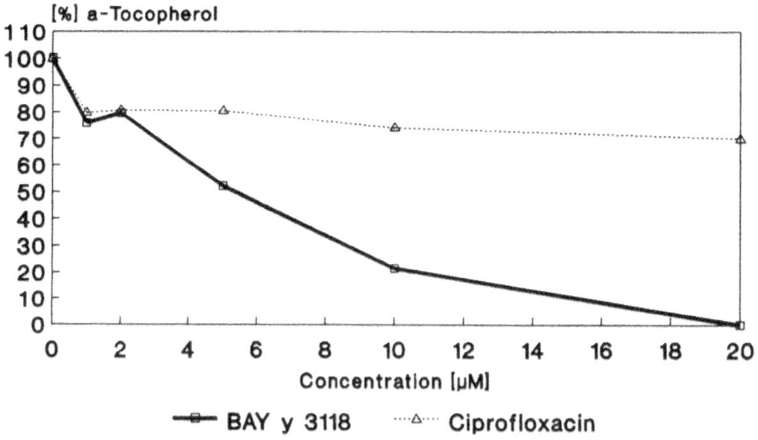

Figure 1. BAY y 3118/ciprofloxacin + α-tocopherol (10 μM). UVA (15-16 mW/cm^2) 30 Min.

Table 1. Determination of BAY y 3118 and BAY y 3114 in rat tissues

Study I			BAY y 3118		BAY y 3114	
Rat	Dose [mg/kg]	Sample after min	Skin [nmol/g]	Plasma [μM]	Skin [nmol/g]	Plasma [μM]
R1	50 + UVA	90	3.9	3.3	0.6	trace
R2	100 + UVA	90	4.0	4.4	1.0	trace
R3	200 + UVA	90	8.8	9.7	1.4	trace
R4	200	90	8.1	6.1	—	—

After irradiation of BAY y 3118 (100 μM) for 30 min., no reaction with α-t took place if the radical scavenger was added 5, 20 or 60 sec. after irradiation was stopped. That means, there is no chemical reaction with photo degradation products.

2.2. Toxicokinetic of Photo Degradation Products of BAY y 3118 (in Vivo)

Regarding the unclear mechanism of the systemic toxicity in guinea pigs and the much lower sensitivity in rats, we started in vivo studies on rats with oral administration of BAY y 3118 and UV-A exposure. Aim of the studies was to detect the formation of photo degradation products in the skin and the systemic availability of these compounds in the blood.

I. In the first experiment 3 rats (dorsal hairs shaved) were administered orally 50, 100 and 200 mg/kg BAY y 3118. After 30 min. the UV-A exposure with 4.5-5 mW/cm^2 for 60 min. was started. Rat 4 got 200 mg/kg without UV-A exposure and was used as control. Then 90 min. after administration of the substance the rats were sacrificed and blood and skin samples were taken.

For the sensitive detection of BAY y 3118 as for degradation products in the plasma or skin extracts we used a HPLC-Method with a photochemical-reactor (Beam Boost) and fluorescence detector (λ 277 nm/418 nm).

In *plasma* the main substance related peaks are BAY y 3118 and the glucuronide of BAY y 3118. Only very small amounts of photo degradation products could be detected. A lot of efforts were needed to find a reasonable extraction method for the skin samples. In *skin* the main substance related peaks are BAY y 3118 and the dechlorinated compound BAY y 3114. So it seems, that the initial photochemical step in the skin is the abstraction of a chlorine radical leading to a highly reactive radical intermediate. But also other photo degradation products could be shown.

Quantitative results for BAY y 3118 and BAY y 3114 (for which reference substances were available) in plasma and skin are given in Table 1. A relatively good concentration/dose response was demonstrated in plasma as in skin.

II. In the essential experiment with guinea pigs it was only possible to administer 100 mg/kg BAY y 3118 because of the higher sensitivity of this species.

The design of the study was similar to the first rat study: by our HPLC-method we found no obvious difference in the pattern of the photo degradation products in plasma as in skin extracts between rat and guinea pig. Also the quantitative results were very similar to the data in rat and revealed no argument for the higher sensitivity of guinea pigs against this phototoxic reaction with BAY y 3118.

3. DISCUSSION

The following mechanism for the photo toxicity induced by BAY y 3118 is proposed. During UV-A irradiation, quinolone in the target tissue generates toxic oxygen radicals which attack biological systems such as the mitochondria [6]. The degree of cell damage is dependent on the balance between the amount of toxic oxygens from quinolone + UV-A and the scavenging activity of vitamin E and possibly other biological defence systems.

4. CONCLUSION

Photolability (in vitro)

- BAY y 3118 is an efficient producer of radicals (singlet oxygen ?) in the presence of UV-A
- It was shown, that α-tocopherol is a very efficient scavenger of these radicals
- BAY y 3114 (dechloro) is the main but not the phototoxic degradation product

Toxicokinetic of photo degradation products (in vivo)

- It was demonstrated, that after oral administration of BAY y 3118 and UV-A exposure, BAY y 3114 and other degradation products were present in plasma and skin
- There was no obvious difference in concentration and degradation pattern between rat and guinea pig

ACKNOWLEDGMENTS

I like to thank Dr. Renhof and Dr. Vohr for the animal experiments and Mrs. Hancke and Mr. Tran for the technical assistance in the analytical procedures.

5. REFERENCES

1. VOHR, H.W., Bayer AG Pharma Report no.22921 (1994)
2. DAYHAW-BARKER, P. and TRUSCOTT, T.G., Direct detection of singlet oxygnen sensitized by nalidixic acid:The effect of pH and melanin, Photochemistry and Photobiology, **47(5)** pp, 765-767 (1988)
3. WAGAI, N. and TAWARA, K., Important role of oxygen metabolites in quinolone antibacterial agent-induced cutaneous phototoxicity in mice, Archives of Toxicology, **65**, 495-499 (1991)
4. ROBERTSON, D.G.; EPLING, G.A.; KIELY, J.S: BAILEY, D.L. and SONG, B., Mechanistic studies of the phototoxic potential of PD 117596, a quinolone antibacterial compound, Toxicology and Applied Pharmacology, **111**, 221-232 (1991)
5. WAGAI, N. and TAWARA, K., Possible direct role of reactive oxygens in the cause of cutaneous phototoxicity induced by five quinolones in mice, Archives of Toxicology, **66**, 392-397 (1992)
6. SHIMODA, K., YOSHIDA, M., WAGAI, N., TAKAYAMA, S. and KATO, M., Phototoxic lesions induced by quinolone antibacterial agents in auricular skin and retina of albino mice, Toxicologic Pathology **21**(6) 554-561 (1993)

4-OXO-RETINOIC ACID IS GENERATED FROM ITS PRECURSOR CANTHAXANTHIN AND ENHANCES GAP JUNCTIONAL COMMUNICATION IN 10T1/2 CELLS

Wilhelm Stahl, Michael Hanusch, and Helmut Sies*

Institut für Physiologische Chemie I and Biologisch-Medizinisches Forschungszentrum
Heinrich-Heine-Universität Düsseldorf
Düsseldorf, Germany

INTRODUCTION

Retinoids and carotenoids inhibit the transformation of murine C3H/10T1/2 fibroblasts by chemical and physical carcinogens (Hossain et al., 1989; Pung et al., 1988). The protective ability of carotenoids was reported to be independent of their provitamin A activity and their ability to inhibit lipid peroxidation (Zhang et al., 1991). One of the most active carotenoids is canthaxanthin, which is not a known provitamin A in mammals (Bertram et al., 1991). The inhibitory effect of both retinoids and carotenoids towards neoplastic transformation correlates with their activity to induce gap junctional communication (Zhang et al., 1991; Bertram et al., 1991).

Gap junctions are transmembrane proteins composed of connexins (Beyer et al., 1990) and one of the major gap junctional protein in many cell types, including 10T1/2 cells, is connexin43 (Cx43). Connexin43 gene expression is induced when 10T1/2 cells are treated with retinoids or carotenoids (Hossain et al., 1989; Pung et al., 1988). While the retinoids may affect gene expression by various nuclear retinoic acid receptors (RARs), specific receptors for carotenoids are not known. It has been speculated, however, that carotenoids could act after conversion to retinoids shown to be active in this system.

In addition to retinoic acid its metabolite 4-oxo-retinoic acid binds to a specific nuclear retinoic acid receptor (RARβ) with high affinity, similar to that of the parent compound (Pijnapple et al., 1993). Since 4-oxo-retinoic acid might be a product also formed

*Address for correspondence: Prof. Dr. Helmut Sies, Institut für Physiologische Chemie I, Heinrich-Heine-Universität, Postfach 101007, D-40001 Düsseldorf, Germany; Phone: +49-211-311-2707; Fax: +49-211-311-3029.

Biological Reactive Intermediates V
Edited by R. Snyder et al., Plenum Press, New York, 1996

from canthaxanthin under cell culture conditions, we have investigated whether 4-oxo-retinoic acid is effective in inducing gap junctional communication in 10T1/2 cells and whether it is formed under cell culture conditions. The present report is based on a recent paper by Hanusch et al (1995).

MATERIALS AND METHODS

Chemicals. Lucifer Yellow CH was purchased from Sigma (Deisenhofen, Germany). 4-Oxo-Retinoic Acid and canthaxanthin were kind gifts from Drs. J. Bausch and H. E. Keller, Hoffmann-La Roche (Basel, Switzerland). The purity of these compounds was checked by HPLC to be > 97%.

Plasmids. pRLCGAP containing rat glyceraldehydephosphate dehydrogenase (GAPDH) cDNA was kindly provided by Dr. R. Wu (Ithaca, NY). pBluescript SK+ containing mouse Connexin43 cDNA was kindly provided by Prof. K. Willecke (Cologne, Germany).

Cells and Culture Conditions. The mouse embryo fibroblast C3H/10T1/2 clone 8 cell line (ATCC, No. CCL 226) was used throughout this study. The cells were cultured in fibroblast growth medium (PromoCell, Heidelberg, Germany) supplemented with 10% fetal calf serum (FCS) (Life Technologies, Eggenstein, Germany). The confluent cells were treated with 13-cis or all-trans 4-oxo-retinoic acid (10^{-6} M) or with the different solutions of canthaxanthin (10^{-5} M) in medium with 3 % FCS. Medium change and cell treatment with 13-cis or all-trans 4-oxo-retinoic acid was every 3 days using acetone as solvent at a final concentration of 0.5 % in the culture medium, for canthaxanthin every 5 days using tetrahydrofuran (THF) as solvent (0.5 %). Control cells were treated with acetone (0.5 %) or THF (0.5 %).

Gap Junctional Communication Assay. Junctional permeability was assayed by microinjection of the fluorescent dye Lucifer Yellow CH (10% in 0.33 M LiCl) into cells of confluent cell cultures by means of a microinjector and micromanipulator (Microinjector 5242 and ECET Micromanipulator 5170, Eppendorf, Hamburg, Germany). The total number of fluorescent neighbors of the injected cells was scored 5 min after the injection and served as an index of junctional communication.

Retinoid and Carotenoid Stability. Stock solutions of retinoids (10^{-3} M) were prepared in ethanol and stored in the dark at -20°C. Stock solutions of canthaxanthin (10^{-2} M) were prepared in chloroform and stored in the dark at -20°C. The concentration of the stock solutions was determined spectrophotometrically.

HPLC Conditions

Analytical HPLC. Reversed-phase HPLC was conducted on a Suplex pKB 100 column (endcapped, 5 μm, 250 x 4,6 mm) from Supelco (Bellefonte, PA, USA) and on a LiChrospher RP-18 column (endcapped, 5 μm, 250 x 4 mm) from Merck (Darmstadt, Germany). The mobile phase for canthaxanthin consisted of acetonitrile/methanol/toluene/water (7/7/1/0.8) (Sundquist et al., 1993a) and for 4-oxo-retinoic acid of acetonitrile/methanol/acetic acid (95/5/0.5), (Sundquist et al., 1993b).

Preparative HPLC. Preparative reversed-phase HPLC was done on a LiChrospher 100 RP-18 endcapped column (10 μm, 250 x 10 mm) from Merck (Darmstadt, Germany); eluent was methanol. The flow rate was 4 ml min^{-1}. Fractions from 3 to 6 min and from 7 to 11 min were collected and combined, designated as oxidation fraction I. Oxidation fraction II was collected from 6 to 7 min, including the retention time of 4-oxo-retinoic acid (6.5 min). Canthaxanthin was purified by collecting from 16 to 18 min. The fractions were evaporated under nitrogen, the residues were dissolved in THF and directly tested in the cell-cell communication assay.

Northern Analysis. Isolation of RNA from C3H/10T1/2 cells was performed by the acid phenol method (Chomszynsky and Sacchi, 1987). Total RNA was separated on a 1% agarose gel containing 0.66 M formaldehyde, blotted and UV-crosslinked to GeneScreen nylon membranes (Dupont, Dreieich, Germany). The probes used for hybridizations were a 0.7 kbp AccI fragment of Connexin43 cDNA and a 1.3 kbp EcoRI fragment of GAPDH cDNA.

Isolation of Decomposition Products of Canthaxanthin

Canthaxanthin was dissolved in THF and added to FGM (3 % FCS) to a final concentration of 100 μM. After incubation for 5 days at 37°C in the dark, 1 ml of the incubation mixture was acidified with 0.2 ml of TCA (5 %) and extracted with 1 ml methanol plus 5 ml hexane/dichloromethane (3/1). The residue was further processed and chromatographed as described by Hanusch et al. (1995). A diode array detector (model 168, Beckman, Munich, Germany) was used for spectral identification.

RESULTS

Induction of Gap Junctional Communication by 4-Oxo-Retinoic Acid and Canthaxanthin

The structures of the compounds tested for their activity in the cell-cell-communication assay are shown in Figure 1. Gap junctional communication markedly increased after addition of all-trans and 13-cis 4-oxo-retinoic acid and canthaxanthin (Fig.2). The increases produced by all-trans- and 13-cis 4-oxo-retinoic acid at their maximally effective concentrations (10^{-6} M) were comparable. Upon treatment with canthaxanthin a similar response was observed but a higher concentration (10^{-5} M) of the carotenoid was required. Treatment of 10T1/2 cells with canthaxanthin and all-trans 4-oxo-retinoic acid also led to increased the levels of Cx43 mRNA. The raise in connexin43 levels was 2.8-fold with canthaxanthin and 4.7-fold with 4-oxo-retinoic acid as compared to the solvent control. Lower concentrations of all-trans 4-oxo-retinoic acid (down to 10^{-8} M) were also active.

Activity of Canthaxanthin Decomposition Products in the Communication Assay

Incubation of cells with purified canthaxanthin was followed by a two-fold induction of gap junctional communication as compared to the control. A significantly higher communication was obtained when a stock solution of canthaxanthin which had been stored for 2 months in chloroform at -20°C ("predissolved") was used. Thus, it was suggested that

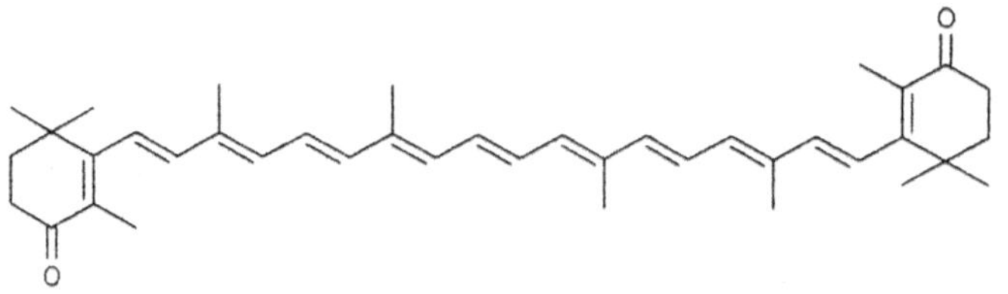

all–trans Canthaxanthin

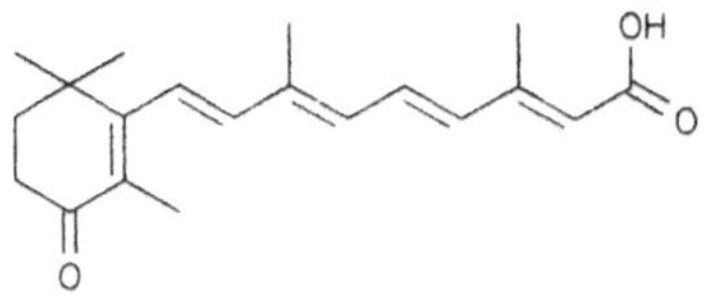

all–trans 4–oxo Retinoic Acid

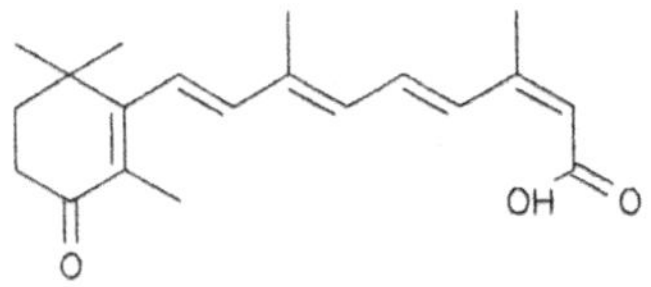

13–cis 4–oxo Retinoic Acid

Figure 1. Structures of the compounds examined in this study.

decomposition products of canthaxanthin are responsible for the inductory effects of this carotenoid in gap junctional communication.

After fractionation of the stock solution by preparative HPLC (Fig.3), the different fractions were tested for activity in the cell-cell communication assay. Oxidation fraction I showed only a slight if any effect, whereas oxidation fraction II induced gap junctional

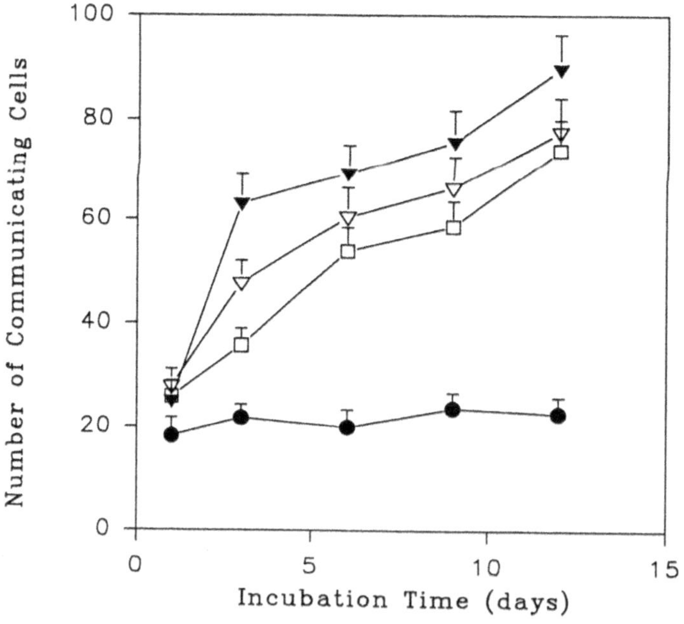

Figure 2. Effect of 4-oxo-retinoic acid and canthaxanthin on gap junctional communication. Confluent cultures of C3H/10T1/2 cells were treated with retinoids every 3 days, with canthaxanthin every 5 days or with THF (0.5 %) as solvent control. Gap junctional communication was assayed as described in Materials and Methods. Data points are the means ± SE of 10 microinjections within one dish. Each experiment was repeated 2-4 times with similar outcome. (●), THF control (0,5 %); (▽) all-trans 4-oxo-retinoic acid (10^{-6} M); (▼) 13-cis 4-oxo-retinoic acid (10^{-6} M); (□) canthaxanthin (10^{-5} M).

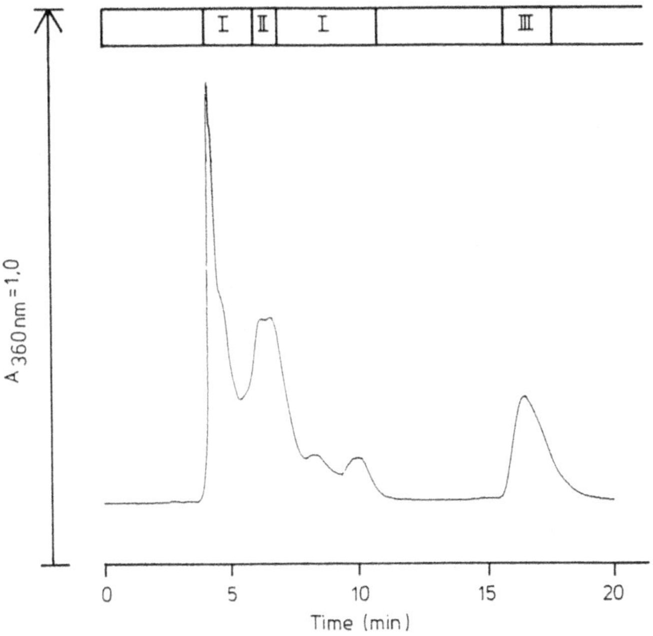

Figure 3. HPLC chromatogram of decomposition products of predissolved canthaxanthin. A two-months-old stock solution of canthaxanthin (10^{-2} M) in chloroform was fractionated by preparative HPLC. Fractions from 3 to 6 min and from 7 to 11 min were collected and combined, designated as oxidation fraction I. Oxidation fraction II was collected from 6 to 7 min, including the retention time of 4-oxo-retinoic acid (6.5 min). Canthaxanthin was purified by collecting from 16 to 18 min.

communication as strongly as canthaxanthin. 4-Oxo-retinoic acid coelutes with oxidation fraction II.

In order to examine products formed from canthaxanthin during preincubation in the culture medium, a 100 μM canthaxanthin solution in fibroblast growth medium was incubated for 5 days and analyzed for decomposition products. All-trans and 13-cis 4-oxo-retinoic acid were identified by HPLC co-chromatography (Fig.4) and by comparison of their UV/VIS spectra with reference compounds. Compared to the control (100 μM canthaxanthin without preincubation) a 9-fold increase in the amount of 4-oxo-retinoic acid was detected in the incubation mixture.

DISCUSSION

Induction of Gap Junctional Communication by 4-Oxo-Retinoic Acid

The all-trans and 13-cis isomer of 4-oxo-retinoic acid, enhanced gap junctional communication in 10T1/2 cells (Fig.2). The activity of both compounds was similar and comparable to the efficiacy of all-trans retinoic acid (data not shown), a compound which was reported to inhibit neoplastic transformation and induce gap junctional communication in 10T1/2 cells (Hossain et al., 1989).

Recently, it has been reported that 4-oxo-retinoic acid modulates positional specification in early embryos and is capable of activating the RARβ receptor (Pijnapple et al. 1993). 4-Oxo-retinoic acid induces cell differentiation in F9 mouse teratocarcinoma cells as effectively as 9-cis and all-trans retinoic acid (Nikawa et al., 1995). Here, we document a further biological activity of the all-trans and 13-cis isomers of 4-oxo-retinoic acid as inducers of gap junctional communication and Cx43 mRNA in 10T1/2 cells.

Canthaxanthin

The nonprovitamin A carotenoid canthaxanthin elevated junctional communication to almost the same level as 4-oxo-retinoic acid, but 10-fold higher concentrations were required to obtain comparable effects. The induction of Cx43 mRNA in 10T1/2 cells was stronger with 4-oxo-retinoic acid (Fig.3) and a slight difference was observed in the time course of induction. Communication induced by 4-oxo-retinoic acid occurred earlier (after 3 days of treatment) as compared to canthaxanthin which required 6 days of incubation to produce similar effects. Such a delayed response to canthaxanthin was also described by Zhang et al. (1991) in comparison to all-trans retinoic acid.

Although purified canthaxanthin induced gap junctional communication within the time period studied, "predissolved" canthaxanthin was more active. This finding suggests that decomposition of canthaxanthin results in the formation of active compounds. Isolated oxidation products of canthaxanthin (oxidation fraction II) enhanced gap junctional communication to a level comparable to the effect of canthaxanthin "predissolved". Two active decomposition products were identified resulting from autoxidation of canthaxanthin: all-trans and 13-cis 4-oxo-retinoic acid (Fig.4). The data support the idea that the activity of canthaxanthin is at least in part due to the formation of active retinoids. To obtain a significant induction of gap junctional communication in 10T1/2 cells, only 0.1 % of 10 μM canthaxanthin need to be converted to active retinoids such as 4-oxo-retinoic acid, which is active at 10 nM.

The stability of carotenoids in biological systems might be an important determinant of their biochemical activities. For instance, β-carotene, less stable than canthaxanthin, accumulates in cells and undergoes degradation spontaneously; e.g., only 58% of 3.5 μM

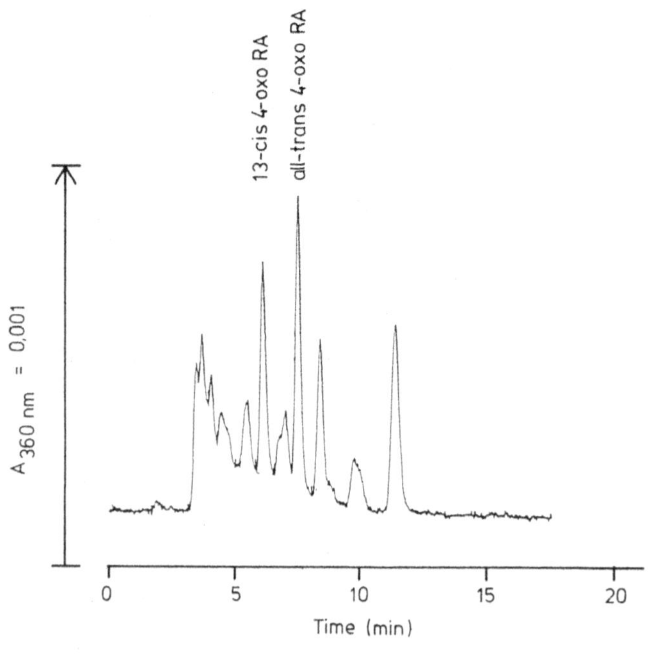

Figure 4. HPLC analysis of canthaxanthin decomposition products in cell culture medium. Canthaxanthin (100 μM) in fibroblast growth medium containing 3% FCS was incubated under cell culture conditions for 5 days, and decomposition products were isolated as described in Materials and Methods.

β-carotene remained after 24 h incubation with FU-5 hepatoma cells (Oarada et al., 1993). As with canthaxanthin several different autoxidation products were identified (Handelman et al. 1991). Furthermore, 9-cis β-carotene is converted to 9-cis retinoic acid in human intestinal mucosa (Wang et al. 1994), and 9-cis retinoic acid is found in human serum (Rushin et al. 1990). For canthaxanthin, our data suggest that its biological activity may be mediated via formation of 4-oxo-retinoic acid.

ACKNOWLEDGMENTS

This work was supported by the National Foundation for Cancer Research, Bethesda, USA, and by the Institut Danone für Ernährung, Rosenheim, Germany.

REFERENCES

Bertram, J.S., Pung, A., Churley, M., Kappock, T.J., Wilkins, L.R. and Cooney, R.V. (1991). Diverse carotenoids protect against chemically induced neoplastic transformation. *Carcinogenesis* 12, 671-678.

Beyer, E.C., Paul, D.L. and Goodenough, D.A. (1990). Connexin family of gap junction proteins. *Membrane Biol.* **116**, 187-194.

Chomszynsky, P. and Sacchi, N. (1987). Single-step method of RNA isolation by acid guanidinium thiocyanate-phenol chloroform extraction. *Anal. Biochem.* **162**, 156-159.

Handelman, G.J., van Kuijk, F.J.G.M., Chatterjee, A. and Krinsky, N.I. (1991). Characterization of products formed during the autoxidation of β-carotene. *Free Rad. Biol. Med.* **10**, 427-437.

Hanusch, M., Stahl, W., Schulz, W., and Sies, H. (1995). Induction of gap junctional communication by 4-oxoretinoic acid generated from its precursor canthaxanthin. *Arch. Biochem. Biophys.* **317**, 423-428

Hossain, M.Z., Wilkens, L.R., Mehta, P.P., Loewenstein, W.R. and Bertram, J.S. (1989). Enhancement of gap junctional communication by retinoids correlates with their ability to inhibit neoplastic transformation. *Carcinogenesis* **10**, 1743-1748.

Levin, A.A., Sturzenbecker, L.J., Kazmer, S., Bosakowski, T., Huselton, C., Allenby, G., Speck, J., Kratzeisen, C., Rosenberger, M., Lovey, A. and Grippo, J.F. (1992). 9-cis Retinoic acid stereoisomer binds and activates the nuclear receptor RXR α. *Nature* **355**, 359-361.

Mehta, P.P., Bertram, J.S., and Loewenstein, W.R. (1989). The action of retinoids on cellular growth correlates with their action on gap junctional communication. *J. Cell Biol.* **108**, 1053-1065.

Nikawa, T., Schulz, W.A., van den Brink, C.E., Hanusch, M., van der Saag, P., Stahl, W. and Sies, H. (1995). Efficacy of all-trans β-carotene, canthaxanthin, and all-trans-, 9-cis, and 4-oxoretinoic acids in inducing differentiation of an F9 embryonal carcinoma RAR-β-lacZ reporter cell line. *Arch. Biochem. Biophys.*, **316**, 665-672.

Oarada, M., Stahl, W. and Sies, H. (1993). Cellular levels of all-trans β-carotene under the influence of 9-cis β-carotene in FU-5 rat hepatoma cells. *Biol. Chem. Hoppe-Seyler* **374**, 1075-1081.

Pijnappel, W.W.M., Hendriks, H.F.J., Folkers, G.E., van den Brink, C.E., Dekker, E.J., Edelenbosch, C., van der Saag, P.T. and Durston, A.J. (1993). The retinoid ligand 4-oxo-retinoic acid is a highly active modulator of positional specification. *Nature* **366**, 340-344.

Pung, A., Rundhaug, J.E., Yoshizawa, C.N. and Bertram, J.S. (1988). β-Carotene and canthaxanthin inhibit chemically- and physically-induced neoplastic transformationin 10T1/2 cells. *Carcinogenesis* **9**, 1533-1539.

Rushin, W.G., Catignani, G.L. and Schwartz, S.J. (1990). Determination of β-carotene and its cis-isomers in serum. *Clin. Chem.* **36**, 1986-1989.

Sundquist, A.R., Hanusch, M., Stahl, W. and Sies, H. (1993a). Cis/trans isomerization of carotenoids by the triplet carbonyl source 3-hydroxymethyl-3,4,4-trimethyl-1,2-dioxetane. *Photochem. Photobiol.* **57**, 785-791

Sundquist, A.R., Stahl, W., Steigel, A. and Sies, H. (1993b). Separation of retinoic acid all-trans, mono-cis, and poly-cis isomers by reversed phase high-performance liquid chromatography. *J. Chromatography* **637**, 201-205.

Wang, X.D., Krinsky, N.I., Benotti, P.N. and Russell, R.M. (1994).Biosynthesis of 9-cis retinoic acid from 9-cis β-carotene in human intestinal mucosa in vitro. *Arch. Biochem. Biophys.* **313**, 150-155.

Zhang, L.X., Cooney, R.V. and Bertram, J.S. (1991). Carotenoids enhance gap junctional communication and inhibit lipid peroxidation in C3H/10T1/2 cells: Relationship to their cancer chemopreventive action. *Carcinogenesis* **12**, *2109-2114.*

HEPATIC EPOXIDE CONCENTRATIONS DURING BIOTRANSFORMATION OF 1,2- AND 1,4-DICHLOROBENZENE*

The Use of *in Vitro* and *in Vivo* Metabolism, Kinetics and PB-PK Modeling

Erna Hissink, Ben van Ommen, Jan J. Bogaards, and Peter J. van Bladeren

TNO Toxicology
P.O. Box 360, 3700 AJ Zeist, The Netherlands

INTRODUCTION

Halogenated benzenes are extensively used as solvents, fumigants and intermediates in the production of pesticides and dyes and therefore represent significant environmental contaminants (Hawley, 1971; US-EPA, 1985). It has been shown that biotransformation is necessary to exhibit toxicity (Brodie et al., 1971). Both 1,2-dichlorobenzene and its 1,4-isomer are metabolized by cytochrome P450 enzymes to an epoxide intermediate. This epoxide may either rearrange to produce phenols, hydrolyse to form dihydrodiols or conjugate with glutathione.

Major differences in toxicity exist between the two isomers, which may be related to the way the epoxide is detoxified. In rats, the 1,2-isomer shows acute liver toxicity, while 1,4-DCB is suspected to be a liver carcinogen (Stine et al., 1991; NTP, 1987). A set of *in vivo* and *in vitro* studies was designed to investigate whether differences in metabolism are responsible for these differences in toxicity.

MATERIALS AND METHODS

Animals

Male Wistar rats, 9-11 weeks old, were used for the *in vivo* kinetics and metabolism studies and for preparation of liver microsomes. Food and tap water was provided *ad libitum.*

* Abbreviations used: DCB = dichlorobenzene; DCP = dichlorophenol; AA = ascorbic acid; GSH = glutathione

Biological Reactive Intermediates V
Edited by R. Snyder et al., Plenum Press, New York, 1996

Chemicals

[^{14}C]-1,2- and [^{14}C]-1,4-dichlorobenzene (DCB) were purchased from Sigma Chemical Company (St. Louis, MO). 1,2-DCB was from Merck (Darmstadt, Gernany) and 1,4-DCB was from Aldrich (Steinheim, Germany). Isoniazid was purchased from Janssen (Geel, Belgium). [^{35}S]-Glutathione (GSH) was purchased from DuPont NEN (The Netherlands). Microsomes of cell lines selectively expressing human CYP1A1, 1A2, 3A4, 2E1 or 2D6 were obtained from Gentest (Woburn, MA) and human microsomes were obtained from Human Biologics Inc. (Phoenix, AR).

Microsomal Incubations

0.4 mM 1,2- or 1,4-DCB was incubated with liver microsomes from rats pretreated with isoniazid (an inducer of cytochrome P450 2E1, Ryan et al., 1984) in the presence of [^{35}S]-GSH. Also incubations were conducted in the presence of rat liver cytosol. Conjugates were quantified by HPLC using radiochemical detection. [^{14}C]-1,2-DCB was incubated with isoniazid microsomes at a concentration of ca. 40 μM, with possible additions of 0.5 mM glutathione or 1 mM ascorbic acid. Products were identified and quantified by HPLC using radiochemical detection. Covalent binding to protein was determined by extensive washing of the protein pellet with solvents.

Human microsomes and microsomes of cell lines selectively expressing the various P450's were incubated with 100 μM 1,2-and 1,4-DCB; the produced phenols were quantified using gas chromatography.

In Vivo Kinetics, Metabolism and Toxicity

Animals were orally dosed with 10, 50 or 250 mg/kg radiolabelled 1,2- or 1,4-DCB. Excretion of radioactivity in urine, feces, bile and blood was measured during one week. Urinary metabolites were identified using LC and MS techniques. For the toxicity study, animals were dosed with 50 or 250 mg/kg 1,2- or 1,4-DCB and sacrificed after 4, 8, 24, 48 and 72 hours. Plasma alanine aminotransferase (ALAT) activity was determined as a measure of liver toxicity. Hepatic and renal glutathione concentrations were determined using a fluorometric method developed by Hissin and Hilf (1976).

PB-PK Modeling

A mathematical model was constructed describing the kinetics and biotransformation of the two dichlorobenzenes in rat and man, based on biological, physical and biochemical parameters (a "physiologically based pharmacokinetic" [PB-PK] model). This model is based on a model for styrene developed by Ramsey and Andersen (1984). The biochemical parameters V_{max} and Km were derived from the model by calculating best fits of the experimental data (blood concentration of parent compound). Apart from the general kinetics and metabolism, this model describes the hepatic epoxide concentration as a resultant of the formation of the epoxide from DCB by the cytochrome P450 isoenzymes involved, which was described by Michaelis Menten kinetics, and the formation of phenols and glutathione conjugates from the epoxide, which were described by first order processes.

RESULTS AND DISCUSSION

In Vivo Kinetics, Metabolism and Toxicity

After oral administration of a single dose of [^{14}C]-1,2-DCB to rats, the radioactivity was excreted predominantly via the urine (70-85%). LC and MS analysis indicated the formation of mercapturic acids (ca. 60%), together with sulfates (ca. 30%) and phenols (ca. 10%). Bile cannulated animals excreted ca. 60% of the dose via the bile, predominantly as glutathione conjugates. Mercapturic acids as well as glutathione conjugates were detected both as aromatic compounds and as dihydro-hydrate conjugates.

Radioactivity derived from [^{14}C]-1,4-DCB was also excreted primarily via the urine, but predominantly as the sulfate (ca. 60%) and glucuronide (ca. 30%) conjugate of its phenol . Mercapturic acids were found only to a very minor extent, while the bile contained 5% (at 10 mg/kg bw) to 30% (at 250 mg/kg bw) of the radioactive dose, identified as glucuronides.

After single oral administration of 1,2-DCB the concentration of glutathione in the liver decreased dose related, followed by a typical overshoot reaction. 1,4-DCB hardly affected the hepatic glutathione concentration. The glutathione depletion coincided with an increase in liver damage, as determined by plasma ALAT.

In Vitro Biotransformation

Human P450's. Figure 1 shows the results of incubations conducted with microsomes of cell lines selectively expressing human CYP1A1, 1A2, 3A4, 2E1 or 2D6 with 1,2- or 1,4-DCB. Furthermore, a rank correlation between catalytic activity towards specific substrates and 1,2-DCB was performed for 22 human liver microsomes, which is shown in

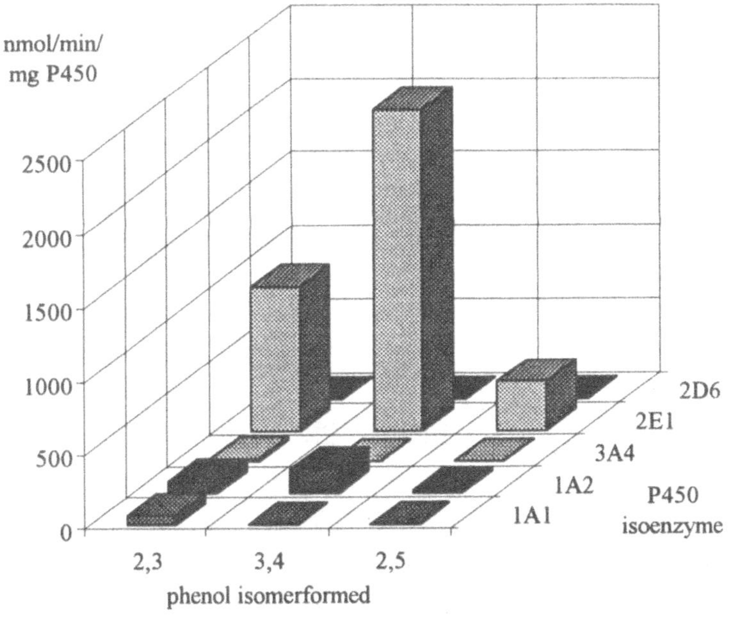

Figure 1. Isoenzyme selectivity in the hydroxylation of dichlorobenzenes. 1,2-DCB is metabolized to 2,3- and 3,4-DCP, 1,4-DCB is metabolized to 2,5-DCP.

Table 1. Rank correlation between catalytic activity towards specific substrates and 1,2-DCB for 22 human liver microsomes

Substrate	Chlorzox. 6-OH	Testoster. 6-OH	2,3-DCP	3,4-DCP
EROD	-0.29	-0.46*	-0.27	-0.17
chlorzoxazone 6-OH		0.27	0.81***	0.90***
testosterone 6-OH			0.02	0.03

Table 1. From these data it is clear that CYP 2E1 most efficiently catalyzes the DCB oxidation.

Rat liver microsomes. Incubations with 1,2- and 1,4-DCB and [^{35}S]-GSH revealed that the epoxides of 1,2-DCB readily react with GSH, while the epoxide of 1,4-DCB hardly conjugates with GSH. Addition of rat liver cytosol resulted for the 1,2-isomer in an increased conjugation of ca. 1.2 fold at high GSH concentration, and an increase of 3.75 fold at low concentration. For 1,4-DCB, addition of cytosol had no effect.

In Table 2 the results are shown of incubations conducted with [^{14}C]-1,2-DCB and possible additions of ascorbic acid (AA) and GSH. Without these compounds, the main product was the dihydrodiol, with highest covalent binding (8.2% of the total conversion). Addition of GSH resulted in formation of its conjugate with the epoxide, with a concomitant decrease in formation of the dihydrodiol; covalent binding was decreased by 50%. Ascorbic acid had minor effects on the products formed, while addition of both agents resulted in the same products and covalent binding as with GSH alone. These results indicate a major role for GSH in the protection of covalent binding vs. reactive epoxides.

PB-PK modeling of the hepatic epoxide concentration. Using the data from *in vivo* studies with 1,2-DCB and 1,4-DCB in control rats and rats pretreated with isoniazid, the kinetics were described in the PB-PK model as outlined above. In Table 3 results are shown of calculations conducted with these models, including the biochemical parameters V_{max} and Km. Comparing the Km values for both substrates, it is clear that 1,2-DCB is oxidized more efficiently than 1,4-DCB. The V_{max} for 1,4-DCB is somewhat higher than for 1,2-DCB and is increased after induction. Concerning the epoxide concentration, it can be concluded that oxidation of 1,2-DCB results in lower concentrations of hepatic epoxides than oxidation of 1,4-DCB. Induction of the enzymatic oxidation of 1,4-DCB results in a much higher C_{max}.

From the presented data, it is clear that the epoxide of the 1,2-isomer easily conjugates with glutathione. The *in vitro* data reflect the *in vivo* biotransformation very well. This

Table 2. Product formation during microsomal oxidation of [^{14}C]-1,2-DCB. Effect of addition of glutathione and /or ascorbic acid

addition	product formation (% conversion)						
	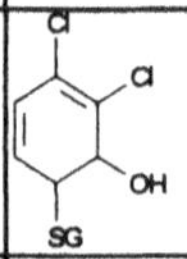	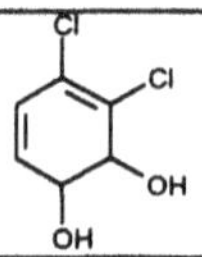	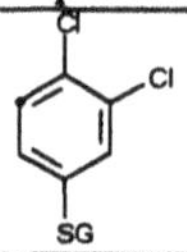	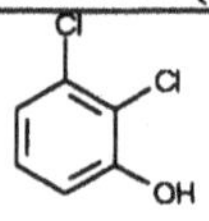	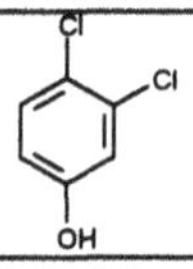	total	covalent binding to protein (% of control)
		1.8	0	0.2	0.5	3.1	100
GSH	1.8	0.4	0.3	0.2	0.5	3.9	48
AA	0	1.9	0	0.3	0.6	3.5	70
AA, GSH	1.8	0.4	0.3	0.2	0.5	3.6	45

Table 3. PB-PK modeling of the hepatic epoxide concentrations and the Michaelis-Menten constants V_{max} and Km. Effect of isomer, dose and induction

parameter	1,2-DCB	1,4-DCB		
induction (isoniazid)	no	no	yes	yes
dose (mg/kg)	250	250	250	50
V_{max} (μmol/hr)	17.5	22	36	32
Km (μmol/l)	1.5	21.4	21.4	21.4
C_{max} [epoxide] (μmol/l)	27	81	175	66

cojugation results in hepatic glutathione depletion at high dose levels, resulting in acute hepatotoxicity. On the other hand, the epoxide of the 1,4-isomer hardly reacts with glutathione, possibly resulting in higher intracellular epoxide concentrations. This compound is not acutely toxic, but is a suspected carcinogen. Model calculations indeed indicate that the epoxide concentration after administration of 1,4-DCB is much higher than after administration of 1,2-DCB. These data suggest that the epoxide concentration and reactivity are important determinants in the type and extent of toxicity.

REFERENCES

1. Brodie, B.B., Reid, W.D., Cho, A., Sipes, I.G., Krishna, G., and Gillette, J.R. (1971). Possible mechanism of liver necrosis caused by aromatic organic compounds. *Proc. Nat. Acad. Sci.* 68, 160-164.
2. Hawley, G.G. (1971). The Condensed Chemical Dictionary, 8th. ed., pp. 283-284. Van Nostrand Reinhold, New York/London.
3. Hissin, P.J. and Hilf, R. (1976). A fluorometric method for determination of oxidized and reduced glutathione in tissues. *Anal. Biochem.* 74, 214-226.
4. National Toxicology Program (1987). Toxicology and Carcinogenesis studies of 1,4-dichlorobenzene (CAS No. 67-72-1) in F344/N rats and B6C3F1 mice (gavage studies) Abstracts from long-term cancer studies, 1976-1992. Environmental Health Perspectives. National Institute of Health, vol. 101, supplement 1 (1993). 185-186.
5. Ramsey, J.C., and Andersen, M.E. (1984). A physiologically based description of the inhalation pharmacokinetics of styrene in rats and humans. *Toxicol. Appl. Pharmacol.* 73, 159-175.
6. Ryan, D.E., Ramanathan, L., Iida, S., Thomas, P.E., Haniu, M., Shively, J.E., Lieber, C.S., and Levin, W. (1985). Characterization of a major form of rat hepatic microsomal cytochrome P-450 induced by isoniazid. *J. Biol. Chem.* 260, 6385-6393.
7. Stine, E.R., Gunawardhana, L., and Sipes, I.G. (1991). The acute hepatotoxicity of the isomers of dichlorobenzene in Fisher-344 and Sprague-Dawley rats: Isomer-specific and strain-specific differential toxicity. *Toxicol. Appl. Pharmacol.* 109, 472-481.
8. US EPA. (1985). Health assessment document for chlorinated benzenes. Final report. Office of health and environment assessment, Washington D.C., U.S. Environmental Protection Agency. Report EPA/600/8-84/015F.

OXYGEN RADICAL FORMATION DUE TO THE EFFECT OF VARYING HYDROGEN ION CONCENTRATIONS ON CYTOCHROME P450-CATALYZED CYCLOSPORINE METABOLISM IN RAT AND HUMAN LIVER MICROSOMES*

S. Sohail Ahmed, Kimberly L. Napoli, and Henry W. Strobel

Department of Biochemistry/Molecular Biology
and Division of Immunology/Organ Transplantation
The University of Texas Health Science Center, Houston, Texas 77225

INTRODUCTION

Cyclosporin A (CyA) is a hydrophobic cyclic undecapeptide whose potent immunosuppressive effects have resulted in its widespread usage for suppressive allograft rejection. However, its metabolism by the cytochrome P450 mixed-function oxidase (MFO) system frequently results in hepatotoxicity as seen by jaundice and altered serum enzyme in hepatic transplant recipients, hyperbilirubinemia, hypoproteinemia, and albuminemia in rats (Whiting et al., 1983). In previous experiments, the sources of the hepatotoxicity associated with CyA metabolism was implicated as being oxygen radicals (Ahmed et al., 1993). Xenobiotics, like CyA or FK-506, lead to the production of free radicals by a redox cycle involving one-electron reductions by oxidoreductase that reduce molecular oxygen to superoxide. Not only can superoxide result in the observed toxicity, but it may, in the presence of transition metals such as iron, generate reactive hydroxy radicals by the Haber-Weiss or the Fenton reaction. Oxidoreductase is a vital element of the cytochrome P450 system which catalyzes the biotransformation of xenobiotics by either oxidation or reduction. Oxidoreductase usually transfers two reducing equivalents from NADPH to the cytochrome P450-substrate complex resulting in substrate oxidation and subsequent product release. This reductase-mediated transfer of electrons is characterized as an electron flux. Certain investigators have measured metabolite ratios of different substrates at varying pH (Grogan et al., 1993 and Hille et al., 1981) and have reported differences in the electron transfer from oxidoreductase

*Supported by Grant No. CA-53191 from the National Cancer Institute DHHW.

to cytochrome P450. These results have prompted us to investigate the possible role of pH in either uncoupling the electron-flux at the level of oxidoreductase and P450 or that of P450 and CyA with the generation of hydroxy radicals as seen by lipid peroxidation.

MATERIALS

Cyclosporine powder was a gift of Sandoz Research Institute, N.J., DL-isocitric acid (trisodium salt), cytochrome c (bovine heart type V), isocitric dehydrogenase (NADP-porcine heart type VI), catalase, magnesium phosphate (dibasic trihydrate), ethylenediaminetetraacetic acid (EDTA-disodium salt dihydrate), 2-thiobarbituric acid, superoxide dismutase, and trizma base were purchased from Sigma Chemical Company. Adrenaline hydrogen tartrate was purchased from BDH Chemical Limited, benzphetamine hydrochloride was obtained from The Upjohn Company, nicotine adenine dinucleotide phosphate (NADPH - reduced form, tetrasodium salt) was purchased from Boehringer Mannheim, carbon tetrachloride (CCl_4) was obtained from Mallinckrodt, and trichloroacetic acid was purchased from Fisher Scientific.

METHODS

Microsomes were prepared from rat and human liver (Sun and Strobel, 1986; Dignam and Strobel, 1977). The experiments were performed in either duplicate (cytochrome c reduction, benzphetamine assay, and adrenochrome reaction) or triplicate -(thiobarbituric acid assay). Chromatography analysis was performed using a Waters Associates HPLC system. Error was represented as standard deviation from the mean, and data were analyzed for significant differences, compared to physiological pH, by the ANOVA program. Significance was reported at a level of probability (*P), 0.05, (**P) < 0.01, and (***P) < 0.001 when compared to values at physiological pH (7.5).

RESULTS AND DISCUSSION

Effect of pH on Oxidoreductase

Cytochrome C Reduction In order to illustrate the effect of hydrogen ion concentration on the transfer of electrons by oxidoreductase (and excluding the effects of P450-substrate interactions), cytochrome c reduction was measured in purified rat and human reductase (Table 1A). Both rat and human oxidoreductase show minimum electron-flux at pH 6.0 (14.22 and 19.95 (μmol/min)/mg, respectively). By shifting the pH by 0.5 unit (pH 6.5), the rate of electron-flux increases significantly (35.54 and 42.62 (μmol/min)/mg, respectively) and at a pH 7.0 there is dramatic activity (101.46 and 99.95 (μmol/min)/mg, respectively). Both preparations showed a plateau in reduction activity with increasing pH that dropped sharply at pH 10.0. This pattern was mirrored identically for rat and human microsomal preparations (Table 1B). Therefore, the electron flux from NADPH to cytochrome c via oxidoreductase is clearly sensitive to high (pH<7.0) and low (pH>9.0) hydrogen ion concentrations, but stable between pH 7.0 to 9.0. As can be seen by the decrease in cytochrome c reduction at high and low hydrogen ion concentrations, the electron-flux can be uncoupled. However, as demonstrated by our results, any change in oxygen radical production cannot be attributed to effect on the electron flux from oxidoreductase to P450

Table 1. Cytochrome c Reduction

	A. Reductase ((μmol/min)/mg)		B. Microsomes ((μmol/min)/ml)	
pH	Rat reductase	Human reductase	Rat microsomes	Human microsomes
6.0	14.22**	19.95**	0.47**	5.73**
6.5	35.54**	42.62**	2.03**	11.07**
7.0	101.46	99.95	7.56	26.44
7.5	105.42	101.20	8.13	27.10
8.0	105.26	96.54	8.41	26.83
8.5	98.83	87.54	7.69	24.26
9.0	96.68	87.45	7.39	23.54
9.5	93.42	79.87	6.88	22.92
10.0	55.15**	42.50**	4.11*	14.88**

because reduction activity is constant between pH 7.0 and 9.5. Therefore, the role of pH in influencing P450-CyA interactions and resulting in lipid peroxidation were explored.

Effect of pH on P450-CyA Interactions

Malondialdehyde Formation. In order to explore the role of pH in affecting P450-CyA interactions and resulting in lipid peroxidation by oxygen radical production, MDA and adrenochrome formation were compared to benzphetamine and CyA metabolism (both of which are substrates of P450 3A). MDA formation is one means to measure lipid peroxidation in microsomal preparations. However, this method does not discriminate between lipid peroxidation due to substrate oxidation by the MFO and peroxidation by "leaked" electrons (due to autooxidation). Therefore, the source of lipid peroxidation shown in Table 2A could be questionable. However, this problem can be resolved by using three other means of assessment; in our case, adrenochrome formation, benzphetamine metabolism, and high performance liquid chromatography of CyA metabolites.

Adrenochrome Formation. The adrenochrome reaction, which specifically measures oxygen radicals, reflects the source of the autooxidation at hydrogen ion concentrations where substrate metabolism is decreased. As seen in the rat system (Table 3) the production of adrenochrome at pH 6.5 (where enzymatic activity is elevated) is low compared to the

Table 2. Malondialdehyde Formation (nmols) and Benzphetamine Metabolism (nmols)

	A. Malondialdehyde		B. Benzphetamine	
pH	CCl_4 (rat/hum)	CyA (rat/hum)	rat micro.	human micro.
6.0	0.14/0.06**	0.04/0.07	0.68	0.68
6.5	0.48**/0.12**	0.04/0.12	4.89**	4.88
7.0	0.42**/0.12**	0.05/0.15	4.83**	4.49
7.5	0.06/0.68	0.03/0.17	0.33	3.06
8.0	0.08/0.62**	0.05/0.45	0.34	5.76
8.5	0.24**/0.46**	0.11*/1.94**	1.19	7.16*
9.0	0.12/3.57**	0.10/3.49**	1.71**	21.29**
9.5	0.24**/2.47**	0.29**/0.45*	1.58*	11.26**
10.0	0.32**/0.13**	0.35**/0.12	0.68	5.57

Table 3. Adrenochrome Reaction (nmols)

pH	Microsomes[†]	CCl_4[†]	CyA =
6.0	3.85**/3.54	3.47/7.33	12.18**/27.23**
6.5	10.75/5.09	4.54/24.88**	20.74*/63.67*
7.0	*8.20/9.20	2.42/22.20**	27.11/51.36
7.5	14.92/5.22	0.74/4.60	28.35/54.34
8.0	16.16/4.85	1.74/15.17**	33.08/52.21
8.5	17.90/2.48	0.99/4.97	16.41**/37.22**
9.0	14.92/4.97	3.11/2.48	23.01/24.46**
9.5	11.44/3.60	3.10/4.97	28.10/24.87**
10.0	6.21**/7.58	1.36/3.23	34.20/29.84**

[†]Rat and human

increased levels of oxygen radicals produced at pH>8.5. Since the metabolism of CyA is increased at pH<6.5 , this would lead to the conclusion that the elevated levels of MDA in Table 2A at pH>8.5 may be attributable to autooxidation resulting from lower buffering ability of rat microsomes which resulted in increased sensitivity to pH and electron "leaking" due to denaturing. In contrast, the human system (Table 3) displays decreased levels of oxygen radicals at pH 9.0 (where metabolism is increased) and elevated levels of oxygen radicals at pH<8.5. Since CyA metabolism is decreased at pH<9.0, this would lead to the conclusion that any lipid peroxidation that occurs is the result of electron "leaking". Peroxidation due to electron "leaking", as seen to occur in the rat at pH>8.5 is not seen in the human at pH<8.5 (Table 3). This may be explained by the ten-fold difference in protein concentration in the human system (20.1 mg/ml) compared to the rat system (1.69 mg/ml). Previous experiments (data not shown) have reflected a decreasing trend of lipid peroxidation as the concentration of protein was increased. This ability for protein to "buffer" the effects of lipid peroxidation would explain the lower electron "leaking"-induced peroxidation at pH<9.0 in the human (Table 2A).

Benzphetamine Metabolism and HPLC of CyA Metabolites. In order to elucidate the effect of hydrogen ion concentrations on enzymatic activity, the benzphetamine assay was used measuring its product formaldehyde. This assay is measuring the direct formation of formaldehyde and is insensitive to peroxidized lipid product that is indiscriminately detected by the TBA assay. Therefore, oxygen radicals formed due to "leaking" will peroxidize the lipid membrane and not be detected by the benzphetamine assay. Therefore, the pH ranges at which substrate turnover is greatest should show increased levels of formaldehyde and those ranges at which turnover is lower should show decreased formaldehyde production. As seen in Table 2B, the rat model showed increased and significant metabolism at pH 6.5 and significantly lower activity at pH values>8.5. The human model showed elevated metabolism at pH 9.0 with decreasing activity at pH<8.5. This higher level

Table 4. HPLC of Cyclosporine Metabolism (% metabolism)

pH	% increase in peak 14	% decrease in parent CyA
6.5	37.7 (rat)/*(human)	58.50 (rat)/*(human)
7.5	32.4 (rat)/1.8(human)	0.75 (rat)/93.5 (human)
9.0	*(rat)/28.6 (human)	*(rat)/95.0 (human)

*= compared to pH=9.0 (rat microsomes) or pH=6.5 (human microsomes)

of activity (which correlates well with increased peak formation as detected by HPLC (Table 4) would explain the elevated peroxidation seen in the human at pH 9.0 and that in the rat system (Table 2A) at pH 6.5 which is not significantly elevated due to the lower electron flux at that hydrogen ion concentration.

CONCLUSION

In conclusion, when the P450 MFO system can metabolize CyA, the superoxide radicals that are produced from metabolism will bind to transition metal (iron) in the lipid membrane and produce hydroxy radicals which result in increased lipid peroxidation. However, electrons that "leak" from autooxidation can, if sufficient in number or due to the release of iron by protein denaturation, be able to carry out lipid peroxidation by the Fenton reaction. The cause of lower substrate turnover has been reported as due to differences in interaction of enzyme with substrate rather than changes in the electron-flux (Grogan et al., 1993) Our findings have led to the similar conclusion that decreased substrate metabolism at higher pH in rats or lower pH in humans is reflective of changes in the interaction of the MFO system with CyA and not because of uncoupling of the electron-flux at the level of oxidoreductase and P450. The fact that earlier investigations have suggested several ionic domains on the surface of P450 (Nelson and Strobel, 1988), that chemical modification studies and site-directed mutagenesis of multiple lysine residues have reduced oxidoreductase-P450 dependent demethylation activity (Shen and Strobel, 1993), and our current finding of reduced substrate metabolism at varying hydrogen ion concentrations provide further support for the involvement of electrostatic interactions in cytochrome P450 metabolism of CyA.

REFERENCES

1. Ahmed, S. S., Strobel, H. W., Napoli, K., and Grevel, J.: Adrenochrome reaction implicates oxygen radicals in metabolism of cyclosporine A and FK-506 in rat and human liver microsomes. J. Pharm. and Exp. Therap. 265(3): 1047-1054..
2. Dignam, J. D. and Strobel, H. W.: NADPH-cytochrome P-450 reductase from rat liver: Purification by affinity chromatography and characterization. Biochemistry. 16: 1116-1123, 1977.
3. Grogan, J., Shou, M., Zhou, D., Chen, S., and Korzekwa, K. R.: Use of aromatase (CYP 19) metabolite ratios to characterize electron transfer from NADPH-cytochrome P450 reductase. Biochem. 32: 12007-12012, 1993.
4. Hille, R., Fee, J. A., and Massey, V.: Equilibrium properties of xanthine oxidase containing FAD analogs of varying oxidation-reduction potential. J. Biol. Chem. 256(17): 8933-8940, 1981.
5. Nelson, D. R. and Strobel, H. W.: On the membrane topology of vertebrate cytochrome P450 proteins.J. Biol. Chem. 263:6038-6050, 1988.
6. Shen, S. and Strobel, H. W.: Role of lysine and arginine residues of cytochrome P450 in the interaction between cytochrome P450 2B1 and NADPH-cytochrome P450 reductase. Arch. Bioch. and Biophysics. 304(1): 257-265, 1993.
7. Whiting, P. H., Simpson, J. G., Davidson, R. J. L. and Thomson, A. W.: Pathological changes in rats receiving cyclosporin A at immunotherapeutic dosages for 7 weeks. Br. J. Exp. Pathol. 64: 437-444, 1983.

NITRIC OXIDE PRODUCTION IN THE LUNG AND LIVER FOLLOWING INHALATION OF THE PULMONARY IRRITANT OZONE*

Jeffrey D. Laskin, Diane E. Heck, and Debra L. Laskin†

Environmental and Occupational Health Sciences Institute
Rutgers University
The University of Medicine and Dentistry of
New Jersey-Robert Wood Johnson Medical School
Piscataway, New Jersey

Following irritation or injury, an array of growth factors and cytokines as well as lipid mediators accumulate in target tissues. It is becoming increasingly apparent that many of these agents can cause both immune and nonimmune cells within the tissue to secrete nitric oxide (Nathan, 1992; Laskin et al., 1994). Nitric oxide is a highly reactive mediator that plays a critical role in a variety of physiological processes including non-specific host defense (Nathan 1992). Nitric oxide also regulates vascular tone, gastric motility, neurotransmission (Moncada et al., 1991; Nathan, 1992), bone marrow cell growth and development (Punjabi et al., 1992, 1994b) as well as wound healing (Heck, et al., 1992). Nitric oxide is generated enzymatically in mammalian cells by the NADPH-dependent enzyme, nitric oxide synthase, from the amino acid l-arginine (Nathan 1992; Nathan and Xie, 1994). Enzyme activity requires flavin mononucleotide, adenine dinucleotide, tetrahydrobiopterin and in some cases, calcium and calmodulin as cofactors. Three different isoforms of nitric oxide synthase have been characterized including two constitutively expressed, relatively low output forms of the enzyme, and one relatively high output cytokine-inducible form of the enzyme (Nathan and Xie, 1994). Constitutive forms of nitric oxide synthase have been identified in endothelial cells and the brain and are known to require calcium and calmodulin for activity. The cytokine inducible form is not calcium-dependent. Although first characterized in macrophages, this form of the enzyme can be induced in many cell types including endothelial cells, epithelial cells, hepatocytes and neurons (Nathan and Xie, 1994).

---

* Abbreviations used: alveolar macrophages (AM); interstitial macrophages (IM); interferon-γ (IFN-γ); lipopolysaccharide (LPS).

† Dr. Debra Laskin, Rutgers University, Department of Pharmacology & Toxicology, 681 Frelinghuysen Rd., Piscataway, NJ 08855-1179; Phone: 908-445-4702; FAX: 908-445-2534.

Biological Reactive Intermediates V
Edited by R. Snyder et al., Plenum Press, New York, 1996

Table 1. Examples of Changes in Cells and Tissues Induced by Nitric Oxide

	References
Altered leukocyte function	Moncada et al. 1991
Altered smooth muscle cell tone	Ignarro, 1990; Moncada, 1991
Altered microvascular permeability	Laskin et al., 1994
DNA damage	Wink et al., 1991
Decreased in synthesis of prostaglandins	Laskin et al., 1994
Altered production of reactive oxygen intermediates	Laskin et al., 1994
Metabolic stress	Nathan, 1992
Altered DNA synthesis	Heck et al., 1992; Punjabi et al., 1992

Excessive nitric oxide production during irritation and inflammation can cause marked changes in the tissue leading to damage (Laskin et al., 1994). This may be due to a direct effect on immune cells activated during the inflammatory response. For example, leukocytes are required for host defense and the resolution of inflammation and nitric oxide can suppress their functioning (Nathan, 1992). Alternatively, nitric oxide alters patterns of blood flow and this may contribute to tissue damage during inflammation (Moncada, 1991). Nitric oxide can also directly inactivate critical metabolic enzymes (Nathan, 1992). As a free radical species, it has a tendency to abstract electrons from electron-rich substrates. Sulfhydryl- iron- and iron-sulfur-containing proteins such as ribonucleotide reductase, ubiquinone oxidoreductase, succinate oxidoreductase and *cis*-aconitase are targets for nitric oxide (Moncada et al., 1991; Nathan, 1992). When these important metabolic enzymes are inhibited, impaired tissue functioning may result. Nitric oxide also activates some enzymes, for example, guanylyl cyclase (Nathan, 1992). Subsequent changes in levels of cyclic guanine nucleotides in cells can alter normal physiological processes. In smooth muscle cells, enhanced production of this cyclic nucleotide causes relaxation (Moncada et al., 1991; Nathan, 1992).

Our laboratories have been interested in the role of nitric oxide in inflammatory reactions in the lung (Pendino et al., 1992; 1993; 1994; Punjabi, 1994a). For our studies, we have used ozone, a highly reactive oxidant present in low concentrations in ambient air (Lippman, 1989), as a model irritant. It is well recognized that inhalation of ozone disrupts epithelial cells in the proximal alveolar regions and the terminal bronchioles of the lung. During this process, there is a marked inflammatory response characterized by the accumulation of granulocytes and macrophages at sites of tissue damage. In previous studies, we found that granulocytes and macrophages isolated from ozone treated animals are activated to produce increased amounts of cytotoxic mediators and inflammatory cytokines including interleukin-1, tumor necrosis factor-a and fibronectin (Pendino et al., 1994). We hypothesize that these mediators contribute to ozone-induced tissue damage. Unexpectedly, during the course of our work, we discovered that type II epithelial cells were also functionally activated to release cytotoxic mediators by brief exposure of rats to ozone (Punjabi et al., 1994a). Taken together, these data indicate that following ozone inhalation, multiple cell types in the lung may contribute to tissue damage.

Since nitric oxide appears to be an important mediator of inflammation and tissue damage, a question arises as to whether the inflammatory reaction induced by ozone was associated with altered ability of lung cells to produce this mediator. Previous studies have demonstrated that alveolar macrophages can be stimulated to produce nitric oxide by inflammatory mediators such as lipopolysaccharide (LPS) and interferon-γ (IFN-γ) (Pendino et al., 1993; Prokhorova, 1994). Similarly, we found that alveolar macrophages as well as interstitial macrophages from control animals produced nitric oxide in response to LPS and

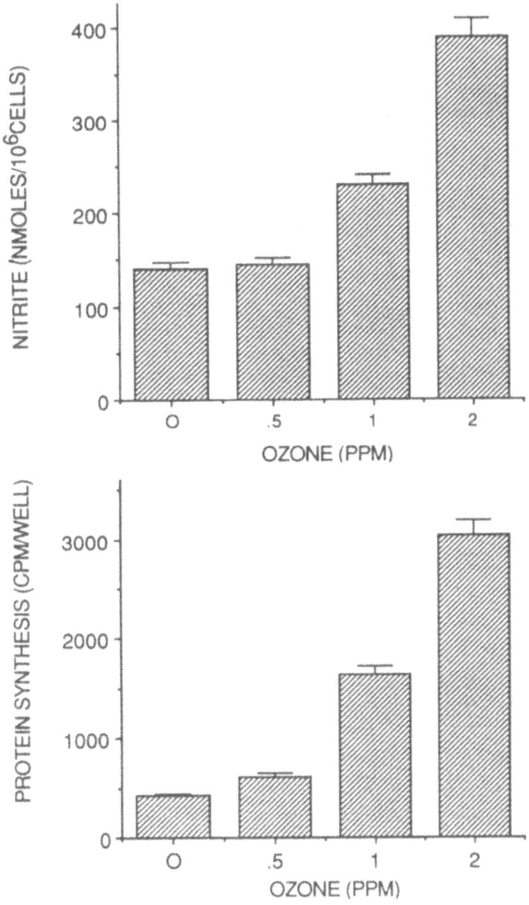

Figure 1. Effects of Inhalation of Ozone on Nitric Oxide Production by Lung Macrophages and Type II cells. Female Sprague-Dawley rats were exposed to ozone (2 ppm), generated from oxygen gas via a UV-light ozone generator, or ultrapure air in 5.5 ft^3 airtight plexiglass chambers for 3 hours. Forty-eight hours later, alveolar macrophages were isolated from perfused lungs by lavage. Interstitial macophages and type II cells were isolated from lavaged lung by differential centrifugation and digestion of the tissue with collagenase or elastase, respectively (Punjabi et al., 1994; Prokhorova et al., 1994). Nitric oxide was quantified by the accumulation of nitrite in the culture medium after 48 hours using a procedure based on the Greiss reaction (Punjabi et al., 1994b) with sodium nitrite as the standard. Cells were cultured in the presences of interferon-γ or lipopolysaccharide (LPS). Each value is the average nmoles/10^6 cells ± SE.

IFN-γ (Figure 1). Nitric oxide production by both cell types was time-dependent reaching maximal levels after 24 to 48 hours (Wizemann et al., 1994) and inhibited by the nitric oxide synthase inhibitor, N^G-monomethyl-l-arginine (Pendino et al., 1993; Prokhorova et al., 1994; Wizemann et al., 1994). We also found that production of nitric oxide by the cells was due to expression of protein for inducible nitric oxide synthase in the cells which resulted from greater amounts of enzyme mRNA (Pendino et al., 1992; 1993). Brief exposure of rats to ozone caused significantly more nitric oxide to be produced spontaneously by both the alveolar and interstitial macrophages (Figure 1). In addition, macrophages from ozone-exposed animals were found to be sensitized to produce nitric oxide in response to inflammatory stimuli (Figure 1). This was due to increased expression of mRNA and protein for inducible nitric oxide synthase in the cells (Pendino et al., 1992; 1993). These data indicate that lung macrophages from ozone treated animals are activated to produce reactive nitrogen intermediates and that this response is due to increased inducible nitric oxide synthase.

Additional studies revealed that type II cells, like lung macrophages, also have the capacity to be activated to produce nitric oxide. Figure 1 shows that type II cells treated with IFN-γ produce nitric oxide, a response which was time- and concentration-dependent (Punjabi et al., 1994a). However, unlike macrophages, type II cells failed to respond to LPS (Figure 1). Following exposure of rats to ozone, type II cells produced significantly greater amounts of nitric oxide spontaneously as well as in response to IFN-γ (Figure 1). In separate

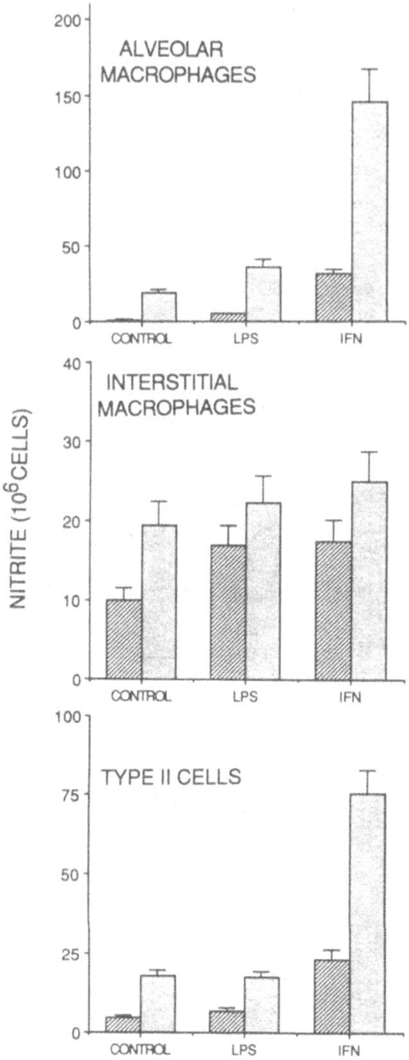

Figure 2. Protein Synthesis and Nitric Oxide Production by Hepatocytes After Acute Inhalation of Ozone by Rats. Animals were exposed to various doses of ozone for three hours. Hepatocytes were isolated from the perfused liver 48 hours later by collagenase digestion and assessed for their ability to produce nitric oxide and synthesize proteins. Nitrite in the culture medium was assayed after 48 hours as indicated in the legend to Figure 1. To quantify protein synthesis, hepatocytes were cultured in 96 well dishes (2 x 10^4 cells well) for 44 hours and then pulse labeled with 1 mCi/well of ^{3}H-leucine for 4 hours. The cells were then lysed and counted for radioactivity. To stimulate nitric oxide production, hepatocytes were cultured in the presence of LPS (5 mg/ml) and interferon-γ (50 U/ml). Each point is the average ± SE from one representative experiment.

experiments, we also found that ozone exposure induced the production of nitric oxide in the lung *in vivo*. Thus, examination of histological sections of stained tissues revealed inducible nitric oxide synthase protein as well as mRNA localized throughout the lung particularly in macrophages and epithelial cells, as well as interstitial cells and fibroblasts lining the alveoli (Pendino et al., 1993). Taken together, these data indicate that expression of inducible nitric oxide synthase may underlie increased spontaneous nitric oxide production in cells isolated from ozone exposed animals.

Inflammatory cytokines produced by lung macrophages following ozone inhalation may be released into the general circulation. When localized in tissues such as the liver, parenchymal cells may functionally respond to these mediators. For example, we have previously reported that ozone treatment of rats causes lung macrophages to release tumor necrosis factor-α (Pendino et al., 1994) which we subsequently identified in hepatocytes. When the hepatocytes were isolated, they were found to produce significantly more nitric oxide in response to LPS and IFN-γ than did cells from control animals (Figure 2). Both

Northern and Western blot analysis revealed that this was due to increased expression of inducible nitric oxide synthase (Laskin et al., 1995). Interestingly, increased protein synthesis was also noted in hepatocytes from animals exposed to ozone (Figure 2). Thus, hepatocytes isolated from ozone-treated animals are activated to produce cytotoxic mediators and also primed for an acute phase response (Kushmer and Mackiewicz, 1987).

It is clear from the present studies that ozone markedly enhances the ability of macrophages as well as alveolar epithelial cells in the lung to produce nitric oxide. In addition, following ozone exposure, hepatocytes are stimulated to produce increased amounts of nitric oxide as well as protein. These latter findings indicate that ozone has distinct extrapulmonary effects that may be important in the response to this toxicant. A general question arises as to whether the extrapulmonary effects of ozone are a sensitive indicator of toxicity, at least in the rat model. At the present time, the relation between nitric oxide production and protein synthesis is not clear. Induction of acute phase protein production following endotoxemia has been reported to be associated with enhanced nitric oxide production (Frederick et al., 1992). It may be that nitric oxide produced in the liver is a mediator of xenobiotic-induced toxicity. However, since nitric oxide can detoxify reactive oxygen intermediates (Laskin, et al., 1994), we cannot rule out the possibility that it acts to protect the tissue from oxidant-induced damage. Further studies are necessary to explore this possibility.

The contribution of nitric oxide to ozone-induced inflammation and injury is not known. Since nitric oxide is vasoactive, it is possible that this mediator is involved with altered blood flow in the lung (Moncada et al., 1991). At high concentrations, one would expect nitric oxide induced toxicity most likely to be due to inhibition of enzymes of intermediatory metabolism and/or respiration (Laskin et al., 1994). Nitric oxide has been reported to mediate a variety of inflammatory reactions including carrageenin-induced increases in epidermal vascular permeability and edema (Ialenti et al., 1992), immune complex-induced injury to rat lung and skin (Mulligan et al., 1991; 1992) and endotoxin-induced hepatic injury (Laskin et al., 1995). Nitric oxide produced by lung macrophages and epithelial cells after ozone exposure may damage surrounding tissue as well as the growth and differentiation of type II cells. This latter possibility may be important during the healing process since type II cells are precursors for the more abundant type I alveolar epithelial cells. Impairment of regenerating type I cells may prolong ozone-induced epithelial cell damage. The precise role of nitric oxide in controlling this process remains to be determined.

ACKNOWLEDGMENTS

This work was supported by NIH grants ES04738 and ES05022.

REFERENCES

1. Frederick JA, Hasselgren PO, Davis S, Higashiguchi T, Jacob TD, Fischer JE. Nitric oxide may upregulate in vivo hepatic protein synthesis during endotoxemia. Arch Surg 128: 152-157, 1992.
2. Heck, DE, Laskin, DL, Gardner, CR, Laskin, JD. Epidermal growth factor suppresses nitric oxide and hydrogen peroxide production by keratinocytes: Potential role for nitric oxide in the regulation of wound healing. J. Biol. Chem. 267: 21277-21280, 1992.
3. Ialenti A, Ianaro A, Moncada S, Di Rosa M. Modulation of acute inflammation by endogenous nitric oxide. Eur. J. Pharmacol. 211: 177-182, 1992.
4. Ignarro, LJ. Biosynthesis and metabolism of endothelium-derived nitric oxide. Annu Rev Pharmacol Toxicol 30:535-560, 1990.
5. Kushmer I, Mackiewicz J. Acute phase proteins as disease markers. Dis Markers 5: 1-15, 1987.

6. Laskin, JD, Heck, DE, Laskin, DL. Multifunctional role of nitric oxide in inflammation. Trends Endocrin. Metabolism 5:377-382, 1994.
7. Laskin, DL, Rodrigues-del Valle, M, Heck, DE, Hwang, SM, Ohnishi, ST, Durham, SK, Goller, NL, Laskin, JD. Hepatic nitric oxide production following acute endotoxemia in rats is mediated by increased inducible nitric oxide synthase gene expression, Hepatology, in press, 1995.
8. Lippman M. Health effects of ozone. A critical review. J Air Poll Control Assoc 39: 672-695 (1989).
9. Moncada, S., Palmer, RMJ, Higgs, EA. Nitric oxide: Physiology, pathophysiology and pharmacology. Pharmacol. Rev. 43:109-142, 1991.
10. Mulligan MS, Havel JM, Marletta MA, Ward PA. Tissue injury caused by deposition of immune complexes is L-arginine dependent. Proc Natl Acad Sci USA 88: 6338-6342, 1991.
11. Mulligan MS, Warren JS, Smith CW, Anderson DA, Yeh CG, Rudolph R, Ward PA. Lung injury after deposition of IgA immune complexes: Requirement for CD18 and L-arginine. J Immunol 148: 3086-3092, 1992.
12. Nathan, CF. Nitric oxide as a secretory product of mammalian cells. FASEB J 6:3051-3064, 1992.
13. Nathan, CF, Xie, QW, Regulation of biosynthesis of nitric oxide. J. Biol. Chem. 268:13,725-13,728, 1994.
14. Pendino K, Punjabi C, Lavnikova N, Gardner C, Laskin J, Laskin D. Inhalation of ozone stimulates nitric oxide production by pulmonary alveolar and interstitial macrophages. Am Rev Respir Dis 145: A650, 1992.
15. Pendino KJ, Laskin JD, Shuler RL, Punjabi CJ, Laskin DL. Enhanced production of nitric oxide by rat alveolar macrophages following inhalation of a pulmonary irritant is associated with increased expression of nitric oxide synthase. J Immunol., 151: 7196-7205, 1993.
16. Pendino KJ, Shuler RL, Laskin JD, Laskin DL. Enhanced production of interleukin-1, tumor necrosis factor-α and fibronectin by rat lung phagocytes following inhalation of a pulmonary irritant. Am. J. Respiratory Cell Mol. Biol. 11:279-286, 1994.
17. Prokhorova S, Lavnikova N, Laskin DL. Characterization of interstitial macrophages and subpopulations of alveolar macrophages from rat lung. J Leuk Biol, 55:141-146, 1994.
18. Punjabi, CJ, Laskin, DL, Heck, D, Laskin, JD. Nitric oxide production by murine bone marrow cells: Inverse correlation with cellular proliferation. J. Immunology 149: 2179-2184, 1992.
19. Punjabi CJ, Pendino KJ, Durham SK, Laskin JD, Laskin DL. Production of nitric oxide by rat type II pneumocytes. Increased expression of inducible nitric oxide synthase following exposure to a pulmonary irritant, Am J Respir Cell Mol Biol. 11:165-172, 1994a.
20. Punjabi, CJ, Laskin, JD, Hwang, SM, MacEachern, L, Laskin, DL. Enhanced production of nitric oxide by bone marrow cells and increased sensitivity to macrophage colony-stimulating factor (CSF) and granulocyte-macrophage stimulating factor (GM-CSF) after benzene treatment of mice. Blood 83:3255-3263, 1994b.
21. Wink, DA, Kasprzak, KS, Maragos, CM, Elespuru, R, Misra, M, Dunams, TM, Cebula, TA, Koch, WH, Andrews, AW, Allen, JS, Keefer, LK. DNA deaminating ability and genotoxicity of nitric oxide and its progenitors. Science 254: 1001-1003, 1991.
22. Wizemann, T, Gardner, CR, Laskin, JD, Quinones, S., Durham, SK, Goller, NL, Ohnishi, ST, Laskin, DL. Production of nitric oxide and peroxynitrite in the lung following acute exposure of rats to endotoxin. J. Leukocyte Biol. 56: 759-768, 1994.

21

THE IMPORTANCE OF SUPEROXIDE IN NITRIC OXIDE-DEPENDENT TOXICITY*

Evidence for Peroxynitrite-Mediated Injury

John P. Crow and Joseph S. Beckman†

Department of Anesthesiology, THT 958
University of Alabama at Birmingham
619 19th Street South
Birmingham, Alabama 35233-1924

INTRODUCTION

The potent vasodilator effects of pharmacological agents such as nitroglycerin have been known for over 100 years. In the last 20 years evidence has accumulated suggesting that nitric oxide is the species responsible for the biological effects of nitro-vasodilators. However, only within the last 10 years has the importance of earlier studies become clear; nitric oxide is an enzymatically-produced second messenger. Work of atmospheric chemists would indicate that nitrogen oxides like nitric oxide are harmful, yet we live out our lives constantly generating nitric oxide in endothelial cells and neurons and pharmacologic doses of nitro-vasodilators are devoid of overt toxicity. Furthermore, at physiologically relevent concentrations, no direct toxicity of nitric oxide itself has been demonstrated. Nontheless, endogenous production of nitric oxide has been implicated in reoxygenation injury following ischemia (1,2), glutamate-mediated neuronal toxicity (2,3), inflammation (4-6) and graft versus host disease (7-9). How then, can nitric oxide be injurious?

We propose that nitric oxide is one component of a biologic binary chemical bomb. The second component is the ubiquitous oxygen radical, superoxide. The bi-radical reaction of nitric oxide with superoxide occurs at a near diffusion-limited rate (10) and yields the potent oxidant and nitrating agent peroxynitrite. Evidence for endogenous peroxynitrite formation has been obtained using an antibody we raised which specifically recognizes nitrotyrosine residues in proteins. This chapter will examine the evidence suggesting that peroxynitrite is primarily responsible for nitric oxide-mediated pathology in a number of inflammatory and neurodegenerative conditions. Data will be presented which directly links

* Invited paper

† (Ph: 205-934-8123; Fax: 205-934-7437)

Biological Reactive Intermediates V
Edited by R. Snyder et al., Plenum Press, New York, 1996

peroxynitrite and mutations in the superoxide dismutase (SOD) gene to neuronal death in the familial form of amyotrophic lateral sclerosis, a fatal motor neuron disease.

I. SOURCES OF NITRIC OXIDE

Nitric oxide is a small, hydrophobic molecule with chemical properties that make it uniquely suited as both an intra- and intercellular messenger. It is a relatively stable, uncharged radical which readily crosses lipid membranes and interacts with specific target molecules like the heme iron prosthetic group of guanylate cyclase (11-13). Binding of nitric oxide to guanylate cyclase results in production of the intracellular messenger cyclic GMP thereby initiating a cascade of downstream cellular events via cGMP-binding proteins such as cGMP-dependent protein kinase.

At least six different isoforms of nitric oxide synthase (NOS) have been purified, cloned, and characterized from many different cell types and the list continues to grow. To date, all NOS isoforms fall into one of two broad categories: 1) constitutively expressed and calcium-regulated, or 2) cytokine-induced and unregulated, i.e., constitutively active. The common enzymatic mechanism of all NOS's involves the utilization of molecular oxygen to oxidize arginine and give citrulline and nitric oxide as products. In addition, all NOS's are heme-containing and require the flavin cofactors FAD and FMN as well as the pterin cofactor, tetrahydrobiopterin (14).

The constitutive and inducible forms of NOS contribute differently to total nitric oxide generation. Calcium-regulated NOS's produce small bursts of nitric oxide in response to calcium influx. Nitric oxide rapidly diffuses to neighboring cells and elicits reponses such as vascular smooth muscle relaxation or neuronal firing. Guanylate cylase in effector cells is maximally stimulated by 10-20 nM of nitric oxide (13,15) and nitric oxide produced for signal transduction is not thought to exceed 0.1 μM under normal physiological conditions (16). Nitric oxide which encounters a capillary will react rapidly with oxyhemoglobin to yield methemoglobin and nitrate.

The inducible isoform of NOS (iNOS) was originally purified from macrophages, however, it has been found in virtually all tissues following stimulation by inflammatory mediators like bacterial lipopolysaccharide (17,18) or cytokines like tumor necrosis factor-alpha or various interleukins (19,20). Although iNOS binds calmodulin and appears to require calcium for nitric oxide production, the binding affinity for calcium is so high that basal levels of intracellular calcium are sufficient for maximal activity (21,22). Thus, once iNOS has been induced it will continue to make nitric oxide until substrate is depleted or the enzyme itself is degraded. Under such conditions, the steady-state concentration of nitric oxide becomes more a function of its diffusion and decomposition than of its production. Actual concentrations of 2-4 μM nitric oxide have been measured in ischemic brain following reoxygenation (23)—conditions which are consistent with nitric oxide production from constitutive NOS isoforms following pathological calcium influx.

II. CHEMISTRY OF NITRIC OXIDE

Studies *in vitro* have implicated nitric oxide in the inhibition of critical iron-sulfur-containing enzymes involved in mitochondrial respiration (24), and in ribonucleotide reductase (25,26), as well as in the initiation of ADP-ribosylation of proteins (4478), and direct DNA damage (27,28). As mentioned earlier, endogenous production of nitric oxide has been implicated in reoxygenation injury, glutamate-mediated neuronal toxicity, inflammation, graft versus host disease, and as a major arm of host defense against viruses, bacteria, and other intracellular parasites (29,30). In many cases *in vivo* or *ex vivo* the involvement of nitric oxide was deduced from the ability of either NOS inhibitors or scavengers like hemoglobin to protect against a particular pathological change or, in the latter case, to inhibit

A $$2\cdot NO + O_2 \xrightarrow{4 \times 10^{6} M^{-2}s^{-1}} 2\cdot NO_2 \text{ (nitrogen dioxide)} \longrightarrow \text{one-electron oxidation reactions}$$

B $$\cdot NO + e^{-} \xrightarrow{+0.39\ v} NO^{-} \text{ (nitroxyl anion)} \longrightarrow ?$$

C $$\cdot NO - e^{-} \xrightarrow[Fe^{+3}; Cu^{+2}]{-1.21\ v} NO^{+} \text{ (nitrosonium ion)} \longrightarrow \text{nitrosation reactions}$$

D $$\cdot NO + O_2^{\cdot -} \xrightarrow{6.7 \times 10^{9} M^{-1}s^{-1}} ONOO^{-} \text{ (peroxynitrite anion)} \longrightarrow \text{nitration, hydroxylation, oxidation reactions}$$

Figure 1. Primary reaction pathways for nitric oxide with pathological implications.

parasite killing by activated macrophages. Often, nitric oxide donor compounds were seen to exacerbate the change, thereby providing further evidence for the apparently straightforward interpretation that nitric oxide per se is toxic. However, when considering toxicities associated with nitric oxide, it is important to note that although a number of reactive secondary products can be formed from nitric oxide, nitric oxide itself is a small, relatively simple molecule. Thus, the mechanisms by which it can directly interact/react with biomolecules are limited to only a few possibilities: one-electron reduction, one-electron oxidation, or radical-radical addition reactions (Fig. 1).

At physiological concentrations, nitric oxide is a relatively stable radical with very limited direct reactivity. As with all radicals, direct attack of nitric oxide on a neutral, non-radical molecule results in a stoichiometrically unbalanced reaction, i.e., another radical will be produced. In a strongly reducing environment, nitric oxide can accept an electron to generate nitroxyl anion (NO^-) (Fig. 1, panel B). Although nitroxyl could easily act as a one-electron reductant and simply regenerate nitric oxide, its chemistry is not well understood; nitroxyl can exist in both triplet and singlet states each with markedly different reactivities. Oxidation of nitric oxide by one electron results in nitrosonium ion (NO^+) (Fig. 1, panel C), a potent nitrosating agent typically generated by the decomposition of nitrous acid (HNO_2). Nitrosonium will directly attack nucleophiles like amines, thiols, and activated aromatic rings to give the corresponding nitroso compounds. Dietary nitrite (NO_2^-) at gastric pH may well be responsible for some formation of carcinogenic nitrosamines as well as biologically active nitrosothiols.

The most favorable direct reaction of nitric oxide is with another radical such as superoxide (Fig. 1, panel D). Peroxynitrite, which is formed from the radical-radical addition of nitric oxide and superoxide, reacts in complex ways with many different biomolecules. The variety of reaction products seen could easily be (mis)interpreted to suggest that multiple reactive species were being formed from nitric oxide.

Standard chemistry textbooks describe the reaction of nitric oxide with oxygen to give nitrogen dioxide (Fig. 1, panel A), the orange-brown gas in smog. Nitrogen dioxide is a strong one-electron oxidant and potent toxin in any biological system. However, it is important to note that formation of nitrogen dioxide from nitric oxide and oxygen is a third-order process and is dependent on the square of nitric oxide concentration. The practical outcome of this relationship is seen at physiological concentrations of nitric oxide; 0.7 hours would be required for 0.1 μM of nitric oxide to decay by one-half to form 0.025 μM of nitrogen dioxide. The biological half-life of nitric oxide is measured in seconds. Neither the half-life nor the toxicity of nitric oxide can be explained on the basis of nitrogen dioxide formation.

Nitrite (NO_2^-) and nitrate (NO_3^-) are stable oxidation products which are often used as indirect measures of endogenous nitric oxide production; inhibition of nitrite and nitrate formation by inhibitors of nitric oxide synthases clearly indicate that nitric oxide is the ultimate source. Again, standard chemistry texts illustrate the formation of nitrite and nitrate from nitric oxide as follows.

$$2\ \cdot NO + O_2 \rightarrow 2\cdot NO_2 \qquad (1)$$

The reaction of nitric oxide with oxygen gives nitrogen dioxide radical which in turn reacts with nitric oxide radical to give dinitrogen trioxide (N_2O_3). Dinitrogen trioxide then nitrosates water to give two moles of nitrite.

$$\cdot NO_2 + \cdot NO \rightarrow N_2O_3 + H_2O \rightarrow 2\ NO_2^- \qquad (2)$$

Nitrate is shown to be formed from the bi-radical reaction of nitrogen dioxide with itself to give dinitrogen tetroxide (N_2O_4) which nitrosates water to give one mole each of nitrite and nitrate.

$$\cdot NO_2 + \cdot NO_2 \rightarrow N_2O_4 + H_2O \rightarrow NO_2^- + NO_3^- \qquad (3)$$

Formation of both nitrite and nitrate via these mechanisms is dependent on nitrogen dioxide formation which, as discussed earlier, may not occur to a significant extent *in vivo*. Even if nitrogen dioxide were formed, it has a much greater probability of encountering and oxidizing nucleophilic biomolecules (to generate nitrite directly) than of colliding with another molecule of nitrogen dioxide or nitric oxide.

Direct one-electron oxidation by nitrogen dioxide yields nitrite but the source of nitrate in biological systems is less clear as it is dependent on nitrogen dioxide reacting with itself. Nitric oxide reacts quite readily with oxyhemoglobin to give methemoglobin and nitrate. In addition, nitrite will react with oxyhemoglobin to give nitrate. Clearly a large fraction of total nitrate in blood can be accounted for by these reactions. However, in blood free systems like cultured cells, nitrate production may occur largely via the formation and decomposition of peroxynitrite as shown below.

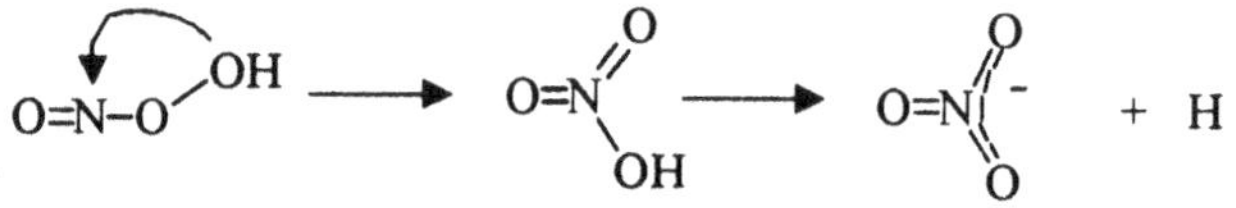

(4)

III. CELLULAR SOURCES OF SUPEROXIDE

The orderly transfer of four electrons to molecular oxygen in mitochondria results in complete reduction to give two moles of water. Because oxygen has has two unpaired electrons it is also a good acceptor of single electrons. Sequential transfer of three single electrons to molecular oxygen gives the anionic radical superoxide, neutral hydrogen peroxide, and the strongly oxidizing hydroxyl radical, respectively.

<u>sequential electron transfer yielding reactive oxygen species</u>

$$O_2 + 1e^- \rightarrow O_2^{\cdot -} + 1e^- \rightarrow H_2O_2 + 1e^- \rightarrow \cdot OH + {}^-OH$$

superoxide — hydrogen peroxide — hydroxyl radical

(5)

The primary cellular source of superoxide appears to be "leakage" of electrons from the mitochondrial flavoprotein NADH dehydrogenase and the ubiquinone-cytochrome *b* segment (31,32). It has been estimated that 1-2% of all oxygen utilized by mitochondria is inadvertently converted to superoxide (33). Other sources of superoxide include the autooxidation/redox cycling of small molecular weight compounds like catechols, flavins, quinones, and pterins as well as enzymatic sources like xanthine oxidase (34). Phagocytic cells utilize NAD(P)H oxidases to specifically generate superoxide as part of their antimicrobial armament. Superoxide production in the vasculature occurs via the autooxidation of iron (II) hemoglobin to methemoglobin which occurs at a rate of up to 3% per day.

The steady-state concentration of superoxide has been estimated to be in the range of 10-100 pM (35). Superoxide is maintained at this relatively low level by the combined action of two intracellular isoforms of superoxide dismutase (SOD). The primary isoform, Cu,Zn SOD, is present at a cytosolic concentration of approximately 10 μM, or 100,000 to 1,000,000 times the concentration of its substrate. SOD catalyzes the transfer of an electron from one superoxide molecule to another generating one-half equivalent of both oxygen and hydrogen peroxide as products. Hydrogen peroxide is then rapidly removed by the action of catalase.

$$2\ O_2^{\cdot -} \xrightarrow[\text{SOD}]{\text{spontaneous dismutation, } H^+} O_2 + H_2O \qquad (6)$$

The term "superoxide" was originally coined to describe the unusual electronic configuration of this anionic radical (36). A common misconception is that "superoxide" is "superoxidizing". At physiological pH superoxide has has little potential to oxidize other molecules since this would require placing a second negative charge on superoxide (37). Neutralization of the first negative charge by protonation facilitates superoxide acting as an oxidant, however, the pK_a for superoxide is 4.7 (38). Thus, superoxide acts as an oxidant only in a more acidic environment or when the molecule under attack is a good one-electron donor such as reduced iron in Fe (II)-sulfur centers. Superoxide is, in fact, more likely to act as a reductant; its ability to carry out the one-electron reduction of iron (III) in cytochrome c serves as the basis for a frequently used assay (39).

The tendency for superoxide to act as a reductant largely determines the mechanisms by which it can be directly toxic. Superoxide has been shown to initiate reductive release of ferric iron from ferritin (40,41) thereby allowing iron to participate in numerous oxidative reaction pathways. The entire field of oxygen radical-mediated pathology has relied to a large extent on explanations of oxidant injury based on the interaction between redox active iron and superoxide. Problems with this conventional view, as well as alternative possibilities, will be discussed later.

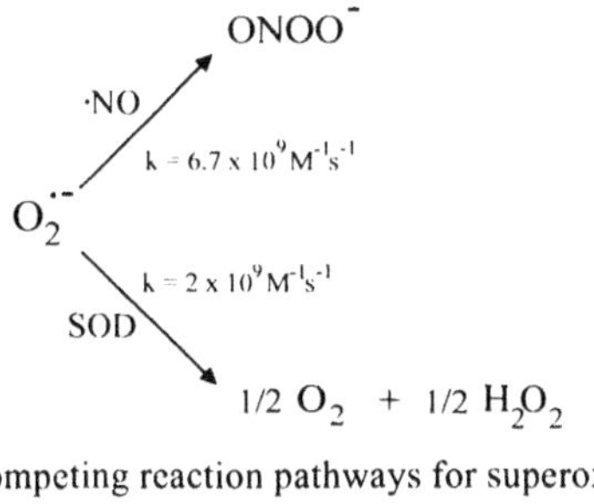

Figure 2. Competing reaction pathways for superoxide *in vivo*.

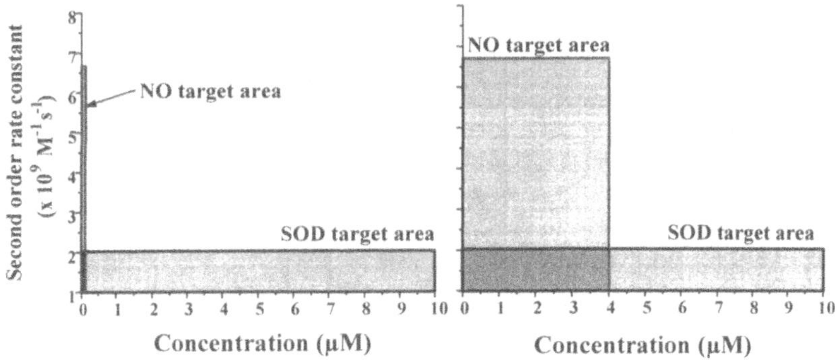

Figure 3. Relative target areas for superoxide. Left panel. Target areas of SOD and nitric oxide under physiological and right panel, pathological conditions. Target areas are determined by multiplying the second order rate constants for reaction with superoxide times the cellular concentration of the reactant.

Dismutation of superoxide by SOD occurs at a rate of 2.3 x 10^9 $M^{-1}s^{-1}$ (Fig. 2) (42). Thus, the only likely reactions of superoxide would be those which occur at least as fast or where the concentration of the other reactant far exceeds that of SOD. The only known reactant which can effectively out-compete SOD for the available superoxide is nitric oxide. The product of this reaction is peroxynitrite—a highly reactive compound which could account for both superoxide- and nitric oxide-dependent toxicity.

Superoxide reacts with nitric oxide three times faster than it reacts with SOD and the concentration of nitric oxide can approach that of SOD under pathological conditions. The relative target areas for superoxide are illustrated in Figure 3.

Target areas are a function of the second order rate constants for the given reaction multiplied by the concentration of the target. The competing targets of nitric oxide and SOD for superoxide are illustrated, however, this type of analysis can be used to determine the probability of attack of any given target by any potential reactant.

IV. CONVENTIONAL MECHANISMS OF SUPEROXIDE-DEPENDENT TOXICITY

SOD is present in all aerobic organisms examined to date and total lack of SOD appears to dramatically decrease the life-span of bacteria, mice, and rats, particularly when

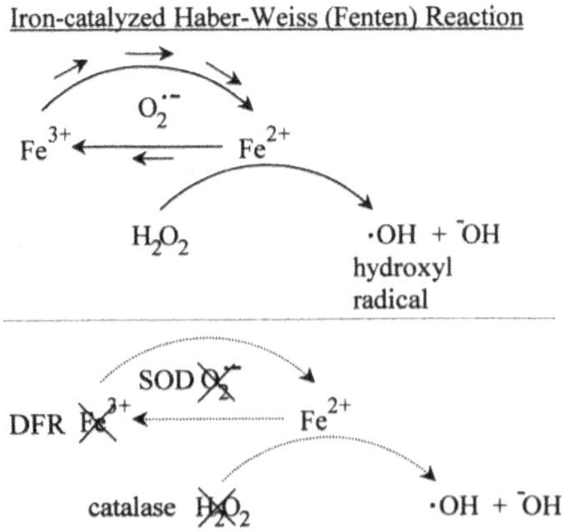

Figure 4. Superoxide-dependent generation of hydroxyl radical via the Fenten Reaction. Superoxide serves as the reductant for iron (III) as well as the source of hydrogen peroxide. Iron (II) reduces hydrogen peroxide to generate hydroxide and hydroxyl radical.

they are exposed to hyperoxic conditions. Also, the ability of exogenously administered SOD to limit or prevent injury under pathological conditions (e.g., hypoxia/reoxygenation) is well-established. The inescapable implication is that superoxide is toxic or that it is required for the production of some secondary toxic species. Superoxide itself reacts with very few biomolecules and, as mentioned earlier, is a better reductant than oxidant at neutral pH. For many the years the accepted explanation for this observed superoxide-dependent toxicity has been hydroxyl radical production via the iron-catalyzed Haber-Weiss reaction (43).

In this scheme, superoxide serves primarily as the reductant for iron. This seems an improbable *in vivo* scenario considering that the steady-state superoxide concentration is in the range of 10-100 pM while other reductants like ascorbate are present in much higher concentrations and reduce Fe^{3+} to Fe^{2+} quite effectively. If, on the other hand, superoxide served primarily as the source of hydrogen peroxide, then SOD would not be protective because hydrogen peroxide is the product. As mentioned earlier, superoxide may serve to increase the pool of free iron by releasing it from ferritin. However, even with a relatively large pool of Fe^{+2}, the rate at which it reduces hydrogen peroxide to produce hydroxyl radical is quite slow (10^3-10^4 $M^{-1}s^{-1}$) (44). Thus, the iron-catalyzed Haber-Weiss reaction requires the cooperative interaction of three species present in low concentrations *in vivo* due to efficient scavenging systems (SOD, catalase, ferritin), making the rate of hydroxyl radical formation dependent on the product of three minuscule concentrations which react at relatively low rates.

Even if hydroxyl radical were efficiently formed *in vivo*, it is simply too reactive to do any specific damage. The ability of hydroxyl radical to react at near diffusion-limited rates with virtually all biological molecules limits its effective diffusion radius to only a few Ångstroms. Thus, hydroxyl radical would have to be formed directly at critical sites (e.g., active sites of enzymes, bases of DNA, within membranes, etc.) to affect significant injury. Many such critical targets are capable of binding iron *in vitro* thereby promoting hydroxyl radical formation in close proximity. However, the rates of hydrogen peroxide reduction by bound iron can be even lower than with free iron due to non-optimal coordination geometries. Overall, the slow rates of reactions leading to hydroxyl radical production and the low concentrations of the reactants suggest that other reactions may be important for understanding superoxide toxicity.

Hydroxyl radical production via the iron-catalyzed Haber-Weiss reaction became widely accepted largely because there was no better explanation at the time for how SOD, catalase, and iron chelators could protect against oxygen radical-mediated injury. Demonstration of endogenous nitric oxide production occurred years later. Had nitric oxide production and subsequent peroxynitrite formation been known at the time, it is quite likely that Haber-Weiss chemistry would never have been invoked.

V. CHEMISTRY AND REACTIVITY OF PEROXYNITRITE

At least four possible mechanisms for *in vivo* peroxynitrite formation exist, however, the direct bi-radical reaction of nitric oxide with superoxide is the most reasonable because the reaction is extremely fast (approaching the diffusion limit) and is not dependent on pH or the presence of transition metals. For *in vitro* use, peroxynitrite is readily synthesized from acidified hydrogen peroxide and nitrous acid; the reaction mixture is rapidly quenched with sodium hydroxide to yield peroxynitrite anion (45). Peroxynitrite exists in the *cis* conformation in alkaline solutions (Fig. 5) and is stable for several weeks at pH 12.

Dilution of the alkaline stock solution into pH 7.4 buffer results in protonation to give peroxynitrous acid (pK_a 6.8). Neutralization of the negative charge lessens the energy barrier associated with isomerization to the *trans* conformer. *Trans* peroxynitrous acid can

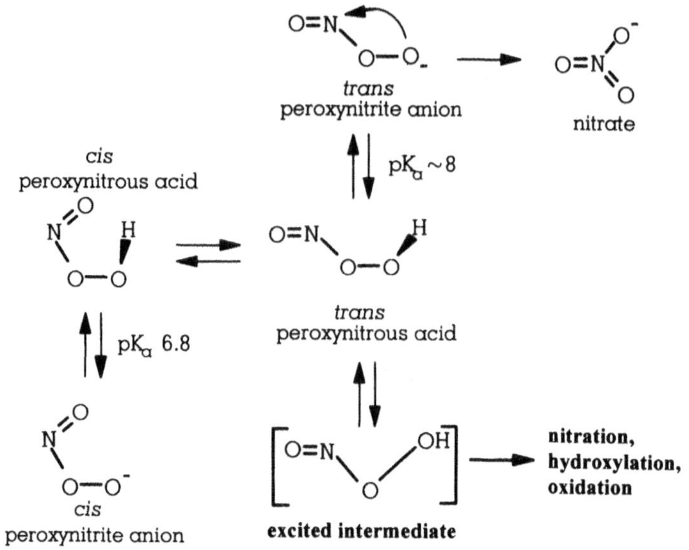

Figure 5. Conformational states of peroxynitrite. Isomerization from cis to trans occurs only after protonation. Trans peroxynitrous acid can either ionize to trans peroxynitrite anion which can rearrange to nitrate or undergo transition to the more energetic intermediate which reacts via several distinct pathways with various biomolecules.

either ionize to *trans* peroxynitrite and undergo internal rearrangement to give nitrate or react with target molecules via formation of the excited intermediate (Fig. 5).

At pH 7.4 and 37°C the half-life of peroxynitrite is 1-2 sec (46). Peroxynitrous acid can react through three major pathways or rearrange to nitrate (see Fig. 6). Reaction of peroxynitrite with target molecules results in products characteristic of both nitrogen dioxide ($^{\bullet}NO_2$) and hydroxyl radical ($^{\bullet}OH$).

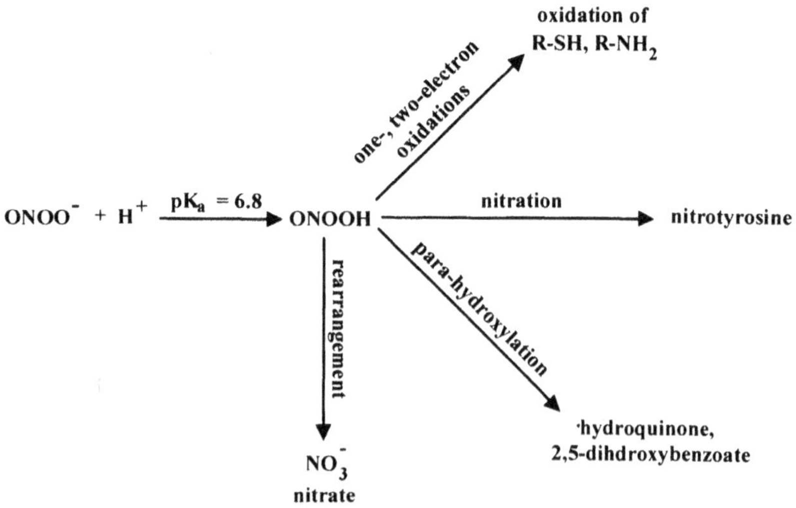

Figure 6. Reaction pathways for peroxynitrous acid.

$$ONOO^- + H^+ \rightarrow ONO\cdots OH \rightarrow \text{"}{\cdot}NO_2 + {\cdot}OH\text{"} \quad (7)$$

However, kinetic data and thermodynamic calculations indicate that homolytic fission of peroxynitrous acid is unfavorable and that no free hydroxyl radical is produced (47). We have proposed that peroxynitrous acid reacts as a vibrationally excited intermediate (Fig. 5) (48).

Decomposition of one molecule of peroxynitrous acid yields the equivalent of two potent one-electron oxidants each capable of participating in a variety of reactions or acting in a concerted fashion to nitrate phenolic rings (49,50) or hydroxylate aromatic rings (Fig. 6) (49,51). Peroxynitrous acid has also been shown to oxidize lipids (52), proteins (53,54), and DNA (55) via attack on nucleophilic centers.

Under physiological conditions, peroxynitrite is sufficiently stable to diffuse several cell diameters to reach critical cellular targets before becoming protonated and decomposing. We have also seen that it can cross lipid membranes, presumably in the protonated (acid) form (Chen and Beckman, unpublished observations). While most oxidative species are derived from peroxynitrous acid, direct reactions of the otherwise stable peroxynitrite anion occur with nucleophiles like thiols . Selective reactions of peroxynitrite anion with thiols and metal thiolates like those found in zinc-finger proteins serves to effectively target its oxidative capabilities to potentially critical sulfhydryl centers (Crow et al., Biochemistry 1995, in press). Thus, peroxynitrite may be considered the "smart bomb" of biological oxidants in that it is formed from two weak oxidants in a metal-independent manner and the anionic form is relatively stable allowing it to deliver multiple reactivites (including hydroxyl radical-like reactivity) over a relatively long distance.

VI. NITRATION BY PEROXYNITRITE AND CATALYSIS BY SOD

Peroxynitrite will nitrate phenolics like tyrosine in the absence of catalysts. The activation energy for non-catalytic or spontaneous phenolic nitration by peroxynitrite is 20 kcal/mole (47). Transition metals like iron (III) and copper (II) or Cu,Zn SOD reduce the activation energy required for nitration to 12 kcal/mole. The attraction of Cu,Zn SOD for peroxynitrite is likely due to peroxynitrite's charge and its structural similarity to superoxide. Unlike spontaneous nitration, which appears to involve two concerted one-electron oxidations, the reaction of peroxynitrite with the copper atom of SOD is thought to result in formation of a reactive nitronium ion (NO_2^+) at the active site of SOD (50).

$$ONOO^- + Cu^{2+}\text{–SOD} \rightarrow NO_2^{+}\cdots O^-\text{–}Cu^{1+}\text{–SOD} + H^+ \rightarrow$$

$$\text{protein–tyr–}NO_2 + HO\text{-}Cu^{1+}\text{–SOD} + H^+ \rightarrow Cu^{2+}\text{–SOD} + H_2O \quad (8)$$

In the simplest sense, Cu/Zn SOD is acting as a chelator for copper when it reacts with peroxynitrite. However, the size of the active site and the charge distribution around it affect both how peroxynitrite binds and what the nitronium ion might subsequently attack. The reaction of peroxynitrite with bovine Cu/Zn SOD occurs at a rate of 10^5 $M^{-1}s^{-1}$ and is limited by the rate of isomeriztion of the C-shaped *cis* conformer of peroxynitrite to the more linear *trans* form (65). This is evidenced by hyperbolic kinetics when the decomposition rate of peroxynitrite is plotted as a function of SOD concentration (65). Conversely, the rate of peroxynitrite reaction with iron (III)-EDTA continues to increase as a function of chelated iron concentration idicating that the iron can react with either conformer.

VII. FUNCTIONAL EFFECTS OF THE SOD MUTATIONS IN ALS

Amyotrophic lateral sclerosis (ALS) or Lou Gehrig's Disease is a degenerative disease which seletively affects motor neurons with an average age of onset of 45 years. The recent association of familial amyotrophic lateral sclerosis (FALS) with 26 distinct and independent mutations at 16 different amino acid positions in copper/zinc superoxide dismutase (SOD) points to an oxygen radical-mediated mechanism of motor neuron injury at least in the inherited form. Transgenic mice expressing any one of several SOD mutants in addition to wild-type become paralyzed. Thus, neuronal injury is best explained in terms of gain of a deleterious function of SOD rather than decreased superoxide scavenging.

The active site of Cu,Zn SOD is a deep hydrophobic pocket that matches the shape of superoxide. The catalytically active copper is buried at the bottom of the active site. Because peroxynitrite is most stable in a *cis* or C-shaped conformation, it can not fit into the active site. In wild-type SOD only the small fraction of peroxynitrite that isomerizes to the *trans* configuration can fit into the active site and react to nitrate phenolic compounds. Our fundamental hypothesis is that the SOD mutants found in ALS patients all have a more accessible active site which allows greater access of peroxynitrite to the copper to produce the nitrating species.

With one exception, all of the familial ALS SOD mutations affect the ß-barrel, which is an extremely stable structure that forms the foundation of the SOD molecule. Banci et al. (REF) used a molecular dynamics program to predict the thermal motion of individual amino acid residues in Cu,Zn SOD. The FALS mutations identified thus far occur in amino acids that have lower thermal mobilities on average than other residues. The presence of new amino acid residues which have higher thermal mobilities would serve to increase the range of thermal motion of the ß-barrel and thereby the dynamic "flex" of the active site. The end result would be a more accessible active site which is better able to accomodate the larger peroxynitrite molecule. This is analogous to a well-worn baseball glove (i.e., the mutant) which can be used to catch a softball (peroxynitrite) much better than a new, stiff glove.

In support of this hypothesis, we have seen increased nitration of a tyrosine analog by peroxynitrite in the presence of any of three SOD mutants relative to wild-type SOD. We have also seen up to 10-fold enhancement of nitration of a protein specifically found in neurons, neurofilament-L (NFL) (Fig 7).

Neurofilaments-L, -M, and -H are low, medium, and high molecular weight cytoskeletal proteins present in neurons. They form long, macromolecular, structural filaments which help determine axonal caliber and maintain rigidity. Neurofilament assembly/disassembly is a dynamic process which is involved in axonal transport and appears to be regulated by phosphorylation. Hyperphosphorylated neurofilament tangles are found in axonal swel-

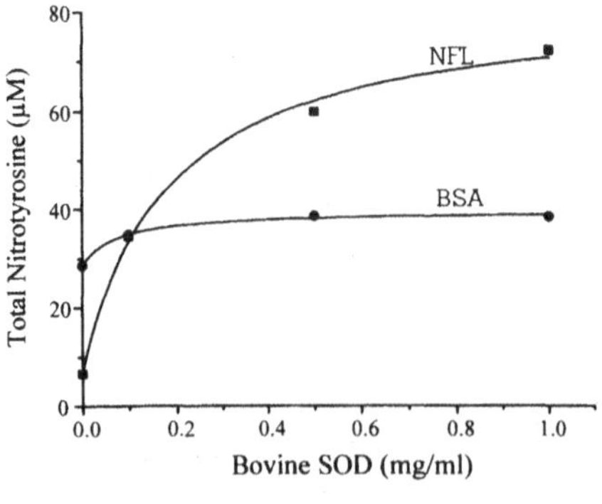

Figure 7. Nitration of neurofilament-L (NFL) and bovine serum albumin (BSA) by peroxynitrite as a function of superoxide dismutase (SOD). Both NFL and BSA are ~ 68 kD proteins containing 20 tyrosine residues. Peroxynitrite (1 mM) was added to protein solutions (0.4 mg/ml) in 50 mM potassi (77) um phosphate, pH 7.4 in the absence and present of increasing amounts of SOD. Nitrotyrosine content of proteins was quantified spectrally at 430 nm (ε_{430nm} = 4,400 $M^{-1}cm^{-1}$) following alkalinization to pH 10.

lings termed spheroids, which are characteristic of ALS. Recently we have obtained evidence of neurofilament nitration within spheroids bodies.

VIII. SELECTIVE KILLING OF MOTOR NEURONS

We propose that SOD mutant are intrinsicly more efficient in catalyzing nitration by peroxynitrite and that neurofilaments represent a neuron-specific target for nitration. Nitration could affect phosphorylation/dephosphorylation or otherwise alter protein function such that proper assembly/disassembly is adversely affected. Under these conditions, improperly assembled neurofilaments could lead to axonal gridlock, swelling, and ultimately death of the neuron.

IX. EVIDENCE OF PEROXYNITRITE FORMATION *IN VIVO*: IMMUNOLOGICAL DETECTION OF NITROTYROSINE

Even before endothelial-derived relaxing factor (EDRF) was identified as nitric oxide, numerous studies suggested that SOD could prolong the biological half-life of either endogenous nitric oxide or authentic nitric oxide added to a tissue bath (66,67). The most reasonable interpretation of this effect is that SOD scavenges superoxide, thereby preventing its reaction with nitric oxide.

Using an anti-nitrotyrosine antibody we raised to nitrated protein, we and others have seen specific staining for nitrated protein in a number of diseases and conditions including the endothelium and plaques in atherosclerotic vessels (68), synovial joints in rheumatoid arthritis, septic lung and heart (Royall and Kooy, unpublished, Univ. of Alabama), in ischemic brain, and spinal cord of patients with amyotrophic lateral sclerosis (M. Strong, Univ of West Ontario, unpublished results). We contend that nitrotyrosine residues in proteins are specific footprints of peroxynitrite based on the demonstrated ability and efficiency of peroxynitrite to act as a nitrating agent both *in vitro* and *ex vivo* (e.g., when added to cells or tissue homogenates) and based on the lack of credible evidence for production of any other reasonable nitrating species *in vivo*.

X. OTHER POTENTIAL MECHANISMS OF PEROXYNITRITE-MEDIATED PATHOLOGY

At present it is not clear whether the nitration seen in various pathological conditions is merely a marker or is fundamentally related to the primary pathology. There are several reasons why the latter may be true. 1) Nitration inhibits tyrosine phosphorylation (69) and may, in some circumstances, mimic phosphorylation in an irreversible manner (Fig. 8).

2) Nitrotyrosine may be recognized by tyrosine phosphatases leading to inhibition of dephosphorylation. 3) By introducing a negative charge onto tyrosine, nitration may alter protein conformation and function (70-75), or 4) "tag" a protein for proteolysis. 5) Nitrotyrosine structurally resembles dinitrophenol, a strongly antigenic compound used for making haptens. Endogenous nitration may, therefore, initiate autoimmune processes. 5) Nitration of cytoskeletal proteins like neurofilaments may alter the dynamics of assembly/disassembly, a process critical to motor neuron survival (76).

Nitration of tyrosine is the peroxynitrite reaction pathway of highest yield, primarily due to the ability of SOD to selectively enhance nitration in complex media such as the milieu

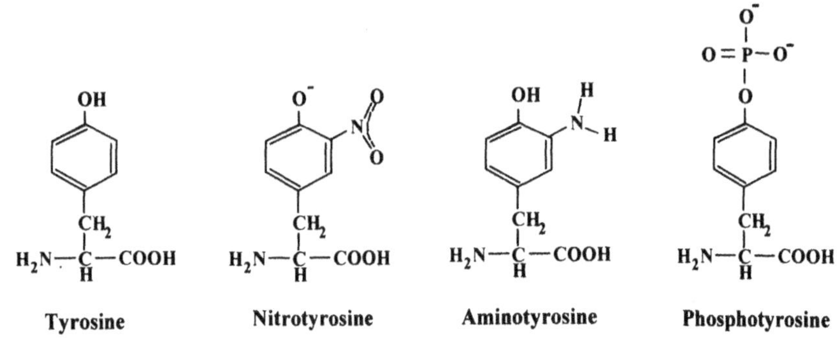

Figure 8. The structure of various tyrosine analogs of biological interest. The nitro group involves a stable (78) carbon-nitrogen bond, unlike phosphotyrosine which is readily hydrolyzed. Nitrotyrosine is readily reduced to aminotyrosine by dithionite at alkaline pH, which is an important control utilized in immunohistochemistry studies.

of a cell. In addition to the nitration pathway, we have recently shown that peroxynitrite reacts quite rapidly ($2 \times 10^5 M^{-1}s^{-1}$) (Crow et al., Biochemistry 1995, in press) with zinc-thiolate centers such as those present in numerous transcription factors. Oxidation of such "zinc fingers" could dramatically alter their ability to recognize and bind DNA. Thus, peroxynitrite could exert significant effects on gene regulation.

REFERENCES

1. Masini, E., Bianchi, S., Mugnai, L., Gambassi, F., Lupini, M., Pistelli, A., and Mannaioni, P.F. 1991. The effect of nitric oxide generators on ischemia reperfusion injury and histamine release in isolated perfused guinea-pig heart. *Agents & Actions* 33:53-56.
2. Cazevieille, C., Muller, A., Meynier, F., and Bonne, C. 1993. Superoxide and nitric oxide cooperation in hypoxia/reoxygenation-induced neuron injury. *Free Radical Biology & Medicine* 14:389-395.
3. Dawson, V. L., Dawson, T.M., Bartley, D.A., Uhl, G.R., and Snyder, S.H. 1993. Mechanisms of nitric oxide-mediated neurotoxicity in primary brain cultures. *Journal of Neuroscience* 13:2651-2661.
4. Green, S. J., Mellouk, S., Hoffman, S.L., Meltzer, M.S., and Nacy, C.A. 1990. Cellular mechanisms of nonspecific immunity to intracellular infection: cytokine-induced synthesis of toxic nitrogen oxides from L-arginine by macrophages and hepatocytes. *Immunology Letters* 25:15-19.
5. Mulligan, M. S., Hevel, J.M., Marletta, M.A., and Ward, P.A. 1991. Tissue injury caused by deposition of immune complexes is L-arginine dependent. *Proceedings of the National Academy of Sciences of the United States of America* 88:6338-6342.
6. Billiar, T. R., Hoffman, R.A., Curran, R.D., Langrehr, J.M., and Simmons, R.L. 1992. A role for inducible nitric oxide biosynthesis in the liver in inflammation and in the allogeneic immune response. [Review]. *Journal of Laboratory & Clinical Medicine* 120:192-197.
7. Hoffman, R. A., Langrehr, J.M., and Simmons, R.L. 1992. The role of inducible nitric oxide synthetase during graft-versus-host disease. *Transplantation Proceedings* 24:2856
8. Garside, P., Hutton, A.K., Severn, A., Liew, F.Y., and Mowat, A.M. 1992. Nitric oxide mediates intestinal pathology in graft-vs.-host disease. *European Journal of Immunology* 22:2141-2145.
9. Langrehr, J. M., Murase, N., Markus, P.M., Cai, X., Neuhaus, P., Schraut, W., Simmons, R.L., and Hoffman, R.A. 1992. Nitric oxide production in host-versus-graft and graft-versus-host reactions in the rat. *Journal of Clinical Investigation* 90:679-683.
10. Huie, R. E. and Padmaja, S. 1993. The reaction of NO with superoxide. *Free Radical Research Communications* 18:195-199.
11. Mittal, C. K. and Murad, F. 1977. Activation of guanylate cyclase by superoxide dismutase and hydroxyl radical: A physiological regulator of guanosine 3',5'-monophosphate formation. *Proc. Natl. Acad. Sci. U. S. A.* 74:4360-4364.

12. Craven, P. A. and DeRubertis, F.R. 1978. Restoration of the responsiveness of purified guanylate cyclase to nitrosoguanidine, nitric oxide, and related activators by heme and hemeproteins. Evidence for involvement of the paramagnetic nitrosyl-heme complex in enzyme activation. *J. Biol. Chem.* 253:8433-8443.
13. Ignarro, L. J., Adams, J.B., Horwitz, P.M., and Wood, K.S. 1986. Activation of soluble guanylate cyclase by NO-hemoproteins involves NO-heme exchange. Comparison of heme-containing and heme-deficient enzyme forms. *J. Biol. Chem.* 261:4997-5002.
14. Sessa, W. C. 1994. The nitric oxide synthase family of proteins. [Review]. *Journal of Vascular Research* 31:131-143.
15. Murad, F., Mittal, C.K., Arnold, W.P., Katsuki, S., and Kimura, H. 1978. Guanylate cyclase: Activation by azide, nitro compounds, nitric oxide, and hydroxyl radical and inhibition by hemoglobin and myoglobin. *Adv. Cyclic. Nucleotide. Res.* 9:145-158.
16. Shibuki, K. 1990. An electrochemical microprobe for detecting nitric oxide release in brain tissue. *Neuroscience Research* 9:69-76.
17. Hauschildt, S., Luckhoff, A., Mulsch, A., Kohler, J., Bessler, W., and Busse, R. 1990. Induction and activity of NO synthase in bone-marrow-derived macrophages are independent of Ca2+. *Biochem. J.* 270:351-356.
18. Knowles, R. G., Merrett, M., Salter, M., and Moncada, S. 1990. Differential induction of brain, lung and liver nitric oxide synthase by endotoxin in the rat. *Biochem. J.* 270:833-836.
19. Busse, R. and Mulsch, A. 1990. Induction of nitric oxide synthase by cytokines in vascular smooth muscle cells. *FEBS. Lett.* 275:87-90.
20. Corbett, J. A., Lancaster, J.R., Jr., Sweetland, M.A., and McDaniel, M.L. 1991. Interleukin-1 beta-induced formation of EPR-detectable iron-nitrosyl complexes in islets of Langerhans. Role of nitric oxide in interleukin-1 beta-induced inhibition of insulin secretion. *J. Biol. Chem.* 266:21351-21354.
21. Stuehr, D. J., Cho, H.J., Kwon, N.S., Weise, M.F., and Nathan, C.F. 1991. Purification and characterization of the cytokine-induced macrophage nitric oxide synthase: An FAD- and FMN-containing flavoprotein. *Proc. Natl. Acad. Sci. U. S. A.* 88:7773-7777.
22. Mollace, V., Colasanti, M., Rodino, P., Massoud, R., Lauro, G.M., and Nistico, G. 1993. Cytokine-induced nitric oxide generation by cultured astrocytoma cells involves Ca(++)-calmodulin-independent NO-synthase. *Biochemical & Biophysical Research Communications* 191:327-334.
23. Malinski, T., Bailey, F., Zhang, Z.G., and Chopp, M. 1993. Nitric oxide measured by a porphyrinic microsensor in rat brain after transient middle cerebral artery occlusion. *Journal of Cerebral Blood Flow & Metabolism* 13:355-358.
24. Welsh, N., Eizirik, D.L., Bendtzen, K., and Sandler, S. 1991. Interleukin-1 beta-induced nitric oxide production in isolated rat pancreatic islets requires gene transcription and may lead to inhibition of the Krebs cycle enzyme aconitase. *Endocrinology* 129:3167-3173.
25. Kwon, N. S., Stuehr, D.J., and Nathan, C.F. 1991. Inhibition of tumor cell ribonucleotide reductase by macrophage-derived nitric oxide. *Journal of Experimental Medicine* 174:761-767.
26. Lepoivre, M., Fieschi, F., Coves, J., Thelander, L., and Fontecave, M. 1991. Inactivation of ribonucleotide reductase by nitric oxide. *Biochemical & Biophysical Research Communications* 179:442-448.
27. Wink, D. A., Kasprzak, K.S., Maragos, C.M., Elespuru, R.K., Misra, M., Dunams, T.M., Cebula, T.A., Koch, W.H., Andrews, A.W., Allen, J.S., et al. 1991. DNA deaminating ability and genotoxicity of nitric oxide and its progenitors. *Science* 254:1001-1003.
28. Nguyen, T., Brunson, D., Crespi, C.L., Penman, B.W., Wishnok, J.S., and Tannenbaum, S.R. 1992. DNA damage and mutation in human cells exposed to nitric oxide in vitro. *Proceedings of the National Academy of Sciences of the United States of America* 89:3030-3034.
29. Green, S. J., Meltzer, M.S., Hibbs, J.B., Jr., and Nacy, C.A. 1990. Activated macrophages destroy intracellular Leishmania major amastigotes by an L-arginine-dependent killing mechanism. *Journal of Immunology* 144:278-283.
30. Oswald, I. P., Wynn, T.A., Sher, A., and James, S.L. 1994. NO as an effector molecule of parasite killing: Modulation of its synthesis by cytokines. [Review]. *Comparative Biochemical Physiological Pharmacological and Toxicological Endocrinology* 108:11-18.
31. Boveris, A., Cadenas, E., and Stoppani, A.O.M. 1976. Role of ubiquinone in the mitochondrial generation of hydrogen peroxide. *Biochem. J.* 156:435-444.
32. Turrens, J. F. and Boveris, A. 1980. Generation of superoxide anion by the NADH dehydrogenase of bovine heart mitochondria. *Biochem. J.* 191:421-427.
33. Boveris, A. and Chance, B. 1973. The mitochondrial generation of hydrogen peroxide: General properties and effect of hyperbaric oxygen. *Biochem. J.* 134:707-716.

34. Cross, A. R. and Jones, O.T.G. 1991. Enzymic mechanisms of superoxide production. *Biochim. Biophys. Acta Bio-Energetics* 1057:281-298.
35. Gardner, P. R. and Fridovich, I. 1991. Superoxide sensitivity of the Escherichia coli 6-phosphogluconate dehydratase. *Journal of Biological Chemistry* 266:1478-1483.
36. Pauling, L. 1979. The discovery of the superoxide radical. *Trends in Biochemical Sciences* 4:N270-N271.
37. Sawyer, D. T. and Valentine, J.S. 1981. How super is superoxide? *Accts. Chem. Res.* 14:393-400.
38. Bielski, B. H. J. 1978. Re-evaluation of the spectral and kinetic properties of HO_2 and O_2^- free radicals. *Photochem. Photobiol.* 28:645-649.
39. Mccord, J. M. and Fridovich, I. 1969. Superoxide dismutase: An enzymic function for erythrocuprein (hemocuprein). *J. Biol. Chem.* 244:6049-6055.
40. Biemond, P., van Eijk, H.G., Swaak, A.J.G., and Koster, J.F. 1984. Iron mobilization from ferritin by superoxide derived from phagocytosing polymorphonuclear leucocytes. Possible mechanism in inflammation diseases. *J. Clin. Invest.* 73:1576-1579.
41. Thomas, C. E., Morehouse, L.A., and Aust, S.D. 1985. Ferritin and superoxide-dependent lipid peroxidation. *J. Biol. Chem.* 260:3275-3280.
42. Klug, D., Rabani, J., and Fridovich, I. 1972. A direct demonstration of the catalytic action of superoxide dismutase through the use of pulse radiolysis. *J. Biol. Chem.* 247:4839-4842.
43. Haber, F. and Weiss, J. 1934. The catalytic decomposition of hydrogen peroxide by iron salts. *Proc.Roy.Soc.* 147:332-351.
44. Gutteridge, J. M. C. 1984. Lipid peroxidation initiated by superoxide-dependent hydroxyl radicals using complexed iron and hydrogen peroxide. *FEBS Lett.* 172:245-249.
45. Wang, Q. J., Giri, S.N., Hyde, D.M., and Li, C.F. 1991. Amelioration of bleomycin-induced pulmonary fibrosis in hamsters by combined treatment with taurine and niacin. *Biochem. Pharmacol.* 42:1115-1122.
46. Beckman, J. S., Beckman, T.W., Chen, J., Marshall, P.A., and Freeman, B.A. 1990. Apparent hydroxyl radical production by peroxynitrite: Implications for endothelial injury from nitric oxide and superoxide. *Proc. Natl. Acad. Sci. U. S. A.* 87:1620-1624.
47. Koppenol, W. H., Moreno, J.J., Pryor, W.A., Ischiropoulos, H., and Beckman, J.S. 1992. Peroxynitrite, a cloaked oxidant formed by nitric oxide and superoxide. *Chemical Research in Toxicology* 5:834-842.
48. Crow, J. P., Spruell, C., Chen, J., Gunn, C., Ischiropoulos, H., Tsai, M., Smith, C.D., Radi, R., Koppenol, W.H., and Beckman, J.S. 1994. On the pH-dependent yield of hydroxyl radical products from peroxynitrite. *Free Radical Biology & Medicine* 16:331-338.
49. Halfpenny, E. and Robinson, P.L. 1952. The nitration and hydroxylation of aromatic compounds by pernitrous acid. *J. Chem. Soc.* 939-946.
50. Ischiropoulos, H., Zhu, L., Chen, J., Tsai, M., Martin, J.C., Smith, C.D., and Beckman, J.S. 1992. Peroxynitrite-mediated tyrosine nitration catalyzed by superoxide dismutase. *Archives of Biochemistry & Biophysics* 298:431-437.
51. van der Vliet, A., O'Neill, C.A., Halliwell, B., Cross, C.E., and Kaur, H. 1994. Aromatic hydroxylation and nitration of phenylalanine and tyrosine by peroxynitrite. Evidence for hydroxyl radical production from peroxynitrite. *FEBS Letters.* 339:89-92.
52. Radi, R., Beckman, J.S., Bush, K.M., and Freeman, B.A. 1991. Peroxynitrite-induced membrane lipid peroxidation: The cytotoxic potential of superoxide and nitric oxide. *Archives of Biochemistry & Biophysics* 288:481-487.
53. Moreno, J. J. and Pryor, W.A. 1992. Inactivation of alpha 1-proteinase inhibitor by peroxynitrite. *Chemical Research in Toxicology* 5:425-431.
54. Bauer, M. L., Beckman, J.S., Bridges, R.J., Fuller, C.M., and Matalon, S. 1992. Peroxynitrite inhibits sodium uptake in rat colonic membrane vesicles. *Biochim. Biophys. Acta.* 1104:87-94.
55. King, P. A., Anderson, V.E., Edwards, J.O., Gustafson, G., Plumb, R.C., and Suggs, J.W. 1992. A stable solid that generates hydroxyl radical upon dissolution in aqueous solution: Reaction with proteins and nucleic acid. *J. Am. Chem. Soc.* 114:5430-5432.
56. Radi, R., Beckman, J.S., Bush, K.M., and Freeman, B.A. 1991. Peroxynitrite oxidation of sulfhydryls. The cytotoxic potential of superoxide and nitric oxide. *J. Biol. Chem.* 266:4244-4250.
57. Rubbo, H., Denicola, A., and Radi, R. 1994. Peroxynitrite inactivates thiol-containing enzymes of Trypanosoma cruzi energetic metabolism and inhibits cell respiration. *Archives of Biochemistry & Biophysics* 308:96-102.
62. Hofman, M. A. 1991. From Here to Eternity. Brain Ageing in an Evolutionary Perspective. *Neurobiol. Aging.* 12:338-340.
65. Beckman, J. S., Ischiropoulos, H., Zhu, L., van der Woerd, M., Smith, C., Chen, J., Harrison, J., Martin, J.C., and Tsai, M. 1992. Kinetics of superoxide dismutase- and iron-catalyzed nitration of phenolics by peroxynitrite. *Archives of Biochemistry & Biophysics* 298:438-445.

66. Bates, J. N., Harrison, D.G., Myers, P.R., and Minor, R.L. 1991. EDRF: Nitrosylated compound or authentic nitric oxide. *Basic Research In Cardiology* 86 Suppl 2:17-26.
67. Chen, F. Y. and Lee, T.J. 1993. Role of nitric oxide in neurogenic vasodilation of porcine cerebral artery. *Journal of Pharmacology & Experimental Therapeutics* 265:339-345.
68. Beckman, J. S., Ye, Y.Z., Anderson, P.G., Chen, J., Accavitti, M.A., Tarpey, M.M., and White, C.R. 1994. Extensive nitration of protein tyrosines in human atherosclerosis detected by immunohistochemistry. *Biological Chemistry Hoppe-Seyler* 375:81-88.
69. Martin, B. L., Wu, D., Jakes, S., and Graves, D.J. 1990. Chemical influences on the specificity of tyrosine phosphorylation. *Journal of Biological Chemistry* 265:7108-7111.
70. Chantler, P. D. and Gratzer, W.B. 1975. Effects of specific chemical modification of actin. *European Journal of Biochemistry* 60:67-72.
71. Feste, A. and Gan, J.C. 1981. Selective loss of elastase inhibitory activity of alpha 1-proteinase inhibitor upon chemical modification of its tyrosyl residues. *Journal of Biological Chemistry* 256:6374-6380.
72. Chacko, G. K. 1985. Modification of human high density lipoprotein (HDL3) with tetranitromethane and the effect on its binding to isolated rat liver plasma membranes. *Journal of Lipid Research* 26:745-754.
73. Deckers-Hebestreit, G., Schmid, R., Kiltz, H.H., and Altendorf, K. 1987. F0 portion of Escherichia coli ATP synthase: Orientation of subunit c in the membrane. *Biochemistry* 26:5486-5492.
74. Lundblad, R. L., Noyes, C.M., Featherstone, G.L., Harrison, J.H., and Jenzano, J.W. 1988. The reaction of bovine alpha-thrombin with tetranitromethane. Characterization of the modified protein. *Journal of Biological Chemistry* 263:3729-3734.
75. Tawfik, D. S., Chap, R., Eshhar, Z., and Green, B.S. 1994. pH on-off switching of antibody-hapten binding by site-specific chemical modification of tyrosine. *Protein Engineering* 7:431-434.
76. Nixon, R. A. 1993. The regulation of neurofilament protein dynamics by phosphorylation: clues to neurofibrillary pathobiology. [Review]. *Brain Pathology* 3:29-38.
77. Stork, G., Suh, H.S., and Kim, G. 1991. The temporary silicon connection method in the control of regiochemistry and stereochemistry: Applications to radical-mediated reactions - The stereospecific synthesis of C-glycosides. *J. Am. Chem. Soc.* 113:7054-7056.
78. Schuchmann, M. N., Schuchmann, H.P., Hess, M., and Vonsonntag, C. 1991. O2.- Addition to ketomalonate leads to decarboxylation: A chain reaction in oxygenated aqueous solution. *J. Am. Chem. Soc.* 113:6934-6937.

UNDERSTANDING THE STRUCTURAL ASPECTS OF NEURONAL NITRIC OXIDE SYNTHASE (NOS) USING MICRODISSECTION BY MOLECULAR CLONING TECHNIQUES

Molecular Dissection of Neuronal NOS

B. S. S. Masters,[1] K. McMillan,[1] J. Nishimura,[1] P. Martasek,[1]
L. J. Roman,[1] E. Sheta,[2] S. S. Gross,[3] and J. Salerno[4]

[1] Department of Biochemistry
The University of Texas Health Science Center at San Antonio
7703 Floyd Curl Drive
San Antonio, Texas 78284-7760
[2] Department of Biochemistry
Faculty of Science
Alexandria University, Alexandria, Egypt
[3] Department of Pharmacology
Cornell University Medical Center
1300 York Avenue, New York, New York 10021
[4] Department of Biology
Rensselaer Polytechnic Institute, Troy, New York 12180

INTRODUCTION

The neuronal isoform (Type I) of nitric oxide synthase (NOS) requires Ca^{+2}/calmodulin to catalyze the formation of NO• and citrulline from L-arginine (1) and molecular oxygen (2) with reducing equivalents from NADPH. The overall reaction is a five-electron process involving two successive monooxygenation steps with the obligatory formation of N-hydroxy-L-arginine as the oxygenated intermediate. The neuronal NOS shares with the other two known isoforms the unusual property, for a mammalian enzyme, of containing iron protoporphyrin IX (3-6), FAD and FMN (3, 7-9), and tetrahydrobiopterin (9). Bredt, *et al.* (10) had previously demonstrated the remarkable (58%) sequence similarity of rat brain NOS to rat liver NADPH-cytochrome P450 reductase and determined that the C-terminal 641 amino acids of NOS display consensus regions for both flavin and nucleotide binding.

Biological Reactive Intermediates V
Edited by R. Snyder et al., Plenum Press, New York, 1996

It was reported that carbon monoxide (CO) inhibited both macrophage and neuronal NOS activity (3-6) and these various preparations exhibited reduced-CO difference spectra with absorbance maxima in the region of 445 nm, a property shared with the members of the cytochrome P450 family. However, the low degree of sequence similarity to any of the more than 200 members of this family and other CO-binding, oxygenating heme proteins, such as chloroperoxidase or *Bacillus megaterium P450* (BM-3), suggests that the homology may end with the sharing of a cysteine thiolate ligand to the heme iron at the fifth axial position. Recent studies have established that this involves $cysteine_{415}$ of the neuronal NOS (11, 12), as suggested originally by McMillan, *et al.* (3), and $cysteine_{184}$ of the endothelial NOS (13) as determined by site-directed mutagenesis of the holoenzyme (11, 13) and the heme domain (12), respectively.

These apparently common properties have led our laboratory and others to seek the basis for similarities and differences among the isoforms since their regulation and expression in various cells and tissues are keys to their functions. The approach we have initiated to determine structural aspects of the neuronal NOS, which may lead to the understanding of its function, is to dissect the protein at sites which will make possible the expression of the domains in question in *E. coli* expression systems. If this tactic is to succeed, it will require the folding of the individual domains into "native structures" and the subsequent binding of their respective prosthetic groups. Characterization of the resulting expressed domains by a variety of biophysical methods, including attempts at crystallography, will determine the ultimate success of this approach.

RESULTS AND DISCUSSION

Background

The entry of our laboratory into the realm of molecular microdissection of neuronal NOS was suggested first by the discovery that the mammalian isoforms of NOS contain both flavin and heme prosthetic groups (3-8) and tetrahydrobiopterin (9) and the observation by Bredt, *et al.* (8) that a calmodulin binding consensus sequence could be located between residues 725-745, which coincidentally divided the protein almost exactly into two major domains. Our first experiments were performed by limited proteolysis, utilizing immobilized trypsin, and showed that the neuronal NOS, in the absence of calmodulin, could be cleaved into heme-binding and flavin-binding domains (14). Sheta, *et al.* (14) showed that calmodulin protected the protein against cleavage into ~89 kDa and ~79 kDa moieties, indicating that the calmodulin binding motif suggested by Bredt, *et al.* (8) was probably correct. Previously, Vorherr, *et al.* (15) demonstrated that a synthetic NOS peptide, corresponding to the calmodulin binding region in the enzyme, bound with a high affinity to calmodulin and the data of Sheta, *et al.* (14) estimated that calmodulin binding to nNOS occurred with a dissociation constant of 1 nM. These results led us to attempt the expression of the individual heme- and flavin-binding domains utilizing molecular cloning techniques (12).

The formation of NO• requires two monooxygenation reactions, which involve the incorporation of an atom from molecular oxygen in each reaction. As can be seen in Figure 1, the intermediate in the formation of citrulline from L-arginine is N-hydroxy-L-arginine, an oxygenated intermediate. Before the discovery of the heme constituency of the NOS isoforms (3-6), it was assumed that BH_4 would serve this function. The reports by Giovanelli, *et al.* (16) that BH_4 appeared not to be involved in an oxidation-reduction function and by Baek, *et al.* (17) that it stabilized the association of inactive monomers to active dimers in cytokine-induced macrophage NOS from RAW 264.7 cells in the overall reaction forming

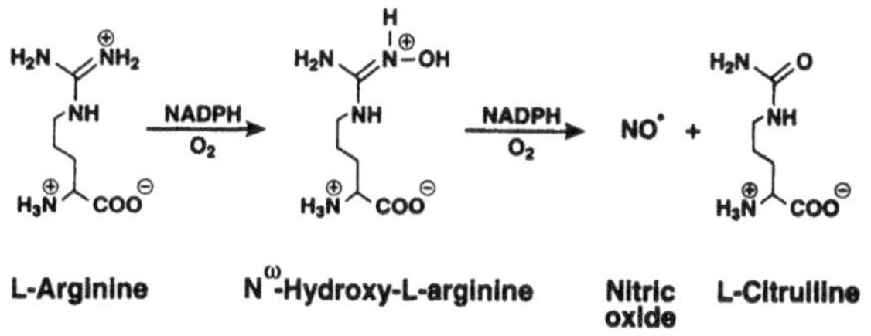

Figure 1. The reaction scheme for the formation of citrulline and NO• from L-arginine.

NO_2^- and NO_3^- (representing NO• production) appeared to settle this controversy. However, recent publications from Mayer's (18) and Kaufman's (19) laboratories have raised the issue again of a redox function for BH_4 in the nitric oxide synthase-mediated reaction sequence. While our recent studies do not address this issue, the employment of microdissection by molecular cloning techniques affords the unique opportunity of addressing these issues by studying the individual domains of nNOS with respect to their individual catalytic, binding, and biophysical properties.

The model which has been chosen for these experiments is the *Bacillus megaterium* BM-3 cytochrome P450 which is a cytochrome P450 that mediates the ω-1, ω-2, and ω-3 hydroxylation of medium and long chain fatty acids (20-22). Following the initial studies of Narhi and Fulco (20), who first showed that trypsin cleavage produced two domains from this FAD-, FMN-, and heme-binding protein, Li, *et al.* (21) and Boddupalli, *et al.* (22) demonstrated that these domains could be prepared by heterologous expression in *E. coli.* Boddupalli, *et al.* (22) and Gonvindaraj, *et al.* (23) demonstrated that the heme domain catalyzed fatty acid hydroxylation supported by the isolated flavin domain and by an artificial reductant (FMN + NADPH), respectively. Accordingly, our laboratory decided to take this approach to the study of neuronal NOS, which contains not only FAD, FMN, and heme but also binding sites for tetrahydrobiopterin and calmodulin, both of which are required for activity.

General Methodology

The expression of the heme domain in *E. coli* was achieved (12) utilizing the expression vector, pCW_{ori+} (24), which had proved useful to our laboratory and others in the expression of several members of the cytochrome P450 family (25, 26). Polymerase chain reaction amplification of nucleotides 349-2490, including NdeI/Hind III restriction sites, produced ample DNA which was recovered by band excision, restricted with NdeI/Hind III, and ligated into similarly restricted pCW_{ori+} . *E. coli* JM109 cells were then transformed with the resultant plasmid construct and plated on LB agar containing ampicillin. After preliminary screening by restriction mapping, positive clones were identified on SDS-PAGE electrophoresis by immunoblot analysis using rabbit anti-rat recombinant neuronal NOS immunoglobulin G. Purification of the expressed non-mutated protein from solubilized cell paste was performed as described (12). Site-directed mutagenesis of the cysteine residue, suggested by McMillan, *et al.*, (3) to act as the cysteinyl thiolate ligand, was performed utilizing a unique 304-bp restriction fragment flanked by NarI and Bst 1107I restriction sites. Coincident amplification of this fragment and incorporation of a mutagenic phosphorylated oligonucleotide were performed using Taq DNA polymerase and Taq ligase in thermal cycling reactions (12). The PCR products were excised, restricted with Nar I/Bst 1107I and ligated into the plasmid containing the heme domain of NOS (pNOS). The clones were screened by searching for the NarI/Bst 1107I insert and a novel Afl III site at nucleotide 1591 introduced by the $C_{415}H$ mutation. Positive clones of the mutated heme domain were subsequently amplified by PCR using a primer designed to introduce a histidine run of four residues into the N-terminus. This permitted the purification of the isolated mutagenized domain by Ni-chelate chromatography (12). All DNA sequences were confirmed by automated dideoxy sequencing using an Applied Biosystems 373A DNA sequencer and fluorescent-labeled dideoxynucleotides.

Since this vector was not suitable for the expression of the flavin-binding domain, a Clonetech plasmid vector, pPROK-1, was employed. The constructs chosen for expression of the flavoprotein domain contained EcoR1 inserts corresponding to nucleotides 2491-4635 and 2593-4635 for the expression of the C1 fragment (containing the calmodulin-binding sequence; residues 715-1429) and the C2 fragment (excluding the calmodulin-binding sequence; residues 749-1429), respectively. Expression levels of the reductase domain, estimated from cyanide-insensitive NADPH-cytochrome c reduction, were 12-25 nmol/liter (1-2 mg/liter). Purification of the flavoprotein domain, which exhibits many properties similar to NADPH-cytochrome P450 reductase, was performed on 2',5'-ADP-Sepharose 4B, an affinity medium first introduced for purification of the reductase by this laboratory (28). Further purification of the calmodulin-binding sequence-containing C1 fragment was effected by chromatography on calmodulin:Sepharose 4B while the C2 fragment was further purified by DEAE-Sepharose CL6B and Sephacryl 200 chromatography, since it did not contain the calmodulin binding sequence (12).

Protein expression was performed in cultures grown in Fernbach flasks at 30° C for the heme protein and 37° C for the flavoprotein and growth was monitored by absorbance at 600 nm. The addition of IPTG turned on the synthesis of protein, and heme and flavin precursors were added to the medium during the induction period with harvesting carried out 17-18 hours and 36 hours post-induction for the heme- and flavin-containing domains, respectively. Details of protein purification are as described previously (12).

Expression of the tetrahydrobiopterin-binding domain has been successfully achieved using the approach of fusing and expressing the cDNAs for glutathione-S-transferase and selected residues of neuronal NOS in *E.* coli, and purifying the resultant fusion protein by glutathione-bound Sepharose. This is particularly important when there are no noteworthy structural attributes to utilize for protein purification. An insert including the nucleotides 2020-2503, which encode residues 558-721 of neuronal NOS, was engineered with BamH1 and XhoI restriction sites for subcloning into the expression vector, pGEX-4T-1 (27; Pharmacia). The resulting expressed protein included glutathione-S-transferase and the putative dihydrofolate reductase domain of NOS which would encompass the pterin binding site.

Enzyme assays were performed as described previously (3; 12). Spectral characteristics were determined as described by McMillan and Masters (12, 29) upon the addition of substrate and various analogs.

Experimental Results and Discussion

Heterologous expression of the heme protein was achieved with residues 1-714 of neuronal NOS being induced in *E. coli* and exhibiting a molecular weight of approximately 80 kDa on SDS-PAGE gels. However, purification of these expressed domains proved laborious and conversion of the ~445 nm ferrous-CO species to a denatured 420 nm species resulted. The design of a construct which incorporated an N-terminal arginine and four histidine residues (His_4 heme protein) made it possible to purify the protein rapidly by Ni-NTA agarose chromatography and yields of ~7.5 nmol/liter (0.6 mg/liter) were obtained, permitting the preparation of sufficient quantities of the heme domain for optical and EPR spectroscopic and other biophysical studies. The ferrous-CO difference spectrum of the partially purified expressed heme protein (without the histidine tract) showed an absorbance maximum at ~445 nm, as with the intact neuronal NOS, and the addition of L-arginine produced a low- to high-spin shift indicated by a "Type I" optical difference spectrum (12, 29). The wild type His_4 hemeproteins, eluted from the Ni-NTA column with 100 mM imidazole, on the other hand, exhibited low spin characteristics with Soret, β, and α transition bands at 428, 544, and 575 nm (12). With the mutant heme protein ($C_{415}H$), absorbance maxima were seen at 412, 530, and ~530 (shoulder) nm, suggesting imidazole ligation at both the proximal and distal axial sites. The latter spectral signature is seen with the bis-imidazole methemoglobin complex (30). Also seen with mutant and wild type species were long wavelength absorbance bands centering around 600 nm, indicative of residual charge-transfer species. In addition, the spectral signature of the cysteine thiolate ligand at 445 nm in the reduced, CO difference spectrum is obliterated in the $C_{415}H$ mutant,which exhibits a low, broad absorbance in the region of 420 nm. Examination of other spectral properties of the wild type heme protein, such as the formation of ferric cyanide complexes, were consistent with the preservation of an intact, native heme pocket in the isolated heme domain.

In other experiments in which the binding of L-arginine to the heme domain was tested in the presence and absence of BH_4, it was shown that BH_4 enhanced the binding of L-arginine up to 3-fold, approaching the changes in absorbance observed in the ferrous-CO difference spectra. Indeed, BH_4 alone elicited low- to high-spin transitions (12). In addition, recent experiments show that the isolated heme domain also binds BH_4. These data strongly suggested that the localization of the pterin-binding site is within the N-terminus, *i.e.*, the heme domain, and that interaction is occurring between the pterin cofactor and the heme prosthetic group.

More recent experiments have been performed on the glutathione-S-transferase fusion protein construct of residues 558-721 of neuronal NOS which encompass the putative dihydrofolate reductase domain. The construct permitted the purification of this domain both as a fusion protein and, after thrombin treatment, of the isolated domain. N^{ω}-nitro-L-arginine binding studies indicate that even this subdomain of the heme domain is capable of participating in substrate binding and further supports our contention that the substrate binding site of neuronal NOS is complex and involves the participation of both the heme- and pterin-binding regions of the protein.

Expression of the two constructs of the flavoprotein domain permitted the determination of the catalytic capabilities of each and the potential of regulation by Ca^{+2}/calmodulin of the flavin-binding module. The C1 flavoprotein domain, effectively purified by calmodulin affinity chromatography, catalyzed the NADPH-mediated reduction of cytochrome c at a rate of 3-4 μmol/min/mg, which is 2-4-fold the rate catalyzed by the intact neuronal NOS (14). Upon the addition of Ca^{+2}/calmodulin, the rate of cytochrome c reduction was stimulated 2-fold. The addition of EGTA or the calmodulin antagonist, trifluoperazine, reduced the activity to the unstimulated rate. The addition of EGTA or trifluoperazine to the C2 flavoprotein had no effect on cytochrome c reductase activity, indicating that the activation of C1 was due to the presence of the calmodulin-binding site. This result raises the possibility that the regulation of neuronal NOS can reside within individual prosthetic group domains as well as between domains (31) and in monomer-dimer interactions. In addition, these data strongly support the identification of the calmodulin binding site as residues 725-745, as originally suggested by Bredt, *et al.* (8).

CONCLUSIONS

The model we have constructed from the results derived from our studies to date is shown below.

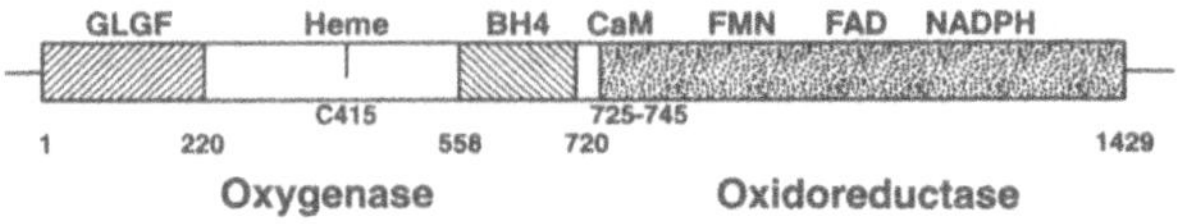

Figure 2. The modular structure of neuronal nitric oxide synthase.

The experimental approach of microdissection using molecular cloning techniques has proved particularly useful in the determination of structure-function relationships in this complex heme-, flavin-, pterin-, and calmodulin-binding protein. The resulting domains appear to retain their structural integrity, which permits characterization by a variety of biophysical techniques, including X-ray crystallography, of these somewhat smaller structures.

REFERENCES

1. Bredt, D.S., and Snyder, S.H., 1990, Isolation of nitric oxide synthetase, a calmodulin-requiring enzyme, *Proc. Natl. Acad. Sci., USA* 87: 682-685.
2. Kwon, N.S., Nathan, C.F., Gilker, C., Griffith, O.W., Matthews, D.E., and Stuehr, D.J., 1990, L-Citrulline production from L-arginine by macrophage nitric oxide synthase. The ureido oxygen derives from dioxygen, *J. Biol. Chem.* 265: 13442-13445.
3. McMillan, K., Bredt, D.S., Hirsch, D.J., Snyder, S.H., Clark, J.E., and Masters, B.S.S., 1992, Cloned, expressed rat cerebellar nitric oxide synthase contains stoichiometric amounts of heme, which binds carbon monoxide, *Proc. Natl. Acad. Sci., USA* 89: 11141-11145.
4. Stuehr, D.J., and Ikeda-Saito, M., 1992, Spectral characterization of brain and macrophage nitric oxide synthases, *J. Biol. Chem.* 267: 20547-20550.
5. Klatt, P., Schmidt, K., and Mayer, B., 1992, Brain nitric oxide synthase is a heme protein, *Biochem. J.* 288: 15-17.
6. White, K.A., and Marletta, M.A., 1992, Nitric oxide synthase is a cytochrome P-450 type hemoprotein, *Biochemistry* 31: 6627-6631.
7. Stuehr, D.J., Kwon, N.S., Nathan, C.F., Griffith, O.W., Feldman, P.L., and Wiseman, J., 1991, N^{ω}-Hydroxy-L-arginine is an intermediate in the biosynthesis of nitric oxide from L-arginine, *J. Biol. Chem.* 266: 6259-6263.
8. Bredt, D.S., Ferris, C.D., and Snyder, S.H., 1992, Nitric oxide synthase regulatory sites. Phosphorylation by cyclic AMP-dependent protein kinase, protein kinase C, and calcium/calmodulin protein kinase; identification of flavin and calmodulin binding sites, *J. Biol. Chem.* 267: 10976-10981.
9. Mayer, B., John, M., Heinzel, B., Werner, E.R., Wachter, H., Schultz, G., and Bohme, E., 1991, Brain nitric oxide synthase is a biopterin- and flavin-containing multi-functional oxido-reductase, *FEBS Lett.* 288: 187-191.
10. Bredt, D.S., Hwang, P.M., Glatt, C.E., Lowenstein, C., Reed, R.R., and Snyder, S.H., 1991, Cloned and expressed nitric oxide synthase structurally resembles cytochrome P-450 reductase, *Nature* 351: 714-718.
11. Richards, M.K., and Marletta, M.A., 1994, Characterization of neuronal nitric oxide synthase and a C415H mutant, purified from a baculovirus overexpression system, *Biochemistry* 33: 14723-14732.
12. McMillan, K., and Masters, B.S.S., 1995, Procaryotic expression of the heme- and flavin-binding domains of rat neuronal nitric oxide synthase as distinct polypeptides: Identification of the heme-binding proximal thiolate ligand as $cysteine_{415}$, *Biochemistry, in press.*
13. Chen, P-F., Tsai, A-L., and Wu, K.K., 1994, Cysteine 184 of endothelial nitric oxide synthase is involved in heme coordination and catalytic activity, *J. Biol. Chem.* 269: 25062-25066.
14. Sheta, E.A., McMillan, K., and Masters, B.S.S., 1994, Evidence for a bidomain structure of constitutive cerebellar nitric oxide synthase, *J. Biol. Chem.* 269: 15147-15153.
15. Vorherr, T., Knopfel, L., Hofmann, F., Mollner, S., Pfeuffer, T., and Carafoli, E., 1993, The calmodulin binding domain of nitric oxide synthase and adenylyl cyclase, *Biochemistry* 32: 6081-6088.
16. Giovanelli, J., Campos, K.L., and Kaufman, S., 1991, Tetrahydrobiopterin, a cofactor for rat cerebellar nitric oxide synthase, does not function as a reactant in the oxygenation of arginine, *Proc. Natl. Acad. Sci. USA* 88: 7091-7095.

17. Baek, K.J., Thiel, B.A., Lucas, S., and Stuehr, D.J., 1993, Macrophage nitric oxide synthase subunits. Purification, characterization, and role of prosthetic groups and substrate in regulating their association into a dimeric enzyme, *J. Biol. Chem.* 268: 21120-21129.
18. Mayer, B., Klatt, P., Werner, E.R., and Schmidt, K., 1995, Kinetics and mechanism of tetrahydrobiopterin-induced oxidation of nitric oxide, *J. Biol. Chem.* 270: 655-659.
19. Campos, K.L., Giovanelli, J., and Kaufman, S., 1995, Characteristics of the nitric oxide synthase-catalyzed conversion of arginine to N-hydroxyarginine, the first oxygenation step in the enzymic synthesis of nitric oxide, *J. Biol. Chem.* 270: 1721-1728.
20. Narhi, L.O., and Fulco, A.J., 1987, Identification and characterization of two functional domains in cytochrome P-450_{BM-3}, a catalytically self-sufficient monooxygenase induced by barbiturates in *Bacillus megaterium*, *J. Biol. Chem.* 262: 6683-6690.
21. Li, H., Darwish, K., and Poulos, T.L., 1991, Characterization of recombinant *Bacillus* megaterium cytochrome P-450_{BM-3} and its two functional domains, *J. Biol. Chem.* 266: 11909-11914.
22. Boddupalli, S.S., Oster, T., Estabrook, R.W., and Peterson, J.A., 1992, Reconstitution of the fatty acid hydroxylation function of cytochrome P-450_{BM-3} utilizing its individual recombinant hemo- and flavoprotein domains, *J. Biol. Chem.* 267: 10375-10380.
23. Gonvindaraj, S., Li, H., and Poulos, T.L., 1994, Flavin supported fatty acid oxidation by the heme domain of *Bacillus megaterium* cytochrome P450BM-3, *Biochem. Biophys. Res. Commun.* 203: 1745-1749.
24. Gegner, J.A., and Dahlquist, F.W., 1991, Signal transduction in bacteria: CheW forms a reversible complex with the protein kinase CheA, *Proc. Natl. Acad. Sci. USA* 88: 750-754.
25. Barnes, H.J., Arlotto, M.P., and Waterman, M.R., 1991, Expression and enzymatic activity of recombinant cytochrome P450 17α-hydroxylase in *Escherichia coli*, *Proc. Natl. Acad. Sci., USA*, 88: 5597-5601.
26. Nishimoto, M., Clark, J.E., and Masters, B.S.S., 1993, Cytochrome P450 4A4: Expression in *Escherichia coli*, Purification, and Characterization of Catalytic Properties, *Biochemistry* 32: 8863-8870.
27. Smith, D.B., and Johnson, K.S., 1988, Single-step purification of polypeptides expressed in *Escherichia coli* as fusions with glutathione-S-transferase, *Gene* 67: 31-40.
28. Yasukochi, Y., and Masters, B.S.S., 1976, Some properties of a detergent-solubilized NADPH-cytochrome c (cytochrome P-450) reductase purified by biospecific affinity chromatography, *J. Biol. Chem.* 251: 5337-5344.
29. McMillan, K., and Masters, B.S.S., 1993, Optical difference spectrophotometry as a probe of rat brain nitric oxide synthase heme-substrate interaction, *Biochemistry* 32: 9875-9880.
30. Brill, A.S., and Williams, R.J.P., 1961, The absorption spectra, magnetic moments and the binding of iron in some haemproteins, *Biochem. J.* 78: 246-253.
31. Abu-Soud, H.M., Yoho, L.L., and Stuehr, D.J., 1994, Calmodulin controls neuronal nitric-oxide synthase by a dual mechanism. Activation of intra- and interdomain electron transfer, *J. Biol. Chem.* 269: 32047-32050.

MOLECULAR MECHANISMS REGULATING NITRIC OXIDE BIOSYNTHESIS

Role of Xenobiotics in Epithelial Inflammation

Diane E. Heck

Department of Pharmacology and Toxicology
Rutgers University, Piscataway New Jersey 08855

The squamous epithelium, which comprises the outermost barrier to the environment, functions as the primary initial site of xenobiotic insult. Following xenobiotic insult, the epithelium responds by producing a complex inflammatory micro-environment. During inflammation, hyperplasia, edema as well as leukocyte infiltration are typically observed (Adams, 1993). Stromal cells, keratinocytes in the case of the skin, resident immune-type dendritic cells, and infiltrating leukocytes release an array of cytokines and inflammatory mediators that regulate the inflammatory process. Initially, these mediators induce increased vascular permeability, which in turn facilitates an influx of serum-derived factors including complement, hormones, leukotrienes and cytokines (Camp, 1990, Gallo, 1989). These mediators then enhance recruitment of additional leukocytes to the site of xenobiotic insult as well as stimulating the proliferation of resident epithelial cells and fibroblasts. This later effect is required for the resolution of inflammation and successful wound repair (Friedman, 1993, Knighton, 1991, Kupper, 1990).

Within the inflammatory micro environment activated leukocytes produce a variety of bioactive mediators including cytokines, hydrolytic enzymes, collagen and lipid degradation products (Adams, 1993). For some time it has been recognized that inflammatory leukocytes also produce short-lived reactive intermediate such as active oxygen species including superoxide anion, hydrogen peroxide and hydroxyl anion. These reactive oxygen intermediates (ROI) are important in non-specific host defense (Flohe, 1985). Recently, immunologically activated leukocytes have been found to produce reactive nitrogen intermediates (Nathan, 1992, Feng, 1990). Nitric oxide, the best recognized of these species also plays an important role in host defense (Marletta, 1993). However peroxynitrite and other transient nitrogen oxides may be important in inflammation (Stamler, 1994). One of the major functions of nitric oxide and its oxidation products during inflammatory reactions is microbial killing. This activity is due to nitric-oxide-mediated inactivation of critical metabolic enzymes necessary for the survival of many invading microorganisms. However, host tissues also respond to nitric oxide formed during an inflammatory response. As a consequence, this may alter physiologic functions within the inflamed tissue. For example, nitric oxide causes relaxation of smooth muscle cells lining the blood vessels, resulting in

Biological Reactive Intermediates V
Edited by R. Snyder et al., Plenum Press, New York, 1996

vasodilation. This presumably occurs because nitric oxide binds to and activates enzymes such as guanylyl cyclase, causing an increase in cellular levels of cGMP, and/or calcium-dependent membrane potassium channels, two important components of signal transduction pathways regulating smooth muscle cell contractility (Marletta, 1992, Feng, 1990). In epithelial tissues such as the skin, subsequent alterations in blood flow may lead to erythema and edema (Adams, 1993).

Exposure of epithelial cells to some inflammatory cytokines such as tumor necrosis factor-a, interleukins 1 and 6, as well as γ-interferon, microbial products including bacterial lipopolysaccharides, and chemical agents such as phorbol esters, induce these cells to become immunologically activated (Lugar, 1991). Activated epithelial cells perform many functions of activated leukocytes including expression of major histocompatibility antigens and leukocyte adhesion molecules, and production of inflammatory cytokines such as interleukins 1 and 6 (Barker, 1991, McKay, 1991). Activated epithelial cells also produce reactive oxygen and nitrogen intermediates (Heck, 1992).

Characterization of cellular effects induced by two intriguingly dissimilar cytokines produced following xenobiotic exposure may provide some insight into the complex interplay of mediators within inflamed epidermal tissues. Both of these cytokines, epidermal growth factor (EGF), a growth-stimulatory cytokine important in stimulating cellular proliferation necessary for tissue repopulation during healing, and γ-interferon (γ-IFN), a cytokine critical for the control of microbial infection which is associated with inhibition of cell growth, are found in increased concentrations following a xenobiotic insult. As appropriate expression of, and cellular responses to, these mediators is required for successful resolution of inflammation and tissue repair, alterations of these effects by xenobiotics may adversely affect the healing process, and so compound the initial insult.

Epidermal growth factor, and its structural analog, transforming growth factor-α are found in high concentrations following injury to epithelial tissues (Adams, 1993, Mc Kay, 1991, McKenzie, 1990). These mediators, which promote repair of epithelial tissues through stimulation of cellular proliferation, have also been found to regulate the production of free radicals by keratinocytes (Heck, 1992). Many cell types respond to γ-interferon by exhibiting anti-viral activity. This is a complex function involving a host of cellular responses which is, in part, mediated by production of nitric oxide and hydrogen peroxide, as well as expression of major histocompatibility antigens (Gewert, 1992). We have found that EGF regulates cellular responses to γ-IFN (Heck, 1992). For example, both basal levels and cytokine-stimulated production of hydrogen peroxide are inhibited by EGF in cultured keratinocytes. We have also shown that EGF is a potent inhibitor of γ-IFN-induced nitric oxide production. In these studies treatment of keratinocytes with increasing concentrations of EGF in the presence of γ-IFN resulted in a dose-related decrease in nitric oxide production. This effect, which was maximal within 24 hours, remained constant throughout the experiment (Heck, 1992). It is recognized that EGF, a peptide growth factor that stimulates tyrosine as well as serine and threonine phosphorylation of its receptor, inhibits nitric oxide production by blocking induction of nitric oxide synthase in pulmonary epithelial cells (Asano, 1994). However, the molecular mechanisms mediating this effect remain uncharacterized. In further studies, using immunohistochemistry in conjunction with flow cytometry we have found that EGF also inhibits γ-IFN-induced expression of major histocompatibility antigens (Fig.1).

The squamous epithelium , exemplified by the skin, is a plastic, continuously renewing tissue, with ongoing proliferation of basal cells, differentiation of maturing cells and finally apoptotic cell death (Holbrook, 1993). Within the complex micro environment produced in response to xenobiotic insult, proliferating as well as differentiating cells are exposed to both growth-stimulatory and growth-inhibitory cytokines (Adams, 1993, Friedman, 1993, Camp, 1990). Primary cultures of keratinocytes provide a useful model for

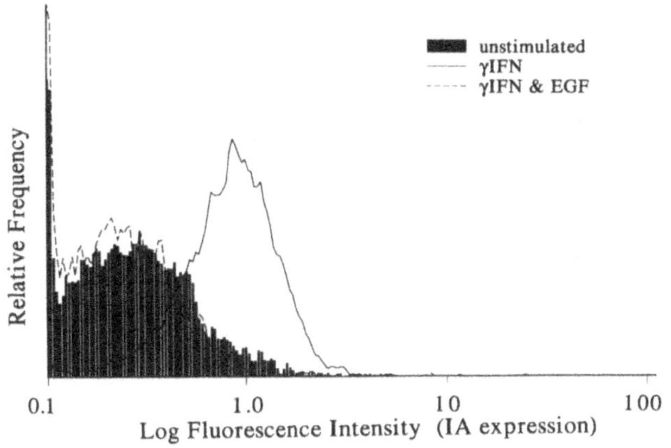

Figure 1. Inhibition of major histocompatability antigen expression (la expression) by EGF. Unstimulated PAM 212 murine keratinocytes (shaded histogram) and interferon-γ stimulated cells (100 μ/mL 48 hr, open histograms) were incubated with a FITC labeled monoclonal anti-la antibody (Sigma). Antigen expression was quantified using flow cytometry. Interferon treatment stimulated expression of la antigen (solid line), an effect which was inhibited by the presence of EGF (100 ng/mL, broken line).

investigating differential cellular responses to inflammatory mediators. These cells proliferate in a low calcium-containing medium, however, the addition of calcium to the cell culture medium causes the cells to differentiate. Differentiating cells take on the characteristics of supra basal cells of the squamous epithelium. In the absence of calcium they are refractile in appearance, and form a monolayer in culture. Exposure to high concentrations causes the cells to flatten and stratify in a manner similar to the regeneration of wounded skin. Treatment of undifferentiated keratinocytes with γ-IFN stimulates the production of low levels of nitric oxide, whereas interferon-treatment of calcium differentiated cells results in the production of copious amounts of nitric oxide (Heck, 1993, Heck, 1994). Similarly, interferon-stimulated hydrogen peroxide production is enhanced following differentiation. In figure 2 hydrogen peroxide production was measured flow cytometrically as the oxidation of dichlorofluorescin-diacetate (Heck, 1992). In both proliferating and calcium differentiated cells enhanced hydrogen peroxide production was observed following γ-IFN stimulation. Interestingly, basal levels of hydrogen peroxide are also increased following calcium-induced differentiation.

Growth-stimulatory polypeptide hormones of the epidermal growth factor family induce cellular effects through interaction with specific membrane-bound receptors. Ligand binding stimulates intrinsic tyrosine kinase activity of these receptors. Activation of these kinases initiates a cascade of intracellular phosphorylation events. Phosphorylation of

Figure 2. Differentiation mediated regulation of hydrogen peroxide production by mouse keratinocytes. Relative hydrogen peroxide concentrations in the cells were quantified by flow cytometry using the hydroperoxide sensitive dye DCFH-DA (Heck, 1992). Proliferating (unshaded bars) and calcium differentiated (1.7 mM Ca^{+2}, 24 hr, shaded bars) cells were stimulated with interferon-γ and incubated with DCFH-DA (5 μM) during the final 30 minutes of each treatment. Fluorescence intensity is directly proportional to hydrogen peroxide production.

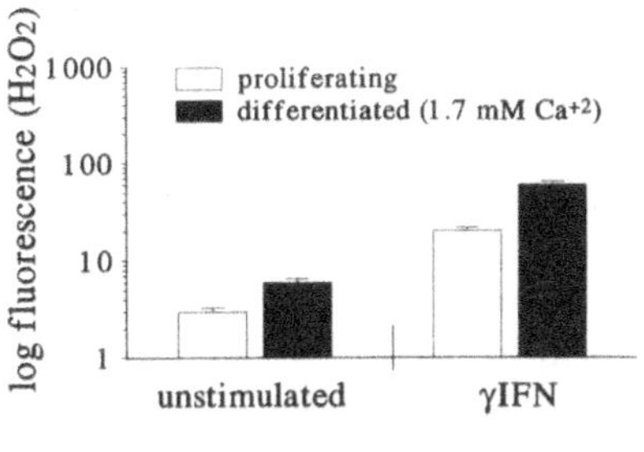

cellular proteins is required for initiation of biological responses by this family of growth factors (Hunter, 1995). Many cytokine receptors, such as those for the interferon family of polypeptide hormones, do not have intrinsic tyrosine kinase activity. These multimeric receptors act through assembly of membrane bound subunits and accessory proteins. Once assembled, active receptor complexes then recruit cytosolic kinases (Pestka, 1992, Stuber, 1992). For example, association of α, β, or γ-interferon with their specific cellular receptors initiates the formation of active multimeric receptor complexes. Following ligand-induced multimerization, cytosolic tyrosine kinases associate with the active membrane-bound receptor complexes. This process, in turn, results in phosphorylation of specific latent cytosolic transcription factors, referred to as signal transduction and activator of transcription (STAT) proteins. Phosphorylated STAT proteins translocate to the nucleus where they form active transcriptional-regulatory complexes (Sadowski, 1993, Shuai, 1993). Interestingly, signals initiated by epidermal growth factor also induce activation of STAT proteins (Fu, 1993, Ruff-Jamison, 1993). In this case, the mechanism(s) by which STAT proteins are activated is less well defined.

Recent studies of this highly conserved family of transcription factors have addressed the perplexing problem of specificity, that is, how can diverse receptors apparently use common transcription factors to regulate expression of receptor specific target genes. Molecular cloning of STAT proteins has provided some insight into this dilemma. These studies have demonstrated that the STAT family of proteins consists of numerous highly homologous proteins. One avenue for preserving receptor-specific discrimination is through selective association of unique STAT proteins with a limited number of receptors (Hill, 1995). Selectivity is further amplified by differential affinity of cytosolic kinases for various receptor-STAT protein complexes (Darnell, 1994). Phosphorylation is followed by STAT protein dimerization, apparently prior to the formation of protein-DNA complexes, variant kinase affinity for receptor-STAT complexes may result in the formation of distinct hetero- or homo-dimeric moieties. Potentially, additional selectivity may be provided by spatial and temporal organization of nuclear transcription complexes. Finally, divergent sequences flanking the DNA-protein binding consensus region may also contribute to selectivity (Hill, 1995). For example, in initial studies it was reported that both EGF and γ-interferon induced activation of STAT-1α, a 91 kDa protein (Fu, 1993, Ruff-Jamison, 1993). In more recent studies, it has become apparent that while γ-IFN does activate STAT-1, EGF preferentially activates highly homologous STAT-3, a discrete 92 kDa protein (Zhong, 1994). Additionally, STAT proteins may form a variety of hetero- and homo- dimers in response to receptor activation, each multimer exhibiting different affinities for unique DNA sequences; interestingly only STAT-3 homodimers are transcriptionally active in response to EGF (Ruff-Jamison, 1994).

We and others have reported that cellular responses to γ-IFN are, in some instances, antagonized by the presence of EGF (Asano, 1994, Heck, 1992). These observations led us to hypothesize that cross-talk between receptors which use the STAT signal transduction system may mediate this effect. In studies designed to investigate this hypothesis we initially observed that treatment of differentiating keratinocytes with γ-IFN in the presence of EGF partially diminished the stimulation of p-91 induced by EGF. However, proliferating cells did not respond to EGF by activation of p-91 (Heck,1994). It is likely that expression of unique STAT proteins during differentiation may favor low affinity receptor-protein interactions. Potentially, EGF may induce activation of p-91 during differentiation, when little or no p-92 is expressed, due to a low affinity interaction between this protein and the EGF receptor. Proliferating cells, which express p-92, respond to EGF treatment by selectively activating p-92, a much higher affinity substrate for the EGF receptor. The recent availability of selective antibodies for these two proteins may facilitate investigations of this later possibility (Darnell, 1994).

Xenobiotics may appear to act silently, producing an array diverse effects through intervention in this signal transduction system. Specific toxins may directly activate, or conversely inhibit, activation of STAT proteins (Heck, 1994). Alternatively, some toxins act upstream of this signal transduction pathway by regulating expression of unique STAT proteins, thus affecting the ability of a signal to be transduced by altering the components available (Heck, 1995, Heck, 1994). Altering cellular responses to endogenous growth factors and cytokines provides a previously unrecognized mechanism by which xenobiotics disrupt critical metabolic processes.

REFERENCES

Adams, R.A., Disorders due to Drugs and Chemical Agents, in: Fitzpatrick, T.B., Eisen, A.Z., Wolff, K., Freedberg, I.M., Austen, K.F., eds. *Dermatology in General Medicine*, pp. 1767-1783. New York, McGraw-Hill Inc., 1993.

Asano, K., Chee, C. B. E., Gaston, B., Lilly, C. M., Gerard, C., Drazen, J., M., Stamler, J. S., Constitutive and Inducible Nitric Oxide Synthase Gene Expression, Regulation, and Activity in Human Lung Epithelial Cells., Proc. Natl. Acad. Sci. USA, 91, 10089-10093, 1994.

Barker, J.N.W.N., Mitra, R.S., Griffiths, C.E.M., Dixit, V.M., Nickoloff, B.J. Keratinocytes as initiators of inflammation. The Lancet, 337: 211-214, 1991.

Clark, R.A., The commonality of cutneous wound repair and lung injury, Chest 99:575-605, 1991.

Camp, R.D.R., Fincham, N.J., Ross, J.S., Bacon, K.B., Gearing, A.J.H., Leukocyte chemoattractant cytokines of the epidermis. J. Invest. Dermatol., 95: 108S-110S, 1990.

Darnell, J. E., Kerr, I. M., Stark, G. R., Jak-STAT Pathways and transcriptional activation in response to IFNs and other extracellular signaling proteins. Science, 264, 1415-1421, 1994.

Feng, Q., Hedner, T., Endothelium derived relaxing factor (EDRF) and nitric oxide (NO). I. physiology, pharmacology and pathophysiological implications. Clin. Physiol., 10: 407-426, 1990.

Flohe, L., Beckman, R., Giertz, H., Loschen, G., Oxygen centered free radicals as mediators of inflammation. in: Sies, H. ed., *Oxidative Stress*, pp.403-428, Orlando Fl., Academic Press, 1985.

Fu, X-Y., Zhang, J-J., Transcription factor p91 interacts with the epidermal growth factor receptor and mediates activation of the c-FOS gene promoter. Cell, 74, 1135-1145, 1993.

Friedmann, P.S., Strickland, I., Memomoa, A.A., Johnson, P.M., Early time course of recruitment of immune surveillance in human skin after chemical provocation. Clin.Exp. Immunol. 91:351-356, 1993.

Gallo, R.L., Staszewski, R., Granstein, R.D., Physiology and pathology of skin photoimmunology, in: Bos, J.D. ed., *Skin and the Immune System*, pp. 381-402. Boca Raton, Fl., CRC Press, 1989.

Gewert, D., Finter, N. B., Antiviral Effects of the interferons: Studies in animals and at the cellular level, in: *Interferon, Principles and Medical Applications*, pp. 129-138, Galveston TX, The University of Galveston Medical Branch at Galveston Department of Microbiology, 1992.

Heck, DE and JD Laskin, Pertussis toxin inhibits induction of p-91 STAT by γ- interferon in A549 pulmonary epithelial cells. Proceedings of the American Association for Cancer Research, 1995 (in Press).

Heck, DE, DL Laskin and JD Laskin, γ-Interferon and inhaled irritants induce production of vasoregulatory mediators by rat alveolar Type II epithelial cells. Am. J. Respir. Crit. Care Med., 149 (2) A551, 1994.

Heck DE, JD Laskin, Interferon dependent regulation of cytoplasmic transcription factors in keratinocytes during wound healing. J. Invest. Dermatol., 102, 528, 1994.

Heck DE, DL Laskin, JD Laskin, J Finkelstein, T Liberati and G Oberdorster, Effects of silicon dioxide on the production of vasoregulatory mediators by rat alveolar type II epithelial cells. The Toxicologist, 14, 200. 1994.

Heck DE, DL Laskin, CR Gardner and JD Laskin, Role of nitric oxide in chemical induced skin injury. The Toxicologist, 13, 183, 1993.

Heck, D.E., and J.D. Laskin, Nitric oxide production by mouse and human keratinocytes is regulated by insulin-like growth factor-1. J. Invest. Dermatol., 100, 511, 1993.

Heck, D. E., Laskin, D.L., Gardner, C.R., Laskin, J.D., Epidermal growth factor supresses nitric oxide and hydrogen peroxide production by keratinocytes. J. Biol. Chem., 267: 21277-21280, 1992.

Hill, C.S., Treisman, C.S., Transcriptional regulation by extracellular signals: Mechanisms and specificity. Cell, 80, 199-212, 1995.

Holbrook, K. A., Wolff, K., The structure and development of skin. in: Fitzpatrick, T.B., Eisen, A.Z., Wolff, K., Freedberg, I.M., Austen, K.F., eds. *Dermatology in General Medicine*, pp. 1767-1783. New York, McGraw-Hill Inc., 1993.

Hunter, T., Protein kinases and phosphatases: The yin and yang of protein phosphorylation and signaling. Cell, 80, 225-236, 1995.

Hunter, T., Cytokine connections. Nature, 366, 114-116, 1993.

Katz, A.M., Rosenthal, D., Sauder, D.N., Cell adhesion molecules. Structure, function and implication in a variety of cutaneous and other pathologic conditions. Int. J. Dermatol., 30,153-160, 1991.

Kingsworth, A.N., Slavin, J., Peptide growth factors and wound healing. Br. J. Surg., 78:1286-1290, 1991.

Knighton, D.R. and Fiegel, V.D., Regulation of cutaneous wound healing by growth factors and the microenvironment, Invest. Radiol. 26:604-611, 1991.

Kupper, T.S., Immune and inflammatory processes in cutaneous tissues. Mechanisms and speculations, J. Clin. Invest. 86:1783-1789, 1990.

Luger, T.A. and Schwarz, T., Evidence for an epidermal cytokine network, J. Invest. Dermatol., 95:100s-104s.

Mariano, T. M., Donnely, R. J., Soh, J., Pestka, S. Structure and function of the Type I interferon receptor, in: *Interferon, Principles and Medical Applications*, pp. 129-138, Galveston TX, The University of Galveston Medical Branch at Galveston Department of Microbiology, 1992.

Marletta, M.A., Mammalian synthesis of nitrite, nitrite, nitric oxide and N-nitrosylating agents. Chem. Res. Toxicol., 1: 249-257, 1993.

McKay, I.A. and Leigh, I.M., Epidermal cytokines and their roles in cutaneous wound healing, Brit. J. Dermatol., 124:513-518, 1991.

McKenzie, R.C. and Sauder, D.N., The role of keratinocyte cytokines in inflammation and immunity, J. Invest. Dermatol., 95:105s-107s, 1990.

Nathan, C., Nitric oxide as a secretory product of mammalian cells. FASEB J., 6: 3051-3064, 1992.

Ruff-Jamison, S., Zhong, Z., Wen, Z., Chen, K., Darnell, J. E., Cohen, S. Epidermal Growth Factor and Lipopolysaccharide Activate STAT 3 Transcription Factor in Mouse Liver, J. Biol. Chem., 269, 21933-21935, 1994.

Ruff-Jamison, S., Chen, K., Cohen, S., Induction by EGF and interferon-γ of tyrosine phosphorylated DNA binding proteins in mouse liver nuclei. Science, 261, 1733-1736, 1993.

Sadowski, H. B., Shuai, K., Darnell, J. E., Gilman, M. Z., A common nuclear signal transduction pathway activated by growth factor and cytokine receptors. Science, 261, 1739-1744, 1993.

Shuai, K., Stark, G. R., Kerr, I. M., Darnell, J. E. A single phosphotyrosine residue of STAT91 required for gene activation by interferon-γ. Science, 261, 1744-1746, 1993.

Stamler, J. S., Redox signaling: Nitrosylation and related target interactions of nitric oxide. Cell, 78, 931-936, 1994.

Stuber, D., Fountoulakis, M., Garotta, G., IFN-g Receptor: Protein structure and function in: *Interferon, Principles and Medical Applications*, pp. 129-138, Galveston TX, The University of Galveston Medical Branch at Galveston Department of Microbiology, 1992.

Zhong, Z., Wen, Z., Darnell, J. E., STAT 3 and STAT 4: Members of the family of signal transducers and activators of transcription, Proc. Natl. Acad. Sci. USA, 91, 4806-4810, 1994.

DNA DAMAGE AND NITRIC OXIDE

Larry K. Keefer and David A. Wink

Chemistry Section
Laboratory of Comparative Carcinogenesis
National Cancer Institute
Frederick Cancer Research and Development Center
Frederick, Maryland 21702

While there is no evidence that nitric oxide (NO) reacts directly with DNA, NO can modify the biopolymer by a variety of indirect pathways. In this review, we describe a selection of the reported mechanisms by which exposure to NO might affect the integrity of DNA, with emphasis on the biological reactive intermediates (BRIs) capable of participating in such transformations.

BRIS FORMED DURING AUTOXIDATION OF NO (NO_2 AND NO_X)

Many of the pathways by which NO can damage DNA are initiated by the reaction of NO with O_2. In the gas phase, this leads to nitrogen dioxide (NO_2) (Butler & Williams, 1993) according to equation 1. NO_2 is an abundant environmental pollutant that has been reported to cause single strand breaks in cellular DNA under conditions in which NO itself does not (Görsdorf et al., 1990).

$$2\,NO + O_2 \longrightarrow 2\,NO_2 \tag{1}$$

In contrast to the gas phase reaction, autoxidation of NO in aqueous solution has recently been found not to involve free NO_2 (Wink et al., 1993a). The rate law (equation 2) is of the same form as that in the gas phase [k values of $8 \times 10^6\ M^{-2}\ s^{-1}$ and $7 \times 10^3\ M^{-2}\ s^{-1}$ for the aqueous (Ford et al., 1993) and gaseous (Olbregts, 1985) reactions at 25 °C, respectively]. Moreover, reactive intermediates capable of oxidizing as well as nitrosating other molecules are produced in both the gaseous and the aqueous reactions. However, the ratio of oxidation to nitrosation observed when all reactants are in water solution can differ dramatically from that seen when NO autoxidation occurs in the gas phase (Wink et al., 1993a). For example, when solutions of the redox-active 2,2′-azinobis(3-ethylbenzothiazoline-6-sulfonate) ion (ABTS) were exposed to bubbles of NO/O_2 in air (which according to equation 1 must be rich in NO_2), extensive conversion of ABTS to its green one-electron oxidation product occurred. This reaction was observed whether azide, a highly efficient scavenger of nitrosating agents

Biological Reactive Intermediates V
Edited by R. Snyder et al., Plenum Press, New York, 1996

(Williams, 1988), was present at $[N_3^-]/[ABTS]=2$ or not (equation 3a). However, when the experiment was repeated using preformed aqueous solutions of NO in water in place of NO/O_2 in air, formation of green $ABTS^+$ could be almost quantitatively quenched in the presence of azide (equation 3b). A similar dichotomy was observed when plasmid DNA was reacted in aerobic buffer with gaseous versus solution-phase sources of nitric oxide (Routledge et al., 1994). We conclude despite continuing controversy regarding this point that, whatever the identity (or identities) of the important BRI(s) produced during autoxidation of NO in neutral aqueous media, free NO_2 contributes little or nothing to the resulting reactivity pattern. We shall refer henceforth to the as-yet-unidentified nitrosating and oxidizing agent(s) produced when NO and O_2 react in aqueous solution as "NO_X".

$$\text{Rate of NO disappearance} = \frac{-d[NO]}{dt} = k[NO]^2[O_2] \qquad (2)$$

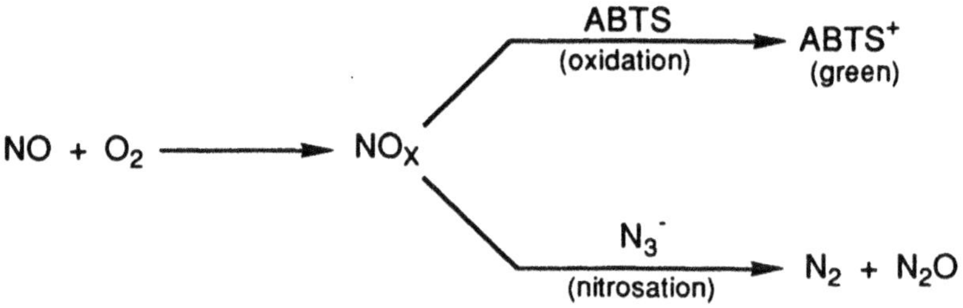

(3)

NO_X: DNA DAMAGE BY NITROSATIVE DEAMINATION OF BASES

Once formed, NO_X can influence the integrity of DNA not only via diverse indirect mechanisms, as detailed below, but also by direct attack on its bases. As mentioned above, NO_X can nitrosate (i.e., it can transfer an NO^+ moiety to) nucleophilic centers. When the nucleophile is a primary amine, deamination according to equation 4 can result. This proved to be the outcome when the primary amino groups of cytosine, guanine, and adenine were exposed to NO_X. By bubbling NO through neutral, aqueous, aerobic solutions of 2′-deoxy-5-methylcytidine, for example, significant deamination to thymidine was observed. We conclude that a nitrosative pathway was responsible for the NO-mediated deamination, as in equation 5. Indeed, NO exposure in neutral, aqueous, aerobic solution has led to deamination of all the bases, nucleosides, and nucleotides studied to date, as well as of calf thymus DNA itself (Nguyen et al., 1992; Wink et al., 1991).

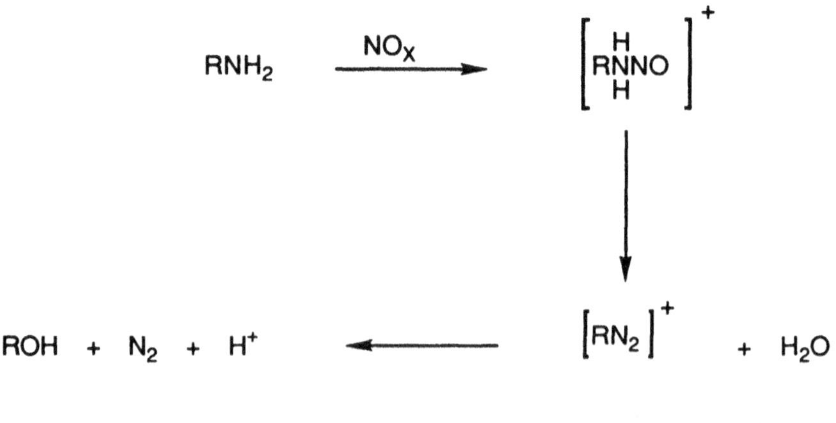

(4)

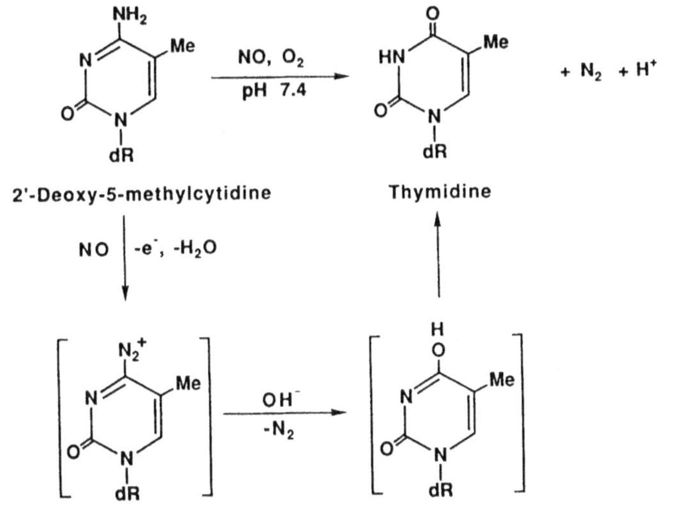

(5)

NO_X: FORMATION OF DNA-ALKYLATING *N*-NITROSO COMPOUNDS

When NO_X reacts with a secondary amine, a stable *N*-nitroso derivative is produced, as in equation 6. It is presumably by this mechanism that formation of a carcinogenic *N*-nitroso compound has been demonstrated in vivo in rats exposed to bacterial lipopolysaccharide whose immune systems were thus stimulated to synthesize greatly increased amounts of NO endogenously (Leaf et al., 1991). Many such *N*-nitroso compounds, including the 2-carbon nitrosamine carcinogen, *N*-nitrosodimethylamine (NDMA), can induce point mutations by alkylating DNA, particularly on the oxygen atom at the 6 position of its guanine residues (Saffhill et al., 1985), as outlined in equation 7. Therefore, it seems likely that NO could damage DNA indirectly via the successive nitrosation and alkylation steps shown in equations 6 and 7 as well as by direct deaminative attack on the bases as described in the previous paragraph.

$$R_2NH \xrightarrow{NO_X} R_2NNO \qquad (6)$$

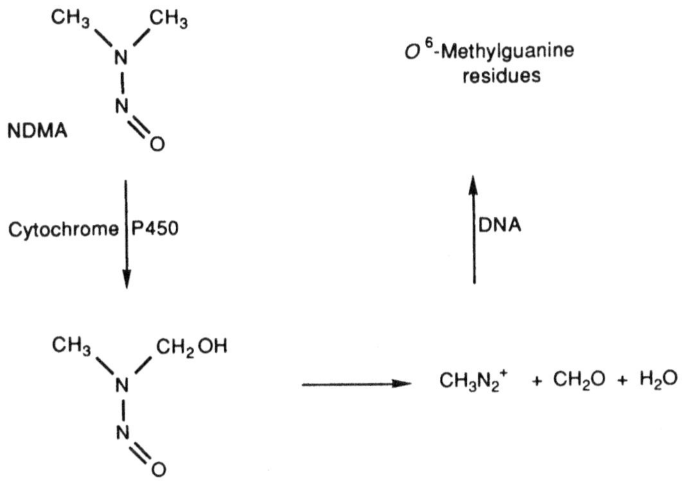

(7)

NO_X: INHIBITION OF DNA REPAIR BY *S*-NITROSATION OF PROTEINS

In addition to inducing alkylation damage by way of *N*-nitroso compound formation as discussed above, NO-derived NO_X can also aggravate the toxic consequences of DNA alkylation by inactivating the protein responsible for repairing it. For example, O^6-alkylguanine-DNA-alkyltransferase contains a thiol function at its active site that normally displaces promutagenic O^6-alkyl groups from the guanine residues, as shown in equation 8a. This thiol function can be *S*-nitrosated by NO_X, however, rendering it incapable of accepting alkyl groups and prohibiting it from acting as a repair protein (equation 8b) (Laval & Wink, 1994).

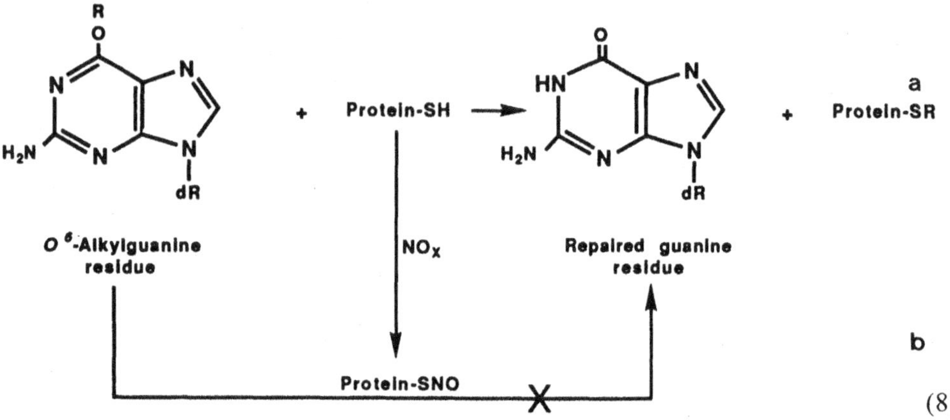

(8)

Another DNA repair protein that is inhibited by NO_X is formamidopyrimidine-DNA-glycosylase (Wink & Laval, 1994). In this case, enzymatic activity depends on the structural integrity of a zinc finger motif, in which a Zn^{2+} center holds four thiol groups in the active spatial array through simultaneous coordination; these thiol groups too can be *S*-nitrosated by NO_X, as shown in equation 9. The resulting *S*-nitrosothiol residue can no longer coordinate the metal, leading to structural degradation of the zinc finger and loss of activity. As a result, oxidatively damaged bases such as 8-oxoguanine cannot be efficiently repaired and the DNA-damaging effects of oxidants are potentiated.

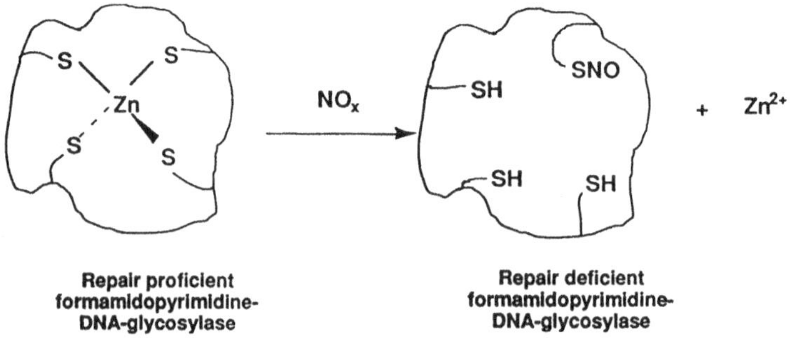

(9)

INHIBITION OF CYTOCHROMES P450

Another enzyme system whose catalytic efficiency can be impaired on exposure to nitric oxide is the cytochrome P450 complex (Khatsenko et al, 1993; Stadler et al, 1994;

Wink et al., 1993c). At least two mechanisms are operative here. One appears to involve direct coordination of NO to the heme iron center in a reversible process that can completely block coordination of O_2 until the NO concentration falls to a level at which dioxygen can again compete effectively for available binding sites.

A second mode of P450 inhibition involves an irreversible effect of exposure to NO on enzymatic activity. In this case, metabolism resumes when NO is depleted but at a slower rate than was observed before the NO was introduced. This mechanism may involve reaction of NO_X with nucleophilic centers that are remote from the active site but that nonetheless affect the kinetics of metabolism.

Cytochromes P450 are important xenobiotic-metabolizing enzymes (Nelson et al., 1993), so NO could significantly influence overall DNA damage by inhibiting their activity. For example, the toxic threat of NDMA might be at least temporarily negated if continued exposure to NO interfered with the ability of cytochrome P4502E1 to metabolize this otherwise inactive nitrosamine to the DNA-alkylating methyldiazonium ion ($CH_3N_2^+$) as shown in the mechanism of equation 7.

NO: INACTIVATION OF FENTON INTERMEDIATES

It is possible that NO can also protect DNA by preventing damage attributable to reactive oxygen intermediates (ROI) such as those produced in the Fenton reaction. It has recently been shown, for example, that NO inhibits the cytotoxicity of hydrogen peroxide in mammalian cell culture (Wink et al., 1993b). NO does not react with H_2O_2 rapidly enough for its protective effect to be rationalized in terms of a direct NO/H_2O_2 reaction in the cited experiments, but it has been reported to interfere with Fenton reactivity according to the mechanism illustrated in equation 10d (Kanner et al., 1991). One might speculate that cytotoxicity in the absence of NO was mediated by Fenton intermediates formed in the reaction of hydrogen peroxide with trace transition metals in the medium, and that inhibition of cytotoxicity resulted from interception of these intermediates by NO. This outcome would be expected whether the given Fenton intermediate was a hypervalent metallo-oxo species (equation 10a), coordinated peroxide ion (equation 10b), or hydroxyl radical (equation 10c). It has been shown that NO can reduce the levels of ROI-mediated strand breaks seen in naked DNA exposed to hydrogen peroxide in neutral solutions of chelated transition metal ions (Pacelli et al., 1994).

It should be noted here that the nitrogenous product of equation 10d is nitrous acid (nitrite ion at pHs above its pK_a of 3.3); nitrite is also produced in the nitrosation of water by the NO_X produced in the NO/O_2 reaction (equation 11) (Ignarro et al., 1993). Nitrous acid and nitrite have long been known as mutagenic species (Zimmermann, 1977) capable of crosslinking (Kirchner & Hopkins, 1991) and deaminating (Shapiro & Chargaff, 1966) DNA in a reaction analogous to equation 5.

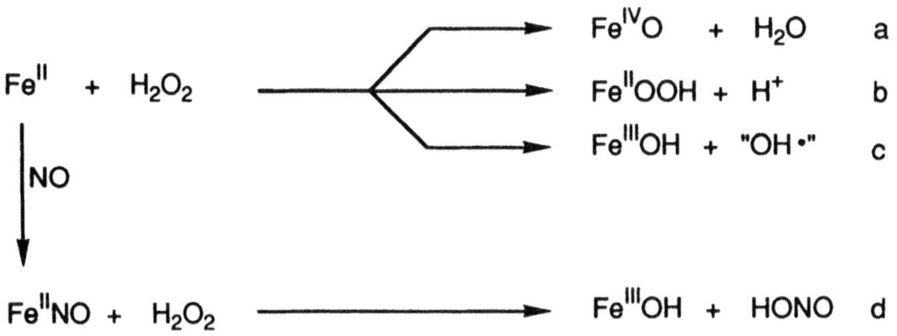

(10)

$$H_2O \xrightarrow{NO_X} HONO \xrightarrow{pH\ 7.4} NO_2^- \qquad (11)$$

NO PLUS SUPEROXIDE

Coupling of the two biologically important radicals, NO and O_2^-, occurs at nearly diffusion controlled rates to produce peroxynitrite ion, as in equation 12 (Huie & Padmaja, 1993). This biological reactive intermediate has been proposed to induce damage in sensitive biomolecules (Koppenol et al., 1992). Despite the potential for harm arising from peroxynitrite-mediated transformations, it may be that the NO/O_2^- reaction results in net protection of DNA. This could be the case because the acid-catalyzed rearrangement of $ONOO^-$ to nitrate (equation 13), a detoxication pathway, is quite rapid even at pH 7.4 (Koppenol et al., 1992). If rearrangement occurs before the peroxynitrite encounters the DNA, covalent modification of the biopolymer would be averted.

$$NO + O_2^- \longrightarrow ONOO^- \qquad (12)$$

$$ONOO^- \xrightarrow[k \approx 1\ s^{-1}]{pH\ 7.4} NO_3^- \qquad (13)$$

NO: INHIBITION OF RIBONUCLEOTIDE REDUCTASE

A second radical coupling reaction in which NO may impact the integrity of DNA is seen in its ability to quench the electron spin resonance signal for the tyrosyl radical of ribonucleotide reductase (Lepoivre et al., 1992). This inhibits the enzyme, which is critical for the conversion of ribonucleotides to deoxyribonucleotides, an essential step in DNA synthesis whose inhibition may interfere with DNA replication and repair. NO might also affect reactions at the active site metal as well as initiate NO_X-mediated attack on the enzyme's sulfhydryl groups critical to the ultimate reduction step. Whatever the mechanism, inhibition of ribonucleotide reductase activity may be an essential component of another pathway by which NO can indirectly protect DNA, suppression of viral replication (Karupiah et al., 1993).

NO IN THE FIXATION OF RADIATION DAMAGE

NO can also couple directly with the radicals formed upon irradiating biopolymers. By thus "fixing" the damage, cells in which such reactions occur may be lethally compromised in their function (Mitchell et al., 1993). This effect can be used to advantage in the radiotherapy of certain tumors containing pockets of poorly vascularized tissue whose cells have acquired the ability to thrive in a hypoxic environment. When the tumor is X-irradiated, radicals formed in normoxic areas can couple with dioxygen (equation 14a) to generate a permanent, covalent change. Radicals formed in hypoxic cells, on the other hand, have a greater opportunity to abstract hydrogen atoms from expendable molecules nearby, restoring the biopolymer's original structure and averting irreversible damage (equation 14b). This outcome has the practical consequence that tumors with significant populations of hypoxic cells can, despite nearly complete destruction of the surrounding tissue during radiotherapy, regenerate rapidly by unfettered expansion of the surviving hypoxic clones. Recent experiments show, however, that treatment of animals bearing hypoxic tumors with NO donor compounds can sensitize these otherwise refractory sites to the radiation, leading to tumor regrowth delays and even cures unseen in animals that were not treated with NO (Liebmann et al., 1994). The damage-fixing pathway inferred to be operative in this case is shown in equation 14c.

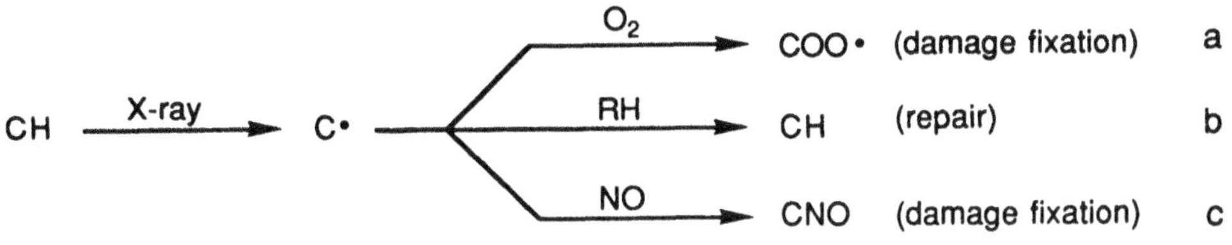

(14)

CONCLUSIONS

In summary, nitric oxide is capable of imposing a complex array of often interrelated and sometimes competing effects on DNA. While there is no evidence for a direct reaction between DNA and NO, the latter's oxidation products (NO_2, peroxynitrite ion, nitrous acid, and the as-yet-uncharacterized "NO_X" species produced on autoxidation of NO in aqueous media) may deaminate, crosslink, and oxidize the bases of DNA. Alkylation of DNA can result from the reaction of NO_X with disubstituted nitrogen species to form carcinogenic alkylating agents of the *N*-nitroso compound series, and the reaction of NO_X with DNA repair proteins can lead to increased persistence of alkylated (as well as oxidized) bases, potentiating the toxic effects thereof. NO and NO_X can inhibit cytochrome P450, decreasing DNA damage by certain *N*-nitroso compounds and other agents that require metabolic activation by this enzyme system. NO can inhibit ribonucleotide reductase, interfering with DNA synthesis, and it can couple with radicals formed on X-irradiation to "fix" the damage. The net effect of exposing DNA to NO may be deleterious or protective, cytotoxic or cytostatic, harmless or mutagenic, depending on the interplay among these various potential pathways in any given set of circumstances.

ACKNOWLEDGMENT

Equation 5 was adapted from "Experimental Tests of the Mutagenicity and Carcinogenicity of Nitric Oxide and Its Progenitors" by Larry K. Keefer, Lucy M. Anderson, Bhalchandra A. Diwan, Craig L. Driver, Diana C. Haines, Chris M. Maragos, David A. Wink, and Jerry M. Rice (in press) with the permission of Academic Press.

REFERENCES

Butler, A.R., and Williams, D.L.H., 1993, The physiological role of nitric oxide, *Chem. Soc. Rev.* 22:233-241.

Ford, P.C., Wink, D.A., and Stanbury, D.M., 1993, Autoxidation kinetics of aqueous nitric oxide, *FEBS Lett.* 326:1-3.

Görsdorf, S., Appel, K.E., Engeholm, C., and Obe, G., 1990, Nitrogen dioxide induces DNA single-strand breaks in cultured Chinese hamster cells, *Carcinogenesis* 11:37-41.

Huie, R.E., and Padmaja, S., 1993, The reaction of NO with superoxide, *Free Radical Res. Commun.* 18:195-199.

Ignarro, L.J., Fukuto, J.M., Griscavage, J.M., Rogers, N.E., and Byrns, R.E., 1993, Oxidation of nitric oxide in aqueous solution to nitrite but not nitrate: Comparison with enzymatically formed nitric oxide from L-arginine, *Proc. Natl. Acad. Sci. (USA)* 90:8103-8107.

Kanner, J., Harel, S., and Granit, R., 1991, Nitric oxide as an antioxidant, *Arch. Biochem. Biophys.* 289:130-136.

Karupiah, G., Xie, Q.-w., Buller, R.M.L., Nathan, C., Duarte, C., and MacMicking, J.D., 1993, Inhibition of viral replication by interferon-γ-induced nitric oxide synthase, *Science* 261:1445-1448.

Khatsenko, O.G., Gross, S.S, Rifkind, A.B., and Vane, J.R., 1993, Nitric oxide is a mediator of the decrease in cytochrome P450-dependent metabolism caused by immunostimulants, *Proc. Natl. Acad. Sci. (USA)* 90:11147-11151.

Kirchner, J.J., and Hopkins, P.B., 1991, Nitrous acid cross-links duplex DNA fragments through deoxyguanosine residues at the sequence 5'-CG, *J. Am. Chem. Soc.* 113:4681-4682.

Koppenol, W.H., Moreno, J.J., Pryor, W.A., Ischiropoulos, H., and Beckman, J.S., 1992, Peroxynitrite, a cloaked oxidant formed by nitric oxide and superoxide, *Chem. Res. Toxicol.* 5:834-842.

Laval, F., and Wink, D.A., 1994, Inhibition by nitric oxide of the repair protein, O^6-methylguanine-DNA-methyltransferase, *Carcinogenesis* 15:443-447.

Leaf, C.D., Wishnok, J.S., and Tannenbaum, S.R., 1991, Endogenous incorporation of nitric oxide from L-arginine into *N*-nitrosomorpholine stimulated by *Escherichia coli* lipopolysaccharide in the rat, *Carcinogenesis* 12:537-539.

Lepoivre, M., Flaman, J.-M., and Henry, Y., 1992, Early loss of the tyrosyl radical in ribonucleotide reductase of adenocarcinoma cells producing nitric oxide, *J. Biol. Chem.* 267:22994-23000.

Liebmann, J., DeLuca, A.M., Coffin, D., Wink, D.A., Keefer, L.K., and Mitchell, J.B., 1994, *In vivo* radioprotection of normal tissue with tumor radiosensitization by nitric oxide, *Proc. Am. Assoc. Cancer Res.* 35:649 (abstract 3868).

Mitchell, J.B., Wink, D.A., DeGraff, W., Gamson, J., Keefer, L.K., and Krishna, M.C., 1993, Hypoxic mammalian cell radiosensitization by nitric oxide, *Cancer Res.* 53:5845-5848.

Nelson, D.R., Kamataki, T., Waxman, D.J., Guengerich, F.P., Estabrook, R.W., Feyereisen, R., Gonzalez, F.J., Coon, M.J., Gunsalus, I.C., Gotoh, O., Okuda, K., and Nebert, D.W., 1993, The P450 superfamily: Update on new sequences, gene mapping, accession numbers, early trivial names of enzymes, and nomenclature, *DNA Cell Biol.* 12:1-51.

Nguyen, T., Brunson, D., Crespi, C.L., Penman, B.W., Wishnok, J.S., and Tannenbaum, S.R., 1992, DNA damage and mutation in human cells exposed to nitric oxide *in vitro*, *Proc. Natl. Acad. Sci. (USA)* 89:3030-3034.

Olbregts, J., 1985, Termolecular reaction of nitrogen monoxide and oxygen: A still unsolved problem, *Int. J. Chem. Kinet.* 17:835-848.

Pacelli, R., Krishna, M.C., Wink, D.A., and Mitchell, J.B., 1994, Nitric oxide protects DNA from hydrogen peroxide-induced double strand cleavage, *Proc. Am. Assoc. Cancer Res.* 35:540 (abstract 3214).

Routledge, M.N., Wink, D.A., Keefer, L.K., and Dipple, A., 1994, DNA sequence changes induced by two nitric oxide donor drugs in the *supF* assay, *Chem. Res. Toxicol.* 7:628-632.

Saffhill, R., Margison, G.P., and O'Connor, P.J., 1985, Mechanisms of carcinogenesis induced by alkylating agents, *Biochim. Biophys. Acta* 823:111-145.

Shapiro, H.S., and Chargaff, E., 1966, Studies on the nucleotide arrangement in deoxyribonucleic acids. XI. Selective removal of cytosine as a tool for the study of the nucleotide arrangement in deoxyribonucleic acid, *Biochemistry* 5:3012-3018.

Stadler, J., Trockfeld, J., Schmalix, W.A., Brill, T., Siewert, J.R., Greim, H., and Doehmer, J., 1994, Inhibition of cytochromes P4501A by nitric oxide, *Proc. Natl. Acad. Sci. (USA)* 91:3559-3563.

Williams, D.L.H., 1988, *Nitrosation*, Cambridge U. Press, Cambridge.

Wink, D.A., and Laval, J., 1994, The Fpg protein, a DNA repair enzyme, is inhibited by the biomediator nitric oxide *in vitro* and *in vivo*, *Carcinogenesis* 15:2125-2129.

Wink, D.A., Kasprzak, K.S., Maragos, C.M., Elespuru, R.K., Misra, M., Dunams, T.M., Cebula, T.A., Koch, W.H., Andrews, A.W., Allen, J.S., and Keefer, L.K., 1991, DNA deaminating ability and genotoxicity of nitric oxide and its progenitors, *Science* 254:1001-1003.

Wink, D.A., Darbyshire, J.F., Nims, R.W., Saavedra, J.E., and Ford, P.C., 1993a, Reactions of the bioregulatory agent nitric oxide in oxygenated aqueous media: Determination of the kinetics for oxidation and nitrosation by intermediates generated in the NO/O_2 reaction, *Chem. Res. Toxicol.* 6:23-27.

Wink, D.A., Hanbauer, I., Krishna, M.C., DeGraff, W., Gamson, J., and Mitchell, J.B., 1993b, Nitric oxide protects against cellular damage and cytotoxicity from reactive oxygen species, *Proc. Natl. Acad. Sci. (USA)* 90:9813-9817.

Wink, D.A., Osawa, Y., Darbyshire, J.F., Jones, C.R., Eshenaur, S.C., and Nims, R.W., 1993c, Inhibition of cytochromes P450 by nitric oxide and a nitric oxide-releasing agent, *Arch. Biochem. Biophys.* 300:115-123.

Zimmermann, F.K., 1977, Genetic effects of nitrous acid, *Mutat. Res.* 39:127-147.

INHIBITION OF CYTOCHROME P450 ENZYMES BY NITRIC OXIDE

J. Stadler,[1] W. A. Schmalix,[2] and J. Doehmer[2]

[1] Chirurgische Klinik und Poliklinik
[2] Institut für Toxikologie und Umwelthygiene
Technische Universität München
Ismaninger Str. 22, 81675 München, Germany

INTRODUCTION

Inflammatory stimulation of the liver causes significant alterations in liver cell metabolism. Recent experimental studies indicate that induction of nitric oxide (NO) biosynthesis may play a major role in the regulation of inflammatory processes and subsequent metabolic changes (23). Under cell culture conditions bacterial toxins and proinflammatory cytokines, such as TNFα, IL-1 and IFNγ, were identified as inducers of NO biosynthesis in parenchymal as well as non-parenchymal liver cells (6). Hepatic NO biosynthesis was also demonstrated in several animal models for local or systemic inflammation including injection of *Corynebacterium parvum* or lipopolysaccharides (4, 10). Finally, human hepatocytes were shown to express inducible nitric oxide synthase (iNOS) activity, which set the basis for cloning and heterologous expression of the human iNOS gene (24, 11). These findings support the idea that NO biosynthesis of the liver is a clinically relevant phenomenon in diseases characterized by local or systemic inflammatory reactions.

THE BIOCHEMICAL BASIS OF METABOLIC EFFECTS OF NO

NO is a highly reactive radical with a half live of only a few seconds within oxygenated biological solutions. Among many other factors, the chemical reactivity of NO depends on the redox-state of the environment where it is generated (39). NO may be interconverted into either nitrosonium or the nitroxyl anion, resulting in considerably different chemical affinities. These processes seem to be specifically interesting for reactions of NO with other radical species. In this context, it was demonstrated that NO is able to either neutralize reactive oxygen intermediates or increase their toxicity by formation of peroxynitrate and subsequent release of the highly toxic hydroxyl anion (2, 15). However, the result of this radical interactions also depends on the relative amounts of reaction partners.

Biological Reactive Intermediates V
Edited by R. Snyder et al., Plenum Press, New York, 1996

Table 1. Nitric oxide biosynthesis in the liver: historical perspective

1987 Palmer et al./Ignarro et al. (26, 14)	NO was identified as a "endothelium derived relaxing factor"
1989 Billiar et al. (5)	NO biosynthesis was demonstrated in stimulated liver cells
1990 Curran et al. (6)	Proinflammatory cytokines and endotoxin were shown to induce NO biosynthesis in rat hepatocytes
1992 Nüssler et al. (24)	Demonstration of inducible NO biosynthesis in human hepatocytes
1993 Geller et al. (11)	Molecular cloning and functional expression of the inducible NO synthase from human hepatocytes

Other important targets for NO are thiol-groups, which allow NO to bind to suitable aminoacids and a whole variety of proteins, including albumin. Being bound to serum proteins, NO may be transported to distant areas of the organism where it could then be released and exert its biological functions (38). In addition, NO-dependent S-nitrosylation also results in modulation of enzyme activity. For example, NO was shown to bind to the Cys-149 of glyceraldehyde-3-phosphate dehydrogenase (GAPDH) (21). S-nitrosylation of this specific site increases the sensitivity of GAPDH for auto ADP-ribosylation which in turn leads to inhibition of enzyme activity (8). Since GAPDH is involved in glycolysis as well as in gluconeogenesis, it remains to be studied whether inhibition of this enzyme is responsible for NO-dependent suppression of hepatocyte glucose output (32). Although NO is highly lipophilic and readily transverses biological membranes, binding to GAPDH seems to play an important role for intracellular transport of NO, too (20).

Analyzing electron paramagnetic resonance (EPR) spectra demonstrated that hepatocyte NO biosynthesis leads to the formation of iron-nitrosyl compounds (33). The EPR signal was found to be almost identical to the one generated by NO-binding to iron-sulphur clusters. However, the signal was only detected in the cytosol of the cells. These findings explain why hepatocytes in contrast to other cell types are not susceptible to NO-dependent inhibition of enzyme complexes of the mitochondrial electron transport chain, which also contain iron-sulphur clusters (40, 35). The cytosolic target for NO, which protects hepatocellular mitochondria and is responsible for the characteristic EPR signal, has not been identified. It is very likely a scavenger molecule such as metallothionein (17). Interpretation of these data must take into consideration that EPR spectra of cellular preparations always represent the summation of many individual signals. Therefore, the predominant signal may be due to binding of NO to iron-sulphur clusters, but it is very well possible that there are other important targets such as heme-groups.

Many biological effects of NO are based on binding to prosthetic heme-groups, which may activate or suppress enzyme activity. The most prominent heme-containing enzyme, which is activated by NO, is the soluble guanylate cyclase (13). NO-dependent cGMP formation is involved in the regulation of vascular tone, aggregation of thrombocytes, neural transmission and several other biological processes (22). In hepatocytes cGMP production is also increased by NO biosynthesis (3). However, exposure to NO also results in inhibition of heme-containing enzymes. This effect was demonstrated for catalase, which catalyzes the detoxication of hydrogen peroxide (19). Furthermore, it was reported that NO suppresses its own production in a feed-back mechanism by inhibiting the NOS, which is a hemoprotein by itself (30). Another heme-containing enzyme, the cyclooxygenase, is influenced by NO in a more complex fashion. Small quantities of NO seem to increase the activity of this enzyme, while large amounts will almost totally block enzymatic activity (34). This variance

Table 2. Effects of NO on hemoproteins

Protein	Metabolic function	Effect of NO
cytosolic guanylate cyclase	formation of cGMP	activity increased
catalase	detoxification of H_2O_2	activity decreased
nitric oxide synthase	production of NO	activity decreased
cyclooxygenase	synthesis of prostaglandins	activity increased or decreased
lipoxygenase	synthesis of leukotrienes	activity decreased
cytochromes P450	mixed-function oxidation	activity decreased
hemoglobin	transport of oxygen	methemoglobin formation

in effectiveness of NO may be associated with different susceptibilities of the two isoforms of cyclooxygenase (42). There is also experimental data suggesting that there is no direct interaction of NO with cyclooxygenase heme groups (41). Nevertheless, the effects of NO on eicosanoid production have a major impact on the development of inflammatory reactions (36). Last but not least, it is worth mentioning that NO reacts with hemoglobin to generate methemoglobin. For adult individuals this reaction was never shown to be of pathologic significance but for infants this effect may be of relevance (12).

THE EFFECT OF NO ON CYTOCHROME P450 ACTIVITY

NO was used for decades as a spin label probe for enzymological investigations of prosthetic heme-groups, specifically in cytochrome P450 (CYP) research (25). However, the first report about the consequence of NO-binding on enzymatic CYP activity was published in 1993 (43). Using microsomal preparations and purified CYP enzymes the authors were able to demonstrate that exposure to NO solutions or socalled NO donors results in concentration-dependent inhibition of enzyme activity. The results of this study also indicated differential susceptibilities of miscellaneous CYP-dependent reactions towards the inhibitory effects of exogenously applied NO. Most interesting, however, was the observation of two distinct phases of CYP inhibition by NO. The first phase was characterized by reversible, but almost complete cessation of catalytic activity, followed by a second irreversible phase characterized by a varying extent of recovery of activity. These findings led to the suggestion that NO may inhibit CYP enzyme activity by two distinct mechanisms.

In our laboratory, the approach to study the effects of NO on CYP activity was based on the use of genetically engineered cell lines, which constitutively express single CYP enzymes. The cell lines are derived from V79 Chinese hamster fibroblasts, which do not exhibit any endogenous CYP activity (9). We tested cell lines expressing rat and human CYP1A1 and CYP1A2 and found that these enzymes were inhibited by different NO donors in a concentration-dependent manner (37). With both rat and human enzymes CYP1A1 was more sensitive towards the inhibitory effect of NO than CYP1A2. In a next step, CYP activity was tested in intact rat hepatocytes, which were stimulated to endogenously produce NO. Under the influence of NO-biosynthesis the turnover of benzo[a]pyrene, which reflects the activity of the CYP1A family, was dramatically reduced to almost unmeasurable levels. With the addition of N^G-monomethyl-L-arginine, a competitive inhibitor of NO synthesis, we were able to significantly counteract cytokine-evoked CYP-inactivation. However, enzymatic activity did not reach that of untreated controls. Western blot analysis demonstrated the relative concentrations of immunodetectable CYPs to be only marginally decreased by NO synthesis, which may be the result of the suppressive effect of NO on total protein synthesis

(7). In contrast, Northern blot analysis revealed a significant decrease of CYP-mRNA steady state levels in cells, which were induced to produce NO. Taking all together, these results indicate an immediate direct effect of NO on CYP catalytic activity and a secondary long term effect which suppresses CYP activity by modulating mRNA levels.

In an experimental model of endotoxemia these *in vitro* studies were confirmed under *in vivo* conditions (18). The authors demonstrated a significant correlation of inhibition of CYP activity and induction of NO biosynthesis in rats treated with i. p. injections of lipopolysaccharide. When these animals received N^{ω}-nitro-L-arginine methyl ester before lipopolysaccharide administration, NO synthesis as well as suppression of CYP activity were mostly prevented. In this study a significant reduction of the CYP content in hepatic microsomal preparations was measured under the influence of NO biosynthesis using the CO binding method. However, as discussed by the authors, this finding may reflect masking of CO binding by NO rather than true changes in CYP concentrations.

There are several lines of evidence that the inhibitory effect of NO is at least in part mediated by binding to the heme moiety of the CYP enzymes. Khatsenko et al. (18) demonstrated characteristic changes in absorption spectra which indicate that NO binds to the ferrous as well as to the ferric state of this specific kind of hemoproteins. Another important point was raised by Wink et al. (43) considering the fact that two different phases of NO-mediated catalytic inhibition were identified. The first, reversible phase most likely represents the binding of NO to the heme-group, which blocks oxygen binding and results in complete cessation of enzymatic activity. It remains unclear what the molecular basis for the second phase of inhibition might be. However, a similar mode of biphasic inhibition was described for NO-mediated inhibition of two other hemoproteins, cyclooxygenase and lipoxygenase (16). Finally, Lancaster et al. (19) have demonstrated microsomal heme-loss and consequently, an increase in heme-oxidase activity in isolated hepatocytes, when NO biosynthesis was induced.

INHIBITION OF DETOXICATION PROCESSES DURING INFLAMMATORY STIMULATION OF THE LIVER

Activation as well as suppression of specifc liver functions are known to occur under conditions of local or systemic inflammation. Metabolic pathways of detoxification are significantly inhibited in the early course of inflammatory diseases, resulting in clinically well known phenomena such as hyperbilirubinemia and decreased degradation of specific drugs (27). Experimental studies have indicated that suppression of CYP activity may be responsible for many of these defects of hepatocellular performance (29) and recently, clinical studies seem to confirm these findings (31). It has also been shown that proinflammatory cytokines are involved in the suppression of CYP activity (28, 1). NO biosynthesis may be the missing link which explains how cytokines and bacterial toxins really act on CYPs.

Table 3. Evidence for heme-binding as a mechanism of NO-mediated CYP inhibition

Characteristic changes in absorption spectra
Microsomal heme-loss
Increase of heme-oxidase activity
Demonstration of two distinct phases of CYP inhibition with complete, but reversible cessation of catalytic activity in the first phase

Development of hepatocellular insufficiency is a serious problem in the managment of critically ill patients. At the moment, there are basically no therapeutic interventions available which allow causative treatment of hepatic dysfunction. Therefore, liver failure significantly contributes to the syndrome of multiple organ dysfunction which is a major killer on intensive care units. We hope that studying NO biosynthesis will provide new insights into the molecular mechanisms of hepatocellular insufficiency and finally result in applicable therapeutic strategies to improve this situation.

REFERENCES

1. Abdel-Razzak, Z., P. Loyer, A. Fautrel, J.-C. Gautier, L. Corcos, B. Turlin, P. Beaune, and A. Guillouzo. Cytokines down-regulate expression of major cytochrome P450 enzymes in adult human hepatocytes in primary culture. *Mol Pharmacol.* 44: 707-715, 1993.
2. Beckman, J. S., T. W. Beckman, J. Chen, P. A. Marshall, and B. A. Freeman. Apparent hydroxyl radical production by peroxynitrite: Implications for endothelial injury from nitric oxide and superoxide. *Proc Natl Acad Sci USA.* 87: 1620-1624, 1990.
3. Billiar, T. R., R. D. Curran, B. G. Harbrecht, J. Stadler, D. L. Williams, J. B. Ochoa, M. Di Silvio, R. L. Simmons, and S. A. Murray. Association between synthesis and release of cGMP and nitric oxide biosynthesis by hepatocytes. *Am J Physiol.* 262: C1077-1082, 1992.
4. Billiar, T. R., R. D. Curran, D. J. Stuehr, J. Stadler, R. L. Simmons, and S. A. Murray. Inducible cytosolic enzyme activity for the production of nitrogen oxides from L-arginine in hepatocytes. *Biochem Biophys Res Commun.* 168: 1034-1040, 1990.
5. Billiar, T. R., R. D. Curran, D. J. Stuehr, M. A. West, B. G. Bentz, and R. L. Simmons. An L-arginine-dependent mechanism mediates Kupffer cell inhibition of hepatocyte protein synthesis in vitro. *J Exp Med.* 169: 1467-1472, 1989.
6. Curran, R. D., T. R. Billiar, D. J. Stuehr, J. B. Ochoa, B. G. Harbrecht, S. G. Flint, and R. L. Simmons. Multiple cytokines are required to induce hepatocyte nitric oxide production and inhibit total protein synthesis. *Ann Surg.* 212: 462-469, 1990.
7. Curran, R. D., F. K. Ferrari, P. H. Kispert, J. Stadler, D. J. Stuehr, R. L. Simmons, and T. R. Billiar. Nitric oxide and nitric oxide-generating compounds inhibit hepatocyte protein synthesis. *FASEB J.* 5: 2085-2092, 1991.
8. Dimmeler, S., F. Lottspeich, and B. Brune. Nitric oxide causes ADP-ribosylation and inhibition of glyceraldehyde-3-phosphate dehydrogenase. *J Biol Chem.* 267: 16771-16774, 1992.
9. Doehmer, J., C. Wölfel, S. Dogra, C. Doehmer, A. Seidel, K. L. Platt, F. Oesch, and H. R. Glatt. Applications of stable V79-derived cell lines expressing rat cytochromes P4501A1, 1A2, and 2B1. *Xenobiotica.* 22: 1093-1099, 1992.
10. Geller, D. A., P. D. Freeswick, D. Nguyen, A. K. Nüssler, M. Di Silvio, R. A. Shapiro, S. C. Wang, R. L. Simmons, and T. R. Billiar. Differential induction of nitric oxide synthase in hepatocytes during endotoxemia and the acute-phase response. *Arch Surg.* 129: 165-171, 1994.
11. Geller, D. A., C. J. Lowenstein, R. A. Shapiro, A. K. Nüssler, M. Di Silvio, S. C. Wang, D. K. Nakayama, R. L. Simmons, S. H. Snyder, and T. R. Billiar. Molecular cloning and expression of inducible nitric oxide synthase from human hepatocytes. *Proc Natl Acad Sci USA.* 90: 3491-3495, 1993.
12. Hegesh, E. and J. Shiloah. Blood nitrates and infantile methemoglobinemia. *Clin Chim Acta.* 125: 107-115, 1982.
13. Ignarro, L. J., J. B. Adams, P. M. Horwitz, and K. S. Wood. Activation of soluble guanylate cyclase by NO-hemoproteins involves NO-heme exchange. Comparison of heme-containing and heme-deficient enzyme forms. *J Biol Chem.* 261: 4997-5002, 1986.
14. Ignarro, L. J., G. M. Buga, K. S. Wood, R. E. Byrns, and G. Chaudhuri. Endothelium-derived relaxing factor produced and released from artery and vein is nitric oxide. *Proc Natl Acad Sci USA.* 84: 9265-9269, 1987.
15. Kanner, J., S. Harel, and R. Granit. Nitric oxide as an antioxidant. *Arch Biochem Biophys.* 289: 130-136, 1991.
16. Kanner, J., S. Harel, and R. Granit. Nitric oxide, an inhibitor of lipid oxidation by lipoxygenase, cyclooxygenase and hemoglobin. *Lipids.* 27: 46-49, 1992.
17. Kennedy, M. C., T. Gan, W. E. Antholine, and D. H. Petering. Metollothionein reacts with Fe2+ and NO to form products with a g = 2.039 ESR signal. *Biochem Biophys Res Commun.* 196: 632-635, 1993.

18. Khatsenko, O. G., S. S. Gross, A. B. Rifkind, and J. R. Vane. Nitric oxide is a mediator of the decrease in cytochrome P450-dependent metabolism caused by immunostimulants. *Proc Natl Acad Sci USA.* 90: 11147-11151, 1993.
19. Lancaster jr, J. R. Diffusion and reactions of nitric oxide in isolated hepatocytes. *First International Conference: Biochemistry and Molecular Biology of Nitric Oxide.* Los Angeles, CA: July 16-21, 1994, 1994.
20. McDonald, B., B. Reep, E. G. Lapetina, and L. Molina y Vedia. Glyceraldehyde-3-phosphate dehydrogenase is required for the transport of nitric oxide in platelets. *Proc Natl Acad Sci USA.* 90: 11122-11126, 1993.
21. Molina y Vedia, L., B. McDonald, B. Reep, B. Brune, M. Di Silvio, T. R. Billiar, and E. G. Lapetina. Nitric oxide-induced S-nitrosylation of glyceraldehyde-3-phosphate dehydrogenase inhibits enzymatic activity and increases endogenous ADP-ribosylation. *J Biol Chem.* 267: 24929-24932, 1992.
22. Moncada, S. and A. Higgs. Mechanisms of disease: The L-arginine-nitric oxide pathway. *N Engl J Med.* 329: 2002-2012, 1993.
23. Nüssler, A. K. and T. R. Billiar. Inflammation, immunoregulation, and inducible nitric oxide synthase. *J Leukoc Biol.* 54: 171-178, 1993.
24. Nüssler, A. K., M. Di Silvio, T. R. Billiar, R. A. Hoffman, D. A. Geller, R. Selby, J. Madariaga, and R. L. Simmons. Stimulation of the nitric oxide synthase pathway in human hepatocytes by cytokines and endotoxin. *J Exp Med.* 176: 261-264, 1992.
25. O'Keefe, D., R. Ebel, and J. Peterson. Studies of the oxygen binding site of cytochrome p-450. *J Biol Chem.* 253: 3509-3516, 1978.
26. Palmer, R. M., A. G. Ferrige, and S. Moncada. Nitric oxide release accounts for the biological activity of endothelium-derived relaxing factor. *Nature.* 327: 524-526, 1987.
27. Park, G. R. Pharmcokinetics and pharmacodynamics in the critically ill patient. *Xenobiotica.* 23: 1195-1230, 1993.
28. Peterson, T. C. and K. W. Renton. Kupffer cell factor mediated depression of hepatic parenchymal cell cytochrome P450. *Biochem Pharmcol.* 35: 1491-1497, 1986.
29. Renton, K. Relationships between the enzymes of detoxication and host defense mechanism. In: Caldwell J, Jakoby W, eds. Biological basis of detoxication. New York; Academic Press, Inc.; 1983: 307-324.
30. Rogers, N. E. and L. J. Ignarro. Constitutive nitric oxide synthase from cerebellum is reversibly inhibited by nitric oxide formed from L-arginine. *Biochem Biophys Res Commun.* 189: 242-249, 1992.
31. Shedlofsky, S. I., B. C. Israel, C. J. McClain, D. B. Hill, and R. A. Blouin. Endotoxin administration to humans inhibits heaptic cytochrome P450-mediated drug metabolism. *J Clin Invest.* 94: 2209-2214, 1994.
32. Stadler, J., D. Barton, H. Beil-Moeller, S. Diekmann, C. Hierholzer, W. Erhard, and C. D. Heidecke. Hepatocyte nitric oxide biosynthesis inhibits glucose output and competes with urea synthesis for L-arginine. *Am J Physiol.* 268: G183-188, 1995.
33. Stadler, J., H. A. Bergonia, M. Di Silvio, M. A. Sweetland, T. R. Billiar, R. L. Simmons, and J. R. Lancaster jr. Nonheme iron-nitrosyl complex formation in rat hepatocytes: detection by electron paramagnetic resonance spectroscopy. *Arch Biochem Biophys.* 302: 4-11, 1993.
34. Stadler, J., T. R. Billiar, R. D. Curran, L. A. McIntyre, H. I. Georgescu, R. L. Simmons, and C. H. Evans. Articular chondrocytes synthesize nitric oxide in response to cytokines and lipopolysaccharide. *J Immunol.* 147: 3915-3920, 1991.
35. Stadler, J., T. R. Billiar, R. D. Curran, D. J. Stuehr, J. B. Ochoa, and R. L. Simmons. Effect of exogenous and endogenous nitric oxide on mitochondrial respiration of rat hepatocytes. *Am J Physiol.* 260: C910-916, 1991.
36. Stadler, J., B. G. Harbrecht, M. Di Silvio, R. D. Curran, M. L. Jordan, R. L. Simmons, and T. R. Billiar. Endogenous nitric oxide inhibits the synthesis of cyclooxygenase products and interleukin-6 by rat Kupffer cells. *J Leukoc Biol.* 53: 165-172, 1993.
37. Stadler, J., J. Trockfeld, W. A. Schmalix, T. Brill, H. Greim, J. R. Siewert, and J. Doehmer. Inhibition of cytochromes P4501A by nitric oxide. *Proc Natl Acad Sci USA.* 91: 3559-3563, 1994.
38. Stamler, J. S., O. Jaraki, J. Osborne, D. I. Simon, J. Keaney, J. Vita, D. Singel, C. R. Valeri, and J. Loscalzo. Nitric oxide circulates in mammalian plasma primarily as an S-nitroso adduct of serum albumin. *Proc Natl Acad Sci USA.* 89: 7674-7677, 1992.
39. Stamler, J. S., D. J. Singel, and J. Loscalzo. Biochemistry of nitric oxide and its redox-activated forms. *Science.* 258: 1898-1902, 1992.
40. Stuehr, D. J. and C. F. Nathan. Nitric oxide. A macrophage product responsible for cytostasis and respiratory inhibition in tumor target cells. *J Exp Med.* 169: 1543-1555, 1989.
41. Tsai, A.-L., C. Wei, and R. J. Kulmacz. Interaction between nitric oxide and prostaglandin H synthase. *Arch Biochem Biophys.* 313: 367-372, 1994.

42. Vane, J. R., J. A. Mitchell, I. Appleton, A. Tomlinson, D. Bishop-Bailey, J. Croxtall, and D. A. Willoughby. Inducible isoforms of cyclooxygenase and nitric oxide synthase in inflammation. *Proc Natl Acad Sci USA*. 91: 2046-2050, 1994.
43. Wink, D. A., Y. Osawa, J. F. Darbyshire, C. R. Jones, S. C. Eshenaur, and R. W. Nims. Inhibition of cytochromes P450 by nitric oxide and a nitric oxide-releasing agent. *Arch Biochem Biophys*. 300: 115-123, 1993.

NO AS A PHYSIOLOGICAL SIGNAL MOLECULE THAT TRIGGERS THYMOCYTE APOPTOSIS

Karin Fehsel, Klaus-Dietrich Kröncke, and Victoria Kolb-Bachofen

The Institute of Immunobiology and Biomedical Research Center
Department of Medicine, Heinrich-Heine-University, Düsseldorf, Germany

ABSTRACT

Nitric oxide (NO) produced at high concentrations by the inducible NO synthase (iNOS) is an important effector molecule involved in immune regulation and defence. In in vitro experiments we could show that NO represents a signal for triggering apoptosis in 30% of thymocytes with a concomittant decrease in $CD4^{+}CD8^{+}$ cells. In addition it protects the remaining cell population from apoptosis induced by glucocorticoids comparable to the protective effect of heat shock.

NO-induced DNA-strand breaks led to increased expression of p53, as detected by PCR analysis 2h after NO donor addition.

Furthermore, in cocultures of thymocytes with NO-producing endothelial cells the rate of thymocyte apoptosis was significantly increased, and this could be completely prevented by inhibiting NO production. Addition of dexamethasone to these cocultures did not lead to a further increase in the percentage of apoptotic thymocytes, underlining the protective effect of NO on dexamethasone-induced apoptosis.

INTRODUCTION

Apoptosis is a mode of cell death defined by morphologic and biochemical criteria. The most common biochemical feature of apoptosis is the fragmentation of DNA into internucleosomal fragments, which can be detected by in situ nick translation. This method stains nuclei of apoptotic cells directly in tissue sections or cell cultures (1). Experimentally, apoptosis in the thymus can be induced by various treatments such as glucocorticoids, anti CD3 antibodies or LPS (2). After LPS injection in mice apoptotic foci are grouped around capillaries (3), suggesting that substances penetrating the vessel wall or mediators secreted from endothelial cells induce programmed cell death. In this study, we report that LPS induces iNOS expression in thymic endothelial cells, which in turn induces apoptosis in thymocytes. Furthermore, we have investigated whether NO chemically generated or released from activated, cultured endothelial cells may serve as an inducing signal for apoptosis in thymocytes.

MATERIALS AND METHODS

Cells

Thymocytes from 30 day old NMRI mice were incubated in the absence or presence of the NO donors SNAP, SNOG and SNOC at a final concentration of $5x10^6$ cells/ml in RPMI 1640/10% FCS on 8-chamber tissue culture slides (Nunc LabTec, Wiesbaden, FRG). At various time points medium was removed and the slides were dried and processed for *in situ* nick translation.

Coculture with Activated ECs

$1.5x10^4$ rat aorta ECs were seeded on 8-chamber tissue culture slides and stimulated by the addition of IFNγ (100U/ml), IL-1ß (200U/ml) and LPS (20µg/ml) in the presence or absence of 0.25 mM NMA. After 30h $3x10^4$ freshly isolated thymocytes were added to these EC cultures and incubated for 3 to 24h. As control experiments thymocytes were incubated with unstimulated ECs or with medium containing cytokines and LPS only. In protection experiments Dex (1µM) was added to the cocultures after 30 min.

In Situ Nick Translation

DNA strand breaks were visualized by incorporating biotin-labeled dUTP at the sites of strand breaks exactly as described elsewhere (1). Cells were then washed in PBS and processed for immunocytochemical detection of incorporated biotin-dUTP by FITC-labeled avidin for FACS analysis or peroxidase-labeled avidin followed by an enzyme reaction using DAB as substrate for light microscopy.

PCR-Analysis of p53 mRNA in Thymocytes

Total RNA was isolated from treated thymocytes and reversely transcribed using oligo(dT) as primer. This cDNA was used as template for PCR primed with p53 specific primers. PCR was carried out according to the manufactorer's instructions. After a total of 35 cycles an aliqout of each reaction was subjected to electrophoresis on a 2% agarose gel. Bands were visualized by ethidium bromide staining under UV light.

RESULTS

NO-Induced Apoptosis in Thymocytes

Chemical NO donors, which spontaneously release NO, induce thymocyte apoptosis time- and concentration-dependently. All NO donors worked in the same concentration range from 0,1 - 1 mM. The kinetics of NO-induced strand breaks in cultured thymocytes is shown in Fig.1. Heat treatment (20 min; 42°C) of thymocytes induced apoptosis in the same percentage and time-dependence as compared to SNAP incubation. Addition of $ZnSO_4$ (0.5 mM) to any of the inducing agents resulted in complete protection and DNA fragmentation was comparable to sham-treated controls. Furthermore, preincubation with SNAP for 30 min

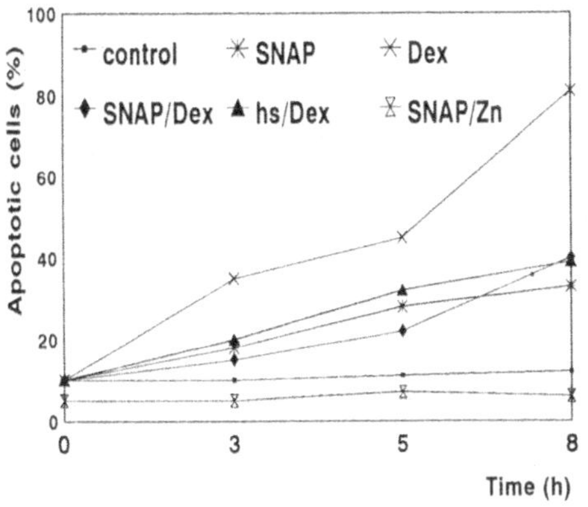

Figure 1. Kinetics of NO induced apoptosis. Thymocytes were incubated as indicated. After 3, 5, 8, 15 h cells were fixed, stained by in situ nick translation and counted. Values are the mean ± SD from 5 individual experiments.

had the same protective, anti-apoptotic effect on Dex-treated thymocytes as was previously described for heat shock (4). Flow cytometric analysis revealed that mainly the number of $CD4^+CD8^+$ cells decreased in response to NO treatment. RT-PCR analysis of cultured thymocytes resulted in a p53 specific amplification product only in SNAP-treated cells.

Coculture of Thymocytes with Endothelial Cells (ECs)

ECs were activated by cytokines and LPS to produce high concentrations of NO. Coculturing thymocytes for 8, 15 or 24h with activated ECs (Fig.2) resulted in a significantly increased rate of thymocyte apoptosis, in contrast to coincubation with resident ECs or activated ECs in the presence of the NOS inhibitor L-NMA. Dex had only minimal effects on the thymocyte apoptosis rate in cultures with activated ECs, but led to a significant increase in the percentage of apoptotic thymocytes in both, in cocultures with resident ECs as well as in cocultures with activated plus NMA incubated ECs.

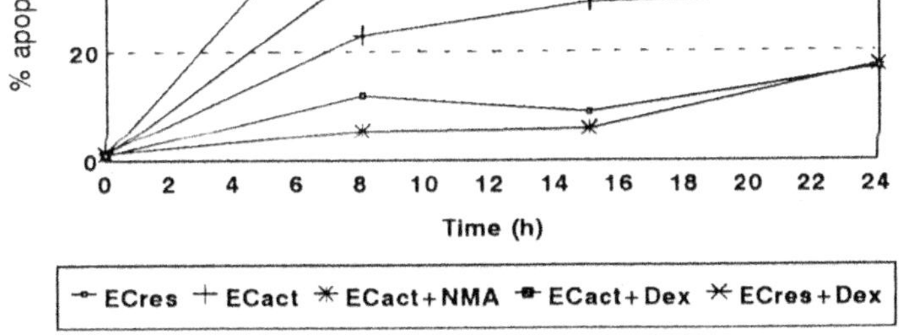

Figure 2. Rate of apoptosis in coculture studies. Thymocytes and aorta endothelial cells were incubated at a target:effector ratio of 1:1. At different time points the amount of apoptotic thymocytes was determined by in situ nick translation.

DISCUSSION

Many substances are known to induce thymic apoptosis, but the physiological inducing signal(s) triggering thymocyte apoptosis as it is observed for instance during inflammatory reactions are largely unknown. In this report, we could clearly show that endothelial cells activated by cytokines and LPS produce NO in high concentrations, which in turn induced apoptosis in thymocytes. It had already been shown that NO can induce apoptosis in murine peritoneal macrophages and splenocytes (5,6). To assess whether this holds true for thymocytes also, we used the chemical NO donors SNAP, SNOG and SNOC and searched for apoptosis. All NO-donors induced apoptosis in a dose-dependent way, probably via inducing DNA strand breaks, since this is paralleled by a strong increase in p53 expression. The time-scale of induction, the amount of apoptotic cells and the affected subpopulation were found to be strikingly similar to heat shock induced programmed cell death (4). Both, heat treatment as well as NO partially protected from Dex-induced apoptosis probably by induction of heat shock protein expression.

Furthermore, NO produced by activated endothelial cells inhibited Dex-induced apoptosis in cocultured thymocytes. In cocultures without the addition of Dex the degree of apoptosis correlated with the nitrite concentrations measured in culture supernatants. In conclusion, we find that NO induces apoptosis in murine thymocytes involving p53 expression. In addition we present evidence, that NO-formation by thymic ECs can be involved in the elimination of thymocyte subsets during thymic atrophy for instance following endotoxin challenge.

Indeed, we found apoptotic foci in thymic cortex in the vicinity of iNOS expressing endothelial cells and histiocytes 18h after LPS injection.

REFERENCES

1. Fehsel K., V. Kolb-Bachofen, and H. Kolb. 1991. Analysis of TNFÂ -induced DNA strand breaks at the single cell level. Am. J. Path. 139: 251.
2. Thomas D.J., and T.C. Caffrey. 1991. Lipopolysaccharide induces double-stranded DNA fragmentation in mouse thymus: protective effect of zinc treatment. Toxicology 68: 327.
3. Fehsel K., K.-D. Kröncke, H. Kolb, and V. Kolb-Bachofen. 1993. In situ nick translation detects focal apoptosis in thymuses of glucocorticoid- and LPS-treated mice. J. Histochem. Cytochem. 42: 613.
4. Migliorati G., I. Nicoletti, F. Crocicchio, C. Pagliacci, F. D'Adamio, and C. Riccardi. 1992. Heat shock induces apoptosis in mouse thymocytes and protects them from glucocorticoid-induced cell death. Cell. Immunol. 143:348
5. Albina J.E., S. Cui, R.B. Mateo, and J.S. Reichner. 1993. Nitric oxide-mediated apoptosis in murine peritoneal macrophages. J. Immunol. 150:5080
6. Cui S., J.S. Reichner, and J.E. Albina. 1994. Nitric oxide (NO) induces apoptosis in ConA-stimulated splenocytes. FASEB J. 8:4470

AMPLIFICATION OF NITRIC OXIDE SYNTHASE EXPRESSION BY NITRIC OXIDE IN INTERLEUKIN 1β-STIMULATED RAT MESANGIAL CELLS

Heiko Mühl and Josef Pfeilschifter*

Department of Pharmacology
Biozentrum, University of Basel
Klingelbergstrasse 70, CH-4056 Basel, Switzerland

INTRODUCTION

Nitric oxide (NO), a free-radical gas produced by many cell types, mediates blood vessel relaxation, functions as a neurotransmitter in the central and peripheral nervous system mediates macrophage cytotoxicity during host defence and leads to tissue injury in some inflammatory and autoimmune diseases. Increasing evidence indicates that NO orchestrates acute and chronic inflammatory processes and contributes to the pathomechanisms of septic shock, destruction of pancreatic islet cells during the development of insulin-dependent type 1 diabetes, adjuvant arthritis, progression of renal failure and neural destruction in vascular stroke and other neurodegenerative diseases (1). The careful control of this extremely reactive molecule is essential for the prevention of deleterious inflammatory reactions. Whereas the activity of the constitutive brain and endothelial cell NO synthases are mainly regulated posttranslationally by cytoplasmic Ca^{2+} levels or phosphorylation by a variety of protein kinases, the inducible NO synthase is regulated primarily at a transcriptional level. Once induced, the latter enzyme synthesizes NO for long periods of hours and days.

We have used renal mesangial cells, a specialized type of vascular smooth muscle cell that takes part in the regulation of the glomerular filtration rate. These cells respond to endothelial-derived NO with increased levels of intracellular cyclic GMP and express a macrophage type of NO synthase when exposed to inflammatory cytokines (2). In the present report we have addressed possible self-regulatory mechanisms modulating the expression of NO synthase in mesangial cells.

* Author to whom correspondence should be addressed: Professor Josef M. Pfeilschifter, Department of Pharmacology, Biozentrum, University of Basel, Klingelbergstrasse 70, CH-4056 Basel, Switzerland; Tel: +41/61 267 22 23; Fax: +41/61 267 22 08; E-Mail: wittker@yogi.urz.unibas.ch

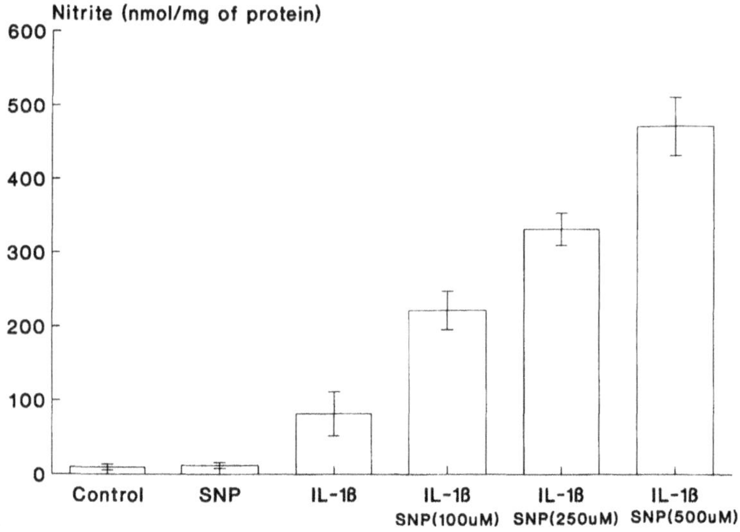

Figure 1. Nitric oxide amplifies IL-1β-stimulated nitrite production in mesangial cells. Confluent mesangial cells were incubated for 16h with vehicle (control), SNP (500 μM), IL-1β (2 nM) or combinations of IL-1β (2 nM) plus the indicated concentrations of SNP. Thereafter, the cells were washed several times to remove all agents and fresh medium with or without IL-1β but without SNP was added for a second incubation period of 6h. Nitrite production during this second incubation period was measured as a read-out for NO synthase activity. Data are means ± S.D. of four experiments.

RESULTS AND DISCUSSION

Fig. 1 shows that sodium nitroprusside (SNP), a NO-generating compound, was unable to induce NO synthase activity on its own. In contrast, when SNP was added to cells together with IL-1β, a cytokine known to induce NO synthase expression in mesangial cells, NO synthase activity and nitrite synthesis were markedly enhanced (Fig. 1). Comparable data were obtained with S-nitroso-N-acetyl-D,L-penicillamine (SNAP), a second NO-generating compound that is structurally not related to SNP. The amplifying effect of SNP is dependent on the concentration of IL-1β used to trigger NO synthase expression and is most pronounced at submaximal concentrations of the cytokine. The increased production of nitrite is paralleled by increased NO synthase protein and mRNA (3) levels as determined by Western blot and Northern blot analyses, respectively.

We have shown recently that inducible NO synthase is expressed in mesangial cells in response to two principal classes of activating signals that interact in a synergistic fashion. These two groups of activators comprise inflammatory cytokines such as IL-1β and tumor necrosis factor α and agents that elevate cellular levels of cyclic AMP (4,5). SNP only amplified the action of IL-1β but not NO synthase induction triggered by cyclic AMP (data not shown). To test the possibility that NO augmented cytokine-induced NO synthase expression by changing the level of cellular cyclic GMP, we treated the cells with a membrane-permeable analogue, dibutyryl cyclic GMP. Addition of dibutyryl cyclic GMP did not induce NO synthase activity and did not alter IL-1β-stimulated nitrite production (data not shown). Taken together, these findings suggest that amplification of NO synthase expression is an action of NO that is not mediated by cyclic GMP and is selective for inflammatory cytokines like interleukin 1β and tumor necrosis factor α (data not shown) without affecting cAMP induction of NO synthase.

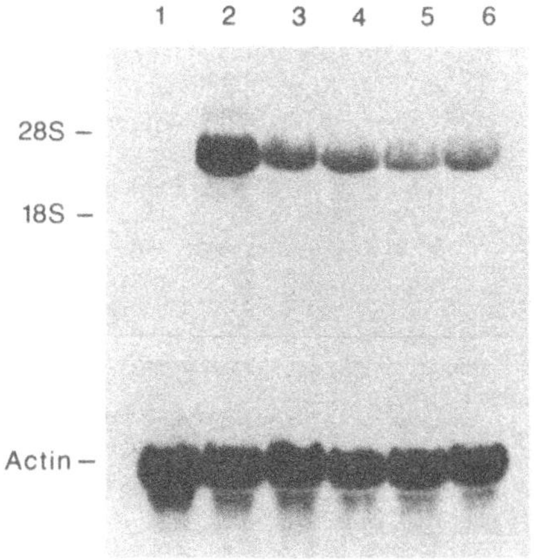

Figure 2. Effects of L-NMMA on IL-1β-stimulated NO synthase mRNA accumulation in mesangial cells. Confluent cells were incubated with vehicle (lane 1), IL-1β (2nM, lane 2) or combinations of IL-1β (2 nM) plus L-NMMA (100 μM, lane 3), 500 μM (lane 4), 1 mM (lane 5) or 5 mM (lane 6) for 20 h. Total cellular RNA was successviely hybridized to 32P-labelled NO synthase and β-actin probes as described in the Methods section. Reprinted with permission from Ref. 3.

To determine whether such a positive feedback loop triggered by NO is physiologically relevant we used different compounds known to modulate NO formation in cytokine-stimulated mesangial cells. Interleukin-1β induction of NO synthase mRNA was substantially reduced by inclusion in the culture media of NG-monomethyl-L-arginine (L-NMMA), a guanidino-N-substituted L-arginine analogue that acts as a competitive inhibitor of NO synthase, as shown in Fig. 2. Conversely, changes of L-arginine concentration markedly modulate NO synthase mRNA levels.

These results suggest that NO acts as an autocrine mediator that upregulates IL-1β-induced NO synthase gene expression in mesangial cells and thus leads to an optimal generation of NO by the cells. This positive feedback mechanism of NO serving to maximally amplify its own production is not restricted to renal mesangial cells, but is also observed in vascular smooth muscle cells derived from a rat aorta (3) and thus may be a general mechanism operative in cells and tissues that are able to express the inducible form of NO synthase.

The physiological role of the described potent amplification mechanism of NO generation may be to rapidly provide injured cells with a powerful host defence mechanism that also may form the basis for the dramatic production of NO in acute and chronic inflammatory diseases. It is tempting to suggest that the regional or systemic use of compounds that block the positive amplification cycle of NO production may be of value in therapy for autoimmune diseases. The careful control of this extremely reactive molecule is essential for the prevention of deleterious inflammatory reactions.

ACKNOWLEDGMENTS

This work was supported by Swiss National Science Foundation Grant 31-33653.92 and by a grant from the Commission of the European Union (Biomed I, BMH1-CT92-1893).

REFERENCES

1. Nathan, C. 1992. Nitric oxide as a secretory product of mammalian cells. *FASEB J.* **6**:3051-3064.
2. Pfeilschifter, J., D. Kunz, and H. Mühl. 1993. Nitric oxide: An inflammatory mediator of glomerular mesangial cells. *Nephron* **64:**518-525.

3. Mühl, H. and Pfeilschifter, J. 1995. Amplification of nitric oxide synthase expression by nitric oxide in interleukin 1β-stimulated rat mesangial cells. *J. Clin. Invest.* (in press).
4. H. Mühl, D. Kunz, J. Pfeilschifter. 1994. Expression of nitric oxide synthase in rat glomerular mesangial cells mediated by cyclic AMP. *Br. J. Pharmacol.* **112,** 1-8.
5. Kunz, D., H. Mühl, G. Walker, and J. Pfeilschifter. 1994. Two distinct signaling pathways trigger the expression of nitric oxide synthase in rat renal mesangial cells. *Proc. Natl. Acad. Sci. USA* **91**:5387-5391.

THE KIDNEY AS A TARGET FOR BIOLOGICAL REACTIVE METABOLITES

Linking Metabolism to Toxicity

Terrence J. Monks, Maria I. Rivera, Jos J. W. M. Mertens,
Melanie M. C. G. Peters, and Serrine S. Lau

Division of Pharmacology and Toxicology
College of Pharmacy
University of Texas at Austin
Austin, Texas

INTRODUCTION: THE KIDNEY AS A TARGET OF POLYPHENOLIC-GLUTATHIONE CONJUGATES

The conjugation of potentially toxic electrophiles with glutathione (GSH), the process of thioether formation, is usually associated with detoxication and excretion (Sies and Ketterer, 1988). Compounds that are conjugated with GSH are usually excreted in urine as their corresponding mercapturic acids, *S*-conjugates of *N*-acetylcysteine. The kidney is susceptible to chemical induced injury, and GSH conjugation seems to play a major role in the bioactivation of many of renal toxicants. Halogenated alkanes and alkenes are targeted to the kidney *via* GSH conjugation, and subsequent bioactivation of the corresponding cysteine conjugate by cysteine conjugate β-lyase (Dekant *et al.*, 1994). Polyphenols also seem to be targeted to the kidney *via* extra-renal conjugation with GSH (Monks and Lau 1994a) and tissue selectivity in this instance appears to be conferred by the high activity of γ-glutamyl transpeptidase on the brush border membrane of renal proximal tubular epithelial cells. Thus, in contrast to the generally accepted role of GSH conjugation serving as a detoxication mechanism conjugation of quinones with GSH results in the formation of potent, and selective, nephrotoxicants (Monks *et al.*, 1985, 1988; Lau *et al.*, 1988, 1990; Mertens *et al.*, 1991), and evidence has accumulated suggesting that GSH conjugation of quinones may be a common mechanism of bioactivation (Monks and Lau, 1992).

4-Aminophenol

For example, 4-aminophenol causes acute renal proximal tubular necrosis following administration to rats (Green *et al.*, 1969; Newton *et al.*, 1982) and depletion of hepatic GSH by pretreatment of animals with buthionine sulfoximine, or cannulation of the bile duct to

Biological Reactive Intermediates V
Edited by R. Snyder et al., Plenum Press, New York, 1996

decrease the delivery of hepatic metabolites to the kidney, afforded partial protection against 4-aminophenol nephrotoxicity (Gartland *et al.*, 1989) suggesting a role for GSH conjugation in the toxicity. Indeed, oxidation of 4-aminophenol to the quinoneimine and reaction with GSH gives rise to several isomeric multi-substituted conjugates (Eckert *et al.*, 1990a). Subsequently, 4-amino-3-(glutathion-*S*-yl)phenol was shown to reproduce 4-aminophenol nephrotoxicity in male Fischer 344 rats at doses 3-4 fold lower than that of 4-aminophenol (Fowler *et al.*, 1991), and 4-amino-2-(glutathion-*S*-yl)phenol, 4-amino-3-(glutathion-*S*-yl)phenol, 4-amino-2,5-(di-glutathion-*S*-yl)phenol, and 4-amino-2,3,5 (or 6)-(triglutathion-*S*-yl)phenol were identified in the bile of Wistar rats following administration of 4-aminophenol (100 mg/kg; ip) (Klos *et al.*, 1992) . The latter three conjugates were all capable of causing cytotoxicity when incubated with rat kidney cortical cells, and the toxicity was prevented by inhibition of γ-glutamyl transpeptidase (γ-GT). Coadminstration of ascorbic acid with 4-aminophenol significantly protected against the toxicity with a concomitant reduction in the concentration of both total and covalently bound radiolabel in the kidney (Fowler *et al.*, 1993), indicating that intrarenal oxidation of 4-aminophenol and/or its metabolites play an important role in the mechanism of toxicity.

Hydroquinone

Renal tubular cell degeneration has been described in the renal cortex of male and female rats (F344/N) receiving 1.82 mmol/kg hydroquinone (13 week gavage studies) (Kari *et al.*, 1992). In addition, the results from long-term studies demonstrated that hydroquinone exhibited "*evidence of carcinogenic activity*" in male Fischer 344/N rats as shown by marked increases in tubular cell adenomas of the kidneys (Kari *et al.*, 1992). These findings were confirmed by Shibata *et al.* (1991), who reported the induction of renal cell tumors in rats and mice following exposure to hydroquinone in the diet (0.8%) for two years. Although the mechanism of hydroquinone-mediated nephrotoxicity and nephrocarcinogenicity in male Fischer 344/N rats is not known, GSH conjugates of hydroquinone may play an important role. For example, 2-(*N*-acetyl-cystein-*S*-yl)hydroquinone and 2-(glutathion-*S*-yl)hydroquinone have been identified as *in vivo* and *in vitro* metabolites of hydroquinone (Tunek *et al.*, 1980; Lunte and Kissinger, 1983; Sawahata and Neal, 1983; Nerland and Pierce, 1990) and chemical reaction of GSH with 1,4-benzoquinone gives rise to 2-(glutathion-*S*-yl)hydroquinone and all of the possible multi-substituted GSH adducts, 2,3-(di-glutathion-*S*-yl)hydroquinone, 2,5-(di-glutathion-*S*-yl)hydro-quinone, 2,6-(diglutathion-*S*-yl)hydroquinone, 2,3,5-(tri-glutathion-*S*-yl)hydroquinone, and 2,3,5,6-(tetra-glutathion-*S*-yl)hydroquinone (Lau *et al.*, 1988; Eckert *et al.*, 1990b). Administration of each of these conjugates to rats demonstrated a direct correlation between the degree of GSH substitution and the extent of proximal tubular necrosis, as determined by elevations in blood urea nitrogen. In particular, 2,3,5-(tri-glutathion-*S*-yl)hydroquinone (10-20 μmol/kg, iv) caused severe renal proximal tubular necrosis in male Sprague-Dawley rats (Lau *et al.*, 1988). Histological examination of the kidneys from 2,3,5-(tri-glutathion-*S*-yl)hydroquinone treated rats (20 μmol/kg,iv) revealed that severe necrosis was localized to the S3 segment of the proximal tubules within the cortico-medullary junction. Thus, GSH conjugates of hydroquinone, in particular, 2,3,5-(tri-glutathion-*S*-yl)hydroquinone may contribute to hydroquinone mediated nephrocarcinogenicity. The identification of 2,3,5-(tri-glutathion-*S*-yl)hydroquinone and other multi-GSH conjugates as *in vivo* metabolites of hydroquinone in the rat in concentrations sufficient to produce nephrotoxicity (Hill *et al.*, 1993) lends further support to this hypothesis.

GSH conjugates of hydroquinone have also been shown to catalyze 8-hydroxydeoxy-guanosine formation in calf-thymus DNA (Canales *et al.*, 1993). Such oxidative DNA damage may be important in addressing the question of whether or not hydroquinone is a genotoxic or non-genotoxic carcinogen. Hydroquinone is non-genotoxic in most short term tests. Current evidence suggests that certain chemicals may induce carcinogenesis by a mechanism involving cytotoxicity, followed by sustained regenerative hyperplasia and ultimately tumor formation. Because the GSH conjugates of hydroquinone are potent nephrotoxicants we examined the effects of 2,3,5-(tri-glutathion-*S*-yl)hydroquinone (7.5 μmol/kg) on renal tubular cell proliferation, detected immunohistochemically *via* the incorporation of bromodeoxyuridine into newly synthesized DNA. Examination of kidney slices revealed the presence of tubular necrosis in the S3M segment of the proximal tubule, extending into the medullary rays. Cell proliferation rates in this region were 2.4, 6.9, 15.3, and 14.3% after 12, 24, 48 and 72 h respectively, compared to 1.6% in vehicle controls (Peters *et al.*, 1995). Hydroquinone (1.8 mmol/kg) was also mildly nephrotoxic, and bromodeoxyuridine incorporation into proximal tubular cells of the S3M region correlated with the degree of toxicity in individual rats, and ranged from 4.3-9.2; 1.2-9.6; 2.9-19.5% after 24, 48 and 72 h, respectively. Together with the metabolic studies, these data indicate a role for thioether metabolites in hydroquinone-mediated nephrotoxicity and nephro-carcinogenicity.

3-*tert*-Butyl-4-hydroxyanisole

3-*tert*-Butyl-4-hydroxyanisole and its desmethyl analog, *tert*-butyl-hydroquinone, are anti-oxidants used in food, and both compounds have been shown to promote kidney and bladder carcinogenesis in the rat. Thus, long-term administration of 3-*tert*-butyl-4-hydroxyanisole increases the formation of preneoplastic and neoplastic foci in rat kidney (Tsuda *et al.*, 1984), and thioether metabolites have been reported in rat urine (Tajima *et al.*, 1991). 2-*Tert*-butylhydroquinone, a metabolite of 3-*tert*-butyl-4-hydroxyanisole, is oxidized in rat liver microsomes, and in the presence of GSH forms 2-*tert*-butyl-5-(glutathion-*S*-yl)hydroquinone and 2-*tert*-butyl-6-(glutathion-*S*-yl)hydroquinone (Tajima *et al.*, 1991). Whether the GSH conjugates are involved in the carcinogenic effects of 3-*tert*-butyl-4-hydroxyanisole remains to be determined. However, exposure of male Fischer 344 rats to 2-*tert*-butyl-3,6-(di-glutathion-*S*-yl)hydroquinone (200 μmol/kg) caused mild damage to renal proximal tubules, as observed histologically and by the significant increases in the urinary excretion of γ-GT, alkaline phosphatase, lactate dehydrogenase, and glucose (Lau *et al.*, 1994). In addition to nephro-toxicity, 2-*tert*-butyl-3,6-(di-glutathion-*S*-yl)hydroquinone also increased bladder wet weight 2-4 fold, and caused severe hemorrhaging of the bladder. Long-term exposure to 3-*tert*-butyl-4-hydroxyanisole may therefore result in the continuous delivery of redox-active GSH conjugates to renal proximal tubules, causing cell damage and a sustained regenerative response that may be sufficient to promote the formation of neoplasia. The role of GSH in 3-*tert*-butyl-4-hydroxyanisole induced cell proliferation is further supported by the report that depletion of GSH with diethylmaleate prevented the forestomach hyperplasia associated with the ingestion of this antioxidant (Hirose *et al*, 1987). Recent studies have shown that 2-*tert*-butylhydroquinone is also capable of inducing oxidative DNA damage in the presence of prostaglandin H synthase (Schilderman *et al.*, 1993). It is not yet known whether the GSH conjugates can also catalyze such DNA damage, but this would appear likely in view of the effects of GSH conjugation on the redox properties of hydroquinones, and on the ability of the quinol-GSH conjugates to catalyze 8-hydroxy-deoxygaunosine formation.

2-Bromo-(diglutathion-*S*-yl)hydroquinone

The nephrotoxicity of bromobenzene is mediated *via* hepatic cytochrome P450 catalyzed oxidation to 2-bromohydroquinone, followed by conjugation with GSH and delivery of the conjugates to the kidney (Monks and Lau, 1994a,b). We have recently addressed the cellular and molecular mechanism of 2-Br-(diglutathion-*S*-yl)hydroquinone-mediated nephrotoxicity.

The Brush Border Membrane. We integrated the time dependent alterations in biochemical parameters and cell morphology to identify subcellular targets of 2-Br-(diglutathion-*S*-yl)hydroquinone and to gain initial insights into the mechanism of 2-Br-(diglutathion-*S*-yl)hydroquinone toxicity *in vivo.* The majority of brush border γ-GT excreted in response to 2-Br-(diglutathion-*S*-yl)hydroquinone occurs in the first or second voiding of the bladder (4-6 hr after treatment) and occurred prior to any sign of overt renal damage (Rivera *et al.*, 1994). Significant elevations in blood urea nitrogen concentrations only occurred 8 hr after treatment (Rivera *et al.*, 1994). The rapid increase in γ-GT activity in the urine is indicative of damage to the brush border membrane, and was confirmed by electron microscopy. As early as 30 min following 2-Br-(diglutathion-*S*-yl)hydro-quinone administration there was extensive loss of the brush border membrane into the tubular lumen (Rivera *et al.*, 1994). The significance of the loss of the brush border membrane in response to such structurally diverse chemicals or hypoxic stress (Reimer *et al.*, 1972) is unclear. However, since loss of the apical membrane occurs in response to diverse stimuli, this suggests a common signalling pathway (stress response ?) may be involved. Alternatively, since the activity of γ-GT is necessary for 2-Br-(diglutathion-*S*-yl)hydroquinone mediated nephrotoxicity, probably by either facilitating the uptake into renal cells of the corresponding cystein-*S*-ylglycine or cystein-*S*-yl conjugate, and/or the oxidation of the quinol moiety (Monks and Lau, 1990, 1994b), it is possible that a direct interaction occurs between reactive metabolites of 2-Br-(diglutathion-*S*-yl)hydroquinone and the brush border membrane. In support of this view, covalently bound radiolabel is associated with the plasma membrane fraction of renal tissue obtained from rats treated with 2-Br-[^{14}C]-HQ (Rivera *et al.*, unpublished data).

The Nucleus. The nucleus is also an early target of 2-Br-(diglutathion-*S*-yl)hydroquinone induced nephrotoxicity, and a potential role for hydrogen peroxide (H_2O_2) and the hydroxyl radical (•OH) in 2-Br-(diglutathion-*S*-yl)hydroquinone mediated cytotoxicity to cultured renal proximal tubular epithelial cells was proposed. An additional early target revealed by morphological studies was the nucleus, which exhibited severe karyolysis (Rivera *et al.*, 1994). During cell necrosis, mitochondrial swelling usually precedes nuclear changes (Wyllie *et al.*, 1980). Our studies contrast with this scenario in that 2-Br-(diglutathion-*S*-yl)hydroquinone induced severe karyolysis prior to any apparent adverse effects on mitochondria (see below). Early changes in nuclear structure, in the presence of intact mitochondria, do occur during apoptotic cell death (Wyllie *et al.*, 1980; Boobis *et al.*, 1989; Bursch *et al.*, 1992) but the decrease in cytosolic density caused by 2-Br-(diglutathion-*S*-yl)hydroquinone contrasts with the increase in cytosolic density that occurs during apoptotic cell death. Thus, 2-Br-(diglutathion-*S*-yl)hydroquinone-mediated cell death exhibits some distinctive morphological features.

Consistent with the *in vivo* data, 2-Br-(diglutathion-*S*-yl)hydroquinone causes the formation of single strand breaks in DNA in renal proximal tubular epithelial cells (LLC-PK_1) (Mertens *et al.*, 1995). DNA damage triggers a variety of cellular responses which, in a coordinated fashion, are designed to repair damage prior to cell division. One such repair response involves the activation of poly(ADP-ribose)polymerase. This enzyme is inhibited by 3-aminobenzamide (Purnell and Whish, 1980), which represses the repair of DNA strand breaks (James and Lehman, 1982; Zwelling *et al.*, 1982; Ahnstrom and Ljungman, 1988). This can either increase (James and Lehman, 1982; Nduka *et al.*, 1980; Cleaver and Morgan, 1985; Shen *et al.*, 1992) or decrease (Seto *et al.*, 1985; Tanizawa *et al.*, 1987) toxicity, depending on the degree of DNA-damage. Incubation of LLC-PK_1 cells with 3-aminobenzamide following treatment with 2-Br-(diglutathion-*S*-yl)hydroquinone partially decreased cytotoxicity (Mertens *et al.*, 1995), suggesting that utilization of cofactors (NAD^+ and ATP) required for efficient DNA repair may adversely affect other cellular processes (Berger, 1985; Schraufstatter *et al.*, 1986; Gaal *et al.*, 1987), and exacerbate toxicity. Consistent with an increased demand for ATP to effect DNA repair is the observation that mitochondrial dehydrogenases appear to be activated within 30 min of 2-Br-(diglutathion-*S*-yl)hydroquinone exposure (Mertens *et al.*, 1995). Indeed, extensive DNA fragmentation, could be observed as early as 15 min after exposure to 400 μM 2-Br-(diglutathion-*S*-yl)hydroquinone, the severity of which decreased following initial exposure, consistent with the activation of DNA repair mechanisms. In addition, cytotoxicity and DNA fragmentation in this cell model appear to be independent of endonuclease activation (Mertens *et al.*, 1995). The finding that the nucleus and DNA are sensitive targets of 2-Br-(diglutathion-*S*-yl)hydroquinone is further supported by the observation that the growth arrest and DNA damage inducible gene, *gadd*153, is activated by relatively low concentrations (50 μM) of 2-Br-(diglutathion-*S*-yl)hydroquinone (Jeong *et al.*, 1994).

2-Br-(diglutathion-S-yl)hydroquinone-Mediated Generation of Reactive Oxygen. Although the mechanism of 2-Br-(diglutathion-*S*-yl)hydroquinone-mediated nuclear damage is unclear, it is known that the GSH and *N*-acetylcysteine conjugates of menadione can redox cycle, with the concomitant formation of reactive oxygen species (Wefers and Sies, 1983; Brown *et al.*, 1991) and 2-methyl-3-(*N*-acetylcystein-*S*-yl)-1,4-naphthoquinone is nephrotoxic when administered to rats (Lau *et al.*, 1990). Semi-quinone free radical formation from 2-methyl-3-(glutathion-*S*-yl)-1,4-naphthoquinone, 3-(glutathion-*S*-yl)-1,4-naphthoquinone, 2,3-(di-glutathion-*S*-yl)-1,4-naphthoquinone and 2-(glutathion-*S*-yl)-1,4-benzoquinone (Takahashi *et al.*, 1987; Rao *et al.*, 1988) and 2,6-dimethoxy-3,5-(di-glutathion-*S*-yl)hydroquinone (Wolf and Spector, 1987) has been demonstrated. We therefore investigated whether reactive oxygen species play a role in 2-Br-(diglutathion-*S*-yl)hydroquinone-mediated-mediated cytotoxicity. Treatment with catalase, and pretreatment with deferoxamine mesylate protected LLC-PK_1 cells against 2-Br-(diglutathion-*S*-yl)hydroquinone-mediated-mediated DNA single strand breaks (unpublished data) and cytotoxicity (Mertens *et al.*, 1995) suggesting that toxicity involves the iron-catalyzed Haber-Weiss reaction, in which $O_2^{\bullet -}$ undergoes dismutation to form H_2O_2, and reduces Fe^{3+} to Fe^{2+}. The H_2O_2 then reacts with Fe^{2+} to generate the hydroxyl radical, which is probably the reactive species responsible for the DNA damage. In support of this scenario, the alkaline elution profile obtained with 2-Br-(diglutathion-*S*-yl)hydroquinone-mediated was non-linear, and similar to elution profiles reported in studies examining DNA single strand breaks induced by H_2O_2 (Djuric *et al.*, 1993) and the phorbol ester tumor promoters (Sun *et al.*, 1992). In addition, the pattern of DNA repair may also reflect the formation of quinone-thioether derived hydroxyl radicals (Mertens *et al.*, 1995) which cause several distinct modified bases, and which are likely to be repaired at different rates. Exposure of bacterial and mammalian cells to H_2O_2 or hydroquinone has also been shown to cause extensive DNA damage (Mello Filho and Meneghini, 1984; Birnboim, 1986; Schraufstatter *et al.*, 1986; Leanderson and Tagesson, 1992). Catalase and deferoxamine mesylate protect against toxicity mediated by H_2O_2 directly (Kvietys *et al.*, 1989), or *via* its generation from hypoxanthine/xanthine oxidase (Kvietys *et al.*, 1989; Andreoli and McAteer, 1990) and glucose/glucose oxidase (Walker and Shah, 1991). The hydroxyl radical can initiate DNA fragmentation *via* base modifications (Dizdaroglu *et al.*, 1991) and/or *via* attack on the sugar phosphate backbone (Imlay and Linn, 1988), eventually resulting in cell death (Mello Filho *et al.*, 1984; Birnboim, 1986;Kvietys *et al.*, 1989). The cellular source of the Fe^{3+}/Fe^{2+} redox couple, and the site of H_2O_2 formation are unclear and are under investigation. Bathocuproine did not protect against 2-Br-(diglutathion-*S*-yl)hydroquinone-mediated-mediated cytotoxicity (Mertens *et al.*, 1995) indicating that Cu^{2+} ions are not required for the generation of reactive oxygen species.

Mitochondria. In contrast to the alterations observed in the nucleus and brush border membrane, biochemical and morphological studies indicate that mitochondria are not primary targets of either 2-Br-(diglutathion-*S*-yl)hydroquinone (Rivera *et al.*, 1994) or 2,3,5-(triglutathion-*S*-yl)hydroquinone (Hill *et al.*, 1992). Thus, mitochondrial respiratory function was not altered 30 min after 2-Br-(diglutathion-*S*-yl)hydroquinone administration and morphological studies indicated the presence of intact mitochondria at this time (Rivera *et al.*, 1994). Changes in mitochondrial morphology observed between 2 and 4 hr were consistent with functional mitochondria. Thus, condensation of the mitochondrial matrix, in addition to intercristal swelling, are considered to be reversible alterations associated with an active State 3 respiration in the presence of an increased ADP to ATP ratio (Tzagoloff, 1982; Hackenbrock, 1966; Trump *et al.*, 1965). Although a significant decrease in the respiratory control ratiowas observed between 2-4 hr after 2-Br-(diglutathion-*S*-yl)hydroquinone (⇑State 4) similar early increases in State 4 respiration after 2,3,5-(triglutathion-*S*-yl)hydroquinone administration were unrelated to the development of cytotoxicity (Hill *et al.*, 1992). Decreases in the respiratory control ratio beyond 8 hr are a consequence of a decrease in State 3 respiration and occurred concomitantly with significant elevations in blood urea nitrogen concentrations (Rivera *et al.*, 1994). The finding that mitochondrial condensation is not followed by secondary high amplitude swelling prior to cell death and necrosis in this mode of chemical-induced nephrotoxicity represents a novel finding that to our knowledge has not been described with any other nephrotoxic agent.

We subsequently investigated the time course of cellular alterations caused by 2-Br-(diglutathion-*S*-yl)hydroquinone in LLC-PK_1 cells. Neutral red accumulation, MTT-formazan formation, and intracellular lactate dehydrogenase activity were used as indicators of lysosomal, mitochondrial, and plasma membrane integrity, respectively. Lysosomal neutral red accumulation was the most sensitive indicator of viability (Mertens *et al.*, 1995), indicative of H^+ loss from lysosomes. Consistent with these findings, intracellular acidification occurs rapidly (<30 min) in LLC-PK_1 exposed to 2-Br-(diglutathion-*S*-yl)hydroquinone (Albuquerque *et al.*, 1995). Viability determined by either MTT-formazan accumulation or intracellular lactate dehydrogenase activity was essentially identical in 2-Br-(diglutathion-*S*-yl)hydroquinone treated cultures suggesting that damage to mitochondria occurs concomitant with the loss of plasma membrane integrity. Thus, the LLC-PK_1 cell model appears to reflect the temporal changes in renal proximal tubules caused by 2-Br-(diglutathion-*S*-yl)hydroquinone *in vivo*.

In summary, the nucleus and brush border membrane appear to be early targets of 2-Br-(diglutathion-*S*-yl)hydroquinone-mediated renal proximal tubular cell necrosis *in vivo*. In addition, the cytotoxicity mediated by 2-Br-(diglutathion-*S*-yl)hydroquinone in cultured renal epithelial cells is probably due to the generation of hydrogen peroxide and subsequent iron catalyzed formation of hydroxyl radicals. A major target of these reactions is the nucleus, since DNA single strand breaks occur rapidly at relatively low concentrations. The molecular mechanisms responsible for the tissue and cellular response to 2-Br-(diglutathion-*S*-yl)hydro-quinone-mediated nephrotoxicity are under investigation.

ACKNOWLEDGMENTS

Portions of the work described in this manuscript were supported by grant number ES 04662 from the National Institute of Environmental Health Sciences (T.J.M.) and GM

39338 from the National Institute of General Medical Sciences (S.S.L.). Maria I. Rivera was supported by a MARC predoctoral fellowship (GM 13270)

REFERENCES

Ahnström, G., and Ljungman, M. (1988) Effects of 3-aminobenzamide on the rejoining of DNA-strand breaks in mammalian cells exposed to methyl methanesulphonate; Role of poly(ADP-ribose) polymerase. *Mutat. Res.* 194, 17-22.

Albuquerque, S.J., Monks, T.J. and Lau, S.S. (1995) Modulation of intracellular pH by 2-bromo-6-glutathion-*S*-yl-hydroquinone. *Toxicologist,* 15, 305 (Abstract).

Andreoli, S., and McAteer, J.A. (1990) Reactive oxygen molecule mediaed injury in endothelial and renal tubular epithelial cells in vitro. *Kidney Int.* 38, 785-794.

Berger, N.A. (1985) Symposium: Cellular response to DNA damage; The role of poly(ADP-ribose). Poly(ADP-ribose) in the cellular response to DNA damage. *Rad.Res.* 101, 4-15.

Birnboim, H.C. (1986) DNA strand breaks in human leukocytes induced by superoxide anion, hydrogen peroxide and tumor promoters are repaired slowly compared to breaks induced by ionizing radiation. *Carcinogenesis* 7, 1511-1517.

Boobis, A.R., Fawthrop, D.J., and Davies, D.S. (1989). Mechanisms of cell death. *TIPS,* 10, 275-280.

Brown, P.C., Dulik, D.M., and Jones, T.W. (1991) The toxicity of menadione (2-methyl-1,4-naphthoquinone) and two thioether conjugates studied with isolated renal epithelial cells. *Arch. Biochem. Biophys.,* 285, 187-196.

Bursch, W., Oberhammer, F., and Schulte-Hermann, R.(1992). Cell death by apoptosis and its protective role against disease. *TIPS.* 13, 245-251.

Canales, P.L., Kleiner, H.E., Monks, T.J., and Lau, S.S. (1993) Formation of 8-hydroxydeoxygaunonsine by quinol-thioethers. *Toxicologist,* 13, 202 (Abstr).

Cleaver, J.E., and Morgan, W.F. (1985) Poly(ADP-ribose) synthesis is involved in the toxic effects of alkylating agents but does not regulate DNA repair. *Mutation Res.* 150, 69-76.

Dekant, W., Vamvakas, S. and Anders, M.W. (1994) Formation and fate of nephrotoxic and cytotoxic glutathione *S*-conjugates: Cysteine conjugate β-lyase pathway. In: *Conjugation-Dependent Carcinogenicity and Toxicity of Foreign Compounds.* (M.W. Anders and W. Dekant, Eds.) pp. 115-162, Academic Press, San Diego CA.

De Mello Filho, A.C., Hoffmann, M.E., and Meneghini, R. (1984) Cell killing and DNA damage by hydrogen peroxide are mediated by intracellular iron. *Biochem. J.* 218, 273-275.

De Mello Filho, A.C. and Meneghini, R. (1984) *In vivo* formation of single-strand breaks in DNA by hydrogen peroxide is mediated by the Haber-Weiss reaction. *Biochim. Biophys. Acta* 781, 56-63.

Dizdaroglu, M., Nackerdien, Z., Chao, B-C., Gajewski, E., and Rao, G. (1991) Chemical nature of *in vivo* DNA base damage in hydrogen peroxide treated mammalian cells. *Arch. Biochem. Biophys.* 285, 388-390.

Djuric, Z., Everett, C.K. and Luongo, D. (1993) Toxicity, single-strand breaks, and 5-hydroxymethyl-2'-deoxyuridine formation in human breast epithelial cells treated with hydrogen peroxide. *Free Rad. Biol. Med.* 14, 541-547.

Eckert, K.-G., Eyer, P., Sonnenbichler, J. and Zetl, I. (1990a) Activation and detoxication of aminophenols. II. Synthesis and structural elucidation of various thiol addition products of 1,4-benzoquinoneimine and *N*-acetyl-1,4-benzoquinoneimine. *Xenobiotica,* 20, 333-350

Eckert, K.-G., Eyer, P., Sonnenbichler, J. and Zetl, I. (1990b) Activation and detoxication of aminophenols. III. Synthesis and structural elucidation of various glutathione addition products to 1,4-benzoquinone. *Xenobiotica,* 20, 351-361.

Fowler, L.M., Moore, R.D., Foster, J.R., and Lock, E.A. (1991) Nephrotoxicity of 4-aminophenol glutathione conjugate. *Human Expt. Toxicol.,* 10, 451-459.

Fowler, L.M., Foster, J.R., and Lock, E.A. (1993) The effect of ascorbic acid, acivicin and probenecid on the nephrotoxicity of 4-aminophenol in the Fischer 344 rat. *Arch. Toxicol.,* 67, 613-621.

Gaal, J.C., Smith, K.R., and Pearson, C.K. (1987) Cellular euthanasia mediated by a nuclear enzyme: A central role for nuclear ADP-ribosylation in cellular metabolism. *Trends in Biochem. Sci.* 12, 129-130.

Gartland, K.P.R., Bonner, F.W., Timbrell, J.A., and Nicholson, J.K. (1989) Biochemical characterization of 4-aminophenol-induced nephrotoxic lesions in the F344 rat. *Arch. Toxicol.,* 63, 97-106, 1989.

Green, C.R., Ham, K.N., and Tange, J.D. (1969) Kidney lesions induced in rats by 4-aminophenol. *Br. Med. J.,* 1, 162-164.

Hackenbrock, C.R. (1966). Ultrastructural basis for metabolically linked mechanical activity in mitochondria. I. Reversible ultrastructural changes with change in metabolic steady state in isolated liver mitochondria. *J. Cell Biol.,* 30, 269-297.

Hirose, M., Inoue, T., Masuda, A., Tsuda, H., and Ito, N. (1987) Effects of simultaneous treatment with various chemicals on BHA-induced development of rat forestomatch hyperplasia. Complete inhibition by diethylmaletate in a 5-week feeding study. *Carcinogenesis*, 8, 1555-1558.

Hill, B.A., Heather, H.K., Ryan, E.A., Dulik, D.M., Monks, T.J. and Lau, S.S. (1993) Identification of multi-S-substituted conjugates of hydroquinone by HPLC-coulometric electrode array analysis and mass spectroscopy. *Chem Res. Toxicol.* 6, 459-469.

Hill, B.A., Monks T.J., and Lau S.S. (1992). The effects of 2,3,5-(triglutathion-*S*-yl)hydro-quinone on renal mitochondrial respiratory function *in vivo* and *in vitro*: Possible role in cytotoxicity. *Toxicol. Appl. Pharmacol.*, 117, 165-171.

Imlay, J.A. and Linn, S. (1988) DNA damage and oxygen radical toxicity. *Science* 240, 1302-1309.

James, M.R. and Lehman, A.R.(1982) Role of poly(adenosine diphosphate ribose) in deoxyribonucleic acid repair in human fibroblasts. *Biochemistry* 21, 4007-4013.

Jeong, J.K., Stevens, J.L., Lau, S.S., and Monks, T.J. (1994) Gene expression in response to quinone-thioethers in LLC-PK1 cells. *Toxicologist*, 14, 180 (Abstract).

Kari, F.W., Bucher, J., Eustis, S.L., Haseman, J.K., and Huff, J.E. (1992) Toxicity and carcinogenicity of hydroquinone in F344/N rats and B6C3F1 mice. *Food Chem. Tox.*, 30, 737-747.

Klos, C., Koob, M., Kramer, C., and Dekant, W. (1992) *p*-Aminophenol nephrotoxicity: Biosynthesis of toxic glutathione conjugates. *Toxicol. Appl. Pharmacol.*, 115, 98-106.

Kvietys, P.R., Inauen, W., Bacon, B.R., and Grisham, M.B. (1989) Xanthine oxidase-induced injury to endothelium: Role of intracellular iron and hydroxyl radical. *Am. J. Physiol.* 257, H1640-H1646.

Lau, S.S., Hill, B.A., Highet, R.J., and Monks, T.J. (1988) Sequential oxidation and glutathione addition to 1,4-benzoquinone: Correlation of toxicity with increased glutathione substitution. *Molec. Pharmacol.* 34, 829-836.

Lau, S.S., Jones, T.W., Highet, R.J., Hill, B.A. and Monks, T.J. (1990) Differences in the localization and extent of the renal proximal tubular necrosis caused by mercapturic acid and glutathione conjugates of menadione and 1,4-naphthoquinone. *Toxicol. Appl. Pharmacol.*, 104, 334-350.

Lau, S.S., Peters, M.M., Meussen, E., Rivera, M.I., Jones, T.W., van Ommen, B., van Bladeren, P.J., and Monks, T.J. (1994) Nephrotoxicity of 2-*tert*-butyl-hydroquinone glutathione conjugates. *Toxicologist*, 14, 180 (Abstract).

Leanderson, P. and Tagesson, C. (1992) Cigarette smoke-induced DNA damage in cultured human lung cells: Role of hydroxyl radicals and endonuclease activation. *Chem. Biol. Int.* 81, 197-208.

Lunte, S.M. and Kissinger, P.T. (1983) Detection and identification of sulfhydryl conjugates of *p*-benzoquinone in microsomal incubations of benzene and phenol. *Chem. Biol. Int.*, 47, 195-212.

Mertens, J.J.W.M., Gibson, N.W., Lau, S.S. and Monks, T.J. Reactive oxygen species and DNA damage in 2-bromo-(glutathion-*S*-yl)hydroquinone mediated cytotoxicity. *Arch. Biochem. Biophys.* In Press, 1995.

Mertens, J.J.W.M., Temmink, J.H.M., van Bladeren, P.J., Jones, T.W., Lo, H.-H., Lau, S.S., and Monks, T.J. (1991) Inhibition of γ-glutamyl transferase potentiates the nephrotoxicity of glutathione conjugated chlorohydroquinones. *Toxicol. Appl. Pharmacol.*, 110, 45-60.

Monks, T.J. and Lau, S.S. (1990) Glutathione conjugation, γ-glutamyl transpeptidase and the mercapturic acid pathway as modulators of 2-bromohydroquinone oxidation. *Toxicol. Appl. Pharmacol.*, 103, 557-563. 1990.

Monks, T.J. and Lau, S.S. (1992) Toxicology of quinone-thioethers. *CRC Crit. Rev. Tox.* 22, 243-270.

Monks, T.J. and Lau, S.S. (1994a) Glutathione conjugation as a mechanism for the transport of reactive metabolites. In: *Conjugation-Dependent Carcinogenicity and Toxicity of Foreign Compounds*. (M.W. Anders and W. Dekant, Eds.) pp. 183-210, Academic Press, San Diego CA.

Monks, T.J. and Lau, S.S. (1994b) Glutathione conjugate mediated toxicities. In: *Handbook of Experimental Pharmacology, Vol. 112: Conjugation-Deconjugation Reactions in Drug Metabolism and Toxicity.* (Kauffman, F.C., Ed.) pp. 459-509, Springer Verlag, Berlin Heidelberg.

Monks, T.J., Lau, S.S., Highet, R.J. and Gillette, J.R. (1985) Glutathione conjugates of 2-bromohydroquinone are nephrotoxic. *Drug Metab. Dispos.* 13, 553-559.

Monks, T.J., Highet, R.J. and Lau, S.S. (1988) 2-Bromo-(diglutathion-S-yl)hydroquinone nephrotoxicity: Physiological, biochemical and electrochemical determinants. *Molec. Pharmacol.* 34, 492-500.

Nduka, N., Skidmore, C.J., and Shall, S. (1980) The enhancement of cytotoxicity of *N*-methyl-*N*-nitrosourea and of γ-irradiation by inhibition of poly(ADP-ribose) polymerase. *Eur. J. Biochem.* 105, 525-530.

Nerland, D.E. and Pierce, W.M. (1990) Identification of *N*-acetyl-*S*-(2,5-dihydroxyphenyl)-L-cysteine as a urinary metabolite of benzene, phenol, and hydroquinone. *Drug Metab. Disp.*, 18, 958-961.

Newton, J.F., Kuo C.-H., Gemborys, M.W., Mudge, G.H., and Hook, J.B. (1982) Nephrotoxicity of 4-aminophenol, a metabolite of acetaminophen in the F344 rat. *Toxicol. Appl. Pharmacol.*, 65, 336-344.

Peters, M.M.C.G., Jones, T.W., Monks, T.J., and Lau, S.S. (1995) Cytotoxicity and cell proliferation induced by the nephrocarcinogen hydroquinone and its tri-glutathionyl metabolite. *Toxicologist,* 15, 231 (Abstract).

Purnell, M. and Whish, W.J.D. (1980) Novel inhibitors of poly(ADP-ribose)synthetase. *Biochem. J.* 185, 775-777.

Rao, D.N.R., Takahashi, N., and Mason, R.P. (1988) Characterization of a glutathione conjugate of the 1,4-benzosemiquinone-free radical formed in rat hepatocytes. *J Biol Chem* 263: 17981-17986.

Reimer, K.A., Ganote, C.E., and Jennings, R.B. (1972). Alterations in renal cortex following ischemic injury. III. Ultrastructure of proximal tubules after ischemia or autolysis. *Lab. Invest.* 26, 347-363.

Rivera, M.I., Jones, T.W., Lau, S.S. and Monks, T.J. (1994) Early morphological and biochemical changes during 2-bromo-(diglutathion-*S*-yl)hydroquinone-induced nephrotoxicity. *Toxicol. Appl. Pharm.*, 128, 239-250.

Sawahata, T. and Neal, R.A. (1983) Biotransformation of phenol to hydroquinone and catechol by rat liver microsomes. *Molec. Pharmacol.*, 23, 453-460.

Schilderman, P.A.E.L., van Maanen, J. M. S., Smeets, E.J., ten Hoor, F., and Kleinjans, J. C. S. (1993) Oxygen radical formation during prostaglandin H synthase-mediated biotransformation of butylated hydroxy-anisole. *Carcinogenesis*, 114, 347-353.

Schraufstatter, I.U., Hinshaw, D.B., Hyslop, P.A., Spragg, R.G., and Cochrane, C.G. (1986) Oxidant injury of cells. DNA strand-breaks activate polyadenosine diphosphate-ribose polymerase and lead to depletion of nicotinamide adenine dinucleotide. *J. Clin. Invest.*, 77, 1312-1320.

Seto, S., Carrera, C.J., Kubota, M., Wasson, D.B., Carson, D.A. (1985) Mechanism of deoxyadenosine and 2-chlorodeoxyadenosine toxicity to nondividing human lymphocytes. *J. Clin. Invest.* 75, 377-383.

Shen, W., Kamendulis, M., Ray, S.D., and Corcoran, G.B. (1992) Acetaminophen-induced cytotoxicity in cultured mouse hepatocytes: Effects of Ca^{2+}-endonuclease, DNA repair, and glutathione depletion inhibitors on DNA fragmentation and cell death. *Toxicol. Appl. Pharmacol.* 112, 32-40.

Shibata, M-A., Hirose, M. Tanaka, H., Asakawa, E., Shirai, T., and Ito, N. (1991) Induction of renal cell tumors in rats and mice, and enhancement of hepatocellular tumor development in mice after long-term hydroquinone treatment, *Jpn. J. Cancer Res.*, 82, 1211-1219.

Sies H, and Ketterer, B. (1988) Glutathione conjugation: Mechanisms and biological significance, Academic Press, San Diego, CA.

Sun, Y. Pommier, Y. Colburn, N.H. (1992) Acquisition of a growth inhibitory response to phorbol esters involves DNA damage. *Cancer Res*. 52, 1907-1915.

Tajima, K., Hashizaki, M., Yamamoto, K., and Mizutani. T. (1991) Identification and structure characterization of *S*-containing metabolites of 3-*tert*-butyl-4-hydroxyanisole in rat urine and liver microsomes. *Drug Metab. Disp.*, 19, 1028-1033.

Takahashi, N., Schreiber, J., Fischer, V., and Mason, R.P. (1987) Formation of glutathione-conjugated semiquinones by the reaction of quinones with glutathione: An ESR study. *Arch Biochem Biophys* 252: 41-48.

Tanizawa, A., Kubota, M., Takimoto, T., Akiyama, Y., Seto, S., Kiriyama, Y., and Mikawa, H. (1987) Prevention of adriamycin-induced interphase death by 3-aminobenzamide and nicotinamide in a human pre-myelocytic leukemia cell line. *Biochem. Biophys. Res. Communications* 144, 1031-1036.

Trump, B.F., Goldblatt, P.J., and Stowell, R.E. (1965). Studies on necrosis of mouse liver *in vitro*: Ultrastructural alterations in the mitochondria of hepatic parenchymal cells. *Lab. Invest.* 14, 343-371.

Tsuda, H., Fukushima, S., Imaida, K., Sakata, T., and Ito, N. (1984) Modification of carcinogenesis by antioxidants and other compounds. *Acta. Pharmacol. Toxicol.*, 55, 125-143.

Tunek, A., Platt, K.L., Pryzbylski, M., and Oesch, F. (1980) Multi-step metabolic activation of benzene. Effect of superoxide dismutase on covalent binding to microsomal macromolecules, and identification of glutathione conjugates using high pressure liquid chromatography and field desorption mass spectrometry. *Chem. Biol. Int.*, 33, 1-17.

Tzagoloff, A. (1982). *Mitochondria.* p.17. Plenum Press, New York.

Walker, P.D. and Shah, S.H. (1991) Hydrogen peroxide cytotoxicity in LLC-PK_1 cells: A role for iron. *Kidney Int.* 40, 891-898.

Wefers, H. and Sies, H. (1983) Hepatic low-level chemiluminesence during redox cycling of menadione and the menadione-glutathione conjugate: Relation to glutathione and NAD(P)H: Quinone reductase (DT-diaphorase) activity. *Arch. Biochem. Biophys.*, 224, 568-578.

Wolf, S.P., and Spector, A. (1987). Pro-oxidant activation of ocular reductants. II. Lens epithelial cell cytotoxicity of a dietary quinone is associated with a stable free radical formed with glutathione *in vitro*. *Exp. Eye Res.* 45, 791-801.

Wyllie, A.H., Kerr, J.F.R., and Currie, A.R. (1980). Cell death: The significance of apoptosis. *Int. Rev. Cytol.* 68, 251-306.

Zwelling, L.A., Kerrigan, D., and Pommier, Y. (1982) Inhibition of poly-(adenosine diphosphoribose) synthesis slows the resealing rate of X-ray induced DNA strand breaks. *Biochem. Biophys. Res. Comm.* 104, 897-902.

REACTIVE INTERMEDIATES OF XENOBIOTICS IN THYROID

Formation and Biological Consequences

U. Andrae

GSF-Forschungszentrum für Umwelt und Gesundheit
Institut für Toxikologie, Neuherberg
D-85764 Oberschleissheim, Germany

INTRODUCTION

The formation of reactive intermediates from xenobiotics in thyroid and their potential toxic effects on the organ have received only very little attention up to now. At first glance this appears surprising, as numerous chemicals have been shown to be capable of inducing thyroid tumours in experimental animals. The lack of interest in the fate of these chemicals in thyroid is largely the consequence of the current theory of thyroid follicular cell carcinogenesis which centers on perturbations of the hormonal regulatory system which are caused by many thyroid carcinogens (Hill et al. 1989, Thomas and Williams 1991).

To understand the way in which administration of xenobiotics may lead to thyroid tumours through a disturbance of thyroid hormone homeostasis, it may be helpful to have a short look on the mechanisms responsible for maintaining homeostasis.

The highest tier of control is exerted by the hypothalamus, which secretes thyrotropin-releasing hormone, TRH. TRH causes the release of thyroid-stimulating hormone (TSH) from the anterior pituitary. TSH stimulates several key stages in thyroid hormone synthesis which results in increased secretion of thyroid hormones into the blood. The circulating thyroid hormones, especially T4, depress TSH release by the pituitary. There is, therefore, an inverse relationship between the levels of circulating thyroid hormones and the level of TSH secretion. This system provides an extremely sensitive feedback mechanism to maintain adequate production of thyroid hormones.

Application of xenobiotics which inhibit thyroid hormone synthesis, so-called "antithyroid" agents, reduces the output of thyroid hormones into the blood, which results in compensatory release of excess TSH from the pituitary. TSH is the main growth factor for the thyroid epithelium, and one effect of a long-term application of these compounds is a chronic growth stimulus to the thyroid which is thought to result in an increased frequency of mutations due to increased cell replication (Cohen and Ellwein 1990). It is assumed that some of the mutated cells aquire the ability to secrete certain growth factors, such as IGF-1

Biological Reactive Intermediates V
Edited by R. Snyder et al., Plenum Press, New York, 1996

(Thomas and Williams 1991), which would result in their clonal expansion. The consequence of these processes can be thyroid hyperplasia and, eventually, cancer.

INTERACTION OF ANTITHYROID CHEMICALS WITH THYROID PEROXIDASE AND FORMATION OF REACTIVE INTERMEDIATES

There are numerous inhibitors of thyroid hormone synthesis which have been shown to cause thyroid tumours in experimental animals (Hill et al. 1989). Mechanistic studies on the action of these compounds on thyroid were almost exclusively performed on xenobiotics known to inhibit thyroid peroxidase (TPO), the key enzyme in thyroid hormone synthesis. TPO is responsible for the oxidation of iodide, the iodination of tyrosine residues in thyroglobulin and the coupling of iodotyrosines to give the thyroid hormones. Several of these studies actually used lactoperoxidase (LPO) as model for TPO, because both enzymes are very similar for many criteria (Ohtaki et al. 1982, 1985) but LPO is much more available than TPO. These experiments, which were aimed at elucidating the mechanism underlying the inhibition, were mainly conducted by Daniel Doerge and his co-workers in Honolulu. They have shown that these agents are metabolized by the TPO, that reactive intermediates are formed during metabolism and that even closely related inhibitors can differ considerably in their mode of action. The following mechanisms resulting in an inhibition of TPO (or LPO) and the formation of reactive metabolites will be briefly discussed using ethylene thiourea, methimazole and related compounds and amitrole as examples.

Ethylene Thiourea

Ethylene thiourea (ETU) has been shown to cause inhibition of thyroid hormone synthesis *via* an interaction with the iodinating species of TPO (Doerge and Takazawa 1990). This species is formed from the porphyrin cation radical form of TPO, compound I, by reaction with iodide. Inhibition is the consequence of a competition of ETU with tyrosines in thyroglobulin for the iodinating species. When ETU is used up, tyrosine iodination continues without any remaining effect on TPO. Thus, TPO inhibition by ETU is a completely reversible process (Doerge and Takazawa 1990).

The interaction of ETU with the TPO in the presence of iodide is accompanied by the oxidative metabolism of ETU. Oxidation leads to the formation of the highly reactive intermediate imidazoline sulfinic acid, which decomposes to yield imidazoline and bisulfite (Doerge and Takazawa 1990).

Methimazole and 1-Methylbenzimidazoline-2-thione

In contrast to the reversible peroxidase inhibition by ETU, inhibition by other thiocarbamides, such as methimazole or 1-methylbenzimidazoline-2-thione, is essentially irreversible (Doerge 1986). These inhibitors do not interact with compound I, the porphyrin cation radical form, but with another ferryl form of the enzyme, where the radical is centered on the enzyme protein. Interaction with the protein radical results in the formation of an intermediate which can lead to a covalent binding between the heme moiety and the thiocarbamide in a 1:1 mode. The result of this binding is the irreversible suicide inactivation of the enzyme. Alternatively, the intermediate can turnover to release the enzyme and oxidized thiocarbamide products (Doerge 1986).

It has been shown that, similar to the oxidation of ETU, the highly reactive sulfenic or sulfinic acid derivatives of the thiocarbamides are produced and that these intermediates are responsible for covalent binding and inactivation of the enzyme. And again, as with ETU, bisulfite is formed as a product of the oxidation (Doerge 1988).

Amitrole

Clues to the mechanism by which the herbicide amitrole (3-amino-1,2,4-triazole) inhibits TPO have been obtained by studies on the interaction of the compound with LPO. Binding of amitrole to the enzyme in the presence of H_2O_2 leads to the formation of a complex which can either turnover to release the peroxidase and oxidized amitrole metabolites or which can result in a covalently modified enzyme which is irreversibly inactivated. However, in contrast to the reaction with the thiocarbamides, binding of the amitrole metabolite to the enzyme occurs at the protein, not at the heme, and 7 amitrole equivalents are bound to specific amino acids of the enzyme molecule (Doerge and Niemczura 1989). The nature of the reactive amitrole intermediate involved in binding is not yet clear. However, the predominant mode of action of peroxidases with arylamines is N-oxidation to yield nitrogen-centered free radicals (Eling et al. 1990), and it has been suggested that formation of a cation radical may be responsible for the inhibition of TPO and thyroid function by amitrole (Doerge and Niemczura 1989).

GENOTOXICITY AND CARCINOGENICITY OF ANTITHYROID CHEMICALS

The majority of compounds known to cause thyroid tumours in experimental animals have been shown capable of activating the thyroid-pituitary regulatory feedback system and causing increased TSH secretion (Hill et al. 1989, Thomas and Williams 1991). The critical role of an elevated TSH level as a promoter in thyroid carcinogenesis by these agents has been demonstrated in numerous experiments (Thomas and Williams 1991). However, the observation that the tumorigenic effect of radiation, which is of course a genotoxic agent, can be inhibited by hypophysectomy (Nadler et al. 1970) or administration of thyroid hormone (Doniach 1974) indicates that a potential genotoxic component of the tumorigenicity of an antithyroid agent may be very difficult to identify. Therefore, it is no wonder that the possibility of an involvement of reactive intermediates of these compounds in thyroid not only in the inhibition of TPO and thyroid function, but also in the induction of genotoxic effects in the organ, has received very little attention. However, there are three lines of evidence suggesting that it may be advisable not to preclude an involvement of genotoxic mechanisms in the induction of thyroid carcinogenesis *a priori*, namely the observations that some of the antithyroid compounds (a) effect TPO inhibition *via* the formation of reactive metabolites, as discussed above, (b) show genotoxicity in certain *in vitro* test systems, and (c) also cause tumours in organs other than thyroid. In the following, these aspects will be briefly addressed using ethylene thiourea and amitrole as examples. (A third example, the carcinogenic antithyroid chemical thiourea, has been recently discussed elsewhere (Andrae and Greim 1992)).

Ethylene Thiourea

ETU not only induces thyroid tumours in rats and mice after prolonged administration (IARC 1974, NTP 1992) but has been also reported to cause kidney tumours in rats (NTP

1992) and liver tumours in mice (Innes et al. 1969, NTP 1992). With the exception of a reproducible, weak mutagenicity in the *Salmonella typhimurium* strain TA1535, *in vitro* studies on the genotoxicity of ETU in bacteria, yeast and mammalian cells yielded mainly negative results (Dearfield 1994). ETU is a substrate for cytochrome P450 (Hui et al. 1988) and the flavin-containing monooxygenase (FMO) (Poulsen et al. 1979, Hui et al. 1988). In mouse liver the compound is preferentially metabolized by the FMO system, which catalyzes the oxidation of ETU to the reactive imidazoline-2-sulfinic acid (Poulsen et al. 1979). Imidazoline-2-sulfinic acid is also produced in the thyroid as an intermediate during oxidation of ETU by the TPO (Doerge and Takazawa 1990). Since (a) oxidative ETU metabolites have been shown to bind to proteins (Hui et al. 1988, Decker and Doerge 1991), (b) the sulfinic acid metabolite of the structurally related thyroid carcinogen thiourea induces DNA damage and gene mutations in cultured mammalian cells (Ziegler-Skylakakis and Andrae 1990), and (c) the sulfinic acid metabolite of ETU undergoes hydrolytic cleavage to imidazoline and bisulfite (Doerge and Takazawa (1990), another genotoxic species (see below), it is tempting to speculate that genotoxic metabolites of ETU may be also formed in thyroid. Whether this is in fact the case remains to be determined.

The apparent inactivity of ETU in most of the *in vitro* test systems for genotoxicity may be attributable to a lack of enzymes required for the activation of ETU to reactive metabolites. The FMO is very sensitive to thermal inactivation in the absence of NADPH (Uehleke 1973), a property which may result in its virtual absence in the rodent liver fractions commonly used as metabolic activation systems in *in vitro* mutagenicity testing (Ziegler 1980). Moreover, peroxidase activity is also generally lacking in both the cells and the metabolizing systems utilized.

Amitrole

Amitrole has been found to produce thyroid and pituitary tumours in rats and thyroid tumours in mice after oral administration (IARC 1986). In addition, it has been reported to induce hepatocellular tumours in mice following administration in the diet at doses of 500 mg/kg (Vesselinovich 1983) or higher (Innes et al. 1969). Similar to ETU, the compound has been tested in a wide variety of *in vitro* short term test systems for genotoxicity. (For an overview, see Hill et al. 1989). It was not mutagenic in various assays utilizing *Salmonella typhimurium* or *E. coli* either in the presence or in the absence of a metabolic activation system from rat liver (De Serres and Ashby 1981). Tests for the induction of DNA damage, gene mutations and chromosomal aberrations in mammalian cells also yielded mostly negative results. However, in cultured Syrian hamster embryo cells, which are distinguished by a high activity of the peroxidase prostaglandin H synthase (PHS) (Degen et al. 1983), amitrole induced gene mutations at two different genetic loci and morphological transformation (Tsutsui et al. 1984).

Studies on the interactions of amitrole with LPO have indicated how the compound may cause genotoxicity in peroxidase-containing cells. Doerge and Niemczura (1989) showed that in parallel with the irreversible inhibition of the enzyme (see above) substantial amounts of reactive metabolites are released which may bind to nucleophilic sites away from the active site. In agreement with this observation are previous findings by Krauss and Eling (1987) who demonstrated that LPO, TPO, and PHS activate amitrole to a reactive intermediate that binds to protein and RNA *in vitro*. It seems possible that formation of reactive amitrole metabolites by TPO or PHS in rat thyroid cells can result in covalent binding to the DNA or the mitotic apparatus of the follicular cells and in mutation induction. This hypothesis is consistent with the mutagenic and transforming activity of amitrole in Syrian hamster embryo cells *in vitro* (Tsutsui et al. 1984). In this context it appears interesting that diethylstilbestrol, another peroxidase substrate which induces cell transformation in Syrian hamster embryo cells (Degen et al. 1983, McLachlan et al. 1982), has been reported to induce thyroid tumours in mice (Greenman et al. 1990).

The hypothesis that amitrole may be genotoxic in rat follicular cells has been recently tested by Mattioli et al. (1994). Rats were given amitrole with the drinking water at a daily dose of approximately 200 mg/kg for twelve successive days. Immediately after treatment, the plasma level of T4 had decreased by 99% and the frequency of follicular cells in the S-phase of the cell cycle had increased by 700%, indicating that the exposure had resulted in a complete inhibition of thyroid hormone synthesis and in follicular cell hyperplasia. No increase in the frequency of DNA single-strand breaks could be detected in rats killed after eight days of treatment. Similarly, no increase in DNA fragmentation was detected in primary cultures of rat and human follicular cells immediately after exposure to 5.6-18 mM amitrole for twenty hours. These results may be taken as an indication that amitrole is not genotoxic in rat follicular cells. However, measurement of strand breaks was conducted at the end of extended exposures to high concentrations of amitrole which completely blocked the synthesis of thyroid hormones. This observation suggests that at least one of the enzymes potentially responsible for a formation of genotoxic metabolites from amitrole, TPO, was not active any more when DNA strand breaks were determined. In addition, it has been shown that large numbers of DNA adducts can be present in DNA without causing a measurable increase in DNA fragmentation (Martelli et al. 1995). Moreover, a possible interaction of reactive amitrole metabolites with critical cellular targets other than DNA, e.g., the mitotic spindle, would not have been detected. It is, therefore, still open to question whether amitrole is capable of causing genotoxic effects in thyroid cells or not.

Bisulfite as Genotoxic Metabolite of Antithyroid Agents

During the peroxidase-mediated oxidation of certain TPO-inhibitors, such as 1-methylbenzimidazolidine-2-thione (Doerge 1988) and ethylene thiourea (Doerge and Takazawa 1990) bisulfite is formed as a metabolite. Mottley et al. (1982) have shown that the PHS oxidizes bisulfite is to the DNA-reactive $^{\bullet}SO_3^-$-radical, and in SHE-cells which contain high activities of PHS, bisulfite causes morpholocical and neoplastic transformation (DiPaolo et al. 1981, Tsutsui and Barrett 1990). It appears likely, although it has not been proven yet, that bisulfite may be also activated in thyroid, as thyroid cells contain PHS activity (Friedman et al. 1975, Takasu et al. 1984, Burch et al. 1986). It remains to be determined whether there is a link between radical formation and cell transformation and whether peroxidase-catalyzed oxidation of bisulfite can also occur in thyroid *in vivo*.

CONCLUSIONS

The current knowledge on the chemical induction of thyroid follicular cell carcinogenesis in experimental animals supports the idea that tumour formation in the gland is governed by non-genotoxic mechanisms. This notion is in line with the observation that many of the chemicals known to inhibit TPO and, thus, thyroid function produce thyroid tumours during prolonged administration. However, the molecular mechanisms by which TPO inhibitors affect the enzyme are just beginning to be understood, and evidence has been obtained that the same or similar mechanisms which alter TPO activity may also lead to irreversible changes in the genetic material of the follicular cell. Metabolism of some of the thyroid carcinogens by the TPO has been shown to give rise to the formation of electrophiles which may act as initiators by causing DNA damage or affecting the mitotic apparatus of the follicular cells. In addition, there is evidence that some thyroid carcinogens or metabolites derived from them may be activated by the PHS in the thyroid. Chronic application of these compounds may thus create a pool of initiated cells which could respond aberrantly to the growth stimulus induced by the hormonal imbalance. This possible bifunctional activity of some of the thyroid carcinogens should receive more attention in order to help better define the importance of reactive intermediates of xenobiotics in thyroid carcinogenesis.

REFERENCES

Andrae, U., and Greim, H., 1992, Initiation and promotion in thyroid carcinogenesis. In "*Tissue-specific Toxicity: Biochemical Mechanisms*" (eds. W. Dekant and H.-G. Neumann), pp. 71-93, Academic Press

Burch, R.M., Luini, A., Mais, D.E., Corda, D., Vander-hoek, J.Y., Kohn, L.D., and Axelrod, J., 1986, 1-Adrenergic stimulation of arachidonic acid release and metabolism in rat thyroid cell line. Mediation of cell replication by prostaglandin E2, *J. Biol. Chem.* 261:11236-11241.

Cohen, S.M., and Ellwein, L.B., 1990, Cell proliferation in carcinogenesis, *Science* 249:1007-1011.

Dearfield, K.L., 1994, Ethylene thiourea (ETU). A review of the genetic toxicity studies, *Mutat. Res.* 317:111-132.

Degen, G.H., Wong, A., Eling, T.E., Barrett, J.C., and McLachlan, J.A., 1983, Involvement of prostaglandin synthetase in peroxidative metabolism of diethylstilbestrol in Syrian hamster embryo fibroblast cell cultures, *Cancer Res.* 43:992-996.

DeSerres, F.J., and Ashby, J., 1981, Evaluation of short-term tests for carcinogens. Report of the international collaborative program, *Progr. Mutat. Res. 1*, Elsevier, Amsterdam

DiPaolo, J.A., DeMarinis, A.J., and Doniger, J., 1981, Transformation of Syrian hamster embryo cells by sodium bisulfite, *Cancer Lett.* 12:203-208.

Doerge, D.R., 1986, Oxygenation of organosulfur compounds by peroxidases: evidence of an electron transfer mechanism for lactoperoxidase, *Arch. Biochem. Biophys.* 244:678-685.

Doerge, D.R., 1988, Mechanism-based inhibition of lactoperoxidase by thiocarbamide goitrogens. Identification of turnover and inactivation pathways, *Biochemistry* 27:3697-3700.

Doerge, D.R., and Niemczura, W.P., 1989, Suicide inactivation of lactoperoxidase by 3-amino-1,2,4-triazole, *Chem. Res. Toxicol.* 2:100-103.

Doerge, D.R., and Takazawa, R.S., 1990, Mechanism of TPO inhibition by ethylene thiourea, *Chem. Res. Toxicol.* 3:98-101.

Doniach, I., 1974, Carcinogenic effect of 100, 200, 250 and 500 rad X-rays on the rat thyroid gland, *Brit. J. Cancer* 30:487-495.

Eling, T.E., Petry, T.W., Hughes, M.F., and Krauss, R.S., 1988, Aromatic amine metabolism catalyzed by prostaglandin H synthase. In "*Carcinogenic and Mutagenic Responses to Aromatic Amines and Nitroarenes*" (eds. C.M. King et al.), pp. 161-172, Elsevier, New York

Friedman, Y., Lang, M., and Burke, G., 1975, Further characterization of bovine thyroid prostaglandin synthase, *Biochim. Biophys. Acta.* 397:331-341.

Greenman, D.L., Highman, B., Chen, J., Sheldon, W., and Gass, G., 1990, Estrogen-induced thyroid follicular cell adenomas in C57BL/6 mice, *J. Toxicol. Environ. Health* 29:269-278.

Hill, R.N., Erdreich, L.S., Paynter, O.E., Roberts, P.A., Rosenthal, S.L., and Wilkinson, C.F., 1989, Thyroid follicular cell carcinogenesis, *Fund. Appl. Toxicol.* 12: 629-697.

Hui, Q.Y., Armstrong, C., Laver, G., and Iverson, F., 1988, Monooxygenase-mediated metabolism and binding of ethylene thiourea to mouse liver microsomal protein, *Tox. Letters* 41:231-237.

IARC, 1974, *IARC Monographs on the evaluation of carcinogenic risk of chemicals to man.* Some antithyroid and related substances, nitrofurans and industrial chemicals, Vol.7, WHO, IARC, Lyon, France

IARC, 1987, *IARC Monographs on the evaluation of carcinogenic risks to humans*, Supplement 7, WHO, IARC, Lyon, France

Innes, J.R.M., Ulland, B.M., Valerio, M.G., Petrucelli, L., Fishbein, L., Hart, E.R., Pallotta, A.J., Bates, R.R., Falk, H.L., Gart, J.J., Klein, M., Mitchell, I., and Peters, J., 1969, Bioassay of pesticides and industrial chemicals for tumorigenicity in mice: A preliminary note, *J. Natl. Cancer Inst.* 42:1101-1114.

Krauss, R.S., and Eling, T.E., 1987, Macromolecular binding of the thyroid carcinogen 3-amino-1,2,4-triazole (amitrole) catalyzed by prostaglandin H synthase, lactoperoxidase and TPO, *Carcinogenesis* 8:659-664.

Martelli, A., Mattioli, F., Fazio, S., Andrae, U., and Brambilla, G., 1995, DNA repair synthesis and DNA fragmentation in primary cultures of human and rat hepatocytes exposed to cyproterone acetate, *Carcinogenesis* (in press)

Mattioli, F., Robbiano, L., Fazzuoli, L., and Baracchini, P., 1994, Studies on the mechanism of the carcinogenic activity of amitrole, *Fund. Appl. Toxicol.* 23:101-106.

McLachlan, J.A., Wong, A., Degen, G.H., and Barrett, J.C., 1982, Morphological and neoplastic transformation of Syrian hamster embryo fibroblasts by diethylstilbestrol and its analogs, *Cancer Res.* 42:3040-3045.

Mottley, C., Mason, R.P., Chignell, C.F., Sivarajah, K., and Eling, T.E., 1982, The formation of sulfur trioxide radical anion during the prostaglandin hydroperoxidase catalyzed oxidation of bisulfite (hydrated sulfur dioxide), *J. Biol. Chem.* 257:5050-5055.

Nadler, N.J., Mandavia, M., and Goldberg, M., 1970, The effect of hypophysectomy on the experimental production of rat thyroid neoplasia, *Cancer Res.* 30:1909-1911.

NTP, 1992, National Toxicology Program: NTP technical report on the perinatal toxicology and carcinogenesis studies of ethylene thiourea (CAS No. 96-45-7) in F334/N rats and B6C3F$_1$ mice (feed studies). NTP technical report Series No. 388. NIH Publication No. 92-2843. U.S. Department of Health and Human Services 1992

Ohtaki, S., Nakagawa, H., Nakamura, M., and Yamazaki, I., 1982, Reactions of purified hog TPO with H_2O_2, tyrosine, and methylmercaptoimidazole (goitrogen) in comparison with bovine lactoperoxidase, *J. Biol. Chem.* 257:761-766.

Ohtaki, S., Nakagawa, H., Nakamura, M., Nakamura, S., and Yamazaki, I., 1985, Characterization of hog TPO, *J. Biol. Chem.* 260:441-448.

Poulsen, L.L., Hyslop, R.M., and Ziegler, D.M., 1979, S-Oxygenation of N-substituted thioureas catalyzed by the pig liver microsomal FAD-containing monooxygenase, *Arch. Biochem. Biophys.* 198:78-88.

Takasu, N., Takahashi, K., Yamada, T., and Sato, S., 1984, Modulation of prostaglandin E_2, $F_{2\alpha}$, I_2 content and synthesis by thyrotropin in cultured porcine thyroid cells and intact rat thyroid glands, *Biochim. Biophys. Acta* 797:51-63.

Thomas, G.A., and Williams, E.D., 1991, Evidence for and possible mechanisms of non-genotoxic carcinogenesis in the rodent thyroid, *Mutat. Res.* 248:357-370.

Tsutsui, T., and Barrett, J.C., 1990, Sodium bisulfite induces morphological transformation of cultured Syrian hamster embryo cells but lacks the ability to induce detectable gene mutations, chromosome mutations or DNA damage, *Carcinogenesis* 11:1869-1873.

Tsutsui, T., Maiuzmi, H., and Barrett, J.C., 1984, Amitrole-induced cell transformation and gene mutations in Syrian hamster embryo cells in culture, *Mutat. Res.* 140:205-207.

Uehleke, H., 1973, The role of cytochrome P-450 in the N-oxidation of individual amines, *Drug Metab. Dispos.* 1:299-313.

Vesselinovich, S.D., 1983, Perinatal hepatocarcinogenesis. *Biol. Res. Pregnancy Perinatology* 4:22-25.

Ziegler, D.M., 1980, Microsomal flavin-containing monooxygenase: oxygenation of nucleophilic nitrogen and sulfur compounds. In "*Enzymatic Basis of Detoxication*" (ed. W.B. Jakoby), vol.1, pp. 201-227, Academic Press, New York

Ziegler-Skylakakis, K., and Andrae, U., 1990, Genotoxicity of formamidine sulfinate, a metabolite of thiourea, in V79 cells, *Mutat. Res.* 234:409-410.

MECHANISMS OF CYTOCHROME P450-MEDIATED FORMATION OF PNEUMOTOXIC ELECTROPHILES

Garold S. Yost

Department of Pharmacology and Toxicology
University of Utah
Salt Lake City, Utah 84112

INTRODUCTION

Toxicity to lung tissues from systemically circulated chemicals is largely mediated by the bioactivation of these agents by cytochrome P450 enzymes. The toxicity to lung cells is generally caused by reactive electrophiles that are formed as a result of oxygenation or desaturation of the starting xenobiotic molecule. Examples of the reactive intermediates include epoxides, quinone methides, methylene imines, and acyl halides. Selective lung toxicants include naphthalene, butylated hydroxytoluene, 4-ipomeanol, and 3-methylindole (3MI). Examples of pneumotoxic electrophiles that are formed by P450-mediated oxidation of the parent molecules are shown in Figure 1. Also shown are the P450 enzymes that have been proposed to be responsible for the bioactivation of each compound.

The participation of P450 enzymes (1) in the selective bioactivation of a large number of pneumotoxicants, including carcinogens, has been demonstrated. Examples include naphthalene (2,3,4), 4-ipomeanol (5), butylated hydroxytoluene (6), eugenol (7), benzo[a]pyrene (8), *N*-nitrosodibutylamine (9), 4-(methylnitrosamino)-1-(3-pyridyl)-1-butanone (a tobacco-specific carcinogen, 10), and styrene (11). Most of these toxicants were studied with animal tissues, but several have included human lung and liver tissues (12,10) or human-derived cDNA clones and expressed enzymes (13,5). Immunochemical studies have shown that several cytochrome P450 enzymes are expressed in human pulmonary tissues (14,15).

Several cytochrome P450 genes are selectively transcribed in lung tissues of animals and man; many of these have also been shown to be translated to active enzymes as well. Cytochrome P450 genes that are selectively transcribed and/or expressed in the lung tissues of animals include: *CYP1A1*, *CYP1A2*, *CYP2A3*, *CYP2B4*, *CYP2E1*, *Cyp2f-2*, *CYP3A2*, *CYP4A4*, and *CYP4B1* (see references 16 and 17 for reviews). Several of these genes are also transcribed and/or expressed in lung tissues of humans: *CYP1A1*, *CYP2B6*, *CYP2E1*, *CYP2F1*, *CYP3A4*, and *CYP4B1* (14,15, and see 18 for review).

Biological Reactive Intermediates V
Edited by R. Snyder et al., Plenum Press, New York, 1996

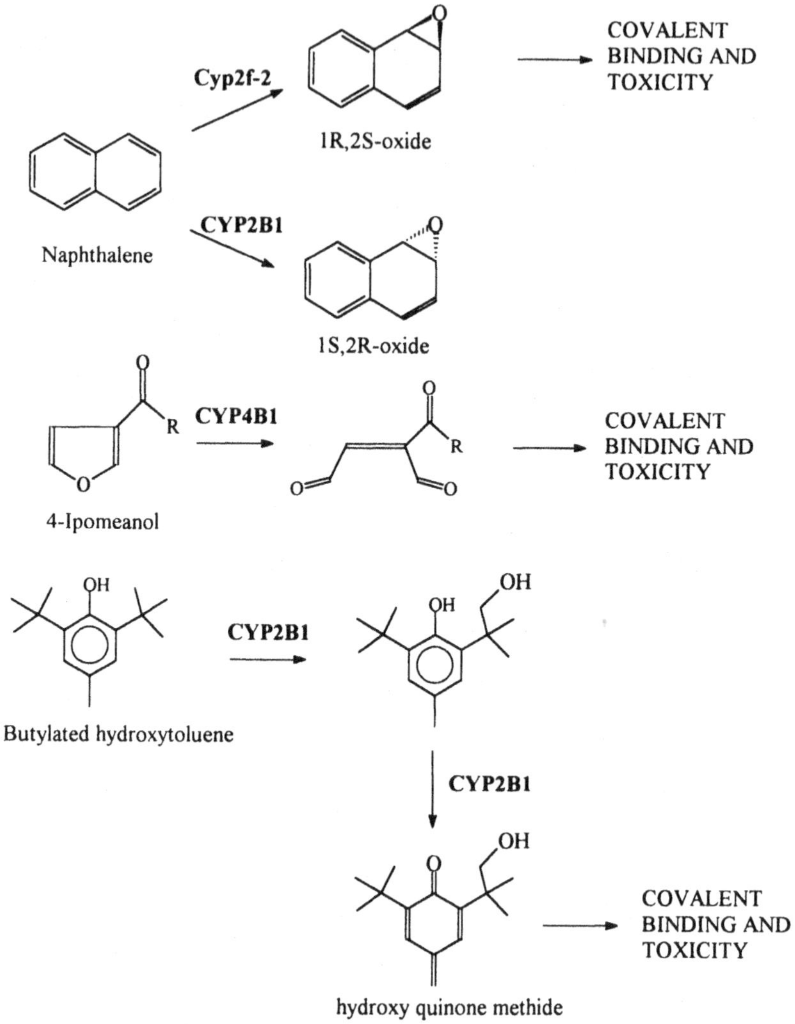

Figure 1. Electrophiles formed from selected pneumotoxic chemicals.

Not only are the P450 enzymes selectively expressed in lung tissues, but they are usually selectively expressed in certain cells within the tissues (17,19). Quite often the selective expression of P450 enzymes in certain cells correlates with the susceptibilities of these cells to toxicants. Examples include the selective damage to Clara cells by 3-methylindole (3MI) (20) and naphthalene (21).

3-Methylindole is an anaerobic fermentation product of tryptophan formed in the large intestine of man (22,23) and in the rumen of cattle, goats and sheep (24). In addition to natural production in the large intestine, cigarette smoke provides another source of exposure to 3MI for man, since 3MI is formed by pyrolysis of tryptophan in burning tobacco (25). Several other sources of human exposures to 3MI are known, including seafood and cheese (26,27). Excess 3MI has also been detected in feces of cancer patients with decreased colonic movement (28). A major metabolite of 3MI in rats, mice and goats, 3-hydroxy-3-methyloxindole (structure **4** of Figure 2) (29) has been isolated from urine from normal human volunteers, although the association of this oxindole with pathological processes or

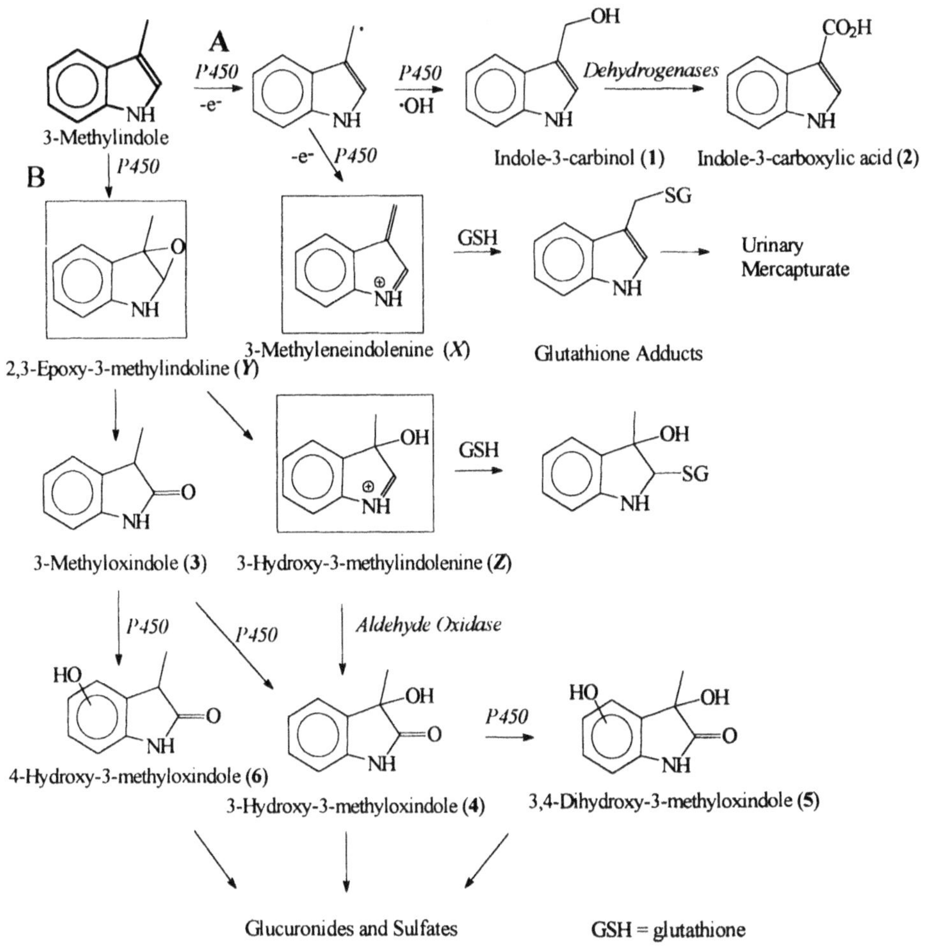

Figure 2. Metabolic pathways for 3-methylindole (proposed electrophilic intermediates are shown in boxes). Pathway A = methyl oxidation. Pathway B = indole oxidation.

the presence of 3MI was not established. Thus, the production of 3MI in humans may be linked to the populations of anaerobic bacteria that are responsible for the synthesis of 3MI, and the toxicity of 3MI under these conditions is unknown.

The urinary metabolites of 3MI in two species, goats and mice, have been described (30) and are summarized in Figure 2. Immediately obvious is the extensive oxidation of 3MI to a number of different products. Some of the metabolites are formed after the parent molecule has been oxidized three successive times, e.g., structure **5**. The excreted metabolites of 3MI were predominately the glucuronides or sulfates of the oxidized molecules.

3-Methyloxindole (structure **3**) was previously presumed to be an excreted metabolite of 3MI (31), but recent work has demonstrated that the only oxindoles that are excreted are multiple oxidation products (structures **4**, **5** and **6**) (30). All of the oxindole metabolites are presumed to be nontoxic compounds because 3-methyloxindole, a precursor to most of the oxindole metabolites, was shown to be nontoxic to goats (32).

The bioactivation of 3MI to reactive intermediates that are requisite steps in the pneumotoxicity of 3MI has been studied in vitro with animal and human microsomes

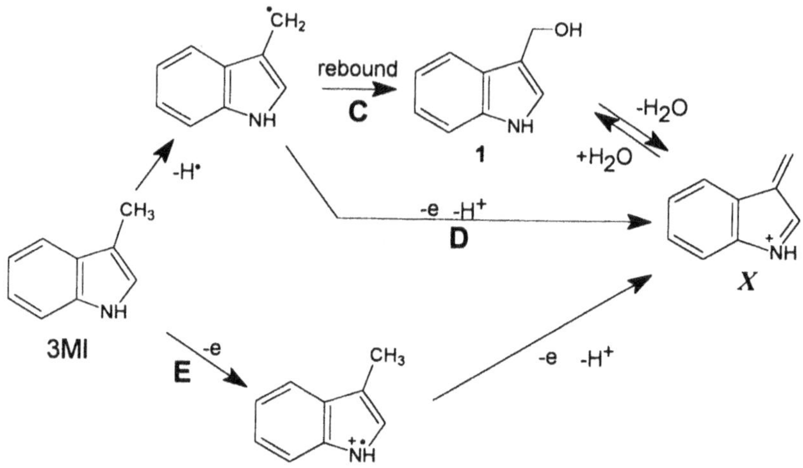

Figure 3. Pathways of cytochrome P450-mediated desaturation and hydroxylation of 3-methylindole.

(33,34,35,12), lung cells (36), a purified goat lung cytochrome P450 enzyme (unpublished data), and cDNA-expressed cytochrome P450 enzymes (13). In total, these studies present a scheme of 3MI metabolism and bioactivation that involves two primary routes of oxidation by cytochrome P450 enzymes (pathways A and B of Scheme I). Pathway A involves methyl oxidation via the hydrogen atom abstraction of 3MI by P450 enzymes that can lead to either the normal hydroxyl rebound product, indole-3-carbinol (**1**), or can undergo a second one-electron oxidation to produce the methylene imine (***X***). Indole-3-carbinol has been shown to be nontoxic when administered to goats via jugular infusion (32). Therefore, formation of indole-3-carbinol leads to detoxication of 3MI while formation of the methylene imine leads to protein alkylation and resultant toxicity. It is the branching (see Figure 3) between these two processes, along with detoxication by other enzymes like glutathione *S*-transferase, that probably controls whether 3MI is toxic to certain cells and organs. Certain P450 enzymes such as the human pulmonary CYP2F1 (13) and a goat pulmonary P450 (possibly 4B2), appear to be highly proficient at the dehydrogenation reaction, producing relatively little of the hydroxylation product.

Pathway B was previously presumed to produce only oxindoles that were nontoxic. However, recent evidence (36) has shown that two additional intermediates may be produced from this indole oxidation pathway. One intermediate (***Y***) is an epoxide that is probably the direct precursor of 3-methyloxindole (**3**) and may bind to DNA. The other intermediate (***Z***) is an imine that is probably the direct precursor of the detoxication product, 3-hydroxy-3-methyloxindole (**4**), after hydration by aldehyde oxidase enzymes. Conversely, ***Z*** is an electrophile that may be responsible for the depletion of glutathione, and therefore it may participate in the toxic process. Thus, the two primary pathways of 3MI oxidation both appear to participate in the bioactivation of 3MI, but also both participate in the detoxication of 3MI. The studies presented here represent the use of stable isotopes and cDNA-expressed P450 enzymes to evaluate the mechanisms of electrophile formation and the enzymes that catalyze these bioactivation events.

RESULTS AND CONCLUSIONS

The use of deuterium-labeled analogues of 3MI to address the mechanisms of metabolism and bioactivation has been highly successful. The fully deuterated analogue

(D3-3MI), the mono-deuterated analogue (D1-3MI), and the di-deuterated analogue (D2-3MI) of 3MI were synthesized and utilized for **inter**molecular and **intra**molecular isotope effect studies (37,38). The results of these studies with goat lung microsomal incubations, demonstrated that essentially no isotope effect ($^{D}V/K=1.1$) could be observed when the bioactivation of D3-3MI to the electrophilic 3-methyleneindolenine was compared with 3MI bioactivation, i.e., no **inter**molecular isotope effect. Thus, there was either no intrinsic isotope effect or the enzyme displayed a high commitment to catalysis that masked the effect. However, when the rate of hydrogen abstraction vs. deuterium abstraction was compared by utilization of D2-3MI, a significant **intra**molecular isotope effect of 5.5 was observed. This result demonstrated that the rate-determining step of 3MI bioactivation to an electrophilic intermediate (Figure 3, structure ***X***) was hydrogen atom abstraction from the methyl group to form a radical (pathway D), rather than nitrogen oxidation to form a cation radical (pathway E). This pathway (D) represents a unique example of initial hydrogen atom abstraction rather than nitrogen oxidation of a heterocyclic amine by P450 enzymes.

Oxygen-18 isotopes of O_2 and H_2O were also employed (38) to evaluate the mechanisms of formation of indole-3-carbinol (ICOH, structure **1**), a product of 3MI metabolism. It was postulated that ICOH was formed by hydration of the methylene imine (pathway C) or that ICOH might be a precursor of the reactive imine. The results demonstrated that ICOH was not readily dehydrated to the imine, but that the opposite reaction, hydration of the imine to ICOH produced approximately 80% of the ICOH formed in goat lung microsomes. When the results from the deuterium isotope effects and the ^{18}O studies were combined it was shown that at least one P450 enzyme in goat lung microsomes is exquisitely efficient at bioactivating 3MI to the methylene imine. In fact, the ratio of dehydrogenation (desaturation) of 3MI to hydroxylation of 3MI was approximately 50:1, a ratio unparalleled in the literature. This means that this enzyme(s) has unique properties as a catalyst for the production of the methylene imine reactive electrophile.

Studies with ^{18}O-labeled H_2O, incubated with 3MI and goat lung microsomes and cytosol, demonstrated that the carbonyl oxygen of 3-hydroxy-3-methyloxindole (Figure 2, structure **4**) was labeled almost exclusively with ^{18}O. The formation of this metabolite was dependent on the presence of cytosol in the incubations. The isotope incorporation was determined by GC/MS analysis of the oxindole. The hydroxyl oxygen was not labeled by ^{18}O-labeled H_2O. These results were interpreted to mean that the reactive intermediate, 3-hydroxy-3-methylindolenine (***Z***), was formed by cytochrome P450 oxidation of 3MI and then was oxidized by aldehyde oxidase using the labeled H_2O to place the label in the carbonyl oxygen. Thus, the pathways of methyl oxidation or ring oxidation proceed through specific reactive intermediates such as *X*, *Y*, or *Z* that are all probably involved in the toxicity of 3MI.

At least one cytochrome P450 in the 2F or 4B subfamilies was presumed to be the enzyme primarily responsible for the uniquely efficient dehydrogenation of 3MI that was identified with the goat lung microsomal studies. It is possible, however, that another enzyme or several P450 enzymes in goat lungs are responsible for these catalytic properties. One important goal of this work has been to identify and characterize the goat lung P450 enzyme(s) that selectively and efficiently dehydrogenates 3MI to the methylene imine intermediate.

Initial studies concerning the organ-selective transcription of cytochrome P450 genes corresponding to the human *2A6*, *2B7*, *2F1*, and *4B1* cDNAs in pulmonary and hepatic mRNA from goats, rabbits, and mice were recently conducted (39). This study demonstrated that transcripts corresponding to the 2F and 4B subfamilies were ubiquitously found in pulmonary tissues but, with only two exceptions (2F1 in mouse liver and 4B1 in rabbit liver), not in hepatic tissues. Thus, it is possible that an enzyme belonging to the 2F or 4B P450 subfamilies could be responsible for the selective bioactivation and toxicity of 3MI to lung tissues in several of these species.

A cDNA library was constructed from goat lung mRNA in the Lambda ZAP®II vector containing a pBluescript® SK⁻ phagemid with multiple cloning sites. The cDNAs were placed in the library and the library was amplified. Screening of the library was accomplished with the human *CYP2F1* and *CYP4B1* cDNAs that had been labeled with ^{32}P by random primer labeling. About 200 positive clones were identified from each screening and several were chosen for restriction mapping and sequencing. The clones with inserts of appropriate sizes, approximately 1.5 kb or larger, were subjected to in vivo excision, and the inserts were sequenced in both directions by the Sanger dideoxy nucleotide method. The deduced amino acid sequences of these two clones were approximately 82% similar to the CYP2F1 or CYP4B1 human sequences and were therefore tentatively named CYP2F3 and CYP4B2, respectively. It seems likely that one or both of these two enzymes catalyze the formation of the methylene imine electrophilic intermediate of 3MI.

Previous work (13) with cDNA-expressed cytochrome P450 enzymes had demonstrated that several human P450 enzymes and two animal P450 enzymes efficiently catalyzed the covalent binding of 3MI to protein. These studies showed that the human 2F1 enzyme was the most active human enzyme in bioactivating 3MI to an electrophilic intermediate, although it was not known whether the intermediate was *X*, *Y*, or *Z* (see Figure 2). Other studies (40) have been carried out to determine the specific metabolites of 3MI that were produced from each of the enzymes. These results demonstrated conclusively that CYP2F1 efficiently oxidized 3MI to the methylene imine intermediate to a much greater extent than any other human enzyme (Table 1). The mouse CYP1A2 and rabbit CYP4B1 were also shown to form the reactive intermediate with high efficiency. The production of 3-methyloxindole was also measured and these results demonstrated that an entirely different human P450, CYP1A2, produced the greatest amounts of this metabolite. These results are important because 3-methyloxindole is probably produced from an epoxide (Figure 2, intermediate ***Y***) that may interact with DNA. The most remarkable observation from these studies is the

Table 1. Metabolic products from incubation of 3-methylindole with cDNA-expressed cytochrome P450 enzymes

	Products formed (pmol/mg protein x hour)		
cDNA*-Expressed P450	3-Methyl-oxindole (**3**)	Indole-3-carbinol (**1**)	N-Aceytl-cysteine Adduct of *X*
control	100	ND	ND
1A2	1310	280	ND
2A6	300	115	50
2B6	125	ND	ND
2C8	ND	ND	ND
2C9	ND	ND	ND
2D6	135	ND	ND
2E1	320	ND	ND
2F1	750	ND	200
3A3	110	ND	ND
3A4	ND	ND	ND
3A5	ND	ND	ND
4B1	130	ND	ND
mouse 1A2	425	525	302
rabbit 4B1	290	560	405

* All cDNAs were human DNA unless otherwise specified
ND = not detected

striking contrast between the human 1A2 and 2F1 enzymes when one compares the production of the methylene imine (catalyzed primarily by 2F1) to the production of the epoxide (catalyzed primarily by 1A2). These data show that 3MI can be used as a highly selective probe to evaluate the catalytic mechanisms of known P450 enzymes, as well as enzymes produced by site-directed mutagenesis or chimeric constructs.

These studies provide strong evidence that human pulmonary P450s metabolize 3MI to electrophilic intermediates, that at least three distinct reactive intermediates are formed by these enzymes, and that all three electrophiles may participate in the toxicity of 3MI. In addition, the reactive intermediates are produced selectively by different P450 enzymes that either catalyze ring oxidation or methyl oxidation.

ACKNOWLEDGMENTS

The author would like to thank Drs. Gary Skiles, Janice Thornton-Manning, and Swayampakula Ramakanth, and Ms. Huifen Wang for their outstanding work on this research. This research was supported by Grant HL13645 from the U.S. Public Health Service, National Institutes of Health.

REFERENCES

1. Nelson, D.R., Kamataki, T., Waxman, D.J., Guengerich, F.P., Estabrook, R.W., Feyereisen, R. Gonzalez, F.J., Coon, M.J., Gunsalus, I.C., Gotoh, O., Okuda, K., and Nebert, D.W. (1993) The P450 superfamily: Update on new sequences, gene mapping, accession numbers, early trivial names of enzymes, and nomenclature. *DNA Cell Biol.* **12**, 1-51.
2. Nagata, K., Martin, B.M., Gillette, J.R., and Sasame, H.A. (1990) Isozymes of cytochrome P-450 that metabolize naphthalene in liver and lung of untreated mice. *Drug Metab. Dispos.* **18**, 557-564.
3. Buckpitt, A., Buonarati, M., Avey, L.B., Chang, A.M., Morin, D., and Plopper, C.G. (1992) Relationship of cytochrome P450 activity to Clara cell cytotoxicity. II. Comparison of stereoselectivity of naphthalene epoxidation in lung and nasal mucosa of mouse, hamster, rat and rhesus monkey. *J. Pharmacol. Exp. Therap.* **261**, 364-372.
4. Buckpitt, A. Chang, A.-M., Weir, A., Van Winkle, L. Duan, X., Philpot, R., and Plopper, C. (1995) Relationship of cytochrome P450 activity to Clara cell cytotoxicity. IV. Metabolism of naphthalene and naphthalene oxide in microdissected airways from mice, rats, and hamsters. *Mol. Pharmacol.* **47**, 74-81.
5. Czerwinski, M., McLemore, T.L., Philpot, R.M., Nhamburo, P.T., Korzekwa, K. Gelboin, H.V., and Gonzalez, F.J. (1991). Metabolic activation of 4-ipomeanol by complementary DNA-expressed human cytochromes P-450: Evidence for species-selective metabolism. *Cancer Res.* **51**, 4636-4638.
6. Bolton, J.L., and Thompson, J.A. (1991) Oxidation of butylated hydroxytoluene to toxic metabolites. Factors influencing hydroxylation and quinone methide formation by hepatic and pulmonary microsomes. *Drug Metabol. Dispos.* **19**, 467-472.
7. Thompson, D., Constantin-Teodosiu, D., Egestak, B., Mickos, H., and Moldeus, P. (1990) Formation of glutathione conjugates during oxidation of eugenol by microsomal fractions of rat liver and lung. *Biochem. Pharmacol.* **39**, 1587-1595.
8. Shimada, T., Yamazaki, H., Mimura, M., and Guengerich, F.P. (1992) Rat pulmonary microsomal cytochrome P-450 enzymes involved in the activation of procarcinogens. *Mutat. Res.* **284**, 233-241.
9. Schulze, J., Richter, E., and Philpot, R.M. (1990) Tissue, species, and substrate concentration differences in the position-selective hydroxylation of *N*-nitrosodibutylamine. Relationship to the distribution of cytochrome P-450 isozymes 2 (IIB) and 5 (IVB). *Drug Metabol. Dispos.* **18**, 398-402.
10. Smith, T.J., Guo, Z., Gonzalez, F.J., Guengerich, F.P., Stoner, G.D., and Yang, C.S. (1992) Metabolism of 4-(methylnitrosamino)-1-(3-pyridyl)-1-butanone in human lung and liver microsomes and cytochromes P-450 expressed in hepatoma cells. *Cancer Res.* **52**, 1757-1763.
11. Nakajima, T. Elovaara, E., Gonzalez, F.J., Gelboin, H.V., Raunio, H., Pelkonen, O., Vainio, H., and Aoyama, T. (1994) Styrene metabolism by cDNA-expressed human hepatic and pulmonary cytochromes P450. *Chem. Res. Toxicol.* **7**, 891-896.

12. Ruangyuttikarn, W., Appleton, M.L., and Yost, G.S. (1991) Metabolism of 3-methylindole in human tissues. *Drug Metab. Dispos.* **19**, 977-984.
13. Thornton-Manning, J.R., Ruangyuttikarn, W., Gonzalez, F.J., and Yost, G.S. (1991) Metabolic activation of the pneumotoxin, 3-methylindole, by vaccinia-expressed cytochrome P450 enzymes, *Biochem. Biophys. Res. Commun.* **181**, 100-107.
14. Anttila, S., Vainio, H., Hietanen, E., Camus, A.-M., Malaveille, C., Brun, G., Husgafvel-Pursiainen, K., Heikkila, L., Karjalainen, A., and Bartsch, H. (1992) Immunohistochemical detection of pulmonary cytochrome P450IA and metabolic activities associated with P450IA1 and P450IA2 isozymes in lung cancer patients. *Environ. Health Perspec.* **98**, 179-182.
15. Wheeler, C.W., Wrighton, S.A., and Guenthner, T.M. (1992) Detection of human lung cytochromes P450 that are immunochemically related to cytochrome P450 IIE1 and cytochrome P450IIIA. *Biochem. Pharmacol.* **44**, 183-186.
16. Guengerich, F.P. (1990) Purification and characterization of xenobiotic-metabolizing enzymes from lung tissue. *Pharmacol. Ther.* **45**, 299-307.
17. Baron, J., and Voigt, J.M. (1990) Localization, distribution, and induction of xenobiotic-metabolizing enzymes and aryl hydrocarbon hydroxylase activity with lung. *Pharmacol. Ther.* **47**, 419-445.
18. Wheeler, C.W., and Guenthner, T.M. (1991) Cytochrome P-450-dependent metabolism of xenobiotics in human lung. *J. Biochem. Toxicol.* **6**, 163-169.
19. Serabjit-Singh, C.J., Nishio, S.J., Philpot, R.M., and Plopper, C.G. (1988) The distribution of cytochrome P-450 monooxygenase in cells of rabbit lung: An ultrastructural immunocytochemical characterization. *Mol. Pharmacol.* **33**, 279-289.
20. Nichols, W.K., Larson, D.N., and Yost, G.S. (1990) Bioactivation of 3-methylindole by isolated rabbit lung cells. *Toxicol. Appl. Pharmacol.* **105**, 264-270.
21. Kanekal, S., Plopper, C. Morin, D. and Buckpitt, A. (1990) Metabolic activation and bronchiolar Clara cell necrosis from naphthalene in the isolated perfused mouse lung. *J. Pharmacol. Exp. Ther.* **252**, 428-437.
22. Fordtran, J.S., Scroggie, W.B., and Potter, D.E. (1964) Colonic absorption of tryptophan metabolites in man. *J. Lab. Clin. Med.* **64**, 125-132.
23. Yokoyama, M.T. and Carlson, J.R. (1979) Microbial metabolites of tryptophan in the intestinal tract with special reference to skatole. *Am. J. Clin. Nutr.* **32**, 173-178.
24. Carlson, J.R. and Dickinson, E.O. (1978) Tryptophan-induced pulmonary edema and emphysema in ruminants. In *Effects of Poisonous Plants on Livestock*, R.F. Keeler, K.R. Van Kampen, L.F. James, eds., Academic Press, New York, p. 261-262.
25. Wynder, E.L., and Hoffman, D. (1967) Certain constituents of tobacco products. In *Tobacco and Tobacco Smoke, Studies in Experimental Carcinogenesis*, Academic Press, New York, pp. 377-379.
26. Kowalewska, J., Zelazowska, H., Babuchowski, A., Hammond, E.G., Glatz, B.A., and Ross, F. (1985) Isolation of aroma-bearing material from Lactobacillus helveticus culture and cheese. *J. Dairy Sci.* **68**, 2165-2171.
27. Schulz H. (1986) Determination of indole and skatole in sea food by high-performance liquid chromatography (HPLC). *Z. Lebensm.-Unters. Forsch.* **183**, 331-334.
28. Karlin, D.A., Mastromarino, A.J., Jones, R.D., Stroehlein, J.R., and Lorentz, O. (1985) Fecal skatole and indole and breath methane and hydrogen in patients with large bowel polyps or cancer. *Cancer Res. Clin. Oncol.* **109**, 135-141.
29. Albrecht, C.F., Chorn, D.J., and Wessels, P.L. (1989) Detection of 3-hydroxy-3-methyloxindole in human urine. *Life Sci.* **45**, 1119-1126.
30. Smith, D.J., Skiles, G.L., Appleton, M.L., Carlson, J.R., and Yost, G.S. (1993) Identification of goat and mouse urinary metabolites of the pneumotoxin, 3-methylindole, *Xenobiotica* **23**, 1025-1044.
31. Hammond, A.C., Carlson, J.R., and Willett, J.D. (1979) The metabolism and disposition of 3-methylindole in goats. *Life Sci.* **25**, 1301-1306.
32. Potchoiba, M.J., Carlson, J.R., and Breeze, R.G. (1982). Metabolism and pneumotoxicity of 3-methyloxindole, indole-3-carbinol and 3-methylindole in goats. *Am. J. Vet. Res.* **43,** 1418-1423.
33. Nocerini, M.R., Carlson, J.R., and Yost, G.S. (1985) Adducts of 3-methylindole and glutathione: species deferences in organ-selective bioactivation. *Toxicol. Lett.* **28**, 79-87
34. Nocerini, M.R., Carlson, J.R., and Yost, G.S. (1985) Glutathione adduct formation with microsomally activated metabolites of the pulmonary alkylating and cytotoxic agent, 3-methylindole. *Toxicol. Appl. Pharmacol.* **81**, 75-84.
35. Huijzer, J.C., Adams, J.D., Jr., and Yost, G.S. (1987). Decreased pneumotoxicity of deuterated 3-methylindole: bioactivation requires methyl C-H bond breakage. *Toxicol. Appl. Pharmacol.* **90**, 60-68.

36. Thornton-Manning, J.R., Nichols, W.K., Manning, B.W., Skiles, G.L., and Yost, G.S. (1993) Metabolism and bioactivation of 3-methylindole by Clara cells, alveolar macrophages and subcellular fractions from rabbit lungs, *Toxicol. Appl. Pharmacol.* **122**, 182-190.
37. Skiles, G.L., and Yost G.S. (1990) Mechanisms of oxidative metabolism of the systemic pneumotoxin 3-methylindole in mice, *Toxicologist*, **10**, 67.
38. Skiles, G.L., and Yost G.S. (1992) Stable-isotope mechanistic studies on the oxidation of 3-methylindole, *Toxicologist*, **12**, 289.
39. Ramakanth, S., Thornton-Manning, J. R., Wang, H. H., Maxwell, H., and Yost, G. S. (1994) Correlation between pulmonary cytochrome P450 transcripts and the organ-selective pneumotoxicity of 3-methylindole, *Toxicol. Lett.* **71**, 77-85.
40. Thornton-Manning, J.R., Gonzalez, F.J., and Yost, G.S. (1992) Metabolism of 3-methylindole by vaccinia-expressed cytochrome P450 enzymes," *Toxicologist*, **12**, 288.

ROLE OF FREE RADICALS IN FAILURE OF FATTY LIVERS FOLLOWING LIVER TRANSPLANTATION AND ALCOHOLIC LIVER INJURY

Ronald G. Thurman,[1] Wenshi Gao,[1] Henry D. Connor,[1*] Yukito Adachi,[1]
Robert F. Stachlewitz,[1] Zhi Zhong,[1] Kathryn T. Knecht,[1]
Blair U. Bradford,[1] Ronald P. Mason,[2] and John J. Lemasters[3]

[1] Laboratory of Hepatobiology and Toxicology
Department of Pharmacology and Curriculum in Toxicology
CB# 7365, FLOB
The University of North Carolina, Chapel Hill, North Carolina 27599
[2] Laboratory of Molecular Biophysics, NIEHS, NIH
Research Triangle Park, North Carolina 27709
[3] Department of Cell Biology and Anatomy
The University of North Carolina, Chapel Hill, North Carolina 27599

INTRODUCTION

A critical factor in the extreme shortage of livers for transplantation is frequent failure due to primary non-function of ethanol-induced fatty livers when employed as donor organs (Starzl et al., 1988). Although fatty livers due to ethanol are frequently available in the donor pool since a major source of liver grafts is brain-dead victims of accidents involving alcohol (Butts & Patetta, 1988), surgeons must sometimes discard these organs because of high lipid content. Thus, an examination of the relationship between alcohol, fatty liver, and graft failure following liver transplantation could lead to a larger donor pool of usable organs. With this as a goal, we examined the connection between Kupffer cells and reperfusion injury in ethanol-induced fatty liver since Kupffer cells, which are activated following cold storage and reperfusion (Thurman, Cowper, Marzi, Currin, & Lemasters, 1988), have been implicated in primary non-function. Kupffer cells, when activated, release toxic mediators including cytokines and eicosanoids (Decker, 1990) which may play a role in reperfusion injury following transplantation.

* Permanent address: Department of Chemistry, Kentucky Wesleyan College, Owensboro, Kentucky 42301.

Biological Reactive Intermediates V
Edited by R. Snyder et al., Plenum Press, New York, 1996

Furthermore, chronic exposure to ethanol has been linked to activation of Kupffer cells. For instance, ethanol affects phagocytosis, bactericidal activity, and cytokine production by Kupffer cells (Martinez, Abril, Earnest, & Watson, 1992; Yamada, Mochida, Ohno, Hirata, Ogata.I., Ohta, & Fujiwara, 1991), and serum of alcoholics have elevated TNF levels (Stahnke, Hill, & Allen, 1991). These observations support the concept that Kupffer cells are activated in patients with alcoholic liver disease since TNF is produced by the monocyte-macrophage lineage, which is largely made up of Kupffer cells (Decker, Lohmann-Matthes, Karck, Peters, & Decker, 1989). In addition, Ca^{2+} activates Kupffer cells (Decker, 1990), and Ca^{2+} channels in Kupffer cells are opened more easily following long-term ethanol exposure (Goto, Lemasters, & Thurman, 1993). Therefore, we investigated whether elimination of Kupffer cells would affect early alcohol-induced liver injury.

Contributing to hepatic injury observed in sepsis is the release of chemical mediators and free radicals by Kupffer cells (Monden, Arii, Itai, Sasaoki, Adachi, Funaki, Higashitsuji, & Tobe, 1991; Monden, Arii, Itai, Sasaoki, Adachi, Funaki, & Tobe, 1991). Kupffer cells are activated in sepsis by endotoxin originating primarily from the cell wall of gram negative bacteria (Keller, West, Cerra, & Simmons, 1985). Endotoxin levels in plasma are raised in rats on an enteral alcohol feeding model as well as in the chronic alcoholic (Bode, Kugler, & Bode, 1987; Fukui, Brauner, Bode, & Bode, 1991), and these values correlate well with pathology (Nanji, Khettry, Sadrzadeh, & Yamanaka, 1993). Therefore, endotoxin derived from intestinal bacteria is most likely involved in alcohol-induced liver injury and activation of Kupffer cells by endotoxin likely plays a role in the mechanism of pathophysiology. To evaluate this hypothesis, endotoxin production was blocked with antibiotic treatment. Additionally, we assessed free radical formation in bile of rats on a continuous enteral feeding protocol. In summary, these data indicate that Kupffer cells are involved in hepatic injury due to alcohol and in reperfusion injury to fatty liver caused by ethanol.

METHODS

Enteral Feeding Model

Male Wistar rats, weighing about 300g, were employed in the enteral feeding model. An intragastric cannula was implanted into the stomach of rats as described by Tsukamoto and French (Tsukamoto, Reiderberger, French, & Largman, 1984), and diet was administered continuously for up to 4 weeks.

A liquid diet was prepared as described by Thompson and Reitz (Thompson & Reitz, 1978). It consisted of corn oil as fat (37% of total calories), protein (23%), carbohydrate (5%), minerals and vitamins, and ethanol or isocaloric maltose dextrin (35%). For each rat, ethanol levels in the diet were adjusted daily based on the urine alcohol concentration. Where indicated, $GdCl_3$ (10 mg/kg) was administered twice weekly via the tail vein beginning on the day of the operation to destroy Kupffer cells (Adachi, Bradford, Gao, Bojes, & Thurman, 1994).

Antibiotic Treatment

The liquid diet of rats on the enteral feeding protocol was supplemented with Polymyxin B and neomycin (Satoh, Guth, & Grossman, 1983) to prevent bacterial growth, the main source of endotoxin in the gastrointestinal tract. Based on the findings of preliminary experiments, 150mg/kg/day of polymyxin B and 450mg/kg/day of neomycin were given via the liquid diet.

Blood Collection and Asparate Aminotransferase (AST) Assay

Blood was collected weekly via the tail vein or, in transplantation experiments, via the vena cava at 0, 15, 30, 60, and 180 min after arterial or venous clamps were removed in a parallel group of rats. Sera were separated by centrifugation and stored in a microtube at -20°C until assayed for AST by standard enzymatic procedures (Bergmeyer, 1988).

Free Radical Detection in Bile

For at least two weeks, male Wistar rats (300-320g) were infused continuously with a high-fat liquid diet. Rats were anesthetized with Nembutal (75 mg/kg). To make ethanol levels of all animals comparable, ethanol was given to some rats fed a high-fat diet or chow control diet and animals were breathalyzed. Blood for later AST analysis was collected from the tail vein, centrifuged, and stored frozen as serum. Bile ducts were cannulated with PE10 tubing, the spin trap POBN (100 mg/kg) was administered, and bile samples were collected at 15 minute intervals into 30 μl of a solution of desferroxamine mesylate (5 mM) to prevent *ex vivo* radical formation. Samples were immediately frozen on dry ice and stored at -70°C prior to EPR analysis. Bile samples were thawed and placed in a quartz flat cell, and a Varian E-109 EPR spectrometer fitted with a TM110 cavity was used. Instrument settings were as follows: 20 mW microwave power, 80-G scan width, 0.53-G modulation amplitude, 16 minute time scan, and 1-s time constant. Data were stored on an IBM-type computer interfaced to the spectrometer. Simulations and double integrations of spectra to calculate amplitude were carried out with a computer program.

Low-Flow, Reflow Perfusion Model

A low-flow, reflow reperfusion model was used to study reperfusion injury to the liver. Livers from female Sprague-Dawley rats (125-150g) were perfused at low flow rates of 1 ml/g/min, causing anoxia in pericentral regions. After normal flow rates (4 ml/g/min) were restored for 40 min, an oxygen-dependent reperfusion injury resulted. Lactate dehydrogenase (LDH) was measured in the effluent perfusate.

STATISTICS

Data are presented as mean ± S.E.M. Statistical analyses were calculated with Students' t-test or ANOVA, as indicated. For all experiments, the criterion for significance was $p < 0.05$.

RESULTS

Role of Kupffer Cells in Reperfusion Injury in Ethanol-Induced Fatty Liver

A low-flow, reflow perfusion model was used to assess the effect of ethanol on reperfusion injury in lipid-loaded livers due to exposure to a modified Lieber-DeCarli diet. During reperfusion, maximal rates of LDH release were 17 IU/g/h in control rats but were elevated to around 37 IU/g/h by ethanol treatment, indicating that ethanol exacerbated reperfusion injury. In livers from control and ethanol-treated rats exposed to the Kupffer cell

Table 1. Effect of $GdCl_3$ on maximal release of LDH during reperfusion in livers from control and ethanol-treated rats

Treatment Group	LDH Release (IU/g/h)
Control	17.2 ± 2.2
Control + $GdCl_3$	7.9 ± 0.9
Ethanol	36.8 ± 3.0
Ethanol + $GdCl_3$	7.5 ± 1.3

Livers were perfused with the low flow, reflow model as described in Methods.
$GdCl_3$ (20 mg/kg) was injected intravenously 24h prior to perfusion experiments.
Values are mean ± SEM.
n = 7. a, $p < 0.05$ compared to low-flow period; b, $p < 0.05$ compared to control group.
Data were analyzed using ANOVA.

toxicant $GdCl_3$, however, release of LDH during reperfusion was only around 8 IU/g/h (Table 1). These results are consistent with the hypothesis that Kupffer cells are involved in hepatic reperfusion injury in rats treated with ethanol.

Livers were perfused with the low flow, reflow model as described in Methods. $GdCl_3$ (20 mg/kg) was injected intravenously 24h prior to perfusion experiments. Values are mean ± SEM. n = 7. a, $p < 0.05$ compared to low-flow period; b, $p < 0.05$ compared to control group. Data were analyzed using ANOVA.

Effect of Inactivation of Kupffer Cells on Early Alcohol-Induced Hepatic Injury

In rats exposed continuously to ethanol with an enteral feeding protocol (Tsukamoto-French model) for 2 and 4 weeks, serum AST levels reached 192 ± 13 and 244 ± 56 IU/L, respectively (Figure 1), versus control values of 88 ± 7 IU/L. To determine whether Kupffer cells are implicated in early ethanol-induced liver injury, rats were treated with $GdCl_3$and alcohol-induced liver injury was almost entirely prevented (Figure 1). Furthermore, $GdCl_3$

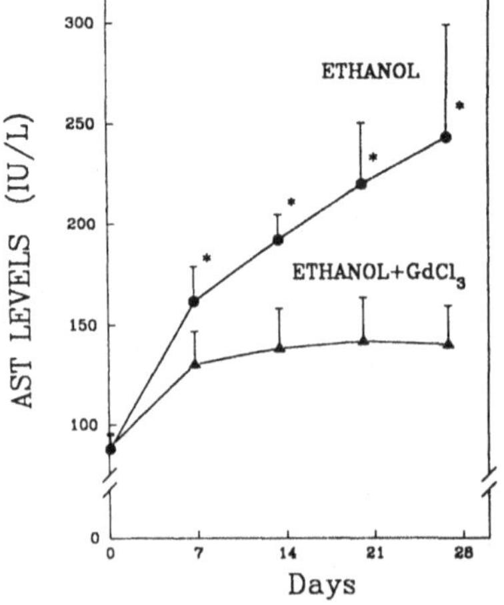

Figure 1. Effect of ethanol and $GdCl_3$ treatment on serum AST levels.

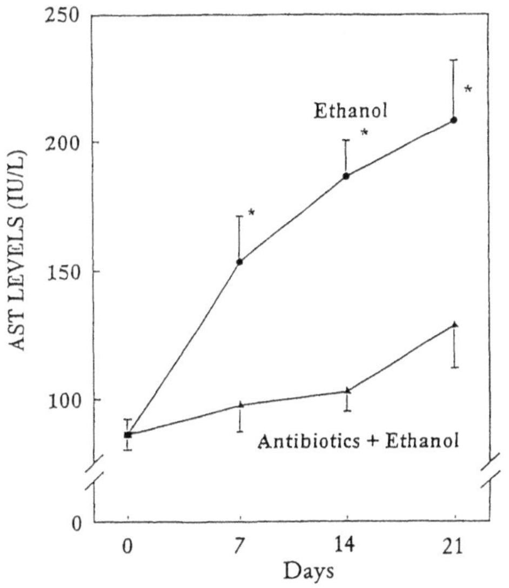

Figure 2. Effect of ethanol and antibiotic treatment on serum AST levels. Blood was collected from the tail vein once a week and AST was determined as described in Methods. Data represent mean ±± SEM (n = 5-10). *, p<0.05 compared with other values.

minimized fatty changes, inflammation and necrosis, as well as elevated rates of ethanol elimination and hepatic hypoxia (Adachi, Bradford, Gao, Bojes, & Thurman, 1994).

Blood was collected from the tail vein once a week and AST was determined as described in Methods. Data represent mean ± SEM (n = 4-8). *, $p < 0.01$ compared to control value.

Effect of Antibiotics on Hepatic Injury

Because endotoxin affects Kupffer cell activation, which triggers the release of toxic mediators, levels of endotoxin produced in the intestine were diminished with polymyxin B and neomycin. These antibiotics eliminated gram negative bacteria in the intestinal tract almost completely. Following 2 and 3 weeks on the Tsukamoto-French protocol, AST levels in the serum of ethanol-fed rats gradually increased to 185 ± 14 and 205 ± 24 IU/L, respectively (Figure 2). These values were considerably higher than values in untreated rats or rats fed a high-fat control diet without ethanol. Polymyxin B and neomycin, however, reduced elevated AST levels to 101 ± 8 and 126 ± 17 IU/L following 2 and 3 weeks of diet, respectively.

Possible Role of Free Radicals in Alcoholic Liver Disease

EPR spectroscopy was used to determine if elevated levels of endotoxin, which activates Kupffer cells to release toxic cytokines, increases deleterious free radicals as well. A free radical was recently detected in bile from rats treated with ethanol via intragastric ethanol administration (Figure 3; Knecht, Adachi, & Thurman, 1993). This radical signal was diminished by over 50% by $GdCl_3$ treatment. Thus, free radicals are formed in the bile of rats treated intragastrically with an ethanol-containing, high-fat diet via mechanisms involving Kupffer cells.

Representative EPR spectra of radical adducts in bile from rats treated for at least two weeks with continuous intragastric infusion of an ethanol-containing, high-fat diet (A) or an ethanol diet administered to $GdCl_3$-treated rats (B). Bile ducts were cannulated under Nembutal anesthesia and the spin trap POBN (100 mg/kg) was administered i.p. Bile samples

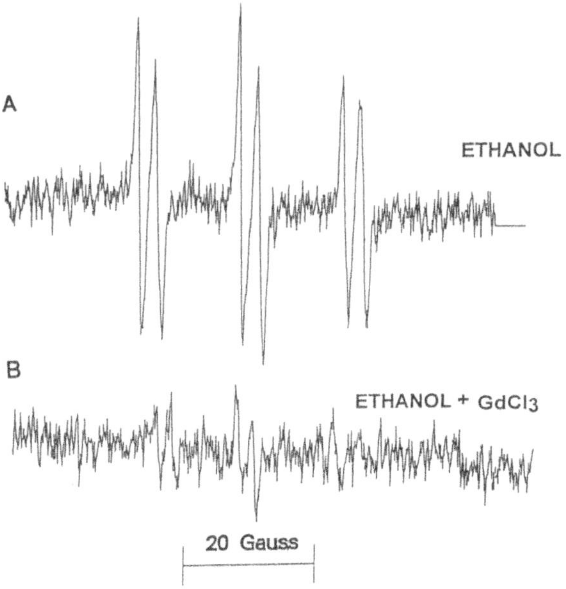

Figure 3. Destruction of Kupffer cells prevents free radical adduct formation due to chronic enteral ethanol exposure.

were collected into vials containing desferal (50 mM) in order to prevent *ex vivo* free radical formation for 3-4 h; they were then frozen on dry ice, and analyzed for free radical adducts with EPR spectroscopy as described in Methods. Typical experiments.

DISCUSSION

Reperfusion Injury following Transplantation in Ethanol-Induced Fatty Liver

Kupffer cells, which are activated following cold storage and reperfusion (Thurman, Cowper, Marzi, Currin, & Lemasters, 1988), release toxic mediators such as tumor necrosis factor (TNF). Our studies demonstrated recently that elevated levels of TNF and IL-6 following liver transplantation under non-survival conditions could be lowered by nisoldipine (Savier, Shedlofsky, Swim, Lemasters, & Thurman, 1992), a calcium channel blocker which reduces liver and lung injury subsequent to transplantation (Takei, Marzi, Kauffman, Currin, Lemasters, & Thurman, 1990). The protective effect of nisoldipine is presumably due to prevention of Kupffer cell activation and thus of TNF release.

Since Kupffer cells are clearly implicated in liver injury following transplantation, we assessed their involvement in reperfusion injury to fatty liver due to ethanol. In perfused livers from lipid-loaded, ethanol-treated rats ethanol treatment exacerbated reperfusion injury; elimination of Kupffer cells with $GdCl_3$ minimized injury (Table 1). These studies demonstrate that Kupffer cells indeed participate in reperfusion injury to fatty livers from ethanol-treated rats.

In this laboratory recent studies in rats fed high-fat, ethanol-containing diets have established that reperfusion injury to fatty livers that were stored and transplanted correlated strongly with decreased survival of the recipients (Gao, Connor, Lemasters, Mason, & Thurman, 1994). The oxygen-dependent reperfusion injury which follows orthotopic liver transplantation (Thurman, Marzi, Seitz, Thies, Lemasters, & Zimmermann, 1988) could be due to lipid loading, which cause grafts to be more vulnerable to swelling, or to accumulation of lipid droplets, which lead to increased lipid peroxidation after reperfusion. In our study, exposure to high-fat and alcohol-containing diets for 3-5 weeks produced lipid loading of 20-30% of the hepatocytes, and following transplantation (i.e., reperfusion), serum enzymes reached peak values in about 1 hr, indicating massive injury. Furthermore, survival decreased from around 60% to less than 10% (Gao, Connor, Lemasters, Mason, & Thurman, 1994). Since increased hepatic lipid content differentiated rats fed the high-fat- and alcohol-containing diets from rats fed the low-fat diet, it was concluded that lipid loading increased reperfusion injury following transplantation (Gao, Connor, Lemasters, Mason, & Thurman, 1994).

The oxygen-dependent hepatic reperfusion injury which occurs after orthotopic liver transplantation (Thurman, Marzi, Seitz, Thies, Lemasters, & Zimmermann, 1988) is possibly due to accumulation of xanthine and hypoxanthine during cold storage (Marzi, Zhong, Zimmermann, Lemasters, & Thurman, 1989), resulting in increased free radical production following reintroduction of oxygen. We detected free radical adducts in the bile of rats treated intragastrically with a high-fat, ethanol-containing diet (Figure 3), and we have also determined that formation of free radicals correlated with graft failure after orthotopic liver transplantation of livers from normal rats (Connor, Gao, Nukina, Lemasters, Mason, & Thurman, 1992). Additionally, Carolina rinse, a new solution containing antioxidants and SOD/catalase, minimized reperfusion injury and significantly improved survival (Gao, Currin, Lemasters, Connor, Mason, & Thurman, 1992). Recently, we attempted to identify the free radicals formed from fatty livers and their sources in order to determine why fatty livers fail more frequently after liver transplantation (Gao, Connor, Lemasters, Mason, & Thurman, 1994). Transplantation and storage increased radical adduct formation, yet similar amounts of radical adduct were observed in rats fed low-fat chow, a high-fat control diet, and a high-fat + ethanol diet. Thus, it seemed that free radical formation correlated with neither lipid content nor survival. However, the lipid and lipid/alcohol transplants were insensitive to infusion of the antioxidants SOD and catalase, while a significant decrease in radical adduct resulted in low-fat controls (Gao, Connor, Lemasters, Mason, & Thurman, 1994; Connor, Gao, Nukina, Lemasters, Mason, & Thurman, 1992). Thus, it is concluded that fatty livers produce antioxidant-insensitive free radicals.

Role of Kupffer Cells in Ethanol-Induced Liver Injury

Elimination of Kupffer cells with $GdCl_3$ prevented early injury to the liver due to ethanol (Figure 1). Kupffer cells are thought to be activated by ethanol exposure (D'Souza, Bagby, Lang, Deaciuc, & Spitzer, 1993), and endotoxin could be implicated. In the peripheral blood of chronic alcoholics with liver disease, endotoxin levels are elevated (Fukui, Brauner, Bode, & Bode, 1991). Furthermore, plasma levels of endotoxin in rats supplied continuously with an ethanol-containing diet increased in 1-2 weeks and plasma endotoxin levels paralleled the severity of liver injury (Nanji, Mendenhall, & French, 1989). Endotoxin activates Kupffer cells, causing the release of a number of potent effectors and cytokines which probably mediate most of the biologic effects of endotoxin. Furthermore, since 90% of endotoxin injected intravenously is scavenged in the liver by Kupffer cells (Arii, Monden, Itai, Sasaoki, Shibagaki, & Tobe, 1988), most of the endotoxin derived from the intestine is removed by the liver. The production of mediators from activated Kupffer cells could cause parenchymal cell injury (Monden, Arii, Itai, Sasaoki, Adachi, Funaki, & Tobe, 1991).

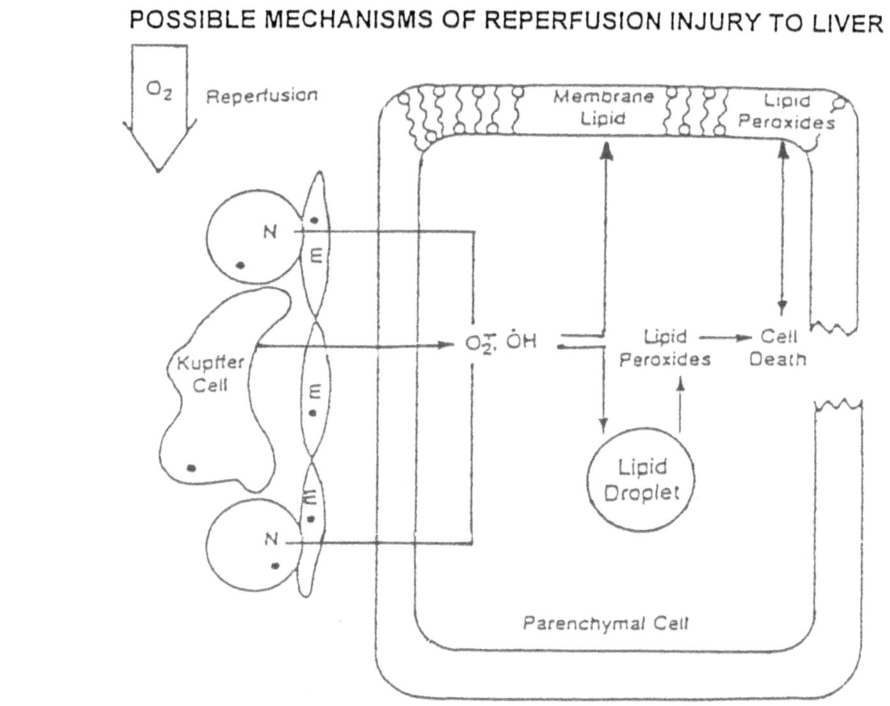

Figure 4. Scheme depicting hypothetical interactions between fatty liver and cell death (reperfusion injury). Abbreviations used: N, neutrophils; E, hepatic endothelium.

Our data demonstrating that antibiotic treatment prevented early liver injury induced by ethanol supports the role of endotoxin in the pathophysiology of ethanol-induced liver injury (Figure 2). Antibiotics prevented elevated AST in serum after exposure to ethanol, and the hepatic pathology score was also reduced significantly under these conditions (Adachi, Moore, Bradford, Gao, & Thurman, 1994). Polymyxin B and neomycin prevent growth of gram negative bacteria and are not absorbed in the intestine (Satoh, Guth, & Grossman, 1983). Further, polymyxin B binds endotoxin avidly, blocking its biological effect (Lopes & Inniss, 1969). Therefore, by diminishing endotoxin, antibiotic treatment prevented ethanol-induced liver injury. Short-term and long-term exposure to ethanol have been reported to be a major cause of endotoxemia, which is thought to participate in alcoholic liver injury (Nanji, Khettry, Sadrzadeh, & Yamanaka, 1993; Bode, 1980; Remmer, 1981).Endotoxin levels in the portal circulation might be elevated by increased gram negative bacterial flora, which is a major source of endotoxin (Bode, Bode, Heidelbach, Durr, & Martini, 1984), as well as by altered permeability of the intestinal membrane.

Possible Mechanisms of Kupffer Cell Participation in Hepatic Injury

Kupffer cells, activated by ethanol, may fuction in hepatic injury through several mechanisms, including hypoxia, formation of free radicals, and release of toxic mediators. Hypoxia has been linked to alcohol-induced liver injury (Videla, Bernstein, & Israel, 1973; Bernstein, Videla, & Israel, 1973; Ji, Lemasters, & Thurman, 1982; Yuki & Thurman, 1980), and in our studies, oxygen tension on the liver surface was significantly diminished by ethanol treatment, a phenomenon prevented by antibiotics (Adachi, Moore, Bradford, Gao,

& Thurman, 1994). Thus, Kupffer cells may participate in ethanol-induced liver injury by contributing to pericentral hypoxia.

Alternatively, Kupffer cells might be activated by ethanol to release mediators directly toxic to liver cells or which attract cytotoxic neutrophils or macrophages. Two mediators known to be released from activated Kupffer cells, TNF and interleukin-1 (Martinez, Abril, Earnest, & Watson, 1992), are cytotoxic, and both stimulate white cell migration and activation as well as protease and oxygen radical release (Thiele, 1989). Cellular infiltration of activated white blood cells, which release oxygen radicals, may enhance the inflammatory response and cause cell injury and death, and inflammatory cell infiltration due to ethanol is blocked by $GdCl_3$ (Adachi, Bradford, Gao, Bojes, & Thurman, 1994). Microcirculatory disturbances caused by vasoconstrictive agents released from Kupffer cells and white blood cells could increase hypoxia and result in a vicious cycle of pathophysiology.

Free radicals have been implicated in hepatic injury after liver transplantation (Connor, Gao, Nukina, Lemasters, Mason, & Thurman, 1992), and their release from Kupffer cells activated following ethanol administration is most likely a factor in the hepatotoxicity of ethanol. The precise pathway responsible for formation of free radicals in ethanol-treated rats is unclear; however, oxygen radical production by the NADPH oxidase system in Kupffer cells and adhering white blood cells is a strong possibility since the EPR signal was reduced by $GdCl_3$. Another possibility is reperfusion injury involving hypoxia and free radical formation via the xanthine-xanthine oxidase system in parenchymal cells. Whatever the exact mechanism, it is clear that toxic free radicals are involved.

ACKNOWLEDGMENTS

This work was supported, in part, by grants AA-09156 and AA-03624 from NIAAA and DK-37034 from NIH. W. Gao is supported by NRSA #5T32ESO7126 from NIEHS.

REFERENCES

Adachi, Y., Bradford, B.U., Gao, W., Bojes, H.K., & Thurman, R.G. (1994). Inactivation of Kupffer cells prevents early alcohol-induced liver injury. *Hepatology*, 20, 453-460.

Adachi, Y., Moore, L.E., Bradford, B.U., Gao, W., & Thurman, R.G. (1995). Antibiotics prevent liver injury following chronic exposure of rats to ethanol. *Gastroenterology*, in press

Arii, S., Monden, K., Itai, S., Sasaoki, T., Shibagaki, M., & Tobe, T. (1988). The three different phases of reticuloendothelial system phagocytic function in rats with liver injury. *J.Surg.Res*. 45, 314-319.

Bergmeyer, H.U. (1988). *Methods of Enzymatic Analysis*. New York: Academic Press.

Bernstein, J., Videla, L., & Israel, Y. (1973). Metabolic alterations produced in the liver by chronic ethanol administration. Changes related to energetic parameters of the cell. *Biochem.J.* 134, 515-521.

Bode, C., Kugler, V., & Bode, J.C. (1987). Endotoxemia in patients with alcoholic and non-alcoholic cirrhosis and in subjects with no evidence of chronic liver disease following acute alcohol excess. *J Hepatol*, 4, 8-14.

Bode, J.C., Bode, C., Heidelbach, R., Durr, H.-K., & Martini, G.A. (1984). Jejunal microflora in patients with chronic alcohol abuse. *Hepato-gastroenterol*, 31, 30-34.

Bode, J.C. (1980). Alcohol and the gastrointestinal tract. In H.P. Frick, G.A. Harnack, G.A. Martini, & A. Prader (Eds.), *Advances in Internal Medicine and Pediatrics* (pp. 1-75). Heidelberg: Springer-Verlag.

Butts, J.D. & Patetta, M. (1988). *North Carolina Medical Examiner System Annual Report 1988*. Chapel Hill, N.C. Office of the Chief Medical Examiner and the Staff of the Division of Statistics and Information Services.

Connor, H.D., Gao, W., Nukina, S., Lemasters, J.J., Mason, R.P., & Thurman, R.G. (1992). Free radicals are involved in graft failure following orthotopic liver transplantation: An EPR spin trapping study. *Transplantation*, 54, 199-204.

Decker, K. (1990). Biologically active products of stimulated liver macrophages (Kupffer cells). *Eur J Biochem*, 192, 245-261.

Decker, T., Lohmann-Matthes, M.L., Karck, U., Peters, T., & Decker, K. (1989). Comparative study of cytotoxicity, tumor necrosis factor, and prostaglandin release after stimulation of rat Kupffer cells, murine Kupffer cells, and murine inflammatory liver macrophages. *J.Leukocyte Biol.* 45, 139-146.

D'Souza, N.B., Bagby, G.J., Lang, C.H., Deaciuc, I.V., & Spitzer, J.J. (1993). Ethanol alters the metabolic response of isolated, perfused rat liver to a phagocytic stimulus. *Alcoholism: Clinical and Experimental Research*, 17, 147-154.

Fukui, H., Brauner, B., Bode, J.C., & Bode, C. (1991). Plasma endotoxin concentrations in patients with alcoholic and non-alcoholic liver disease: Reevaluation with an improved chromogenic assay. *Hepatology*, 12, 162-169.

Gao, W., Connor, H.D., Lemasters, J.J., Mason, R.P., & Thurman, R.G. (1994). Primary non-function of fatty livers produced by alcohol is associated with a new, antioxidant-insensitive free radical species. *Transplantation*, in press

Gao, W., Currin, R.T., Lemasters, J.J., Connor, H.D., Mason, R.P., & Thurman, R.G. (1992). Reperfusion rather than storage injury predominates following long-term (48 hrs) cold storage of grafts in UW solution: Studies with Carolina Rinse in transplanted rat liver. *Transplant International*, 5, S329-S335.

Goto, M., Lemasters, J.J., & Thurman, R.G. (1993). Activation of voltage-dependent calcium channels in Kupffer cells by chronic treatment with alcohol in the rat. *J. of Pharmacol. and Exp. Ther*, 267, 1264-1268.

Ji, S., Lemasters, J.J., & Thurman, R.G. (1982). Intralobular hepatic pyridine nucleotide fluorescence: Evaluation of the hypothesis that chronic treatment with ethanol produces pericentral hypoxia. *Proc Natl Acad Sci*, 80, 5415-5419.

Keller, G.A., West, M.A., Cerra, F.B., & Simmons, R.L. (1985). Multiple system organ failure modulation of hepatocyte protein synthesis by endotoxin activated Kupffer cells. *Ann.Surg.* 201, 87

Knecht, K.T., Adachi, Y., & Thurman, R.G. (1993). Detection of free radical adducts in the bile of rats treated chronically with intragastric alcohol. *Hepatology*, 18, 270A. (Abstract)

Lopes, J. & Inniss, W.E. (1969). Electron microscopy of effect of polymyxin B on E. coli lipopolysaccharide. *J Bacteriol*, 100, 1128-1130.

Martinez, F., Abril, E.R., Earnest, D.L., & Watson, R.R. (1992). Ethanol and cytokine secretion. *Alcohol*, 9, 455-458.

Marzi, I., Zhong, Z., Zimmermann, F.A., Lemasters, J.J., & Thurman, R.G. (1989). Xanthine and hypoxanthine accumulation during storage may contribute to reperfusion injury following liver transplantation in the rat. *Transplantation Proceedings*, 21, 1319-1320.(Abstract)

Monden, K., Arii, S., Itai, S., Sasaoki, T., Adachi, Y., Funaki, N., Higashitsuji, H., & Tobe, T. (1991). Enhancement and hepatocyte-modulating effect of chemical mediators and monokines produced by hepatic macrophages in rats with induced sepsis. *Research in Experimental Medicine*, 191, 177-187.

Monden, K., Arii, S., Itai, S., Sasaoki, T., Adachi, Y., Funaki, N., & Tobe, T. (1991). Enhancement of hepatic macrophages in septic rats and their inhibitory effect on hepatocyte function. *J Surg Res*, 50, 72-76.

Nanji, A.A., Khettry, U., Sadrzadeh, S.M.H., & Yamanaka, T. (1993). Severity of liver injury in experimental alcoholic liver disease. Correlation with plasma endotoxin, prostaglandin E2, leukotriene B4, and thromboxane B2. *Am.J.Pathol.* 142, 367-373.

Nanji, A.A., Mendenhall, C.L., & French, S.W. (1989). Beef fat prevents alcoholic liver disease in the rat. *Alcohol.Clin.Exp.Res.* 13, 15-19.

Remmer, H. (1981). *Die Wirkungen des Alkohols*. Alkoholwirkungen, 17, 1-11.

Satoh, H., Guth, P.H., & Grossman, M.I. (1983). Role of bacteria in gastric ulceration produced by indomethacin in the rat: Cytoprotective action of antibiotics. *Gastroenterology*, 84, 483-489.

Savier, E., Shedlofsky, S.I., Swim, A.T., Lemasters, J.J., & Thurman, R.G. (1992). The calcium channel blocker nisoldipine minimizes the release of tumor necrosis factor and interleukin-6 following rat liver transplantation. *Transplant International*, 5, S398-S402.

Stahnke, L.L., Hill, D.B., & Allen, J.I. (1991). TNFα and IL-6 in alcoholic liver disease. In E. Wisse, D.L. Knook, & R.S. McCuskey (Eds.), *Cells of the hepatic sinusoid* (pp. 472-475). Leiden,The Netherlands : Kupffer cell foundation.

Starzl, T.E., van Thiel, D., Tzakis, A.G., Iwatsuki, S., Todo, S., Marsh, J.W., Koneru, B., Staschak, S., Stieber, A., & Gordon, R.D. (1988). Orthotopic liver transplantation for alcoholic cirrhosis. *JAMA*, 260, 2542-2544.

Takei, Y., Marzi, I., Kauffman, F.C., Currin, R.T., Lemasters, J.J., & Thurman, R.G. (1990). Increase in survival time of liver transplants by protease inhibitors and a calcium channel blocker, nisoldipine. *Transplantation*, 50, 14-20.

Thiele, D.L. (1989). Tumor necrosis factor, the acute phase response and the pathogenesis of alcoholic liver disease. *Hepatology*, 9, 497-499.

Thompson, J.A. & Reitz, R.C. (1978). Effects of ethanol ingestion and dietary fat levels on mitochondrial lipids in male and female rats. *Lipids*, 13, 540-550.

Thurman, R.G., Cowper, K.B., Marzi, I., Currin, R.T., & Lemasters, J.J. (1988). Activation of Kupffer cells by storage of the liver in Euro-Collins solution. *Hepatology*, 8, 261. (Abstract)

Thurman, R.G., Marzi, I., Seitz, G., Thies, J., Lemasters, J.J., & Zimmermann, F.A. (1988). Hepatic reperfusion injury following orthotopic liver transplantation in the rat. *Transplantation*, 46, 502-506.

Tsukamoto, H., Reiderberger, R.D., French, S.W., & Largman, C. (1984). Long-term cannulation model for blood sampling and intragastric infusion in the rat. *Am.J.Physiol.* 247, R595-R599.

Videla, L., Bernstein, J., & Israel, Y. (1973). Metabolic alteration produced in the liver by chronic alcohol administration. Increased oxidative capacity. *Biochem.J.* 134, 507-514.

Yamada, S., Mochida, S., Ohno, A., Hirata, K., Ogata.I., Ohta, Y., & Fujiwara, K. (1991). Evidence for enhanced secretory function of hepatic macrophages after long-term ethanol feeding in rats. *Liver*, 11, 220-224.

Yuki, T. & Thurman, R.G. (1980). The swift increase in alcohol metabolism. Time course for the increase in hepatic oxygen uptake and the involvement of glycolysis. *Biochem.J.* 186, 119-126.

THE LIVER AS ORIGIN AND TARGET OF REACTIVE INTERMEDIATES EXEMPLIFIED BY THE PROGESTERONE DERIVATIVE, CYPROTERONE ACETATE

L. R. Schwarz,*[1] S. Werner,[1] J. Topinka,[2] U. Andrae,[1] I. Neumann,[1] and T. Wolff[1]

[1] GSF-Forschungszentrum für Umwelt und Gesundheit
Institut für Toxikologie, D-85764 Neuherberg, Germany
[2] Regional Hygiene Institute of Central Bohemia
Videnska 1083, 14220 Prague 4, Czech Republic

INTRODUCTION

The Liver as Origin and Target of Reactive Intermediates

Since the middle of the last century it is known that chemicals can cause liver injury (cf. Zimmerman, 1978). In 1860, a severe fatty liver in man was ascribed to phosphorous intoxication. During the first half of the 20th century several hepatotoxins have been extensively studied and in the second half of this century it became clear that in many cases enzymatically 'activated' metabolites of xenobiotics are responsible for the toxic effects observed (cf. Zimmerman, 1978; Hinson et al., 1994). The liver was found to express the highest concentrations and greatest variety of xenobiotic metabolizing enzymes in the body (cf. Zimmerman, 1978; Watkins, 1990). This high metabolic capacity turned out to be a primary cause of the liver specificity of toxic xenobiotics. Chemically induced liver damage may have three main manifestations (Fig. 1).

1. Toxic chemicals may cause disturbance of intermediary metabolism and cellular functions and ultimately cell death (Plaa, 1991). Classical toxins of the parenchymal liver cells which are metabolically activated are bromobenzene, acetaminophen and CCl_4 (Lau and Monks, 1988; Vermeulen et al., 1992; Brattin et al., 1985). If necrosis of parenchymal cells is massive or occurs frequently, liver fibrosis and cirrhosis may develop. Also nonparenchymal liver cells may become targets of toxic chemicals (Steinberg and Oesch, 1992).

* To whom correspondence should be addressed

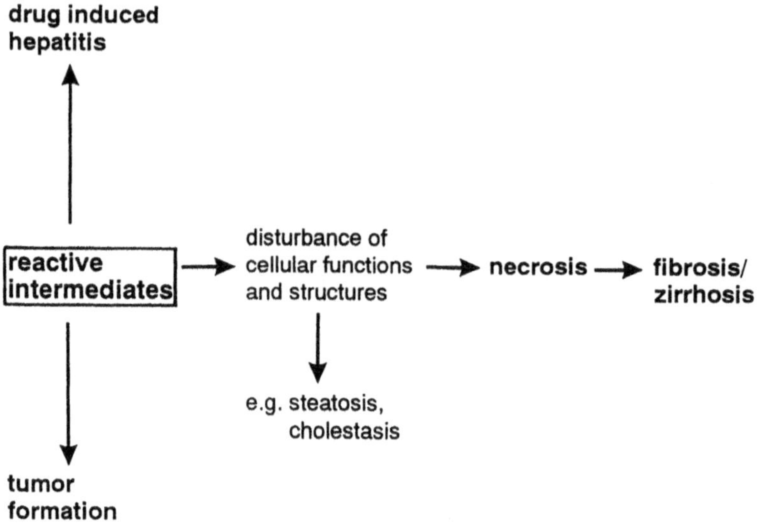

Figure 1. Toxic effects of reactive intermediates in the liver.

2. Drugs may induce the formation of haptens and thus provoke an autoimmune reaction in the liver, which manifests as a hepatitis. Several drugs have been shown to form neoantigens by binding covalently to proteins of the liver cell. The most prominent example of a drug causing a fatal hepatitis is the anesthetic agent halothane (Martin et al., 1993).
3. A great variety of carcinogens has been shown to react with hepatic DNA and to give rise to tumor formation in the liver. Examples are certain nitrosamines, arylamines, polycyclic aromatic hydrocarbons and aflatoxins (Pitot and Dragan, 1994). Tumor formation in the liver may not only be due to the genotoxic activity but also to the tumor promoting activity of chemicals (Schwarz and Greim, 1986). The latter mechanism may, for example, cause the tumorigenicity of several synthetic steroids in the liver, such as the estrogen ethinyl estradiol (Porter et al.,1987; Yager et al.,1991; Kemp et al.,1989).

Cyproterone Acetate

Similar to other synthetic steroids, the progesterone derivative cyproterone acetate (CPA) (Fig.2) has been shown to increase tumor formation in rat liver (Schuppler and Günzel, 1979). Several lines of evidence suggested that tumorigenicity of CPA is due to tumor promotion (Schulte-Hermann et al., 1983): Like other hepatic tumor promoters, CPA stimulated growth of preneoplastic liver cells, induced cytochrome P450-dependent enzyme activities and decreased apoptosis (Schulte-Hermann et al., 1981 ; Schulte-Hermann et al., 1988; Bursch et al., 1984). Moreover, CPA showed no genotoxicity in frequently used test systems. It did not induce gene mutations in Salmonella typhimurium and V79 cells in the absence and presence of metabolic activation and did not induce micronuclei in bone marrow cells in the Rhesus monkey (Lang and Redmann, 1979; Lang and Reimann, 1993). In this presentation, however, we show that CPA is clearly genotoxic in the liver of rats and in primary cultures of hepatocytes of rat and man. In view of the wide spread use of CPA in human therapy, these new findings raise questions about the safety of the drug CPA. CPA has found a particularly wide-spread use as an antiacnegenic drug with contraceptive activity (Diane^{R} and Diane-35^{R}).

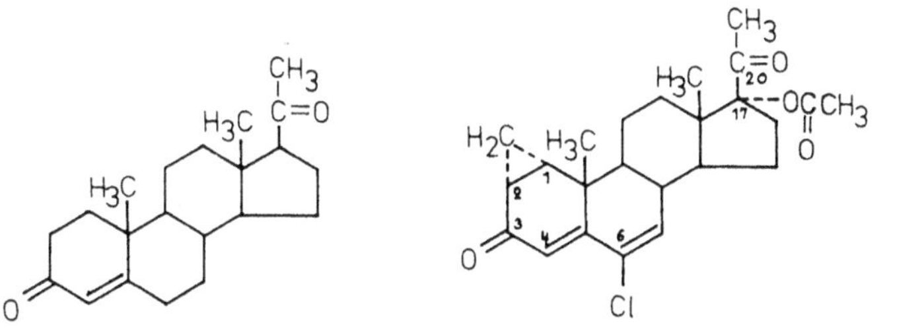

Figure 2. Structures of the 'natural' and synthetic progestins progesterone and cyproterone acetate (CPA).

METHODS

Animals. Male and female Wistar rats (8 to 10 weeks old, inbred strain, Neuherberg) were fed a standard diet (Altromin, Lage) and had free access to tap water.

Isolation of Hepatocytes. Rat hepatocytes were isolated by collagenase perfusion of the liver as described recently (Schwarz and Watkins, 1992). Human hepatocytes were isolated from liver pieces which were removed during surgery of secondary tumors. The patients had no known liver disease or history of pharmacotherapy. Cell isolation was performed basically according to Strom et al., (1982). Following perfusion of the liver specimen, the dissected tissue was incubated with collagenase/dispase. More than 80% of the hepatocytes excluded the dye trypan blue. Hepatocytes were cultured as described recently (Topinka et al,. 1993).

Measurement of DNA Repair Synthesis. DNA repair synthesis was determined by means of the BrdUrd density-shift method (Andrae et al., 1988). Hepatocytes were preincubated in William's medium E containing 40 μM FdUrd and 200 μM BrdUrd for 1h and subsequently incubated in fresh medium with the test compounds in the presence of FdUrd, BrdUrd and 10 μCi [^{3}H]dCyd/ml for 20 h. Following isolation of DNA, unreplicated DNA strands were separated from replicated, density-labeled strands by equilibrium centrifugation in alkaline $CsCl/Cs_2SO_4$ gradients. Repair synthesis was determined by measuring the incorporation of radioactivity in unreplicated DNA strands and expressed as cpm/μg DNA.

Determination of DNA Adducts. DNA was isolated and purified by RNAse/proteinase K/phenol treatment according to Gupta (1984). DNA adducts were identified and quantified by the ^{32}P-postlabeling technique using the butanol extraction procedure of Gupta for enrichment of DNA adducts (1985). Modified nucleotides were separated by PEI-cellulose thin layer chromatography as decribed recently (Topinka et al. 1993).

RESULTS AND DISCUSSION

Induction of DNA Repair Synthesis

The first indication that the synthetic steroid CPA is genotoxic was obtained fortuitously. In our initial experiments we were interested in the mitogenic effects of the steroidal drug on cultured hepatocytes obtained from female rats (Neumann et al. 1992). We noticed

Table 1. Effect of CPA on DNA repair synthesis in hepatocytes of female and male rats

	cpm /μg DNA [a]	
CPA (μM)	female	male
0	99.7 ± 11	95.3 ± 13
2	154.0 ± 23	62.1 ± 3
5	301.1 ± 11	78.5 ± 4
10	368.1 ± 29	84.1 ± 24
20	405.9 ± 8	86.8 ± 24
50	303.6 ± 15	66.4 ± 3

[a] Results represent the mean and range of two independent cell preparations

that CPA greatly stimulated DNA synthesis in the cells, as indicated by increases in the labelling index and enhanced incorporation of [^{3}H]thymidine into bulk cellular DNA. In order to characterize the effects of CPA on DNA synthesis in hepatocytes in more detail, studies on the effect of CPA on DNA repair synthesis were performed. Using the bromodeoxyuridine density-shift method, we could show that CPA strongly induces DNA repair synthesis (Tab. 1) (Neumann et al. 1992). Induction of repair was already clearly detectable at 2 μM CPA and maximal effects were obtained at about 20 μM. In contrast, CPA did not stimulate DNA repair in hepatocytes of male rats.

Formation of CPA-DNA Adducts in Rat Liver and Extrahepatic Tissues

Using the ^{32}P-postlabeling technique, we studied whether CPA induces the formation of DNA adducts in rat liver. Male and female rats were treated with a single dose of CPA p.o., and 24h later DNA binding was analysed. Exposure to CPA resulted in the formation of several DNA adducts in rat liver as shown by the autoradiograms obtained after ^{32}P-postlabeling of the adducts (Fig. 3). The four adducts which were detectable in the fingerprints of the liver DNA of male and female rats were termed 'A', 'B', 'C' and 'D'. Formation of the CPA-DNA adducts was dose-dependent (Fig. 4). The relative levels of the four adducts differed in the two genders. In male liver, B was the predominant adduct up to a dose of

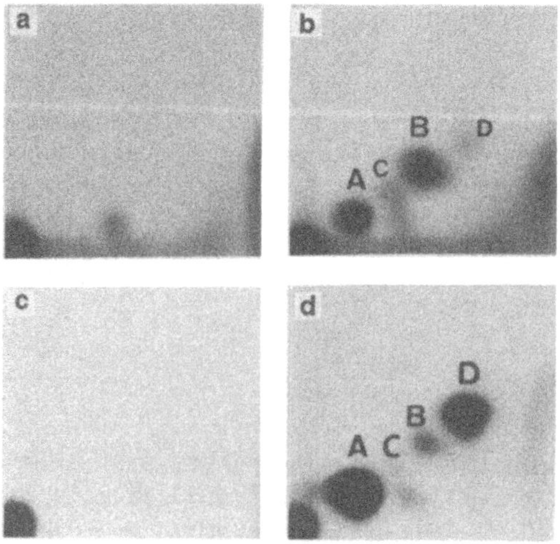

Figure 3. Adduct fingerprints of hepatic DNA from male (b) and female (d) rats treated with a single oral dose of 30 mg and 10 mg CPA/kg b.wt., respectively, and killed 24 h later. Autoradiograms (a) and (c) are derived from DMSO-treated male (a) and female (c) controls. Exposure of the autoradiograms was for 72 h (a,b) and 24 h (c, d) (From Topinka et al., 1993).

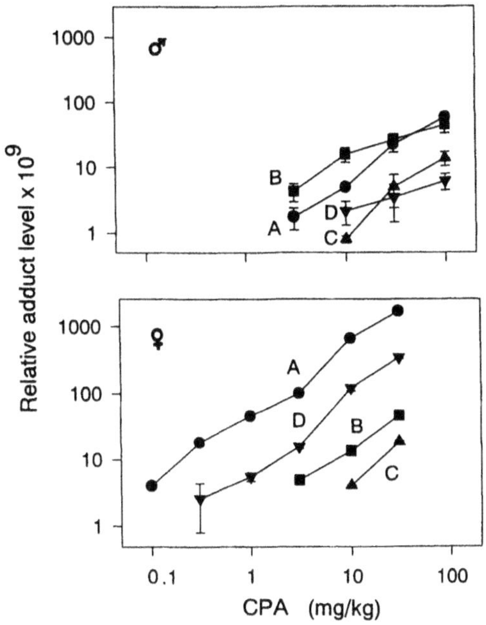

Figure 4. Formation of CPA-induced DNA adducts in the liver of male and female rats. Rats received single oral doses of CPA in 300 μl DMSO and were killed 24 h later. Results represent the mean ± SD of 3 animals. A = adduct A; B = adduct B + adduct B_1; C = adduct C; D = adduct D. When no error bars are given, SD was less than the size of the symbol (From Topinka et al., 1993).

10 mg CPA/kg b.wt. At higher doses, the relative amount of A increased, and at a dose of 100 mg/kg b.wt. A was the predominant adduct in the liver of male rats; C and D represented minor adducts. In female liver, A is the main adduct at all doses used and adduct D, a minor adduct in males, was the second frequent DNA adduct. Formation of the CPA-DNA adducts differed not only qualitatively but also quantitatively in the two sexes. Total adduct levels were up to 35 times higher in female compared to male rats. In the livers of female rats, DNA adducts could be detected already at 100 μg CPA/kg b.wt. (the lowest dose used) and high DNA adduct levels of more than 2000 adducts/10^9 nucleotides were reached at 30 mg CPA/kg b.wt.

In contrast to the results obtained for the liver, formation of CPA-DNA adducts was very low in several extrahepatic tissues/cells, such as white blood cells, ovary, small intestine, uterus and kidney (Tab. 2).

Table 2. Formation of CPA-induced DNA adducts in different organs of rats

	relative adduct level x 10^9 [a]	
tissue	female (20 mg CPA/kg)	male (100 mg CPA/kg)[b]
white blood cells	2 ± 0.3	n.d.[c]
ovary	2 ± 0.7	—
small intestine	1 ± 1	< 1
uterus	4 ; 2[d]	—
kidney	7 ± 5	< 1
liver	2387 ± 628	60,25 [d]

[a] mean ± SD (n = 3)
[b] single dose of CPA, treatment for 24 h
[c] n.d. = not detectable
[d] values of two animals

The present results show that CPA is genotoxic in cultured hepatocytes and in the liver of rats, but not in extrahepatic tissues and that genotoxicity exhibits a clear sex dependence in the rat. These findings suggest that CPA is activated by liver-specific drug metabolizing enzymes, which are predominantly expressed in female rat liver, and that the supposed reactive metabolites of CPA are unstable or do not leave the hepatocyte.

Role of Cytochrome P450 in Metabolic Activation of CPA

Preliminary experiments indicate that cytochrome P450 is not involved in the activation of the steroid. Pretreatment of female rats with ß-naphtoflavone, phenobarbital (PB), isopropanol and dexamethasone, known inducers of cytochromes P450 1A1/2, 2B1/2, 2E1 and 3A, respectively, did not increase the formation of hepatic CPA-DNA adducts. Induction of cytochromes P450 3A (dexamethasone, PB) and possibly 2B1/2 (PB) even decreased DNA binding significantly, suggesting an inactivating role of these cytochrome P450 isoenzymes. The levels of the main adducts A and D decreased by 87% and 95% in dexamethasone-treated animals and by 45% and 30% in PB-treated animals. In this context it is important to note that cytochrome P450 3A1 has been shown to catalyse the 15ß-hydroxylation of the steroid testosterone (Waxman, 1988) and that the 15ß-OH-derivative is the main metabolite of CPA (Hümpel et al., 1979).

Role of Sulfate Conjugation in Metabolic Activation of CPA

In view of the negative results of in vitro mutagenicity tests (Lang and Redmann, 1979) and the finding that phase I metabolism by several cytochrome P450 isoenzymes was not involved in the activation of CPA, we addressed the question whether the synthetic steroid may be activated by a conjugation reaction, in particular, by sulfation. To test this hypothesis, we studied the formation of CPA-DNA adducts in hepatocytes of female rats in the absence and presence of inorganic sulfate in the culture medium. As shown previously, sulfation by cultured hepatocytes greatly depends on the presence of sulfate which stems from the oxidative metabolism of the amino acids methionine and cysteine and from inorganic sulfate of the medium (Mulder, 1981; Schwarz, 1980, 1984). Sulfate is required for the biosynthesis of the cosubstrate of sulfation, 3′-phosphoadenosine-5′-phosphosulfate (PAPS). Formation of CPA-DNA adducts was almost completely suppressed using a medium free of sulfate and methionine/cysteine (Tab. 3). This finding applies to all DNA adducts formed by CPA. Thus, we put forward the hypothesis that CPA is reduced to the 3-OH-derivative and subsequently sulfated at this position, giving rise to a labile sulfate ester. Following the spontaneous elimination of the sulfate group, a reactive carbon cation will be

Table 3. Effect of sulfate and dehydroepiandrosterone on the formation of CPA-induced DNA adducts in cultured hepatocytes from female rats

sulfate (µM)	dehydroepiandrosterone (µM)	relative adduct [a,b] levels x 10^9
0	0	594 ± 110
800	—	11003 ± 686
800	100	590 ± 158

[a] Hepatocytes were incubated for 6 h in the presence of 30 µM CPA

[b] Results represent the mean and range of two independent cell preparations

formed. Of the various sulfotransferase isoenzymes catalysing sulfation of endogenous and foreign molecules, hydroxysteroid-sulfotransferases may possibly mediate sulfation of CPA (Hobkirk, 1985). We therefore studied whether dehydroisoepiandrosterone, a substrate of hydroxysteroid-sulfotransferases, inhibits activation of CPA; in addition to the parent compound, the sulfate conjugate of dehydroisoepiandrosterone, which is formed in hepatocytes, is also a known selective inhibitor of these enzymes (Okuda et al., 1989). Dehydroisoepiandrosterone strongly inhibited the formation of CPA-DNA adducts in hepatocytes (Tab. 3).

In rats hydroxysteroid-sulfotransferases have been shown to be almost selectively expressed in the liver and in particular in the liver of females. This is consistent with the present finding that genotoxicity of CPA has been primarily observed in the liver of female rats (Fig. 4).

Formation of DNA Adducts by CPA in Human Hepatocytes

Hydroxysteroid-sulfotransferases, which are likely to activate CPA, are also expressed in human liver (Hobkirk, 1985; Aksoy et al., 1993). When we studied whether CPA-DNA adducts are also formed in human hepatocyte cultures, we found high levels of a DNA adduct which was chromatographically identical to the adduct A observed in rat liver. As shown in Fig. 5, formation of this DNA adduct was concentration-dependent and maximal DNA binding was observed at a CPA concentration of about 10 μM. Similar levels of the CPA-DNA adduct were determined in human hepatocytes isolated from further liver specimens of female and male liver donors (see contribution of Werner et al., in this book). This finding is in line with the observation that expression of hydroxysteroid-sulfotransferase does not exhibit a significant sex difference in man (Aksoy et al., 1993).

Mutagenicity and Tumor Initiating Activity of CPA

No information is available at present on the mutagenic potency of the CPA-DNA adducts. Since cultured liver cells have only a very limited life-time and a very low capacity to proliferate in culture, it is hardly possible to determine mutagenic effects in this cell system (Cao et al. 1993). When isolated rat hepatocytes were co-cultured with V79 Chinese hamster

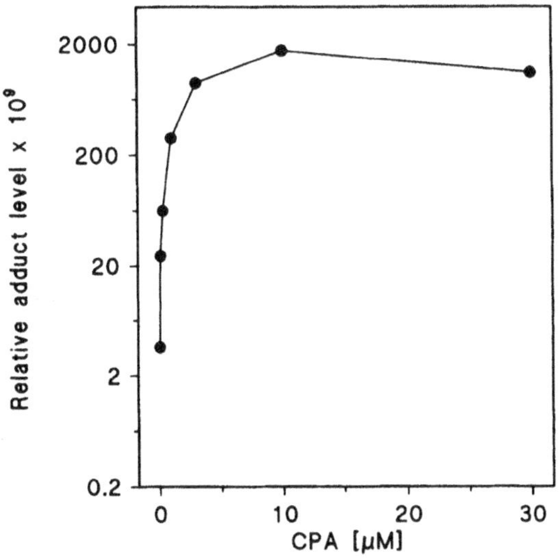

Figure 5. Formation of CPA-induced DNA adducts in cultures of human hepatocytes. Hepatocytes of a man were incubated for 20 h with the CPA concentrations indicated.

cells in the presence of CPA, no induction of mutations was observed in the V79 'indicator' cells (Kasper et al., 1993). The lack of mutagenicity of CPA in this test system is probably due to the fact that the CPA sulfate esters are too short-lived and unable to leave the hepatocyte. Only recently V79 cells have been constructed which express heterologous hydroxysteroid-sulfotransferase (Czich et al., 1994). In the future it will be possible to test the mutagenicity of CPA in this cell system. Nevertheless, there are already indications that CPA has not only genotoxic but also tumor initiating activity in female rats. When tested in the 'rat liver foci bioassay', the synthetic steroid initiated the formation of preneoplastic hepatocytes (Deml et al., 1993).

In Conclusion. The data presented clearly show that CPA is genotoxic in liver cells. Induction of DNA damage by CPA is most likely the consequence of the formation of reactive sulfate esters. In view of the fact that the CPA-DNA adducts are (1) formed in human liver cells, (2) remarkably stable (see contribution of Werner et al., in this book) and (3) exhibit tumor initiating activity in a rat liver, the therapeutic use of the steroid may entail a carcinogenic risk.

REFERENCES

Aksoy, I.A., Sochorová V. and Weinshilboum, R., 1993, Human liver dehydroepiandrosterone sulfotransferase: Nature and extent of individual variation. *Clin. Pharmacol. Ther.* 54:498-506.

Andrae, U., Homfeldt, H., Vogl, L., Lichtmannegger, J., and Summer, K.H., 1988, 2-Nitropropane induces DNA repair synthesis in rat hepatocytes in vitro and in vivo, *Carcinogenesis* 9:811-815.

Brattin, W.J., Glende, E.A. and Recknagel, R.O., 1985, Pathological mechanisms in carbon tetrachloride hepatotoxicity, *J. Free Radic. Biol. Med.* 1:27-38.

Bursch, W., Lauer, B., Timmermann-Trosiener, I., Barthel, G., Schuppler, J., and Schulte-Hermann, R., 1984, Controlled death (apoptosis) of normal and putative preneoplastic cells in rat liver following withdrawal of tumor promoters, *Carcinogenesis* 5:453-458.

Cao, J., Leibold E., Beisker, W., Schranner T., Nüsse M. and Schwarz, L.R., 1993, Flow cytometric analysis of in vitro micronucleus induction in hepatocytes treated with carcinogens, *Toxic. in Vitro* 7:447-451.

Czich, A., Bartsch, I., Dogra, S., Hornhardt, S., and Glatt, H.R., 1994, Stable heterologous expression of hydroxysteroid sulphotransferase in Chinese hamster V79 cells and their use for toxicological investigations, *Chemico-Biological Interaction* 92:119-128.

Deml, E., Schwarz L.R. and Oesterle, D., 1993, Initiation of enzyme-altered foci by the synthetic steriod cyproterone acetate in rat liver foci bioassay, *Carcinogenesis* 14:1229-1231.

Gupta, R.C., 1984, Non-random binding of the carcinogen N-hydroxy-2-acetylaminofluorene to repetitive sequences of rat liver DNA in vivo, *Proc. Natl. Acad. Sci. USA* 81:6943-6947.

Gupta, R.C.,1985, Enhanced sensitivity of ^{32}P-postlabeling analysis of aromatic carcinogen adducts, *Cancer Res.* 45:5656-5662.

Hinson, J.A., Pumford, N.J., and Nelson, S.D., 1994, The role of metabolic activation in drug toxicity, *Drug Metabol. Rev.* 26:395-412.

Hobkirk, R., 1985, Steroid sulfotransferases and steroid sulfate sulfatases: Characteristics and biological roles, *Can. J. Biochem. Cell Biol.* 63:1127-1144.

Hümpel, M., Nieuweboer, B., Düsterberg, B. and Wendt, H., 1979, Die Pharmakokinetik von Cyproteronacetat beim Menschen, In *Androgenisierungserscheinungen bei der Frau* (J. Hammerstein et al., Eds.) Exerpta Medica.

Kasper, P., Gerhard, A., Kaufmann, G., Madle, H. and Müller, L., 1993, Further investigations into the gentoxicity of cyproterone acetate. Abstract of a poster presented at the 23rd Ann. Meeting of the European Environmental Mutagen Society (EEMS) in Barcelona, 27. Sept. - 2. Oct. 1993.

Kemp, C.J., Leary, C.N., and Drinkwater, N.R., 1989, Promotion of murine hepatocarcinogenesis by testosterone is androgen receptor-dependent but not cell autonomous, *Proc. Natl. Acad. Sci.* 86:7505-7509.

Lang, R. and Redmann, U., 1979, Non-mutagenicity of some sex hormones in the Ames Salmonella/microsome mutagenicity test. *Mutation Res.* 67:361-365.

Lang, R. and Reimann, R., 1993, Studies for a genotoxic potential of some endogenous and exogenous sex steroids. I. Communication: Examination for the induction of gene mutations using the Ames Salmonella / Microsome Test and the HGPRT Test in V79 cells, *Environ. Molec. Mutag.* 21:272-304.

Lau, S.S., and Monks, T.J., 1988, The contribution of bromobenzene to our current understanding of chemically-induced toxicities, *Life Sci.* 42:1259-1269.

Martin, J.L., Kenna, J.G., Martin, B.M., Thomassen, D., Reed, G.F., and Pohl, L.R., 1993, Halothane patients have serum antibodies that react with protein disulfide isomerase, *Hepatology* 18:858-863.

Mulder, G.J., 1981, Sulfate availability in vivo, In *Sulfation of Drugs and Related Compounds* (G.J. Mulder, Ed.) pp. 31-52. CRC Press, Boca Raton.

Okuda, H., Nojima, H., Watanabe, N., and Watabe, T., 1989, Sulphotransferase-mediated activation of the carcinogen 5-hydroxymethyl-chrysene, *Biochem. Pharmacol.* 38:3003-3009.

Pitot, H.C., and Dragan, Y.P., 1994, Chemical induction of hepatic neoplasia, In *Liver: Biology and Pathobiology, 3rd edition* (I.M. Arias, Ed.), pp. 1467-1498. Raven Press, New York.

Plaa, G.L., 1991, Toxic responses of the liver, In *Casarett and Doull's Toxicology* (C.D. Klaassen, M.O. Amdur, and J. Doull, Eds.), pp. 286-309. Macmillan Publishing Company, New York.

Porter, L.E., van Thiel, D.H., and Eagon, P.K., 1987, Estrogens and progestins as tumor inducers, *Seminars in Liver Disease* 7:24-31.

Schulte-Hermann, R., Ohde, G., Schuppler, J., and Timmermann-Trosiener, I., 1981, Enhanced proliferation of putative preneoplastic cells in rat liver following treatment with the tumor promoters phenobarbital, hexachlorocyclohexane, steroid compounds, and nafenopin, *Cancer Res.* 41:2556-2562.

Schulte-Hermann, R., Timmermann-Trosiener, I., and Schuppler, J., 1983, Promotion of spontaneous preneoplastic cells in rat liver as a possible explanation of tumor production by nonmutagenic compounds, *Cancer Res.* 43:839-844.

Schulte-Hermann, R., Ochs, H., Bursch, W., and Parzefall, W., 1988, Quantitative structure-activity studies on effects of sixteen different steroids on growth and monooxygenases of rat liver, *Cancer Res.* 48:2462-2468.

Schuppler, J., and Günzel, P., 1979, Liver tumors and steroid hormones in rats and mice, *Arch. Toxicol.* 2:181-195.

Schwarz, L.R., 1980, Modulation of sulfation and glucuronidation of 1-naphthol in isolated rat liver cells, *Arch. Toxicol.* 44:137-145.

Schwarz, L.R., 1984, Sulfation of 1-naphthol in isolated rat hepatocytes, *Hoppe-Seyler's Z. Physiol. Chem.* 365:43-48.

Schwarz, L.R., and Greim, H., 1986, Environmental chemicals in hepatocarcinogenesis: The mechanism of tumor promoters, In *Progress in Liver Diseases*, vol. **VIII** (H. Popper and F. Schaffner Eds.), pp. 581-595. Grune & Stratton, Orlando, New York.

Steinberg, P., and Oesch, F., 1992, Liver cell specific toxicity of xenobiotics, *Tissue Specific Toxicity: Biochemical Mechanisms* (W. Dekant and H.-G. Neumann, Eds.), pp. 117-137. Academic Press, London.

Strom, S.C., Jirtle, R.L., Jones, R.S., Novicki, D.L., Rosenberg, M.R., Novotny, A., Irons, G., McLain, J.R., and Michalopoulos, G., 1982, Isolation, culture and transplantation of human hepatocytes, *JNCI, J. Natl. Cancer Inst.* 68:771-778.

Topinka, J., Andrae, U, Schwarz, L.R., and Wolff, T., 1993, Cyproterone acetate generates DNA adducts in rat liver and in primary rat hepatocyte cultures, *Carcinogenesis* 14:423-427.

Vermeulen, N.P.E., Bessems, J.G.M, and van de Straat, R., 1992, Molecular aspects of paracetamol-induced hepatotoxicity and its mechanism-based prevention, *Drug Metabolism Reviews* 24:367-407.

Watkins III, J.B., Thierau, D., and Schwarz, L.R., 1992, Biotransformation in carcinogen-induced diploid and polyploid hepatocytes separated by centrifugal elutriation, *Cancer Res.* 52:1149-1154.

Yager, J.D., Zurlo, J., and Ni, N., 1991, Sex hormones and tumor promotion in liver, *Proc. Soc. Exp. Biol. Med.* 198:667-674.

Watkins, P.B., 1990, Role of cytochromes P450 in drug metabolism and hepatotoxicity, *Semin. Liver Dis.* 10:235-250.

Waxman, D.J., 1988, Interactions of hepatic cytochromes P450 with steroid hormones: Regioselectivity and stereoselectivity of steroid metabolism and hormonal regulation of rat P450-enzyme expression, *Biochem. Pharmacol.* 37:71-84.

Zimmermann, H.J., 1978, The adverse effects of drugs and other chemicals on the liver,. In *Hepatotoxicity*. Appleton-Century-Crofts, New York.

STUDIES ON THE FORMATION OF HEPATIC DNA ADDUCTS BY THE ANTIANDROGENIC AND GESTAGENIC DRUG, CYPROTERONE ACETATE

1. Adduct Levels in Various Species Including Man and 2. Persistence and Accumulation in the Rat

S. Werner,[1] J. Topinka,[2] S. Kunz,[1] T. Beckurts,[3] C.-D. Heidecke,[3] L. R. Schwarz,[1] and T. Wolff[1]*

[1] GSF-Forschungszentrum für Umwelt und Gesundheit
Institut für Toxikologie, D-85764 Oberschleißheim, Germany
[2] Regional Hygiene Institute of Central Bohemia
Videnska 1083, 14220 Prague 4, Czech Republic
[3] Chirurgische Klinik u. Polyklinik der Technischen Universität München

INTRODUCTION

Cyproterone acetate (CPA) [Fig. 1] is a sex steroid with strong antiandrogenic and gestagenic action, which is widely used in human therapy. Upon long term feeding CPA causes liver tumors in rats [1]; this activity has been attributed to tumor promotion. The safety of the therapeutic use of the steroid has been questioned recently, since findings from our laboratories indicate that CPA exhibits genotoxic activity. CPA induces DNA repair synthesis in cultured hepatocytes from female rats [2] and the formation of CPA-derived DNA adducts in rat liver and in hepatocytes [3].

To estimate the health risk associated with the long term intake of CPA, it is important to know whether and to what extent DNA adducts are formed in human liver. Therefore, we have studied DNA adduct formation in human hepatocytes which were isolated from liver biopsies removed during surgery. The isolation procedure of hepatocytes from liver biopsies was first worked out using liver pieces from pigs. Pig and human hepatocytes were cultured as monolayers and the formation of DNA adducts was studied at various CPA concentrations using the ^{32}P-DNA postlabeling technique.

* To whom all correspondence should be sent.

Biological Reactive Intermediates V
Edited by R. Snyder et al., Plenum Press, New York, 1996

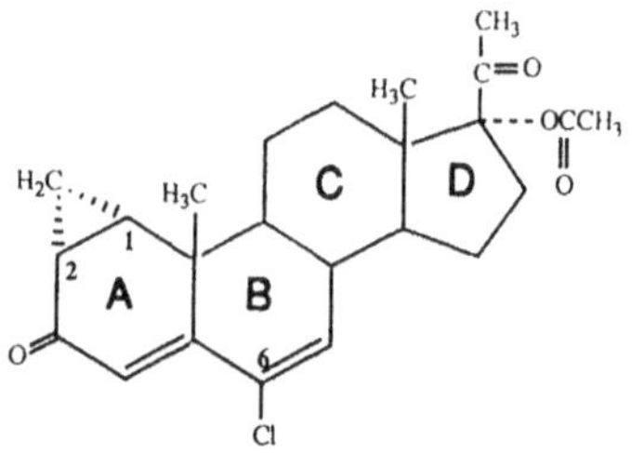

Figure 1. Structure of cyproterone acetate.

To further assess the health risk for CPA, the question needs to be answered whether the adducts formed by CPA are persistent and consequently will accumulate upon repeated exposure. In the second part of the present paper we present results on the persistence and accumulation of hepatic DNA adducts in the rat, the animal model in which the genotoxic and tumorigenic effects of CPA had been detected.

METHODS

Animal Treatment

To study the accumulation of DNA adducts induced by CPA, male and female Wistar rats received 50 μg CPA/kg b.wt. daily by gavage for 42 days. CPA was dissolved in olive oil. In the persistence experiment male and female rats received a single oral dose of 100 and 10 mg CPA/kg b.wt. dissolved in DMSO, respectively; untreated controls received the vehicles only. Mice [C57Bl/6, (102xC3H)F_1] were treated by intragastric intubation of either 10 mg or 35 mg CPA dissolved in olive oil.

Isolation of Hepatocytes and Incubation of Hepatocyte Cultures with CPA

Human and pig liver cells were isolated by two step collagenase perfusion, according to Strom et al. [4] with some modifications, followed by incubation of the dissected liver tissue with collagenase/dispase. Liver cells were cultured as reported recently [3]. Hepatocyte monolayers were incubated for 6 hours with various concentrations of CPA dissolved in DMSO (0.25 % v/v).

Determination of DNA Adducts

DNA was isolated and purified by RNAse/proteinase K/phenol treatment according to Gupta [5]. DNA adducts were identified and quantified by the ^{32}P-postlabeling technique using the butanol enrichment procedure of Gupta [6].

RESULTS AND DISCUSSION

Formation of DNA Adducts in Hepatocytes of Man, Pig and Rat

Incubation of hepatocytes from pigs and man with CPA for 6 hours caused the formation of one major DNA adduct. As revealed by co-chromatography on PEI-cellulose

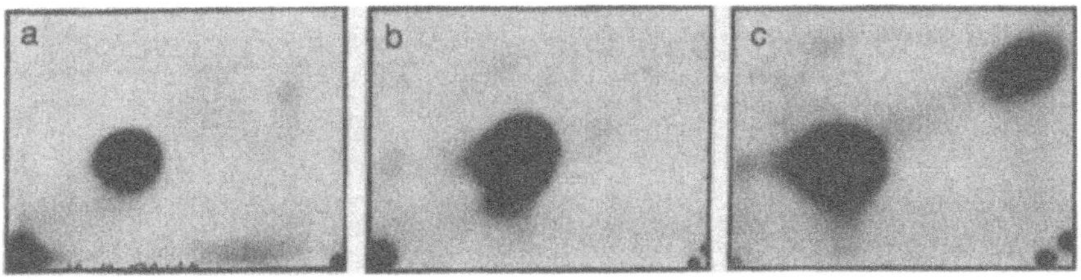

Figure 2. Fingerprints of CPA-induced DNA adducts in hepatocytes from various animal species. Cultured hepatocytes from female pig and a female human liver donator were incubated in the presence of 10 μM CPA for 6 hours, those from female rat for 20 hours. After isolation and digestion of the DNA, the DNA adducts were extracted from the digests into butanol, labelled with [y-^{32}P]-ATP and separated by PEI-cellulose-TLC. The autoradiograms shown are derived from pig (a), human (b) and rat (c) hepatocytes. Autoradiograms a, b and c were exposed for 24, 6 and 1.5 hours, respectively. Control incubations in the absence of CPA performed with rat and pig hepatocytes did not indicate the presence of any adduct.

tlc, this adduct was identical to the major adduct observed in DNA adduct fingerprints of rats termed adduct "A" [2]. Rats showed an additional minor adduct spot not detectable in the other species. Adduct "D" observed in females differed from adduct "B" detected in males (Fig. 2).

The levels of CPA-induced hepatic DNA adducts largely varied among the 3 species, when tested at a concentration of 10 μM CPA. Hepatocytes from pigs developed only low levels of about 20 adducts/10^9 nucleotides (Tab. 1). The adduct levels did not significantly differ between males and females.

High adduct levels were found in cultured human hepatocytes of both genders, comparable to those observed in hepatocytes of female rats (Tab. 1) Between 500 and 2700 adducts/10^9 nucleotides were determined in hepatocytes of 4 liver donors. The variation observed among the four samples was not related to the sex of the donors but seems to reflect individual variations of the activity of the activating enzymes.

Formation of Hepatic DNA Adducts in Mice Treated with CPA in Vivo

Treatment of mice with CPA resulted in the formation DNA adducts at a very low level (Tab. 2). Female mice of 2 different strains had adduct levels between 3 and 12 adducts/10^9 nucleotides, presumably of adduct type A. No adducts were detectable in males of both strains.

Table 1. DNA adduct levels in hepatocyte cultures of various species induced by CPA

species	sex	relative adduct level* [total adducts/10^9 nucleotides]
human	female	816, 1071
	male	475, 2676
pig	female	15, 19
	male	15, 27
rat	female	1050, 1211
	male	50, 52

*hepatocytes were incubated in the presence of 10 μM CPA for 6 hours

Table 2. Hepatic DNA adduct levels in mice pretreated with CPA

strain	CPA-dose [mg/kg]	duration [days]	rel. adduct level x 10^{9*} male	female
C57 BL/6	1 x 35	1	<1	12, 5
$(102xC3H)F_1$	1 x 10	1	<1	<1, 3
		8	<1	5, 4

*presumably adduct "A"

The finding that adduct A was formed in rat, pig, and man and probably also in female mice suggests that CPA was converted to the DNA binding species by a common activation mechanism. The activity of the enzymes involved, however, seems to differ between female rat and humans on the one hand, and pig, male rats and female mice on the other.

Persistence of Adducts in Rat Liver after a Single CPA Dose

To determine the lifetime of the CPA-induced DNA adducts, rats were treated with a single oral CPA dose and the adduct levels were determined at various time intervals after the treatment. Four DNA adducts were detected. In female rats CPA induced two major adducts termed "A" and "D" amounting to about 90 % and two minor adducts, "B" and "C". Males developed predominantly adducts A and B with C and D as minor adducts [2, L. R. Schwarz et al., this issue]. In females treated with 10 mg CPA/kg b.wt. the total adduct levels increased during the first days reaching a maximum level of 3397 adducts/10^9 nucleotides one week after the administration of CPA. Eleven weeks later, 40% of the DNA adducts were still detectable. Male rats that had received a single oral dose of 100 mg CPA/kg b.wt. reached a maximum level of about 100 adducts/10^9 nucleotides 2 weeks after dosing. About 40% of the maximal adduct level were still detectable six weeks after the beginning of the experiment. From these data a half-life of the adducts of about 6-7 weeks in female and 2-3 weeks in male rats can be estimated.

Accumulation of DNA Adducts in Rat Liver after Daily Administration of CPA

In view of the high persistence of the CPA induced DNA adducts observed, a considerable accumulation can be expected, when CPA is administered repeatedly. To examine the accumulation at conditions of antiacnegenic therapy, female rats were treated daily with low CPA doses of 50 μg/kg b.wt. for 6 weeks. The total level of hepatic DNA adducts showed a continuous increase. Adduct levels of 265 and 380 adducts/10^9 nucleotides were found after 3 and 6 weeks of treatment, respectively. When the fractionated CPA dose administered for 3 weeks was given as a single dose of 1.05 mg (21 x 50 μg) CPA/kg b.wt., maximal levels of 299 adducts/10^9 nucleotides were determined 6 days after dosing, the time of maximal DNA adduct levels. DNA adduct formation was much lower in male than in female rats. After 6 weeks of daily treatment with 50 μg CPA/kg b. wt. only 6 adducts/10^9 nucleotides were found.

According to the adduct half-life calculated for females of 6-7 weeks and 2-3 weeks for males, the time period required to reach the maximum level under steady state conditions can be estimated, since pharmacokinetic considerations reveal that 94 % of the steady level of a continously applied drug is reached after 4 half-life times [7]. In females this time period amounts to about 6 months; for males the corresponding value can be roughly estimated to be 2 months.

CONCLUSIONS

Human hepatocytes from both genders formed high levels of DNA-adducts upon treatment with CPA. We assume that quite high DNA adduct levels are also formed in human liver during therapeutic treatment with CPA. This notion is supported by our finding that DNA adducts strongly accumulated in female rats upon treatment with low CPA-doses. We anticipate that CPA-DNA adducts will also largely accumulate in women during antiacnegenic therapy. The biological significance of the adducts however, remains to be established.

REFERENCES

1. Schuppler, J. and Günzel, P., 1979, Liver tumors and steroid hormones in rats and mice, Arch. Toxicol. 2:181-195.
2. Neumann, I., Thierau, D., Andrae, U., Greim, H. and Schwarz, L.R., 1992, Cyproterone acetate induces DNA damage in cultured rat hepatocytes and preferentially stimulates DNA synthesis in gamma-glutamyltranspeptidase-positive cells, Carcinogenesis, 13:373-378.
3. Topinka, J., Andrae, U., Schwarz, L.R. and Wolff, T., 1993, Cyproterone acetate generates DNA adducts in rat liver and in primary hepatocyte cultures, Carcinogenesis, 14:423-427.
4. Strom, S. C., Jirtle, R. L., Jones, R. S., Novicki, D. L., Rosenberg, M. R., Novotny, A., Irons, G., McLain, J. R., Michalopoulos, G., 1982, Isolation, culture, and transplantation of human hepatocytes, J. Natl. Cancer Inst. 68:771-778.
5. Gupta, R.C., 1984, Non-random binding of the carcinogen N-hydroxy-2-acetylaminofluorene to repetitive sequences of rat liver DNA in vivo. Proc. Natl. Acad. Sci. U.S.A., 81:6943-6947.
6. Gupta, R.C., 1985, Enhanced sensitivity of ^{32}P-postlabelling analysis of aromatic carcinogen adducts, Cancer Res., 45:5656-5662.
7. Forth W., Henschler D., Rummel W., 1988, Pharmakologie und Toxikologie, Wissenschaftsverlag Mannheim, 68

MECHANISTIC STUDIES OF BENZENE TOXICITY – IMPLICATIONS FOR RISK ASSESSMENT

Martyn T. Smith

Division of Environmental Health Sciences
School of Public Health
140 Earl Warren Hall
University of California
Berkeley, California 94720-7360

INTRODUCTION

Benzene has been an established human carcinogen for some time now. Its use in the scientific laboratory, its disposal and use in the workplace is therefore tightly regulated in developed nations. It continues to be used, however, as a solvent in developing countries resulting in extremely high exposures. Further, with the introduction of unleaded gasoline in many countries, environmental and occupational exposure to benzene has increased. Most unleaded gasoline contains approximately 1% benzene, but some varieties contain 5% or more. The annual production of benzene in the US exceeds 1.2 billion gallons, accounting for 30% of worldwide production. Approximately 165,000 metric tons are released annually into the air in the US. Benzene is therefore a ubiquitous environmental pollutant. Smoking is another source of benzene exposure. A one pack/day smoker takes in 1800 μg benzene each day, while non-smokers are exposed to 180 - 1300 μg per day (1).

Benzene is an established cause of acute myelogenous leukemia (AML) in humans (2). It's ability to cause other types of leukemia, lymphomas and epithelial cancers remains highly controversial. Studies in China have suggested that benzene can cause lymphocytic leukemias and lung cancer in addition to AML (3). Benzene is certainly a multi-site carcinogen in animals (4), but as yet there is no good animal model of myelogenous leukemia development. The controversy surrounding the epidemiological studies and the lack of an animal model has made the risk assessment process for benzene difficult. Especially problematic has been assessment of the risk posed by low doses of benzene. What insight then can mechanistic research throw on this? In this article, I will argue that the latest mechanistic insights suggest that benzene is a genotoxic carcinogen which poses a risk even at low doses. This conclusion is admittedly controversial and I know that others will interpret the data differently. However, I present my opinions here with the purpose of stimulating debate.

Biological Reactive Intermediates V
Edited by R. Snyder et al., Plenum Press, New York, 1996

METABOLIC ACTIVATION OF BENZENE

The metabolism of benzene has been described in detail elsewhere (5). I will therefore present only a summary. Benzene is converted to benzene oxide mainly via cytochrome P4502E1(6). It can then rearrange non-enzymatically to phenol, the major metabolite, or be converted by epoxide hydrolase to benzene dihydrodiol. A variety of secondary metabolites can be formed from the dihydrodiol, including epoxides, catechol and the ring-opened products trans, trans-muconaldehyde (MUC) and trans, trans-muconic acid (7). MUC has been shown to have a variety of toxicological properties including genotoxicity (8) and has been proposed as an important toxic metabolite of benzene (9). I remain skeptical, however, that sufficient amounts of this metabolite reach the bone marrow to cause toxicity.

The major metabolite of benzene, phenol, is either conjugated, primarily via sulfation in humans, or further hydroxylated to hydroquinone, catechol and 1,2,4-benzenetriol. These compounds are all excellent substrates for peroxidase enzymes and are converted by them to toxic quinone species via their respective semiquinone free radicals (10). Phenol is also a good substrate for peroxidases forming diphenoquinones and polyquinone radicals (11, 12). For as yet unknown reasons, the phenolic metabolites of benzene accumulate in bone marrow. We have proposed that they undergo a secondary activation in the bone marrow via the enzyme myeloperoxidase. This yields the quinone and semiquinone products described above and is proposed to play a key role in benzene toxicity (13). Myeloperoxidase is present at high concentrations in myeloid progenitor cells including multipotent stem cells (14). Other peroxidases are also present, namely eosinophil peroxidase and prostaglandin synthetase. These may also play some role, and indeed some researchers (15) consider the latter to play a major role.

We have further shown that the peroxidase-dependent metabolism of hydroquinone is stimulated by other metabolites, including phenol and catechol (16, 17). This led us to suggest that multiple metabolites are involved in benzene toxicity (16). Data supporting this hypothesis has been accumulating ever since (Table 1). Initially, we showed that multiple doses of phenol and hydroquinone produced a hematopoietic toxicity in mice similar to benzene. Others confirmed this finding using different methods (18). Next, Barale and co-workers (19, 20) showed that phenol and hydroquinone produced a synergistic induction of micronuclei in mouse bone marrow. Eastmond and co-workers recently confirmed this finding and further showed that the type of micronuclei formed were very similar to those formed by benzene (21). Eastmond and Chen (22) have further postulated that this is because the phenol metabolite 4,4′-diphenoquinone inhibits topoisomerase II and enhances the clastogenic effects of hydroquinone .

Table 1. Support for Multi-Metabolite Hypothesis

Finding	Reference
PH* + HQ is toxic to mouse bone marrow	Eastmond *et al*, 1987(16) Guy *et al* (18)
PH + HQ is genotoxic to mouse bone marrow	Barale *et al*, 1990(19) Chen *et al* 1995(22)
PH enhances covalent binding of HQ in marrow	Subrahmanyam *et al*, 1990(23)
Combinations enhance DNA adduct formation	Levay *et al*, 1992(34)
Combinations enhance oxidative DNA damage	Kolachana *et al*, 1993 (24)
PH metabolites inhibit topoisomerase II and enhances the genotoxic effects of HQ	Chen and Eastmond, 1995 (21,22)

*PH = Phenol; HQ = Hydroquinone

Table 2. The Phenolic and Quinonoid Metabolites of Benzene are Genotoxic

Endpoint	System	Reference
Chromosome aberrations	cell cultures	Dean, 1985 (28)
Micronuclei	mouse	Tice *et al*, 1980 (46)
		Barale *et al* 1990 (19)
	human cells	Yager *et al*, 1990 (31)
SCE's	human cells	Erexson *et al*, 1985 (29)
	mouse	Tice *et al*, 1980 (46)
Gene mutations	V79 cells	Glatt *et al*, 1989 (32)
Chromosome-specific aneuploidy	human cells	Zhang *et al*, 1994
		Rupa *et al* 1994 (47)
DNA adducts	HL60 cells	Levay *et al*, 1992 (34)
	mouse	Yin *et al*, 1992 (36)
		Bodell *et al*, 1995 (37)
8-hydroxy-2'-deoxyguanosine	HL60 cells	Kolachana *et al*, 1993 (24)
	mouse	

We followed up our initial finding of combined metabolite toxicity by showing that phenol administration enhanced 14 C-hydroquinone binding in the blood and marrow (23). The combination of phenol and hydroquinone also produced oxidative DNA damage in mouse bone marrow similar to that produced by benzene, as did the minor metabolite, benzenetriol (24). One can, in fact, make the case that all the phenolic metabolites of benzene play a role in benzene toxicity and it is the combination that is toxic (25).

THE PHENOLIC AND QUINONOID METABOLITES OF BENZENE ARE GENOTOXIC

Benzene is a somewhat unusual carcinogen in that its metabolites are not genotoxic in simple mutation assays, such as the Ames Salmonella test. They do produce chromosome damage, however, and gene mutations in more complex mammalian systems (Table 2). Benzene and its metabolites have long been known to cause classical chromosome aberrations in cell cultures and exposed humans (26-28). They also cause sister chromatid exchanges (SCEs) (29, 30) and we have compared their potency in causing micronucleus and aneuploidy formation in human cells (31). Glatt *et al* (32) compared their ability to cause SCE's, gene mutations and micronuclei in rodent V79 cells.

In general hydroquinone, its oxidation product 1,4-benzoquinone, and the minor metabolite 1,2,4-benzenetriol are the most potent genotoxic metabolites. Hydroquinone and 1,4-benzoquinone, have also been shown to form DNA adducts (33-35). Interestingly, this adduct formation is stimulated by other phenolic metabolites (34). More recently, Chinese researchers (36) and Bodell and co-workers (37)have shown that benzene causes DNA adduct formation in rodent blood and bone marrow. Turteltaub and coworkers (38) have further shown that chromatin-adduct formation is linear down to very low doses using accelerator mass spectrometry.

One of our most recent findings is that benzene and its metabolites produce oxidative DNA damage in mouse bone marrow in vivo and in human HL60 cells in vitro (24). Benzene produced a 2 to 5 fold increase in 8 - hydroxy-2′-deoxyguanosine (8-oxodG) in mouse bone marrow one hour after administration. The phenolic metabolites, most notably 1,2,4-benzenetriol and the combination of phenol plus hydroquinone, also produced increases in 8-oxodG formation both in vivo and in vitro.

Thus, benzene and its phenolic metabolites produce many different types of genetic damage. The question is which, if any, of these types of genetic damage are involved in benzene-induced leukemia in humans? In order to attempt to address this and other unanswered questions we have recently embarked on studies of humans exposed to benzene in China.

COLLABORATIVE STUDIES ON CHINESE WORKERS EXPOSED TO BENZENE

In 1992 together with researchers from the National Cancer Institute, the California EPA and China, we began to study a group of highly exposed Chinese workers using a variety of biological markers. In a total of 168 individuals we have measured exposure by questionnaire, personal monitoring and urinary analysis, and a variety of susceptibility and effect markers (39). These include micronuclei, chromosome aberrations, hematotoxicity, cytochrome P4502E1 activity using chlorzoxazone, and mutations in the glycophorin A (GPA) gene. The latter has produced extremely interesting results which throw light on the type of mutations benzene produces in human bone marrow.

The GPA assay (40) is highly suited to the study of benzene genotoxicity in humans because it measures mutational events in bone marrow progenitor cells. It uses antibodies and flow cytometry to measure changes in the forms of GPA on the surface of red blood cells. The GPA gene has two alleles M and N. In the assay loss of the M allele is monitored. Two types of mutation are detectable. The first, called N0, arise from gene-inactivating mutations such as point mutations or small deletions. The second, called NN, result from gene-duplicating events such as mitotic recombination and chromosome loss with reduplication.

We found that benzene exposure produced a 2-fold increase in the GPA mutant frequency (41, 42). Interestingly, all of this increase was caused by a highly significant increase in NN gene-duplicating mutations. This finding is consistent with the genetic toxicology of benzene and its metabolites described above. To the extent that benzene metabolites cause similar changes in key genes involved in leukemogenesis, our results suggest that benzene is leukemogenic by producing chromosomal damage and mitotic recombination.

ROLE OF GENETIC AND EPIGENETIC FACTORS IN BENZENE-INDUCED LEUKEMIA

It is clear that benzene is a genotoxic carcinogen. However, other epigenetic phenomena may play a role in benzene-induced hematotoxicity. For example, Kalf and coworkers have reported that benzene toxicity to the bone marrow can be prevented by IL-1 administration (43). They postulate that hydroquinone is oxidized to benzoquinone which then inhibits the conversion of pre-IL-1 to the active form, IL-1. Further, this effect occurs within the stromal cells of the bone marrow and prevents the normal maturation of progenitor cells. Kalf believes this plays an important role in benzene-induced aplastic anemia. Irons and co-workers (44)have also shown an interesting effect of hydroquinone pre-treatment on the granulocyte-macrophage lineage of hematopoetic progenitor cells. His group has shown that hydroquinone pre-treatment increases the number of colonies formed by recombinant

GM-CSF-induced CFU-GM. At concentrations as low as 10^{-9} molar, hydroquinone produces this stimulatory effect. This would suggest that hydroquinone causes the recruitment of hematopoetic progenitor cells into the granulocyte-macrophage pathway. In theory, this recruitment could provide more target for the genotoxic effects of hydroquinone and other metabolites. Further, it has been shown that this phenomena is specific to chemicals which cause myeloid leukemias. Leukemia may therefore result from benzene inducing both altered differentiation and genotoxic damage leading to a selective growth advantage for immature cells.

IMPLICATIONS FOR RISK ASSESSMENT

One of the major questions for risk assessment is "does benzene have a threshold?" If, as I have suggested above, benzene is indeed a genotoxic carcinogen, then current dogma would dictate that it does not have a threshold and that one molecule is sufficient to cause one hit and increase a person's risk of cancer. If, however, more than one metabolite is necessary for benzene toxicity and more than one type of genetic and epigenetic change must occur for the development of leukemia, then it would appear that a threshold for benzene-induced leukemia seems likely. For the development of leukemia, it is clear that multiple molecules of benzene metabolites will be needed and that these metabolites must also target a stem cell in the bone marrow. This would imply that benzene has a non-linear dose response curve in the low-dose region.

One must also consider that there is a background level of exposure to benzene and that the risk of additional exposures is the one we are really concerned about. The average person is exposed to 180 μg daily. If all of this benzene were absorbed and distributed to the bone marrow, more than 1 million benzene molecules would be available to each bone marrow cell, including the stem cells. This is obviously an overestimate. However, if we assume that approximately 1% of the benzene is absorbed and distributed in metabolite form to the bone marrow then there are still 10,000 molecules available to each bone marrow cell. This is more than enough molecules to overcome the need for multiple metabolites and for multiple targets to be hit. It also suggests that background levels of benzene will form adducts in the bone marrow and produce some background level of risk. For a smoker taking in 1800 μg of benzene the number of molecules available could theoretically exceed 10 million molecules per cell. There are also numerous sources of the benzene metabolites, phenol, hydroquinone and catechol in our environment. These sources include the diet, cigarette smoke and pharmaceuticals. The number of benzene metabolite molecules reaching the human bone marrow in a typical human is therefore likely to measure in the tens of thousands. Thus, additional benzene exposure, such as occupational exposure, would contribute additional risk upon this environmental background and is likely to be linear with dose.

If these concepts are correct, one would expect to detect a high level of benzene metabolite adducts in the bone marrow and that additional benzene exposures would increase this adduct level in an approximately linear fashion. Recent work by Rappaport and co-workers (45) strongly supports this notion. As shown in Figure 1, their data shows that a high background of 1,4-benzoquinone protein adducts occur in mouse bone marrow protein, presumably derived from hydroquinone oxidation. After exposure of mice to benzene this level of protein adducts increases linearly on the high background. Thus, additional molecules of benzene produce metabolites in the bone marrow which bind covalently to macromolecules in a linear fashion. This data suggest that background exposure to benzene and its metabolites in the normal human environment contributes to the background level of leukemia. Further, it suggests that additional benzene exposures at the work place and elsewhere would cause a linear increase in leukemia risk on this background. In conclusion,

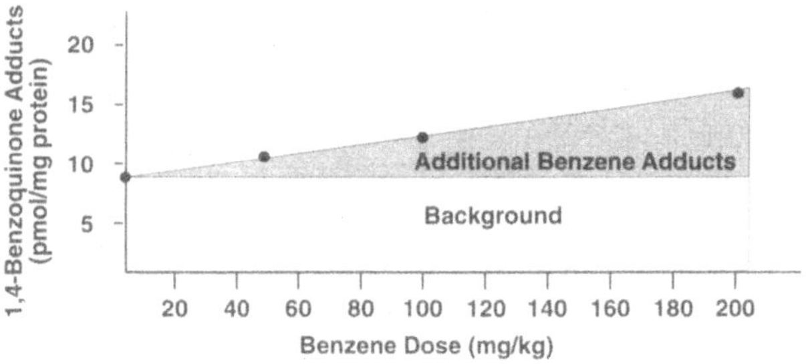

Figure 1. Adducts formed between 1,4-benzoquinone and bone marrow protein in mice exposed to different doses of benzene. Data from Reference (45).

the most recent mechanistic data on benzene, suggest that benzene is a genotoxic carcinogen that poses a risk, albeit small, even at low doses. Further, the most recent data suggest that the risk posed by additional benzene exposures in the low dose region is most likely linear.

ACKNOWLEDGMENTS

Supported by grants RO1 ES06721, P42 ES04705 and P30 ES01896 from the National Institute of Environmental Health Sciences. I thank Drs. D. Eastmond, N. Rothman, S-N Yin and L. Zhang for their open collaboration and helpful discussion. I thank Maura Storace, Adam Delu and Elinor Fanning for help in preparing this manuscript.

REFERENCES

1. Wallace LA. The exposure of the general population to benzene. Princeton, New Jersey: Princeton Scientific Publishing, 1989. (Mehlman MA, ed. Advances in Modern Environmental Toxicology Volume XVI - Benzene: Occupational and Environmental Hazards - Scientific Update;
2. IARC. IARC Monographs on the Evaluation of Carcinogenic Risks to Humans. 1987;Suppl. 6:91-95.
3. Yin SN, Li GL, Tain FD, et al. Leukaemia in benzene workers: A retrospective cohort study. Br J Ind Med 1987;44(2):124-128.
4. Huff JE, Haseman JK, DeMarini DM, et al. Multiple-site carcinogenicity of benzene in Fischer 344 rats and B6C3F1 mice. Environ. Health Perspect. 1989;82:125-163.
5. Snyder R, Kalf GF. A perspective on benzene leukemogenesis. Crit Rev Toxicol 1994;24:177-209.
6. Koop DR, Laethem CL, Schnier GC. Identification of ethanol-inducible P450 isozyme 3a(P450IIE1) as a benzene and phenol hydroxylase. Toxicol Appl Pharmocol 1989;98:278-288.
7. Witz G, Maniara W, Mylavarapu V, Goldstein B. Comparative metabolism of benzene and trans, trans-muconaldehyde to trans, trans-muconic acid in DBA/2N and C57BL/6 Mice. Biochem Pharmacol 1990;40:1275 - 1280.
8. Witz G, Gad S, Tice R, Oshiro Y, Piper C, Goldstein B. Genetic toxicity of the benzene metabolite trans, trans-muconaldehyde in mammalian and bacterial cells. Mutat Res 1990;240:295 - 306.
9. Goldstein BD, Witz G, Javid J, Amoruso MA, Rossman T. Muconaldehyde, a potential toxic intermediate of benzene metabolism. Adv Exp Med 1982;136A:331-339.
10. Smith MT, Yager JW, Steinmetz KM, Eastmond DA. Peroxidase-dependent metabolism of benzene's phenolic metabolites and its potential role in benzene toxicity and carcinogenicity. Environ Health Perspect 1989;82:23-29.
11. Eastmond DA, Smith MT, Ruzo LO, Ross D. Metabolic activation of phenol by human myeloperoxidase and horseradish peroxidase. Mol Pharmacol 1986;30:674-679.
12. Subrahmanyam VV, O'Brien PJ. Phenol oxidation product(s), formed by a peroxidase reaction, that bind to DNA. Xenobiotica 1985;15(10):873-885.

13. Subrahmanyam VV, Ross D, Eastmond DA, Smith MT. Potential role of free radicals in benzene-induced myelotoxicity and leukemia. Free Rad Biol Med 1991;11:495-515.
14. Schattenberg DG, Stillman WS, Gruntmeir JJ, Helm KM, Irons RD, Ross D. Peroxidase activity in murine and human hematopoietic progenitor cells: Potential relevance to benzene-induced toxicity. Mol Pharmacol 1994;46:346-351.
15. Schlosser MJ, Shurina RD, Kalf GF. Prostaglandin H synthase catalyzed oxidation of hydroquinone to a sulfhydryl-binding and DNA-damaging metabolite. Chem Res Toxicol 1990;3(4):333-339.
16. Eastmond DA, Smith MT, Irons RD. An interaction of benzene metabolites reproduces the myelotoxicity observed with benzene exposure. Toxicol Appl Pharmacol 1987;91:85-95.
17. Subrahmanyam VV, Kolachana P, Smith MT. Metabolism of hydroquinone by human myeloperoxidase: Mechanisms of stimulation by other phenolic compounds. Arch Biochem Biophys 1991;286:76-84.
18. Guy R, Dimitriadis E, Hu P, Cooper K, Snyder R. Interactive inhibition of erythroid 59Fe utilization by benzene metabolites in female mice. Chem. -Biol. Interact. 1990;74:55 - 62.
19. Barale R, Marrazzini A, Betti C, Vangelisti V, Loprieno N, Barrai I. Genotoxicity of two metabolites of benzene: Phenol and hydroquinone show strong synergistic effects in vivo. Mutat Res 1990;244:15-20.
20. Marrazzini A, Chelotti L, Barrai I, Loprieno N, Barale R. In vivo genotoxic interactions among three phenolic benzene metabolites. Mutat Res 1994;341:29-46.
21. Chen H, Eastmond D. Syngergistic increase in chromosomal breakage within the euchromatin induced by an interaction of the benzene metabolites phenol and hydroquinone in mice. Carcinogenesis 1995;in press.
22. Chen H, Eastmond D. Inhibition of human topoisomerase II by benzene metabolites (Abstract). The Toxicologist 1995;15(1):221.
23. Subrahmanyam V, Doane-Setzer P, Steinmetz K, Ross D, Smith M. Phenol-induced stimulation of hydroquinone bioactivation in mouse bone marrow in vivo: Possible implications in benzene myelotoxicity. Toxicology 1990;62:107 - 116.
24. Kolachana P, Subrahmanyam VV, Meyer KB, Zhang L, _Smith MT. Benzene and its phenolic metabolites produce oxidative DNA damage in HL60 cells in vitro and in the bone marrow in vivo. Cancer Res 1993;53:1023-1026.
25. Zhang L, Smith M, Bandy B, Tamaki S, AJ D. Role of quinones, active oxygen species and metals in the genotoxicity of 1,2,4-benzenetriol, a metabolite of benzene. London: Richelieu Press, 1995. (Nohl H, Esterbauer H, Rice-Evans C, eds. *Free radicals in the Environment, and Toxicology*;
26. Sasiadek M. Nonrandom distribution of breakpoints in the karyotypes of workers occupationally exposed to benzene. Environ Health Perspect 1992;97:255-257.
27. Forni A, Pacifico E, Limonta A. Chromosome studies in workers exposed to benzene or toluene or both. Arch Environ Health 1971;22:373-378.
28. Dean BJ. Recent findings on the genetic toxicology of benzene, toluene, xylenes and phenols. Mutat Res 1985;154:153-181.
29. Erexson GL, Wilmer JL, Kligerman AD. Sister chromatid exchange induction in human lymphocytes exposed to benzene and its metabolites in vitro. Cancer Res ;45:2471-2477.
30. Morimoto K, Wolff S, Koizumi A. Induction of sister-chromatid exchanges in human lymphocytes by microsomal activation of benzene metabolites. Mutat Res 1983;119:355-360.
31. Yager JW, Eastmond DA, Robertson ML, Paradisin WM, Smith MT. Characterization of micronuclei induced in human lymphocytes by benzene metabolites. Cancer Res 1990;50:393-399.
32. Glatt H, Padykula R, Berchtold GA, et al. Multiple activation pathways of benzene leading to products with varying genotoxic characteristics. Environ Health Perspect 1989;82:81-89.
33. Pongracz K, Kaur S, Burlingame A, Bodell W. Detection of (3'-hydroxy)-3,N4-benzetheno-2'-deoxycytidine-3'-phosphate by 32P-postlabeling of DNA reacted with p-benzoquinone. Carcinogenesis 1990;11(No. 9):1469 - 1472.
34. Levay G, Bodell WJ. Potentiation of DNA adduct formation in HL-60 cells by combinations of benzene metabolites. Proc Natl Acad Sci 1992;89:7105-7109.
35. Jowa L, Kalf GF, Witz G, Snyder R. DNA or nucleoside adducts with hydroquinone (HQ) and benzoquinone (BQ). Toxicologist 1985;5(1):146.
36. Yin W, Li G, Yin S. DNA adduct formation in rodents exposed to benzene. Toxicologist 1992;12:250.
37. Levay G, Pathak D, Bodell W. Detection of benzene-DNA adducts in the white blood cells of mice treated with benzene. 86th annual meeting American Association for Cancer Research. Toronto, Canada, 1995:111, abstract 658.
38. Creek M, Vogel J, Turtletaub K. Extremely low dose benzene pharmacokinetics and macromolecular binding in B6C3F1 male mice. Abstract 1703. The Toxicologist 1994;14(1):430.

39. Rothman N, Li G-L, Dosemeci M, et al. Hematoxicity among Chinese workers heavily exposed to benzene. American Journal of Industustrial Medicine 1995 (in press).
40. Langlois R, Nisbet B, Bigbee W, Ridinger D, Jensen R. An improved cytometric assay for somatic mutations at the glycophorin-A locus in humans. Cytometry 1990;11:513-521.
41. Smith MT, Rothman N, Holland NT, et al. Biomarkers of genetic damage in humans exposed to benzene. EMS Abstracts 1994;23 (Suppl 23):63.
42. Rothman N, Haas R, Hayes R, et al. Benzene induces gene-duplicating but not gene-inactivating mutations at the glycophorin A locus in exposed humans. Proceedings of the National Academy of Science USA 1995 (in press).
43. Renz JF, Kalf GF. Role for interleukin-1 (IL-1) in benzene-induced hematotoxicity: Inhibition of conversion of Pre-IL-1a to mature cytokine in murine macrophages by hydroquinone and prevention of benzene-induced hematotoxicity. Blood 1991;78:938-944.
44. Irons RD, Stillman WS, Colagiovanni DB, Henry VA. Synergistic action of the benzene metabolite hydroquinone on myelopoietic stimulating activity of granulocyte/macrophage colony-stimulating factor in vitro. Proc Natl Acad Sci 1992;89:3691-3695.
45. McDonald T, Yeowell-O'Connell K, Rappaport S. Comparison of protein adducts of benzene oxide and benzoquinone in the blood and bone marrow of rats and mice exposed to [$^{14}C/^{13}C_6$]benzene. Cancer Research 1994;54:4907-4914.
46. Tice RR, Costa DL, Drew RT. Cytogenetic effects of inhaled benzene in murine bone marrow: Induction of sister chromatid exchanges, chromosomal aberrations, and cellular proliferation inhibition in DBA/2 mice. Proc Natl Acad Sci 1980;77:2148-2152.
47. Eastmond DA, Rupa DS, Hasegawa LS. Detection of hyperdiploidy and chromosome breakage in interhpase lymphocytes following exposure to the benzene metabolite hydroquinone using multicolor fluorescence in situ hybridization with DNA probes. Mutat Res 1994; 322:9-20.

LINKING THE METABOLISM OF HYDROQUINONE TO ITS NEPHROTOXICITY AND NEPHROCARCINOGENICITY*

Serrine S. Lau,† Melanie M. C. G. Peters, Heather E. Kleiner, Patricia L. Canales, and Terrence J. Monks

Division of Pharmacology and Toxicology, College of Pharmacy
The University of Texas at Austin, Austin, Texas 78712

INTRODUCTION

Hydroquinone (HQ) is nephrocarcinogenic in male Fischer-344 (F344) rats. Kari *et al.* (1992) described renal tubular cell degeneration in the renal cortex of male F344 rats receiving 1.8 mmol/kg HQ (13 week gavage study) and results from long-term studies demonstrated that HQ exhibited "*evidence of carcinogenic activity*" in male F344 rats as shown by marked increases in tubular cell adenomas of the kidneys. These findings were confirmed by Shibata *et al.* (1991), who reported the induction of renal cell tumors in rats following exposure to HQ in the diet (0.8%) for two years. Although the mechanism of HQ-mediated nephrocarcinogencity in male F344 rats is not known, GSH conjugates of HQ may play an important role. We have previously shown that multi-substituted glutathione (GSH) conjugates of HQ cause selective necrosis of renal proximal tubule cells in Sprague-Dawley rats when administered by intravenous injection, with 2,3,5-(triGSyl)HQ being the most potent nephrotoxicant (Lau *et al.*, 1988). Furthermore, HQ-GSH conjugates have been identified as both *in vitro* and *in vivo* metabolites of HQ in the Sprague-Dawley rat, in quantities sufficient to propose a role for GSH conjugation in the nephrotoxicity of HQ (Hill *et al.*, 1993). In the present study we investigated the metabolism of [^{14}C]-HQ to HQ thioethers in the F344 rat, the rat strain used in the carcinogenicity studies, and also examined the effect of subchronic administration on the elimination and metabolic profile of [^{14}C]-HQ. We propose that the formation of nephrotoxic metabolites may play a role in HQ-induced nephrocarcinogenicity by a mechanism that involves the generation of reactive oxygen species, oxidative DNA damage, and cytotoxicity followed by sustained hyperplasia.

* This work was supported in part by an award from the NIGMS (GM 39338) to SSL.

† All correspondence should be addressed to: Dr. Serrine S. Lau, Division of Pharmacology and Toxicology, College of Pharmacy, University of Texas at Austin, Austin, Texas 78712, Tel: (512) 471-5190; FAX: (512) 471-5002; email: slau@uts. cc. utexas. edu

RESULTS AND DISCUSSION

Male F344 rats (150-200 g) received either a single oral dose of 1.8 mmol/kg [^{14}C]-HQ in corn oil or 14 daily doses of HQ (1.8 mmol/kg in corn oil), followed by a single oral dose of [^{14}C]-HQ (1.8 mmol/kg) on day 15. Urine was collected over dry ice and in the dark for 72 h and analyzed for the recovery of total radioactivity. Following a single gavage dose of [^{14}C]-HQ, 21% of the dose was recovered in the 0-5 h urine, with an additional 35% excreted in the next 19h, indicating that [^{14}C]-HQ is rapidly eliminated, predominantly *via* urine. After subchronic administration of HQ, the elimination of [^{14}C]-HQ was significantly faster, with 46% of the radiolabel recovered in the 0-5 h urine and 31% in the following 19 h urine. Urinary metabolites of HQ were determined by HPLC coupled to UV (280 nm) and electrochemical (EC) detection. HQ was extensively metabolized, with less than 2% of dose excreted unchanged. The major metabolites identified in the 0-24 h urine samples were HQ-glucuronide (21%) , HQ-sulfate (15%), and HQ-mercapturate (13%). A change in the metabolic profile was observed after subchronic administration of HQ, with increased formation of the HQ-glucuronide (2 fold) and the HQ-mercapturate (1.4 fold), while the percentage of dose that underwent conjugation with sulfate remained constant. These results are in agreement with the general finding of the activation of the glucuronidation pathway by a variety of phenolic substances, such as *tert*-butylhydroquinone and butylated hydroxyanisole (Verhagen *et al.*, 1991), which may be due to either increased UDPGA-glucuronsyltransferase activity and/or an increase in cofactor supply (UDPGA) upon subchronic administration. The findings also support our hypothesis that subchronic administration of HQ increases the rate and extent by which HQ is metabolized to its nephrotoxic thioethers. When HQ-mercapturic acid excretion is considered as a measure of internal exposure, it is evident that the amounts formed are sufficient to propose a role for these metabolites in the chronic toxicity of HQ. In addition, 2-(GSyl)HQ, 2,5-(diGSyl)HQ, 2,6-(diGSyl)HQ and 2,3,5-(triGSyl)HQ were identified as metabolites of HQ in bile after oral administration to F344 rats.

HQ is cytotoxic, and exhibits clear clastogenic effects in a variety of assay systems (reviewed in Kari *et al.*, 1992); this has been associated with its ability to redox cycle and create an oxidative stress and/or to react directly with cellular nucleophiles, particularly protein and non-protein sulfhydryls, such as GSH (Schlosser *et al.*, 1990; Nakagawa and Moldeus, 1992; Hanzlik *et al.*, 1994). Conjugation of redox-active compounds with GSH frequently has little effect on their redox behavior, and in some instances may even facilitate oxidation of the quinol (Monks and Lau, 1994) and therefore, quinone thioethers can possess as much, if not more biological reactivity than the parent quinones. For instance, we have shown that GSH conjugates of HQ exhibit oxidation potentials similar to that of HQ (Hill *et al.*, 1993). In addition to the well known modification of proteins by oxidants (Stadtman and Oliver, 1991; Davies, 1987) oxidants mediate DNA damage *via* both DNA strand breaks and the formation of oxidatively modified DNA bases. Cantoni *et al.*, (1991) have shown that DNA strand breaks (measured *via* the alkaline elution assay) occurred in CHO cells exposed to low (20-50 μM) concentrations of menadione, and that at least a portion of this damage occurred as a consequence of H_2O_2 production. Menadione redox cycling was also shown to cause both single and double-strand DNA breaks in human MCF-7 cells (Nutter *et al.*, 1992). ESR studies provided evidence that the damage was mediated via the •OH radical and suggested that menadione-induced cytotoxicity was a consequence of DNA damage. Thus, exogenously added catalase reversed menadione cytotoxicity and inhibited •OH radical formation. Consistent with these findings, we have also shown that the cytotoxicity of 2-Br-(glutathion-S-yl)hydroquinone in renal epithelial cell culture is a consequence of reactive oxygen species generation, which catalyzes the formation of single strand breaks in DNA (Mertens *et al.,* 1995). Solveig Walles (1992) suggested that DNA single strand breaks induced by HQ in isolated hepatocytes were due to arylation of DNA bases by 1,4-benzoquinone, and HQ has been shown to bind to deoxyguanosine (Jowa *et al.*, 1990). However, DNA adducts were not found in kidney after HQ administration *in vivo* (English *et al.*, 1994).

There is an increased awareness of the importance of oxidative, as opposed to covalent DNA damage (Loeb *et al.*, 1988), which may be mediated primarily by reactive oxygen species, in particular •OH. Hydrated thymine derivatives, including 5,6-dihydroxy-dihydrothymine (thymine glycol) are recognized as important products of ionizing radiation and are released from bacterial and mammalian DNA *in vivo* after exposure to various oxidative agents (Cerutti, 1978).The C-8 position of deoxyguanosine residues in DNA are also susceptible to hydroxylation. 8-Hydroxydeoxyguanosine (8-OH-dG) formation has been demonstrated *in vitro* by incubating deoxyguanosine in the presence of various oxygen

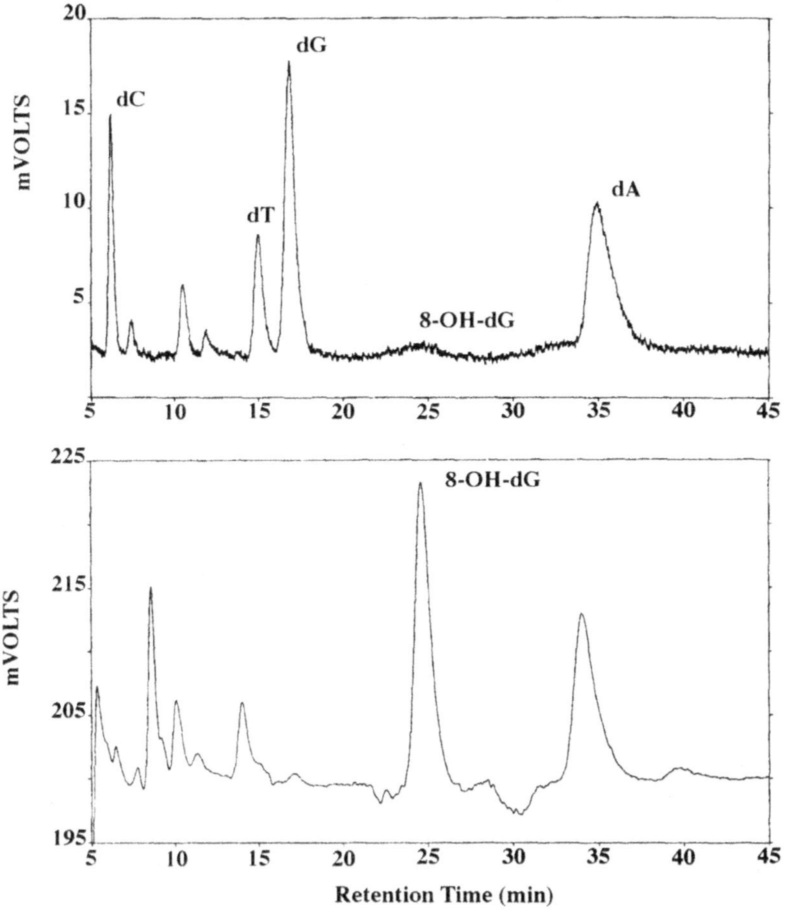

Figure 1. HPLC-UV (top panel) and HPLC-EC (bottom panel) profile of 2-(GSyl)HQ catalyzed formation of 8-OH-dG in calf thymus DNA. DNA hydrolysates (40 μl) were analyzed for 8-OH-dG and dG by HPLC (Shimadzu LC-6A) UV detection (254 nm, Shimadzu SPD-6AV) and EC detection (detector 1, 0V; detector 2, +0.28V; ESA Coulochem, model 5100 A), using a Partisil 5 μm ODS-3 reverse phase analytical column (Whatman, Clifton, NJ) and a mobile phase containing 4 mM citric acid, 8 mM ammonium acetate, 5% methanol, and 20 mg/L EDTA (pH 4.0) at a flow-rate of 1 ml/min.

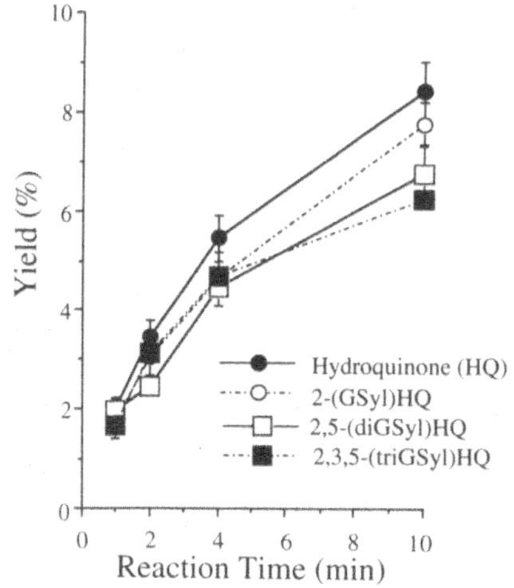

Figure 2. Hydroquinone (HQ) and HQ-GSH conjugate catalyzed formation of 8-OH-dG. The formation of 8-OH-dG was expresssed as the percentage initial dG. Reaction mixtures contained 1.3 mM dG, 1 mM HQ or HQ-GSH conjugate (dissolved in water), 5 mM H_2O_2, 0.1 mM $FeCl_3$, 0.5 mM EDTA, and 0.1 M phosphate buffer (pH 7.4) in a total volume of 2 ml. After a 2 min preincubation, the reactions were initiated by the addition of H_2O_2. Incubations were performed in the dark at 37°C, aliquots were removed at appropriate times, and placed in 2 mM deferoxamine on ice to terminate the reaction. dG and 8-OH-dG were assayed by HPLC-UV/EC as described in the legend of figure 1. Values represent mean ± SE (n=3).

radical-producing agents, such as asbestos plus hydrogen peroxide (H_2O_2) (Kasai and Nishimura, 1984a) and polyphenol(s) in the presence of H_2O_2 and ferric ion (Kasai and Nishimura, 1984b). More importantly, Kasai *et al.* (1986) have shown that 8-OH-dG is formed in DNA *in vivo* following exposure of mice or rats to oxygen radical producing agents. In particular, oral administration of the renal carcinogen, potassium bromate, specifically produced 8-OH-dG in rat target organ (kidney) DNA (Kasai *et al.*, 1987). The detection of 8-OH-dG in human lymphocyte DNA has also been reported (Kiyosawa *et al.*, 1988). The functional significance of oxidative DNA damage was emphasized when it was demonstrated that introduction of an 8-OH-dG residue into DNA caused misreading at the site of the 8-OH-dG moiety during *in vitro* DNA synthesis (Kuchino *et al.*, 1987).

Previous studies have shown that HQ catalyzes the *in vitro* formation of 8-OH-dG (Kasai and Nishimura, 1984b; Leanderson and Tagesson, 1990). The aim of the present study was to determine the relative ability of HQ and HQ-GSH conjugates to catalyze the formation of 8-OH-dG from deoxyguanosine (dG), and from calf thymus DNA. The 8-OH-dG was determined by HPLC coupled to UV (280 nm) and electrochemical (EC) detection (Figure 1). Incubations of dG with HQ or HQ-GSH conjugates in the presence of $FeCl_3$, EDTA, and H_2O_2 resulted in the formation of 8-OH-dG (Figure 2). The initial rate of 8-OH-dG formation from HQ, 2-(GSyl)HQ, 2,5-(diGSyl)HQ, and 2,3,5-(triGSyl)HQ was 52 ± 6, 45 ± 8, 51 ± 6, and 43 ± 1 nmol/min (mean ± SE), respectively. In addition, HQ, 2-(GSyl)HQ, 2,5-(diGSyl)HQ and 2,3,5-(triGSyl)HQ increased the amount of 8-OH-dG formed in calf thymus DNA over 30-fold compared to control incubations in the absence of HQ or HQ-thioethers (Figures 3). The occurrence of this reaction *in vivo* and the importance of oxidative DNA damage in HQ-mediated toxicity and carcinogenicity remains to be determined. However, our data indicate that the redox-activity of HQ and its ability to form 8-OH-dG are not impaired by conjugation with GSH. Studies in our laboratory indicate that the toxicity of 2-Br-(glutathion-S-yl)hydroquinone in renal proximal epithelial cells is in part mediated by the generation of •OH radicals, which lead to the formation of DNA single strand breaks (Mertens *et al.*, 1995). The formation of •OH radicals by HQ-thioethers may not only contribute to the acute nephrotoxicity of HQ, but also to its nephrocarcinogenicity, by causing heritable changes in the DNA of sublethally injured cells. In addition, proliferating cells may be more susceptible to oxidative DNA damage (Adachi *et al.*, 1994).

To further delineate the role of metabolism-dependent cytotoxicity in HQ-mediated carcinogenicity we determined the potential of HQ and 2,3,5-(triGSyl)HQ to induce cell

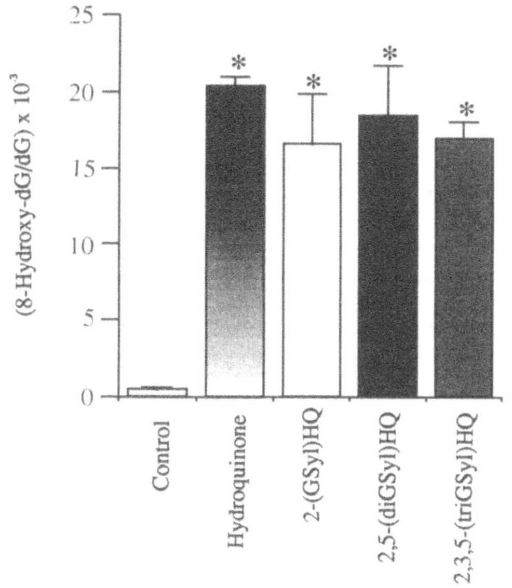

Figure 3. The formation of 8-OH-dG in calf thymus DNA following treatment with 1 mM hydroquinone (HQ) or HQ-GSH conjugates. Calf thymus DNA (1 mg/ml) was incubated with 0.1 mM $FeCl_3$, 0.5 mM EDTA, 1 mM HQ or HQ-GSH conjugate, and 5 mM H_2O_2 in 0.1 M sodium phosphate buffer (pH 7.4), in a total volume of 0.5 ml, in the dark at 37°C and terminated by the addition of 1 ml ice-cold absolute ethanol. Samples were centrifuged at 16,000xg for 10 min at 4°C and washed once with ice-cold 70% ethanol. DNA samples were dried by vacuum centrifugation and DNA was resuspended in 100 µl of 20 mM sodium acetate buffer (pH 4.8), boiled for 10 min, immediately placed on ice, and digested in the dark at 37°C with nuclease P1 (10 units) for 30 min and alkaline phosphatase (8 units) for 1 h. Values represent mean ± SE (n=3). * Significantly different from control ($p<0.01$).

proliferation and DNA synthesis in vivo. Current evidence suggests that certain chemicals may induce carcinogenesis by a mechanism involving cytotoxicity, followed by sustained regenerative hyperplasia, and ultimately tumor formation. Since the GSH conjugates of HQ are potent nephrotoxicants in the Sprague-Dawley rat, we examined the effects of HQ (200 mg/kg, po) and 2,3,5-(triGSyl)HQ (7.5 µmol/kg, i.v.) on nephrotoxicity in the F344 rat. 2,3,5-(TriGSyl)HQ caused significant increases in blood urea nitrogen, urinary γ-glutamyl transpeptidase, alkaline phosphatase, GSH-S-transferase and glucose 24 h after dosing (Peters *et al.*, 1995). By 72 h, most of these parameters had returned to control levels. Light microscopy of kidney preparations stained with hematoxylin and eosin showed that both HQ and 2,3,5-(triGSyl)HQ were selectively toxic to proximal tubular cells in the S_3M segment in the outer stripe of the outer medulla of the kidney. As early as 24 h after dosing regeneration was visible in this area, with maximal regeneration 72 h after dosing. Both HQ and 2,3,5-(triGSyl)HQ increased DNA-synthesis in proximal tubular cells in the outer stripe of the outer medulla. Following 2,3,5-(triGSyl)HQ administration cell proliferation rates, detected immunohistochemically via the incorporation of bromodeoxyuridine into newly synthesized DNA were 2.4, 6.9, 15.3 and 14.3% after 12, 24, 48 and 72 h, respectively, compared to 1.6% in vehicle controls (Peters *et al.*, 1995). The time-course and localization of the toxicity and DNA synthesis in the S_3 region of the kidney indicate that cell proliferation could not be uncoupled from cytolethality. Proliferation was not observed in regions of the kidney distal to the site of toxicity.

In conclusion, these results support a role for thioether metabolites of HQ in the HQ-induced nephrocarcinogenicity via a mechanism involving initial cytotoxicity and compensatory cell-proliferation, while oxidative DNA damage may contribute both to the acute nephrotoxicity and the carcinogenicity following chronic administration of HQ in the male F344 rat.

REFERENCES

Adachi, S., Kawamura, K., and Takemoto, K. (1994). Increased susceptibility to oxidative DNA damage in regenerating liver. *Carcinogenesis* **15**: 539-543.

Cantoni, O., Fiorani, M., Cattaberi, F. and Bellomo, G. (1991). DNA breakage caused by hydrogen peroxide produced during the metabolism of 2-methyl-1,4-naphthoquinone (menadione) does not contribute to the cytotoxic action of the quinone. *Biochem. Pharmacol.* **42**: 5220-5222.

Cerutti, P. (1978). Repairable damage in DNA. In: *DNA Repair Mechanisms*, (Hanawalt, P., Friedberg, F. and Fox, C., Eds.) Academic Press, N.Y.: 1-14.

Davies, K.J.A. (1987). Protein damage and degradation by oxygen radicals. I. General Aspects *J. Biol. Chem.*, **262**: 9895-9901, (see also Part III pp 9908-9913 and Part IV pp 9914-9920).

English, J.C., Hill, T., O'Donoghue, J.L. and Reddy, M.V. (1994). Measurement of nuclear DNA modification by ^{32}P-postlabeling in the kidneys of male and female Fischer 344 rats after multiple gavage doses of hydroquinone. *Fund. Appl. Toxicol.* **23**: 391-396.

Hanzlik, R.P., Harriman, S.P. and Frauenhoff, M.M. (1994). Covalent binding of benzoquinone to reduced ribonuclease. Adduct structures and stoichiometry. *Chem. Res. Toxicol.* **7**: 177-184.

Hill, B.A., Kleiner, H.E., Ryan, E.A., Dulik, D.M., Monks, T.J. and Lau, S.S. (1993). Identification of multi-S-substituted conjugates of hydroquinone by HPLC-coulometric electrode array analysis and mass spectrometry. *Chem. Res. Toxicol.* **6**: 459-469.

Jowa, L., Witz, G. and Snyder, R. (1990). Synthesis and characterization of deoxyguanosine-benzoquinone adducts. *J. Appl. Toxicol.* **10**: 47-54.

Kari, F.W., Bucher, J., Eustis, S.L., Haseman, J.K. and Huff, J.E. (1992). Toxicity and carcinogenicity of hydroquinone in F344/N rats and B6C3F$_1$ mice. *Fd Chem Toxicol.* **30**: 737-747.

Kasai, H. and Nishimura, S. (1984). DNA damage induced by asbestos in the presence of H_2O_2. *Gann* **75**: 841-844.

Kasai, H. and Nishimura, S. (1984). Hydroxylation of deoxyguanosine at the C-8 postion by polyphenols and aminophenols in the presence of hydrogen peroxide and ferric ion. *Gann* **75**: 565-566.

Kasai, H., Crain, P.F., Kuchino, Y., Nishimura, S., Ootsuyama, A. and Tanooka, H. (1986). Formation of 8-hydroxyguanine in cellular DNA by agents producing oxygen radicals and evidence for its repair. *Carcinogenesis* **7**: 1849-1851.

Kasai, H., Nishimura, S., Kurokawa, Y. and Hayashi, Y. (1987) Oral administration of the renal carcinogen, potassium bromate, specifically produces 8-hydroxydeoxyguanosine in rat target organ DNA. *Carcinogenesis* **8**: 1959-1961.

Kiyosawa, H., Aota, M., Inoue, H. and Makasawa, K. (1988) Detection of 8-hydroxy-deoxygaunosine in human lymphocyte DNA. *Int. Conf. on Med. Biochem. Chem. Aspects of Free Radicals.* Prog. Abstr.: 94.

Kuchino, Y., Mori, F., Kasai, H., Inoue, H., Iwai, S., Iwai, K., Ohtsuka, E. and Nishimura, S. (1987) Misreading of DNA templates containing 8-hydroxydeoxyguanosine at the modified base and at adjacent residues. *Nature* **327**: 77-79.

Lau, S.S., Hill, B.A., Highet, R.J. and Monks, T.J. (1988). Sequential oxidation and glutathione addition to 1,4-benzoquinone: Correlation of toxicity with increased glutathione substitution. *Mol. Pharmacol.* **34**: 829-836.

Leanderson, P. and Tagesson, C. (1990). Hydroquinone and catechol in the formation of the oxidative DNA-adduct, 8-hydroxydeoxyguanosine.*Chem. Biol. Interact.* **75**: 71-81.

Loeb, L.A., James, E.A., Waltersdorph, A.M. & Klebanoff, S.J. (1988). Mutagenesis by the autooxidation of iron with isolated DNA. *Proc. Natl. Acad. Sci.* **85**: 3918-3922.

Mertens, J.J.W.M., Gibson, N.W., Lau, S.S. and Monks, T.J. (1995). Reactive oxygen species and DNA damage in 2-bromo-(glutathio-S-yl)hydroquinone-mediated cytotoxcity. *Arch. Biochem. Biophys.* **320** In Press.

Monks, T.J.and Lau, S.S. (1994). Glutathione conjugate-mediated toxicities. In: *Conjugation- deconjugation reactions in drug metabolism and toxicity* FC Kauffman Ed., Springer-Verlag, Berlin: 459-510.

Nakagawa, Y. and Moldeus, P. (1992). Cytotoxic effects of phenyl-hydroquinone and some hydroquinones on isolated rat hepatocytes. *Biochem. Pharmacol.* **44**: 1059-1065.

Nutter, L.M., Ngo, E.O., Fisher, G.R. and Gutierrez, P.L. (1992). DNA strand scission and free radical production in menadione-treated cells. Correlation with cytotoxicity and role of NADPH quinone acceptor oxidoreductase. *J. Biol. Chem.* **267**: 2472-2479.

Peters, M.M.C.G., Jones, T.W., Monks, T.J. and Lau, S.S. (1995). Cytotoxicity and cell proliferation induced by the nephrocarcinogen hydroquinone and its tri-glutathionyl metabolite.*The Toxicologist* **15**: 231

Schlosser, M.J., Shurina, R.D. and Kalf, G.F. (1990). Prostaglandin H synthase catalyzed oxidation of hydroquinone to a sulfhydryl-binding and DNA-damaging metabolite. *Chem. Res. Toxicol.* **3**: 333-339.

Shibata, M.A., Hirose, M., Tanaka, H., Asakawa, A., Shirai, T. and Nobuyuki, I. (1991). Induction of renal cell tumors in rats and mice, and enhancement of hepatocellular tumor development in mice after long-term hydroquinone treatment. *Jpn. J. Cancer Res.* **82**: 1211-1219.

Solveig Walles, S.A. (1992). Mechanisms of DNA damage induced in rat hepatocytes by quinones *Cancer Lett.* **63**: 47-52.

Stadtman, E.R. and Oliver, C.N. (1991) Metal-ion catalyzed oxidation of proteins. Physiological consequences. *J. Biol. Chem.*, **266**: 2005-2008.

Verhagen H., Schilderman, P.A.E.L. and Kleinjans, J.C.S. (1991). Butylated hydroxyanisole in perspective. *Chem. Biol. Interactions* **80**: 109-134.

ETHYLENE OXIDE AS A BIOLOGICAL REACTIVE INTERMEDIATE OF ENDOGENOUS ORIGIN

Margareta Törnqvist

Department of Environmental Chemistry
Stockholm University
S-106 91 Stockholm, Sweden

ABSTRACT

Reactive intermediates can be monitored in vivo through their reaction products (adducts) with macromolecules. Sensitive methods, based on gas chromatography/mass spectrometry which permit structural identification and quantification, have been developed for the determination of adducts to hemoglobin (Hb). In studies of exposed animals, occupationally exposed workers and smokers a number of Hb adducts have been observed in unexposed control individuals. Methods for the determination of Hb adducts have, particularly through the use of tandem mass spectrometry, reached a sensitivity permitting studies of adducts from reactive intermediates due to "background exposure". *N*-(2-Hydroxethyl)valine (HOEtVal) is one of the background Hb adducts observed. This adduct has been used as a model in studies of sources of background adducts and determinants of their levels.

From the rate of exhalation of ethene by humans and application of a pharmacokinetic model it was concluded that about 70 % of the background HOEtVal in non-smokers originates from ethylene oxide (EO) as the metabolite of endogenously produced ethene. Contributions from ethene in urban air or environmental tobacco smoke are relatively small. Smoking of 2 cig./day approximately doubles the background. In twin studies it was shown that the variations in adduct level are partly due to hereditary factors and family traditions. Animal experiments demonstrate a role of intestinal flora and diet (unsaturated fatty acids, selenium), partly due to influences on metabolic rate. Furthermore, an interaction of dietary fat and intestinal flora is indicated.

From the adduct levels measured and knowledge of the identity of the causative reactive intermediate and of the rate constant for adduct formation, the dose of the reactive intermediate could be calculated. The associated cancer risk could then be estimated through the radiation-dose equivalence of the chemical dose determined in mutation experiments in vitro and application of the cancer risk coefficient for radiation. The risk contribution from the endogenous production of EO has been estimated by this approach to be of approximately the same magnitude as the risk due to background radiation.

INTRODUCTION

For the in vivo dose monitoring of reactive compounds or metabolites hemoglobin (Hb) adducts are useful, partly due to the high sensitivity achievable through the accumulation of adducts over the relatively long life span of the erythrocytes (Ehrenberg and Osterman-Golkar, 1980). The four terminal nitrogens are major nucleophilic sites in Hb and a sensitive method, the N-alkyl Edman method, was developed for mass spectrometrical determination of adducts to the N-terminal valines in Hb (Törnqvist et al., 1986).

The N-alkyl Edman method has been used for in vivo dose monitoring of low-molecular weight compounds in studies of occupational exposures, tobacco smoking and in animal experiments (Törnqvist, 1991; 1993). For a majority of studied low-molecular weight valine adducts background levels have been observed in the unexposed control individuals (animals and humans) (Törnqvist and Kautiainen, 1993). This concerns, for instance, *N*-(2-hydroxyethyl)valine (HOEtVal), the adduct formed by 2-hydroxyethylating agents such as ethylene oxide (EO). Background Hb adducts from high-molecular weight compounds have been observed by other methods (reviewed by Törnqvist, 1993).

A major source of background HOEtVal is endogenous and exogenous ethene, known to be metabolized to EO as has been demonstrated in vitro with rat liver microsomes (Schmiedel et al., 1983) and in vivo in mice (Ehrenberg et al., 1977) and rats (Filser and Bolt, 1983). In model studies in humans and animals the factors that are determinants of the background HOEtVal levels have been investigated. These studies are reviewed in the present paper, partly with reference to papers in preparation.

ANALYTICAL METHODS

Hemogobin adducts could be used for different purposes:

a. Determination of in vivo doses in exposed individuals as a basis for cancer risk estimation.
b. Studies of metabolites and metabolic rates in exposed individuals (also part of a).
c. Monitoring of exposure for the purpose of hygienic surveillance in occupational settings.
d. Identification of unknown adducts from reactive intermediates formed as metabolites from exogenous exposure or produced endogenously.
e. Studies of variations of "background adducts" with the purpose of identification of causative reactive intermediates and their sources.

The development of the N-alkyl Edman method for specific cleavage of alkylated *N*-terminal valines led to a useful tool for these applications (Törnqvist et al., 1986; Törnqvist, 1994). The method is carried out by a rather mild, simple and fast work-up procedure. A high sensitivity in the analysis was reached by using a fluorinated reagent, pentafluorophenyl isothiocyanate (Lequin and Niall, 1972), for the cleavage and derivatization. The formed derivatives, pentafluorophenylthiohydantoins of *N*-alkylated valines, give a high response in analysis by gas chromatography/mass spectrometry, chemical ionization, negative ions (Törnqvist et al., 1986). A high reproducibility in the analysis was achieved i.a. by using a globin alkylated with a deuterated analogue as internal standard (Törnqvist et al., 1992a).

The analytical sensitivity has been sufficient for studies of adducts from several electrophiles/reactive intermediates with the purposes of points a-c above, that means, e.g., after occupational exposure to ethene. Studies of background levels of the same adducts (points d and e above) have also been carried out, but have been difficult as the levels have been close to the detection level of the method/instrument. The introduction of tandem mass spectrometry (MS/MS) for the analysis of Hb adducts by the N-alkyl Edman method has increased the resolving power and made it possible to study variations of background levels with high reproducibility, as revealed in model studies of the HOEtVal adduct (to be published).

Especially in studies of background adducts there are problems to prove that observed adducts are true adducts formed in vivo and not adducts formed as artefacts during preparation of samples or introduced through contamination. The N-alkyl Edman method has several advantages in this respect; for instance, the procedure is mild which reduces the formation of artefacts during work-up of samples, and the residual "tag", valine, from Hb in the derivative to be determined gives a possibility of discriminating from contaminants. In earlier work it was shown that artefact formation of HOEtVal during storage of samples before derivatization caused problems in the analysis of low HOEtVal levels (Törnqvist, 1990). Precautionary measures for the handling of samples were worked out and have now made it possible to control the problem with artefact formation and to study small differences in background levels of HOEtVal.

MAGNITUDE OF OBSERVED HOEtVal LEVELS

Table 1 (column 1) summarizes observed steady-state levels of HOEtVal originating from different exposures. Due to the zero-order kinetics of disappearance of erythrocytes, continuous or intermittent exposures lead to the establishment of steady-state levels (A_{ss}) of Hb adducts equal to one-half of the cumulative levels formed during the life span of erythrocytes (t_{er}) (Granath et al., 1992):

$$A_{ss} = a \cdot \frac{t_{er}}{2} \qquad (1)$$

where a denotes the adduct level increment per unit of time.

The increment of the level of HOEtVal per exposure dose of EO in occupational exposures is uncertain mainly because of difficulties, due to varying exposure concentration, of correctly determining the exposure dose. The value (2400 pmol/g Hb) given in Table 1 is statistically uncertain by a factor of 3. Pharmacokinetic considerations indicate that the most likely value is about 4000 pmol/g Hb per ppm 40 h/week (Ehrenberg and Törnqvist, 1995).

In studies of smokers the value given in Table 1 (85 pmol/g Hb per 10 cig./day) corresponds to 2 % of inhaled ethene giving rise to a dose of EO in the body. This value was confirmed by measurements of the amount of ethene inhaled per cigarette (Granath et al., 1994). In a study of ethene-exposed workers, with intermittent peak exposures, where the exposure dose was determined using a statistical model, the factor for conversion to EO was about four times lower than the value determined for cigarette smokers (to be published). This seems to be due to activation of a first-pass detoxification of EO above a certain concentration of ethene (Filser and Bolt, 1984). A similar effect at high exposure levels of ethene was observed in inhalation studies with human subjects (coll. with Westerholm et al., to be published) and also in experiments with rats (Eide et al., 1995).

Table 1. Steady-state levels of adduct to N-terminal valine in hemoglobin from ethylene oxide (HOEtVal) and corresponding annual doses of ethylene oxide associated with different exposures

Exposure	HOEtVal steady-state level (pmol/g)	Annual dose of ethylene oxide (mMh)
Ethylene oxide, 1 ppm, 40 h/week	2400[a] (4000[b])	$280 \cdot 10^{-3}$
Ethene, 1 ppm, 40 h/week[c]	~ 50 - 100[d]	$12 \cdot 10^{-3}$
Ethene from smoking 10 cig./day[c,e]	85	$10 \cdot 10^{-3}$
Ethene in urban air, 10 ppb, 168 h/week or moderate passive smoking[e]	~ 4[f]	$0.5 \cdot 10^{-3}$
Background (non-exposed)[e,g]	20 (6-25)	$2.3 \cdot 10^{-3}$

[a]Duus et al., 1989; [b]Ehrenberg and Törnqvist, 1995; [c]Granath et al., 1994; [d]At higher exposure levels, a lower adduct level per ppm is indicated (to be published); [e]Törnqvist, 1989; [f]Not detected, estimated; [g]Törnqvist and Kautiainen, 1993.

In all studies background HOEtVal levels in the range of 6-25 pmol/g globin were observed in unexposed individuals (Törnqvist and Kautiainen, 1993). In rodents similar values have been observed.

ENDOGENOUS AND EXOGENOUS CONTRIBUTIONS TO BACKGROUND LEVEL OF HOEtVal

Contribution from Endogenous Ethene

It has long been known that exhaled air contains small amounts of ethene (review in Filser et al., 1992). The rate of exhalation of ethene and the levels of background HOEtVal in Hb were measured in 5 non-smokers (Filser et al., 1992). Application of a pharmacokinetic model to these data led to the conclusion that a major part (about 70 %) of the background level originated from endogenously produced ethene in these persons (the concentration of ethene in the environment was about 15 ppb).

Contribution from Exogenous Ethene

Due to the occurrence of varying background levels the HOEtVal adduct increment due to ethene as an air pollutant can hardly be measured in groups of limited size. In studies of taxi drivers (Hemminki et al., 1994) and parking guards (in coll. with L. Nilsson, to be published) adduct increments compared with control groups could not have been detected with statistical significance. At 5 ppb ethene, a likely value in large cities, the increment is expected to be about 10 % of the background.

Environmental tobacco smoke may occasionally reach considerable levels of ethene (Persson et al., 1989). However, in various studies of adduct levels in humans no significant correlation of the HOEtVal level with recorded passive smoking has been found (Törnqvist, 1989).

Influence of Family-Bound Factors in the Variation of Background HOEtVal Levels

In twin studies significant between-pair variations in the HOEtVal level have been observed in both monozygotic and heterozygotic twins (coll. with M. Svartengren, to be published). A similar between-pair variation was earlier found with respect to *N*-methylvaline in Hb (Törnqvist et al., 1992b). This indicates an involvement of hereditary factors and/or tradition in the inter-individual variation of background levels. A hereditary variation is expected to occur due to polymorphism in detoxification enzymes such as glutathione-S transferase θ (GSTθ) (Hallier et al., 1993), which could result in a high adduct level in deficient individuals (cf. Törnqvist et al., 1988; Törnqvist, 1988).

Animal Studies of the Endogenous Origin of Background Levels

Several mechanisms of production of endogenous ethene have been discussed: peroxidation of lipids (Frank et al., 1980; Sagai and Ichinose, 1980), proteins, and amino acids (particularly methionine), (Lieberman and Mapson, 1964; Clemens et al., 1983; Kessler and Remmer, 1990) and metabolism of enteric bacteria (Primrose and Dilworth, 1976).

In initial studies with mice it was shown that germ-free animals had lower background HOEtVal levels and that unsaturated fatty acids in the lipid component of the diet led to higher values (Törnqvist et al., 1989). The influence of the intestinal flora, also on other adducts, has been confirmed in later studies, carried out with the improved analysis of valine adducts by MS/MS (Kautiainen et al., 1993).

In further experiments detoxification rate and interaction of intestinal flora with dietary fat were studied in germ-free and normal mice by measurement of Hb adducts (collab. with T. Midtvedt, J. de Pierre and A. Kautiainen, to be published). The effect of the intestinal flora on the fate of EO showed a twice higher rate of detoxification in the "normal" mice, in the presence of bacteria, compared to germ-free mice. This was also confirmed by a higher GST activity in the "normal" animals with bacteria. This faster detoxification of EO observed, together with the doubled background level of HOEtVal in normal mice, indicates that the production of ethene is about four times larger in the normal than in the germ-free animals. Furthermore, ongoing experiments indicate that the effect (increase of HOEtVal) of unsaturatedness of lipids is reduced in germ-free animals.

Rats fed a selenium deficient diet, which was expected to increase the lipid peroxidation, were also studied with regard to background levels of some Hb adducts and with regard to detoxification rate of EO (coll. with U. Olsson and A. Kautiainen, to be published). In rats on a selenium deficient diet a role of lipid peroxidation in the induction of GST was indicated (Olsson et al., 1992), leading to a faster detoxification of EO. In this case this was associated with a net decrease of the background HOEtVal level in selenium deficient rats. No significant increase in ethene production (observed as increased level of HOEtVal) due to increased lipid peroxidation could therefore be seen or deduced in selenium deficient rats.

The animal experiments have further indicated an influence of sex on background HOEtVal levels in mice, the values being about 25 % lower in females than in males.

The various endogenous factors influencing the background level are summarized in Table 2.

CANCER RISK ASSOCIATED WITH BACKGROUND HOEtVal

Cancer risks from chemical mutagens have been preliminarily estimated from the radiation dose equivalents of chemical doses, dose being defined as the integral over time of concentration. Chemical dose has thus the dimension concentration · time, e.g., mmol · $L^{-1} \cdot h$ ($mmhL^{-1}$). From equation (1) the daily increment a to the background level of HOEtVal (A_{ss} = 20 pmol/g) is calculated, the value of t_{er} being about 126 days in humans (Berlin, 1964):

Table 2. Endogenous factors having an influence on the background level of adduct from ethene/ethylene oxide to N-terminal valine in hemoglobin

Source	Magnitude of influence
Family-bound (hereditary or acquired) (humans) (monozygotic and dizygotic twinpairs)	± 2
Dietary components (animal experiments)	
Increased saturatedness of dietary fat (mice)	~ 1.5
Selenium deficiency in diet (rats)	
Selenium deficient compared to normal rats	1.6
Induction of detoxification of EO in selenium deficient rats	1.6
Production of precursor in selenium deficient compared to normal rats	(± 0)[a]
Intestinal flora/germ-free condition (mice)	
Normal compared to germ-free mice	+ 2.5
Induction of detoxification of EO in normal compared to germ-free mice	2
Production of precursor in normal compared to germ-free mice	(+ 5)[a]
Interaction of intestinal flora with dietary fat	± 1.2
Sex	
male/female (mice)	+ 1.25

[a] Deduced

$$a = \frac{A_{SS} \cdot 2}{t_{er}} = \frac{20 \cdot 10^{-12}\,(\mathrm{mol\,g^{-1}}) \cdot 2}{126\,(\mathrm{d})} = 0.32 \cdot 10^{-12}\,(\mathrm{mol\,g^{-1}\,d^{-1}}) \tag{2}$$

The relationship between adduct level increment and dose is:

$$a = k \cdot D_d \tag{3}$$

where k is the second-order rate constant for adduct formation. For the reaction of EO with N-terminal valine, $k = 5 \cdot 10^{-5}\ \mathrm{L\,g^{-1}\,h^{-1}}$ (Segerbäck, 1985). The daily dose D_d is then:

$$D_d = \frac{a}{k} = \frac{0.32 \cdot 10^{-12}\,(\mathrm{mol\,g^{-1}\,d^{-1}})}{5 \cdot 10^{-5}(\mathrm{L\,g^{-1}\,h^{-1}})} = 0.64 \cdot 10^{-8}\ \mathrm{Mh\,d^{-1}} \tag{4}$$

The annual dose will hence be $0.64 \cdot 10^{-8} \cdot 365\ \mathrm{Mh\,y^{-1}} = 2.3 \cdot 10^{-3}\,\mathrm{mMh\,y^{-1}}$. This value corresponds to the dose in blood. In Table 1 the annual doses of EO, corresponding to the exemplified exposures, are given. From model studies in the mouse, with a corresponding determination of the dose in different organs from DNA adduct levels from injected EO, it was concluded that the dose of EO is equally distributed over relevant target organs (Segerbäck 1983, 1985).

From a reevaluation and validation of the rad-equivalence of EO it has been concluded that 1 mMh of this compound has the same carcinogenic potency as 40 (20-80) rad of acute gamma-radiation (Segerbäck et al., 1994; Ehrenberg et al., 1995). Accordingly the background HOEtVal level, 20 pmol/ g globin which corresponds to an annual dose of $2.3 \cdot 10^{-3}$ mMh of EO, would be associated with the cancer risk due to about 0.1 rad gamma-radiation per year. This radiation dose, 1 mSv, corresponds to the average annual background radiation dose, the dose from indoor Rn being excluded. In a population of one million, lifelong (70 years) exposure to this annual dose is expected to give rise to a few thousand cancer cases (NRC, 1990; ICRP, 1991) with an uncertainty due to lack of knowledge of the risk at very low dose rates.

CONCLUSION

The reported results pertain to doses of EO and production of its precursor ethene. Partly the study serves as a model for background adducts in general with respect to sources and risk contribution. Similar observations have been made with regard to other low molecular-weight compounds such as propene/propylene oxide. The results indicate that an observed net effect may be due to interaction between changed precursor production and changed detoxification effectiveness. This shows, for instance, that not only heritable polymorphism but also the induction status of detoxification enzymes may be of importance to the dose and the associated cancer risk.

ACKNOWLEDGMENT

I want to express my sincere thank to collaborators mentioned in the text. Furthermore I want to thank Anna-Lena Magnusson, Eden Tareke and Vlado Zorcec for contributions in still unpublished work. The work has been supported financially by the Swedish Cancer Fund and the Swedish Environmental Protection Agency.

REFERENCES

Berlin, N.I., 1964, Life-span of the red cell, In: Bishop, C. and Surgenor, D.M. (eds) *The Red Blood Cell*, Academic Press, New York and London, pp. 423-450.

Clemens, M.R., Einsele, H., Frank, H., Remmer, H. and Waller, H.D., 1983, Volatile hydrocarbons from hydrogen peroxide-induced lipid peroxidation of erythrocytes and their cell components, *Biochem. Pharmacol.* 32:3877-3878.

Duus, U., Osterman-Golkar, S., Törnqvist, M., Mowrer, J., Holm, S. and Ehrenberg, L., 1989, Studies of determinants of tissue dose and cancer risk from ethylene oxide exposure, In: Freij, L. (ed), *Proceedings: Management of Risk from Genotoxic Substances in the Environment*, Swedish National Chemicals Inspectorate, Solna, pp. 141-153.

Ehrenberg, L. and Osterman-Golkar, S., 1980, Alkylation of macromolecules for detecting mutagenic agents. *Teratogen. Carcinogen. Mutagen.* 1:105-127.

Ehrenberg, L. and Törnqvist, M., 1995, The research background for risk assessment of ethylene oxide, *Mutat. Res.* (in press).

Ehrenberg, L., Osterman-Golkar, S., Segerbäck, D., Svensson, K. and Calleman, C.J., 1977, Evaluation of genetic risks of alkylating agents. III. Alkylation of haemoglobin after metabolic conversion of ethene to ethene oxide in vivo. *Mutat. Res.* 45:175-184.

Ehrenberg, L., Granath, F., Hedenskog, M., Magnusson, A.-L., Natarajan, A.T., Tates, A., Törnqvist, M. and Wright, A., 1995, Comparisons of ethylene oxide and low-LET radiation. (ms.)

Eide, I., Hagemann, R., Zahlsen, K., Tareke, E., Törnqvist, M., Kumar, R., Vodicka, P. and Hemminki, K., 1995, Uptake, distribution, and formation of hemoglobin and DNA adducts after inhalation of C2-C8 1-alkenes (olefins) in the rat, *Carcinogenesis* (in press).

Filser, J.G. and Bolt, H.M., 1983, Inhalation pharmacokinetics based on gas uptake studies IV. The endogenous production of volatile compounds. *Arch. Toxicol.* 52:123-133.

Filser, J.G. and Bolt, H.M., 1984, Inhalation pharmacokinetics based on gas uptake studies VI. Comparative evaluation of ethylene oxide and butadiene monooxide as exhaled reactive metabolites of ethylene and 1,3-butadiene in rats, *Arch. Toxicol.* 55:219-223.

Filser, J.G., Denk, B., Törnqvist, M., Kessler, W. and Ehrenberg, L., 1992, Pharmacokinetics of ethylene in man: Body burden with ethylene oxide and hydroxyethylation of hemoglobin due to endogenous and environmental ethylene, *Arch. Toxicol.* 66: 157-163.

Frank, H., Hintze, T. and Remmer, H., 1980, Volatile hydrocarbons in breath, an indication for peroxidative degradation of lipids. In: Kolb, B. (ed.) *Applied Headspace Gas Chromatography*, Heyden & Son, London, pp. 155-164.

Granath, F., Ehrenberg, L. and Törnqvist, M., 1992, Degree of alkylation of macromolecules in vivo from variable exposure, *Mutat. Res.* 284: 297-306.

Granath, F., Westerholm, R., Peterson, A., Törnqvist, M. and Ehrenberg, L., 1994, Uptake and metabolism of ethene studied in smokers, *Mutat. Res.* 313:285-291.

Hallier, E., Langhof, T., Dannappel, D., Leutbecher, M., Schröder, K., Georgens, H.W., Müller, A. and Bolt, H.M., 1993, Polymorphism of glutathione conjugation of methyl bromide, ethylene oxide and dichloromethane in human blood: Influence on the induction of sister chromatid exchanges (SCE) in lymphocytes, *Arch. Toxicol.* 67:173-178.

Hemminki, K., Zhang, L., Krüger, J., Autrup, H. and Törnqvist, M., 1994, Exposure of bus and taxi drivers to urban air pollutants as measured by DNA and protein adducts, *Tox. Lett.* 72:171-174.

ICRP, 1991, *1990 Recommendations of the International Commission on Radiological Protection*, ICRP Publication 60, vol. 21, Pergamon Press, Oxford.

Kautiainen, A., Midtvedt, T. and Törnqvist, M., 1993, Intestinal bacteria and endogenous production of malonaldehyde and alkylators in mice, *Carcinogenesis* 14:2633-2636.

Kessler, W. and Remmer, H., 1990, Generation of volatile hydrocarbons from amino acids and proteins by iron/ascorbate/GSH system, *Biochem. Pharmacol.* 39:1347-1351.

Lequin, R.M. and Niall, H.D., 1972, The application of a fluorinated isothiocyanate as a coupling agent in the Edman degradation, *Biochim. Biophys. Acta.* 257:76-82.

Lieberman, M. and Mapson, L.W., 1964, Genesis and biogenesis of ethylene, *Nature* 4956:343-345.

NRC, 1990, National Research Council, Committee on the Biological Effects of Ionizing Radiations, *Health effects of exposure to low levels of ionizing radiation*, BEIR V Report, Washington, National Academy Press.

Olsson, U., Lundgren, B., Segura-Aguilar, J., Messing-Eriksson, A., Andersson, K., Becedas, L. and de Pierre J.W., 1992, Effects of selenium deficiency on xenobiotic-metabolizing and other enzymes in rat liver, *Internat. J. Vit. Nutr. Res.* 63:27-30,

Persson, K.-A., Berg, S., Törnqvist, M., Scalia-Tomba, G.-P. and Ehrenberg, L., 1988, Note on ethene and other low-molecular weight hydrocarbons in environmental tobacco smoke, *Acta Chem. Scand.* B 42:690-696.

Primrose, S.B. and Dilworth, M.J., 1976, Ethylene production by bacteria, *J. General Microbiology* 93:177-181.

Sagai, M. and Ichinose T., 1980, Age-related changes in lipid peroxidation as measured by ethane, ethylene, butane, and pentane in respired gases of rats, *Life Sci.* 27:731-738.

Schmiedel G., Filser, J.G. and Bolt, H.M., 1983, Rat liver microsomal transformation of ethene to oxirane in vitro, *Toxicol. Lett.* 19:293-297.

Segerbäck, D., 1983, Alkylation of DNA and hemoglobin in the mouse following exposure to ethene and ethene oxide. *Chem.-Biol. Interact.* 45:139-151.

Segerbäck, D., 1985, *In Vivo Dosimetry of Some Alkylating Agents as a Basis for Risk Estimation*, Ph.D. Thesis, Stockholm University, Stockholm.

Segerbäck, D., Rönnbäck, C., Bierke, P., Granath, F. and Ehrenberg, L., 1994, Comparative two-stage cancer tests of ethylene oxide, *N*-(2-hydroxyethyl)-*N*-nitrosourea and X-rays, *Mutat. Res.* 307:387-393.

Törnqvist, M., 1988, Search for unknown adducts: Increase of sensitivity through preselection by biochemical parameters, In: Bartsch, H., Hemminki, K. and O'Neill, I.K. (eds), *Methods for Detecting DNA Damaging Agents in Humans: Applications in Cancer Epidemiology and Prevention*, International Agency for Research on Cancer, Scientific Publication 89, Lyon, pp. 378-383.

Törnqvist, M., 1989, *Monitoring and Cancer Risk Assessment of Carcinogens Particularly Alkenes in Urban Air*, Ph.D. Thesis, Stockholm University, Stockholm.

Törnqvist, M., 1990, Formation of reactive species that lead to hemoglobin adducts during storage of blood samples. *Carcinogenesis* 11:51-54.

Törnqvist, M., 1991, The N-alkyl Edman method for haemoglobin adduct measurement: Updating and applications to humans. In: Garner, R.C., Farmer, P.B., Steel, G.T. and Wright, A.S. (eds), *Human Carcinogen Exposure: Biomonitoring and Risk Assessment*, Oxford University Press, Oxford, pp. 411-419.

Törnqvist, M., 1993, Current research on hemoglobin adducts and cancer risks: An overview. In: Travis, C. (ed.) *Use of Biomarkers in Assessing Health and Environmental Impacts of Chemical Pollutants*, Plenum press, New York, pp. 17-30.

Törnqvist, M., 1994, Epoxide adducts to N-terminal valines, In: *Methods in Enzymology*, vol. 231, pp. 650-657.

Törnqvist, M. and Kautiainen, A., 1993, Adducted proteins for identification of endogenous electrophiles, *Environ. Health Perspect.* 99:39-44.

Törnqvist, M., Mowrer, J., Jensen, S. and Ehrenberg, L., 1986, Monitoring of environmental cancer initiators through hemoglobin adducts by a modified Edman degradation method, *Anal. Biochem.* 154: 255-266.

Törnqvist, M., Osterman-Golkar, S., Kautiainen, A., Näslund, M., Calleman, C.J. and Ehrenberg, L., 1988, Methylations in human hemoglobin, *Mutat. Res.* 204:521-529.

Törnqvist, M., Gustafsson, B., Kautiainen, A., Harms-Ringdahl, M., Granath, F. and Ehrenberg, L., 1989, Unsaturated lipids and intestinal bacteria as sources of endogenous production of ethene and ethylene oxide. *Carcinogenesis* 10:39-41.

Törnqvist, M., Magnusson, A.-L., Farmer, P.B., Tang, Y.-S., Jeffrey, A.M., Wazneh, L., Beulink, G.D.T., van der Waal, H. and van Sittert, N.J., 1992a, Ring test for low levels of *N*-(2-hydroxyethyl)valine in human hemoglobin, *Anal. Biochem.* 203:357-360.

Törnqvist, M., Svartengren, M. and Ericsson, C.H., 1992b, Methylations of hemoglobin from monozygotic twins discordant for cigarette smoking: Hereditary and tobacco-related factors, *Chem.-Biol. Interact.* 82:91-98.

ESTROGEN METABOLITES AS BIOREACTIVE MODULATORS OF TUMOR INITIATORS AND PROMOTERS

Leon Bradlow, Nitin T. Telang, and Michael P. Osborn

Strang Cornell Cancer Research Laboratory
New York, New York

A role for estrogens in the initiation and promotion of breast and endometrial cancer has been clear for more than 100 years since Beatson demonstrated that oophorectomy induced remissions in breast cancer (1). The details of these responses are still not clear and are under further study.

A. SYNTHESIS AND METABOLISM OF ESTRADIOL AND RELATED COMPOUNDS

The formation of all of estrogens derives ultimately from the action of Cyp450 enzymes on precursor androgens. Estradiol, itself, is formed by the action of aromatase on testosterone in a linked series of 3 hydroxylations which occurs primarily in the ovary and also in peripheral fat (Figure 1) (2). The activity of this enzyme complex is regulated by various hormones (FSH, cortisol and others). The metabolism of estradiol in humans involves an initial oxidation to estrone followed by selective oxidation by different selective CyP450s as described below to C-2, C-4 and C-16α hydroxylated metabolites (Figure 2) (3). The relative formation of these metabolites is regulated by the action of various bioreactive molecules. These estrogenic metabolites in turn serve as proximate bioreactive molecules altering cell growth and responses. These responses play a critical role in carcinogenesis in hormone sensitive tissues.

I. Specific Pattern of the Metabolism of Estradiol to Its Metabolites

i. 2-Hydroxyestrone. 2-Hydroxyestrone is quantitatively the most abundant metabolite of estradiol in human subjects and is formed primarily by the action of CyP4501A1 or 1A2 although other cytochromes may also play a role. The enzyme is highly specific and for example 2-hydroxylation of 17α-ethinylestradiol is carried out by a different enzyme in the CyPIII family (4). The activity of CyP1A1(5) is readily modulated in both directions by drugs (6), diet (7), bioreactive molecules (8), micronutrients (9,10), and body composition

Biological Reactive Intermediates V
Edited by R. Snyder et al., Plenum Press, New York, 1996

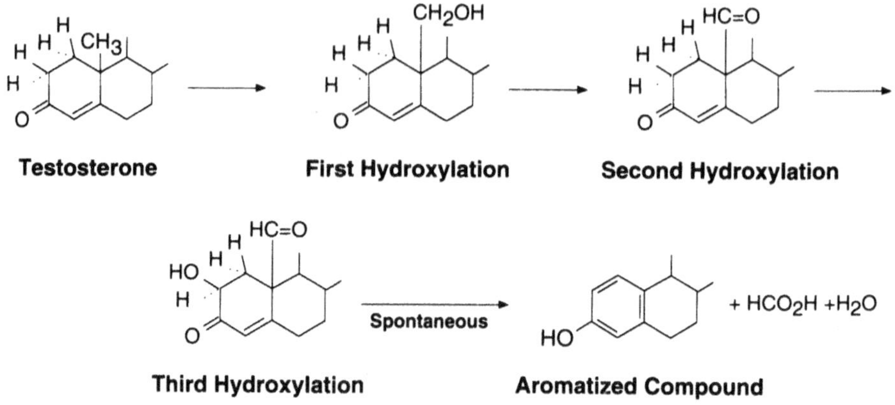

Figure 1. Pathway of aromatization to form estrogens from androgens.

(11). The compound is weakly active and in some circumstances antiestrogenic (12,13). Increases in the extent of this pathway appears to play a role in decreasing tumor risk. The compound is very rapidly methylated to the 2-methoxy compound by red blood cell catechol O-methyl transferase (14). Direct infusion of 2-OHE1 failed to raise its blood level because of the speed of this reaction.

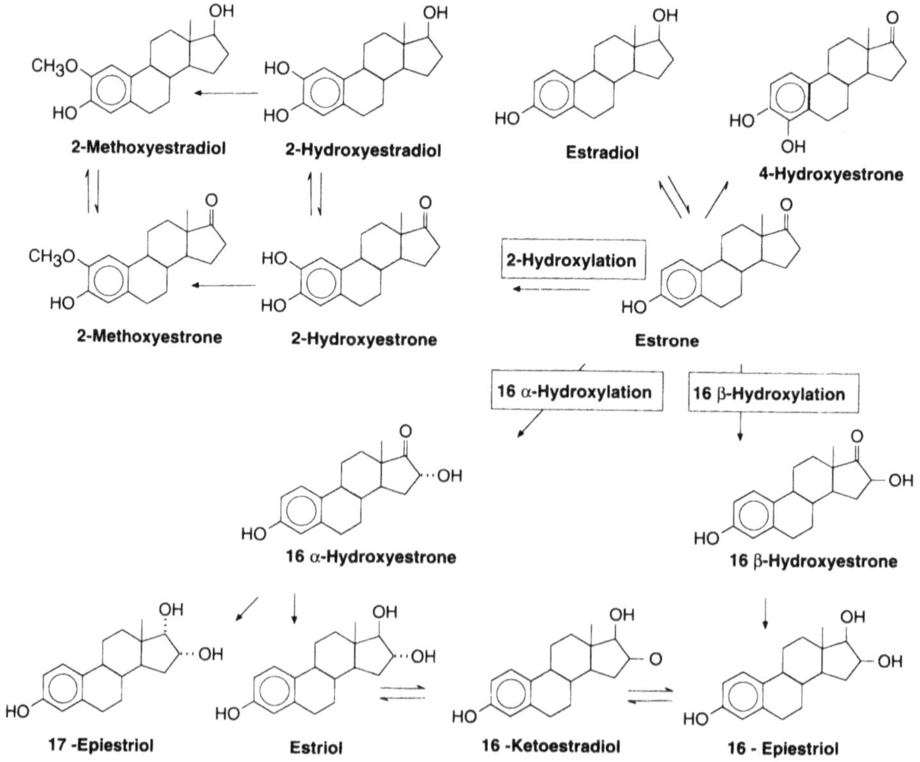

Figure 2. Pattern of estrogen metabolism to its principal metabolites.

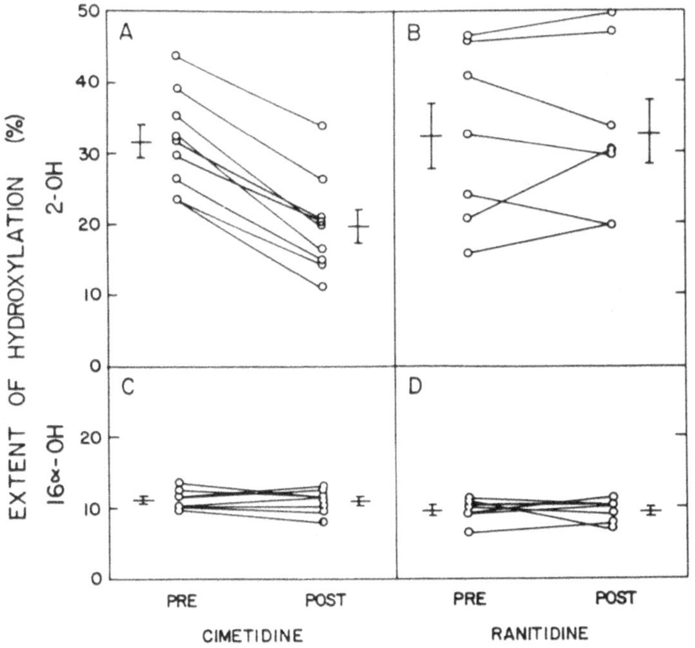

Figure 3. Effect of Cimetidine on estrogen metabolism in human subjects.

ii. 4-Hydroxy Estradiol. This is a metabolite which is the object of considerable interest because of its carcinogenic properties (15). It is formed by the action of an as yet unidentified CyP450 although recent data suggests that Cyp4501B1 may be the specific enzyme involved but not by the action of CyP4501A1 or 1A2 (16). Its formation is minimal

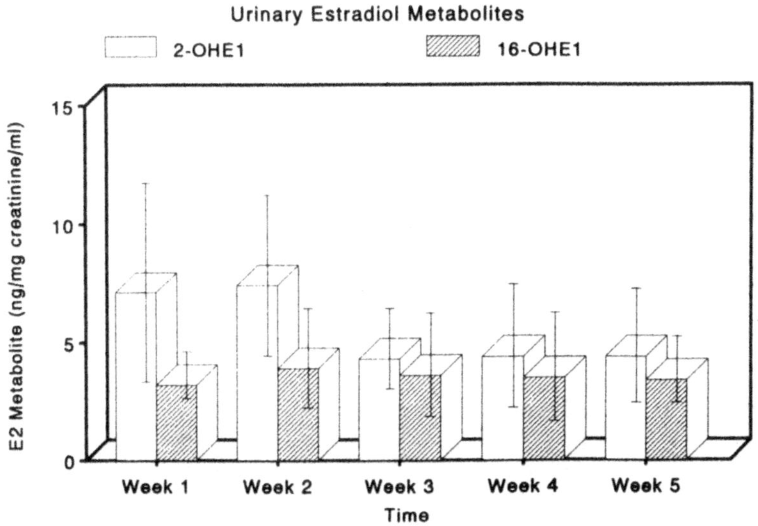

Figure 4. Nutritional intervention study with a decrease in 2-hydroxyestrone formation as a result of semi-synthetic diet containing no inducing agents.

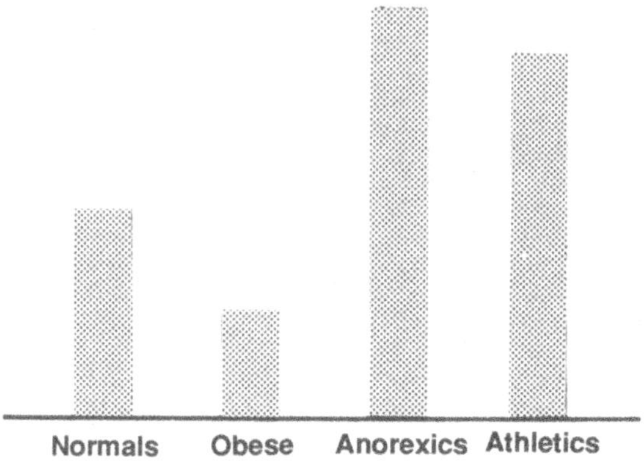

Figure 5. Effect of obesity and exercise on 2-hyroxylation.

in human subjects (3-4%) (17) but much greater in the male Syrian hamster kidney though not in the liver (18). Its formation is increased by dioxins, tetrachlordibenzofuranes, β-naphthoflavone, and other flavones. Methylation of 4-OHE1 occurs somewhat more slowly than for the 2,3 catechol estrogen. Reaction of the 3,4 catechols with glutathione is more rapid than for the 2,3 catechol estrogens (19).

iii. 16α-Hydroxyestrone. As in the case of 4-hydroxylation the enzyme responsible for 16α-hydroxylation has not yet been established. The enzyme appears to be constituitive and it is difficult to directly alter the extent of this reaction. The reaction can be inhibited by high thyroid levels or large amounts of ω-3 fatty acids but neither treatment proved to be practical or well tolerated by patients. Baboons placed upon a very high fat diet (65%) (20) did show an increase in this reaction. It can be stimulated by bioreactive molecules like benz-(α)pyrene or 7,12-dibenzanthracene (21) and by viruses like HPV and MMTV (22,23), as well as by oncogenes like RAS (24) and MYC (25) (Figure 6). The compound is partially reduced to estriol following *in vivo* human administration of labeled 16α-OHE1.

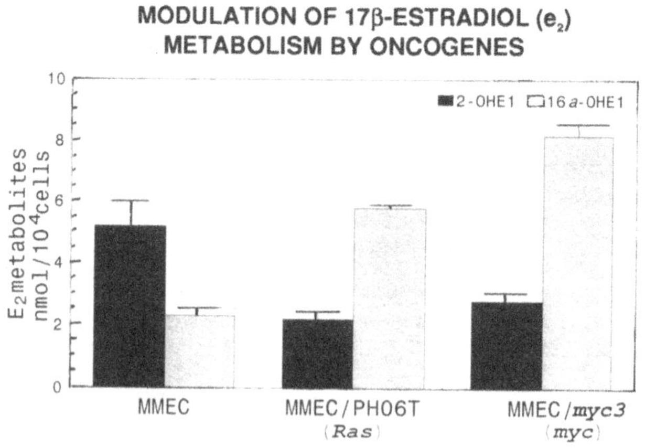

Figure 6. Effect of transfection of normal mammary epithelial cells with the Myc or Ras oncogenes; 2-hydroxylation is depressed and 16α-hydroxylation is enhanced.

I. Biochemical and Physiological Properties of the Metabolites

i. Catechol Estrogens. The biochemical and carcinogenic properties of 2 and 4 catechol estrogens are quite different from each other and no generalized conclusions about the properties of catechol estrogens as a class can be made. Thus 2-hydroxyestrone, the principal catechol estrogen formed in humans, is at best weakly estrogenic and and in some respects antiestrogenic (11,12). In addition Liehr demonstrated that this compound did not cause oxidative attack on guanosine to yield 8-oxyguanosine and does not form DNA adducts (26). Further studies with 2-hydroxyestrone in a multipronged genotoxic assay by Suto showed that 2-OHE1 did not exhibit exhibit genotoxic behavior as measured by unscheduled DNA synthesis, proliferation, or anchorage independent growth (27). Recent studies by Cavalieri on the action of 2,3 and 3,4 quinones with DNA showed that the 3,4 quinones bind covalently at position 7 on guanosine or adenosine resulting in cleavage from the deoxysugar and the formation of an apurinic site (Figure 7). During the subsequent DNA repair process this vacant site is occupied by thymine resulting in a G to T or A to T mutation (Scheme 1). The 2,3 quinone does not react at C-7 but only on the amino groups on G and A with no depurination (Figure 8) (28).

Thus the 2,3-quinone and by extension the 2,3 catechol estrogen are not mutagenic. Direct measurements suggest that 2-hydroxyestrone is at best only weakly estrogenic and antiproliferative. 4-Hydroxyestradiol, on the other hand is a very potent estrogen and does

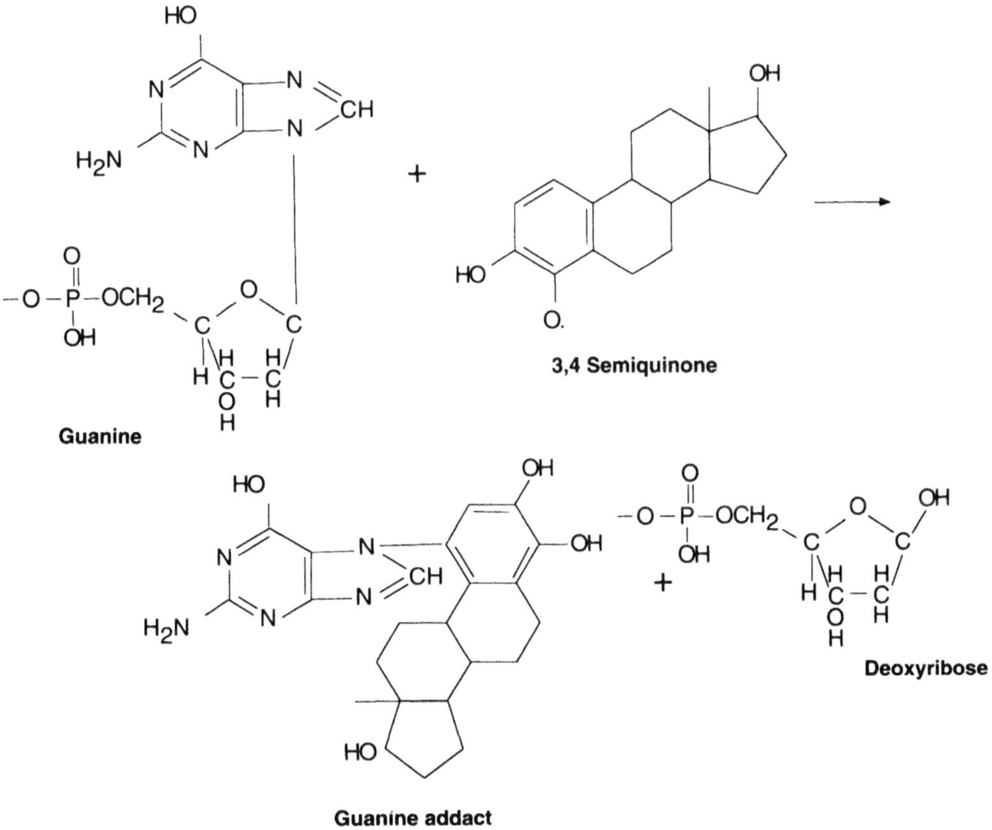

Figure 7. Effect of 3,4 steroid semiquinones residues attached to DNA.

ATCTGCTT ——— Original

ATCT CTT ——— Apurinic site

ATCTTCTT ——— Post Repair

Scheme 1

cause substantial formation of 8-hydroxyguanosine as well as adduct formation in the male Syrian hamster kidney model (29). It has been proposed that formation of the semiquinone containing a one electron system is responsible for the genotoxic effects and the oxidative damage (27) Figure 9.

Since this oxidation-reduction is a reversible reaction between the catechol, the semiquinone, and the quinone, it was proposed that this could generate continuing DNA damage. In actual fact in tissues the semiquinone is likely to be rapidly consumed by irreversible reaction at C-1 with glutathione which circulates at substantial levels. In addition

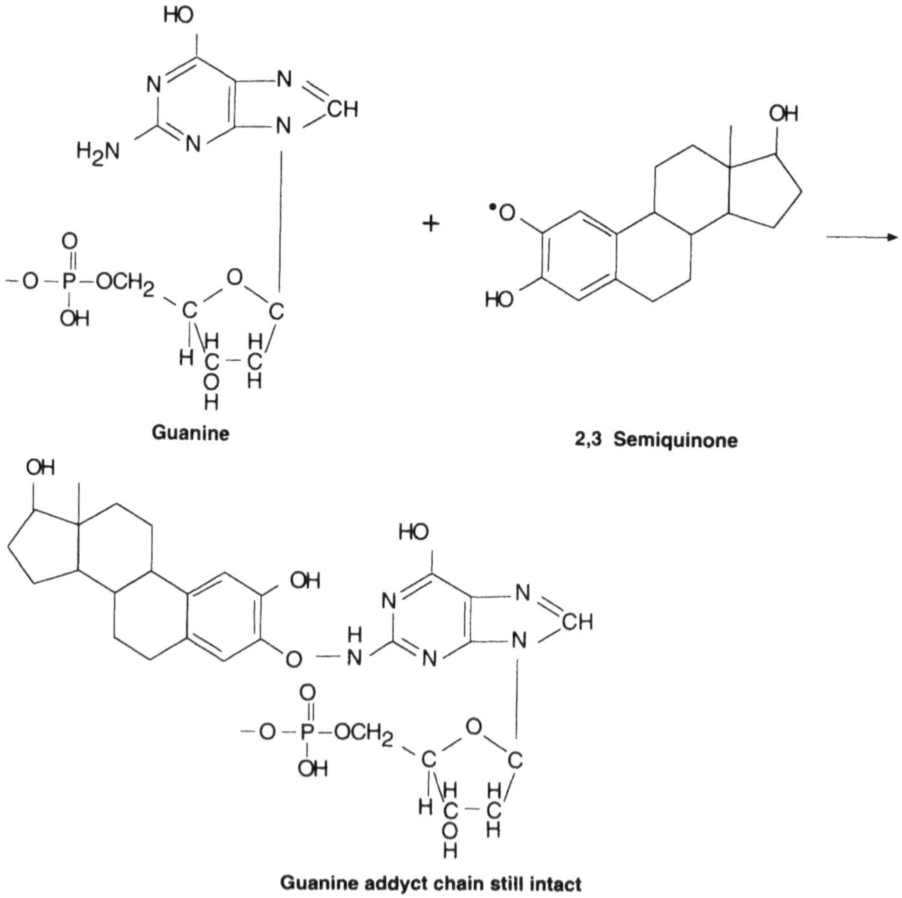

Figure 8. Failure of 2,3 estrogen semiquinones to generate apurinic sites in DNA.

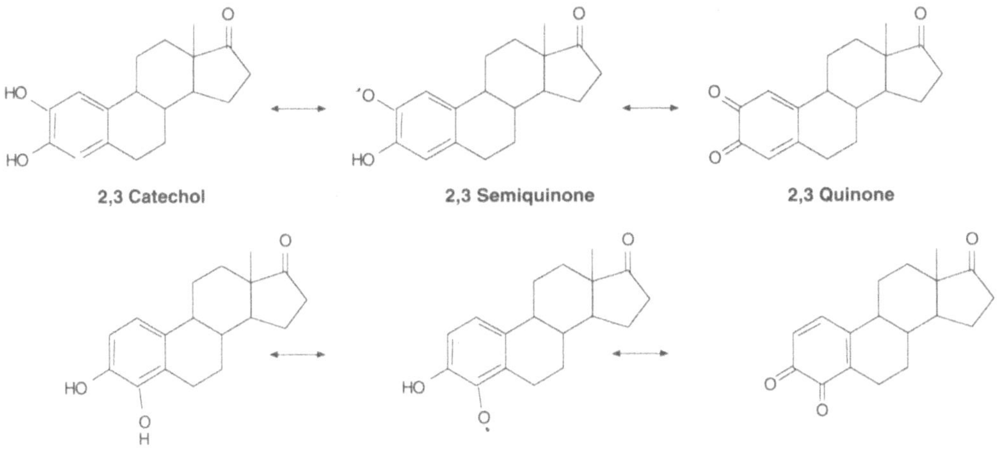

Figure 9. Redox equilibrium between the cathecol estrogens and the estrogen quinones.

Cavalieri has shown that attack on DNA results in irreversible binding of the catechol to the DNA so that repeated damage could not occur by a cycling process. It should also be noted that increasing 2-hydroxylation by the administration of indole-3-carbinol results in a decreased incidence of mammary tumors in a variety of animal models (Figure 10) (30-33).

In patients with laryngeal papillomas increasing 2-hydroxylation of estradiol is associated with inhibition of the growth of these papillomas (34). Similar data has been observed in cell culture systems (22). In sum, with the exception of the male Syrian hamster model where 4-hydroxyestradiol plays a key role there is no animal or human data supporting a role for 2-hydroxyestrone as a carcinogen and considerable evidence for a protective role.

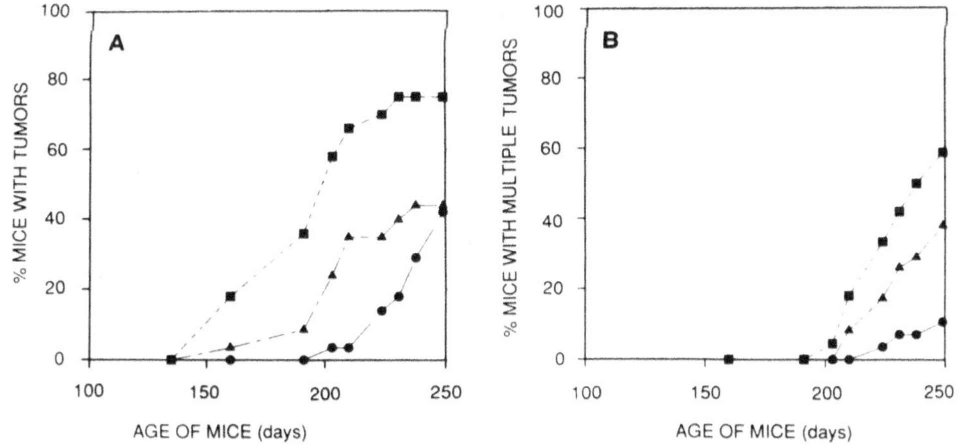

Figure 10. Decrease in mammary tumor formation following induction of 2-hydroxyestrone by feeding indole-3-carbinol to C3H/OuJ mice. The upper panel illustrates tumor incidence and latency; the lower panel represents tumor multiplicity in the same animals. ∞ = Control. ▲ = 500 ppm of Indole-3-carbinol in the diet. ● = 2000 ppm of Indole-3-carbinol in the diet.

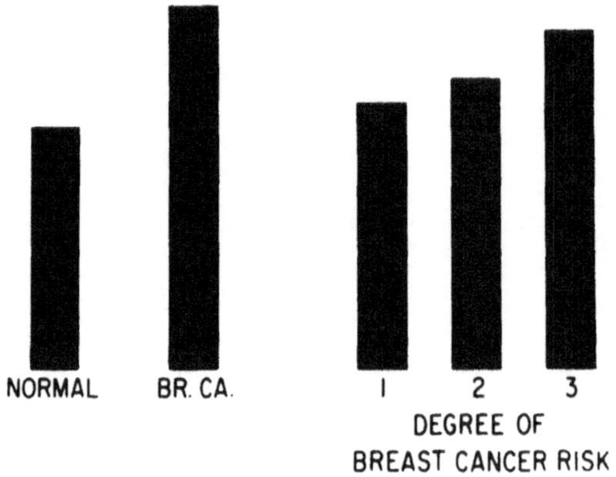

Figure 11. The left side represents the difference in 16 hydroxylation between newly diagnosed cancer patients matched and normal controls. The right side represents the difference in 16α-hydroxylation between women at low risk for breast cancer and high risk for breast cancers.

iii. 16α-Hydroxyestrone. On the other hand we have described a substantial body of data on the role of 16α-hydroxyestrone as a tumor initiator and promoter. These observations started with the finding that 16α-hydroxylation of estradiol was elevated in women with breast cancer (37) and in women at high risk for breast cancer (36) (Figure 11).

Immunohistochemical staining of breast tissue containing tumors showed extensive staining for 16α-hydroxyestrone in the tissue around the tumor but not for 2-hydroxyestrone. It was also elevated in mice at high risk for breast cancer (37). Studies in breast tissue obtained from women undergoing mammoplasty (TDLU-LR) versus tissue obtained from a breast containing a tumor but distal to the tumor (TDLU-HR) showed that latter tissue exhibited higher levels of 16α-hydroxylation and lower levels of 2-hydroxylation. The extent of ras P-21 protein expression was also elevated in this latter tissue (Figure 12) (38).

In addition we have shown that the tissue from the TDLU-HR was hyperresponsive with respect to treatment with BP and linoleic acid responding with a marked increase in 16α-hydroxylation and ras P-21 expression and a decrease in 2-hydroxylation (Figure 13) (39).

Similar responses were observed in breast cancer cell lines treated with BP and DMBA (40). Although PAHs are supposed to elevate P4501A1 and P4501A2 which should increase estradiol 2-hydroxylation we have never seen this response in breast tissue, instead we only get an elevation of 2-hydroxylation.Transfection of normal mammary cells with the *ras* or *myc*oncogenes also increased 16α-hydroxylation and decreased reaction at C-2 at the

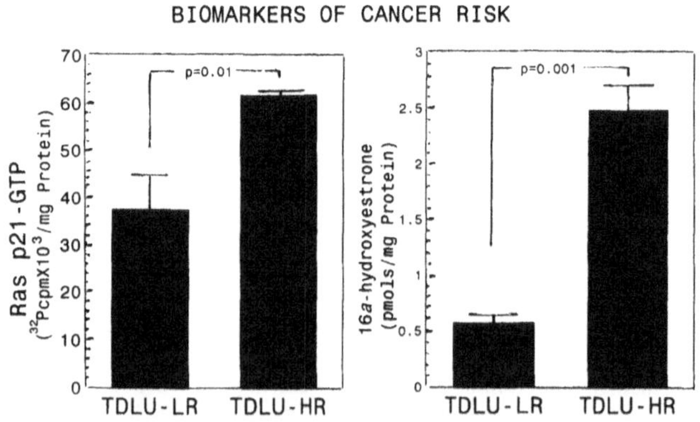

Figure 12. Increased Ras expression and 16α-hydroxylation in TDLU taken from a breast containing a tumor vs. control taken from a mammoplasty specimen.

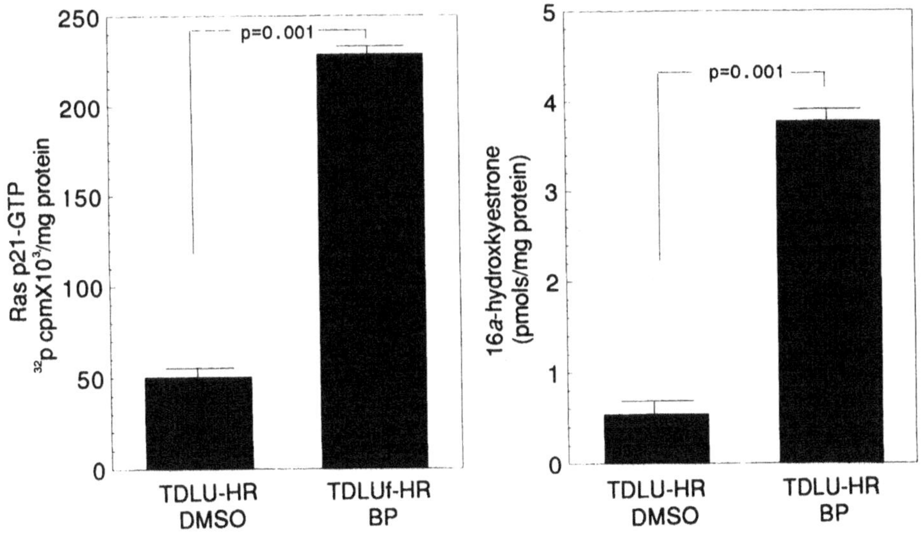

Figure 13. Effect of exposure BP on Ras expression and 16α-hydroxylation TDLU HR vs. TDLU LR.

same time that tumorigenicity was increased (24,25). This change in metabolism which was originally demonstrated radiometrically has now been confirmed by GC-MS analysis of the cell culture media.

A specific role for 16α-hydroxyestrone as a genotoxic compound was demonstrated by Telang et al. (41) using a variety of parameters including increased unscheduled DNA synthesis, hyperproliferation and increased anchoarage independent growth. None of these parameters were exhibited by estriol, estrone, or 2-hydroxyestrone. The latter compound was in fact somewhat inhibitory.

Additional support for the role of 16α-hydroxyestrone as a promotional agent for tumor growth comes from recent observations by Suto (24). When MCF-7 cells are implanted into the flank of nude mice it has long been observed that additional estrogen must be supplied as a sustained release pellet or an Alza pump. When estradiol, 16α-hydroxyestrone, or 2-hydroxyestrone were compared, 16α-hydroxyestrone proved to be the most potent of these compounds at promoting tumor growth with 2-hydroxyestrone being the least active.

CLINICAL APPLICATIONS OF BIOREACTIVE MOLECULES

In light of this data showing the risk relationship of 16α-OHE1 for breast cancer we sought an approach to safely lower 16α-hydroxylation in women at risk for breast cancer . Attempts to decrease the activity of the 16α-hydroxylase directly through the use of thyroid hormone or ω-3 fatty acids all failed to achieve significant reductions in reaction at C-16. Reaction at C-16 appeared to be constitutive and not readily modulated. Since our preliminary data suggested some degree of reciprocity with decreases in 2-hydroxylation being accompanied by increases in the alternative pathway, we turned to exploring the possibility of increasing C-2 hydroxylation as a method of decreasing the expression of the alternative pathway. As we discussed above, C-2 hydroxylation was readily modulated by diet (high protein (43), cruciferous vegetables (46), low body fat (44), vigorous exercise (45), smoking (46), and dioxin exposure (specific for this chlorinated compound) (4\7) which all lead to a marked increase in 2-hydroxylation which we believe to be a protective factor. Since these

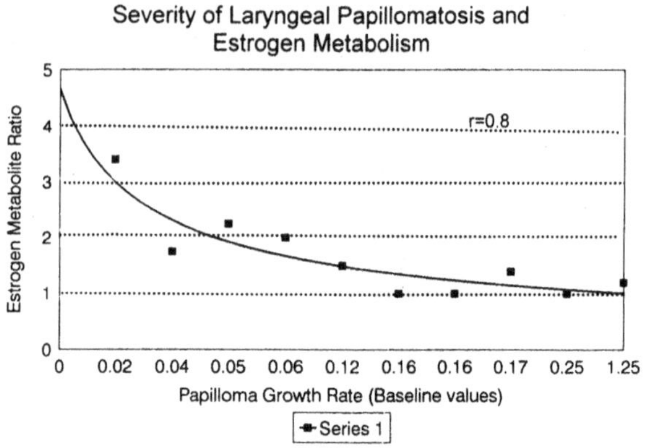

Figure 14. Inverse relationship of the 2/16αmetabolite ratio and rate of growth of HPV induced laryngeal papillomas.

do not appear to be practical approaches we turned to exploring the application of specific compounds present in various vegetables which were bioreactive and capable of modulating estrogen metabolism. We initially studied the use of a variety of indoles and flavones as

10. Sepkovic, D.W., Bradlow, H.L., Michnovicz, J.J., Murtezani, S., Levy, I. and Osborne, M.P. Catechol estrogen production in rat microsomes after treatment with indole-3-carbinol, ascorbigen, or β-napthoflavone: A comparison of stable isotope dilution gas chromatography-mass spectrometry and radiometric methods. Steroids 59: 318-323, 1994.
11. Lustig, R., Hershcopf, R.J. and Bradlow, H.L. The impacts of body weight and diet on estrogen metabolism and estrogen-dependent disease. In Adipose Tissue and Reproduction ed. Frisch, R. Karger Press, Zurich, pp. 119-132, 1989.
12. Gordon J, Cantrall WP, Albers HJ, Mauer S, and Stolar SM. Steroids and lipid metabolism. The hypocholesteremic effect of estrogen metabolites. Steroids 4: 267-91 (1964).
13. Schneider, J., Huh, M.M., Bradlow, H.L. and Fishman, J. Antiestrogen action of 2-hydroxyestrone on MCF-7 human breast cancer cells. J Biol Chem 259: 4840-4845, 1984.
14. Zhu, BT, Ezelb, EL Liehr, JG. Catechol O-methyl transferase catalyzes O-methylation of mutagenic flavonoids: Metabolic inactivation as a possible reason for their lack of carcinogenicity in vivo J. Biol. Chem. 269: 292-299 1994.
15. Li, S.A., Li, J.J., Metabolism of moxestrol in the hamster kidney: significance for estrogen carcinogenesis in Hormonal Carcinogenesis, Li, J.J., Nandi, S., and Li, S.A., eds. Springer-Verlag New york, p 110 1992.
16. Shen, Z, Lie, J., Wells, R.L., and Elkind, M.M. CDNA cloning, sequence analysis, and induction by aryl hydrocarbons of a murine cytochroome P450 gene Cyp1B1 DNA Cell Biol 12: 763-9 1994
17. Bradlow, HL Unpublished observations from this laboratory'
18. Zhu, B.T., Bui, Q.D., Weisz, J, Liehr, J.C. Conversion of estrone to 2- and 4-hydroxyestrone by hamster liver and kidney microsomes: Implications for the mechanism of estrogen induced carcinogenesis. Endocrinol. 135: 1772-1779 1994.
19. Jellinck P.H. and Bradlow H.L. Peroxidase-catalysed displacement of tritium from regio-specifically labeled estradiol and 2.-hydroxyestradiol. J Steroid Biochem 35: 705-710, 1990.
20. Musey, P.I., Collins, D.C., Bradlow, H.L., Gould, K.G., and Preedy, J.K.R. Effect of diet on oxidation of 17β-estradiol in vivo. J. Clin. Endo. Metab 65 792-796 1987.
21. Osborne, M.P., Bradlow, H.L., Wong, G.Y. and Telang, N.T. Increase in the extent of estradiol 16α-hydroxylation in human breast tissue: A potential biomarker of breast cancer risk. J Natl Cancer Inst 85: 1917-1920, 1993.
22. Auborn K.J., Woodworth,C., DiPaolo, J.A. and Bradlow, H.L. The interaction between HPV infection and estrogen metabolism in cervical carcinogenesis Inst. J. Cancer 49: 867-869, 1991.
23. Bradlow, H.L., Hershcopf, R.J.,Martucci, C.P., and Fishman, J. Estradiol 16α-hydroxylation in the mouse correlates with mammary tumor incidence and presence of MMTV: a possible model for hormonal etiology of breast cancer in humans. Proc. Natl. Acad. Sci. USA 82 6295-6299 1985.
24. Suto, A., Bradlow, H.L., Wong, G.Y., Osborne, M.P. and Telang, N.T. Persistent estrogen responsiveness of Ras oncogene-transformed mouse mammary epithelial cells. Steroids 57: 262-268, 1992.
25. Telang, N.T., Arcuri, F., Granata, O.M., Bradlow, H.L., Osborne, M.P. and Castagnetta, L. Alteration of estradiol metabolism in myc oncogene-transfected mouse mammary epithelial cells. Breast Cancer Res Treat, 1994 in press.
26. Han, X, Liehr, JC 8-Hydroxylation of guanine bases in kidney and liver DNA of hamsters treated with estradiol: Role of free radicals in estrogen induced carcinogenesis Cancer Res. 54: 5515-5517 1994.
27. Suto, A., Bradlow, H.L., Wong, G.Y., Osborne, M.P., and Telang, N.T. Experimental down-regulation of intermediate biomarkers of carcinogenesis in mouse mammary epithelial cells. Breast Cancer Res Treat 27: 193-202 1993.
28. Dwivedy, I., Devaneson, P., Cremonesi, P., Rogan. E., and Cavalieri, E. Synthesis and Characterization of Estrogen 2,3- and 3,4-quinones. Comparison of DNA adducts formed by the quinones versus Horseradish peroxidase-activated catechol estrogens. Chem. Res Toxicol 9: 828-833 1992.
29. Li, JJ, Li, SA, Oberley, TG, and Parsons, JA 1995, Carcinogenic Activities of Various Steroidal and Nonsteroidal Estrogens in the Hamster Kidney: Relation to Hormonal Activity and Cell Proliferation. Cancer Res 55 4347-4351.
30. Bradlow HL, Michnovicz JJ, Telang NT, et al: Diet, oncogenes and tumor viruses as modulators of estrogen metabolism in vivo and in vitro. Cancer Detect Prev S16: 35-42, 1992
31. Kojima T, Tanaka T, Mori M: Chemoprevention of spontaneous endometrial cancer in female Donryku rats by dietary indole-3-carbinol Cancer Res. 54:1446-1449, 1994
32. Malloy, V, Bradlow, HL, Matias, J, Orentreich, N: Further studies on chemoprevention of mammary tumors with indole-3-carbinol Terra Symposium on Estrogens 1991. Abst #24
33. Grubbs C, Steele VE, Casebolt T, et al: Chemoprevention of chemically induced mammary carcinogenesis by indole-3-carbinol AACR Meeting 1994 Abst #1305

34. Neufield L, Goldsmith A, Bradlow HL, Auborn K: Estrogen metabolism and human papilloma virus-induced tumors of the larynx: Chemoprophylaxis with indole-3-carbinol. Anticancer Res. 13: 227-234 1993.
35. Schneider, J., Kinne, D., Fracchia, A., Pierce, V., Anderson, K.E., Bradlow, H.L. and Fishman, J. Abnormal oxidative metabolism of estradiol in women with breast cancer. Proc Natl Acad Sci USA, 79: 3047-3051, 1982.
36. Osborne M.P., Karmali R.A., Hershcopf R.J., Bradlow H.L., Kourides I.A., Williams W.R., Rosen P.P. and Fishman J. Omega-3 fatty acids: Modulation of estrogen metabolism and potential for breast cancer prevention. Cancer Invest 8: 629-631, 1988.
37. Bradlow, H.L., Hershcopf, R.J., Martucci, C.P. and Fishman, J. Estradiol 16α-hydroxylation in the mouse correlates with mammary tumor incidence and presence of murine mammary tumor virus: A possible model for the hormonal etiology of breast cancer in humans. Proc Natl Acad Sci USA, 82: 6295-6299, 1985.
38. Telang, N.T., Narayanan.R., Bradlow. H.L. and Osborne, M.P. Coordinated expression of intermediate biomarkers for tumorigenic transformation in Ras-transfected mouse mammary epithelial cells. Breast Cancer Res Treat 18: 155-163, 1991.
39. Telang, N.T., Kurihara, H., Wong, G.T.C., Bradlow, H.L., Osborne, M.P. Preneoplastic transformation in mouse mammary tissue: Identification and validation of intermediate biomarkers for chemoprevention Anticancer Res. 11: 1021-1028 1991
40. Telang, N.T., Bradlow, H.L. and Osborne, M.P. Molecular and endocrine biomarkers in noninvolved breast: Relevance to cancer chemoprevention. J Cell Biochem S 16 G: 161-169, 1992.
41. Telang, N.T., Suto, A., Wong, G.Y., Osborne, M.P. and Bradlow, H.L. Estrogen metabolite 16α-hydroxyestrone induces genotoxic damage and aberrant cell proliferation in mouse mammary epithelial cells in culture. J Natl Cancer Inst 84: 634-638, 1992.
42. Private communication from Dr. Suto.
43. Anderson, K.E., Kappas, A., Conney, A.H., Bradlow, H.L. and Fishman, J. The influence of dietary protein and carbohydrate on the principal oxidative biotransformations of estradiol in normal subjects. J Clin Endocrinol Metab 59: 103-107, 1984.
44. Bradlow, H.L., Michnovicz, J.J., Wong, G.Y.C., Halper, M.P., Miller, D. and Osborne, M.P. Long term responses of women to indole-3-carbinol or a high fiber diet. Cancer Epidemiol Biomarkers Prevention 3 591-595, 1994.
45. Lustig, R.H., Bradlow, H.L. and Fishman, J. Estrogen metabolism in disorders of nutrition and dietary composition. In: The Menstrual Cycle and Its Disorders, eds. H.M. Pirke, J. Wuttke and U. Schweiger, Springer-Verlag, Berlin, Heidelberg. pp 119-132, 1989.
46. Snow, R., Barbieri, R.,and Frisch, R., Estrogen 2-hydroxylase oxidation and menstrual function among elite oarswomen J. Clin.Endocrinol. and Metab., 69: 369-376 1991.
47. Michnovicz, J.J., Hershcopf, R.J., Naganuma, H., Bradlow, H.L. and Fishman, J. Increased 2-hydroxylation of estradiol as a possible mechanism for the antiestrogenic effect of cigarette smoking. N Engl J Med 315: 1305-1309, 1986.
48. Gierthy, J.F., Lincoln, D.W. II, Kampcik, S.J., Dickerman,H.W., Bradlow, H.L., Niwa, T. and Swaneck, G.E. Induction of human breast cancer cells by 2,3,7,8-tetrachlorodibenzo-*p*-dioxin. Biochem Biophys Res Commun 157: 50-55 1988.
49. Niwa, T., Swaneck, G. and Bradlow, H.L. Alterations in estradiol metabolism in MCF-7 cells induced by treatment with indole-3-carbinol and related compounds. Steroids 59 523-527 1994.
50. Bradlow, H.L., Michnovicz, J.J., Telang, N.T. and Osborne, M.P. Effect of dietary indole-3-carbinol on estradiol metabolism and spontaneous mammary tumors in mice. Carcinogenesis 12: 1571-1574, 1991.
51. Michnovicz, JJ and Bradlow, HL. Induction of estradiol metabolism by indole-3-carbinol in humans J. Natl Cancer Instit. 82: 947-951 1990.
52. Michnovicz, J.J. and Bradlow, H.L. Altered estrogen metabolism and excretion in humans following consumption of indole-3-carbinol. Nutr Cancer 16: 59-66, 1991.

BIOSYNTHESIS AND CELLULAR EFFECTS OF TOXIC GLUTATHIONE *S*-CONJUGATES

Wolfgang Dekant*

Institut für Toxikologie und Pharmakologie
Universität Würzburg
Versbacher Str. 9
97078 Würzburg, Germany

ABSTRACT

Glutathione conjugation has been identified as an important detoxication reaction. However, several glutathione-dependent bioactivation reactions have been identified. Current knowledge on the mechanisms and the possible biological importance of these reactions is discussed in this article. Dichloromethane is metabolized by glutathione conjugation to formaldehyde via *S*-(chloromethyl)glutathione. Both compounds are reactive intermediates and may be responsible for the dichloromethane-induced tumorigenesis in sensitive species. Vicinal dihaloalkanes are transformed by glutathione *S*-transferase-catalyzed reactions to mutagenic and nephrotoxic *S*-(2-haloethyl)glutathione *S*-conjugates. Electrophilic episulphonium ions are the ultimate reactive intermediates formed and interact with nucleic acids. Several polychlorinated alkenes are bioactivated in a complex, glutathione-dependent pathway. The first step is hepatic glutathione *S*-conjugate formation followed by cleavage to the corresponding cysteine *S*-conjugates, and, after translocation to the kidney, metabolism by renal cystein conjugate ß-lyase. ß-Lyase-dependent metabolism of halovinyl cysteine *S*-conjugates yields electrophilic thioketenes, whose covalent binding to cellular macromolecules is likely responsible for the observed nephrotoxicity of the parent compounds. Finally, hepatic glutathione conjugate formation with hydroquinones and aminophenols yields conjugates that are directed to γ-glutamyltransferase-rich tissues, such as the kidney, where they undergo alkylation or redox cycling reactions, or both, that cause organ-selective damage.

INTRODUCTION

Glutathione (γ-glutamyl-cysteinylglycine) is a major low molecular weight peptide in mammalian cells and participates in a variety of cellular reactions (Meister, 1988; 1992).

*Tel.: (0931) 201 3449; Fax: (0931) 201 3446.

Biological Reactive Intermediates V
Edited by R. Snyder et al., Plenum Press, New York, 1996

Due to the nucleophilicity of the sulphur atom and the antioxidant properties of glutathione, this tripeptide is an important factor in the detoxication of xenobiotics and oxidants (Boyland and Chasseaud, 1969). The formation of glutathione *S*-conjugates from xenobiotics and their electrophilic metabolites has long been associated with detoxication. Recent evidence, however, indicated that this generalisation is not always true and that some glutathione *S*-conjugates may be toxic (Anders et al., 1988; Dekant et al., 1989; Monks et al., 1990; Anders, 1991). At least three types of toxic glutathione *S*-conjugates have been identified. The objective of this review is to summarize current knowledge on the biosynthesis of toxic glutathione *S*-conjugates, the reactions of these *S*-conjugates in cellular systems and their association with toxicity. Several recent reviews on this topic have appeared (Monks and Lau, 1987; 1989; Lock, 1988; 1989; Dekant et al., 1990a, b; Koob and Dekant, 1991; Dekant and Vamvakas, 1992).

BIOSYNTHESIS AND CELLULAR REACTIONS OF TOXIC GLUTATHIONE *S*-CONJUGATES

Toxic glutathione *S*-conjugates are biosynthesized from three different types of compounds, haloalkanes, haloalkenes and from hydroquinones and aminophenol.

Toxic Glutathione *S*-Conjugates Derived from Halogenated Alkanes

Dichloromethane. Dichloromethane is metabolized by cytochromes P450 via the intermediate formyl chloride to carbon monoxide and HCl. Glutathione *S*-transferases catalyse the transformation of dichloromethane to formaldehyde and HCl. Recent studies implicate the newly described θ-class human hepatic glutathione *S*-transferase and rat hepatic transferase isoforms 5-5 and 12-12 in the biotransformation of dichloromethane to formaldehyde.

In the first step of the glutathione-dependent reaction, *S*-(chloromethyl)-glutathione is formed in a glutathione *S*-transferase-catalyzed reaction. *S*-(Chloromethyl)glutathione is a reactive electrophile and is rapidly hydrolysed to *S*-(hydroxymethyl)glutathione which finally decomposes to formaldehyde and glutathione (Ahmed and Anders, 1976; 1978) (Fig. 1).

Dichloromethane has been reported to be carcinogenic in mice but not in hamsters or rats and pharmacokinetic studies implicate glutathione-dependent bioactivation in the observed tumorigenicity of dichloromethane (Green, 1983; Andersen et al., 1991). The rate of metabolism of dichloromethane by glutathione conjugation is much higher in mice than in other species. Clear evidence for the formation of DNA adducts in dichloromethane-treated animals is lacking. Formaldehyde, formed as a metabolite of dichloromethane, may participate in the production of DNA-protein crosslinks (Heck and Casanova-Schmitz, 1983). A role for the glutathione-dependent bioactivation of dichloromethane and the electrophilic glutathione *S*-conjugate in the observed species differences in dichloromethane

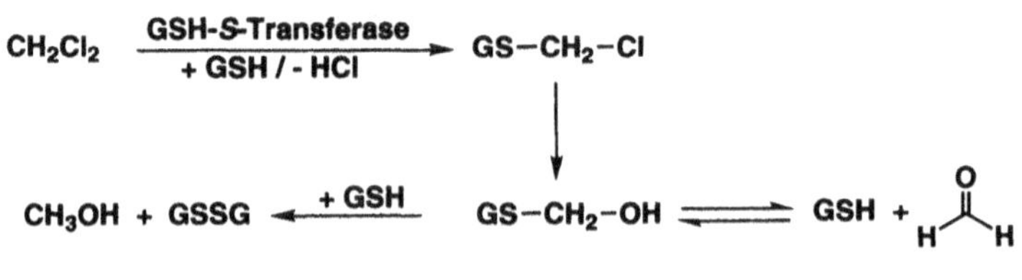

Figure 1. Bioactivation of dichloromethane by glutathione conjugation.

tumorigenicity was also indicated in bacteria and freshly isolated hepatocytes (Graves et al., 1994a, b). Semicarbazide, a formaldehyde trapping reagent, prevented formaldehyde induced mutagenicity, but did not influence dichloromethane mutagenicity (Green et al., 1988a, b; Reitz et al., 1989; Kermani et al., 1990; Blumbach et al., 1993).

Dihaloalkanes

1,2-dichloro-, 1,2-dibromo- and 1-bromo-2-chloroethane. The vicinal dihaloalkanes 1,2-dibromoethane, 1,2-dichloroethane and 1-bromo-2-chloroethane are toxic and carcinogenic (Spencer et al., 1951; Weisburger, 1977). Their metabolism involves two pathways, cytochrome P450-dependent oxidation and glutathione *S*-conjugate formation (Guengerich et al., 1980). P450-dependent oxidation results in formation of chloroacetaldehyde from 1,2-dichloroethane, and bromoacetaldehyde from 1,2-dibromoethane; these reactive aldehydes are thought to be responsible for the covalent binding of 1,2-dibromoethane- and 1,2-dichloroethane-metabolites to proteins (Guengerich et al., 1980). 1,2-Dibromoethane is metabolized to *S*-(2-bromoethyl)glutathione by the rat α class glutathione *S*-transferase 2-2 and, to a lesser extent, by the μ class enzyme 3-3. Human α class glutathione *S*-transferases catalyze the conjugation of 1,2-dibromoethane with glutathione (Inskeep and Guengerich, 1984; Kim and Guengerich, 1989; 1990; 1992; Humphreys et al., 1990; Ozawa and Tsukioka, 1990; Cmarik et al., 1990). 1,2-Dichloroethane and 1-bromo-2-chloroethane are metabolized to *S*-(2-chloroethyl)glutathione (Fig. 2).

The biosynthetic sulphur half mustards are strong alkylating agents and react with nucleic acids (Jean and Reed, 1989) and their electrophilicity is attributable to neighbouring group assistance in nucleophilic displacement (Dohn and Casida, 1987). The concept of episulphonium ions as ultimate intermediates in the metabolism of vicinal dihaloalethanes is supported by structure/activity relationships on the mutagenicity of several derivatives of *N*-acetyl-*S*-(2-haloalkyl)-l-cysteine (Van Bladeren et al., 1979), by the stereochemistry of the DNA-adducts (Peterson et al., 1988) formed and by NMR-data (Dohn and Casida, 1987). The glutathione-dependent formation of these sulphur half-mustards is responsible for the binding of 1,2-dibromoethane- and 1,2-dichloroethane-metabolites to DNA and for 1,2-dibromoethane and 1,2-dichloroethane induced genotoxicity (Van Bladeren et al., 1980). *S*-[2-(N^7-guanyl)ethyl]glutathione is the major DNA-adduct formed *in vivo* (Ozawa and Guengerich, 1983), and the corresponding mercapturic acid *N*-acetyl-*S*-[2-N^7-guanyl)ethyl]-l-cysteine is a urinary metabolite of 1,2-dibromoethane (Kim and Guengerich, 1989). *S*-(2-chloroethyl)glutathione also reacts with guanosine to give *S*-[2-N^7-guanyl)ethyl]glutathione (Foureman and Reed, 1987). *S*-[2-(N^1-adenyl)ethyl]glutathione has been found as a minor DNA-adduct formed from 1,2-dibromoethane. *S*-[2(N^7-guanyl)ethyl]glutathione is

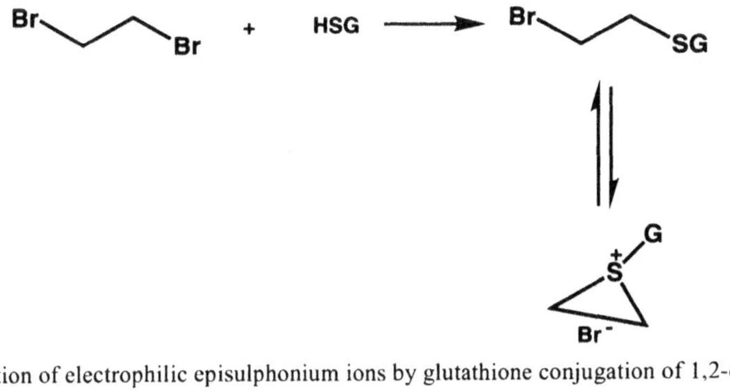

Figure 2. Formation of electrophilic episulphonium ions by glutathione conjugation of 1,2-dihaloalkanes.

a mutagenic DNA-adduct inducing primarily base substitutions of which G:C to A:T transitions accounted for 75 %. *S*-(2-chloroethyl)glutathione and *S*-(2-chloroethyl)-L-cysteine are nephrotoxic metabolites in rats. *S*-(2-chloroethyl)-L-cysteine is also a potent mutagen in *Salmonella typhimurium* TA 100 and induces high rates of DNA repair in cultured cells (Vamvakas et al., 1988a; 1989a).

Glutathione conjugation and the oxidative metabolism of vicinal dihaloethanes give rise to an identical excretory end product of metabolism, *N*-acetyl-*S*-(2-hydroxyethyl)-L-cysteine; therefore, it is difficult to assess the relative contribution of both pathways to 1,2-dihaloethane toxicity and mutagenicity. Studies with radiolabelled and stable isotope-labelled 1,2-dihaloethane suggest that the majority of the radiolabel in DNA is derived from glutathione-dependent reactions and that glutathione conjugation contributes to a larger extent to the excretion of *N*-acetyl-*S*-(2-hydroxyethyl)-L-cysteine.

1,2-dibromo-3-chloropropane. The nematocide 1,2-dibromo-3-chloropropane has been widely used as a soil fumigant. The acute toxicity of 1,2-dibromo-3-chloropropane is characterized by necrosis of the renal proximal tubules, testicular atrophy, and occasional liver damage (Torkelson et al., 1961).

1,2-Dibromo-3-chloropropane toxicity and mutagenicity is dependent on metabolic activation. 1,2-Dibromo-3-chloropropane is metabolized to a variety of polar metabolites excreted with urine and faeces which are largely derived from glutathione conjugates (Jones et al., 1979). The structures of metabolites formed and the result of mechanistic studies on

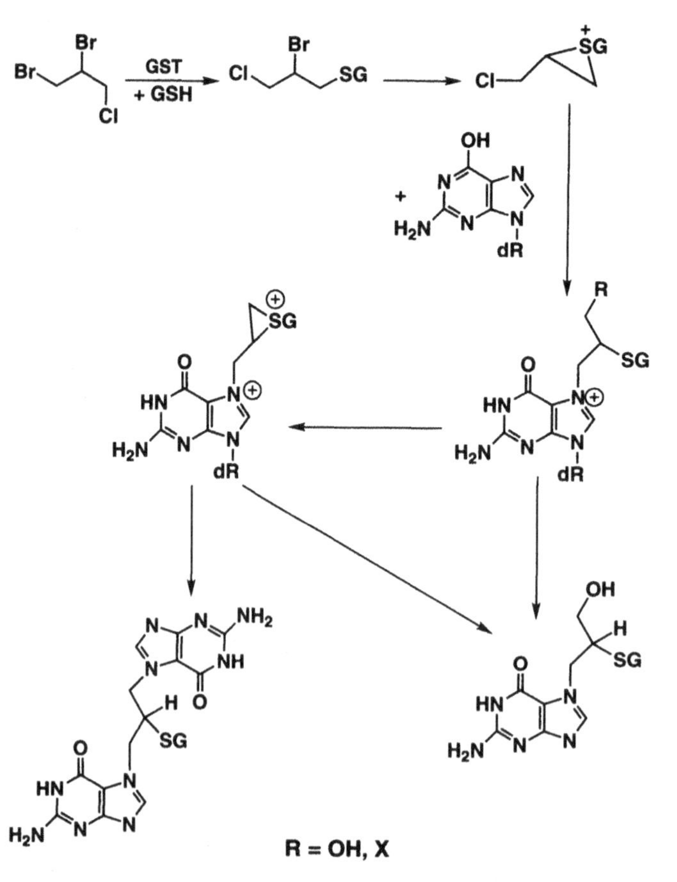

Figure 3. Bioactivation of 1,2-dibromo-3-chloropropane by glutathione conjugation and structures of DNA-adducts identified (modified from Humphreys *et al.* 1991).

1,2-dibromo-3-chloropropane-biotransformation *in vivo* and *in vitro* indicate that *S*-(3-chloro-2-bromopropyl)glutathione is an intermediate in 1,2-dibromo-3-chloropropane-metabolism (Fig. 3) (Pearson et al., 1990); it spontaneously cyclizes to an episulphonium ion and its hydrolysis accounts for a major part of the biliary excreted *S*-(3-chloro-2-hydroxypropyl)glutathione and *S*-(2,3-hydroxypropyl)-glutathione (Fig. 3).

Formation of toxic glutathione conjugates is most likely responsible for the renal and testicular toxicity of 1,2-dibromo-3-chloropropane. Based on the absence of significant isotope effects with perdeutero-1,2-dibromo-3-chloropropane, it appears that the breaking of a carbon-hydrogen bond is not the rate-limiting step in 1,2-dibromo-3-chloropropane-induced renal and and testicular toxicity *in vivo* and *in vitro*. Moreover, renal and testicular necrosis and the ability of 1,2-dibromo-3-chloropropane to induce DNA-damage in these organs are independent of cytochrome P450 and display a requirement for glutathione (Lag et al., 1989a, b; Holme et al., 1991). These observations support the hypothesis that the toxicity of 1,2-dibromo-3-chloropropane is induced by a glutathione-dependent pathway under formation of a reactive episulphonium ion. This episulphonium ion reacts with DNA and two isomers of *S*-[1-(hydroxymethyl)-2-(N^7-guanyl)ethyl]glutathione and *S*-[bis(N^7-guanyl)methyl]glutathione have been identified as major 1,2-dibromo-3-chloropropane-derived adducts found in DNA (Humphreys et al., 1991). Metabolic episulphonium formation by glutathione conjugation by an identical mechanism also seems to occur during the biotransformation on the structural analogue 1,2,3-trichloropropane (Weber and Sipes, 1990; 1992; Mahmood et al., 1991; Winter et al., 1992).

Tris(2,3-dibromopropyl)phosphate. The flame retardant tris(2,3-dibromopropyl) phosphate is nephrotoxic in animals after acute dosage (Osterberg et al., 1977; Elliot et al., 1982; Söderlund et al., 1984) and a selective renal carcinogen after long-term administration (IARC, 1984). Tris(2,3-dibromopropyl)phosphate is metabolized to bis(2,3-dibromopropyl)phosphate, 2,3-dibromopropanol, 2-bromoacrolein and several polar metabolites (Lynn et al., 1982; Söderlund et al., 1984). Radioactivity from ^{14}C-tris(2,3-dibromopropyl)phosphate was covalently bound to proteins in rats *in vivo*; binding to kidney proteins was 5-times higher than to liver proteins. The mechanisms underlying tris(2,3-dibromopropyl)phosphate-nephrotoxicity and carcinogenicity are still unclear; however, formation of reactive glutathione conjugates semms to play an important role in the renal toxicity of tris(2,3-dibromopropyl)phosphate. Incubation of tris(2,3-dibromopropyl)phosphate with cytosolic enzymes, ^{35}*S*-glutathione and DNA resulted in ^{35}*S*-glutathione-binding to DNA (Inskeep and Guengerich, 1984). The mutagenicity of tris(2,3-dibromopropyl)phosphate was also markedly increased in *Salmonella typhimurium* expressing human glutathione *S*-transferases (Simula et al., 1993). The structures of the glutathione conjugates identified in rats given tris-(2,3-dibromopropyl)phosphate also supported the involvement of episulphonium ion intermediate formation, probably by glutathione *S*-transferases, in tris-(2,3-dibromopropyl)phosphate toxicity (Pearson et al., 1993) (Fig 4).

Halogenated Alkenes

The halogenated alkenes hexachlorobutadiene, perfluoropropene, chlorotrifluoroethene, and the alkyne dichloroacetylene, are selectively nephrotoxic in rats and induce proximal tubular damage (Reichert et al., 1975; Potter et al., 1981; Ishmael et al., 1982). Moreover, the widely used solvents trichloroethene (NCI, 1986b) and tetrachloroethene, and also dichloroacetylene (NCI, 1986a) and hexachlorobutadiene (Kociba et al., 1977), induced chronic nephrotoxicity and/or carcinomas of the proximal tubules in rats after giving high doses.

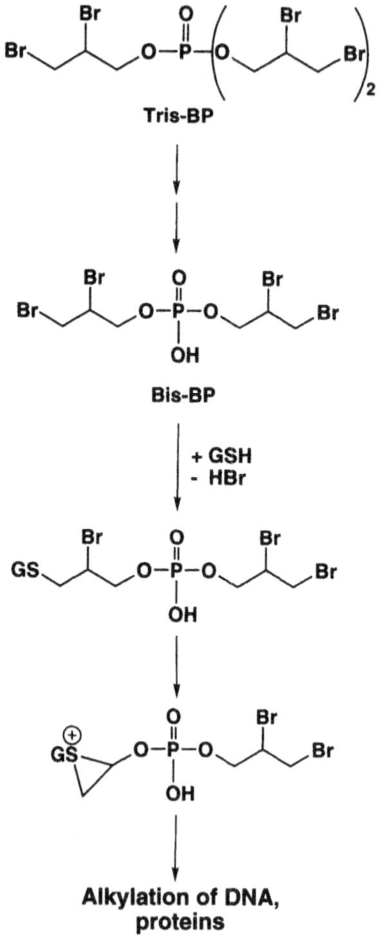

Figure 4. Bioactivation of tris-(2,3-dibromopropyl)phosphate by glutathione conjugation and formation of an electrophilic episulfonium ion.

Glutathione-dependent pathways have been implicated in the renal toxicity of these compounds.

Nephrotoxic haloalkenes are metabolized to glutathione *S*-conjugates by microsomal and cytosolic glutathione *S*-transferases; the microsomal fraction of rat liver generally exhibits a 2- to 10-fold higher activity toward haloalkenes than does the cytosolic fraction.

Hexachlorobutadiene (Wolf et al., 1984; Dekant et al., 1988a, b), 1,1,2-trichloro-3,3,3-trifluoropropene (Vamvakas et al., 1989b), trichloroethene (Dekant et al., 1986a; 1990c) and tetrachloroethene (Dekant et al., 1986a; 1987b) are metabolized by soluble and membrane-bound glutathione *S*-transferases from rat liver by an addition-elimination reaction to give exclusively *S*-(haloalkenyl)glutathione conjugates (Fig. 5).

Formation of these glutathione *S*-conjugates has also been observed from bile obtained in isolated rat livers perfused with the haloalkenes. Glutathione conjugate formation from hexachlorobutadiene has been observed in mouse liver (Dekant et al., 1988a) and in human liver; it has been calculated that glutathione-dependent metabolism of hexachlorobutadiene in intact liver cells is mainly catalyzed by microsomal glutathione *S*-transferase (Wallin et al., 1988; Oesch and Wolf, 1989). In rats *in vivo*, metabolites indicative of glutathione conjugation reactions were also found: the bile of rats given hexachlorobutadiene (Nash et al., 1984) 1,1,2-trichloro-3,3,3-trifluoropropene (Vamvakas et al., 1989b), trichlo-

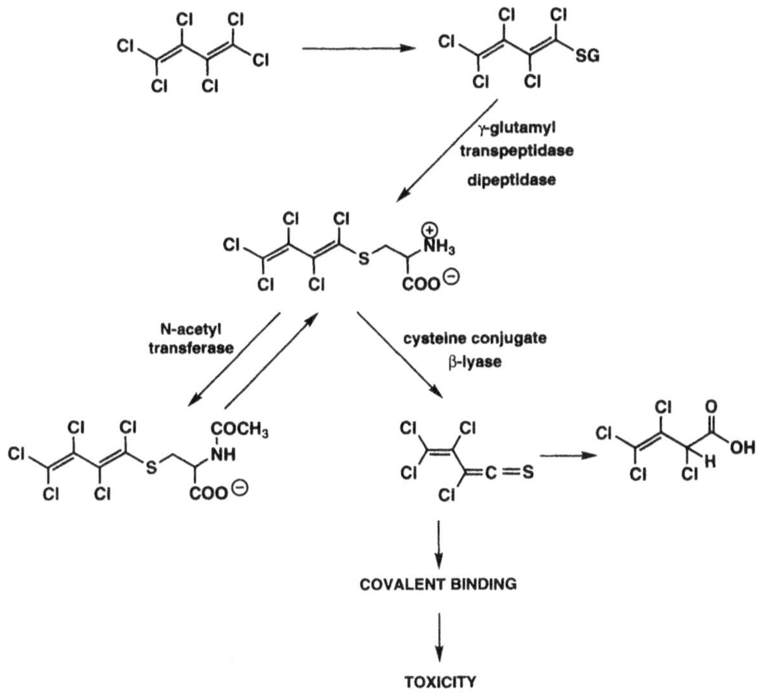

Figure 5. Biosynthesis of glutathione conjugates from hexachlorobutadiene as a representative compound for polychlorinated alkenes, renal processing and bioactivation of cysteine *S*-conjugates by cysteine conjugate ß-lyase.

roethene (Dekant et al., 1990c), or tetrachloroethene (Vamvakas et al., 1989c) contains glutathione *S*-conjugates identical to those formed in liver microsomes and the corresponding mercapturic acids are urinary metabolites (Dekant et al., 1986a, b; Reichert and Schuetz, 1986; 1990; Vamvakas et al., 1989b). Hexachlorobutadiene seems to be metabolized *in vivo* exclusively by glutathione conjugate formation (Wallin et al., 1988), in contrast, both trichloroethene and tetrachloroethene are mainly metabolized by cytochrome P450 (Dekant et al., 1984). Metabolites whose formation is accountable to *S*-conjugate formation and processing are only minor excretory products.

The highly nephrotoxic alkyne dichloroacetylene (Reichert et al., 1975) is metabolized by addition of glutathione to give *S*-(1,2-dichlorovinyl)-glutathione; in rats, *N*-acetyl-*S*-(1,2-dichlorovinyl)-l-cysteine is a major urinary metabolite of dichloroacetylene (Kanhai et al., 1989; 1991).

In contrast to the vinylic *S*-conjugates formed from chloroalkenes, fluoroalkenes are generally metabolized by glutathione *S*-transferases to *S*-(fluoroalkyl)glutathione conjugates. For example, chlorotrifluoroethene and tetrafluoroethene are metabolized by the glutathione *S*-transferases to *S*-(1-chloro-1,1,2-trifluoroethyl)glutathione (Dohn and Anders, 1982) and *S*-(1,1,2,2-tetrafluoroethyl)glutathione (Odum and Green, 1984), respectively.

The enzymic reaction of glutathione with perfluoropropene yields both *S*-(1,1,2,3,3,3-hexafluoropropyl)glutathione and *S*-(1,2,3,3,3-pentafluoro-propenyl)glutathione as products (Koob and Dekant, 1990). The nephrotoxic chlorofluoroalkene 1,1-dichloro-2,2-difluoroethene is metabolized to *N*-acetyl-*S*-(1,1-dichloro-2,2-difluoroethyl)-l-cysteine *in vivo* (Commandeur et al., 1987).

Toxicity of Halovinyl S-Conjugates. The *S*-conjugates formed are toxic metabolites of the parent haloalkenes, which are accumulated in the kidney. Many studies have been performed to investigate the nephrotoxicity, cytotoxicity and genotoxicity of synthetic halovinyl *S*-conjugates (Lock, 1988; 1989). For detailed reviews, see (Anders et al., 1988; Dekant et al., 1989; Koob and Dekant, 1991).

The halovinyl *S*-conjugates *S*-(1,2-dichlorovinyl)-L-cysteine, *S*-(1,2,2-trichlorovinyl)-L-cysteine and *S*-(1,2,3,4,4-pentachlorobutadienyl)-L-cysteine are mutagenic in *Salmonella typhimurium* with (Green and Odum, 1985; Commandeur et al., 1991) and without the addition of exogenous activating systems (Dekant et al., 1986c; Vamvakas et al., 1988a, b). Incubation of these *S*-conjugates with *Salmonella typhimurium* homogenates results in time- and dose-dependent production of pyruvate, which is formed by ß-lyase-catalyzed cleavage of the *S*-conjugates in equimolar amounts with the presumed mutagenic intermediates. Both the mutagenicity and the pyruvate production are decreased in the presence of aminooxyacetic acid, an inhibitor of the ß-lyase. Haloalkyl *S*-conjugates from fluoroalkenes are also cleaved to pyruvate by *Salmonella typhimurium* homogenates, however they are not mutagenic in the Ames preincubation assay (Vamvakas et al., 1988b).

Halovinyl *S*-conjugates induce also ß-lyase-dependent DNA repair in LLC-PK_1 cells, a porcine kidney cell line (Vamvakas et al., 1989a; d). The extent of genotoxicity observed is very low and the concentration range over which DNA repair occurs in the absence of cytotoxicity and cell death is very small.

Studies with fluorescence digital imaging microscopy have shown that *S*-(1,2-dichlorovinyl)-L-cysteine increases cytosolic Ca^{2+}-concentrations prior to the onset of cell death. The increase is associated with impaired ability of the mitochondria to sequester cytosolic Ca^{2+} and precedes the collapse of the mitochondrial membrane potential (Vamvakas et al., 1990). This effect is common to some tumour promotors, such as hydroperoxides, that induce oxidative stress (Cerutti, 1985). *S*-(1,2-Dichlorovinyl)-L-cysteine also induces oxidation of pyridine nucleotides in incubations with kidney mitochondria and promotes dimethyl-nitrosamine-initiated renal tubule carcinomas in rats (Meadows et al., 1988; Vamvakas et al., 1992). The increased Ca^{2+} concentrations activate Ca^{2+}- and Mg^{2+}-dependent endonucleases. This results in increased formation of DNA-double-strand breaks followed by enhanced poly(ADP-ribosyl)ation of nuclear proteins in LLC-PK_1 cells treated with *S*-(1,2-dichlorovinyl)-L-cysteine. Finally, *S*-(1,2-dichlorovinyl)-L-cysteine induces the expression of the proto-oncogenes c-*fos* and c-*myc* in LLC-PK_1 cells (Vamvakas and Köster, 1993). Both the direct interaction of *S*-(1,2-dichlorovinyl)-L-cysteine derived reactive intermediates with DNA and the modification of DNA structure by increased poly(ADP-ribosyl)ation may be involved in the effects of *S*-(1,2-dichlorovinyl)-L-cysteine on the of c-*fos* and c-*myc* expression.

Reactive Intermediates Formed from Haloalkyl and Halovinyl S-Conjugates. Recently, the structures of the final reactive metabolites have been elucidated. Processing of the biosynthetic glutathione *S*-conjugates to the cysteine *S*-conjugates is required for the expression of toxicity. Fluoroalkyl and chloroalkenyl cysteine *S*-conjugates are metabolized by the pyridoxal phosphate-dependent ß-lyase to unstable thiols that yield reactive electrophiles (Dekant et al., 1987a; 1988c, d). Thioketenes are formed from the enethiols produced by ß-lyase-mediated cleavage of halovinyl cysteine *S*-conjugates (Dekant et al., 1991); fluoroalkyl cysteine *S*-conjugates are transformed to reactive thioacyl halides. Both thionoacyl fluorides and thioketenes are potent acylating agents and react with nucleophiles such as lysine in proteins. The role of thioacylating agents in *S*-conjugate-induced toxicity and mutagenicity has been confirmed by structure-activity studies: Only *S*-conjugates that form acylating agents are mutagenic in bacteria and cytotoxic in kidney cells (Vamvakas et al., 1988b; 1989e).

Hydroquinones and Aminophenol

Bromohydroquinone is a major toxic metabolite of bromobenzene and is easily converted to bromoquinone (Lau et al., 1984). Bromobenzene-derived covalently bound radioactivity in liver is mainly due to bromohydroquinone (Zheng and Hanzlik, 1992). Glutathione-dependent reactions have been implicated in bromohydroquinone induced nephrotoxicity.

Bromohydroquinone is oxidized to bromoquinone which reacts with glutathione (Monks et al., 1985). In microsomes and in rats *in vivo*, several isomeric mono- and bisglutathione substituted derivatives of bromohydroquinone have been identified as metabolites (Monks et al., 1985; Lau and Monks, 1990).

These glutathione *S*-conjugates are nephrotoxic in rats and cytotoxic to renal cells (Lau et al., 1988a). These results indicate that bromohydroquinone nephrotoxicity is likely due to the biosynthetic formation of glutathione conjugates derived from bromohydroquinone (Fig. 6).

Benzoquinone is also metabolized in rats to mono-, di- and multisubstituted glutathione conjugates which are nephrotoxic (Hill et al., 1993). Glutathione *S*-conjugates derived from benzoquinone, menadione and chloroquinones are also nephrotoxic in rats and toxic to kidney cells and renal mitochondria (Mertens et al., 1991; Redegeld et al., 1991; Hill et al., 1992). The biosynthetic glutathione *S*-conjugates (Kleiner et al., 1992) are accummulated by the kidney in a γ-glutamyltranspeptidase-dependent pathway, their toxicity is diminished or increased by inhibition of γ-glutamyltranspeptidase but not by inhibition of ß-lyase (Monks et al., 1988; Monks and Lau, 1990). These experiments suggest that ß-lyase-mediated cleavage does not play a role in the toxicity of these hydroquinone *S*-conjugates. The glutathione conjugates of hydroquinones and quinones may serve as transport forms to γ-glutamyl-transpeptidase rich organs to be accumulated there. In the rat, γ-glutamyl-transpeptidase is almost exclusively present in the kidney. Moreover, substitution of bromohydroquinones with cysteine lowers the redox potential and makes these bromohydroquinones more prone to oxidation to toxic quinones. The reduction of nephrotoxicity of bromohydroquinone glutathione *S*-conjugates by simultaneous administration of ascorbic

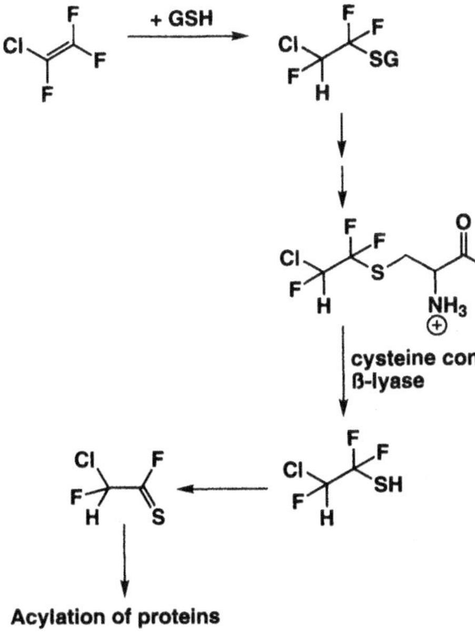

Figure 6. Biosynthesis of glutathione conjugates from chlorotrifluoroethene as a representative compound for polyfluorinated alkenes, renal processing and bioactivation of cysteine *S*-conjugates by cysteine conjugate ß-lyase.

acid suggests that processes involving oxidation to a reactive quinone and, presumably, peroxidative mechanisms are involved in the nephrotoxicity of these compounds (Lau et al., 1988a, b; 1990; Monks and Lau, 1990).

p-Aminophenol is an acute nephrotoxin causing necrosis of the pars recta of the proximal tubules in rats (Gartland et al., 1989a, b). p-Aminophenol is oxidized enzymically to the benzoquinone imine which is a reactive α,ß-unsaturated compound (Crowe et al., 1979; Newton et al., 1982; Josephy et al., 1983). Its interaction with cellular macromolecules when formed in the kidney may cause renal toxicity. However, several recent experimental findings support the assumption that p-aminophenol toxicity is mediated by glutathione-dependent mechanisms (Fig. 6). Synthetic quinoneimine reacts non-enzymically with glutathione (Eckert et al., 1990). Both direct and indirect evidence for the involvement of glutathione conjugates as a transport form for p-aminophenol metabolites has been obtained. Depletion of glutathione by buthioninsulfoximine, which inhibits glutathione synthesis, completely protected rats against p-aminophenol-induced nephrotoxicity (Gartland et al., 1990). Moreover, biliary cannulation partially protected rats from p-aminophenol-induced nephrotoxicity. Glutathione conjugates were identified in bile of rats given p-aminophenol including 1-amino-3-(glutathione-*S*-yl)-4-hydroxybenzene and 1-amino-2-(glutathione-*S*-yl)-4-hydroxybenzene. These metabolites are nephrotoxic in rats and cytotoxic in rat renal epithelial cells. Their toxicity is dependent on γ-glutamyltranspeptidase, but not on ß-lyase which is in line with the effects of hydroquinone *S*-conjugates (Fowler et al., 1991; Klos et al., 1992). These results suggest that nephrotoxins are formed by glutathione conjugation of p-aminophenol metabolites in the liver and are translocated to the kidney.

CONCLUSIONS

The results discussed in this review show that glutathione-dependent bioactivation reactions occur with several classes of compounds. Most experiments on glutathione-dependent bioactivation have been performed in rodents; however, humans also have the enzymatic capacity to synthetize toxic glutathione *S*-conjugates. Hexachlorobutadiene is metabolized to the toxic metabolite *S*-(pentachlorobutadienyl)glutathione (Oesch and Wolf, 1989), and *N*-acetyl-*S*-(1,2-dichlorovinyl)-L-cysteine is excreted as a urinary metabolite of trichloroethene after occupational exposure (Birner et al., 1993). Human tissue samples also express glutathione *S*-transferases 5 - 5 and this may have the capacity to bioactivate dichloromethane to the toxic glutathione *S*-conjugate (Meyer et al., 1991). Human α class glutathione *S*-transferases catalyze the formation of an electrophilic sulphur-mustard from 1,2-dibromoethane indicating that this bioactivation reaction may also occur with this class of compounds (Cmarik et al., 1990).

REFERENCES

Ahmed, A. E., and Anders, M. W., 1976, Metabolism of dihalomethanes to formaldehyde and inorganic halide. I. *In vitro* studies. *Drug Metab. Dispos.* 4:357-361.

Ahmed, A. E., and Anders, M. W., 1978, Metabolism of dihalomethanes to formaldehyde and inorganic halide. II. Studies on the mechanism of the reaction. *Biochem. Pharmacol.* 27:2021-2025.

Anders, M. W., Lash, L. H., Dekant, W., Elfarra, A. A., and Dohn, D. R., 1988, Biosynthesis and biotransformation of glutathione *S*-conjugates to toxic metabolites. *CRC Crit. Rev. Toxicol.* 18:311-342.

Anders, M. W., 1991, Glutathione-dependent bioactivation of xenobiotics. *FASEB J.* 4:87-92.

Andersen, M. E., Clewell, H. J. I., Gargas, M. L., MacNaughton, M. G., Reitz, R. H., Nolan, R. J., and McKenna, M. J., 1991, Physiologically based pharmacokinetic modeling with dichloromethane, its metabolite, carbon monoxide, and blood carboxyhemoglobin in rats and humans. *Toxicol. Appl. Pharmacol.* 108:14-27.

Birner, G., Vamvakas, S., Dekant, W., and Henschler, D., 1993, Nephrotoxic and genotoxic *N*-acetyl-*S*-dichlorovinyl-L-cysteine is a urinary metabolite after occupational 1,1,2-trichloroethene exposure in humans: Implications for the risk of trichloroethene exposure. *Environ. Health Perspect.* 99:281-284.

Blumbach, S., Hashmi, M., Anders, M. W., and Dekant, W., 1993, Bioactivation of dichloromethane: The role of toxic glutathione conjugates in mutagenicity. *ISSX Proceedings* 4:150.

Boyland, E., and Chasseaud, L. F., 1969, Role of glutathione and glutathione *S*-transferases in mercapturic acid biosynthesis. *Adv. Enzymol.* 32:173.

Cerutti, P. A., 1985, Prooxidant states and tumor promotion. *Science* 227:375-381.

Cmarik, J. L., Inskeep, P. B., Meredith, M. J., Meyer, D. J., Ketterer, B., and Guengerich, F. P., 1990, Selectivity of rat and human glutathione *S*-transferases in activation of ethylene dibromide by glutathione conjugation and DNA binding and induction of unscheduled DNA synthesis in human hepatocytes. *Cancer Res.* 50:2747-2752.

Cmarik, J. L., Humphreys, W. G., Bruner, K. L., Lloyd, R. S., Tibbetts, C., and Guengerich, F. P., 1992, Mutation spectrum and sequence alkylation selectivity resulting from modification of bacteriophage M13mp18 DNA with *S*-(2-chloroethyl)glutathione. *J. Biol. Chem.* 267:6672-6679.

Commandeur, J. N. M., Oostendorp, R. A. J., Schoofs, P. R., Xu, B., and Vermeulen, N. P. E., 1987, Nephrotoxicity and hepatotoxicity of 1,1-dichloro-2,2-difluoroethylene in the rat. *Biochem. Pharmacol.* 36:4229-4237.

Commandeur, J. N. M., Boogard, P. J., Mulder, G. J., and Vermeulen, N. P. E., 1991, Mutagenicity and cytotoxicity of two regioisomeric mercapturic acids and cysteine *S*-conjugates of trichloroethylene. *Arch. Toxicol.* 65:373-380.

Crowe, C. A., Yong, A. C., Calder, I. C., Ham, K. N., and Tange, J. D., 1979, The nephrotoxicity of p-aminophenol. I. The effect on microsomal cytochromes, glutathione and covalent binding in kidney and liver. *Chem.-Biol. Interact.* 27:235-243.

Dekant, W., Metzler, M., and Henschler, D., 1984, Novel metabolites of trichloroethylene through dechlorination reactions in rats, mice and humans. *Biochem. Pharmacol.* 33:2021-2027.

Dekant, W., Metzler, M., and Henschler, D., 1986a, Identification of *S*-1,2,2-trichlorovinyl-N-acetylcysteine as a urinary metabolite of tetrachloroethylene: Bioactivation through glutathione conjugation as a possible explanation of its nephrocarcinogenicity. *J. Biochem. Toxicol.* 1:57-72.

Dekant, W., Metzler, M., and Henschler, D., 1986b, Identification of *S*-1,2-dichlorovinyl-N-acetyl-cysteine as a urinary metabolite of trichloroethylene: A possible explanation for its nephrocarcinogenicity in male rats. *Biochem. Pharmacol.* 35:2455-2458.

Dekant, W., Vamvakas, S., Berthold, K., Schmidt, S., Wild, D., and Henschler, D., 1986c, Bacterial ß-lyase mediated cleavage and mutagenicity of cysteine conjugates derived from the nephrocarcinogenic alkenes trichloroethylene, tetrachloroethylene and hexachlorobutadiene. *Chem.-Biol. Interactions* 60:31-45.

Dekant, W., Lash, L. H., and Anders, M. W., 1987a, Bioactivation mechanism of the cytotoxic and nephrotoxic *S*-conjugate *S*-(2-chloro-1,1,2-trifluoroethyl)-L-cysteine. *Proc. Natl. Acad. Sci. USA* 84:7443-7447.

Dekant, W., Martens, G., Vamvakas, S., Metzler, M., and Henschler, D., 1987b, Bioactivation of tetrachloroethylene: Role of glutathione *S*-transferase-catalyzed conjugation versus cytochrome P-450-dependent phospholipid alkylation. *Drug Metab. Dispos.* 15:702-709.

Dekant, W., Schrenk, D., Vamvakas, S., and Henschler, D., 1988a, Metabolism of hexachloro-1,3-butadiene in mice: *in vivo* and *in vitro* evidence for activation by glutathione conjugation. *Xenobiotica* 18:803-816.

Dekant, W., Vamvakas, S., Henschler, D., and Anders, M. W., 1988b, Enzymatic conjugation of hexachloro-1,3-butadiene with glutathione: Formation of 1-(glutathion-*S*-yl)-1,2,3,4,4-pentachlorobuta-1,3-diene and 1,4-bis(glutathion-*S*-yl)-1,2,3,4- tetrachlorobuta-1,3-diene. *Drug. Metab. Dispos.* 16:701-706.

Dekant, W., Berthold, K., Vamvakas, S., Henschler, D., and Anders, M. W., 1988c, Thioacylating intermediates as metabolites of *S*-(1,2-dichlorovinyl)-L-cysteine and *S*-(1,2,2-trichlorovinyl)-L-cysteine formed by cysteine conjugate ß-lyase. *Chem. Res. Toxicol.* 1:175-178.

Dekant, W., Vamvakas, S., and Anders, M. W., 1989, Bioactivation of nephrotoxic haloalkenes by glutathione conjugation: Formation of toxic and mutagenic intermediates by cysteine conjugate ß-lyase. *Drug Metab. Rev.* 20:43-83.

Dekant, W., Vamvakas, S., and Anders, M. W., 1990a, Biosynthesis, bioactivation, and mutagenicity of *S*-conjugates. *Tox. Lett.* 53:53-58.

Dekant, W., Vamvakas, S., Koob, M., Köchling, A., Kanhai, W., Müller, D., and Henschler, D., 1990b, A mechanism of haloalkene-induced renal carcinogenesis. *Environ. Health Perspect.* 88:107-110.

Dekant, W., Koob, M., and Henschler, D., 1990c, Metabolism of trichloroethene: *in vivo* and *in vitro* evidence for activation by glutathione conjugation. *Chem.-Biol. Interact.* 73:89-101.

Dekant, W., Urban, G., Görsman, C., and Anders, M. W., 1991, Thioketene formation from α-haloalkenyl 2-nitrophenyl disulfides: Models for biological reactive intermediates of cytotoxic *S*-conjugates. *J. Amer. Chem. Soc.* 113:5120-5122.

Dekant, W., and Vamvakas, S., 1992, Mechanisms of xenobiotic-induced renal carcinogenicity. *Adv. Pharmacol.* 23:297-337.

Dohn, D. R., and Anders, M. W., 1982, The enzymatic reaction of chlorotrifluoroethylene with glutathione. *Biochem. Biophys. Res. Commun.* 109:1339-1345.

Dohn, D. R., and Casida, J. E., 1987, Thiiranium ion intermediates in the formation and reactions of *S*-(2-haloethyl)-L-cysteines. *Bioorganic Chemistry* 15:115-124.

Eckert, K.-G., Eyer, P., Sonnenbichler, J., and Zetl, I., 1990, Activation and detoxication of aminophenols. II. Synthesis and structural elucidation of various thiol addition products of 1,4-benzoquinoneimine and *N*-acetyl-1,4-benzoquinoneimine. *Xenobiotica* 20:333-350.

Elliot, W. C., Lynn, R. K., Hougton, D. C., Kennish, J. M., and Bennett, W. M., 1982, Nephrotoxicity of the flame retardant tris(2,3-dibromopropyl)phosphate, and its metabolites. *Toxicol. Appl. Pharmacol.* 63:179-182.

Foureman, G. L., and Reed, D. J., 1987, Formation of *S*-(2-(N^7-guanyl)ethyl) adducts by the postulated *S*-(2-chloroethyl)cysteinyl and *S*-(2-chloroethyl)glutathionyl conjugates of 1,2-dichloroethane. *Biochemistry* 26:2028-2033.

Fowler, L. M., Moore, R. B., Foster, J. R., and Lock, E. A., 1991, Nephrotoxicity of 4-aminophenol glutathione conjugate. *Human Exper. Toxicol.* 10:451-459.

Gartland, K. P. R., Bonner, F. W., and Nicholson, J. K., 1989a, Investigations into the biochemical effects of region-specific nephrotoxins. *Mol. Pharmacol.* 35:242-250.

Gartland, K. P. R., Bonner, F. W., Timbrell, J. A., and Nicholson, J. K., 1989b, Biochemical characterisation of *para*-aminophenol-induced nephrotoxic lesions in the F334 rat. *Arch. Toxicol.* 63:97-106.

Gartland, K. P. R., Eason, C. T., Bonner, F. W., and Nicholson, J. K., 1990, Effects of biliary cannulation and buthionine sulphoximine pretreatment on the nephrotoxicity of para-aminophenol in the Fisher 334 rat. *Arch. Toxicol.* 64:14-25.

Graves, R. J., Callender, R. D., and Green, T., 1994a, The role of formaldehyde and *S*-chloromethylglutathione in the bacterial mutagenicity of methylene chloride. *Mutat. Res.* 320:235-243.

Graves, R. J., Coutts, C., Eyton-Jones, H., and Green, T., 1994b, Relationship between hepatic DNA damage and methylene chloride-induced hepatocarcinogenicity in B6C3F1 mice. *Carcinogenesis* 15:991-996.

Green, T., 1983, The metabolic activation of dichloromethane and chlorofluoromethane in a bacterial mutation assay using Salmonella typhimurium. *Mutat. Res.* 118:277-288.

Green, T., and Odum, J., 1985, Structure/activity studies of the nephrotoxic and mutagenic action of cysteine conjugates of chloro- and fluoroalkenes. *Chem.-Biol. Interact.* 54:15-31.

Green, T., Provan, W. M., Collinge, D. C., and Guest, A. E., 1988a, Macromolecular interactions of inhaled methylene chloride in rats and mice. *Toxicol. Appl. Pharmacol.* 93:1-10.

Green, T., Provan, W. M., Dugard, P. H., and Cook, S. K., 1988b, Methylene chloride, dichloromethan: Human risk assessment using experimental animal data. *ECETOC Technical Report* 32:1-62.

Guengerich, F. P., Crawford, W. M. J., Domoradzki, J. Y., Macdonald, T. L., and Watanabe, P. G., 1980, *In vitro* activation of 1,2-dichloroethane by microsomal and cytosolic enzymes. *Toxicol. Appl. Pharmacol.* 55:303-317.

Heck, H. d., and Casanova-Schmitz, M., 1983, Biochemical toxicology of formaldehyde.

Hill, B. A., Monks, T. J., and Lau, S. S., 1992, The effects of 2,3,5-(triglutathion-*S*-yl)hydroquinone on renal mitochondrial respiratory function *in vivo* and *in vitro*: Possible role in cytotoxicity. *Toxicol. Appl. Pharmacol.* 117:165-171.

Hill, B. A., Kleiner, H. E., Ryan, E. A., Dulik, D. M., Monks, T. J., and Lau, S. S., 1993, Identification of multi-*S*-substituted conjugates of hydroquinones by HPLC-coulometric electrode array analysis and mass spectroscopy. *Chem. Res. Toxicol.* 6:459-469.

Holme, J. A., Soderlund, E. J., Brunborg, G., Lag, M., Nelsdon, S. D., and Dybing, E., 1991, DNA damage and cell death induced by 1,2-dibromo-3-chloropropane (DBCP) and structural analogs in monolayer culture of rat hepatocytes: 3-Aminobenzamide inhibits the toxicity of DBCP. *Cell Biol. Toxicol.* 7:413-432.

Humphreys, W. G., Kim, D. H., Cmarik, J. L., Shimada, T., and Guengerich, F. P., 1990, Comparison of the DNA-alkylating properties and mutagenic responses of a series of *S*-(2-haloethyl)-substituted cysteine and glutathione derivatives. *Biochemistry* 29:10342-10350.

Humphreys, W. G., Kim, D. H., and Guengerich, F. P., 1991, Isolation and characterization of N7-guanyl adducts derived from 1,2-dibromo-3-chloropropane. *Chem. Res. Toxicol.* 4:445-453.

IARC, 1984, *Models, mechanisms and etiology of tumour promotion. Proceedings of a symposium organized by the Hungarian Cancer Society and the IARC Budapest, 16-18 May 1983.* IARC Sci. Publ., Lyon.

Inskeep, P. B., and Guengerich, F. P., 1984, Glutathione-mediated binding of dibromoalkanes to DNA: Specificity of rat glutathione-*S*-transferases and dibromoalkane structure. *Carcinogenesis* 5:805-808.

Ishmael, J., Pratt, I., and Lock, E. A., 1982, Necrosis of the pars recta, S3 segment) of the rat kidney produced by hexachloro-1:3-butadiene. *J. Pathol.* 138:99-113.

Jean, P. A., and Reed, D. J., 1989, *In vitro* dipeptide, nucleoside, and glutathione alkylation by *S*-(2-chloroethyl)glutathione and *S*-(2-chloroethyl)-L-cysteine. *Chem. Res. Toxicol.* 2:455-460.

Jones, A. R., Fakhouri, G., and Gadiel, P., 1979, The metabolism of the soil fumigant 1,2-dibromo-3-chloropropane. *Experienta* 35:1432-1434.

Josephy, P. D., Eling, T. E., and Mason, R. P., 1983, Oxidation of p-aminophenol catalyzed by horseradish peroxidase and prostaglandin synthase. *Mol. Pharmacol.* 23:461-466.

Kanhai, W., Dekant, W., and Henschler, D., 1989, Metabolism of the nephrotoxin dichloroacetylene by glutathione conjugation. *Chem. Res. Toxicol.* 2:51-56.

Kanhai, W., Koob, M., Dekant, W., and Henschler, D., 1991, Metabolism of ^{14}C-dichloroetyne in rats. *Xenobiotica* 21:905-916.

Kermani, H. R. S., Sloane, R. A., Moorman, M. P., Yang, R. S. H., Ray, C., and Reitz, R. H., 1990, Enzyme kinetics of methylene chloride: MFO and GST activities in female B6C3F1 mice liver and lung in relation to aging and chronic dosing. *Toxicologist* 10:186-.

Kim, D.-H., and Guengerich, F. P., 1989, Excretion of the mercapturic acid *S*-[2-(N7-guanyl)ethyl]-N-acetylcysteine in urine following administration of ethylene dibromide to rats. *Cancer Res.* 499:5843-5847.

Kim, D.-H., and Guengerich, F. P., 1990, Formation of the DNA adduct *S*-[2-(N7-guanyl)ethyl]glutathione from ethylene dibromide: Effects of modulation of glutathione and glutathione *S*-transferase levels and lack of a role for sulfation. *Carcinogenesis* 11:419-424.

Kleiner, H. E., Hill, B. A., Monks, T. J., and Lau, S. S., 1992, *In vivo* and *in vitro* formation of several *S*-conjugates of hydroquinone. *Toxicologist* 12:1350-1350.

Klos, C., Koob, M., Kramer, C., and Dekant, W., 1992, p-Aminophenol nephrotoxicity: Biosynthesis of toxic glutathione conjugates. *Toxicol. Appl. Pharmacol.* 115:98-106.

Kociba, R. J., Keyes, D. G., Jersey, G. C., Ballard, J. J., Dittenber, D. A., Quast, J. F., Wade, L. E., Humiston, C. G., and Schwetz, B. A., 1977, Results of a two year chronic toxicity study with hexachlorobutadiene in rats. *Am. Ind. Hyg. Assoc. J.* 38:589-602.

Koob, M., and Dekant, W., 1990, Metabolism of hexafluoroproene - Evidence for bioactivation by glutathione conjugate formation in the kidney. *Drug. Metab. Dispos.* 18:911-916.

Koob, M., and Dekant, W., 1991, Bioactivation of xenobiotics by formation of toxic glutathione conjugates. *Chem.-Biol. Interact.* 77:107-136.

Lag, M., Omichinski, J. G., Soderlund, E. J., Brunborg, G., Holme, J. A., Dahl, J. E., Nelson, S. D., and Dybing, E., 1989a, Role of P-450 activity and glutathione levels in 1,2-dibromo-3-chloropropane tissue distribution, renal necrosis and *in vivo* DNA damage. *Toxicology* 56:273-288.

Lag, M., Soderlund, E. J., Brunborg, G., Dahl, J. E., Holme, J. A., Omichinski, J. G., Nelson, S. D., and Dybing, E., 1989b, Species differences in testicular necrosis and DNA damage, distribution and metabolism of 1,2-dibromo-3-chloropropane (DBCP). *Toxicology* 58:133-144.

Lau, S. S., Monks, T. J., and Gillette, J. R., 1984, Identification of 2-bromohydroquinone as a metabolite of bromobenzene and o-bromophenol: Implications for bromobenzene-induced nephrotoxicity. *J. Pharmacol. Exp. Ther.* 230:360-366.

Lau, S. S., McMenamin, M. G., and Monks, T. J., 1988a, Differential uptake of isomeric 2-bromohydroquinone-glutathione conjugates into kidney slices. *Biochem. Biophys. Res. Commun.* 152:223-230.

Lau, S. S., Hill, B. A., Highet, R. J., and Monks, T. J., 1988b, Sequential oxidation and glutathione addition to 1,4-benzoquinone: Correlation of toxicity with increased glutathione substitution. *Mol. Pharmacol.* 34:829-836.

Lau, S. S., Jones, T. W., Highet, R. J., Hill, B., and Monks, T. J., 1990, Differences in the localization and extent of the renal proximal tubular necrosis caused by mercapturic acid and glutathione conjugates of 1,4-naphthoquinone and menadione. *Toxicol. Appl. Pharmacol.* 104:334-350.

Lau, S. S., and Monks, T. J., 1990, The *in vivo* disposition of 2-bromo-[14C]hydroquinone and the effect of γ-glutamyl transpeptidase inhibition. *Toxicol. Appl. Pharmacol.* 103:121-132.

Lock, E. A., 1988, Studies on the mechanism of nephrotoxicity and nephrocarcinogenicity of halogenated alkenes. *CRC Crit. Rev. Toxicol.* 19, 23-42.

Lock, E. A., 1989, Mechanism of nephrotoxic action due to organohalogenated compounds. *Tox. Lett.* 46:93-106.

Lynn, R. K., Garvie-Gould, C., Wong, K., and Kennish, J. M., 1982, Metabolism, distribution, and excretion of the flame retardant tris(2,3-dibromopropyl)phosphate (Tris-BP) in the rat: Identification of mutagenic and nephrotoxic metabolites. *Toxicol. Appl. Pharmacol.* 63:105-119.

Mahmood, N. A., Overstreet, D., and Burka, L. T., 1991, Comparative disposition and metabolism of 1,2,3-trichloropropane in rats and mice. *Drug Metab. Dispos.* 19:411-418.

Meadows, S. D., Gandolfi, A. J., Nagle, R. B., and Shively, J. W., 1988, Enhancement of DMN-induced kidney tumors by 1,2-dichlorovinylcysteine in Swiss-Webster mice. *Drug Chem. Toxicol.* 11:307-318.

Meister, A., 1988, Glutathione metabolism and its selective modification. *J. Biol. Chem.* 263:17205-17208.

Meister, A., 1992, Commentary: On the antioxidant effects of ascorbic acid and glutathione. *Biochem. Pharmacol.* 44:1905-1915.

Mertens, J. J., Temmink, J. H., Bladeren, P. J., Jones, T. W., Lo, H. H., Lau, S. S., and Monks, T. J., 1991, Inhibition of γ-glutamyl transpeptidase potentiates the nephrotoxicity of glutathione-conjugated chlorohydroquinones. *Toxicol. Appl. Pharmacol.* 110:45-60.

Meyer, D. J., Coles, B., Pemple, S. E., Gilmore, K. S., Fraser, G. M., and Ketterer, B., 1991, Theta, a new class of glutathione transferases purified from rat and man. *Biochem. J.* 274:409-414.

Monks, T. J., Lau, S. S., Highet, R. J., and Gillette, J. R., 1985, Glutathione conjugates of 2-bromohydroquinone are nephrotoxic. *Drug Metab. Dispos.* 13:553-559.

Monks, T. J., and Lau, S. S., 1987, Commentary: Renal transport processes and glutathione conjugate-mediated nephrotoxicity. *Drug Metab. Dispos.* 15:437-441.

Monks, T. J., Highet, R. J., and Lau, S. S., 1988, 2-Bromo-(diglutathion-*S*-yl)hydroquinone nephrotoxicity: Physiological, biochemical, and electrochemical determinants. *Mol. Pharmacol.* 34:492-500.

Monks, T. J., and Lau, S. S., 1989, Sulphur conjugate-mediated toxicity. *Rev. Biochem. Toxicol.* 10:41-90.

Monks, T. J., Anders, M. W., Dekant, W., Stevens, J. L., Lau, S. S., and Bladeren, P. J., 1990, Glutathione conjugate mediated toxicities. *Toxicol. Appl. Pharmacol.* 106:1-19.

Monks, T. J., and Lau, S. S., 1990, Glutathione, γ-glutamyl transpeptidase, and the mercapturic acid pathway as modulators of 2-bromohydroquinone oxidation. *Toxicol. Appl. Pharmacol.* 103:557-563.

Nash, J. A., King, L. J., Lock, E. A., and Green, T., 1984, The metabolism and disposition of hexachloro-1:3-butadiene in the rat and its relevance to nephrotoxicity. *Toxicol. Appl. Pharmacol.* 73:124-137.

National Cancer Institute NCI, 1986a, Carcinogenesis bioassay of tetrachloroethylene. *National Toxicology Program* Technical Report 232.

National Cancer Institute NCI, 1986b, Carcinogenesis bioassay of trichloroethylene. *National Toxicology Program Technical Report* 311.

Newton, J. F., Kuo, C.-H., Gemborys, M. W., Mudge, G. H., and Hook, J. B., 1982, Nephrotoxicity of p-aminophenol, a metabolite of acetaminophen, in the Fischer 344 rat. *Toxicol. Appl. Pharmacol.* 65:336-344.

Odum, J., and Green, T., 1984, The metabolism and nephrotoxicity of tetrafluoroethylene in the rat. *Toxicol. Appl. Pharmacol.* 76:306-318.

Oesch, F., and Wolf, C. R., 1989, Properties of the microsomal and cytosolic glutathione transferases involved in hexachloro-1:3-butadiene conjugation. *Biochem. Pharmacol.* 38:353-359.

Osterberg, R. E., Bierbower, G. W., and Hehir, R. M., 1977, Renal and testicular damage following dermal application of the flame retardant tris(2,3-dibromopropyl)phosphate. *J. Toxicol. Environ. Health* 3:979-987.

Ozawa, N., and Guengerich, F. P., 1983, Evidence for formation of an *S*-[2-(N7-guanyl)ethyl]glutathione adduct in glutathione mediated binding of the carcinogen 1,2-dibromoethane to DNA. *Proc. Natl. Acad. Sci. USA* 80:5266-5270.

Ozawa, H., and Tsukioka, T., 1990, Gas chromatographic separation and determination of chloroacetic acids in water by a difluoroanilide derivatisation method. *Analyst* 115:1343-1347.

Pearson, P. G., Soderlund, E. J., Dybing, E., and Nelson, S. D., 1990, Metabolic activation of 1,2-dibromo-3-chloropropane: Evidence for the formation of reactive episulfonium ion intermediates. *Biochemistry* 29:4971-4977.

Pearson, P. G., Omichinski, J. G., Holme, J. A., McClanahan, R. H., Brunborg, G., Soderlund, E. J., Dybing, E., and Nelson, S. D., 1993, Metabolic activation of tris(2,3-dibromopropyl)phosphate to reactive intermediates. II. Covalent binding, reactive metabolite formation, and differential metabolite-specific DNA damage *in vivo*. *Toxicol. Appl. Pharmacol.* 118:186-195.

Peterson, L. A., Harris, T. M., and Guengerich, F. P., 1988, Evidence for an episulfonium ion intermediate in the formation of *S*-[2-(N7-guanyl)ethyl]glutathione in DNA. *J. Am. Chem. Soc.* 110:3284-3291.

Potter, C. L., Gandolfi, A. J., Nagle, R., and Clayton, J. W., 1981, Effects of inhaled chlorotrifluoroethylene and hexafluoropropene on the rat kidney. *Toxicol. Appl. Pharmacol.* 59:431-440.

Redegeld, F. A. M., Hofman, G. A., Loo, P. G. F., Koster, A. S., and Noordhoek, J., 1991, Nephrotoxicity of glutathione conjugate of menadione (2-methyl-1,4-naphthoquinone) in the isolated perfused rat kidney. Role of metabolism by γ-glutamyltranspeptidase and probenecid-sensitive transport. *J. Pharmacol. Exptl. Therap.* 256:665-669.

Reichert, D., Ewald, D., and Henschler, D., 1975, Generation and inhalation toxicity of dichloroacetylene. *Fd Cosmet. Toxicol.* 13:511-515.

Reichert, D., and Schuetz, S., 1986, Mercapturic acid formation is an activation and intermediary step in the metabolism of hexachlorobutadiene. *Biochem. Pharmacol.* 35:1271-1275.

Reitz, R. H., Mendrala, A. L., and Gurengerich, F. P., 1989, *In vitro* metabolism of methylene chloride in human and animal tissues: Use in physiologically based pharmacokinetic models. *Toxicol. Appl. Pharmacol.* 97, 230-246.

Simula, T. P., Glancey, M. J., Söderlund, E. J., Dybing, E., and Wolf, C. R., 1993, Increased mutagenicity of 1,2-dibromo-3-chloropropane and tris(2,3-dibromo-propyl)phosphate in *Salmonella* TA100 expressing human glutathione *S*-transferases. *Carcinogenesis* 14:2303-2307.

Söderlund, E. J., Gordon, W. P., Nelson, S. D., Omichinski, J. G., and Dybing, E., 1984, Metabolism *in vitro* of tris(2,3-dibromopropyl)-phosphate: oxidative debromination and bis(2,3-dibromopropyl)phosphate formation as correlates of mutagenicity and covalent protein binding. *Biochem. Pharmacol.* 33:4017-4023.

Spencer, H. C., Rowe, V. K., Adams, E. M., McCollister, D. D., and Irish, D. D., 1951, Vapor toxicity of ethylene dichloride determined by experiments on laboratory animals. *Arch. Ind. Hyg. Occup. Med.* 4:482-493.

Torkelson, T. R., Sadek, S. E., Rowe, V. K., Kodama, I. K., Anderson, H. H., Loquvam, G. S., and Hine, C. H., 1961, Toxicological investigation of 1,2-dibromo-3-chloropropane. *Toxicol. Appl. Pharmacol.* 3:545-553.

Vamvakas, S., Berthold, K., Dekant, W., and Henschler, D., 1988a, Bacterial cysteine conjugate ß-lyase and the metabolism of cysteine *S*-conjugates: Structural requirements for the cleavage of *S*-conjugates and the formation of reactive intermediates. *Chem.-Biol. Interact.* 65:59-71.

Vamvakas, S., Elfarra, A. A., Dekant, W., Henschler, D., and Anders, M. W., 1988b, Mutagenicity of amino acid and glutathione *S*-conjugates in the Ames test. *Mutat. Res.* 206:83-90.

Vamvakas, S., Dekant, W., and Henschler, D., 1989a, Assessment of unscheduled DNA synthesis in a cultured line of renal epithelial cells exposed to cysteine *S*-conjugates of haloalkenes and haloalkanes. *Mutat. Res.* 222:329-335.

Vamvakas, S., Kremling, E., and W., D., 1989b, Metabolic activation of the nephrotoxic haloalkene 1,1,2-trichloro-3,3,3-trifluoro-1-propene by glutathione conjugation. *Biochem. Pharmacol.* 38:2297-2304.

Vamvakas, S., Herkenhoff, M., Dekant, W., and Henschler, D., 1989c, Mutagenicity of tetrachloroethylene in the Ames-test: Metabolic activation by conjugation with glutathione. *J. Biochem. Toxicol.* 4:21-27.

Vamvakas, S., Dekant, W., and Henschler, D., 1989d, Genotoxicity of haloalkene and haloalkane glutathione *S*-conjugates in porcine kidney cells. *Toxicol. in vitro* 3:151-156.

Vamvakas, S., Köchling, A., Berthold, K., and Dekant, W., 1989e, Cytotoxicity of cysteine *S*-conjugates: structure-activity relationships. *Chem.-Biol. Interact.* 71:79-90.

Vamvakas, S., Sharma, V. K., Shen, S.-S., and Anders, M. W., 1990, Perturbations of intracellular calcium distribution in kidney cells by nephrotoxic haloalkenyl cysteine *S*-conjugates. *Molec. Pharmacol.* 38:455-461.

Vamvakas, S., Bittner, D., Dekant, W., and Anders, M. W., 1992, Events that precede and that follow *S*-(1,2-dichlorovinyl)-L-cysteine-induced release of mitochondrial Ca^{2+} and their association with cytotoxicity to renal cells. *Biochem. Pharmacol.* 44:1131-1138.

Vamvakas, S., and Köster, U., 1993, The nephrotoxin dichlorovinylcysteine induces expression of the protooncogenes *c-fos* and *c-myc* in LLC-PK_1 cells: A comparative investigation with growth factors and 12-*O*-tetradecanoylphorbolacetate. *Cell Biol. Toxicol.* 9:1-13.

Van Bladeren, P. J., Gen, A., Breimer, D. D., and Mohn, G. R., 1979, Stereoselective activation of vicinal dihalogen compounds to mutagens by glutathione conjugation. *Biochem. Pharmacol.* 28:2521-2524.

Van Bladeren, P. J., Breimer, D. D., Rotteveel-Smijs, G. M. T., Jong, R. A. W., Buijs, W., Gen, A., and Mohn, G. R., 1980, The role of glutathione conjugation in the mutagenicity of 1,2-dibromoethane. *Biochem. Pharmacol.* 29:2975-2982.

Wallin, A., Gerdes, R. G., Morgenstern, R., Jones, T. W., and Ormstad, K., 1988, Features of microsomal and cytosolic glutathione conjugation of hexachlorobutadiene in rat liver. *Chem.-Biol. Interact.* 68:1-11.

Weber, G. L., and Sipes, I. G., 1990, Covalent interactions of 1,2,3-trichloropropane with hepatic macromolecules: Studies in the male F-344 rat. *Toxicol. Appl. Pharmacol.* 104:395-402.

Weber, G. L., and Sipes, I. G., 1992, *In vitro* metabolism and bioactivation of 1,2,3-trichloropropane. *Toxicol. Appl. Pharmacol.* 113:152-158.

Weisburger, E. K., 1977, Carcinogenicity studies on halogenated hydrocarbons. *Environ. Health Perspect.* 21:7-16.

Winter, S. M., Weber, G. L., Gooley, P. R., Mackenzie, N. E., and Sipes, I. G., 1992, Identification and comparison of the urinary metabolites of [1,2,3-13C3] acrylic acid and [1,2,3-$^{13}C_3$]propionic acid in the rat by homonuclear ^{13}C nuclear magnetic resonance spectroscopy. *Drug Metab. Dispos.* 20:665-672.

Wolf, C. R., Berry, P. N., Nash, J. A., Green, T., and Lock, E. A., 1984, Role of microsomal and cytosolic glutathione *S*-transferases in the conjugation of hexachloro-1:3-butadiene and its possible relevance to toxicity. *J. Pharmacol. Exp. Ther.* 228:202-208.

Zheng, J., and Hanzlik, R. P., 1992, Dihydroxylated mercapturic acid metabolites of bromobenzene. *Chem. Res. Toxicol.* 5:561-567.

ACYL GLUCURONIDES

Covalent Binding and Its Potential Relevance

Hildegard Spahn-Langguth,* Monika Dahms, and Andreas Hermening

Department of Pharmacology
Johann Wolfgang Goethe-University Frankfurt/M.
Biocenter Niederursel
Marie-Curie-Straße 9, Geb. N260, D-60439 Frankfurt/M., Germany

INTRODUCTION

In general, phase-II metabolites of drugs, such as sulfates, O- or N-glucuronides, were anticipated to be rapidly excreted into urine and bile without exhibiting significant biological/pharmacological activity. However, a few years ago various conjugates were proven to have considerable affinity to pharmacological receptors or to contribute to toxic effects. Today, it is largely accepted that they may be active and/or reactive [Kroemer and Klotz, 1992; Spahn-Langguth and Benet, 1992a, b].

With respect to phase-II metabolism of carboxylic acid xenobiotics, conjugation with D-glucuronic acid represents the major route for elimination and detoxification. As opposed to the unreactive role previously assigned to acyl glucuronides (ester glucuronides), they are, in fact, potentially reactive intermediates. Acyl glucuronides were found to be electrophiles, reacting with sulfhydryl and hydroxyl groups [Stogniew and Fenselau, 1982]. This reactivity is unique for ester glucuronides and - due to the involved mechanisms - neither observed for other O- nor for N-, S- or C-glucuronides.

The electrophilic acyl glucuronide conjugates may undergo typical reactions such as hydrolysis, formation of β-glucuronidase-resistant isomers (acyl migration) and, in addition, covalent binding to endogenous compounds, which appears to result from the unstable nature of acyl glucuronides [Faed 1984, Spahn-Langguth and Benet, 1992b]. The phenomenon of acyl migration as well as covalent binding is now recognized as a common occurrence with ester glucuronides, although their extent varies considerably between compounds.

* Address for correspondence and present address: H. Spahn-Langguth, Ph. D. , Assoc. Professor, Pharmaceutics, School of Pharmacy, Martin-Luther-University Halle-Wittenberg, Weinbergweg 15, 06120 Halle/Saale.

THE BIOCHEMICAL PROCESS OF GLUCURONIDE FORMATION, ITS LOCATION AND THE FATE OF THE PRODUCTS

When a glucuronide is formed, the endogenous glucuronic acid enters the uridinediphospho glucuronosyltransferase-catalyzed (UDPGT-catalyzed) reaction as uridine diphospho glucuronic acid (UDPGA). With respect to glucuronic acid and its activated form stereochemical as well as nomenclature aspects have been summarized by Dudley [1985]. During the conjugation reaction an anomeric inversion at the C-1 atom of glucuronic acid takes place converting the α-D-glucuronic acid in UDPGA to β-D-glucuronic acid in the conjugate. No interconversion is possible between the two glucuronic acid anomers, neither in UDPGA nor in the enzymatically formed glucuronide.

Usually liver microsomes are used to study glucuronidation processes [Fournel-Gigleux et al., 1988; Spahn et al., 1989]. However, glucuronide formation may occur in other organs as well, such as the kidneys and the gut, although the extent of glucuronide formation is usually smaller in these and other organs [el Mouelhi et al., 1991; Vree et al., 1992]. Significant differences in the extent of glucuronide formation and its stereoselectivity were detected in different mammalian species [el Mouelhi et al., 1988].

The kinetic properties of the glucuronide conjugates differ from those of the respective aglycones. The acidic glucuronic acid conjugates are more water-soluble than their precursors at physiological pH. The distribution volume is usually much smaller than that of the more lipophilic aglycone. Hence, high concentrations may result in the systemic circulation, although the total fraction of the dose present in the organism as glucuronide may be fairly low. The total renal clearance (CL_{ren}) of the glucuronides may exceed the glomerular filtration rate (GFR) significantly, at least, when plasma protein binding is negligible. For several glucuronides with a CL_{ren} smaller than GFR significant plasma protein binding was detected, i.e., the unbound renal clearance calculated from these data is much higher. It may hence be assumed that renal excretion occurs preferentially via carrier-mediated tubular secretion. The carrier system responsible for this transport should be the organic anion transporter located in the basolateral membrane of cells in the proximal tubule of the kidney [Ott and Giacomini, 1993]. Another excretion pathway whose extent is species-dependent may be the hepatic route, i.e., via biliary secretion. While significant amounts of glucuronides may be detected in the bile of rats, biliary secretion appears to be of minor relevance in humans.

HYDROLYSIS, ACYL MIGRATION AND OTHER POSSIBILITIES OF ISOMERISM WITHIN THE GLUCURONIC ACID MOIETY

Unlike other glucuronides, ester glucuronides may be subject to acyl migration and hydrolysis. Both reactions are observed under physiological pH and temperature conditions and may occur in vitro and in vivo. Hydrolysis of an acyl glucuronide leads to regeneration of the aglycone. Rates of hydrolysis are extremely dependent upon pH and temperature with more rapid degradation of the enzymatically formed β-1-O-acyl glucuronide at higher (also at physiological) pH than at a more acidic level. Acyl migration is the migration of the acyl residue from the C-1 position to the C-2-, C-3- and C-4 positions with formation of regioisomers. Backformation of the C-3- from the C-4 isomer as well as of the C-2- from the C-3 isomer is possible. However, the originally formed C-1 isomer cannot result from its regioisomers via acyl migration, i.e., the high energy 1-O-acyl bond is not regenerated under in-vitro conditions (Fig. 1). The studies of Bradow et al. [1989] indicated that there is no evidence for rearrangements beyond nearest-neighbor hydroxyl groups.

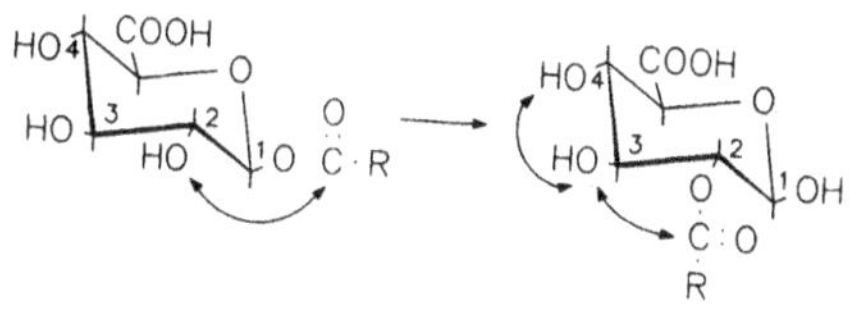

Figure 1. After formation of the C1-O-acyl glucuronide from a carboxylic acid, the acyl residue can "migrate" within the glucuronic acid molecule forming β-glucuronidase-resistant positional isomers at C2, C3, and C4. The rearrangement is reversible with the exception of the enzymatically formed C1-isomer, which cannot be regenerated via acyl migration.

While no interconversion between the two glucuronic acid anomers is possible for the enzymatically formed β-1-O acyl glucuronide, its regioisomers were found to occur as mixtures of the α- and β-anomers [Smith and Benet 1986]. Furthermore, isomerization within the glucuronic acid moiety may result in furanose structures, open-chain and lactone forms as described for bilirubin glucuronides in the early work of Blanckaert et al., [1978].

Acyl migration with formation of β-glucuronidase-resistant isomers is the predominating reaction in vitro, while in vivo, hydrolysis is more relevant.

Like hydrolysis rates, the rates of acyl migration are pH-dependent. Hence, with respect to both in-vitro enzyme kinetic studies and pharmacokinetic studies sample treatment is of crucial importance and may represent a prerequisite for the correct estimation of aglycone and glucuronide concentrations. To avoid post-sampling degradation samples are usually immediately cooled and pH-stabilized (pH 3-4) [Hasegawa et al., 1982]. Earlier studies not employing correct sample stabilization procedures yielded inaccurate measures (Scheme A) of the pharmacokinetics of the carboxylic acid drugs as well as their glucuronides, which differ considerably in their distribution behaviour.

When investigating different acyl glucuronides under similar conditions (150 mM pH 7.4 phosphate buffer; 37 °C) stabilities were found to vary considerably between compounds. The degradation half-lives of the acyl glucuronides formed from beclobric acid, the major and active metabolite of the chiral lipid-regulating agent beclobrate (withdrawn from the market in 1992), were between 20 and 30 h with no relevant difference between the two diastereomers, while those of various 2-arylpropionic acids were considerably shorter and exhibited significant stereoselectivity as summarized in Table 1.

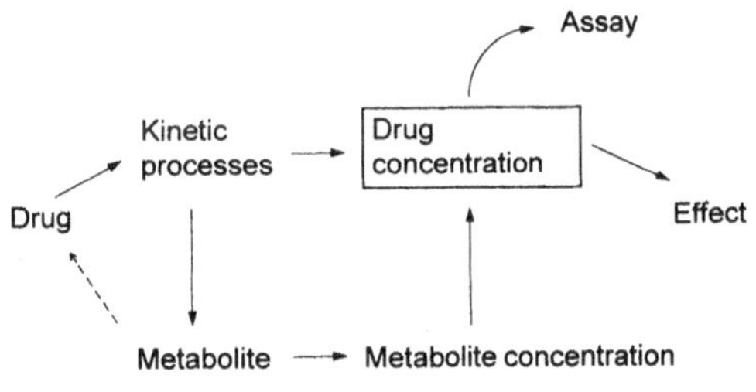

Scheme A. In this situation a drug metabolite, which is pharmacologically inactive, is formed in vivo and the metabolic step is reversible in vivo and in vitro (as observed for various acyl glucuronides), and samples are taken for the analysis of the drug with the aim of correlating kinetics and dynamics. When sample treatment does not inhibit the regeneration of drug from metabolite after sampling, then the measured drug concentrations may contain more or less error, dependent on the velocity of back-formation of drug from metabolite and the drug/metabolite concentration ratio. For acyl glucuronides sample stabilization is possible by immediate cooling and acidification following sampling.

Table 1. Apparent first-order degradation rate constants, k, for β-1-O-acyl glucuronides of several carboxylic acids (in 150 mM pH 7.4 phosphate buffer at 37 °C) and the respective half-lives

	k (h^{-1})	Half-life (h)
(+)Beclobric acid	0.027	25.7
(-)Beclobric acid	0.031	22.4
Furosemide	0.158	4.39
S(+)Carprofen	0.224	3.09
R(-)Carprofen	0.400	1.73
S(+)Fenoprofen	0.360	1.93
R(-)Fenoprofen	0.710	0.98
Zomepirac	1.540	0.45
Tolmetin	1.780	0.39

POTENTIAL CLINICAL IMPLICATIONS OF THE IN-VIVO "INSTABILITY" OF ESTER GLUCURONIDES

With respect to the concentration-time profiles of glucuronide conjugates (e.g., zomepirac, tolmetin, carprofen, naproxen and desmethylnaproxen, gemfibrozil and its phase-II metabolites as well as diflunisal) it becomes obvious that - even though the respective acyl glucuronides are present in plasma at concentrations of different orders of magnitude, the profiles always parallel those of the respective parent compounds. Hence, elimination of the aglycones is the rate-limiting step.

As mentioned before acyl glucuronides are unstable at physiological pH, undergoing alkaline and esterase-catalyzed hydrolysis. Because the acyl glucuronide may hydrolyze in vivo and then be remetabolized, the dispositional relationship may be described by reversible metabolism [as reviewed by Spahn-Langguth and Benet, 1992a, b], where the true clearance for both parent compound and the glucuronide can only be calculated from the AUC values of drug and metabolite, if drug and metabolite are dosed intravenously and both are measured.

The kinetics of most drugs, which are eliminated outside the kidney, are similar for patients with normal and those with impaired renal function, although the kinetics may differ for the respective metabolites, as soon as they are renally excreted. In contrast, when an acyl glucuronide occurs, decreased renal function will block the excretion of this metabolite, which can then be hydrolyzed back to the aglycone in vivo. Consequently, the apparent metabolic clearance is decreased in renal failure, and higher aglycone as well as glucuronide concentrations are maintained over a prolonged period of time. Gugler and coworkers [Gugler et al., 1979] characterized this phenomenon as "futile cycling," which is unique for acyl glucuronides. It appears to be a general phenomenon when ester glucuronides accumulate in plasma due to decreased renal clearance of the glucuronide metabolite as observed for renal failure, elderly patients as well as with respect to drug-drug interactions (e.g., with probenecid) [Hayball et al., 1991; Rowe and Meffin, 1984; Smith et al., 1985; Spahn et al., 1989; see review articles for further refernces on this topic].

FORMATION OF COVALENT ADDUCTS WITH ENDOGENOUS COMPOUNDS

A third reaction typically observed with acyl glucuronides is covalent binding to plasma proteins. A correlation between reversible and irreversible binding of acyl glucuronides to plasma proteins, with reversible binding acting as preliminary or intermediate step, has been discussed by Wells et al. [1987]. Measurements of reversible plasma protein binding of glucuronide conjugates, however, are rare. Stereoselective protein binding of the ether glucuronides of oxazepam has been reported with the glucuronide of S-oxazepam exhibiting a higher binding to HSA than that of R-oxazepam [Boudinot et al., 1985]. In the case of ester glucuronides, only four studies describe a reversible binding to plasma proteins [Wells et al., 1987; Hayball et al., 1992; Iwakawa et al., 1990; Rubin et al., 1972]. With respect to acyl glucuronides, most probably the lack of data results from major experimental problems associated with such binding studies, since the studies need to be carried out at physiological pH, where instabilities occur. As for carprofen [Iwakawa et al., 1990] rapid ultrafiltration was employed to study the binding of zomepirac and tolmetin to human serum albumin (HSA) [Ojingwa et al., 1994]. Interestingly, extensive binding to HSA was found for the aglycones, the β-1-O-acyl glucuronides as well as for the α/β-3-regioisomers of tolmetin and zomepirac (Table 2), where the glucuronides of both compounds bind to a smaller extent than their parent aglycones. Interestingly, the isomeric conjugates showed much stronger reversible binding than the β1-conjugates.

Presumably, reversible binding of the glucuronide accompanies or precedes irreversible binding. Covalent binding to proteins has been demonstrated for numerous acyl glucuronides. Its extent was found to be clearly time- and pH-dependent in in vitro binding studies with HSA as acceptor protein. The isomeric conjugates, but not the aglycone, were found to bind covalently to HSA as well [Smith et al., 1986].

As depicted in Figs. 2A and 2B (and as summarized by Benet and Spahn [1988]), covalent binding may occur via two different mechanisms: direct nucleophilic displacement is possible and leads to formation of an acylated protein and release of glucuronic acid; the second mechanism includes an imine formation. For this reaction acyl migration is a prerequisite. Subsequently, imine formation is possible between the free aldehyde of the open-chain glucuronic acid and a nucleophile, possibly lysine, in the albumin molecule with glucuronic acid being part of the adduct. Imine formation is reversible, but may be followed by an Amadori rearrangement of the imino sugar to the more stable 1-amino-2-keto product. The relevance of the second mechanism was confirmed by studies of different groups [Smith

Table 2. Estimated binding parameters for the interaction of tolmetin, zomepirac and their glucuronides with human serum albumin [according to Ojingwa et al., 1994]

Compound/conjugate	K_{d1} [μM]	K_{d2} [μM]	NS [μM^{-1}]
Tolmetin	0.6	29.4	—
TG	4.42	625.0	—
TG-α/β3	2.22	145.0	—
Zomepirac	0.86	27.8	101.0
ZG	22.2	417.0	—
ZG-α/β3	5.70	196.0	—

(TG, ZG = tolmetin glucuronide, zomepirac glucuronide, TG- and ZG-α/β3 = tolmetin- and zomepirac glucuronide C3-isomers (as α and β anomers)

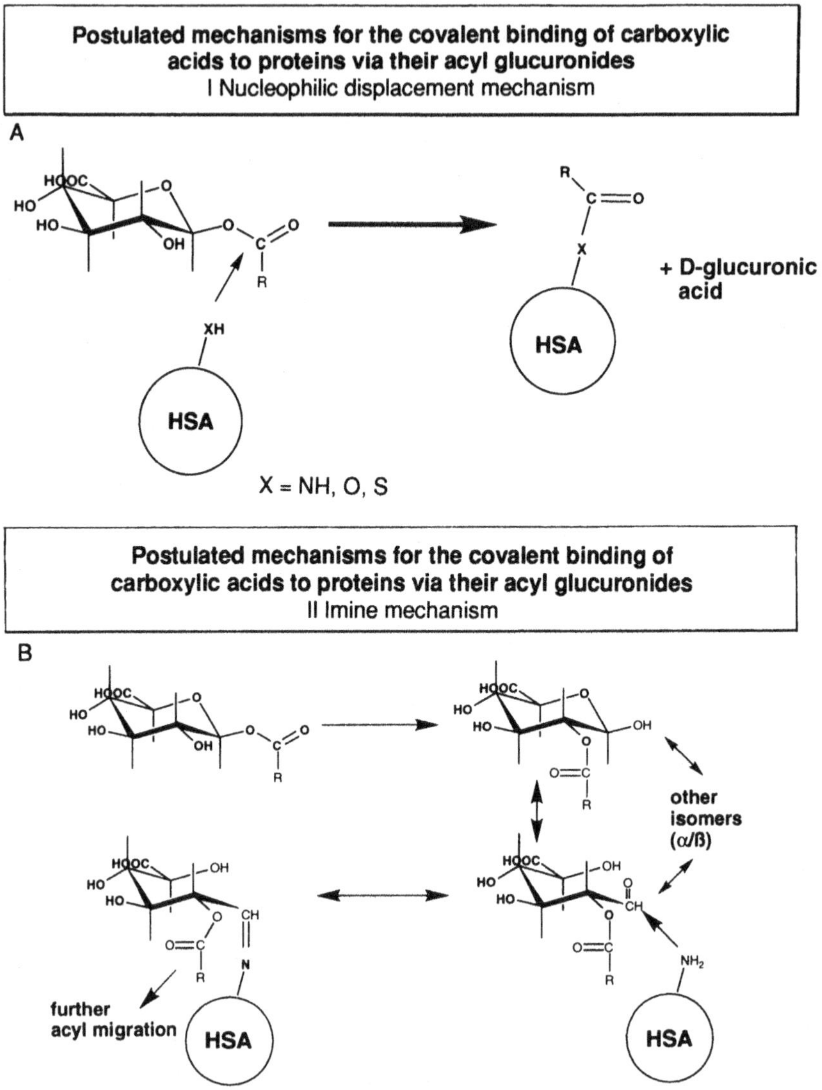

Figure 2. Two mechanisms have been proposed to describe the covalent binding of acyl glucuronides to proteins: I. The first is a displacement mechanism (Fig. 2A), where a nucleophilic group (NH_2, OH, SH) of the protein attacks the carbonyl group of the aglycone. This leads to a regeneration of glucuronic acid and formation of an amide, ester, or thioester bond between the protein and the drug. II. The second is a condensation with an NH_2 group (Fig. 2B), generating an imine (-CN-) which could undergo rearrangement to a more stable 1-amino-keto product [as summarized by Benet and Spahn, 1988 and Zia-Amirhosseini et al., 1994]. In this case the glucuronic acid is retained in the adduct. Both mechanisms may occur in vivo. Recent tandem mass spectrometric analyses of tolmetin protein adducts formed in vitro confirm that both the Schiff base and the nucleophilic displacement mechanism can occur simultaneously, but the relative importance of each route depends on the acyl glucuronide/protein concentration ratio, as summarized by Zia-Amirhosseini et al. [1994].

et al., 1990; Ding et al., 1993; Grubb et al., 1993]. In spite of the presence of glucuronic acid in other types of glucuronides (N-, S-, C-, or ether glucuronides), they do not undergo acyl migration, and hence no ring-opening and aldehyde formation are possible.

Studies performed by Dubois et al., [1993] as well as in our laboratory suggest that HSA is the major binding protein with respect to covalent binding. No explanation has been provided so far for the discrepancy between the higher extent of covalent binding to plasma proteins as compared to HSA, because no covalent binding was detected with fibrinogen and gamma globulins, and only 0.14 % of ketoprofen was bound to globulins after 3 h incubations in the studies of Dubois et al., [1993].

In recent studies performed by Ding et al. [1993] Lys-199 (a lysine ε-amino group located in the hydrophilic pocket of subdomain II-A, which also plays a role in reversible binding to HSA) and Lys-525 were identified as the most prominent covalent binding sites for tolmetin glucuronide on the human serum albumin molecule.

Generally, adducts are quantified following alkaline hydrolysis and release of the aglycone. In the first step protein is precipitated by addition of ispopropanol and acidified acetonitrile or of acetonitrile/ethanol mixtures. The protein pellet obtained after centrifugation is washed several times with methanol/diethylether to remove reversibly bound aglycone and conjugates from the protein binding sites. Adducts are incubated with 0.2-0.5 M potassium hydroxide solution at 70-80 °C for 30-120 min. Adduct concentrations are then defined as aglycone liberated from precipitated and extensively washed protein. Approaches for a direct measurement included tryptic digests followed by HPLC separation of the fragments as well as electrophoresis (SDS-PAGE) and blotting of adducts formed from the fluorescent benoxaprofen and flunoxaprofen acyl glucuronides [van Breemen and Fenselau, 1985; Spahn et al., 1990; Dahms and Spahn-Langguth, 1995].

In-vitro studies with blood constituents showed that plasma protein binding accounts for most of the observed covalent binding in blood. Apparently acyl glucuronides do not permeate red blood cell membranes and hence do not usually have access to the intracellular hemoglobin, for which significant covalent binding of tolmetin was detected in in vitro studies when hemoglobin or hemolyzed erythrocytes were exposed to tolmetin glucuronide [Ojingwa et al., 1994].

In in vitro studies with tissue homogenates as well as in vivo studies significant covalent binding to tissue proteins was detected [Ojingwa et al., 1994; Dickinson and King, 1993; King and Dickinson, 1993]. Incubations of glucuronides with subcellular fractions revealed that covalent binding is negligible for the cytosolic fraction, but considerable for membrane-bound proteins [Ojingwa et al, 1994; Spahn-Langguth et al., 1994; Hargus et al., 1995].

In vitro studies performed with various acyl glucuronides showed that covalent adducts are always detectable [e.g., Smith et al., 1986; Smith et al., 1992; Spahn-Langguth et al., 1994a, b]. Studies with chiral compounds revealed that the less stable diastereomer shows a higher extent of covalent binding. When R- and S-carprofen glucuronides were incubated with HSA (fatty acid-free) at 37 °C for 24 h, the extent of covalent HSA binding was initially (at 1 h) higher for the S- than for the R-glucuronide, while after 6 and 24 h covalent binding was significantly higher for R-carprofen incubations [Iwakawa et al., 1988]. After incubation of beclobric acid β-1-O-acyl glucuronides with pooled plasma and HSA in pH 7.4 buffer no significant difference between the two enantiomers was detected with respect to the magnitude of in-vitro covalent binding [Mayer et al., 1993a, b].

In summary, the in vitro studies showed that the extent of covalent binding to a protein is not only clearly pH- and time- but also glucuronide concentration-dependent.

PROTEIN ADDUCT KINETICS: CONCENTRATION-TIME PROFILES IN VIVO

Covalent plasma protein adducts are usually detected in preclinical and clinical studies when acyl glucuronides occur in plasma (the glucuronides' detectability may depend upon sample treatment). The extent of formation and/or their detectability in samples may be species-dependent, because of the differences in aglycone and glucuronide elimination pathways and activity in plasma esterases [Smith et al., 1990; Iwakawa et al., 1989; Iwakawa et al., 1991].

Only few data are available about in vivo adduct kinetics. However, for all investigated compounds the persistence of covalent protein adducts in the systemic circulation was considerably higher than that of its precursors [McKinnon et al., 1989; Benet et al., 1992], as depicted in Fig. 3 for the two 2-arylpropionic acids R/S-benoxaprofen and S-flunoxapro-

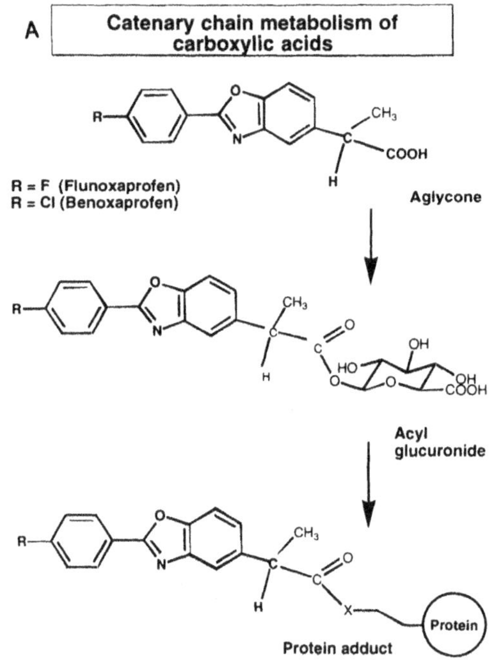

Figure 3. The metabolic pathway for carboxylic acids (the 2-arylpropionic acid NSAIDs flunoxaprofen and benoxaprofen), from which acyl glucuronides and covalent protein adducts are formed consecutively may be regarded as catenary chain and be treated accordingly with respect to pharmacokinetic analyses (Fig. 3A). A representative example for the respective plasma concentration-time profiles obtained in a patient following p.o. dosage of S-flunoxaprofen (100 mg) are depicted in Fig. 3B. Obviously, the terminal half-life is considerably longer for the S-flunoxaprofen adduct than for its precursors, indicating a prolonged persistence of the adduct in the blood. Fig. 3C depicts the respective concentration-times curves obtained in a volunteer for the two benoxaprofen stereoisomers (R and S) following dosage of benoxaprofen racemate (600 mg). The plasma concentrations of S-benoxaprofen exceed those of R-benoxaprofen because of the stereoinversion observed for R-benoxaprofen. As for flunoxaprofen, the concentration-time curves of aglycone and glucuronide are almost parallel in the terminal log-linear part of the curves, however, the half-lives for benoxaprofen isomers and their glucuronides are considerably longer than for flunoxaprofen, while the respective adduct half-lives are in a similar range (Table 3).(All concentrations are given as aglycone equivalents and were determined following hydrolysis of the glucuronides and adducts. Abbreviations are as follows: S-F = S-flunoxaprofen, S-B = S-benoxaprofen, R-B = R-benoxaprofen, Gluc = acyl glucuronide).

fen, both of which have been withdrawn from the market in the meantime [Dahms and Spahn-Langguth, 1995].

Interestingly, covalent binding was also detected for the phase-I metabolites of gemfibrozil, which are further metabolized to their respective acyl glucuronides. For all four phase-I products, the corresponding acyl glucuronides were present in the plasma, however, significant covalent binding was found only for gemfibrozil and one of the four detectable phase-I metabolites (M3) [Gräfe, 1994; Spahn-Langguth et al., 1994].

While the kinetics of an acyl glucuronide is usually rate-limited by its formation and the terminal half-lives of aglycone and glucuronide are similar, the half-lives and residence times of adducts are much higher and kinetics is rate-limited by adduct elimination [Spahn-Langguth and Benet, 1992a]. The half-lives detected for the adducts of different glucuronides vary widely (range: 2 - 14 days) with the highest values approaching that for albumin turn-over. An explanation for these differences between adducts has not been provided yet, but it was hypothesized that different adduct types may be involved. Using standard analytical procedures for adducts the different adduct types are not discriminated, since only the total amount of aglycone that is released from protein via hydrolysis is measured.

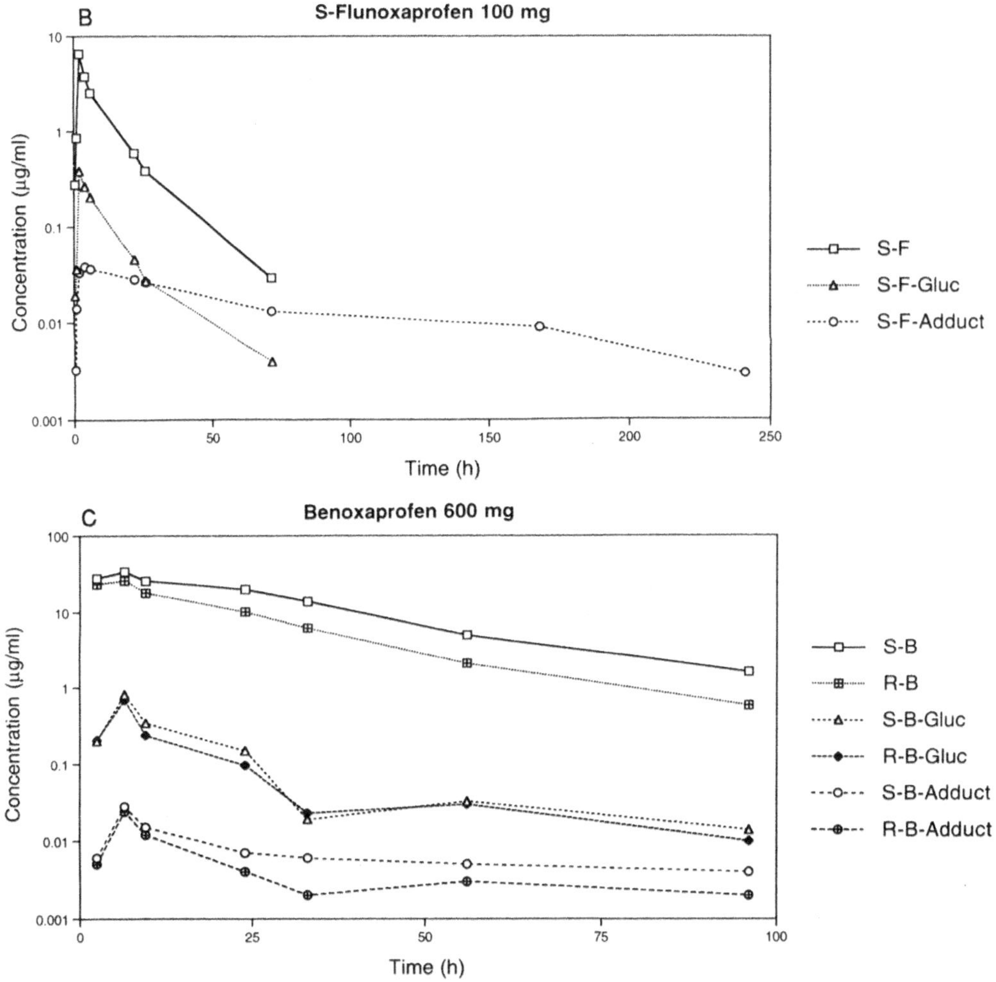

Table 3. Pharmacokinetic parameters of S-flunoxaprofen following a single p.o. dose of 100 mg S-flunoxaprofen and of S- and R-benoxaprofen following a single p.o. dose of 600 mg R/S-benoxaprofen (each number represents arithmetical means of 2 patients) [Dahms and Spahn-Langguth, 1995]

Parameter	Aglycone	Glucuronide	Adduct
S-Flunoxaprofen			
C_{max} [μg/ml]	7.26	0.395	0.0295
t_{max} [h]	2.31	1.75	2.53
AUC_{0-oo} [μg ml^{-1} h]	57.7	3.06	2.44
$t_{1/2}$ [h]	8.13	8.36	56.0
MRT_{tot} [h]	11.5	12.0	81.6
S-Benoxaprofen			
C_{max} [μg/ml]	29.9	0.83	0.028
t_{max} [h]	3.55	4.25	5.65
AUC_{0-oo} [μg ml^{-1} h]	1125	14.1	2.18
$t_{1/2}$ [h]	23.8	42.4	99
MRT_{tot} [h]	36.0	41.5	136.5
R-Benoxaprofen			
C_{max} [μg/ml]	24.7	0.599	0.225
t_{max} [h]	3.30	4.52	6.89
AUC_{0-oo} [μg ml^{-1} h]	602	9.91	1.2
$t_{1/2}$ [h]	21.8	41.8	99
MRT_{tot} [h]	31.4	41.8	122.5

Because of their long half-lives adducts accumulate in the blood (and probably also in tissues) following chronic drug administration [Zia-Amirhosseini et al., 1994]. The "in vivo stability" of tolmetin-plasma protein adducts was characterized in a study in six healthy volunteers after a 400 mg single dose and after a multiple-dose regimen of 400 mg tolmetin every 12 hours for 10 days. The average maximum bound concentration was only 2.7 ng/ml after a single dose, but it was almost an order of magnitude higher after multiple dosing. The protein adduct exhibited an average half-life of 4.8 days in contrast to the much shorter 5-hour half-lives for tolmetin and its glucuronide.

In three male volunteers the extent of covalent binding of beclobric acid enantiomers was studied after single and multiple oral doses of racemic beclobrate (100 mg once daily) and covalent binding observed in all volunteers. The average maximum adduct densities for (-)- and (+)-beclobric acid were 0.147×10^{-4} and 0.177×10^{-4} mol/mol protein for a single

Table 4. Rate constants [h^{-1}] for the in vivo generation and elimination of beclobric acid-plasma protein adducts determined using the compartmental approach (Fig. 7) for single- and multiple-dose data (arithmetic means ± SD; n = 3) as well as rate constants [h^{-1}] for the generation and cleavage of beclobric acid glucuronide as obtained via the compartmental approach from single dose and multiple dose data with beclobrate [Mayer et al., 1993; Hermening et al., 1995]

	(+)	(-)	(+)/(-) Ratio
Aglycone			
Generation	0.34 ± 0.084	0.56 ± 0.056	0.60
Elimination	0.010 ± 0.0006	0.014 ± 0.0040	0.71
Glucuronide			
Formation	0.44 + 0.020	0.49 + 0.050	0.90
Aglyconebackformation	0.018 ± 0.003	0.018 ± 0.008	1.08

beclobrate dose. From preliminary half-life calculations following single and multiple dosage the terminal adduct half-life was found to be in the range of 2 - 3 days and was always shorter for the adduct of levorotatory beclobric acid [Mayer et al., 1993].

Half-lives in the range of 2.5 - 4 days were found for the adducts of S-flunoxaprofen and R- and S-benoxaprofen (Tab. 3). For beclobrate as gemfibrozil aglycone-, glucuronide- and adduct data a multi-compartmental model was used for kinetic modelling, into which a reversible metabolic step was included, since aglycone back-formation may occur under physiological conditions [Hermening et al., 1995]. The obtained rate constants for glucuronide formation and back-formation to aglycone and for protein adduct formation and elimination are given in Table 4.

In general, when the extent of exposure to the glucuronide is increased (as with higher doses or reduced glucuronide clearance due to renal failure, drug-drug interactions as well as prolonged treatment) adduct concentrations are elevated as well. This explains the data obtained with clofibric acid for which significant covalent binding was detected in vivo in laboratory animals as well as in patients. Covalently bound clofibric acid-protein adducts were detected in all 14 investigated patients (daily dose; 0.5 - 2.0 g), even in one subject in whom there was no measurable plasma clofibric acid [Sallustio et al., 1991]. Adduct concentrations appeared to correlate with renal function in the patients. This may be understood on the basis of the scheme depicting reversible metabolism and covalent binding for acyl glucuronides (Scheme B), where the glucuronide can accumulate in plasma when its excretion is inhibited and covalent binding increases because of increased exposure to the reactive metabolite. Furthermore, in rats, protein adducts were present in liver homogenates, and concentrations increased with increasing duration of treatment.

FACTORS AFFECTING THE EXTENT OF COVALENT BINDING AND ITS PREDICTABILITY

As mentioned before, in in vitro studies the extent of covalent binding for one particular compound correlates with the extent of exposure to the reactive acyl glucuronide. Any process that leads to increased glucuronide levels in the organism will hence increase covalent binding. Most relevant are reduced glucuronide clearance, which may be observed with renal failure, for drug-drug interactions, and multiple drug dosages.

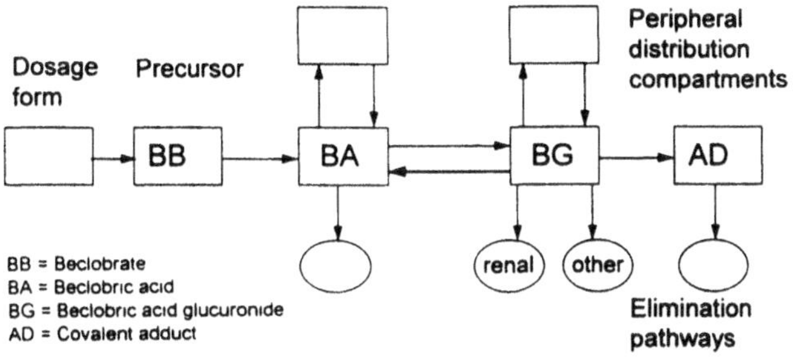

Scheme B. Compartmental approach including acyl glucuronide as well as adduct data: Multi-compartmental model for carboxylic acids forming acyl glucuronides [modified from Hermening et al., 1995]. This metabolic step is reversible, and a covalent adduct may result due to the reactivity of the respective acyl glucuronide. (For beclobric acid, where the two stereoisomers were treated independently because of the lack of stereoinversion, backformation of aglycone was negligible. As for the aglycone enantiomers a two-compartment disposition model yielded better fitting criteria than a one- or more-compartment disposition model for the acyl glucuronides, while a one-compartment model was appropriate for the adducts. This appears reasonable, since the high molecular weight adducts should not permeate through membranes.)

As was expected on the basis of the data obtained by Smith et al. [1985] with zomepirac, the extent of covalent binding of carprofen was increased when probenecid was coadministered as summarized recently [Spahn-Langguth et al., 1994], an observation that was also made with benoxaprofen. Stereochemical aspects of covalent binding have largely been neglected so far. However, there are few exceptions. While probenecid did not significantly affect the S/R ratio of covalent binding in the case of carprofen, it was enhanced for benoxaprofen because of an increased stereoinversion [Spahn et al., 1987].

When tolmetin 400 mg was dosed in patients twice daily for 10 days, its adduct levels were about 10 times higher than they were following a single dose [Zia-Amirhosseini et al., 1994].

In the case of beclobric acid multiple dosing increased covalent binding 3- to 4-fold [Mayer et al., 1993].

Following chronic dosage of carprofen (100 mg, 19 doses) maximum total adduct levels amounted to 4.3 ng/mg protein, whereas total adduct after single dose of enantiomer or racemate (25 or 50 mg, respectively) was in the range of 0.46 - 1.3 ng/mg protein [Iwakawa et al., to be submitted].

However, when trying to compare different acyl glucuronides and their covalent binding to proteins in vitro or in vivo, the relationship becomes increasingly complex. When evaluating the in-vitro covalent binding of the enantiomers of 3 chiral drugs via their glucuronides, carprofen, fenoprofen and beclobric acid, it was found that long degradation half-lives of the β-1-O-acyl glucuronide appeared to correlate with little covalent binding (Table 5). By inclusion of the covalent binding data of acyl glucuronides of achiral compounds into the data pool available in our laboratories, an excellent correlation was obtained between the moles of drug covalently bound per mole of protein of the adduct versus the degradation rate constant for the β-1-O-acyl glucuronide conjugates of each drug or enantiomer [Benet et al., 1993]. With respect to epimeric glucuronides of carboxylic acids, the least stable stereoisomer usually exhibits the highest extent of covalent binding.

Table 5. Apparent first-order degradation rate constants, k, for β-1-O-acyl glucuronides of several carboxylic acids and in vitro "epitope densities," i.e., amounts of carboxylic acid bound covalently per mole protein (in moles aglycone per mole protein after 6 h incubation of 1 μM β-1-O-acyl glucuronides with 0.5 mM human serum albumin at pH 7.4 and 37 °C) [Spahn-Langguth and Benet, 1992; Benet et al., 1993]

	k (h^{-1})	moles carboxylic acid per mole protein x 10^3
(+)-Beclobric acid	0.027	0.065
(-)-Beclobric acid	0.031	0.090
Furosemide	0.158	0.500
S-(+)-Carprofen	0.224	1.000
R-(-)-Carprofen	0.400	1.500
S-(+)-Fenoprofen	0.360	1.510
R-(-)-Fenoprofen	0.710	2.446
Zomepirac	1.540	5.750
Tolmetin	1.780	6.200

It may be expected that the relationship for in vivo reversible binding would be much more complex. However, the degree of covalent binding to plasma proteins should depend, at least, on the plasma concentrations of the acyl glucuronides and the degradation rate of each conjugate. The glucuronide concentrations depend upon the rates of formation, degradation and elimination as well as the administered dose. In vivo data available from different studies with achiral and chiral compounds showed a 25-fold variation in maximum adduct concentration, while AUCs for the glucuronide metabolites showed a 30-fold variation [Benet et al., 1993]. Since for each drug there is a linear relationship between amount covalently bound and the extent of glucuronide present (AUC), bound drug was normalized to AUC for comparison with in vitro degradation rates, yielding a significant linear correlation. Furthermore, there are indications that a structural relationshipcan be related to the facility with which covalent binding occurs. The degree of substitution adjacent to the carboxyl group appears to determine the extent of covalent binding, with low binding for fully substituted compounds (beclobric acid). However, this relationship needs to be further investigated, since some data available in the literature are not consistent with the obtained correlation [Benet et al., 1993].

POTENTIAL TOXICOLOGICAL IMPLICATIONS OF THE REACTIVITY OF EPIMERIC ACYL GLUCURONIDES

There is increasing evidence that the process of acyl glucuronide formation may contribute to undesired effects observed after dosage of carboxylic acids or their precursors. Interestingly, of the 26 drugs withdrawn from the US and British market during the past 30 years because of organ (liver and kidney) toxicity, six were carboxylic acid nonsteroidal anti-inflammatory drugs; these are mainly metabolized via formation of acyl glucuronides, i.e., reactive intermediates [Bakke et al., 1984]. Examples of recent withdrawals in Europe are the 2-arylpropionic acid S-flunoxaprofen and the lipid-regulating agent beclobrate, which is rapidly hydrolyzed to the respective ester in vivo. Other carboxylic acid drugs are still on the market, although it is known that hypersensitivity reactions may occur.

As reviewed and hypothesized earlier the mechanism of toxic side-effects may be of a different nature [Faed, 1984; Spahn-Langguth and Benet, 1992]. For nonsteroidal anti-inflammatory drugs one well-known hypothesis states that hypersensitivity reactions (anaphylactoid reactions as well as organotoxic reactions) are based on the pharmacologic profile of the compound. Inhibition of prostaglandin synthesis leads to an increased production of leukotrienes. Therefore, any compound that inhibits cyclooxygenase would then lead to such side-effects in predisposed patients.

However, other mechanisms that include the occurrence of acyl glucuronides are feasible as well. In this case, determinants of a potential toxicity of carboxylic acid drugs in vivo may be the ability of acyl glucuronides to lead to covalent binding with proteins, the plasma- or tissue concentration-time profile of the reactive glucuronide and the residence time of the respective adduct in the organism [Benet and Spahn, 1988].

From studies reported in the literature there is evidence that antibodies occur in the blood of patients at low levels following dosage of acetylsalicylic acid and valproic acid [Amos et al., 1971; Williams et al., 1992]. An immunological mechanism may be the basis for anaphylactic reactions as well as organ toxicity, since covalent binding to tissue proteins may be followed by subsequent antigen-antibody complex formation. Hence, the third hypothesis includes the theory that acyl glucuronides act as haptens, which, upon covalent binding to proteins, become immunogens and induce antibody production, as shown in the work of Zia-Amirhosseini [1994]. Anti-adduct antibodies formed in mice following adminstration of adducts with mouse serum albumin appeared to be specific for the aglycone, but not for the glucuronic acid part of the molecule. Some cross-reactivity was observed for structurally related aglycones.

Other endogenous molecules with nucleophilic groups may also be reacting with acyl glucuronides. The early studies of Stogniew and Fenselau [1982] on the reactivity of acyl glucuronides towards smaller endogenous nucleophiles (e.g., glutathione) were the basis for an alternative toxicity hypothesis, in which organ toxicities are explained by glutathione depletion. Furthermore, the resulting thioesters may also be reactive.

Very recent studies from our laboratory demonstrate that in vitro significant covalent binding of tolmetin to DNA occurs following incubation of the respective acyl glucuronide with plasmid DNA. The in-vivo relevance of this reaction is currently under investigation.

ACKNOWLEDGMENTS

Studies performed at facilities of the School of Pharmacy, University of California in San Francisco, as well as of the Department of Pharmacology in Frankfurt/M. were in part supported by a grant of the Deutsche Forschungsgemeinschaft (to H.S.-L.) and the Graduiertenförderung des Landes Hessen (to S. Mayer).

REFERENCES

Amos HE, Wilson DV, Taussig MJ, Carlton SJ "Hypersensitivity reactions to acetylsalicylic acid" Clin Exp Immunol 8, 563-572 (1971)

Bakke OM, Wardell WM, Lasagna L "Drug discontinuations in the United Kingdom and the United States: 1964-1983" Clin Pharmacol Ther 35, 559-567 (1984)

Benet LZ, Spahn H "Acyl migration and covalent binding of drug glucuronides - potential toxicity mediators" In: Siest G, Magdalou J, Burchell B (eds) "Molecular and Cellular Aspects of Glucuronidation", Colloques INSERM/John Libbey Eurotext, Vol. 173, London, 305-309 (1988)

Benet LZ, Spahn-Langguth H, Iwakawa S, Volland C, Mizuma T, Mayer S, Mutschler E, Lin ET "Prediction of covalent binding of acidic drugs in man" Life Sci 53, PL141-146 (1993)

Benet LZ, Bisher A, Volland C, Spahn-Langguth H "The pharmaco- and toxicokinetics of reactive and active phase II metabolites" In: Crommelin DJA, Midha KK (eds.) "Topics in Pharmaceutical Sciences" Med Pharm Scientific Publishers, Stuttgart, 533-544 (1992)

Blanckaert N, Compernolle F, Leroy P, Van Hautte R, Fevery J, Heirwegh KPM "The fate of bilirubin-IXα glucuronide in cholestasis and during storage in vitro" Biochem J 171, 203-214 (1978)

Boudinot FD, Homon CA, Jusko WJ, Ruelius HW "Protein binding of oxazepam and its glucuronide conjugates to human serum albumin" Biochem Pharmacol 39, 2115-2121 (1985)

Bradow G, Khan L, Fenselau C "Studies of intramolecular rearrangements of acyl-linked glucuronides using salicylic acid, flufenamic acid, and (S)- and (R)-benoxaprofen and confirmation of isomerization in acyl-linked δ^9-11-carboxytetrahydrocannabinol glucuronide" Chem Res Toxicol 2, 316-324 (1989)

van Breemen RB, Fenselau C "Acylation of albumin by 1-O-acyl glucuronides" Drug Metab Dispos 13, 318-320 (1985)

Dahms M, Spahn-Langgguth H "Acyl glucuronides as reactive intermediates: Detection of flunoxaprofen protein adducts in biological material" Naunyn Schmiedeberg's Arch Pharmacol, Suppl. (1995)

Dickinson RG, King AR "Studies on the reactivity of acyl glucuronides - V. Glucuronide-derived covalent binding of diflunisal to bladder tissue of rats and its modulation by urinary pH and β-glucuronidase" Biochem Pharmacol 46, 1175-1182 (1993)

Ding A, Ojingwa JC, McDonagh AF, Burlingame AL, Benet LZ "Evidence for covalent binding of acyl glucuronides to serum albumin via an imine mechanism as revealed by tandem mass spectrometry" Proc Nat Acad Sci USA 90, 3797-3801 (1993)

Dubois N, Lapicque F, Maurice MH, Pritchard M, Fournel-Gigleux S, Magdalou J, Abiteboul M, Siest G, Netter P "In vitro irreversible binding of ketoprofen glucuronide to plasma proteins" Drug Metab Dispos 21, 617-623 (1993)

Dudley KH "Commentary: Stereochemical formulas of β-D-glucuronides" Drug Metab Dispos 13, 524-528 (1985)

el Mouelhi M, Ruelius HW, Fenselau C, Dulik DM "Species-dependent enantioselective glucuronidation of three 2-arylpropionic acids: naproxen, ibuprofen and benoxaprofen" Drug Metab Dispos 16, 627-634 (1988)

el Mouelhi M, Schwenk M "Stereoselective glucuronidation of naproxen in isolated cells from liver, stomach, intestine, and colon of the guinea pig" Drug Metab Dispos 19, 844-845 (1991)

Faed EM "Properties of acyl glucuronides: Implications for studies of the pharmacokinetics and metabolism of acidic drugs" Drug Metab Rev 15, 1213-1249 (1984)

Fournel-Gigleux S, Hamar-Hansen C, Motassim N, Antoine B, Mothe O, Decolin D, Caldwell J, Siest G "Substrate-specificity and enantioselectivity of arylcarboxylic acid glucuronidation" Drug Metab Dispos 16, 627-634 (1988)

Gräfe AK "PK/PD Correlation model for the triglyceride-lowering effect of lipid-regulating agents and the effect of the patients' age: Gemfibrozil and its metabolites" PhD Thesis, Johann Wolfgang Goethe-University, Frankfurt/M., 1994

Grubb N, Weil A, Caldwell J "Studies on the in vitro reactivity of clofibryl and fenofibryl glucuronides. Evidence for protein binding via a Schiff*s base mechanism" Pharmacology 46, 357-364 (1993)

Gugler R, Kurten JW, Jensen CJ, Klehr U, Hartlapp J "Clofibrate disposition in renal failure and acute and chronic liver disease" Eur J Clin Pharmacol 230, 237-241 (1979)

Hargus SJ, Martin BM, Pohl LR "Covalent modification of rat liver dipeptidyl peptidase IV by diclofenac metabolites" The Fifth International Symposium on Biological Reactive Intermediates, Munich, Jan 1995, Abstract Volume, p. 74 (1995)

Hasegawa J, Smith PC, Benet LZ "Apparent intramolecular acyl migration of zomepirac glucuronide" Drug Metab Dispos 10, 469-473 (1982)

Hayball PJ, Nation RL, Bochner F "Stereoselective interactions of ketoprofen glucuronides with human plasma protein and serum albumin" Biochem Pharmacol 44, 291-299 (1992)

Hayball PJ, Nation RL, Bochner F, Sansom LN, Ahern MJ. Smith MD "The influence of renal function on the enantioselective pharmacokinetics and pharmacodynamics of ketoprofen in patients with rheumatoid arthritis" Br J clin Pharmac 31, 546-550 (1991)

Hermening A, Gräfe AK, Mayer S, Mutschler E, Spahn-Langguth H "Pharmacokinetics of fibrates and their acyl glucuronides: Inclusion of a reversible metabolic step and cvalent binding into kinetic modelling" Naunyn Schmiedeberg's Arch Pharmacol, Suppl (1995)

Iwakawa S, Suganuma T, Lee SF, Spahn H, Benet LY, Lin ET "Direct determination of diastereomeric carprofen glucuronides in human plasma and urine and preliminary measurements of stereoselective metabolic and renal elimination after oral administration of carprofen in man" Drug Metab Dispos 17, 474-480 (1989)

Iwakawa S, Spahn H, Benet LZ, Lin ET "Stereoselective disposition of carprofen, flunoxaprofen, and naproxen in rats" Drug Metab Dispos 19, 853-857 (1991)

Iwakawa S, Spahn H, Benet LZ, Lin ET "Carprofen glucuronides: Stereoselective degradation and interaction with human serum albumin" Pharmaceut Res 5 (Suppl), 5214 (1988)

King AR, Dickinson RG "Studies on the reactivity of acyl glucuronides - IV. Covalent binding of diflunisal to tissues of the rat" Biochem Pharmacol 45, 10431047 (1993)

Kroemer HK, Klotz U "Glucuronidation of drugs. A re-evaluation of the pharmacological significance of the conjugates and modulating factors" Clin Pharmacokinet 23, 292-310 (1992)

Mayer S, Spahn-Langguth H, Gikalov I, Mutschler E "Pharmacokinetics of beclobric acid enantiomers and their glucuronides after single and multiple p.o. dosage of rac-beclobrate" Arzneim Forsch/Drug Res 42-2, 1354-1358 (1993)

Mayer S, Mutschler E, Benet LY, Spahn-Langguth H "In vitro and in vivo irreversible plasma protein binding of beclobric acid enantiomers" Chirality 5, 120-125 (1993)

McKinnon GE, Dickinson RG "Covalent binding of diflunisal and probenecid to plasma protein in humans: Persistence of the adducts in the circulation" Res Commun Chem Pathol Pharmacol 66, 339-354 (1989)

Ojingwa J, Spahn-Langguth H, Benet LZ "Reversible binding of tolmetin, zomepirac and their glucuronide conjugates to human serum albumin" J Pharmacokin Biopharm 22, 19-40 (1994)

Ojingwa J, Spahn-Langguth H, Benet LZ "Irreversible binding of tolmetin to macromolecules via its glucuronide: Bindng to blood constituents, tissue homogenates and subcellular fractions in vitro" Xenobiotica 24, 495-506 (1994)

Ott RJ, Giacomini KM "Stereoselective transport of drugs across epithelia" In: Wainer IW (ed) "Drug stereochemistry: Analytical methods and pharmacology" Marcel Dekker Inc. New York, Basel, Hong Kong, 281-314 (1993)

Rowe BJ, Meffin PJ "Diisopropylfluorophosphate increases clofibric acid clearance: Supporting evidence for a futile cycle" J Pharmacol Exp Ther 230, 237-241 (1984)

Rubin A, Warrick RL, Wolen SM, Chernish SM, Ridolfo AS, Gruber CM "Physiological disposition of fenoprofen in man. III. Metabolism and protein binding of fenoprofen" J Pharmacol Exp Ther 183, 449-457 (1972)

Sallustio BC, Knights KM, Roberts BJ, Zacest R "In vivo covalent binding of clofibric acid to human plasma proteins and rat liver proteins" Biochem Pharmacol 42, 1421-1425 (1991)

Smith PC, Langendijk PNJ, Hasegawa J, Benet LZ "Effect of probenecid on the formation and elimination of acyl glucuronides: Studies with zomepirac" Clin Pharmacol Ther 38, 121-127 (1985)

Smith PC, Benet LZ "Characterization of the isomeric esters of zomepirac glucuronide by proton NMR" Drug Metab Dispos 14, 503-505 (1986)

Smith PC, Benet LZ, McDonagh A "Covalent binding of zomepirac glucuronide to proteins: Evidence for a Schiff base mechanism" Drug Metab Dispos 18, 639-644 (1990a)

Smith PC, McDonagh AF, Benet LZ "Irreversible binding of zomepirac to plasma protein in vitro and in vivo" J Clin Invest 77, 934-939 (1986)

Smith PC, McDonagh AF, Benet LZ "Effect of an esterase inhibitor on the disposition of zomepirac and its covalent binding to plasma proteins in the guinea pig" J Pharmacol Exp Ther 230, 218-224 (1990b)

Smith PC, Song WQ, Rodriguez RJ "Covalent binding of etodolac acyl glucuronide to albumin in vitro" Drug Metab Dispos 20, 962-965 (1992)

Spahn H, Iwakawa S, Lin ET, Benet LZ "Influence of probenecid on the urinary excretion rates of the diastereomeric benoxaprofen glucuronides" Eur J Drug Metab Pharmacokinet 12, 233-239 (1987)

Spahn H, Iwakawa S, Lin ET, Benet LZ "Procedures to properly characterize formation of acyl glucuronides: Studies with benoxaprofen" Pharm Res 6, 125-132 (1989)

Spahn H, Zia-Amirhosseini P, Näthke I, Mohri K, Benet LZ "Characterization of drug adducts - formed with proteins via acyl glucuronides - by electrophoresis and blotting" Pharm Res 7 (Suppl), S-257 (1990)

Spahn H, Spahn I, Benet LZ "Probenecid-induced changes in the clearance of carprofen enantiomers - A preliminary study" Clin Pharmacol Ther 45, 500-505 (1989)

Spahn-Langguth H, Benet LZ "Active and reactive phase-II metabolites: The glucuronides pathway" In: Crommelin DJA, Midha KK (eds.) "Topics in Pharmaceutical Sciences" Med Pharm Scientific Publishers, Stuttgart, 533-544 (1992a)

Spahn-Langguth H, Benet LZ "Acyl glucuronides revisited: Is the glucuronidation process a toxification as well as a detoxification mechanism?" Drug Metab Rev 24, 5 - 48 (1992b)

Spahn-Langguth H, Zia-Amirhosseini P, Iwakawa S, Ojingwa J, Büschges R, Benet LZ "Aktive und reaktive Glucuronide: Reversible und irreversible Interaktionen mit Proteinen" In: Dengler HJ, Mutschler E (eds) "Fremdstoffmetabolismus und Klinische Pharmakologie, Gustav Fischer Verlag, Stuttgart, 29-50 (1994)

Spahn-Langguth H, Gräfe AK, Mayer S, Büschges R, Benet LZ "Fibrates: The occurrence of active and reactive metabolites" In: Reid E, Hill HM, Wilson ID (eds) "Biofluid and tissue analysis for drugs including hyperlipidaemics." Methodological Surveys in Bioanalysis of Drugs, Vol. 23, The Royal Society of Chemistry, Cambridge, 87-102 (1994)

Stogniew M, Fenselau C "Electrophilic reactions of acyl-linked glucuronides: Formation of clofibrate mercapturate in humans" Drug Metab Dispos 10, 609-613 (1982)

Wells DS, Janssen FW, Ruelius HW "Interactions between oxaprozin glucuronide and human serum albumin" Xenobiotica 17, 1437-1449 (1987)

Williams AM, Worrall S, DeJersey J, Dickinson RG "Studies on the reactivity of acyl glucuronides: Glucuronide-derived adducts of valproic acid and plasma proteins and anti-adduct antibodies in humans" Biochem Pharmacol 43, 745-755 (1992)

Vree TB, Hekster YA, Anderson PG "Contribution of the human kidney to the metabolic clearance of drugs" Ann Pharmacother 26, 1421-1428 (1992)

Zia-Amirhosseini P "Hypersensitivity to nonsteroidal anti-inflammatory drugs: Exploration of a theory" Ph.D. Thesis, School of Pharmacy, University of California, San Francisco, 1994

Zia-Amirhosseini P, Ojingwa J, Spahn-Langguth H, McDonagh AF, Benet LZ "Enhanced covalent binding of tolmetin to proteins in humans after multiple dosing" Clin Pharmacol Ther 55, 21-27 (1994)

BENZENE-INDUCED BONE MARROW CELL DEPRESSION CAUSED BY INHIBITION OF THE CONVERSION OF PRE-INTERLEUKINS-1α AND -1β TO ACTIVE CYTOKINES BY HYDROQUINONE, A BIOLOGICAL REACTIVE METABOLITE OF BENZENE

Rodica Niculescu,[1] John F. Renz,[2] and George F. Kalf[1]

[1] Department of Biochemistry and Molecular Biology
Jefferson Medical College of Thomas Jefferson University
Philadelphia, Pennsylvania
[2] Department of Surgery
University of California
San Francisco, California

INTRODUCTION

Benzene (BZ), a widely used industrial chemical and a ubiquitous environmental pollutant, is a hematotoxin that causes bone marrow (BM) cell depression in experimental animals and aplastic anemia in humans that are chronically exposed[1-3]. Marrow stromal macrophage (SØ) dysfunction and deficient interleukin-1(IL-1) production have been demonstrated in aplastic anemia. Markedly depressed IL-1 production was observed in adherence-separated, LPS-stimulated monocytes from 75 to 80 percent of patients with severe aplastic anemia compared with normal control subjects[4,5]. The SØ is involved in hematopoietic regulation[6-8] through the synthesis of several cytokines including IL-1, which synergizes with IL-3 to promote the development of the pluripotent stem cell to myeloid and lymphoid stem cells[9]. IL-1 is also involved in lymphocyte development and in the induction of cytokine production by stromal fibroblasts [8,10,11]. Inhibition of the production of active IL-1 in SØ could result in a lack of cytokines, increased apoptosis of hematopoietic progenitor cells and thus aplastic anemia. Two IL-1 cytokines, IL-1α and IL-1β, are the products of distinct genes located on chromosome 2[12]. The transcript of each gene is translated as a precursor protein of approximately 34 kDa that is converted by a specific protease to a biologically active cytokine of 17 kDa[13]. The processing of pre-IL-1α to mature cytokine is catalyzed by the

Biological Reactive Intermediates V
Edited by R. Snyder et al., Plenum Press, New York, 1996

calcium-activated, cysteine protease, calpain[14,15]. The release of IL-1β from the cell is also associated with the cleavage of its precursor form by a sulfhydryl-dependent protease referred to as IL-1β convertase (ICE)[16-18].

The BM SØ is the target of BZ hematotoxicity[19-21]. Hydroquinone (HQ), an hepatic metabolite of BZ, accumulates in the BM[22] where it undergoes peroxidase-mediated oxidation in the SØ to para-benzoquinone (BQ)[23], a direct-acting biologically reactive electrophile that can interact with the cysteine sulfhydryl groups of cellular proteins[23]. BZ-induced aplastic anemia may result from interference of BQ in IL-1 production, processing and/or secretion by SØ. LPS-induced secretion of IL-1 by murine SØ[24] and P388D$_1$ macrophage-like cells[25] is decreased in vitro after exposure to HQ. A 30 to 40% decrease in chymotrypsin-like activity in extracts of P388D$_1$ cells was observed concomitant with a decrease in the release of IL-1 into the conditioned medium[25] suggesting that HQ decreases IL-1 release by inhibiting proteolytic conversion of the molecule from its membrane-bound precursor form.

We report here that HQ has no effect on the transcription or translation of mRNAs for the 34 kDa pre-IL-1α or pre-IL-1β, but prevents the proteolytic conversion of the pre-IL-1 forms to the 17 kDa active cytokines in murine SØ or human B1 cells respectively. In addition, SØ from BZ-treated mice produce the 34 kDa pre-IL-1α when stimulated in culture with LPS[26], but cannot convert the precursor to IL-1α. These results suggest that BQ, derived from the oxidation of HQ, a marrow metabolite of BZ, also prevents the conversion *in vivo* and thus may be responsible for BZ-induced BM cell depression. In this connection, BZ-induced BM cell depression can be prevented by the concomitant administration with BZ of native, but not heat-inactivated, recombinant IL-1α , thus bypassing the inability of the HQ-inhibited SØ to produce IL-1α[26]. We also show that BQ is an excellent inhibitor of the proteolytic activity of both highly purified calpain and ICE.

RESULTS AND DISCUSSION

The Ability of Recombinant Murine IL-1α to Prevent BZ-Induced Depression of BM Cellularity

If our premise is correct that BZ-induced BM cell depression in a mouse model results from an inhibition of the processing of pre-IL-1α by calpain, then BZ-induced BM hypocellularity should be prevented by the coadministration of IL-1α with BZ. Fig 1. presents a representative of five experiments, all of which showed similar results. Seven groups (n=4) of C57Bl/6J mice were treated as follows: One group was treated with vehicle (corn oil) only and served as control. One group received BZ (600 mg/Kg body weight ip twice/day for 2 days) and another only rMuIL-1α (2000U/mL). Three groups were given rMuIL-1α (500, 1000, or 2000 U/animal) 18 hr prior to the first daily BZ injection. The remaining group was administered heat-inactivated rMuIL-1α (2000U/animal) prior to BZ treatment. Administration of BZ to mice decreased the nucleated cells in the BM (Fig.1) to 25 per cent of control measured 17 hr after the last BZ injection (Day 3). Native rMuIL-1α provided a dose-dependent protection against the depressive effects of BZ (Fig. 1, bars 3-5), with complete protection occurring at 2000 U/animal (100U/g body weight). Heat-inactivated rMuIL-1α did not protect (bar 6) and pretreatment with rMuIL-1α alone (bar 7) did not affect BM cellularity in three days time. rMuIL-1α provided similar protection when the dose of BZ administered was 800 mg/Kg body weight. Taken together, these results suggest that the depression of BM cellularity in mice (the mouse model of aplastic anemia) may result from an inability of the SØ to process pre-IL-1α, the major form in the mouse, to biologically active cytokine required for the induction in stromal fibroblasts of colony-stimulating factors

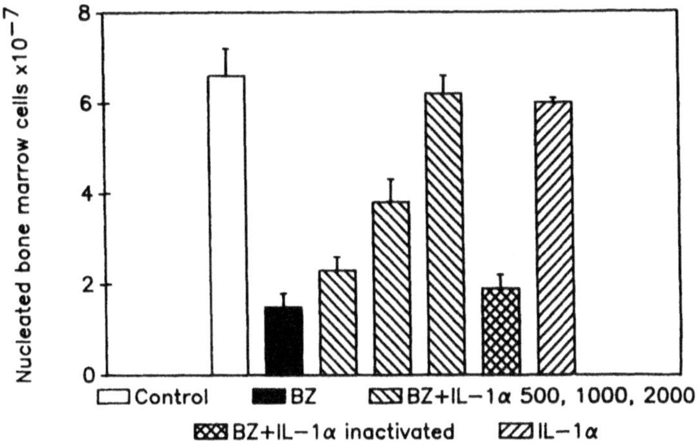

Figure 1. BZ-induced depression of BM cellularity and its prevention by IL-1α. Seven groups (n=4) of male C57Bl/6J mice were established. One group served as the control and received only corn oil and PBS. Five groups were injected ip with BZ (600 mg/kg body weight) in corn oil twice per day, 7 hr apart, for 2 days. One of these groups received BZ only, the others were also injected with rMuIL-1α (500, 1000 or 2000U/animal) or heat-inactivated rMuIL-1α administered in PBS containing 0.2% BSA. A seventh group of animals received only rMuIL-1α (2000U/animal) . Eighteen hours after the final injection, the animals were killed by cervical dislocation, their femurs removed, and the nucleated BM cells obtained and counted. Data are expressed as the mean ± SD. Where there were differences between groups, they were significant at $p < 0.01$level . Figure was taken from Niculescu and Kalf, Arch. Toxicol. with permission.

essential for the survival of hematopoietic progenitor cells. Additional support for this hypothesis is the demonstration that BZ-induced myelotoxicity is completely prevented when exogenous IL-1α obviates the lack of IL-1a in the SØ of BZ-or HQ-treated animals.

Inhibition of the Conversion of Pre-IL-1α to Mature Cytokine in SØ from BZ-Treated Mice

The inhibition of conversion of precursor to cytokine was demonstrated in SØ of mice treated with BZ under conditions which cause myelotoxicity measured as severe depression of BM cellularity. Three groups (n=4) of C57Bl/6J mice were established. One group received the vehicle only. The second was treated with 800 mg/Kg BZ ip twice daily for 2 days. Indomethacin (2 mg/Kg) , a prostaglandin H synthase-peroxidase inhibitor was concomitantly administered to group three. Indomethacin has been shown to prevent BZ-induced BM cell depression and genotoxicity[27]. It has no affect on the hepatic metabolism of BZ at this dosage but does prevent oxidation of HQ to reactive species in the SØ. Eighteen hr after the final BZ administration, femoral BM was obtained and the number of nucleated marrow cells determined. BM cellularity of BZ-treated mice was depressed to 45 percent of control while coadministration of indomethacin prevented depression (data not shown). The ability of SØ from BZ-treated animals to convert 34 kDa pre-IL-1α to 17 kDa cytokine was assessed by placing BM cells from the three experimental groups into culture to establish an adherent layer consisting predominately of SØ and fibroblasts. IL-1α production by the SØ was stimulated by incubation with LPS for 18 hr. Conversion of the precursor to the 17 kDa cytokine was analyzed in cell lysate protein by Western immunoblotting. As can be seen in Fig. 2, SØ from control animals stimulated with LPS produced both a 34 kDa pre-IL-1α and a 17 kDa mature cytokine (lane 2). The precursor was produced in

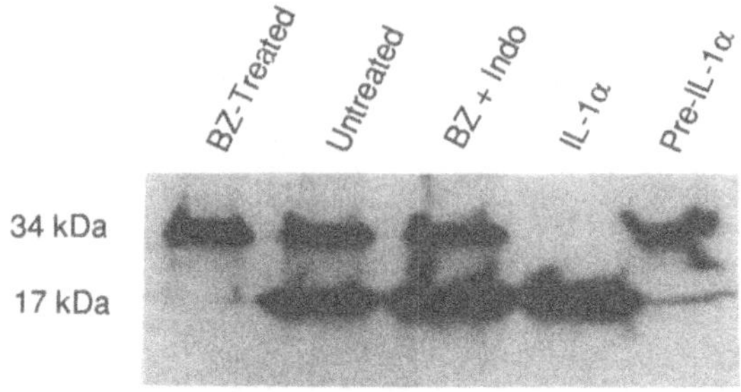

Figure 2. Inhibition of conversion of pre-IL-la to the mature cytokine in stromal macrophages from BZ-treated mice. Mice received BZ (800 mg/Kg body weight in corn oil) ip, twice per day, 7 hr apart, for 2 days. Indomethacin (2 mg/Kg) was administered concomitantly with BZ. Eighteen hr after the last BZ injection, the animals were killed and the femoral BM plug was extruded with RPMI 1640/10 mM Hepes, pH 7.4/10U heparin. BM cells were incubated at 37° C, 5% CO_2 for 2 hr to yield an adherent layer of SØ. IL-1α production was induced by incubating the adherent layer in media containing 10% FBS and 40 μg/mL LPS for 18 hr. The cells were washed, lysed and the cell protein analyzed for the presence of IL-1α by Western immunoblotting. Separation of lysate protein was performed using SDS-PAGE (15% , 0.8-mm thick, 14X14, 150 V). Equal amounts of protein were added to each well. The proteins were electroblotted, to nitrocellulose and the blot probed with specific polyclonal rabbit anti-MuIL-1α antibody followed by ^{125}I-labeled anti-rabbit IgG (2X10^6 cpm) and autoradiography. Lane 1, BZ-treated; lane 2, controls; lane 3. BZ+ indomethacin; lane 4, IL-1α marker; lane 5, pre-IL-1α marker.

LPS-treated SØ from animals administered BZ indicating that transcription and translation were not affected, but did not convert the precursor to the 17 kDa mature cytokine (lane 1). Indomethacin administered concomitantly with BZ prevented the inhibition of precursor processing (lane 3) that results from the oxidation of HQ to BQ by peroxidase.

Inhibition of Conversion of Pre-IL1α to Cytokine by HQ in Mouse SØ

SØ were obtained by culturing femoral BM cells with recombinant macrophage colony-stimulating factor (rM-CSF) for 7 days, which resulted in a population of greater than 95% SØ, as determined by morphological and biochemical analysis (data not shown).

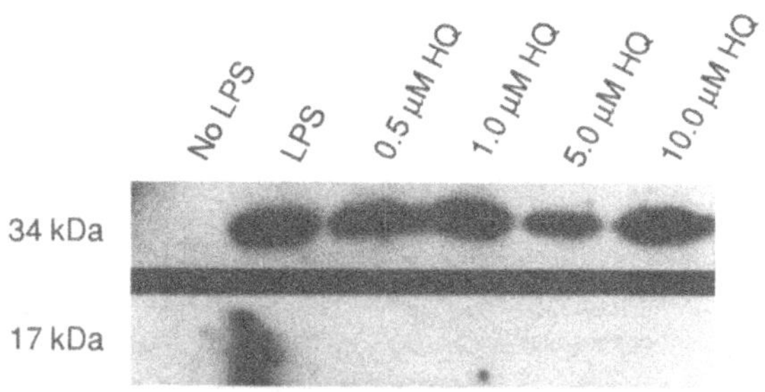

Figure 3. Inhibition of conversion of pre-IL-1α to mature cytokine by HQ in mouse SØ. SØ were grown to confluence in RPMI 1640/ 750 U/mL rM-CSF. SØ were treated with HQ (0.5 to 10 μM final concentration) for 6 hr at 37° C, 5% CO_2 (lanes 3-6). Controls (lanes 1 and 2) did not receive HQ. IL-1α production was stimulated in all cultures with 40 μg LPS except that in lane A. After 24 hr in LPS the cells were collected and lysed and the presence of IL-1α was identified by Western immunoblot as described in the legend for Fig. 2. Lane 1, no LPS, no HQ; lane 2, LPS, no HQ; lanes 3-6, LPS, HQ-0.5-10 μM respectively. Figure was taken from Renz and Kalf, Blood, 78:938, 1991, with permission.

Confluent SØ were treated with HQ (0.5 to 10.0 μM) for 6 hr followed by stimulation with LPS for 24 hr. Immunoblot analysis of cell lysate protein indicated that SØ do not express IL-1α in the absence of LPS stimulation (Fig. 3, lane 1), whereas in the presence of LPS, precursor is produced and converted to mature cytokine (lane 2). Exposure to HQ does not affect transcription or translation of pre-IL-1α at 10 μM, but 0.5 μM completely inhibited the conversion of pre-IL-1α to cytokine (Fig 3, lanes 3-6). HQ is not cytotoxic to SØ as evidenced by a lack of effect on cell viability, as well as on DNA or protein synthesis (data not shown). Thus, SØ from BZ-treated mice and those from treated in culture with HQ, are incapable of producing active cytokine because BQ has inhibited the protease responsible for cleaving the 34 kDa precursor to the 17 kDa active molecule.

Inhibition of Calpain, the Pre-IL-1α Processing Enzyme, by HQ

Processing of the membrane-associated 34 kDa pre-IL-1α to the 17 kDa biologically active cytokine is catalyzed by the membrane -bound sulfhydryl-dependent, calcium-activated neutral protease, calpain[14,15]. BQ, produced from HQ in the cytosol of the SØ, may inactivate calpain by forming a covalent adduct with the SH group of the essential cysteine residue at the active site[28]. We have demonstrated previously that macrophages oxidize radiolabeled HQ to BQ which covalently binds to proteins[23]. Binding is prevented in the presence of excess cysteine by the formation of a monocysteine-BQ conjugate[23]. As can be seen in Fig. 4, BQ causes a concentration-dependent inhibition of the activity of purified human platelet calpain with an IC_{50} of 3 μM, making BQ one of the most potent inhibitors of this protease. For these reasons we postulate that BQ inhibits calpain by covalently binding to the essential SH group of cysteine 108 at the active site[28].

HQ Inhibition of Pre-IL-1β Conversion to Mature Secreted IL-1β

In the human cell, IL-1β represents the major isoform. IL-1β undergoes slow release from the cell which is associated with cleavage of pre-IL-1β by the cysteine protease, ICE. In order to determine the effects of HQ on the processing of pre-IL-1β , we turned to the B1

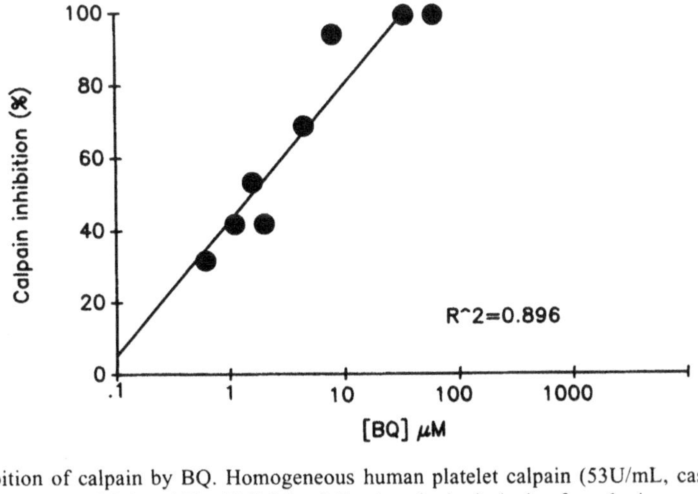

Figure 4. Inhibition of calpain by BQ. Homogeneous human platelet calpain (53U/mL, casein assay) was assayed in the presence of 0.1 to 100 μM BQ by following the hydrolysis of a calpain-specific fluorescent peptide substrate (1 mM) in 60 mM Tris/HCL, pH 7.5, 2.5 % DMSO and 5 mM $CaCl_2$ using an absorbance excitation maximum of 380 nm and measuring an emission response at 440 nm.

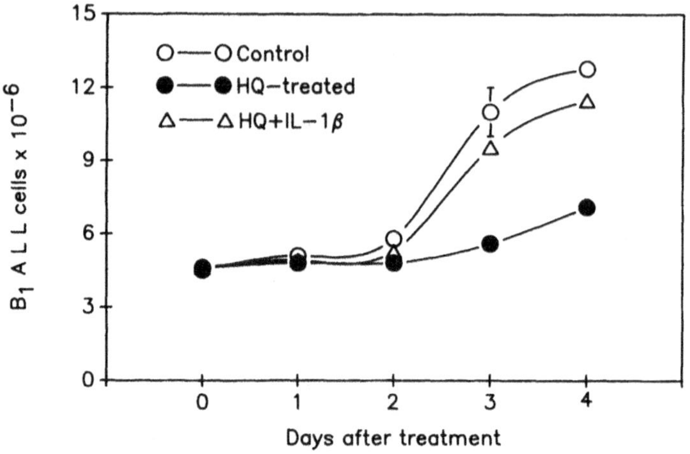

Figure 5. HQ inhibition of IL-1β autocrine stimulation of B1 cell growth. A culture of B1 cells ($4.5X10^6$/mL) was divided into three treatment groups, each set up in triplicate; untreated, treated with a final concentration of 2 μM HQ; treated with 2 μM HQ and 25 pg/mL of rHuIL-1β. IL-1β was added to the HQ-treated cells 4 hr after the addition of HQ to allow time for HQ to enter the cells and be metabolized to BQ which must then bind ICE. Control cultures received PBS. The cells were cultured in a MEM/10 % FBS for 5 days; cell counts and viability studies were performed daily. Viability after treatment. was always greater than 95%. Data points represent the mean number of cells from 3 replicates ± SD. Where no error bars are seen, the SD was too small to plot. The experiment presented is representative of 3 experiments which gave virtually identical results. HQ-treated sample at Day 4 was significantly different from the control group at $p \leq 0.001$.

cell line[29,30] as a model. This cell line exhibits density-dependent growth and secretes an autostimulatory growth factor that has been identified as IL-1β. B1 cells constitutively express mRNA for pre-IL-1β and the IL-1 receptor and secrete the active cytokine. An inhibition of ICE by treatment of the cells with HQ should prevent the: (1) conversion of pre-IL-1β; (2) secretion of IL-1β into the growth medium and; (3) autocrine stimulation of the cells. As can be seen from a representative (Fig. 5) of 3 experiments, all of which gave virtually identical results, untreated B1 cells increase approximately 7-fold in number over the 5 day test period due to autocrine stimulation by IL-1β. HQ (2 μM) significantly inhibited the proliferation of B_1 cells at Day 4 ($p \leq 0.001$) (Fig. 5), presumably because the BQ-inactivated ICE is incapable of converting pre-IL-1β to secretable, biologically-active cytokine. This experiment could not be carried out longer than 5 days because of the cells need for refeeding. The failure of HQ-treated cells to increase in number because of loss of autocrine stimulation by IL-1β is supported by the fact that the growth curve of HQ-treated supplemented with rHuIL-1β, approached that of control cells. The lack of IL-1β secretion by the HQ-treated cells was confirmed by the failure to demonstrate the presence of the 17 kDa IL-1β in the conditioned medium on Day 4 by Western immunoblot with monoclonal anti-IL-1β antibody (Fig. 6, lane 3). A distinct band indicative of a protein of 17 kDa, which migrated to the same position as the IL-1β standard (lane 1), was observed in the culture media of control cells (Fig 6., lane 2) and cells treated with both HQ and IL-1β (lane 4). However, in the presence of HQ, there was no observable band at 17 kDa (lane 3) indicative of the inability of HQ-treated cells to convert pre-IL-1β to the secretable cytokine.

Taken together, these results indicate that B1 cells, whose proliferation in culture is dependent on the autocrine secretion of IL-1β, are prevented by HQ from processing of pre-IL-1β to active secreted cytokine and thus from autocrine growth. As in the case of

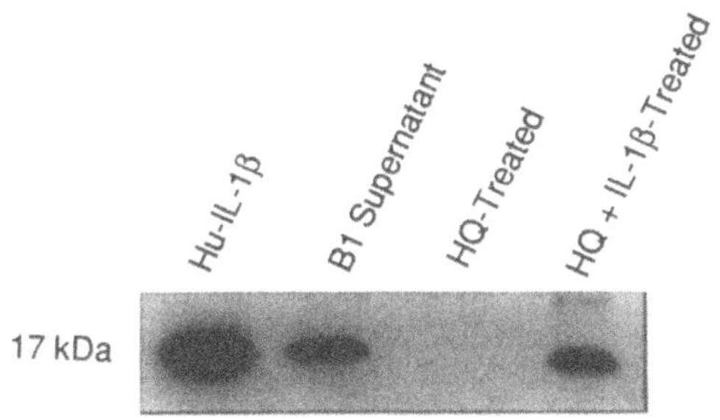

Figure 6. Western immunoblot analysis for the presence or absence of IL-1β in culture supernatants of B1 cells after treatment with HQ and HQ/IL-1β. Supernatant samples from Day 4 cultures (Fig. 5) were mixed with an equal volume of SDS sample solvent and boiled for 5 min. Equal amounts of protein (2 μg) were added to each well of a polyacrylamide mini-gel. PAGE was carried out using a precast 4-15 % gradient polyacrylamide minigel run at 200V for 45 min. Western immunoblotting for the identification of IL-1β was carried out by electroblotting onto a nitrocellulose membrane, blocking and probing the blot with mouse monoclonal anti-human IL-1β (5 mg/mL) for 1 hr. The blot was washed, incubated for 1 hr with horseradish peroxidase-conjugated sheep anti-mouse IgG (1:1000) and developed by incubation with ECL peroxidase developing reagent and exposed to X-ray film. Lane 1, rHuIL-1β (300 ng); lane 2, control B1 supernatant; lane 3, HQ-treated B1 cell supernatant; lane 4, HQ + IL-1β- treated B1 cell supernatant. The experiment was repeated 2 additional times with identical results.

calpain, HQ, by oxidation to BQ in the B1 cells, probably bound to the essential SH group on ICE and inactivated the protease.

Inhibition of ICE by BQ

The ability of BQ to inactivate ICE was assayed by monitoring the conversion of the 34 kDa recombinant pre-IL-1β to the 17 kDa IL-1β by Western immunoblot with a

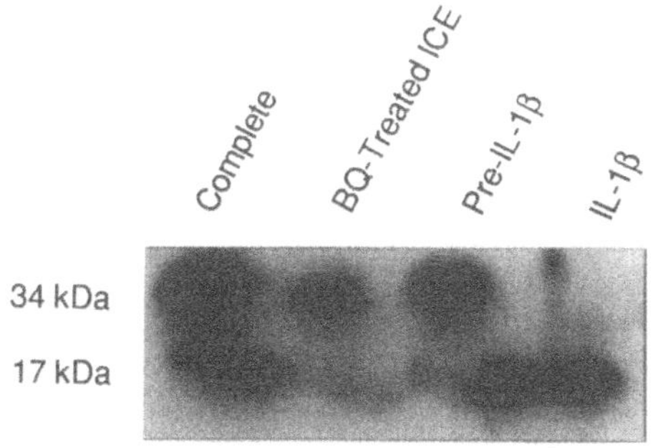

Figure 7. Western immunoblot analysis of the inhibition of pre-IL-1β conversion by pre-treatment of ICE with BQ. Proteolytic activity of homogeneous ICE was measured in the presence or absence of BQ by Western immunoblot using pre-IL-1β as the substrate. Monoclonal anti-IL-1β antibody was used to identify both the precursor and the mature 17 kDa cytokine. The 34 kDa recombinant pre-IL-1β (20 ng) was incubated with ICE (0.6 ng) at 37°C in the presence or absence of 3 μM BQ in 10 mM Tris/HCl, pH 8.1 in a total volume of 10 μL for 3 hr at 37°C. BQ was preincubated with ICE for 10 min before starting the reaction. Pre-IL-1β and ICE were the kind gifts of Dr. Roy Black (Immunex). Pre-IL-1β marker was added at 20 ng and IL-1β marker at 200 ng. The reaction was stopped by boiling samples for 5 min with an equal volume of SDS sample solvent. The conversion of pre-IL-1β to the mature form was analyzed by PAGE and Western immunoblot as described in the legend for Fig. 6. Lane 1, pre-IL-1β + ICE; lane 2, pre-IL-1β + BQ-treated ICE; lane 3, pre-IL-1β marker; lane 4, IL-1β marker. This experiment was repeated 2 additional times with identical results.

monoclonal anti-IL-1β antibody (Fig. 7). Treatment of ICE for 10 min with 3 μM BQ prior to incubation with the precursor resulted in complete inhibition of the conversion of pre-IL-1β to 17 kDa cytokine (Fig. 7, lane 2 compared with lane 1). This experiment was repeated 3 times. The inhibition of ICE protease activity by BQ was equal to inhibition by iodoacetic acid, which is known to alkylate free SH groups and to inhibit ICE (data not presented).

In summary, HQ, by oxidation to the potent electrophile BQ in the SØ, or the myeloid B1 cell, can inactivate the SH-dependent proteases responsible for converting pre-interleukins-1α and -1β to biologically active cytokines. The results with IL-1α support our observations in the the mouse that SØ of mice suffering from BZ-induced BM hypocellularity are incapable of converting pre-IL-1α to active IL-1α. If the results with the mouse model can be extrapolated to humans, the occurrence of aplastic anemia in individuals chronically exposed to BZ may result from apoptosis of hematopoietic progenitor cells brought about by a lack of essential cytokines subsequent to SØ dysfunction and deficient IL-1 production caused by BQ inactivation of the protease responsible for processing pre-IL-1.

ACKNOWLEDGMENTS

This work was supported by EPA Grant R819301. We are indebted to Robert Coleman and Harlan Bradford for performing the experiment presented as Fig.4.

REFERENCES

1. Askoy, M., 1985, Malignancies due to occupational exposure to benzene, *Am. J. Ind. Med.* 7:395-402.
2. Goldstein, B.D., 1988, Benzene toxicity, *Occup. Med.* 3:541-554.
3. Kalf, G. F., 1987, Recent advances in the metabolism and toxicity of benzene, *CRC Crit. Rev. Toxicol.* 18:141-159.
4. Gascon, P., and Scala, G., 1988, Decreased interleukin-1 production in aplastic anemia, JAMA, 85:668-67.
5. Nakao, N., Matsushima, K., and Yoiung, N., 1989, Decreased interleukin-1 production in aplastic anemia, Br. *J. Haematol.* 71:431-436.
6. Bagby, G. C., 1987, Production of multi-lineage growth factors by hematopoietic stromal cells: An intercellular network involving mononuclear phagocytes and interleukin-1, *Blood Cells* 13:147-159.
7. Brandwein, S. R., 1986, Regulation of interleukin-1 production by mouse peritoneal macrophages. Effects of arachidonic acid metabolites and interferons, *J. Biol. Chem.* 261:8624-8632.
8. Fibbe, W. E., Van Damme, J., Billiau, A., Duinkerken, N., Lurvink, E., Ralph, P., Altrock, B. W., Kaushansky, K., Willemeze, R., and Falkenburg, J. H.F., 1988, Human fibroblasts produce granulocyte-CSF, macrophage-CSF, and granulocyte-macrophage-CSF following stimulation by interleukin-1 and poly(rI)·poly(rC), *Blood* 72:860-866.
9. Stanley, R.E., Bartocci, A., Patinkin, D., Rosendahl, M ., and Bradley, T. R., 1986, Regulation of very primitive multipotent hematopoietic cells by hematopoietin-1, *Cell* 45:667-674.
10. Yang, Y-C., Tsai, S., Wong, G. G., and Clark, S. C., 1988, Interleukin-1 regulation of hematopoietic growth factor production by human stromal fibroblasts, *J. Cell Physiol.* 134: 292296.
11. Zucali, J. R., Dinarello, C. A., Oblon, D. J., Gross, M. A., Anderson, L., Weiner, R. S., 1986, Interleukin-1 stimulates fibroblasts to produce granulocyte-macrophage colony-stimulating activity and prostaglandin E_2, *Am. Soc. Clin. Invest.*, 77:1857-1863.
12. Dinarello, C. A., 1988, Interleukin-1, *Ann. NY Acad.* 546:122-132.
13. Kobayashi, Y., Matsushima, K., and Oppenheim, J., 1989, Differential gene expression, synthesis, processing and release of interleukin-1α and interleukin-1β. In Bomford and Henderson (eds): *Interleukin-1, Inflammation and Disease.* New York, NY, Elsevier Science Publishers, pp. 47-62.

14. Kobayashi, Y., Ysamamoto, K., Saido, T., Kawasaki, H., Oppenheim, J., and Matsushima, K., 1990, Identification of calcium-activated neutral protease as a processing enzyme of interleukin-1α. Proc. Natl. Acad. Sci. USA 87:5548-5552.
15. Carruth, L.M., Demczuk, S., and Mizel, S.B.,1991, Involvement of a calpain-like protease in the processing of the murine interleukin-1α precursor. *J. Biol. Chem.* 266:12162-
16. Black, R. A., Kronheim, S. R.T., and Sleath, P.R., 1989, Activation of interleukin-1β by a co-induced protease. *FEBS Lett* . 247:386-390.
17. Ceretti, N. P., Kozlosky, B., Mosley, N., Nelson, K., Van Ness, K., Greenstreet, T. A., et al., 1992, Molecular cloning of the interleukin-1β converting enzyme. *Science* 256:97-99.
18. Thornberry, N. A., Bull, H. G., Calaycay, J. R., Chapman, K. T., Howard, A. D., Kostura, M. J., et al., 1992, A novel heterodimeric cysteine protease is required for interleukin-1β processing in monocytes. *Nature* 356:768-774.
19. Lewis, J. G., Odom, B., and Adams, D. O., 1988, Toxic effects of benzene and benzene metabolites on mononuclear phagocytes. *Toxicol. Appl. Pharmacol.* 92:246-254.
20. Laskin, D., MacEachern, L., and Snyder, R., 1988, Activation of bone marrow macrophages following benzene treatment of mice, *Environ. Health Perspect.* 82:75-79,
21. Thomas, D. J., Reasor, M. J., and Wierda, D., 1989, Macrophage regulation of myelopoiesis is altered by exposure to the benzene metabolite, hydroquinone, *Toxicol.. Appl. Pharmacol.* 97:440-453.
22. Rickert, D. E., Baker, T. S., Bus, J. S., Barrow, C. S., and Irons, R. D., 1979, Benzene disposition in the rat after exposure by inhalation. *Toxicol. Appl. Pharmacol.* 49:417-423.
23. Schlosser, M. J., and Kalf, G. F., 1989, Metabolic activation of hydroquinone by macrophage peroxidase. *Chem. Biol. Interact.* 72:191-207.
24. Thomas, D. J., and Weirda, D., 1988, Bone marrow stromal macrophage production of interleukin-1 activity is altered by benzene metabolites. *Toxicologist* 8:72.
25. Reasor, M. J., Cutler, D. R., Strobl, J. S., Weirda, D., and Landreth, K. S., 1990, Hydroquinone (HQ)-induced toxicity to macrophages. *Toxicologist* 10:59.
26. Renz, J. F., and Kalf, G. F., 1991, Role for interleukin-1 (IL-1) in benzene-induced hematotoxicity: Inhibition of conversion of pre-IL-1α to mature cytokine in murine macrophages by hydroquinone and prevention of benzene-induced hematotoxicity in mice by IL-1α *Blood* 78:938-944.
27. Pirrozi, S. J., Renz, J. F., and Kalf, G. F., 1989, The prevention of benzene-induced genotoxicity in mice by indomethacin, *Mutat. Res.* 222:291-298.
28. Suzuki, K., Hayashi, H., Hayashi, T., and Iwai, K., 1983, Amino acid sequence around the active site cysteine residue of calcium-activated neutral protease (CANP). *FEBS Lett.* 152:67-70.
29. Cohen, A., Grunberger, T., Vanek, W., and Freedman, M. H., 1993, Role of cytokines in growth control of acute lymphoblastic leukemia cell lines. In Guigon et al., (eds): *The Negative Regulation of Hematopoiesis.* Colloque INSERM John Libbey Eurotext Ltd v229 pp341-348.
30. Cohen, A., Petsche, D., Grunberger, T., and Freedman, M. J., 1992, Interleukin-6 induces myeloid differentiation of a human biphenotypic cell line. *Leukemia Res.* 16:751-756.

SULFOTRANSFERASE-MEDIATED ACTIVATION OF SOME BENZYLIC AND ALLYLIC ALCOHOLS

Young-Joon Surh

Division of Environmental Health Sciences
Department of Epidemiology and Public Health
Yale University School of Medicine
New Haven, Connecticut 06520-8034

Sulfo-conjugation in general renders the lipophilic xenobiotics water-soluble, thereby facilitating their removal from the body through excretion in urine or bile. Under certain circumstances, however, sulfonation can also contribute to toxification of some classes of chemicals (recently reviewed by Miller, 1994; Miller and Surh, 1994). Since the first discovery of 2-acetylaminofluorene-*N*-sulfate as an electrophilic metabolite of *N*-hydroxy-2-acetylaminofluorene (Debaun et al., 1968; King and Phillips, 1968), there has been accumulated evidence for the association between sulfonation and induction of tumors by this carcinogen and others. Although the "bay-region" theory is widely accepted as a general concept in determining the tumorigenicity of polycyclic aromatic compounds (PAH) in general, recent studies have shown that PAHs bearing benzylic hydroxyl functional groups can also be activated via sulfuric acid esterification (*vide infra*). This report briefly reviews our recent findings on metabolic activation of some benzylic alcohols to electrophilic, mutagenic, or carcinogenic sulfate esters. The paper also concerns the sulfate ester of the heterocyclic allylic alcohol, 5-hydroxymethylfurfural, as a novel example of electrophilic species formed by sulfotransferase activity.

SULFOTRANSFERASE-DEPENDENT ACTIVATION OF HYDROXYMETHYL POLYARENES

Our recent studies (for review, see Surh and Miller, 1994) and those by others (Watabe et al., 1989; Glatt et al., 1990, 1994a; Falany et al., 1992) have shown that some hydroxymethyl PAHs undergo metabolic activation to electrophilic sulfate esters. These include hydroxymethyl derivatives of 7,12-dimethylbenz[*a*]anthracene, 7-hydroxymethylbenz[*a*]anthracene, 6-hydroxymethylbenzo[*a*]pyrene, 5-hydroxymethylchrysene, 1-hydroxymethylpyrene, and 9-hydroxymethyl-10-methylanthracene, which formed covalently bound adducts with calf thymus DNA and also induced His^+ reversion in Ames *Salmonella* assay in the presence of rodent liver cytosol enriched with the sulfo-group donor, 3'-phosphoadenosine-5'-phosphosulfate (PAPS). Some of these hydrocarbons also afforded the same aralkyl DNA adducts in rat or mouse liver *in vivo* as those produced by cytosolic sulfotransferase activity *in vitro*. Dehydroepiandrosterone (DHEA), a typical substrate for hydroxysteroid sulfotransferases, significantly inhibited the cytosol- and PAPS-mediated DNA binding and mutagenicity of above hydroxymethyl polyarenes (Surh et al., 1987, 1989, 1990a, 1990b, 1991a). Pretreatment of rats with DHEA also attenuated the formation of hepatic aralkyl DNA adducts from 7-hydroxymethyl-12-methylbenz[*a*]anthracene and 6-hydroxymethylbenzo[*a*]pyrene (Surh et al., 1987, 1989, 1991a).

The chemically synthesized sulfuric acid esters of hydroxymethyl PAHs were directly mutagenic without metabolic activation, and produced much higher levels of aralkyl nucleic acid adducts than did their parent hydrocarbons in rodent liver *in vivo* as well as in reactions *in vitro* (Surh et al, 1987, 1989, 1990a). The intrinsic mutagenicity of some of these reactive esters was significantly repressed by glutathione (GSH) and GSH *S*-transferase activity present in rat hepatic cytosol (Surh et al., 1990a, 1990b; Watabe et al., 1987). Ascorbic acid also protects against 6-sulfooxymethylbenzo[*a*]pyrene by forming an inactive adduct with this electrophilic ester (Surh et al., 1994a). When injected intraperitoneally in infant male B6C3F1 mice, 6-sulfooxymethylbenzo[*a*]pyrene exhibited extremely potent hepatocarcinogenicity (Surh et al., 1990b). Single topical application of this reactive ester to female CD-1 mice followed by promotion with 12-*O*-tetradecanoylphorbol-13-acetate produced a higher yield of papillomas than did the parent hydroxymethyl hydrocarbon (Surh et al., 1991b). Some of other sulfuric acid esters of aforementioned hydroxymethyl hydrocarbons had only weak carcinogenic activity compared to their parent compounds (Surh et al., 1990a, 1991c).

METABOLIC ACTIVATION OF SECONDARY BENZYLIC HYDROXYL AROMATIC HYDROCARBONS THROUGH SULFONATION

Benzylic Hydroxyl Derivatives Cyclopenta- and Methylene-Bridged Aromatic Compounds

Besides hydroxymethyl polyarenes which are primary alcohols, certain aromatic hydrocarbons with secondary benzylic hydroxyl functionality, such as monohydroxy and 3,4-dihydroxy derivatives of cyclopenta[*cd*]pyrene (structures shown in Figure 1), also formed highly reactive and mutagenic sulfuric acid ester metabolites by sulfotransferase activity (Surh et al., 1993; Surh and Tannenbaum, 1994a; Glatt et al., 1994b). Another class of secondary benzylic alcohols activated through sulfonation includes methylene-bridged polyarenols (Figure 1) such as 4*H*-cyclopenta[*def*]chrysen-4-ol which exerts relatively potent mutagenicity in *Salmonella typhimurium* TA98 when incubated with rat liver cytosol and PAPS (Glatt, et al., 1993). In contrast to these findings, benzylic hydroxyl derivatives of other methylene-bridged PAHs, for instance, 4*H*-cyclopenta[*def*]phenanthren-4-ol and fluoren-9-ol, did not exhibit appreciable mutagenicity in the presence of rodent sulfotransferase activity (Surh, et al., unpublished observation; Glatt, et al., 1994b). Moreover, the intrinsic mutagenicities of sulfuric acid esters of the latter hydrocarbons were not pronounced compared with that of 1-sulfooxymethylpyrene while their benzylic chloro analogues were found to be extremely potent mutagens (Table 1). Thus, it seems likely that the lack of sulfotransferase-dependent mutagenicity of 4*H*-cyclopenta[*def*]phenanthren-4-ol and fluoren-9-ol is in part associated with the relatively weak inherent mutagenic potency of their metabolically formed sulfuric acid esters.

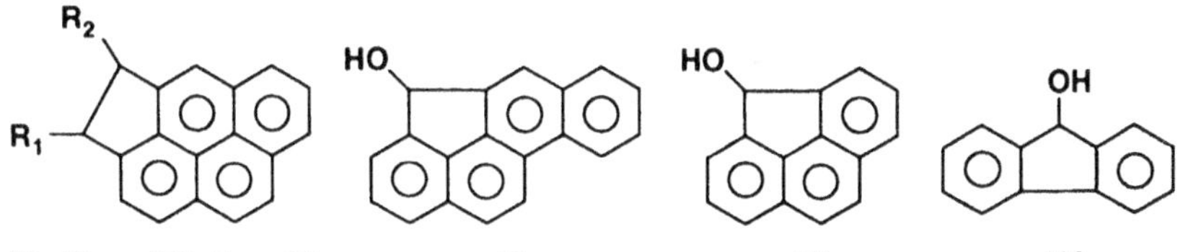

IA (R_1 = OH; R_2 = H) II III IV
IB (R_1 = H; R_2 = OH)
IC (R_1 = R_2 = OH)

Figure 1. The structuctural formulae of some secondary benzylic hydroxyl PAHs which have been tested for sulfotransferase-dependent activation. 3-Hydroxy- (**IA**), 4-hydroxy- (**IB**), and 3,4-dihydroxy-3,4-dihydro (**IC**) derivatives of cyclopenta[*cd*]pyrene; 4*H*-cyclopenta[*def*]chrysen-4-ol (**II**); 4*H*-cyclopenta[*def*]phenanthren-4-ol (**III**); and fluoren-9-ol (**IV**).

Dihydrodiol and Tetraol Derivatives of Benzo[a]pyrene

Benzo-ring dihydrodiol and tetraol derivatives of benzo[*a*]pyrene possess the secondary benzylic hydroxyl functional group(s) at *C*-7 and/or *C*-10 positions. If these primary metabolites of benzo[*a*]pyrene would form reactive esters via benzylic sulfonation, this could represent a potential metabolic activation pathway for benzo[*a*]pyrene alternative to the generally accepted "bay-region epoxidation".

Our preliminary study shows that the model benzo-ring reduced benzyl alcohol, 7-hydroxy-7,8,9,10-tetrahydrobenzo[*a*]pyrene, exerts bacterial mutagenicity and covalently binds to calf thymus DNA (Table 2) when incubated with rat liver cytosol plus PAPS, and that its synthetic sulfuric acid ester is a direct-acting mutagen (Surh and Tannenbaum, manuscript submitted). However, neither the benzo-ring dihydrodiol nor the tetraol derivative of benzo[*a*]pyrene was further activated to electrophilic or mutagenic species in the presence of hepatic cytosol and PAPS (Figure 2). The 3,4-diol derivative derived from benzo[*a*]pyrene K-region oxide was also inactive in terms of sulfotransferase-dependent activation (Figure 2). Thus, it seems likely that the appearance of extra hydroxyl functional

Table 1. Comparative mutagenic activities of benzylic sulfooxy and chloro derivatives of 4*H*-cyclopenta[*def*]phenanthrene (CPP)

Compound	Dose (nmol)	Mutagenicity (No. of His$^+$ revertants/plate) TA98	TA100
DMSO	—	17 (16,17)*	65 (63,68)
4-SO_4Na-CPP	100	14 (16,11)	81 (79,82)
	500	42 (39,44)	126 (103,149)
	1,000	67 (61,72)	216 (189,243)
4-Cl-CPP	0.1	51 (57,45)	184 (169,198)
	1	625 (650,600)	> 1,250
	10	> 2,000	> 2,000
1-CH_2SO_4Na-pyrene	5	~ 1,500	> 2,000

Mutagenicity was assessed according to the liquid-preincubation method as described by Maron and Ames (1983).
*Average of two values obtained from duplicate incubations. Individual values are shown in parentheses.

Table 2. Sulfotransferase-dependent covalent DNA binding of 7-hydroxy-7,8,9,10-tetrahydrobenzo[*a*]pyrene

	Incubation conditions	Covalent DNA binding (pmol/mg DNA)
Experiment 1	Complete	4.7 (3.6, 5.7)
	- PAPS	0.5 (0.6, 0.4)
	Heat-inactivation of cytosol	1.1 (0.3, 1.9)
Experiment 2	Complete	8.6 (8.8, 8.4)
	+ DCNP	1.2 (1.3, 1.1)
	+ DHEA	2.9 (2.7, 3.2)
	+ DCNP & DHEA	0.6 (0.6, 0.5)

Tritium labeled 7-hydroxy-7,8,9,10-tetrahydrobenzo[*a*]pyrene was incubated at 37°C for 30 min (exp. 1) or 2 h (exp. 2) with calf thymus DNA (2 mg) and liver cytosol (2 mg protein) from young adult female Sprague-Dawley rats, in a total volume of 2 ml Tris-HCl buffer (50 mM, pH 7.4) containing 3 mM $MgCl_2$ and 0.1 mM PAPS. The concentrations of the radioactive hydrocarbon used for experiments 1 and 2 were 25 μM (sp. act., 0.156 Ci/mmol) and 5 μM (sp. act. 0.236 Ci/mmol), respectively. The concentration of DHEA and DCNP was 100 μM.

group(s) in benzo[*a*]pyrene dihydrodiols and tetraol hinders the sulfonation of neighbouring benzylic hydroxyl group(s) in these molecules.

ALLYLIC SULFURIC ACID ESTERIFICATION AS A POSSIBLE METABOLIC ACTIVATION PATHWAY FOR 5-HYDROXYMETHYLFURFURAL

In addition to benzylic sulfonation, sulfuric acid esterification of an allylic hydroxyl functional group of a certain chemical may also lead to the formation of a reactive sulfuric acid ester. An example is 5-hydroxymethylfurfural, a ubiquitous food contaminant which is formed as a result of Maillard reactions during cooking or heat-sterilization (Ulbricht et al., 1984). Mutagenic and tumorigenic activities of this aldehyde have been reported (Omura et al., 1983; Schoental et al., 1971; Zhang et al., 1993). Sulfonation of

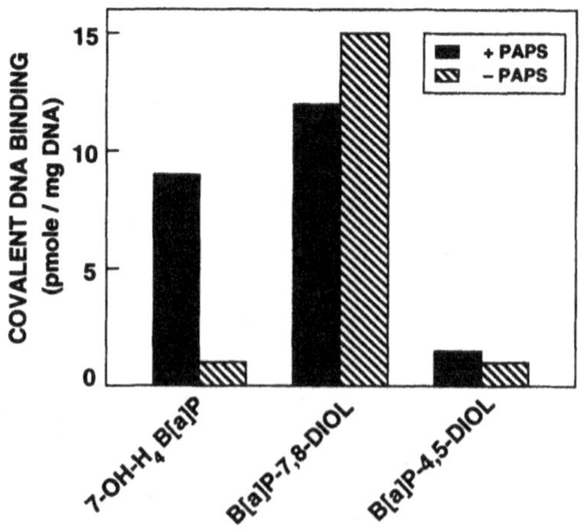

Figure 2. Comparison of covalent DNA binding of 7-hydroxy-7,8,9,10-tetrahydrobenzo[*a*]pyrene, benzo[*a*]pyrene 4,5-diol, and benzo[*a*]pyrene 7,8-dihydrodiol. Tritium-labeled hydrocarbons (12.5 μM) were incubated with calf thymus DNA (2 mg) and hepatic cytosol from female Sprague-Dawley rats (24-day-old) in a total volume of 2 ml Tris-HCl buffer (50 mM, pH 7.4) containing 3 mM $MgCl_2$ at 37°C for 2 h with and without PAPS (0.1 mM).

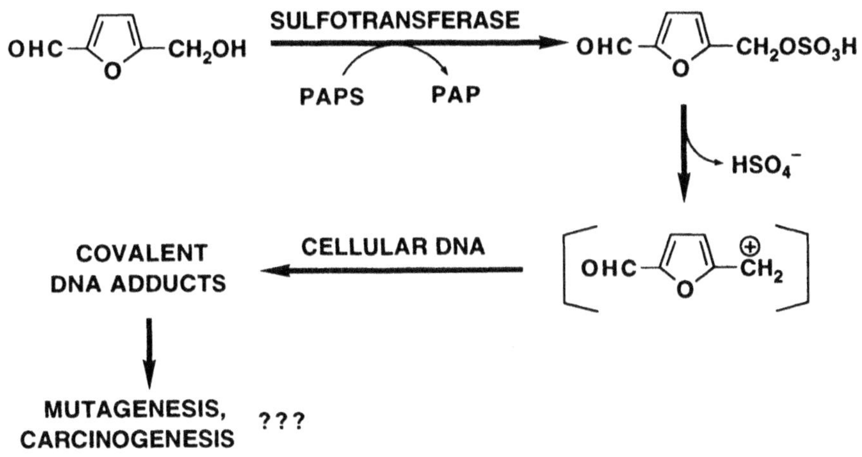

Figure 3. A proposed pathway for metabolic activation of 5-hydroxymethylfurfural via an electrophilic sulfuric acid ester.

5-hydroxymethylfurfural with subsequent departure of the leaving group from the formed reactive sulfuric acid ester would be expected to generate a highly reactive allylic carbonium ion as proposed in Figure 3. The chemically synthesized sulfuric acid ester, 5-sulfooxymethylfurfural, was found to be directly mutagenic in both bacterial and human lymphoblast cells in culture (Surh and Tannenbaum, 1994b; Surh et al., 1994b; Lee et al., 1995), lending further support to the above presumption. The intrinsic bacterial mutagenicity of this electrophilic ester was markedly inhibited by reduced GSH in the presence of dialyzed rat liver cytosol (Lee et al., 1995). 5-Sulfooxymethylfurfural also showed stronger tumor initiating activity than did HMF when topically applied to the mouse skin (Surh et al., 1994b). In our recent investigation, metabolic formation of the allylic sulfate ester of 5-hydroxymethylfurfural has been confirmed by sulfotransferase-dependent enhancement of mutagenicity of the parent compound (Table 3), which is repressed by both DHEA and 2,6-dichloro-4-nitrophenol (DCNP), an inhibitor of aryl-sulfotransferase IV (Table 4). The nonspecific sulfotransferase inhibitor, 3'-phosphoadenosine-5-phosphate (PAP), also attenuated the bacterial mutagenicity induced by 5-hydroxymethylfurfural in the presence of hepatic cytosol and PAPS.

Table 3. Sulfotransferase-dependent bacterial mutagenicity of 5-hydroxymethylfurfural

5-CH_2OH-furfural (10 μmol)	Cytosol	PAPS	No. of His$^+$ revertants/plate
–	–	–	68 (72,63)
–	+	–	124 (133,115)
–	+	+	196 (181,210)
+	+	–	193 ± 9.4
+	–	+	59 ± 2.1
+	+	+	347 ± 29.5

5-Hydroxymethylfurfural (10 μmol) was incubated with *Salmonella typhimurium* TA104 and dialyzed male Sprague-Dawley rat liver cytosol fortified with PAPS (0.25 μmol) at 37° C for 2 h in a total volume of 0.6 ml potassium phosphate buffer (80 mM, pH 7.4).

Table 4. Effect of sulfotransferase inhibitors on cytosol- and PAPS-dependent mutagenicity of 5-hydroxymethylfurfural in *Salmonella typhimurium* TA104

Inhibitor	Conc. (μM)	No. of His^+ revertants/plate
None	–	297 ± 3.1
DCNP	100	115 ± 3.5*
DHEA	100	180 ± 3.9*
PAP	100	152 ± 1.7*

The incubation conditions were same as those described in Table 3 except that 2.5 μmol of the mutagen was used.
*Significantly different from the control ($P < 0.0001$).

ACKNOWLEDGEMENTS

This work was supported in part by Swebilius Cancer Research Grant and American Cancer Society Institutional Grant IN31-34.

REFERENCES

DeBaun, J.R., Rowley, J.Y., Miller, E.C. and Miller, J.A. (1968). Sulfotransferase-activation of *N*-hydroxy-2-acetylaminofluorene in rodent livers susceptible and resistant to this carcinogen. *Proc. Soc. Exp. Biol. Med.*, 129, 268-273.

Falany, C.N., Wheeler, J., Coward, L., Keehan, D., Falany, J.L. and Barnes, S. (1992). Bioactivation of 7-hydroxymethyl-12-methylbenz[*a*]anthracene by rat liver bile acid sulfotransferase I. *J. Biochem. Toxicol.*, 7, 241-248.

Glatt, H.R., Henschler, R., Phillips, D.H., Blake, J.W., Steinberg, P., Seidel, A. and Oesch, F. (1990). Sulfotransferase-mediated chlorination of 1-hydroxymethylpyrene to a mutagen capable of penetrating indicator cells. *Environ. Hlth. Perspect.*, 88, 43-48.

Glatt, H.R., Henschler, R., Frank, H., Seidel, A., Yang, C., Abu-Shqara, E. and Harvey, R.G. (1993). Sulfotransferase-mediated mutagenicity of 1-hydroxymethylpyrene and 4*H*-cyclopenta[*def*]chrysen-4-ol and its enhancement by chloride anions. *Carcinogenesis*, 14, 599-602.

Glatt, H.R., Werle-Schneider, G., Enders, N., Monnerjahan, S., Pudil, J., Czich, A., Seidel, A. and Schwarz, M. (1994a). 1-Hydroxymethylpyrene and its sulfuric acid ester: toxicological effects *in vitro* and *in vivo*, and metabolic aspects. *Chem.-Biol. Interact.*, 92, 305-319.

Glatt, H., Pauly, K., Frank, H., Seidel, A., Oesch, F., Harvey, R.G. and Werle-Schneider, G. (1994b). Substrate-dependent sex differences in the activation of benzylic alcohols to mutagens by hepatic sulfotransferases of the rat. *Carcinogenesis*, 15, 2605-2611.

King, C.M. and Phillips, B. (1968). Enzyme-catalyzed reactions of the carcinogen *N*-hydroxy-2-fluorenylacetamide with nucleic acid. *Science*, 159, 1351-1353.

Lee, Y.-C., Shlyankevich, M. and Surh, Y.-J. (1995). Sulfotransferase-mediated activation of 5-hydroxymethylfurfural to a mutagen. *Proc. Am. Assoc. Cancer Res.*, 36, in press.

Maron, D.M. and Ames, B.R. (1983) Revised methods for the Salmonella mutagenicity test. *Mutat. Res.*, 290, 111-118.

Miller, J.A. (1994). Sulfonation in chemical carcinogenesis– history and present status. *Chem.-Biol. Interact.*, 92, 329-341.

Miller, J.A. and Surh, Y.-J. (1994). Sulfonation in chemical carcinogenesis. In *Conjugation-Deconjugation Reactions in Drug Metabolism and Toxicity* (F.C. Kauffman Ed.), pp. 459-509. Springer-Verlag, Heidelberg.

Omura, H., Jahan, N., Shinohara, K. and Murakami, H. (1983). Formation of mutagens by the Maillard reaction. In *The Maillard reaction in Foods and Nutrition* (G.R. Waller and M.S. Feather, Eds.), pp. 537-563, American Chemical Society, Washington, D.C.

Schoental, R., Hard, G.C. and Gibbard, S. (1971). Histopathology of renal lipomatous tumours in rats treated with the natural products, pyrrolizidine alkaloids and alpha,beta-unsaturated aldehydes. I. *J. Natl Cancer Inst.*, 47, 1037-1044.

Surh, Y.-J., Lai, C.-C., Miller, J.A. and Miller, E.C. (1987). Hepatic DNA and RNA adduct formation from the carcinogen 7-hydroxymethyl-12-methylbenz[*a*]anthracene and its electrophilic sulfuric acid ester metabolite in preweanling rats and mice. *Biochem. Biophys. Res. Commun.*, 144, 576-582.

Surh, Y.-J., Liem, A., Miller, E.C. and Miller, J.A. (1989). Metabolic activation of the carcinogen 6-hydroxymethylbenzo[*a*]pyrene: formation of an electrophilic sulfuric acid ester and benzylic DNA adducts in rat liver *in vivo* and in reactions *in vitro*. *Carcinogenesis*, 10, 1519-1528.

Surh, Y.-J., Blomquist, J.C., Liem, A. and Miller, J.A. (1990a). Metabolic activation of 9-hydroxymethyl-10-methylanthracene and 1-hydroxymethylpyrene to electrophilic, mutagenic, and tumorigenic sulfuric acid esters by rat hepatic sulfotransferase activity. *Carcinogenesis*, 11: 1451-1460.

Surh, Y.-J., Liem, A., Miller, E.C. and Miller, J.A. (1990b). The strong hepatocarcinogenicity of the electrophilic and mutagenic metabolite 6-sulfooxymethylbenzo[*a*]pyrene and its formation of benzylic DNA adducts in the livers of infant male B6C3F$_1$ mice. *Biochem. Biophys. Res. Commun.*, 172, 85-91.

Surh, Y.-J., Liem, A., Miller, E.C. and Miller, J.A. (1991a). Age- and sex-related differences in activation of the carcinogen, 7-hydroxymethyl-12-methylbenz[*a*]anthracene to an electrophilic sulfuric acid ester metabolite in rats: possible involvement of hydroxysteroid sulfotransferase activity. *Biochem. Pharmacol.*, 41, 213-221.

Surh, Y.-J., Liem, A., Miller, E.C. and Miller, J.A. (1991b). Activation and detoxication of 6-hydroxymethylbenzo[*a*]pyrene via its electrophilic and carcinogenic sulfuric acid ester metabolite. *Proc. Am. Assoc. Cancer Res.*, 32, 726.

Surh, Y.-J., Liem, A., Miller, E.C. and Miller, J.A. (1991c). 7-Sulfooxymethyl-12-methyl-benz[*a*]anthracene is an electrophilic mutagen, but does not appear to play a role in carcinogenesis by 7,12-dimethylbenz[*a*]anthracene or 7-hydroxymethyl-12-methyl-benz[*a*]anthracene. *Carcinogenesis*, 12, 339-347.

Surh, Y.-J., Blomquist, J.C. and Miller, J.A. (1991d). Activation of 1-hydroxymethyl- pyrene to an electrophilic and mutagenic metabolite by rat hepatic sulfotransferase activity. *Adv. Exp. Med. Biol.*, 283, 383-391.

Surh, Y.-J., Kwon, H. and Tannenbaum, S.R. (1993). Sulfotransferase-mediated activation of 4-hydroxy- and 3,4-dihydroxy-3,4-dihydrocyclopenta[*cd*]pyrene, the major secondary metabolites of cyclopenta[*cd*]pyrene. *Cancer Res.*, 53, 1017-1022.

Surh, Y.-J. and Miller, J.A. (1994). Roles of electrophilic sulfuric acid ester metabolites in mutagenesis and carcinogenesis by some polynuclear aromatic hydrocarbons. *Chem.-Biol. Interact.*, 92, 351-362.

Surh, Y.-J. and Tannenbaum, S.R. (1994a). Bioactivation of cyclopenta- and cyclohexa-fused polycyclic aromatic hydrocarbons via the formation of benzylic sulfuric acid esters. *Polycyclic Aromatic Compounds.*, 7, 83-90.

Surh, Y.-J. and Tannenbaum, S.R. (1994b). Activation of the Maillard reaction product, 5-hydroxymethylfurfural via allylic sulfonation and chlorination. *Chem. Res. Toxicol.*, 7, 313-318.

Surh, Y.-J., Park, K. K., and Miller, J.A. (1994a). Inhibitory effect of vitamin C on the mutagenicity and covalent DNA binding of the electrophilic and carcinogenic metabolite, 6-sulfooxymethylbenzo[*a*]pyrene. *Carcinogenesis*, 15, 917-920.

Surh, Y.-J., Liem, A., Miller, J.A. and Tannenbaum, S.R. (1994b). 5-Sulfooxymethylfurfural as a possible ultimate mutagenic and carcinogenic metabolite of the Maillard reaction product, 5-hydroxymethylfurfural. *Carcinogenesis*, 15, 2375-2377.

Ulbricht, R.J., Northup, S.J. and Thomas, J.A. (1984). A review of 5-hydroxymethylfurfural (HMF) in parenteral solutions. *Fund. Appl. Toxicol.*, 4, 843-853.

Watabe, T., Hiratsuka, A. and Ogura, K. (1987). Sulfotransferase-mediated covalent binding of the carcinogen 7,12-dihydroxymethylbenz[*a*]anthracene to calf thymus DNA and its inhibition by glutathione transferase. *Carcinogenesis*, 8, 445-453.

Watabe, T., Ogura, K., Okuda, H. and Hiraszuka, A. (1989). Hydroxymethyl sulfate esters as reactive metabolites of the carcinogens, 7-methyl- and 7,12-dihydroxymethyl- benz[*a*]anthracenes and 5-methylchrysene. In *Xenobiotic Metabolism and Disposition* (R. Kato, R.W. Estabrook and M.N. Cayen, Eds.), pp. 393-400. Taylor and Francis, London.

Zhang, X.-M., Chan, C.C., Stamp, D., Minkin, S., Archer, M.C. and Bruce, W.R. (1993). Initiation and promotion of colonic aberrant crypt foci in rats by 5-hydroxymethyl-2-furaldehyde in thermolyzed sucrose. *Carcinogenesis*, 14, 773-775.

APPLICATION OF COMPUTATIONAL CHEMISTRY IN THE STUDY OF BIOLOGICALLY REACTIVE INTERMEDIATES

M. W. Anders, Hequn Yin, and Jeffrey P. Jones

Department of Pharmacology
University of Rochester
601 Elmwood Avenue
Rochester, New York 14642

INTRODUCTION

The toxicity of most xenobiotics is associated with their enzymatic conversion to toxic metabolites, a process termed bioactivation. Although stable, but toxic, metabolites may be formed, as in the biotransformation of dichloromethane to carbon monoxide, most bioactivation reactions afford electrophilic, reactive intermediates. The reactivity of these intermediates usually prevents their direct observation and characterization. Hence strategies that permit the experimentalist to gain insight into the formation and fate of reactive intermediates is of much value in understanding bioactivation reactions. The objective of this review is to point out the utility of computational chemistry in studying the formation and fate of toxic metabolites.

INTRODUCTION TO COMPUTATIONAL CHEMISTRY

Computational chemistry simulates the structures and reactions of chemicals numerically based on the fundamental laws of physics. Some computational chemistry methods can be used to model not only stable molecules, but also transition states, short-lived, unstable intermediates (hence the application to biological reactive intermediates), and even compounds that have not been synthesized.

Computational chemistry can be divided into two broad areas: quantum chemistry and molecular mechanics. Both methods perform calculations on the energy of a particular molecule or of a molecular system and locate the optimized geometry with lowest energy. The two methods do, however, differ from each other in three ways: (1) Quantum chemistry employs the laws of quantum mechanics and molecular mechanics uses the laws of classical physics as the bases for their calculations. (2) Quantum chemistry describes explicitly the motion of electrons in the field of nuclei (after some reasonable approximations), whereas molecular mechanics computations are based on interactions among the nuclei and the electronic effects are implicitly included in the parameterizations of the force fields. (3) Quantum chemical calculations can only be made with relatively small systems, whereas molecular mechanics calculations can be done on large systems, such as proteins.

Molecular Mechanics

As indicated above, molecular mechanics deals with the potential energy of a molecular system as a function of its components and uses the laws of classical physics. Each molecular mechanics method is characterized by its particular force field. Parameterization of a force field is achieved by correlating experimental data with atom types and equations that define how the potential energy of a molecule varies with the spatial locations of its component atoms. The molecular mechanics method has been used, for example, in structure-based drug design (1) and in the analysis of peptide structures (2). A later chapter will deal with the use of molecular mechanics in more detail (See Jones et al., this volume).

Quantum Chemistry

Quantum chemistry deals with the interactions among nuclei and electrons in a molecular system. Quantum chemistry states that the energy and other related properties of a molecule may be obtained by solving the Schrödinger equation:

$$H\Psi = E\Psi$$

where H is the molecular Hamiltonian, Ψ is a wave function, and E is the energy corresponding to the state of a molecule described by Ψ. The molecular Hamiltonian contains operators for the kinetic energy of the nuclei and electrons, for nuclear-nuclear and electron-electron repulsion, and for the attraction between nuclei and electrons. Although it is tempting to seek exact solutions to this equation, such solutions are not practical, except for small and totally symmetric systems (*e.g.*, H_2). Depending on how the approximations were used in the calculations, quantum chemistry can be divided into two classes: ab initio and semiempirical.

Ab Initio Method. The ab initio method uses only the values of physical constants to make the calculations: the speed of light, Planck's constant, and the masses and charges of the electrons and nuclei.

There are different *levels of theory* for dealing with the relationships among electrons and orbitals. HF, or Hartree-Fock theory, is the lowest level. It is a single-determinant wave function that does not take electron correlation (*e.g.*, no account of electron-electron interactions of electrons with opposite spins) into account. Methods that go beyond Hartree-Fock theory in attempting to treat electron correlation are known as *post-SCF* methods, *e.g.*, CI (Configuration Interaction) and MP (Møller-Plesset Perturbation Theory).

Different *basis sets* that describe the composition of orbitals are used. The basis set can be interpreted as restricting each electron to a particular region of space. A larger basis set imposes fewer constraints on electrons and more accurately approximates exact molecular orbitals. A minimum basis set contains the minimum number of basis functions to describe an atom, *e.g.*, H has 1s orbital; C has 1s, 2s, $2p_x$, $2p_y$, $2p_z$ orbitals. A split-valence basis set contains two or more basis functions for each orbital, *e.g.*, H has 1s and 1s' orbitals; C has 1s, 1s', 2s, 2s', etc., orbitals. Other larger basis sets include polarized, diffuse functions. For example, the 3-21G basis set is a split-valence basis set and has 3 Gaussians for the 1s orbital of carbon, 2 Gaussians for the "inner" 2s and 1 Gaussian for the "outer" 2s orbital. A Gaussian is a type of basis function that is used due to its computational efficiency.

Semiempirical Method. The semiempirical method uses parameters derived from experimental data to simplify the computations. Some relatively unimportant terms in the Schrödinger equation were left out and the resulting error was compensated for by adjusting the coefficients (parameters) of the remaining terms. Different ways of generating such approximations give different models, such as AM1, MNDO, and PM3.

The trade-offs between ab initio and semiempirical methods are computational cost and accuracy of results. Although semiempirical methods generally, but not necessarily, produce larger errors, the relative rank order of energies and the energy differences among compounds are reasonably dependable, particularly for systems where good parameter sets exist.

RELATIONSHIP BETWEEN EXPERIMENTAL AND COMPUTATIONAL CHEMISTRY

The validity of computational chemical models must ultimately be tested by experiment. But valid experimental results may be difficult to obtain, particularly for experiments on short-lived, reactive intermediates. The determination of the geometry of triplet, ground-state carbene is a well-documented story of the interplay between computation and experiment (3,4). In 1959, Foster et al. predicted a bent geometry (H-C-H bond angle = 129°) based on their computer calculations. Experimental observations by Herzberg in 1961 suggested a linear molecule. The determination of the geometry of carbene is nontrivial because it is very short-lived species. In 1970, computations by Bender and Schaefer again supported a bent structure (H-C-H bond angle = 135°). Subsequent experimental and theoretical investigations have confirmed the model proposed by Bender and Schaefer.

Computational chemistry can allow determinations that cannot be done by experiment, such as the characterization of transition states. For example, experiments can determine the energies of the transition states by measurement of activation energies, but cannot give geometries. Questions may still remain about the validity of such calculations, because experiments cannot be done to validate the calculations.

APPLICATIONS OF QUANTUM CHEMISTRY

The increasing application of computational chemistry in the study of biological problems is evidenced by the number of relevant publications that have appeared over the past three decades.

Table 1. Number of publications in Medline database that applied computational chemistry methods

Years	Ab initio	AM1	MNDO
1966-1976	20	29	0
1976-1984	80	36	5
1985-1989	110	37	25
1990-1994	133	120	48

Although both ab initio and semiempirical quantum chemical methods have seen consistently increasing applications, semiempirical methods, particularly the AM1 model, have been widely used in the past decade. Applications of quantum chemical methods are largely limited to chemical systems or to chemical models that can be extracted from biological systems.

Overview

Although there are many applications of computational chemistry, all are based on the minimization of the energy of a molecular system. Often, however, computational chemistry can help explain and help understand such systems, *e.g.*, whether a process is thermodynamically possible and kinetically feasible. In cases where the mechanism of reaction or mechanism of toxicity is known, computational chemistry can even help make predictions. The documented applications in Table 1 can be categorized as follows: (1) Prediction (or correlation) of reaction (or biotransformation) rates based on calculation of activation energies (kinetics); (2) Regioselectivity interpretations or predictions based on calculation of either heats of reaction for thermodynamically controlled, multiple-pathway reactions or based on activation energies for kinetically controlled, multiple-pathway reactions; (3) Location of preferred conformers, *e.g.*, chair and boat (thermodynamics); (4) Location of preferred configurations, *e.g.*, *E* and *Z* isomers (thermodynamics); (5) Correlation and prediction of pK_as (thermodynamics); (6) Correlation and prediction of solubilities of compounds as a function of molecular properties, such as dipole moment, charge distribution, and geometrical parameters. A few examples of such applications are discussed below.

Designing Safer Chemicals: Prediction of Cytochrome P450-Dependent Biotransformation Rates of Halogenated Alkanes

HCFCs (hydrochlorofluorocarbons) are being developed as replacements for ozone-depleting CFCs (chlorofluorocarbons) and contain C-H bonds that are oxidized by cytochromes P450 (5). Such oxidations may afford acyl halides or aldehydes, or both, as reactive intermediates, and, for structurally similar compounds that share common bioactivation pathways, toxicity may be a function of their P450-dependent oxidation rates. A recent study in our laboratory compared the experimentally determined *in vitro* rates of cytochrome P450-dependent biotransformation of HCFCs with the calculated activation energies (AM1 Hamiltonian) for the hydrogen-atom abstraction step (the first step in the P450-dependent oxidation of a C-H bond) (6). Studies with HCFC-121 (1-fluoro-1,1,2,2-tetrachloroethane), HCFC-122 (1,1-difluoro-1,2,2-trichloroethane), HCFC-123 (2,2-dichloro-1,1,1-trifluoroethane), HCFC-124 (2-chloro-1,1,1,2-tetrafluoroethane), HCFC-125 (1,1,1,2,2-pentafluoroethane), and halothane (2-bromo-2-chloro-1,1,1-trifluoroethane) gave these correlations (H. Yin et al., unpublished observations):

$$\text{Ln Rate (rat microsomes)} = 44.99 - 1.79\ (\Delta H^{\ddagger}) \qquad r^2 = 0.86$$
$$\text{Ln Rate (human CYP2E1)} = 46.99 - 1.77\ (\Delta H^{\ddagger}) \qquad r^2 = 0.97$$

The expressed isozyme CYP2E1 was used because it is believed to be the predominant isozyme that catalyzes the turnover of HCFCs (7,8); the results are illustrated in Figure 1.

Cross-validation of the models indicated that they provide good predictions (cross-validated $r^2 = 0.71$ and 0.91, respectively). Such relationships may find utility in the design and selection of safer chemicals and may reduce our dependence on animal models.

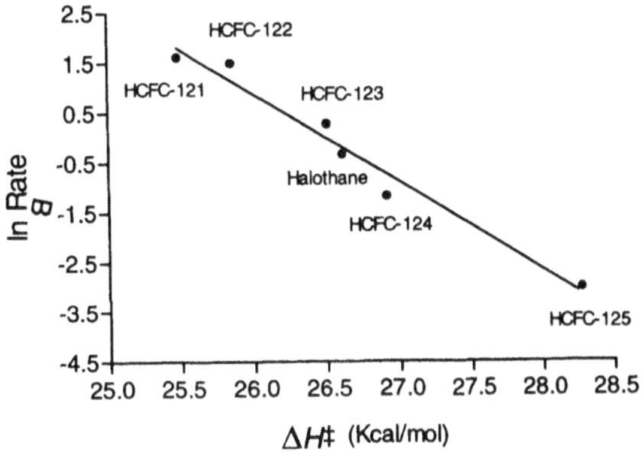

Figure 1. Correlation between measured metabolic rates in expressed human CYP2E1 and the predicted activation energies.

Quantum chemical methods cannot deal with complicated enzymatic systems. They were useful in this study because the chemical process involved (hydrogen-atom abstraction) is the key step in the overall biotransformation rate of HCFCs; therefore the biological problem can be described by a key chemical reaction.

Stereoselectivity in Fluorocitrate Formation

The key step in the bioactivation of the highly toxic fluoroacetate is its citrate synthase-catalyzed conversion to (2*R*,3*R*)-3-fluorocitrate. High stereoselectivity for the pro-S hydrogen abstraction (>100:1 selectivity, corresponding to >3 Kcal/mol energy difference) over pro-R hydrogen abstraction was observed in the citrate synthase-catalyzed conversion of fluoroacetyl CoA to fluorocitrate (9). The authors sought to determine the origin of the observed stereoselectivity. Because fluorine is scarcely larger than a proton (Van der Waals radii = 1.35 Å and 1.2 Å, respectively), steric effects are not likely to be prominent. Although it has been proposed that F—H hydrogen bonds between fluoroacety-CoA and the enzyme may be the source of the energy difference between two transition states, this is unlikely because the F—H hydrogen bond is weak and not sufficient to account for the observed selectivity. Furthermore, kinetic isotope effects ($k_H/k_D = 1.94$) indicated that the rate-limiting step is proton abstraction from fluoroacetyl-CoA. Hence, O'Hagan and Rzepa (9) hypothesized that prochiral hydrogen discrimination by the enzyme must be related to the relative energies of the syn- or anti-orientations of the C-F bond with respect to the oxygen atom of the forming neutral enol or enolate anion intermediate. Ab initio calculations [level of theory: MP2 or B-LYP; basis set: 6-31(3d), or 6-31+(3d)] were carried out on both enols and enolates. The results showed that, first, the *E* enol is more stable than the *Z* enol by 4.3 Kcal/mol; second, that the *E* enolate is more stable than the *Z* enolate by 1.1-1.9 Kcal/mol; and third, that enols are preferred over enolates in either *E* or *Z* isomers. Based on these calculations several conclusions were reached: (1) The degree of protonation of enolates, which is influenced by enzyme, modulates the energy difference between the *E* and *Z* geometries; (2) stabilization of the *E*-form (pro-S hydrogen abstraction) is not due to hydrogen bonding but is rather an example of the cis effect (thermodynamics).

Recent experimental results (10) also showed that an enolic intermediate is the nucleophilic species in the citrate synthetase reaction.

These studies provide a good example of the application of quantum chemical calculations to unravel the mechanistic origin of the stereoselectivity observed in an enzymatic reaction.

Structure-Genotoxicity Relationships of Allylbenzenes and Propenylbenzenes

Allylbenzenes and propenylbenzenes are flavor constituents found in variety of essential oils (11). Similar pathways of bioactivation may apply to these compounds: initial P450-dependent oxidation may lead to the formation of carbon-centered radicals and by oxygen rebound to allyl alcohols, which are conjugated with sulfate and subsequently converted to allyl carbonium ions. Either the intermediate radicals or carbonium ions may be mutagenic. An interesting phenomenon is that some of these compounds (*e.g.*, estragole, methyleugenol, safrole, α-asarone, β-asarone, and elemicin) are genotoxic, whereas others (*e.g.*, allylbenzene, eugenol, *trans*-anethole, isosafrole, and myristicin) are not. The objective of the research was to determine the factors responsible for the differences in mutagenicity. Because these compounds share a common bioactivation pathway, computational calculations (AM1 Hamiltonian) were carried out to compare the stability of metabolic intermediates and rates of reaction. The calculations showed: (1) The heats of reaction for the P450-dependent radical formation for the first group of compounds (14.8 ± 2.9 Kcal/mol) are identical to those of second group (14.6 ± 2.6 Kcal/mol); (2) There are no significant differences in the activation energies of P450-dependent radical formation between the two groups (18.2 ± 0.4 Kcal/mol for active compounds and 18.6 ± 0.6 Kcal/mol for inactive compounds) and, therefore, there is no difference in the rates of radical formation; (3) The heats of reaction for the formation of carbonium ions from alcohols are significantly different (all are less than 231.0 Kcal/mol, mean 227.7 ± 2.2Kcal/mol, for first group and all are more than 231.0 Kcal/mol, mean 233.9 ± 3.2 Kcal/mol, for second group, except for eugenol). The authors concluded that "This difference, while not considerable, is nevertheless genuine and is seen as suggesting that the stability of the carbonium ions may be one of the determinants for the genotoxicity of allyl and propenylbenzenes. Thus, the non-genotoxicity of some allyl- and propenylbenzenes could be due to a greater difficulty of forming its carbonium species rather than to a particular inertness of hydroxylation."

In this example, computational chemistry calculations helped to solve a biologically important problem where experimental approaches are difficult or nearly impossible to apply.

Modeling Cyanide Release from Nitriles: Prediction of Cytochrome P450-Dependent Acute Nitrile Toxicity

The acute toxicity associated with nitriles is believed to be the result of cyanide release (12), a process catalyzed by cytochromes P450. Cytochrome P450-dependent hydrogen-atom abstraction from the α-position of nitriles is the key step in the bioactivation reaction. Oxygen rebound affords cyanohydrins that eliminate cyanide. The relative rate constants ($k_{\alpha corr}$) of this step can be calculated by a computer model (AM1 Hamiltonian) (6). Comparison of the *in vivo* LD_{50} values for 26 nitriles reported in the literature with $k_{\alpha corr}$ and octanol-water partition coefficients (log P) revealed the following correlation (13):

$$\log (1/LD_{50}) = -0.12 (\log P)^2 + 0.08 (\log P) + 0.13 (\ln k_{\alpha corr}) + 7.08 \quad (r^2 = 0.90)$$

This study determined the molecular parameters useful for predicting the acute toxicity of structurally diverse nitriles and provides an example of how computational chemical calculations can be used to model the *in vivo* toxicity of a group of chemicals that are bioactivated by a common mechanism.

CONCLUSIONS

The examples provided in this brief review demonstrate the usefulness of computational chemistry in the study of biological reactive intermediates. Indeed, the ability to gain insight into the chemical behavior of unstable, short-lived intermediates will undoubtedly find considerable application in understanding the formation and fate of biological reactive intermediates. Finally, these tools may be used to design safer chemicals.

ACKNOWLEDGMENTS

This research was supported by National Institute of Environmental Health Sciences grants ES05407 (M.W.A.) and ES06062 (J.P.J.) and by the Pharmaceutical Research and Manufacturers of America Foundation, Inc. (H.Y.). The authors thank Sandra E. Morgan for her assistance in the preparation of the manuscript.

REFERENCES

1. Bowen, J. P., Charifson, P. S., Fox, P. C., Kontoyianni, M., Miller, A. B. Schnur, D., Stewart, E. L., Van Dyke, C., 1993, Computer-assisted molecular modeling: Indispensable tools for molecular pharmacology, *J. Clin. Pharmacol.*, **33**, 1149-1164.
2. Andrianov, A. M., and Akhrem, A. A., 1993, Model of the spatial structure of peptide T, *Molekuliarnai Biologiia* **27**, 934-946.
3. Goddard III, W. A., 1985, Theoretical chemistry comes alive: Full partner with experiment, *Science* **227**, 917-923.
4. Schaefer III, H. F., 1986, Methylene: A paradigm for computational quantum chemistry, *Science* **231**, 1100-1107.
5. Anders, M. W., 1991, Metabolism and toxicity of hydrochlorofluorocarbons: Current knowledge and needs for the future, *Environ. Health Perspect.* **96**, 185-191
6. Korzekwa, K. R., Jones, J. P., and Gillette, J. R., 1990, Theoretical studies on cytochrome P-450 mediated hydroxylation: A predictive model for hydrogen atom abstractions, *J. Am. Chem. Soc.* **112**, 7042-7046.
7. Olson, M. J., Kim, S. G., Reidy, C. A., Johnson, J. T., and Novak, R. F., 1991, Oxidation of 1,1,1,2-tetrafluoroethane (R-134a) in rat liver microsomes is catalyzed primarily by cytochrome P450IIE1, *Drug Metab. Dispos.* **19**, 298-303.
8. Herbst, J., Köster, U., Kerssebaum, R., and Dekant, W., 1994, Role of P4502E1 in the metabolism of 1,1,2,2-tetrafluoro-1-(2,2,2-trifluoroethoxy)-ethane, *Xenobiotica* **24**, 507-516.
9. O'Hagan, D., and Rzepa, H. S., 1994, Stereospecific control of the citrate synthase mediated synthesis of (2*R*,3*R*)-3-fluorocitrate by the relative stabilities of the intermediate fluoroenolates, *J. Chem. Soc., Chem. Commun.* **1994**, 2029-2030.
10. Martin, D. P., Bibart, R. T., and Drueckhammer, D. G., 1994, Synthesis of novel analogs of acetyl coenzyme A: Mimics of enzyme reaction intermediates, *J. Am. Chem. Soc.* **116**, 4660-4668.
11. Tsai, R.-S., Carrupt, P.-A., Testa, B., and Caldwell, J., 1994, Structure-genotoxicity relationships of allylbenzenes and propenylbenzenes: A quantum chemical study, *Chem. Res. Toxicol.* **7**, 73-76.
12. Hartung, R., 1982, Cyanide and nitriles, In *Patty's Industrial Hygiene and Toxicology* (Patty, F. A., Clayton, G. D., and Clayton, F. E., Eds.) pp. 4845-4900, Interscience, New York.
13. Grogan, J., DeVito, S. C., Pearlman, R. S., and Korzekwa, K. R., 1992, Modeling cyanide release from nitriles: Prediction of cytochrome P450 mediated acute nitrile toxicity, *Chem. Res. Toxicol.* **5**, 548-552.

PREDICTING THE REGIOSELECTIVITY AND STEREOSELECTIVITY OF CYTOCHROME P450-MEDIATED REACTIONS

Structural Models for Bioactivation Reactions

J. P. Jones, M. Shou, and K. R. Korzekwa

Department of Pharmacology
University of Rochester, Rochester, New York 14642

INTRODUCTION

Recently an initiative to design safer chemicals was made by the Evironmental Protection Agency (1). The design of safer chemicals must be a multi-discplinary effort merging chemistry and biology. One of the more important areas in this effort is understanding what governs the rate of bioactivation of xenobiotics. This area is of particular importance since many chemicals are protoxins and must be bioactivated to the ultimate toxin (2). This chapter will outline our efforts toward predicting the stereochemistry and regioselectivity of bioactivation reactions, mediated by cytochrome P450, using computational methods. The following chapter will provide some examples of how the electronic factors can be predicted using computational methods. Overall, it is our hope that these methods can be expanded and incorporated into a general program for designing safer chemicals.

To understand how protein structure influences the regioselectivity and stereoselectivity of a reaction, the protein-substrate interactions must be modeled. Molecular mechanics is an area of computational chemistry which uses the classic laws of motion and simple force fields to approximate the structure and dynamics of a molecule. Since the method can be used for very large systems it is the method of choice for studying protein-substrate interactions. Thus, large systems, including solvent, can be modeled using this method. However, the characteristics that allow for the relatively rapid approximation of the structure of large systems makes this method unable to calculate reaction rates, since bonds cannot be made or broken.

Stereoselectivity and regioselectivity are dependent on the tertiary structure of the enzyme. This means that for us to be able to predict the binding of a molecule to an enzyme we must have knowledge of the structure of the enzyme. For the cytochrome P450 family the only known tertiary structures are for the bacterial P450s. Thus, any structural factors important in mammalian P450s can only be obtained by comparison of mammalian and

Biological Reactive Intermediates V
Edited by R. Snyder et al., Plenum Press, New York, 1996

bacterial enzymes. A number of efforts have been made to construct homology models of mammalian P450s from the known crystal structures (3). This approach suffers from the fact that the bacterial and mammalian P450 enzymes share only low sequence homology and the alignments of the primary sequence are certainly in error. Furthermore, while the known tertiary structure of the 3 bacterial enzymes closely resemble each other, there is no direct evidence that the mammalian enzymes have a similar tertiary structure.

Another approach is to compare the metabolic profiles of the bacterial and mammalian isoforms to determine if they give the same results. This approach allows us to establish which bacterial crystal structure should be used to model the tertiary interactions of the mammalian isoforms. Until very recently it was believed that these enzymes only metabolized the natural substrates camphor (P450cam), fatty acids (P450BM3) and terpine (P450terp) and some closely related substrate analogs. The first study to be published on a nonnatural substrate-P450cam interactions was that of Collins et al., in a study of styrene epoxidation. Since then a group of diverse compounds have been identified as P450cam substrates including, styrene(4) , ethylbenzene(5), [R]- and [S]-nicotine(6), benzo[a]pyrene(7), para-substituted thioanisoles (8) and valproic acid (9). A brief discussion of each of these studies follows.

METHYLSTYRENE

The product stereoselectivity of cis-β-methylstyrene metabolism by P450cam was analyzed with in vitro methods and found to give a product ratio of 89% 1S,2R epoxide and 11% 1R,2S epoxide (4). Molecular mechanics simulations were conducted with this substrate and the number of times a given face of the pi system was within 4 angstroms was determined in 2 X 125 ps simulations. The predicted stereoselectivity was 84 % 1S,2R and 16% 1R,2S, in good agreement with experiment. No analysis of what components in the active site may be responsible for the resulting stereoselectivity was given and the results were not compared with the stereoselectivty of mammalian enzymes.

ETHYLBENZENE

The stereospecificity of the biotransformation of ethylbenzene was studied by molecular dynamics and by in vitro metabolism with purified P450cam by Filipovic et al., (5). The stereoselectivity for P450cam was predicted using 5 X 40 ps molecular mechanics simulations and 3 different starting orientations. A given hydrogen atom was assumed to be abstracted if the hydrogen atom came within 3 angstroms of the active iron-oxygen species and the C-H-O bond angle was between 135 and 225 degrees. The predicted stereoselectivity was 76% R and 24% S, the experimental strereoselectivity was 73% R and 27% S. This excellent agreement between theory and experiment is more striking since the substrate is small and can bind in multiple orientations. No detailed analysis was made of the specific interactions that were responsible for the stereoselectivity observed in this system. However, the figure given in the text supports the speculation that the I-helix, and in particular Thr 252, provides a steric environment that makes methyl group rotation to give S hydrogen atom abstraction unfavorable. These authors were also able to predict the amount of coupling between product formation and NADPH consumption. This analysis was performed by counting the number of times any hydrogen atom was within 3.5 angstroms of the active oxygen. It was found that 42 of 805 times a hydrogen atom was abstractable. This predicts that only 5% of the time product would be formed. The experimental result was 6%. The authors conclude that their work "... expands the general validity of computational methods in characterizing and predicting the binding and metabolism of small substrates to P450cam."

VALPROIC ACID

In an ambitous study the stereoelectronic factors responsible for valproic acid regioselectivity was predicted by Collins et al.(9) by a combination of molecular mechanics and molecular orbital theory. Based on the heat of reaction of radical formation the predicted regioselectivity is 2-OH>4-OH>3-OH>5-OH. The results were modified based on molecular dynamics results and the combined results predict a regioselectivity of 3-OH>5-OH>=4-OH. The stereoselectivity at the 4 position was predicted to lead to the R 4-OH compound only. No experimental results were presented. Recent reports test the validity of these calculations (10). The experimental regioselectivity and stereochemistry was determined for the P450cam mediated hydroxylation of valproic acid to be 4-OH>>5-OH and no 3-OH was formed. The stereoselectivity was determined to be 2R4S>2S4R>2R4R>2S4S. A number of new theoretical calculations were performed but none could predict either the in vitro regioselectivity or stereochemistry.

THIOANISOLE

The stereoselective oxidation of thioanisole and 4-methylthioanisole has also been modeled in P450cam (10). The in vitro stereoselectivity for the P450cam mediated reaction was reported to be 72% R and 28% S and 48% R and 52% S for thioanisole and 4-methylthioanisole respectively. Two calculation methods were used for each substrate with differing charges on the substrate. The predicted stereoselectivity for thioanisole was 57% R and 65% R. The predicted stereoselectivity for 4-methylthioanisole was 22% R and 20% R. Different starting orientations made a significant impact on the results for both substrates, with the predicted range for different orientations of each substrate for R sulfoxide formation of 0 to 94% for 4-methylanisole and 1 to 79% for thioanisole. Thus, for this system theory did a very poor job of predicting the experimental result. The authors could not offer an explanation for this result.

NICOTINE

We have performed simulations on the stereoselectivity of binding of the two enantiomers of nicotine and predicted what residues are important in the binding of nicotine to P450cam (6). Free energy perturbation molecular dynamics methods were used to predict the difference in binding energy between R- and S-nicotine. The result of these simulations predicts a binding energy difference of 520 calories, with the R enantiomer favored over the S enantiomer. The experimental difference in binding energy was determined to be 333 calories from the binding spectra. The energy difference was determined to be 200 calories when the product ratios were measured after incubation of the racemic mixture. Thus, the theoretical values were found to closely aproximate the experimental values. These results are of interest since this is the first experiment to show that we can predict the relative affinity of two substrates for a given P450 isoform. Furthermore, these results give a quantitative test of the force field used in these simulations.

The experimental regioselectivity and isotope effect for S-nicotine were determined in P450cam, CYP1A1, 2B1, 2B4 and human microsomes. For each of these enzymes the 5′

oxidation was favored over methyl group oxidation, the 5′ hydrogen trans to the pyridine ring was oxidized in preference to the 5′-cis hydrogen, and no isotope effect was observed at the 5′ position, while very similar isotope effects of ca. 2.4 were found in each system. The regioselectivity for the reaction was predicted based on molecular dynamics calculations to prefer 5′ oxidation to methyl group oxidation and that the 5′-trans hydrogen would be abstracted in favor of the 5′-cis, which is consistent with the experimental results. Since non-enzymatic systems and theoretical calculations indicate that loss of the cis-5′-hydrogen is energetically favored over loss of the trans-5′-hydrogen (11), the preferential loss of the trans-5′-hydrogen by all of the P-450 enzymes studied suggests that the binding of nicotine with the various enzymes has some common feature that lowers the energy barrier for trans-5′-hydrogen loss and converts it to the energetically favored pathway. The molecular modeling studies with CYP101 revealed three primary interaction sites with nicotine, namely the I-helix, the heme and Tyr 96. The I-helix and the heme act to hold the pyrrolidine ring with either the trans-5′- or the cis-5′-hydrogen close to the iron-oxygen species. These orientations are further augmented by a hydrogen bond between Tyr 96 and the pyridine nitrogen of nicotine. When (S)-nicotine is bound in this fashion, placement of the trans-5′-hydrogen in the immediate vicinity of the perferryl oxygen atom provides a minimum energy structure for the enzyme-substrate complex. The model further suggests that the I-helix severely hinders interconversion between the two orientations, which would result in a slow rate of equilibration and thus a masking of the isotope effect.

The tyrosine residue that aligns with Tyr 96 of CYP101 is conserved for CYP2B1 and CYP2B4 (3) but not for CYP1A1. (In CYP1A1 this residue is a glutamine which is still capable of hydrogen bonding with the pyridine ring.) The I-helix is believed to be conserved in all three purified forms used in this study since it is required to bind the heme and since a threonine residue in this helix aids heterolytic cleavage of the dioxygen complex to generate the active oxygen species. Thus, consideration of critical active site elements, which are likely to be common to all the P-450 enzymes studied, is in agreement with the fact that the binding modes of (S)-nicotine with the P-450 enzymes studied must be similar enough to dictate that trans-5′- hydrogen removal be consistently favored, despite the fact that in model chemical studies this stereochemical mode of reaction is not favored energetically.

BENZO[a]PYRENE

One of the interesting riddles in xenobiotic metabolism is why benzo[a]pyrene (B[a]P) is metabolized to the most carcinogenic diol epoxides. Work in our laboratory has established a structural model that provides an answer for this riddle. In vitro studies with 12 different rodent and human P450 enzymes established that all of these enzymes showed very similar stereoselectivity in the formation of the 7,8- and 9,10-epoxides of B[a]P. This was established by chiral phase HPLC analysis of the resulting diols after the epoxide was stereospecifically opened by epoxide hydrolase. These results prompted us to speculate that some conserved feature in the active site of all of the studied P450 enzymes dictates the stereochemistry. Since no 3-dimensional structure was available for any of these enzymes no specific structural feature could be determined. In an attempt to answer this question we incubated B[a]P with P450cam. To our surprise the bacterial P450 catalyzed the reaction, and with the same stereochemistry as the mammalian enzymes. Thus, it was established that an enzyme with known 3-dimensional structure could catalyze the reaction of interest with the same stereochemistry as the mammalian systems.

To determine what structural features were important in determining the stereochemical outcome, we docked B[a]P in the active site of the P450cam crystal structure. B[a]P could be docked in P450cam in 4 distinct orientations. These orientations were not interchangeable without completely removing the substrate from the active site. Of the 4 orientations only two would lead to epoxidation at the 7,8- or 9,10-position. Each orientation would give epoxidation on a different face of the B[a]P molecule leading to a different stereochemistry of the formed epoxide. Furthermore, the observed epoxides at the 7,8- and 9,10-position must arise from epoxidation of opposite faces of the B[a]P molecule. Thus, it appeared that one orientation would give the observed 7,8-diol but not the 9,10-diol and conversely the other orientation would only give the 9,10 epoxide. Molecular dynamics was performed to predict the stereochemistry of epoxidation. The predicted stereochemistry was in agreement with the experimental stereochemistry. These results support the use of the docking experiments in establishing the specific active site interactions that dictate the stereochemistry of epoxidation.

B[a]P fits very tightly into the active site of P450cam in both orientations that could lead to 7,8- or 9,10-epoxidation. Close Van der Waals contacts restrict the B[a]P molecule such that only small fluctuations can be observed during dynamics runs. The residues that have close contact with B[a]P are Pro86, Phe87, Tyr96, Thr101, Leu244, Val247, Gly248, Thr252, Ile395 and Val396. Residues 86, 87 are part of the B′ helix. Tyr96, the amino acid that hydrogen bonds to the carbonyl oxygen of camphor, forms the top of the B[a]P binding site along with Phe86. Residue 101 acts as main contact point of the enzyme on the edge of the B[a]P opposite the oxidation site. The Van der Waals contact with Ile395 and Val396 that comprise the turn in the beta 5 sheet are also in close proximity to B[a]P. Residues 244, 247, 248 and 252 are in the I-helix and act to hold the B[a]P molecule close to the heme-oxygen. Residue 252 is particularly important in that it prevents the translation of the B[a]P molecule in the orientation that would give the 9,10-oxide with the incorrect stereochemistry. Thr 252 also prevents the translation of the B[a]P molecule in the other orientation to a position that would give the 7,8-oxide with the incorrect stereochemistry. Thus, the interaction with Thr 252, and the I helix in general, determines the stereochemistry of epoxidations.

Since Thr 252 is important in the heterolytic cleavage of molecular oxygen to the active oxygen species, and is likely to be conserved across P450 enzymes, it is tempting to say that interaction with this conserved residue is the reason all of the P450 enzymes tested give the same stereochemistry. To test this hypothesis we incubated B[a]P with P450BM-3, another enzyme of known 3-dimensional structure. Again, the stereochemistry was very similar to that observed for the mammalian systems. Thus, B[a]P was docked into the active site of P450BM-3 to establish what residues were important in determing the stereochemistry of the product.

Two orientations, similar to those hypothesized for 7,8- and 9,10-epoxidation in P450cam, were studied for P450BM-3. For each orientation the B[a]P was docked and a molecular dynamic calculation was performed. After the dynamics run a minimization was performed. In this way two orientations, analogous to the P450cam orientations were generated. Very similar results were observed when B[a]P was docked in the active site of P450BM-3 as compared with P450cam. When docked in the P450BM-3 active site, both of the orientations that were studied had close steric interactions with Ser 72, Leu 75, Phe 87, Ala 264, Thr 268, and Leu 437. These amino acids correspond very closely with the regions found to be important in P450cam binding. Amino acids 72, 75, and 87 are in the B′ helical area, amino acids 264 and 268 are in the I helix and amino acids 437 is in a turn in a beta sheet in the same spatial orientation as the residues 395 and 396 of P450cam. Similarly, the steric interactions with Thr 268, which is in the I-helix of P450BM-3 and aligns with Thr 252 of P450cam, prevents the translation of the B[a]P molecule to positions that would favor formation of the epoxide that is not observed at either the 7,8- or 9,10-position. Finally, if constraints were placed on either orientation, such that it would pull the less favored positions into close proximity to the Fe-O species, a large increase (>50 kcal) in energy was observed. It would appear that as with P450cam, P450BM-3 dictates the stereoselective metabolism of B[a]P by precluding the formation of the 7S,8R- and 9R,10S-epoxides. For both isoforms unfavorable steric interactions with the I-helix make formation of the these epoxides, which lead to the S,S-diol products, difficult.

In conclusion, a number of successful applications of molecular mechanics to predicting the stereoslectivity and regioselectivity of xenobiotic metabolism by P450 have been performed. In all but one instance these calculations have used the crystal structure of P450cam. Since no crystal structure is available for mammalian P450 enzymes the steric factors responsible for the interactions with substrate can only be hypothesized based on the known bacterial structures. The validity of these comparisons is dependent on the structural homology between the bacterial and mammalian systems. For both nicotine and B[a]P metabolism we established that the bacterial and mammalian systems have functional homology. In general, it appears likely that since the I helix is important in the catalytic mechanism of P450 that it is likely that the position of this helix is similar in all P450 enzymes and the structural interactions will be conserved. If other areas of the enzyme are of overiding importance, functional and structural homology is less likely. For example, a single amino acid change in CYP2A4 has been found to change the substate specificity from a coumarin hydroxylase to a steroid hydroxylase. Obviously these subtle interactions cannot be modeled by homology with bacterial enzymes. Thus, the methodology stands poised to help us understand specific protein-substrate interactions but knowledge of the 3-dimensional structure of mammalian P450 enzymes is required before this methodology can be applied with confidence.

REFERENCES

1. Flam, F. (1994) EPA Campaigns for Safer Chemicals. *Science* **265**, 1519.
2. Anders, M. W., Ed. (1985) *Bioactivation of Foreign Compounds*, Academic Press, New York.
3. Korzekwa, K. R., and Jones, J. P. (1993) Predicting the cytochrome P450 mediated metabolism of xenobiotics. *Pharmacogenetics* **3**, 1-18.
4. Ortiz de Montellano, P. R., Fruetel, J. A., Collins, J. R., Camper, D. L., and Loew, G. H. (1991) Theoretical and Experimental Analysis of the Absolute Stereochemistry of cis-B-Methylstyrene Epoxidation by Cytochrome P450cam. *J. Am. Chem. Soc.* **113**, 3195-3196.
5. Filipovic, D., Paulsen, M. D., Loida, P. J., Sligar, S. G., and Ornstein, R. L. (1992) Ethylbenzene Hydroxylation By Cytochrome P450cam. *Biochem. Biophys. Res. Commun.* **189**, 488-495.
6. Jones, J. P., Trager, W. F., and Carlson, T. J. (1993) The binding and regioselectivity of reaction of (R)- and (S)-nicotine with cytochrome P-450cam: Parallel experimental and theoretical studies. *J. Am. Chem. Soc.* **115**, 381-387.
7. Jones, J. P., Shou, M., and Korzekwa, K. R. (1995) Stereospecific Activation of the Procarcinogen Benzo[a]pyrene: A Probe for the Active Sites of the Cytochrome P405 Superfamily. *Biochemistry* **34**, In Press.
8. Fruetel, J., Chang, Y., Collins, J., Loew, G., and Ortiz de Montellano, P. R. (1994) Thioanisole Sulfoxidation by Cytochrome P450cam (CYP101): Experimental and Calculated Absolute Stereochemistries. *J. Am. Chem. Soc.* **116**, 11643-11648.
9. Collins, J. R., Camper, D. L., and Loew, G. H. (1991) Valproic Acid Metabolism by Cytochrome P450: A Theoretical Study of the Stereoelectronic Modulators of Product Distribution. *J. Am. Chem. Soc.* **113**, 2736-2743.
10. Chang, Y., Loew, G. H., Rettie, A. E., Baillie, T. A., Sheffels, P. R., and Ortiz de Montellano, P. R. (1993) Binding of Flexible Ligands to Proteins: Valproic Acid and its interactions with Cytochrome P450cam. *Int. J. Quant. Chem. Quant. Biol. Sym.* **20**, 161-180.
11. Peterson, L. A., and Castagnoli, N. (1988) Regio- and Stereochemical Studies on the α-carbon Oxidation of (S)-nicotine by Cytochrome P450 Model Systems. *J. Med. Chem.* **31**, 637-640.

ELECTRONIC MODELS FOR CYTOCHROME P450 OXIDATIONS

Kenneth R. Korzekwa,[1] James Grogan,[2] Steven DeVito,[3] and Jeffrey P. Jones[4]

[1] Center for Clinical Pharmacology, University of Pittsburgh Medical Center 623 Scaife Hall, Pittsburgh, Pennsylvania 15261
[2] Laboratory of Molecular Carcinogenesis, National Cancer Institute National Institutes of Health, Bethesda, Maryland 20892
[3] Office of Pollution Prevention and Toxics (7406) United States Environmental Protection Agency, Washington, DC 20460
[4] Department of Pharmacology, University of Rochester 601 Elmwood Ave, Rochester, New York 14642

INTRODUCTION

The cytochrome P450 enzymes metabolize a wide variety of hydrophobic endogenous (1,2) and exogenous compounds (3). Their involvement in drug metabolism and xenobiotic activation has made them one of the most studied families of enzymes. A part of our efforts in the area of drug metabolism and toxicology has been to develop models to predict cytochrome P450 oxidations. In general, a complete model for an enzymatic reaction should include both protein-substrate interactions and the appropriate chemical transformations. Unfortunately, the mammalian enzymes are membrane bound, and crystal structures are not available. Therefore, exact interactions with the protein will be difficult, if not impossible to predict. For many enzymes and receptors, a general description of the binding site can be developed based on the structures of substrates and inhibitors (or agonists and antagonists). This information is used to construct a pharmacophore, a representation of the characteristics of the binding site of the protein. While pharmacophores may be developed for some of the cytochrome P450 enzymes (4,5), this may not be possible for others. Some of the interactions between drug metabolizing P450 enzymes and their substrates are likely to be nonspecific. This is expected since many P450 enzymes show very broad substrate and regioselectivity. Whereas the nonspecific nature of the substrate-P450 interactions prevents the development of structural models for these enzymes, this characteristic allows for the development of electronic models. For some small hydrophobic substrates, rotation within the P450 active site is fast, relative to the substrate oxidation step. This has been shown and described by isotope effect studies, where slowing the rate of oxidation of one position by deuterium substitution causes an increase in the rate of metabolism of another position (6,7).

Biological Reactive Intermediates V
Edited by R. Snyder et al., Plenum Press, New York, 1996

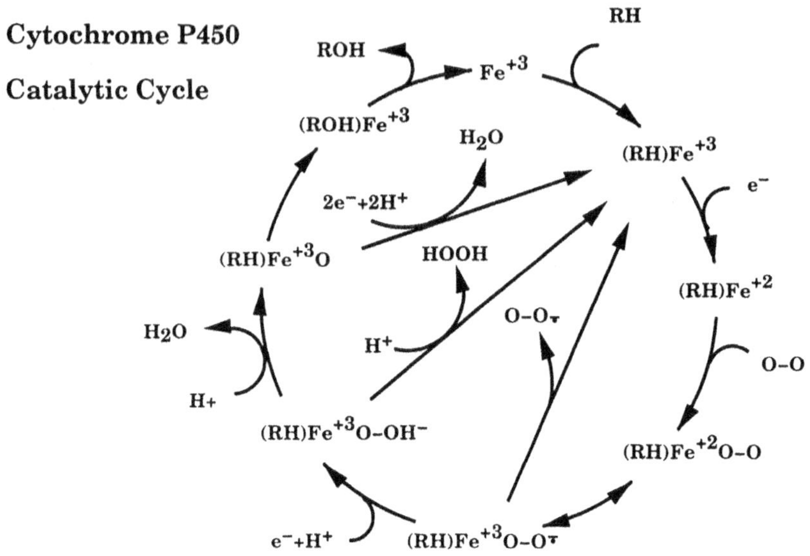

Figure 1. The catalytic cycle for the cytochrome P450 enzymes.

This can only occur if the rate of exchange of the different positions within the active site is as fast or faster than the rate of the oxidation step. For a substrate that can rotate freely in the active site, the regioselectivity will be primarily determined by the reactivity of the various positions on the molecule. Thus, it may be possible to predict the regioselectivities of molecules that rotate freely in the active sites of the enzymes that metabolize them. Another characteristic of these enzymes that can allow the development of electronic models is the competition between substrate oxidation and water formation. The catalytic cycle for the cytochrome P450 enzymes is shown in Figure 1.

It has been shown that the rate limiting steps for the catalytic cycle are associated with reduction of the enzyme by the cytochrome P450 reductase (8,9). This would suggest that the rate of oxidation is not affected by the structure of the substrate. However, if substrate oxidation is slow, the active oxygenating intermediate can be further reduced to form another molecule of water (10-13). This alternate pathway, is in competition with substrate oxidation. Therefore, differences in substrate reactivity can be unmasked in the same way that isotope effects are unmasked (6). The relative tendencies for abstracting hydrogen atoms by the cytochrome P450 enzymes generally follow the stabilities of the radicals formed, as seen in Figure 2.

For example, N-dealkyation is one of the most frequent P450 mediated biotransformations, whereas ω-hydroxylations are very infrequent. This general trend suggests that there may be a Bronsted relationship (14) for P450 mediated hydrogen atom abstraction reactions. A Bronsted relationship occurs when the activation energy of a reaction is proportional to the heat of reaction. This can be envisioned as part of the factors that stabilize or destabilize the products (radicals) are manifest in the transition state. These relationships are common within sets of closely related reactions and are the basis for most linear free energy relationships.

p-NITROSOPHENOXY RADICAL MODEL

If a linear free energy relationship can be developed for cytochrome P450 mediated hydrogen atom abstraction reactions, the relative stability of the radicals can be used to

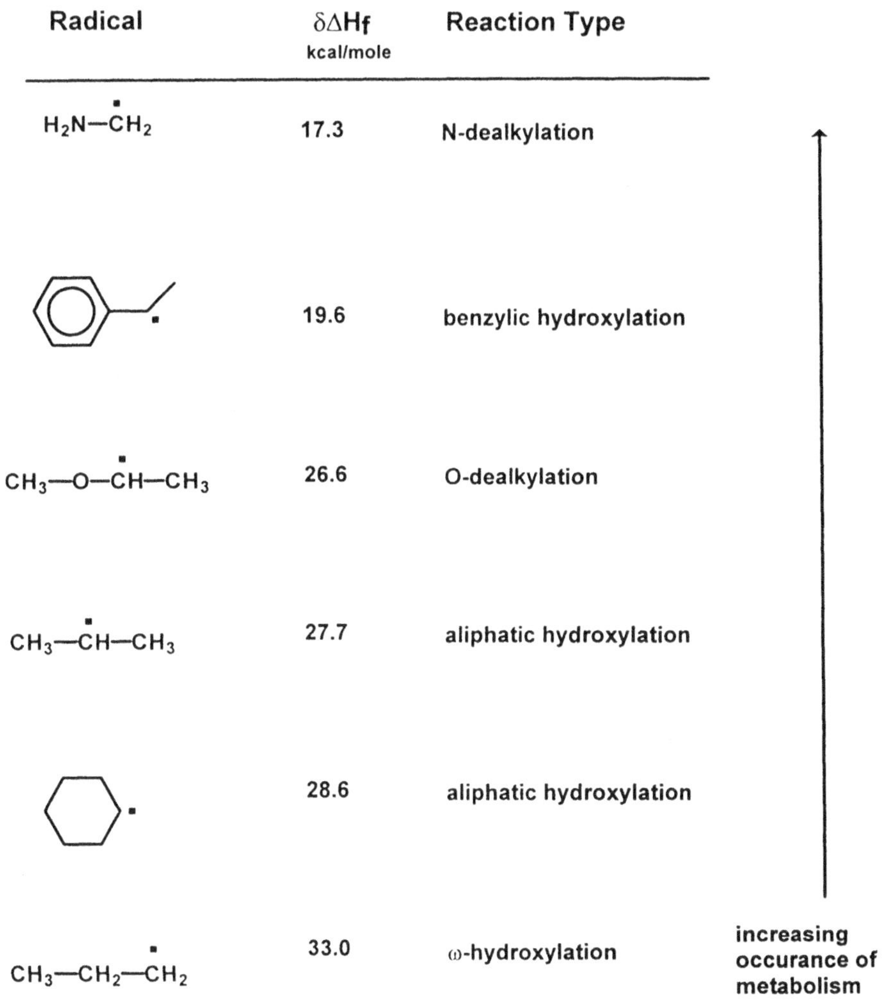

Figure 2. General relationship between radical stability and tendency for oxidation.

determine the electronic tendency for oxidation. Therefore, we used quantum chemical calculations to investigate hydrogen atom abstraction reactions. Although the ideal model for the P450 active oxygenating species would be a heme-oxygen-thiolate molecule, the size limitation of quantum chemical calculations required that we find a smaller model oxidant. Ideally, this model would generate potential energy surfaces similar to those for P450 mediated reactions. The only experimental data available on these surfaces have been provided by isotope effect experiments. Studies on the metabolism of octane suggest that the P450 mediated -hydroxylation of alkanes has a symmetrical reaction coordinate, i.e., the reactants and products have similar energies (15). These experiments also suggested that all P450s have similar energetics (16). We chose the p-nitrosophenoxy radical (PNO) as our model, since abstraction of a hydrogen from ethane (using the AM1 method) has a $H_{R} \approx 0$. We generated reaction coordinates and optimized transition states for PNO abstraction of hydrogen atoms from 20 substrates. These substrates are shown in Figure 3.

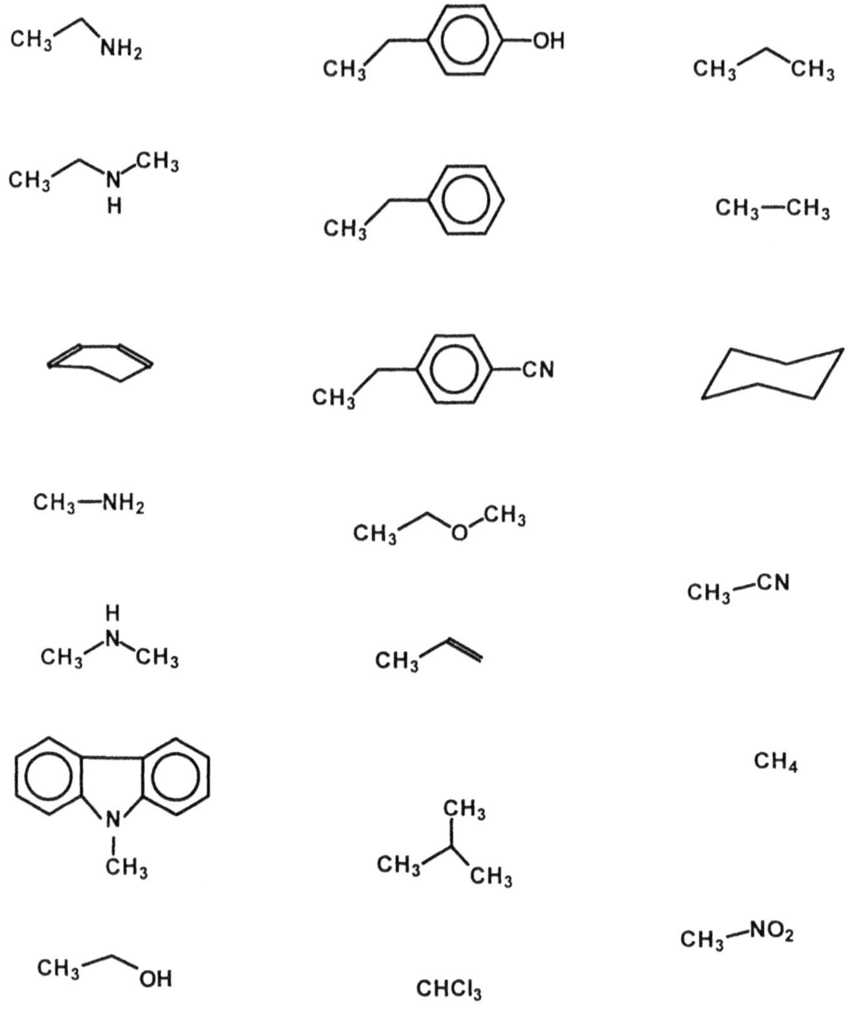

Figure 3. Substrates used to develop the PNO model.

The Bronsted plot for these reactions showed a significant correlation (R^2=0.77) between $\Delta H\ddagger$ and ΔH_R (17). This correlation could be improved if a measure of resonance was included. As a measure of resonance effects, we used the ionization potential of the radical, since this property is generated by the calculations. Inclusion of this term in the regression analysis gave an improved correlation (R^2=0.94). This data suggested that calculating the relative stability and ionization potentials of the product radicals might be used to predict the electronic tendency for hydrogen atom abstractions.

NITRILE METABOLISM

Although the general tendency for metabolism appeared to resemble the predicted PNO activation energies, experimental validation of the model was necessary. We applied the PNO model to the development of a predictive model for the toxicity of nitriles. The acute toxicity

$CH_3—CH_2—CH_2—CN$

γ β α

$NH_2—CH_2—CH_2—CH_2—CN$

$$k\alpha corr = k\alpha\left(\frac{k\alpha}{k\alpha + k\beta + k\gamma}\right)$$

Figure 4. Correction for metabolism at nontoxic positions.

of nitriles is known to be due to cyanide release as a result of oxidation by cytochrome P450. In nitriles, hydroxylation of the carbon atom adjacent (α) to the cyano group gives a cyanohydrin intermediate that chemically decomposes to give the correspinding carbonyl deravitive and cyanide. As with the metabolism of many compounds, both detoxification and activation pathways for nitriles can be catalyzed by the cytochrome P450 enzymes. Therefore, all possible pathways of metabolism must be considered when predicting toxicity of a compound. As an example, butyronitrile and 4-aminobutyronitrile are shown in Figure 4.

Although the reactivities of the α-positions are very similar, the toxicities are likely to be different. The amino group on 4-aminobutyronitrile activates the γ-position, and directs metabolism to this position. This will be expected to decrease the fraction of the dose that goes through the toxic pathway, oxidation at the α-position. In order to include the different metabolic pathways, the rate constants for the toxic pathways (k) were weighted by the fraction of the dose expected to go through that pathway (kα/[kα+kβ+kγ...] (18)). This corrected rate constant (kαcorr) was used to develop a predictive model for the toxicity of nitriles. Tani and Hashimoto published the LD50s for a series of 26 nitriles (19,20). Regression of the $_{LD50}$s of these nitriles on both k corr and hydrophobicity gave a significant linear correlation (R^2=0.73) whereas the actual versus predicted toxicities (1/LD_{50}) for the model based on hydrophobicity alone gave a poor correlation (18). Although these correlations include all 26 nitriles, some of the toxicities are apparently not due to cytochrome P450 metabolism. With the elimination of five compounds that are not expected to fit our paradigm, we obtain a better correlation (R^2=0.81). We have expanded our data set by determining the LD_{50}s of ten additional nitriles. In addition, we now use a different method to compute the hydrophobicities (Phillip Howard and William Meylan, Syracuse Research Corp., Environmental Sciences

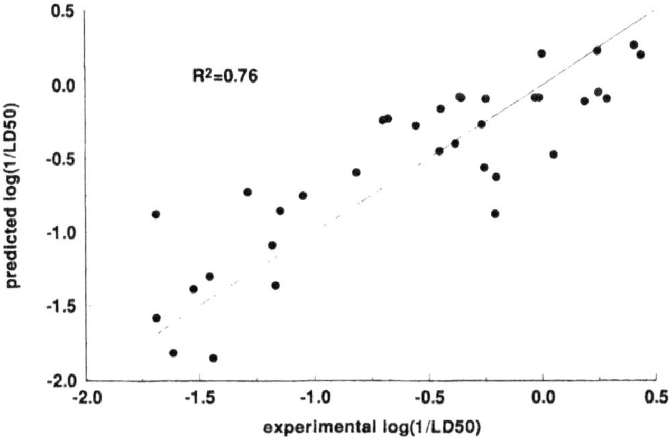

Figure 5. Cross validated regression analysis of 36 nitriles.

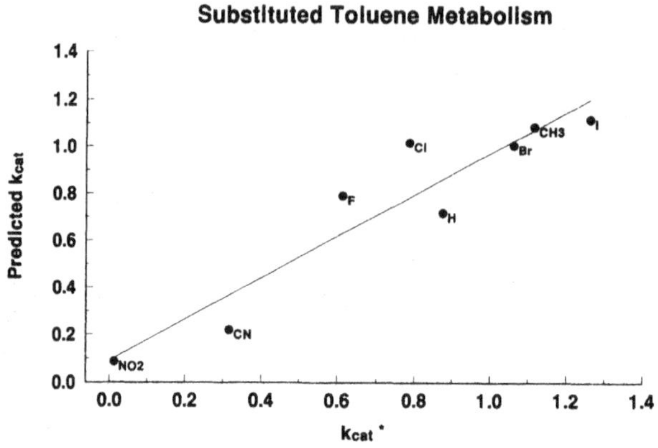

Figure 6. Predicted rates of benzylic oxidation *in vitro* using the PNO model.

Center, Merrill Ln., Syracuse, NY 13210). The cross validated fit for all 36 nitriles gives an R^2=0.76, and is shown in Figure 5. The correlation for these studies is striking, given that a computational model is predicting *in vivo* results.

In another example, the PNO model can successfully predict reaction rates *in vitro*. White and McCarthy (21) published a Hammett linear free energy correlation for the rates of metabolism of substituted toluenes. Using hydrophobicities and predicted PNO activation energies we obtained an excellent predictive model for the rate of benzylic oxidation. The correlation (R^2=0.86) is shown in Figure 6.

Finally, excellent correlations for the rates of metabolism of halogenated hydrocarbons *in vitro* and inhalation anesthetics *in vivo* have been obtained with the PNO model (22).

NEW MODELS

Semiempirical Models

Although the PNO model is useful in predicting hydrogen abstraction reactions, attempts to apply the model to aromatic and olefinic oxidations have been unsuccessful. Addition of the PNO molecule to aromatic systems results in crossed potential energy surfaces (within the UHF formalism). Therefore, several other less conjugated molecules

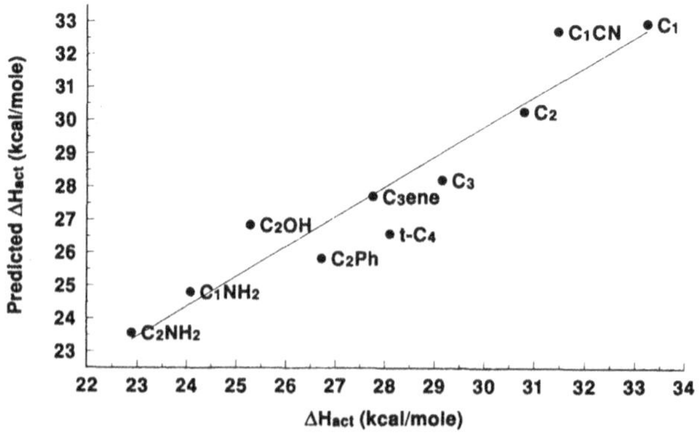

Figure 7. Hydrogen abstraction using the HSO radical.

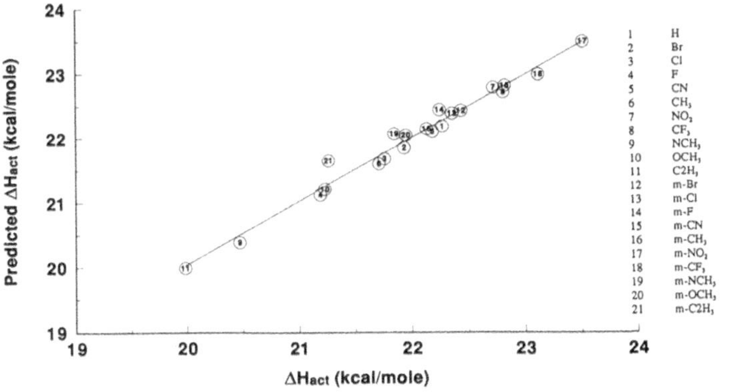

Figure 8. HSO radical addition to aromatic compounds.

were used for abstraction and addition reactions. Two molecules that were used extensively were the HSO radical and the FO radical. These molecules remained on the same potential energy surface for both abstraction and addition reactions. Excellent correlations were obtained for both reactions, as shown in Figures 7 and 8.

Unfortunately, spin contamination (<S2> greater than 0.75) was excessive for the aromatic addition reactions. Spin contamination was also a function of the number of conjugated aromatic rings. Although good correlations were obtained for a series of substituted molecules, the benzenes, naphthalene, and phenanthrene reactions were separated in energies. Increasing the number of rings results in greater spin contamination (larger $\langle S^2 \rangle$ values) and lower observed activation energies. Since there were no methods to correct semiempirical UHF calculations for spin contamination, we turned to ab initio calculations.

Ab Initio Models

The Gaussian package of quantum chemical programs uses spin projection algorithms to remove spin contaminations. Although initial hydrogen abstraction calculations were performed with the FO radical at the MP2/6-31G* level of theory, recent experimental data suggests that alkoxy radicals may closely represent the P450 active oxygen species. The isotope effects for the dealkylation of a series of substituted dimethylanilines were identical for both cytochrome P450 oxidations and the chemical oxidation of these compounds with *t*-butoxy radical (23). This suggests that the potential energy surfaces for these two reactions are similar. Therefore, the t-butoxy radical may be an ideal model for P450 oxidations. Preliminary calculations at the MP2/6-31G* level have been completed for the abstraction of hydrogen atoms from methane, fluoromethane, methanol, methylamine, ethane, propene and cyanomethane. In addition, the same reactions were performed using the methoxy radical. Similar activation energies were obtained for both oxygen radicals, as shown in Figure 9.

The similar stabilities of these methoxy and butoxy radicals are supported by the similar experimental bond dissociation energies of methanol and butanol (24). We are presently performing hydrogen abstraction and aromatic addition reactions at different levels of theory. Hopefully, these calculations will allow us to expand our predictive abilities to include all classes of P450 mediated oxidations.

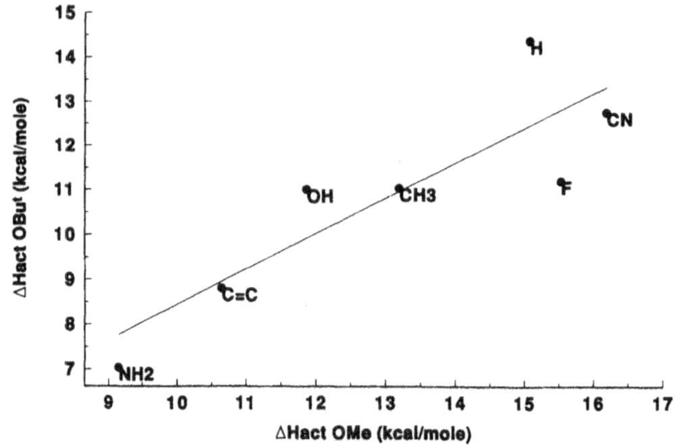

Figure 9. Hydrogen atom abstraction energies for the methoxy and *tert*-butoxy radicals.

SUMMARY

In summary, we have developed predictive models for certain groups of hydrogen abstraction reactions, and are expanding the models to include other P450 oxidations. However, the successful correlations are associated with groups of relatively small, hydrophobic molecules. It would be expected that larger molecules will have more constraints within the P450 active sites. In order to successfully predict the metabolism of these molecules, it will be necessary to include features of the P450 active sites. Calculations which combine quantum chemical and molecular mechanics calculations have been reported (25). When accurate three dimensional representations of the active sites of the cytochrome P450 enzymes become available, these electronic models can be incorporated. The result may be a model that can accurately predict how a cytochrome P450 enzyme will metabolize any real or hypothesized molecule.

REFERENCES

1. Masters, B.S.S., Muerhoff, A.S., and Okita, R.T.,1987, Enzymology of Extrahepatic Cytochromes P450. In: F.P. Guengerich (ed.), *Mammalian Cytochromes P-450*, pp. 107-131, Boca Raton: CRC Press.
2. Jefcoate, C., 1986, Cytochrome P-450 Enzymes in Sterol Biosynthesis and Metabolism. In: P.R. Ortiz de Montellano (ed.), *Cytochrome P-450*, pp. 387-428, New York: Plenum Press.
3. Gonzalez, F.J., 1990, Molecular genetics of the P-450 superfamily. *Pharmacol. Ther.*, 45: 1-38.
4. Jerina, D.M., Michaud, D.P., Feldmann, R.J., Armstrong, R.N., Vyas, K.P., Thakker, D.R., Yagi, H., Thomas, P.E., Ryan, D.E., and Levin, W., 1982, Stereochemical Modeling of the Catalytic Site of Cytochrome P450c. In: R. Sato and R. Kato (eds.), *Microsomes, Drug Oxidations, and Drug Toxicity*, pp. 195-201, Tokyo: Japan Scientific Societies Press.
5. Wolff, T., Distlerath, L.M., Worthington, M.T., Groopman, J.D., Hammons, G.J., Kadlubar, F.F., Prough, R.A., Martin, M.V., and Guengerich, F.P., 1985, Substrate specificity of human liver cytochrome P-450 debrisoquine 4-hydroxylase probed using immunochemical inhibition and chemical modeling. *Cancer Res.*, 45: 2116-2122.
6. Korzekwa, K.R., Trager, W.F., and Gillette, J.R., 1989, Theory for the observed isotope effects from enzymatic systems that form multiple products via branched reaction pathways: cytochrome P-450. *Biochemistry*, 28: 9012-9018.
7. Jones, J.P., Korzekwa, K.R., Rettie, A.E., and Trager, W.F., 1986, Isotopically sensitive branching and its effect on the observed intramolecular isotope effects in cytochrome P-450 catalyzed reactions: a new method for the estimation of intrinsic isotope effects. *J. Am. Chem. Soc.*, 108: 7074-7078.

8. White, R.E. and Coon, M.J., 1980, Oxygen activation by cytochrome P-450. *Annu. Rev. Biochem.*, 49: 315-356.
9. Grogan, J., Shou, M., Zhou, D., Chen, S., and Korzekwa, K.R., 1993, Use of aromatase (CYP19) metabolite ratios to characterize electron transfer from NADPH-cytochrome P450 reductase. *Biochemistry*, 32: 12007-12012.
10. Gillette, J.R., Brodie, B.B., and La Du, B.N., 1957, The Oxidation of Drugs by Liver Microsomes: On the Role of TPNH and Oxygen. *J. Pharmacol. Exp. Ther.*, 119: 532-540.
11. Nordblom, G.D. and Coon, M.J. 1982, Hydrogen Peroxide Formation and Stoichiometry of Hydroxylation Reactions Catalyzed by Highly Purified Liver Microsomal Cytochrome P-450. *Arch. Biochem. Biophys.*, 180: 343-347,.
12. Morgan, E.T., Koop, D.R., and Coon, M.J., 1982, Catalytic Activity of Cytochrome P-450 Isozyme 3a Isolated from Liver Microsomes of Ethanol-treated Rabbits. *J. Biol. Chem.*, 257: 13591-13597.
13. Gorsky, L.D., Koop, D.R., and Coon, M.J., 1982, On the Stoichiometry of the Oxidase and Monooxygenase Reactions Catalyzed by Liver Microsomal Cytochrome P-450. *J. Biol. Chem.*, 259: 6812-6817.
14. Kresge, A.J., 1973, The Bronsted relation - recent developments. *Chem. Soc. Rev.*, 2: 475-503.
15. Jones, J.P. and Trager, W.F., 1987, The Separation of the Intramolecular Isotope Effect for the Cytochrome P-450 Catalyzed Hydroxylation of n-Octane into its Primary and Secondary Components. *J. Am. Chem.* Soc., 109: 2171-2173.
16. Jones, J.P., Rettie, A.E., and Trager, W.F., 1990, Intrinsic isotope effects suggest that the reaction coordinate symmetry for the cytochrome P-450 catalyzed hydroxylation of octane is isozyme independent. *J. Med. Chem.*, 33: 1242-1246.
17. Korzekwa, K.R., Jones, J.P., and Gillette, J.R. Theoretical studies on cytochrome P-450 mediated hydroxylation: a predictive model for hydrogen atom abstractions. J. Am. Chem. Soc., 112: 7042-7046, 1990.
18. Grogan, J., DeVito, S.C., Pearlman, R.S., and Korzekwa, K.R., 1992, Modeling Cyanide Release from Nitriles: Prediction of Cytochrome P450 Mediated Acute Nitrile Toxicity. *Chem. Res. Toxicol.*, 5: 548-552.
19. Tanii, H. and Hashimoto, K., 1984, Studies on the mechanism of acute toxicity of nitriles in mice. *Arch. Toxicol.*, 55: 47-54.
20. Tanii, H. and Hashimoto, K., 1985, Structure-acute toxicity relationship of dinitriles in mice. *Arch. Toxicol.*, 57: 88-93.
21. White, R.E. and McCarthy, M.B., 1986, Active site mechanics of liver microsomal cytochrome P-450. Arch. Biochem. Biophys., 246: 19-32.
22. Anders,M.W., 1995, Previous chapter, this issue.
23. Karki, S.B., Dinnocenzo, J.P., Jones, J.P., and Korzekwa, K.R., 1995, Mechanism of oxidative amine dealkylation of substituted *N,N*-dimethylanilines by cytochrome P-450: Application of isotope effect profiles. *J. Am. Chem. Soc.*(in press).
24. McMillen, D.F. and Golden, D.M., 1982, Hydrocarbon bond dissociation energies. *Ann. Rev. Phys. Chem.*, 33: 493-532.
25. Arad, D., Langridge, R., and Kollman, P.A. A Simulation of the Sulfur Attack in the Catalytic Pathway of Papain Using Molecular Mechanics and Semiemperical Quantum Mechanics. J. Am. Chem. Soc., *112*: 491-502, 1990.

SPECIES DIFFERENCES IN METABOLISM OF 1,3-BUTADIENE

Rogene F. Henderson

Inhalation Toxicology Research Institute
Lovelace Biomedical and Environmental Research Institute
Albuquerque, New Mexico

INTRODUCTION

1,3-Butadiene (BD) is a 4-carbon gaseous compound with two double bonds. Used in high tonnage to make styrene-butadiene polymers in the rubber industry, BD is one of the top 20 chemicals in amount manufactured in the USA (Agency of Toxic Substances and Disease Registry, 1992). BD has also been detected in ppb levels in urban air, gasoline exhaust, and cigarette smoke. Because of large amounts used, BD was tested for toxicity in 2-year inhalation exposures of both Sprague-Dawley rats (Owen et al., 1987) and B6C3F$_1$ mice (Huff et al., 1985; Melnick et al., 1990).

The results of the two-species studies were dramatically different. In the initial study in mice (Huff et al., 1985), BD was shown to be a potent multiple-site carcinogen at exposure levels of 625 and 1250 ppm. There were increased incidences of neoplasia in the heart, lung, mammary gland, and ovary; malignant lymphomas resulted in early deaths of the mice so that the planned 2-year study was stopped after only 61 weeks of exposure. The second study in mice (Melnick et al., 1990) was conducted at much lower exposure concentrations (6.25, 20, 62.5, 200, and 625 ppm) and lasted 104 weeks. Increased incidences of hemangiosarcomas of the heart and lung neoplasia were observed in males exposed to 62.5 ppm BD, while females had increased lung neoplasia even at the 6.25 ppm exposure level. Early deaths from lymphomas were again observed at the high exposure concentration (625 ppm). A noncancer toxicity observed in mice was a macrocytic, megaloblastic anemia.

In contrast, the results of exposures of rats to 1000 or 8000 ppm BD over a 2-year period indicated that BD is a weak carcinogen (Owen et al., 1987). Some increase in mammary tumors was observed in females exposed to 1000 ppm BD; after exposure to 8000 ppm BD, there were also increases in thyroid and testicular tumors. Neither lymphomas nor anemia were observed. The species differences in response to BD caused several laboratories to begin studies on species differences in metabolism of BD as a potential cause for the different responses.

Biological Reactive Intermediates V
Edited by R. Snyder et al., Plenum Press, New York, 1996

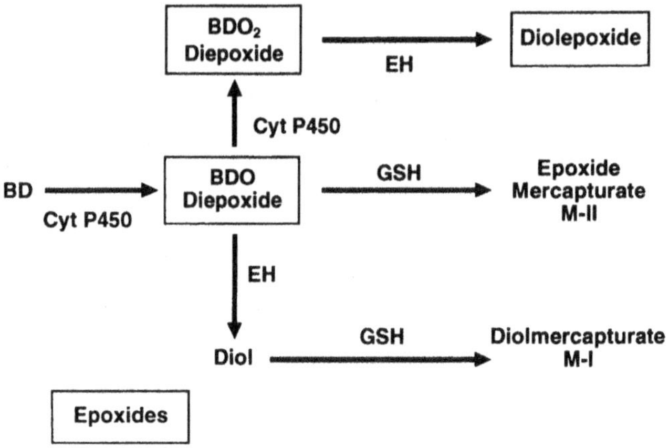

Figure 1. Major metabolic pathways of BD. Cyt P450 = cytochrome P_{450}; EH = epoxide hydrolase; GSH = glutathione; M-I = 1,2-dihydroxy-4-(N-acetylcysteinyl-S-)butane; M-II = 1-hydroxy-2-(N-acetylcysteinyl-S-)-3-butene. The three epoxide metabolites are enclosed in rectangles.

PATHWAYS OF METABOLISM OF BD

A simplified summary of the metabolism of BD to reactive intermediates is shown in Figure 1. Three epoxides can be formed: the monoepoxide (BDO), which is the initial metabolic product; the diepoxide (BDO_2), which can be formed from the monoepoxide; and the diolepoxide, which can be formed from either the partial hydrolysis of the diepoxide or the oxidation of the butene diol formed from hydrolysis of BDO. The mutagenicity of these epoxides has been studied in several laboratories, but recently Cochrane and Skopek (1994) compared the mutagenicity of all three epoxides in the same two test systems. They used TK6 human lymphoblastoid cells and assayed for mutations at the *tk* and *hprt* loci. BDO_2 was an approximately 100-fold more potent mutagen than the other two epoxides. At the *hprt* locus, half of the mutations induced by the diepoxide involved loss of wild-type *hprt* restriction fragments, suggesting cross-linking. Such cross-linking is a characteristic expected from a molecule with two reactive sites, like the diepoxide. Such deletions were not observed in mutations induced by the monoepoxide.

IN VITRO STUDIES OF SPECIES DIFFERENCES IN METABOLISM OF BD

Early studies by Schmidt and Loeser (1986) compared the metabolism of BD and BDO in liver S-9 preparations from mice, rats, monkeys, and humans. The conversion of BD to BDO was greatest in the mouse and lowest in the monkey, with a 5-fold difference between the two species. Rat and human liver were approximately equal and intermediate between the mouse and human preparations. The hydrolysis of BDO to form the butene diol was high in the two primate species and more than 10-fold higher than in the rodents.

More extensive studies have been conducted by Csanàdy et al. (1992, 1993) in which the metabolism of BD and BDO was measured in lung and liver homogenates from mice, rats, and humans; the conversions of BDO to BDO_2 and to the glutathione (GSH) conjugate were measured in addition to BDO hydrolysis. The conversion of BD to BDO was highest in the mouse liver, as was observed in the earlier studies, and lowest in the rat (4-fold lower than mouse). However, the metabolism of BD to BDO in the lung tissue was 10-fold higher in the mouse than in the human or rat tissue and equaled the metabolism found in mouse liver. The conversion of BDO to BDO_2 was detected only in the mouse liver. As was observed in the studies of Schmidt and Loeser (1986), the hydrolysis of BDO to the diol was 10 times higher in human liver preparations than in either rodent samples. On the other hand, the conversion of BDO to a GSH conjugate was 10-fold higher in the mouse preparations than in the human. The data suggest that the operative pathway for eliminating BDO in the mouse is by GSH conjugation; in primates, it is the hydrolysis to the butene diol that is important.

IN VIVO STUDIES OF SPECIES DIFFERENCES IN BD METABOLISM

Concurrent with the *in vitro* studies on BD metabolism, studies were conducted on metabolism of BD in whole animals. In a series of studies at the Inhalation Toxicology Research Institute, $B6C3F_1$ mice, Sprague-Dawley rats, and cynomolgus monkeys were exposed to varying levels of ^{14}C-BD, and metabolites in blood and urine were measured (Bond et al., 1986; Dahl et al., 1991; Sabourin et al., 1992). The studies addressed the question of species differences in uptake of inhaled BD, excretion routes, and blood and urine levels of BD metabolites. At low exposure concentrations, the percent of inhaled BD that was retained in the body was 20% in mice and 3–4% in rats or monkeys (Bond et al., 1986; Dahl et al., 1991). This 5-fold difference between the uptake in mice versus the other species is similar to the 5-fold difference in the metabolism of BD to BDO in mouse liver preparations compared to liver preparations from other species. Excretion patterns were similar in mice and rats with half of the internal dose of BD being excreted in the urine, 15–20% being exhaled as volatile metabolites or small amounts of BD, 15–10% retained bound in the body, and small fractions excreted in feces or exhaled as CO_2. The monkey exhaled almost twice as much of the internal dose of BD as CO_2 as was excreted in urine. The high conversion of BD to CO_2 in the monkey may reflect the high level of epoxide hydrolase activity in primate livers resulting in conversion of BD to the tetrol and further oxidation to CO_2. Further experimental work is required to support this hypothesis. The ^{14}C content of the blood of the exposed mice, rats, and monkeys indicated that for equivalent exposure concentrations, the mouse has the highest amount of BD-derived ^{14}C, rats have half the amount in mice, and monkeys have less than 7% of that of mice. Vacuum distillation of the blood followed by trapping of volatile metabolites by cold traps at the appropriate temperatures suggested the presence of BDO in mouse blood at 5-fold higher levels than in the rat and at 15-fold higher levels than in the monkey (Dahl et al., 1991).

Characterization of the urine from mice, rats, or monkeys exposed to ^{14}C-BD indicated two major urinary metabolites, both mercapturic acids (Sabourin et al., 1992; Bechtold et al., 1994). One mercapturic acid (called M-II) was formed from the GSH conjugation with BDO and was highest in amount relative to the total urinary metabolites in the mouse compared to the rat and the monkey. The other mercapturic acid (called M-I) was identified as 1,2-dihydroxy-4-(N-acetylcysteinyl)butane, the mercapturic acid of butane-1,2-diol, the hydrolysis product of BDO. If fully deuterated BD was administered to rodents, the M-I metabolite lost one deuterium atom, suggesting that an oxidation/reduction reaction occurred in the process (Fig. 2). M-I was the major urinary metabolite in the urine of exposed monkeys with only trace amounts of M-II present. In analyses of urine from humans exposed to low levels of BD in the workplace, using nonradioactive methods of analysis, only M-I, and not M-II, could be detected (Bechtold et al., 1994). The rat urine had approximately equal amounts of M-I and M-II. The ratio of M-I to the total M-I + M-II in

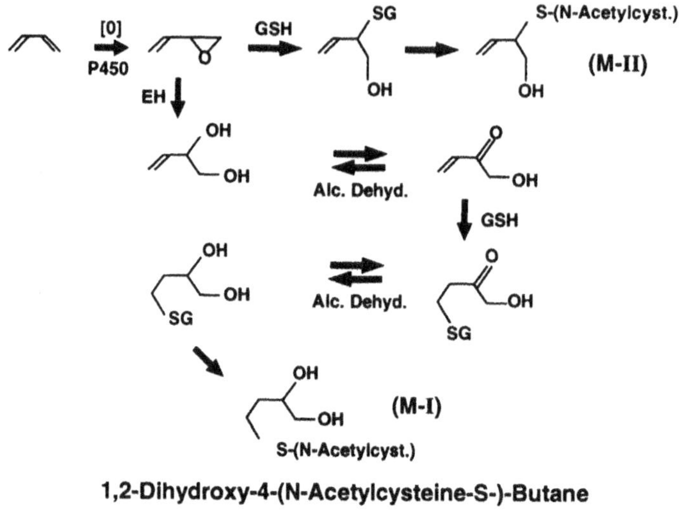

Figure 2. Possible metabolic pathway to form M-I. Used by permission from Bechtold et al., 1994.

the urine of the different species was proportional to the epoxide hydrolase activity (measured using BDO as a substrate) in livers of the different species (Sabourin et al., 1992; Bechtold et al., 1994). The data are in agreement with the *in vitro* data and suggest that a major pathway for removal of BDO in the mouse is via conjugation with GSH, while a major route in primates is by hydrolysis to the butenediol followed by conjugation with GSH.

Recently, there has been increased interest in BDO_2 as the potential potent carcinogen in mice. Two laboratories reported finding no BDO_2 in the blood of rats exposed to BD, while the mouse had blood BDO_2 levels approximately equal to that of BDO (Table 1) (Himmelstein et al., 1994; Bechtold et al., 1995). Subsequently, a method was developed to increase the ability to detect low levels of metabolites (Bechtold et al., 1995). The method used multidimensional gas chromatography/gas chromatography/mass spectroscopy (GC/GC/MS) in which an initial, large-bore GC column was used to increase the volume of samples that could be applied, a cryofocusing unit was used to collect peaks coming off the large column, and the peaks were then reinjected on a small-bore column for resolution. The peaks were identified and quantitated by isotope dilution MS. Even this method did not result in detection of BDO_2 in rat blood. However, when blood or tissues from three rats were combined, and the volatile metabolites (including BDO_2) were vacuum distilled into cold traps, BDO_2 could be detected in rat blood, albeit at levels 100-fold less than in the mouse (J. R. Thornton-Manning, personal communication).

DISCUSSION

The data from both *in vitro* and *in vivo* studies suggest that species differences in the metabolism of BD, particularly in the formation of the BDO_2, can help explain the dramatic species differences in the toxicity of inhaled BD. The qualitative nature of metabolism of BD is similar in the different species, to the extent the pathways have been studied. The quantitative nature of the metabolism along the different pathways is, however, quite different in the different species. All species studied thus far form BDO as the primary metabolite of BD. The major differences lie in how the different species further metabolize the BDO, to form BDO_2 or to clear BDO via excretion as one of two types of mercapturic acid.

Table 1. Blood levels of BD metabolites (nmole/ml; $\overline{X} \pm SE$)

	62.5 ppm BD, 6 hr CIIT[a]	100 ppm BD, 4 hr ITRI[b]
Rat		
BDO	0.07 ± 0.01	0.1 ± 0.06
BDO_2	ND	ND
Mouse		
BDO	0.56 ± 0.04	0.38 ± 0.14
BDO_2	0.65 ± 0.10	0.33 ± 0.19

[a]Himmelstein et al., *Carcinogenesis* 15:1479, 1994.
[b]Bechtold et al., *Chem. Res. Toxicol.* (1995, in press).

The quantitative differences in clearance via the excretion pathways to form either M-I or M-II parallel the quantitative differences in epoxide hydrolase activities in the livers of the different species. In the mouse, more of the BDO is cleared by conjugation of the BDO with GSH, followed by excretion of the mercapturic acid (M-II) in the urine than is cleared by hydrolysis of BDO followed by conjugation with GSH. Knowing that 50% of the internal dose of BD in the mouse is excreted in the urine (Bond et al., 1986), and 62% or 185 of the urinary metabolites are M-II or M-I, respectively (Sabourin et al., 1992; Bechtold et al., 1994), one can calculate that approximately one-third of the internal dose of BD is cleared via direct conjugation of BDO with GSH and only 9% by hydrolysis followed by conjugation with GSH.

In the rat, the amount of M-I and M-II excreted in the urine suggests that approximately 40% of the internal dose of BD is also excreted via the pathways leading to formation of M-I and M-II, but about 20% is excreted by each pathway. However, either very little BDO_2 is formed in the rat, or what is formed is rapidly removed. In primates, in which 31% of the internal dose is excreted in urine, 53% of urinary metabolites is M-I, and 4% is M-II. This suggests that 16% of the BDO is cleared via hydrolysis, then conjugation with GSH, and 1% by direct conjugation with GSH.

The greatest quantitative differences among the species in the metabolism of BD are the formation and clearance of the highly mutagenic BDO_2. In the mouse, a large fraction of BDO is converted to the highly mutagenic BDO_2, based on the findings of two laboratories that blood levels of BDO_2 equal those of BDO. In the rat, levels of BDO_2 are barely detectable. Either the rat does not make much BDO_2, or the rat can rapidly clear the compound. No toxicity studies have been reported in the monkey, so the blood levels of BDO_2 following BD exposure are unknown.

In conclusion, by far the major quantitative difference observed between metabolic products of BD in the sensitive species (mouse) compared to the nonsensitive species (rat) is the 100-fold higher blood concentrations of BDO_2 in exposed mice compared to rats. The evidence suggests that BDO_2 may be of primary importance in the carcinogenicity of BD in mice.

ACKNOWLEDGMENT

The author wishes to recognize the excellent team of investigators who conducted the research on butadiene at ITRI: W. E. Bechtold, J. A. Bond, A. R. Dahl, P. J. Sabourin, and J. R. Thornton-Manning. This research has been supported by the National Institute of Environmental Health Sciences (IAA 222-ES-20092), the Chemical Manufacturers' Association (FIA DE-FIOY-91AL66351), and the Office of Health and Environmental Research of the U.S. Department of Energy under Contract No. DE-AC04-76EV01013.

REFERENCES

Agency for Toxic Substances and Disease Registry, 1992, Toxicological profile for 1,3-butadiene, TP-91/07, Available NTIS, U.S. Department of Commerce, Springfield, VA, USA.

Bechtold, W. E., Strunk, M. R., Chang, I. Y., Ward, J. B. Jr., and Henderson, R. F., 1994, Species differences in urinary butadiene metabolites: Comparisons of metabolite ratios between mice, rats and humans, *Toxicol. Appl. Pharmacol.* 127:44–49.

Bechtold, W. E., Strunk, M. R., Thornton-Manning, J. R., and Henderson, R. F., 1995, Analysis of butadiene, butadiene monoxide and butadiene dioxide in blood by gas chromatography/gas chromatography/mass spectroscopy, *Chem. Res. Toxicol.* (in press).

Bond, J. A., Dahl, A. R., Henderson, R. F., Dutcher, J. S., Mauderly, J. L., and Birnbaum, L. S., 1986, Species differences in the disposition of inhaled 1,3-butadiene, *Toxicol. Appl. Pharmacol.* 84:617–627.

Cochrane, J. E., and Skopek, T. R., 1994, Mutagenicity of butadiene and its epoxide metabolites. I. Mutagenic potential of 1,2-epoxybutene, 1,2,3,4-diepoxybutane and 3,4-epoxy-1,2-butanediol in cultured human lymphoblasts, *Carcinogenesis* 15:713–717.

Csanàdy, G. A., Guengerich, F. P., and Bond, J. A., 1992, Comparison of the biotransformation of 1,3-butadiene and its metabolite, 1,1,3-butadiene monoepoxide, by hepatic and pulmonary tissues from humans, rats and mice, *Carcinogenesis* 13:1143–1153.

Csanàdy, G. A., Guengerich, F. P., and Bond J. A., 1993, Erratum, *Carcinogenesis* 14:784.

Dahl, A. R., Sun, J. D., Birnbaum, L. S., Bond, J. A., Griffith, W. C. Jr., Mauderly, J. L., Muggenburg, B. A., Sabourin, P. J., and Henderson, R. F., 1991, Toxicokinetics of inhaled 1,3-butadiene in monkeys: Comparison to toxicokinetics in rats and mice, *Toxicol. Appl. Pharmacol.* 110:9–19.

Himmelstein, M. W., Turner, M. J., Asgharian, B., and Bond, J. A., 1994, Comparison of blood concentrations of 1,3-butadiene and butadiene epoxides in mice and rats exposed to 1,3-butadiene by inhalation, *Carcinogenesis* 15:1479–1486.

Huff, J. E., Melnick, R. L., Solleveld, H. A., Haseman, J. K., Powers, M., and Miller, R. A., 1985, Multiple organ carcinogenicity of 1,3-butadiene in $B6C3F_1$ mice after 60 weeks of inhalation exposure, *Science* 227:548–549.

Melnick, R. L., Huff, J., Chou, B. J., and Miller, R., 1990, Carcinogenicity of 1,3-butadiene in C57BL/6 × C3H F_1 mice at low exposure concentrations, *Cancer Res.* 50:6592–6599.

Owen, P. E., Glaister, J. R., Gaunt, I. F., and Pullinger, D. H., 1987, Inhalation toxicity studies with 1,3-butadiene. 3. Two-year toxicity/carcinogenicity studies in rats, *Am. Ind. Hyg. Assoc. J.* 48:407–413.

Sabourin, P. J., Burka, L. T., Bechtold, W. E., Dahl, A. R., Hoover, M. D., Chang, I. Y., and Henderson, R. F., 1992, Species differences in urinary metabolites; identification of 1,2-dihydroxy-4-(N-acetylcysteinyl)butane, a novel metabolite of butadiene, *Carcinogenesis* 13:1633–1638.

Schmidt, U., and Loeser, E., 1986, Epoxidation of 1,3-butadiene in liver and lung tissue of mouse, rat, monkey and man, *Adv. Exp. Med. Biol.* 197:951–957.

OVARIAN TOXICITY AND METABOLISM OF 4-VINYLCYCLOHEXENE AND ANALOGUES IN B6C3F$_1$ MICE

Structure-Activity Study of 4-Vinylcyclohexene and Analogues

Julie K. Doerr and I. Glenn Sipes

University of Arizona
Department of Pharmacology and Toxicology
College of Pharmacy
1703 East Mabel
Tucson, Arizona 85721

INTRODUCTION

The industrial chemical, 4-vinylcyclohexene (VCH), is an ovarian toxicant in mice. Administration of VCH for thirty days results in depletion of ovarian follicles, including primordial follicles (Smith et al., 1990a; Hooser et al., 1994). Destruction of these non-dividing follicles produces an irreversible response that leads to the development of premature ovarian failure (Hooser et al., 1994). Since no effect in cyclicity, hormonal levels, and fertility are observed with VCH treatment until the development of ovarian failure (Hooser et al., 1994; Grizzle et al., 1994), early warning signs of ovarian toxicity are not apparent. In addition, ovarian failure is believed to be associated with the development of ovarian neoplasms (Hooser et al., 1994). An increased incidence in the occurrence of uncommon ovarian neoplasms; including mixed benign tumors, granulosa cell tumors, and granulosa cell carcinomas, is observed following chronic administration of VCH (NTP, 1986).

The processes involved in the chemical-mediated ovotoxicity of VCH are not known. Previous studies have established that bioactivation of VCH to epoxides is required for its ovotoxicity (Figure 1; Smith et al., 1990a). In addition, these epoxide metabolites are more potent ovotoxicants than VCH, depleting small ovarian follicles at 4 to 13.5-fold lower doses. The most potent ovotoxicant is VCD. Whether the monoepoxides of VCH themselves are ovotoxic, or require metabolism to the diepoxide to exert ovotoxicity is not known. Thus, structure-activity studies were initiated to determine the individual role of these epoxides in VCH-induced ovotoxicity. It is hypothesized that VCD, which is a potent ovotoxicant, as well as a direct-acting mutagen and carcinogen (NTP, 1989), is the ultimate ovotoxic metabolite of VCH.

Biological Reactive Intermediates V
Edited by R. Snyder et al., Plenum Press, New York, 1996

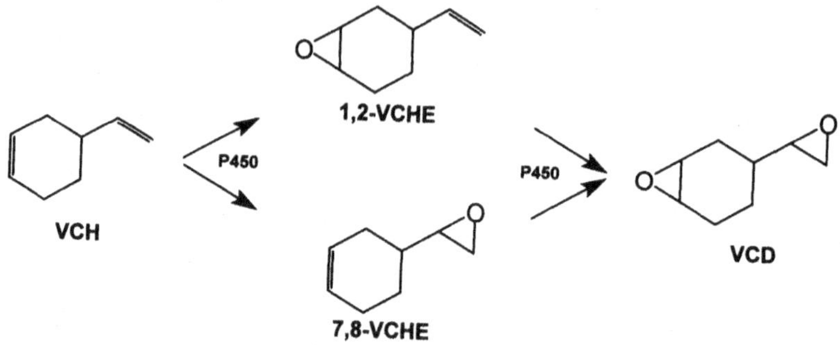

Figure 1. Proposed mechanism of oxidation of 4-vinylcyclohexene. 1,2-VCHE/vinylcyclohexene 1,2-monoepoxide, 7,8-VCHE/vinylcyclohexene 7,8-monoepoxide, VCD/vinylcyclohexene diepoxide.

MATERIALS AND METHODS

Chemicals

Ethylcyclohexene was a gift of Dr. Gerhard Nowack of Phillips Petroleum Company, Bartlesville, OK. Ethylcyclohexene oxide and vinylcyclohexane oxide were custom synthesized in the laboratory of Dr. Eugene Mash of the Department of Chemistry of the University of Arizona, Tucson, AZ. Identity and purity (99%) of the epoxides were established by NMR and IR. Other test compounds were purchased from Aldrich Chemical Company (Milwaukee, WI) with a purity of 98-99%. 3,3,3-Trichloropropene oxide was obtained from Chem Service (West Chester, PA). Glucose-6-phosphate (monosodium salt), glucose-6-phosphate dehydrogenase (Type XV), and β-$NADP^+$ (sodium salt) were purchased from Sigma Chemical Company (St. Louis, MO). Other chemicals were of reagent grade.

Animals

Twenty-one day old female $B6C3F_1$ mice were acquired from Harlan Sprague-Dawley Incorporated (Indianapolis, IN). The mice were housed 5 per cage in a biohazard hood, provided food (Harlan Teklad 4% Mouse/Rat Diet, Madison, WI) and water *ad libitum*, and maintained on a 12 h light/dark cycle in a controlled temperature of 22° ± 2°C. The animals were acclimated to this environment for 7 days prior to use in studies.

Structure-Activity Study

The structure-activity study was designed according to previous ovarian toxicity experiments conducted in our laboratory (Smith et al., 1990a). Treatment groups consisted of VCH (positive control), sesame seed oil (vehicle control), ethylcyclohexene, vinylcyclohexane, and cyclohexene (n = 10/group; Table 1). Twenty-eight day old mice were administered sesame seed oil (2.5 mL/kg) or test compound (7.5 mmol/kg) ip daily for 30 days. Following day thirty, mice were killed by CO_2 inhalation on the first day of diestrus of their cycle. The stage of estrus was determined by vaginal cytology (Allen, 1922). Tissues were removed and weighed. One ovary was randomly selected and further processed for histological examination. The ovary was fixed in Bouin's solution (24 h), transferred to 70% ethanol, serially sectioned (6 μm), and stained with hematoxylin and eosin. Oocytes contained in small and growing pre-antral follicles were counted microscopically at every

Table 1. Structures of the VCH analogues

Test Compound	Structure
Ethylcyclohexene ECHE	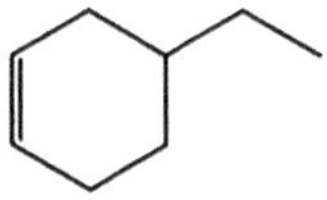
Vinylcyclohexane VCHA	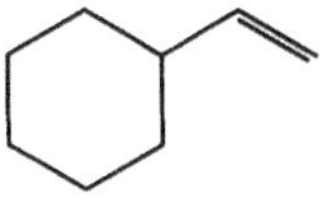
Cyclohexene CHE	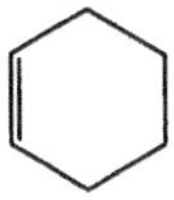

twentieth section. Follicles were classified by the method of Pedersen and Peters (1968). A small follicle (primordial) consisted of an oocyte surrounded by a single layer of noncuboidal granulosa cells. A growing follicle (primary to pre-antral) consisted of an oocyte surrounded by a single or multiple layers of cuboidal granulosa cells. Follicle counts at every twentieth section were summed to calculate the follicle number for each ovary.

Metabolism Studies

Metabolism of VCH and the VCH analogues was examined using hepatic microsomal preparations from 46 day old female mice (16-19 g). Livers were removed and microsomes isolated by differential centrifugation as described by Guengerich (1989). Protein concentration was determined utilizing the Coomassie Plus Protein Assay (Pierce, Rockford, IL). Microsomal cytochrome P450 content was determined by the method of Omura and Sato (1964). Metabolism studies were conducted similar to procedures previously described in our laboratory (Smith et al., 1990b). Samples consisted of test compound (1 mM, VCH or VCH analogue) in methanol (1% v/v), microsomes (0.1 to 1.0 mg/mL), 2 mM 3,3,3-trichloropropene oxide, 0.5 mM $NADP^+$, 10 mM glucose-6-phosphate, 1 unit glucose-6-phosphate dehydrogenase, 50 mM HEPES buffer (pH 7.6), 0.1 mM EDTA, and 15 mM $MgCl_2$ to a final volume of 1 mL. 3,3,3-Trichloropropene oxide, a potent epoxide hydrolase inhibitor (Oesch et al., 1971), was added to the incubations to prevent hydrolysis of epoxides formed. Samples were pre-incubated in a shaking water bath at 37°C for 3 min. Reactions were initiated with the addition of glucose-6-phosphate and incubated at 37°C for an additional 2 to 15 min. Glucose-6-phosphate was absent from blank reactions. Reactions were terminated with the addition of 5 M NaOH (200 µL) and placed in ice. Samples were extracted with 240 µL of hexane containing 4-methyl-1-cyclohexene (0.01 M, internal standard), vortexed, shaken for 10 min, and phases separated by centrifugation (10 min, 3000 rpm). The organic layer was removed and analyzed for epoxide by GC. The extraction efficiencies of the epoxide metabolites were as follows: vinylcyclohexene 1,2-monoepoxide/92%, ethylcyclohexene oxide/70%, vinylcyclohexane oxide/65%, and cyclohexene oxide/55%.

GC Analysis

A single analytical method was developed for quantification of the VCH and VCH analogue monoepoxides. Analyses were done using a Hewlett-Packard HP 5890 Series II GC equipped with a DB624 fused silica column (30 m x 0.255 mm, J and W Scientific, Folsom, CA) and a flame ionization detector. Splitless technique was utilized to introduce 1 μL of the hexane extract with the split vent open at 1 min and a nitrogen carrier gas flow rate of 1 mL/min (10 psi). The initial oven temperature was held at 50°C for 10 min, and than ramped to 230°C at a rate of 12°C/min and held at the final temperature for 5 min. The injector and detector temperatures were 200 and 250°C, respectively. VCH and VCH analogue monoepoxides were quantified by comparing the peak area ratio of the analyte and 4-methyl-1-cyclohexene (internal standard) to an analyte standard curve. The retention times for the analytes were as follows: 4-methyl-1-cyclohexene/12 min, VCH/16 min, vinylcyclohexene 1,2-monoepoxide/21 min, ethylcyclohexene/17 min, ethylcyclohexene oxide/22 min, vinylcyclohexane/16.5, vinylcyclohexane oxide/22.5 min, cyclohexene/10.5 min, and cyclohexene oxide/18 min.

Statistical Analysis

Comparisons between groups were done using a one-way analysis of variance. When appropriate, significance was determined using Student-Newman-Keuls. Data were considered significantly different at $p < 0.05$.

RESULTS

To discern if the monoepoxides of VCH are inherently ovotoxic or require further oxidation to VCD for ovarian toxicity, structural analogues of VCH were tested (Table 1).

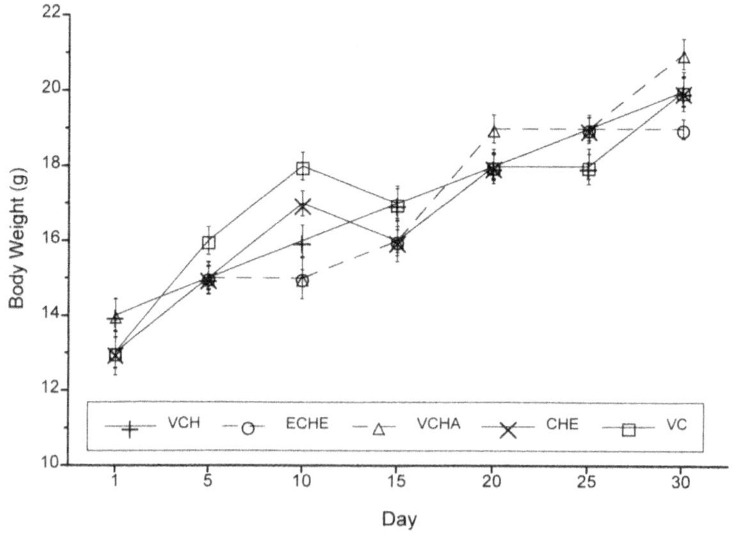

Figure 2. Growth during treatment period with VCH and the VCH analogues. Female B6C3F$_1$ mice were administered test compound (7.5 mmol/kg) or sesame seed oil (VC = vehicle control) ip daily for thirty days. Data represent the mean ± SEM (n = 10/time point). ECHE/ethylcyclohexene, VCHA/vinylcyclohexane, and CHE/cyclohexene.

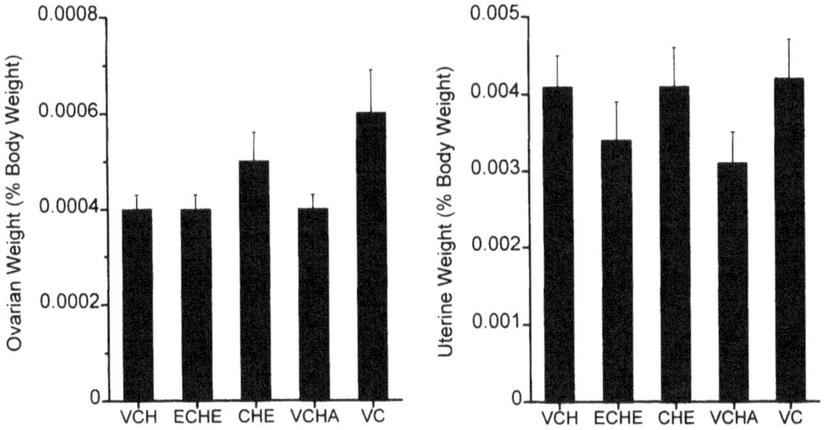

Figure 3. Tissue weights as a percentage of body weight in mice following treatment with VCH or the VCH analogues (see figure 2). Data represent the mean ± SEM (n = 10/group). ECHE/ethylcyclohexene, VCHA/vinylcyclohexane, and CHE/cyclohexene.

General toxicity, as well as ovotoxicity of these compounds was determined. No alteration in growth was observed in the treatment groups over the thirty day dosing period (Figure 2). In addition, ovarian and uterine weights were similar to that of the vehicle control group (Figure 3). Follicular counts revealed that the analogues of VCH did not deplete small follicles (Figure 4). This is in direct contrast to VCH, which consistent with previous studies (Smith et al., 1990a; Hooser et al., 1994) depleted 86% ± 2% of this follicular population at an equimolar dose (Figure 4). A similar effect was observed with the growing follicle population, with the exception of ethylcyclohexene (ECHE). ECHE treatment depleted growing follicles (37% ± 4%), but not to the same extent observed following administration of VCH (72% ± 4%, Figure 4).

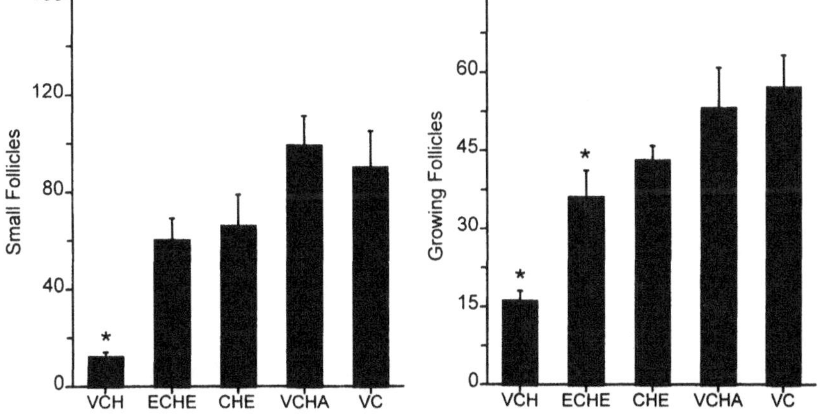

Figure 4. Ovarian follicular counts in mice treated with VCH and the VCH analogues (see figure 2). Mean follicle counts for each treatment group were calculated by summing the counts determined at every twentieth section per mouse and using these totals to obtain an average ± SEM (n = 10/group). *Statistically significant at $p < 0.05$ as determined by ANOVA and Newman-Keuls test. ECHE/ethylcyclohexene, VCHA/vinylcyclohexane, CHE/cyclohexene, and VC/vehicle control.

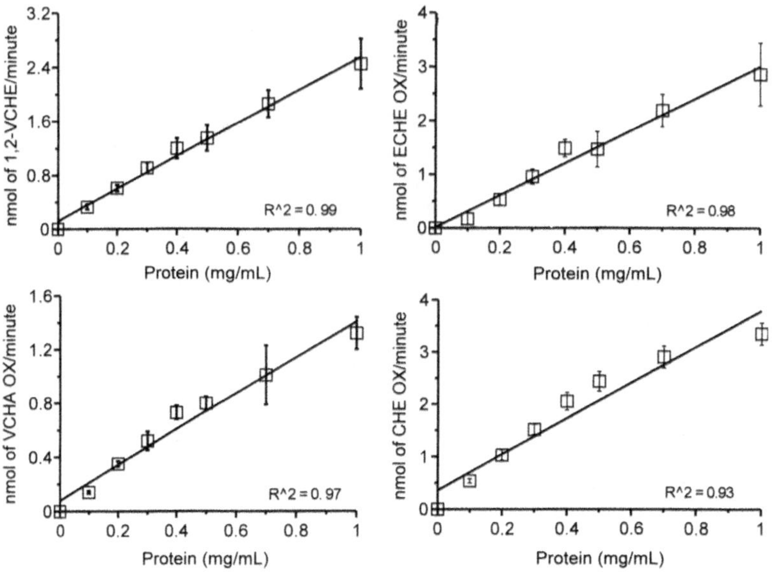

Figure 5. Female mouse hepatic microsomal metabolism of VCH and the VCH analogues. Incubations contained 1 mM of substrate (VCH, ECHE, VCHA, or CHE) and 2 mM 3,3,3-trichloropropene oxide with 0.1 to 1.0 mg of protein for 10 minutes. Hexane extracts of the microsomal preparations were analyzed by GC. Data represent the mean ± SD (n = 2-3). 1,2-VCHE/vinylcyclohexene 1,2-monoepoxide, ECHE OX/ethylcyclohexene oxide, VCHA OX/vinylcyclohexane oxide, and CHE OX/cyclohexene oxide.

The structure-activity study indicated that analogues of VCH, which possess a single unsaturated site and thus can only form monoepoxides, are not ovotoxic. Critical to interpreting these results was the characterization of the metabolism of the VCH analogues, since biotransformation of these compounds has not been previously reported. ECHE, vinylcyclohexane (VCHA), and cyclohexene (CHE) were metabolized to monoepoxides in murine hepatic microsomal preparations (Figure 5). Metabolite production was linear within protein concentrations of 0.1 to 1.0 mg (Figure 5), and between 2 to 15 min (data not shown). Rates of epoxidation for ECHE, VCHA, and CHE were as follows: 3.61 ± 0.33, 1.86 ± 0.12, and 5.34 ± 0.42 nmol/min/mg, respectively. These rates of epoxidation were similar to that of VCH (2.79 ± 0.40 nmol/min/mg).

DISCUSSION

Previous studies have established that metabolism of VCH is necessary for its ovotoxicity in rodents (Smith et al., 1990a). Both monoepoxides of VCH (1,2-VCHE and 7,8-VCHE), as well as the diepoxide (VCD), have been shown to be ovotoxic. Determination of the ED50 values (i.e., dose that reduces 50% of ovarian follicles) showed that the monoepoxides and the diepoxide were 4 to 5.5-fold and 13.5-fold more potent than VCH, respectively (Smith et al., 1990a). An impaired ability of the rat, as compared to the mouse, to bioactivate VCH to these epoxides is associated with the resistance of the rat to the ovotoxic effects of VCH (Smith et al., 1990b). Whether the monoepoxides of VCH are inherently ovotoxic, or require further oxidation *in vivo* to form VCD for VCH-induced ovotoxicity in mice is not known. To address this question, analogues of VCH were tested which contain a single unsaturated site corresponding to either the 1,2-position (ECHE and

CHE) or the 7,8-position (VCHA) of VCH. Thus, it could be determined if epoxidation at one of these sites results in an ovotoxic compound, or if epoxidation at both unsaturated sites is necessary for the ovotoxic effects of VCH.

No alteration in uterine and ovarian weight or small follicle number was observed following treatment with the VCH analogues. However, at an equimolar dose VCH significantly depleted ovarian follicles. This lack of ovotoxicity of the VCH analogues was not due to an inability of these compounds to be biotransformed to monoepoxides, since it was demonstrated that they formed monoepoxide metabolites *in vitro*. Hepatic microsomal oxidation of the analogues occurred at a similar rate to VCH. In addition, circulating levels of the VCH analogue monoepoxides have been observed *in vivo* following administration of a single dose of the parent compound (7.5 mmol/kg, ip; data not shown). Taken together, these results indicate that epoxidation of VCH at a single site does not result in the formation of an ovotoxic metabolite. Rather epoxidation at both unsaturated sites to form the diepoxide metabolite seems essential.

A related olefin, 1,3-butadiene, is ovotoxic in mice causing severe ovarian atrophy and the development of ovarian neoplasms (Melnick et al., 1992). Similar to VCH, an enhanced capacity of mice to metabolize 1,3-butadiene to its epoxide metabolites (butadiene monoepoxide and butadiene diepoxide) results in this species sensitivity to the compound (Himmelstein et al., 1994; Csanady et al., 1992). Of particular interest to our laboratory, is determining if a diepoxide is also critical in 1,3-butadiene-induced ovotoxicity, as well as the ovotoxic potential of its structural analogue, isoprene. Preliminary studies indicate that isoprene and the epoxide metabolites of butadiene are ovotoxic. These compounds depleted ovarian follicle populations to a similar extent as observed with VCH treatment. However, the butadiene epoxides were much more potent. In addition, epoxybutane, administered at an equimolar dose as butadiene monoepoxide, did not deplete ovarian follicles. Since epoxybutane cannot form a diepoxide, these results indicate that a diepoxide metabolite may also be critical in butadiene-induced ovotoxicity.

Future studies in our laboratory will continue to focus on determining the ultimate ovotoxic metabolite of VCH and related olefins. Structure-activity studies will be extended to examining the ovotoxicity of the epoxide-diol metabolites of VCH and butadiene. It is believed that these metabolites are detoxified forms of the diepoxides. Characterization of the ovotoxic metabolite(s) of these compounds will aid in determining the mechanism by which these compounds target the ovary. The ability of hepatic and ovarian tissue to form the diepoxides and/or detoxify these metabolites may be critical in the sensitivity of the ovary to these compounds.

ACKNOWLEDGMENTS

The authors thank Dr. Eugene Mash and Timothy Gregg for the synthesis and characterization of VCHA OX and ECHE OX, and Drs. Jennifer Galvin and Gerhard Nowack and Carrol Kirwin of Phillips Petroleum Company for the gift of ECHE. This work was supported in part by the Center for Toxicology/Flinn Research Fellowship, March of Dimes 15-FY93-0672, NIEHS Center Grant IP30 ES06694, and NO1-ES 85230.

REFERENCES

Allen, E. (1922) The oestrus cycle in the mouse. *Am. J. Anat.* **30**, 297-371.

Csanady, G. A., Guengerich, F. P., and Bond, J. A. (1992) Comparison of the biotransformation of 1,3-butadiene and its metabolite, butadiene monoepoxide, by hepatic and pulmonary tissues from humans, rats and mice. *Carcinogenesis* **13,** 1143-1153.

Grizzle, T. B., George, J. D., Fail, P. A., Seely, J. C., and Heindel, J. J. (1994) Reproductive effects of 4-vinylcyclohexene in Swiss mice assessed by a continuous breeding protocol. *Fundam. Appl. Toxicol.* **22,** 122-129.

Guengerich, F.P. (1989) Analysis and characterization of enzymes. In *Principles and Methods of Toxicology* (Hayes, A. W. Ed.) pp 777-814, Raven Press, Ltd., New York.

Himmelstein, M. W., Turner, M. J., Asgharian, B., and Bond, J. A. (1994) Comparison of blood concentrations of 1,3-butadiene and butadiene epoxides in mice and rats exposed to 1,3-butadiene by inhalation. *Carcinogenesis* **15,** 1479-1486.

Hooser, S. B., Douds, D. P., DeMerell, D. G., Hoyer, P. B., and Sipes, I. G. (1994) Long-term ovarian and gonadotropin changes in mice exposed to 4-vinylcyclohexene. *Reprod. Toxicol.* **8,** 315-323.

Melnick, R. L., and Huff, J. (1992). 1,3-Butadiene: Toxicity and carcinogenicity in laboratory animals and in humans. *Reviews of Environmental Contamination and Toxicology* **124**, 111-144.

National Toxicology Program (1986) Toxicology and carcinogenesis studies of 4-vinylcyclohexene in F344/N rats and $B6C3F_1$ mice (gavage studies). U.S. Department of Health and Human Services, Public Health Service, National Institutes of Health, Research Triangle Park, NC, Technical Report No. 303.

National Toxicology Program (1989) Toxicology and carcinogenesis studies of 4-vinylcyclohexene diepoxide in F344/N rats and $B6C3F_1$ mice (dermal studies). U.S. Department of Health and Human Services, Public Health Service, National Institutes of Health, Research Triangle Park, NC, Technical Report No. 362.

Oesch, F., Kaubisch, N., Jerina, D. M., and Daly, J. W. (1971) Hepatic epoxide hydrase: Structure-activity relationships for substrates and inhibitors. *Biochemistry* **10,** 4858-4866.

Omura, T., and Sato, R. (1964) The carbon monoxide-binding pigment of liver microsomes. *J. Biol. Chem.* **239,** 2370-2385.

Pedersen, T., and Peters, H. (1968) Proposal for a classification of oocytes and follicles in the mouse ovary. *J. Reprod. Fert.* **17,** 555-557.

Smith, B. J., Mattison, D. R., and Sipes, I. G. (1990a) The role of epoxidation in 4-vinylcyclohexene-induced ovarian toxicity. *Toxicol. Appl. Pharmacol.* **105,** 372-381.

Smith, B. J., Carter, D. E., and Sipes, I. G. (1990b) Comparison of the disposition and *in vitro* metabolism of 4-vinylcyclohexene in the female mouse and rat. *Toxicol. Appl. Pharmacol.* **105,** 364-371.

MECHANISMS OF NITROSAMINE BIOACTIVATION AND CARCINOGENESIS*

Chung S. Yang† and Theresa J. Smith

Laboratory for Cancer Research
College of Pharmacy, Rutgers University
Piscataway, New Jersey 08855-0789

It was discovered in the mid-1950s by Magee and Barnes that *N*-nitrosodimethylamine (NDMA) was hepatotoxic. The carcinogenicity of this compound and many other nitrosamines has since been tested by Magee, Druckrey, Preussmann, Schmähl, and others; the results of these pioneer works were summarized in two comprehensive reviews (1, 2). The biochemical mechanisms of the carcinogenicity and the structure-activity relationship of the nitrosamines have since been studied by many investigators and the results have been summarized in three recent symposia (3-5). In this chapter, we will discuss the occurrence and carcinogenicity of nitrosamines, the enzyme mechanisms and the organ site of the activation, and the dose-response relationship in carcinogenesis.

OCCURRENCE AND CARCINOGENICITY OF NITROSAMINES

The structures of several commonly occurring nitrosamines are shown in Figure 1. The occurrence and health effects of nitrosamines were reviewed by the National Research Council (6). NDMA occurs widely in our environment and is present in cigarette smoke, beer, new car interiors, and a variety of food items. However, exposure to high concentrations of NDMA occurs in the rubber, leather tanning, and iron casting industries. *N*-Nitrosodiethylamine (NDEA) and *N*-nitrosomorpholine (NMOR) have also been detected in some of the above sources. *N*-Nitrosodiethanolamine (NDELA) is present in cosmetics, lotions, shampoos, and synthetic cutting fluid, whereas *N*-nitrosopyrrolidine (NPYR) is found in cooked nitrite-cured bacon. In the United States, the general exposure to these volatile nitrosamines

* Abbreviations: NDMA, *N*-nitrosodimethylamine; NDEA, *N*-nitrosodiethylamine; NMOR, *N*-nitrosomorpholine; NDELA, *N*-nitrosodiethanolamine; NPYR, *N*-nitrosopyrrolidine; NNK, 4-(methylnitrosamino)-1-(3-pyridyl)-1-butanone; NNN, *N'*-nitrosonornicotine; NMBzA, *N*-nitrosomethylbenzylamine.

† Send correspondence to: Dr. Chung S. Yang, Laboratory for Cancer Research,.College of Pharmacy, Rutgers University, Piscataway, New Jersey 08855-0789, Tel: (908) 445-5360; Fax: (908) 445-5767.

Biological Reactive Intermediates V
Edited by R. Snyder et al., Plenum Press, New York, 1996

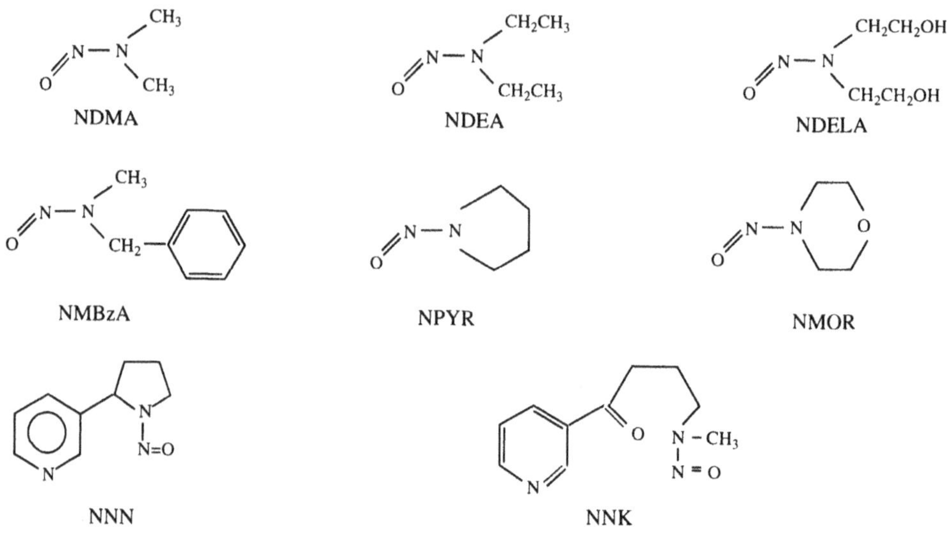

Figure 1. Structures of nitrosamines.

is estimated to be at most a few μg/person/day from our diet, cosmetics and car interiors. Occupational exposures, such as in the leather tannery and rubber factory, however, can be much higher. The largest nonoccupational exposure to nitrosamines are experienced by tobacco users due to the tobacco-specific nitrosamines which are formed from nicotine during the processing of tobacco or during cigarette smoking. The amounts of 4-(methylnitrosamino)-1-(3-pyridyl)-1-butanone (NNK) and *N'*-nitrosonornicotine (NNN) derived from one cigarette are estimated to be 0.12-0.44 and 0.2-3.7 μg, respectively.

Nitrosamines have been shown to be carcinogenic in more than 40 animal species tested and are very likely human carcinogens. Tobacco-specific nitrosamines are believed to be important causative factors for human lung and oral cancers (7). The high incidence of gastric and esophageal cancers in many third world countries are associated with dietary nitrosamines, as well as precursors of nitrosamines and nitrosamides (8). A remarkable feature of nitrosamine carcinogenicity is the organ specificity. For example, NDMA and NDEA are hepatocarcinogens in most species and lung carcinogens in the mouse. *N*-Nitrosomethylbenzylamine (NMBzA), however, induces esophageal and forestomach cancers but not liver cancer in rats. NNK is a strong lung carcinogen in mice and rats, whereas NNN induces esophageal and nasal cavity cancer but not lung cancer in rats. Such specificity can be explained in part by the organ site specific metabolic activation of these nitrosamines, which will be discussed in subsequent sections.

GENERAL MECHANISMS FOR THE METABOLIC ACTIVATION OF NITROSAMINES

It has long been recognized that α-hydroxylation is the key metabolic pathway leading to the formation of carcinogenic alkylating agents (1). In most cases, this pathway is due to oxygenation of the α-carbon, catalyzed by cytochrome P450 enzymes (9). The general pathways are illustrated in Fig. 2. The resulting hydroxylated intermediates (α-nitrosamino alcohols) are not stable and readily rearrange to form an aldehyde and alkyldiazohydroxide. In the specific example shown in Figure 2, the reaction leads to the formation of a methylating agent which can react with DNA and other cellular molecules, causing carcinogenesis and

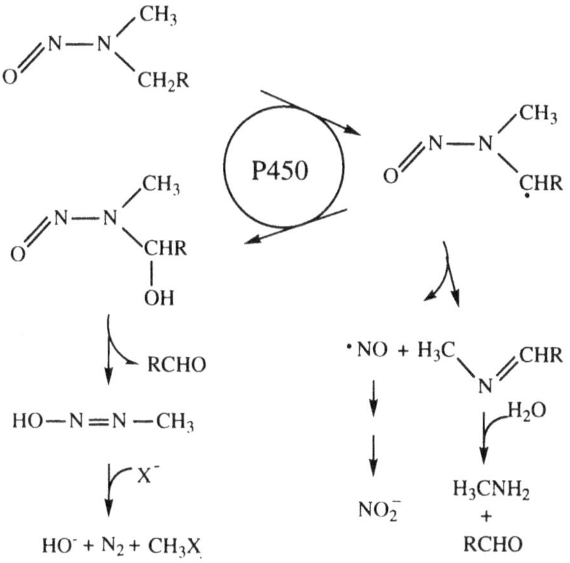

Figure 2. A general scheme for the metabolism of nitrosamines.

cytotoxicity. In this dealkylation pathway for the activation of nitrosamines, the initial oxidation step is believed to form an α-nitrosamino radical and an •OH at the P450 active site. These two radical species can recombine, in a process known as oxygen rebound, to produce the α-nitrosamino alcohol intermediate. Alternatively, the α-nitrosamino radical may fragment into nitric oxide and an imine which is readily hydrolyzed to an aldehyde and methylamine. The nitric oxide is oxidized to nitrite, possibly involving O_2 as an oxidizing agent (10). Since this denitrosation pathway is not known to produce alkylating agents, it is generally considered to be a detoxification pathway. Via a similar mechanism, the other alkyl group (methyl group in the examples shown) can also undergo α-oxidation, leading to similar demethylation and denitrosation pathways. The evidence generated in our laboratory indicated that the denitrosation pathway, sharing an α-nitrosamino radical intermediate with the dealkylation pathway, is an oxidative pathway. This is different from the view of another group of authors suggesting that denitrosation is a reductive pathway (11).

The rates and perhaps also the mechanisms for the denitrosation reactions depend upon the structures of the nitrosamines and P450 enzymes involved for the catalysis. For example, with NDMA, the ratio of the rates of demethylation (formaldehyde formation) to denitrosation (nitrite formation) is about 8 : 1. In the metabolism of *N*-nitrosomethylpentylamine, catalyzed by P450 2B1, the rates of the depentylation, demethylation, and denitrosation are at ratios of 100:50:25. In addition, the carbons of the pentyl moiety group can also be oxidized, forming 2-OH, 3-OH, 4-OH, and 5-OH derivatives at rate ratios of 0.6:1.5:26:2.4, which is much lower than the depentylation rate of 100 (12).

ENZYME SPECIFICITY IN THE METABOLISM OF NITROSAMINES

The vital role of P450 2E1 in the metabolism of NDMA is supported by several lines of evidence. Among all P450 enzymes isolated from rabbits, rats, and humans, P450 2E1 enzymes show the lowest K_m and highest k_{cat}/K_m in catalyzing the oxidation of NDMA (13, 14, and unpublished results). With purified P450 2E1 in the reconstituted system, a K_m value of 3.5 mM has been observed (14), much higher than the value of 20 μM observed in rat and human liver microsomes. This is probably due to the presence of glycerol and possibly other unknown inhibitors in the preparations of purified P450 2E1 and NADPH:P450 oxidoreductase. Recent work with heterologously expressed human P450 2E1 in Hep G2 cells indicates that the P450 2E1 containing microsomes displays a low K_m of 20 μM in the NDMA demethylase assay (15), confirming that P450 2E1 is responsible for the low K_m form of NDMA demethylase. Because the carcinogenic dosages and the human exposure levels of NDMA are not very high, we believe this low K_m form of enzyme is important in the activation of this carcinogen. In comparison to other P450 forms, P450 2E1 is also more efficient in activating NDMA to a mutagen for the V79 cells (16). The predominant role of P450 2E1 in the metabolism of NDMA is demonstrated by the results that antibodies against P450 2E1 inhibited more than 80% of the NDMA demethylase and denitrosation activities in rat and human liver microsomes (unpublished results and17).

P450 2E1 is also very effective in the metabolism of NDEA. Antibodies against P450 2E1 inhibit NDEA deethylase in rat and human microsomes by 50 and 60%, respectively, suggesting that this enzyme also plays an important role in the metabolism of NDEA, but not as dominant as in the metabolism of NDMA (unpublished results and17). In addition to P450 2E1, P450 2A6 activates NDEA in human liver microsomes (18). Recent studies indicate that P450 2A6 is the low K_m enzyme form for the activation of NDEA (personal communication).

The enzyme specificity and alkyl group selectivity in the metabolism of nitrosamines have been investigated. It appears that the structures of the alkyl groups are more important than the nitroso group in determining the enzymes specificity in the metabolism of dialkylnitrosamines. Our tentative conclusion is that P450 2E1 is more effective in oxidizing the α-methyl and α-ethyl groups, whereas other P450 forms, such as P450 2B1, are more effective in catalyzing the dealkylation of larger hydrocarbon chains (19). However, the alkyl group selectivity is an event which is dependent on the concentration of the substrate. When more sensitive methods become available for assaying the enzyme activity at lower substrate concentrations, this tentative conclusion may change. In comparison to most other P450 enzymes, P450 2E1 is known to have relatively small active sites (20). With nitrosamines of large molecular weight, such as NNK, a large K_m value (in the mM range) is observed. Activities with such high K_m values may not have any importance *in vivo*. Therefore, the general statement that P450 2E1 is responsible for the activation of nitrosamines is incorrect.

METABOLIC ACTIVATION OF *N*-NITROSOMETHYLBENZYLAMINE

NMBzA is a potent and rather specific esophageal carcinogen in rats. The molecular basis for the high specificity toward esophagus has attracted the attention of many investigators (21, 22). NMBzA is thought to be very efficiently activated in the esophagus. It was demonstrated that NMBzA is oxidized to benzaldehyde and benzoic acid, an activation pathway leading to the formation of a methylating agent, by an enzyme (or enzymes) displaying apparent K_m values of 3-5 μM in rat esophageal microsomes (unpublished results). This K_m value is about one tenth of the K_m value observed in liver microsomes (in the formation of benzaldehyde and benzyl alcohol); the V_{max} for the esophageal microsomes is approximately one fifth that of the liver microsomes. Again P450 2E1 is an important enzyme in the metabolism of micromolar concentrations of NMBzA in the liver microsomes;

the activity is inducible by pretreatment of rats with acetone, an inducer of P450 2E1 (unpublished results). The presence of P450 2E1 in the esophagus is supported by immunoblot analysis, NDMA demethylase activity, and the presence of CYP 2E1 sequence in clones from a cDNA library of the rat esophagus. This enzyme is only present at low levels in the esophagus and plays a limited role in the activation of NMBzA. In rat esophagus, NMBzA is mainly metabolized by a low K_m P450 enzyme and its identity remains to be established. NMBzA is also oxidized by human liver microsomes, with P450 2E1 again playing an important role (unpublished results). With human esophageal microsomes, NMBzA is metabolized at a much lower rate than in rat esophageal microsomes. NMBzA has been an important experimental tool for the induction of esophageal cancer. Although the presence of NMBzA in the human diet has been reported (23), the results still remain to be confirmed and the role of NMBzA in the causation of human esophageal cancer remains to be investigated.

ENZYMOLOGY OF THE METABOLISM OF NNK AND NNN

In the activation of the potent tobacco-specific carcinogen NNK, the activation pathways are through the oxidation of the α-methyl and α-methylene carbons, leading to the formation of a pyridyloxobutylating agent and a methylating agent, respectively (7) (Figure 3). The detectable products for these pathways are keto alcohol [4-hydroxy-1-(3-pyridyl)-1-butanone] and keto aldehyde [4-oxo-1-(3-pyridyl)-1- butanone], respectively. In rat liver, lung, and nasal mucosa microsomes, mouse liver and lung microsomes, and human liver microsomes, NNK is mainly activated by P450 enzymes (24-29). In rat liver microsomes, the reduction of NNK to NNAL [4-(methylnitrosamino)-1-(3-pyridyl)-1-butanol],

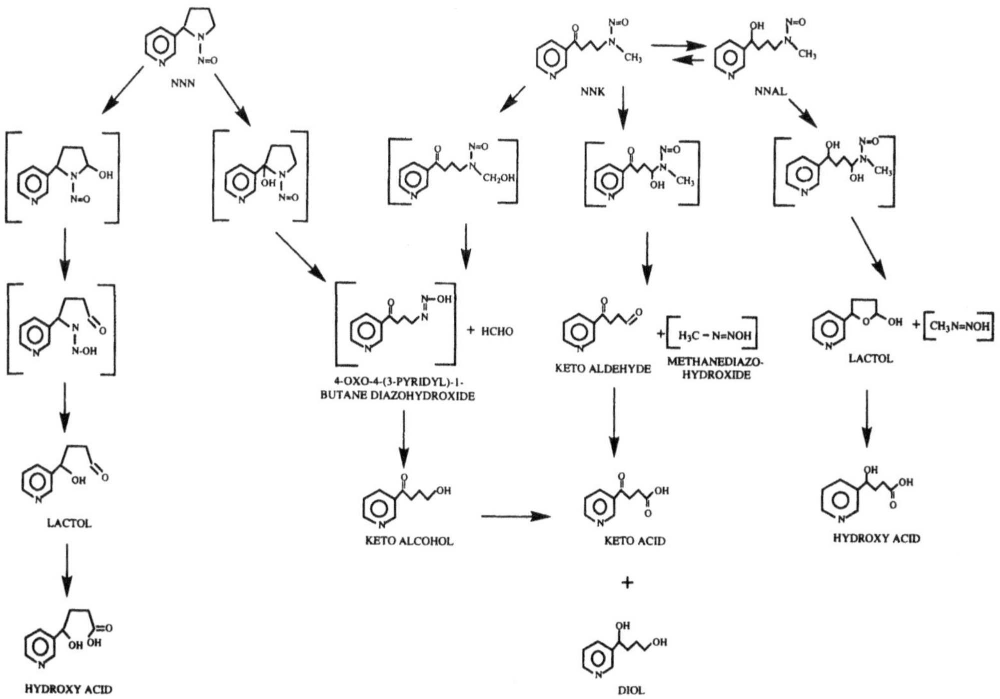

Figure 3. Metabolic pathways for NNK and NNN (modified from reference 7).

catalyzed by carbonyl reductase, is the major route of metabolism whose rate is about 10 times higher than the oxidative pathways. Many P450 forms are capable of catalyzing the oxidization of NNK to different products displaying K_m values of 100-250 μM in rat liver microsomes. The activity is inducible by commonly used P450 inducers such as phenobarbital, 3-methylcholanthrene, Aroclor 1254, pregnenolone 16α-carbonitrile, safrole, and isosafrole, but not by acetone (29). In the rat and mouse lung microsomes, however, the K_m values are much lower (3-30 μM). Antibodies against P450s 2A1 and 2B1 inhibit most of the activities for the formation of keto aldehyde. In addition, antibodies against P450 1A2 also significantly inhibit the oxidation of NNK to keto aldehyde in rat lung microsomes. The results suggest that these three P450 forms or immunochemically-related forms are important in catalyzing the activation of NNK to a DNA methylating agent.

In comparison to other rodent tissues, nasal mucosa microsomes possess the highest activity in catalyzing the oxidation of NNK (28, 30). With rat nasal mucosa microsomes, K_m values of approximately 10 μM and V_{max} values of approximately 3 nmol/min/mg protein have been observed for the formation of both keto aldehyde and keto alcohol (28). Immunoinhibition studies suggest the involvement of P450s 1A2, 2A1, and 3A in this pathway (28). Microsomes from rat olfactory nasal mucosa are 4- to 5-fold more active than those from the respiratory mucosa in catalyzing the oxidation of NNK to keto aldehyde and keto alcohol. With rabbit nasal microsomes, olfactory microsomes are also more active than respiratory microsomes but the difference is less than 2-fold (31). In addition, NNK-*N*-oxide is formed in rabbit nasal mucosa microsomes but not in rat nasal mucosa microsomes. Purified P450 NMa, a major constitutive P450 in rabbit nasal microsomes, shows low K_m values (9-15 μM) and high V_{max} values (1.3 nmol/min/nmol P450) in catalyzing the formation of keto aldehyde and keto alcohol. The olfactory specific P450 NMb has a higher K_m (180 μM) and low activity in catalyzing NNK oxidation (31).

These studies indicate that in nonhepatic tissues of rodents, there are a few low K_m enzymes which catalyze the activation of NNK. In the liver, many P450 enzymes have the ability to catalyze the oxidation of NNK. In addition, the reduction of NNK is the major pathway in hepatic metabolism. The resulting NNAL can be oxidized in a manner similar to the oxidation of NNK, or perhaps more readily, be converted to a glucuronide conjugate and excreted. These metabolic properties may account for the lower susceptibility of the liver than the lung to NNK carcinogenesis.

Among the 12 human P450 forms expressed in Hep G2 cells, P450 1A2 has the highest activity, and P450s 2A6, 2B6, 2E1, 2F1, and 3A5 also have measurable activities in catalyzing the formation of keto alcohol. P450s 2C8, 2C9, 2D6, 3A3, 3A4, and 4B1 in Hep G2 cells have no detectable activity (27). In a reconstituted system, human P450 1A2 catalyzed the formation of keto alcohol exhibiting a K_m value of 383 μM and k_{cat} of 1.7 nmol/min/nmol P450 (unpublished results). Recent studies using the baculovirus expression system demonstrated that human P450s 2A6 (low K_m) and 3A4 (high K_m) can catalyze the formation of keto aldehyde, whereas, P450s 2E1 and 2D6 catalyze the formation of keto alcohol with high K_m values. Low activities for the formation of keto aldehyde and keto alcohol are also detected in P450 2D6 expressed in a B-lymphoblastoid cell line (32). The contribution of each of these forms in human tissues is dependent on the abundance and the activity of each enzyme. In one human liver microsomal sample that we studied, keto alcohol was the major oxidative product formed. Immunoinhibition study suggests that P450s 1A2 and 2E1 each account for approximately 50% of the activity, whereas P450s 2C8, 2D1, 2D6, and 3A4 are not involved (27). With human liver microsomes, carbon monoxide inhibition study suggests that P450 enzymes contribute to 90% and 50% of the activities in the formation of keto alcohol and keto aldehyde, respectively. On the other hand, in human lung microsomes, keto aldehyde is the major oxidative metabolite and the activity was only inhibited 10-40% by carbon monoxide, dependent on the specific lung samples investigated

(27). The results suggest the importance of P450-independent pathways in the activation of NNK in human lung. The identities of the enzymes in catalyzing these pathways is being investigated. Because of this significant species difference, caution has to be applied in extrapolating conclusions obtained from studies with animals to humans.

The activation of NNN involves the 2′- or 5′-hydroxylation of NNN leading to a pyridyloxobutyldiazohydroxide or a diazohydroxide, respectively (7) (Figure 3). DNA adducts have only been detected for the 2′-hydroxylation of NNN. Studies have shown NNN is metabolized by rat liver, nasal mucosa, esophagus and oral tissue in which the 2′-hydroxylation of NNN is greater in the target tissues, whereas, in the rat liver, 5′-hydroxylation of NNN predominates (7, 33-35). This selectivity for 2′-hydroxylation of NNN in the target tissues may account for the carcinogenicity of NNN in the esophagus and nasal mucosa. The activation of NNN is catalyzed by P450. In rat liver microsomes, the 2′-hydroxylation of NNN is induced by aroclor, phenobarbital and 3-methylcholanthrene (35, 36). The specific P450 forms involved in the activation of NNN in the target tissues of rats are presently not known. Human P450 2A6 had the highest activity in catalyzing the formation of keto alcohol (2′-hydroxylation product) from NNN exhibiting a low K_m of 2 μM (unpublished results). This low K_m enzyme form should be important in the activation of NNN in target tissues of humans exposed to low levels of NNN. In human esophageal microsomes, NNN was activated to keto acid (2′-hydroxylation product) and hydroxy acid (5′-hydroxylation product), with keto acid being formed at a higher rate (unpublished results). Comparison of the activation of NNN and NNK in human esophageal microsomes revealed the lack of α-hydroxylation products with NNK. Since the α-hydroxylation pathway leading to the formation of keto alcohol with further oxidation to keto acid is the common pathway shared by NNK and NNN, the lack of α-hydroxylation products with NNK suggests different enzymes are responsible for the activation of NNK and NNN in the esophagus. This may account for the susceptibility of the esophagus to tumor formation induced by NNN.

ORGAN SITES OF NITROSAMINE METABOLIC ACTIVATION AND SPECIES DIFFERENCES

Previous studies in animal models have demonstrated that carcinogens are activated at the target sites. For example, NNK, a potent lung carcinogen in the A/J mouse is most likely to be activated at the target site because of the presence of low K_m forms of activating enzymes (24, 37). NMBzA is efficiently activated in the rat esophagus, and there is evidence to support the thesis that this esophageal carcinogen is activated at the target site (38). Such high activities in the activation of NNK and NMBzA were not observed in human lung and esophageal samples. The possibility that these carcinogens are activated in the liver and then transported to the target sites requires serious investigation.

The organ site of carcinogen activation is an important issue in human cancer susceptibility assessment. The possible use of P450 polymorphisms as parameters for the assessment of human cancer susceptibility has recently received a great deal of attention. In most studies, the approach is mainly correlative without a clear mechanistic link. For example, the possible correlation between P450 2D6 (debrisoquine metabolism) polymorphism and human lung cancer risk is based on the assumption that P450 2D6 is involved in the activation of NNK (39). The available data suggest, however, that this enzyme does not play an appreciable role in the activation of NNK (18, 27), even though a low rate of keto alcohol formation by P450 2D6 can be demonstrated (32) displaying a K_m of 5.5 mM (personal communication). A role for P450 2D6 in the activation of polycyclic hydrocarbons has not been demonstrated either. Non-invasive probes have been developed for assessing the phenotype of different P450 polymorphisms. These approaches, however, mainly reflect the P450 enzyme activities in the liver. In assessing the cancer risk of nonhepatic tissues, there is not enough information to predict their abilities in carcinogen activation based on genetic polymorphism or phenotype due to liver metabolism.

The species differences in drug and carcinogen metabolism are well recognized. This information is vital for the extrapolation of results from animals to humans. For example, phenethyl isothiocyanate and diallyl sulfide are very effective inhibitors of NNK-induced lung tumorigenesis in the A/J mouse model because they are inhibitors and inactivators of the low K_m enzymes responsible for the metabolic activation of NNK (31, 37). Such low K_m enzymes have not been found in the human lung (27); therefore, these compounds cannot be predicted to be effective inhibitors of human lung tumorigenesis.

DOSE-RESPONSE RELATIONSHIP IN NITROSAMINE-INDUCED CARCINOGENESIS

The pioneer work of Druckrey and coworkers (2) established the relationship between the dose of a nitrosamine and the median time to death from cancer in the equation (dose rate) x $(\text{median})^n$ = constant. This equation has been found to be applicable to describe the dose response relationship in many subsequent carcinogenesis studies. By far the most comprehensive study on the dose-response of carcinogenesis was carried out in the early 1980's at the British Industrial Biological Research Association by Brantom and analyzed at the ICRF Cancer Study Unit under the direction of Peto (40). In this study, 4080 inbred rats were maintained from weaning on 15 different concentrations (from 0.033 to 16.896 ppm) of NDEA and NDMA in drinking water for their life time. The highlights of the results are shown in Figure 4. At high doses, the relationship [(dose rate) x $(\text{median})^n$ = constant] was observed, with n = 2.3 for NDEA-induced esophageal and liver tumors and NDMA-induced bile duct neoplasms; and n = 1, for NDMA-induced liver tumors. For doses of NDEA and NDMA sufficiently low for longevity to be nearly normal, the number of liver neoplasms induced by treatment was simply proportional to the dose rate without any "threshold." On

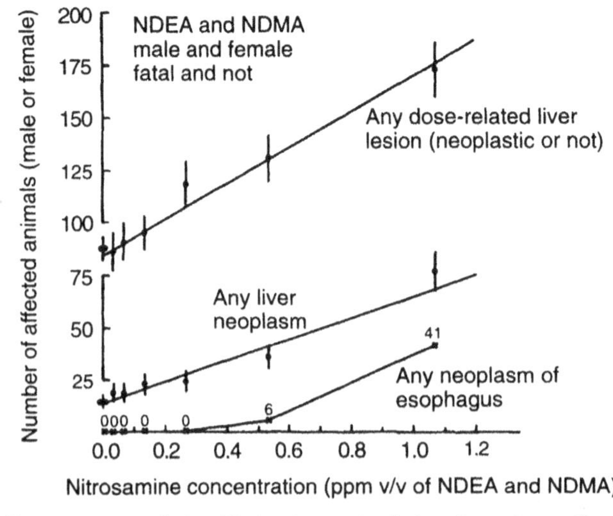

Figure 4. Dose-response relationship in nitrosamine-induced neoplasms (from reference).

the other hand, such a linear dose-response was not observed in NDEA-induced esophageal neoplasms; tumors were observed at 0.528 ppm but not at 0.264 ppm or lower concentrations of NDEA. The molecular basis for this difference is not known. The study illustrates the difficulties in making the proper assumptions for human risk assessment.

REFERENCES

1. Magee, P. N., and Barnes, J. M., 1967, Carcinogenic nitroso compounds, *Adv. Cancer Res.*, 10:163-246.
2. Druckrey, H., Preussmann, R., Ivankovic, S., and Schmahl, D., 1967, Organotrope carcinogene Wirkung bei 65 verschiedenen *N*-Nitroso-Verbindungen an BD-Ratten, *S. Krebsforsch.*, 69:103-201.
3. Loeppky, R. N., and Michejda, C. J. (eds), 1994, *The Chemistry and Biochemistry of Nitrosamines and Related N-Nitroso Compounds*. ACS Symposium Series 553, Washington, D.C.
4. O'Neill, I. K., Chen, J. C., and Bartsch, H. (eds), 1991, *Relevance to Human Cancer of N-Nitroso Compounds, Tobacco Smoke and Mycotoxins*. IARC Publications, Lyon, France.
5. Bartsch, H., O'Neill, I. K., and Schulte-Hermann, R. (eds), 1987, *The Relevance of N-Nitroso Compounds to Human Cancer: Exposures and Mechanisms*. IARC Publications, Lyon, France.
6. National Research Council, N. R., 1981, *The Health Effects of Nitrate, Nitrite, and N-Nitroso Compounds*. National Academy Press, Washington, D.C.
7. Hecht, S. S., 1994, Metabolic activation and detoxification of tobacco-specific nitrosamines- a model for cancer prevention strategies, *Drug Metab. Rev.*, 26:373-390.
8. Bartsch, H., 1991, *N*-Nitroso compounds and human cancer: Where do we stand? In O'Neill, I. K., Chen, J., and Bartsch, H. (eds.) *Relevance to Human Cancer of N-Nitroso Compounds, Tobacco and Mycotoxins*. IARC Publications, Lyon, France, 105, pp. 1-10.
9. Yang, C. S., Smith, T., Ishizaki, H., and Hong, J.-Y., 1991, Enzyme mechanisms in the metabolism of nitrosamines. In O'Neill, I. K., Chen, J. C., and Bartsch, H. (eds.) *Relevance to Human Cancer of N-Nitroso Compounds, Tobacco Smoke, and Mycotoxins*. IARC Publications, Lyon, France, pp. 265-274.
10. Wade, D., Yang, C. S., Metral, C. J., Roman, J. M., Hrabie, J. A., Riggs, C. W., Anjo, T., Keefer, L. K., and Mico, B. A., 1987, Deuterium isotope effect on denitrosation and demethylation of *N*-nitrosodimethyl-amine by rat liver microsomes, *Cancer Res.*, 47:3373-3377.
11. Appel, K. E., Ruhl, C. S., and Hildebrandt, A. G., 1985, Oxidative dealkylation and reductive denitrosation of nitrosomethylaniline *in vitro*, *Chem. Biol. Interactions*, 53:69-76.
12. Ji, C., Mirvish, S. S., Nickols, J., Ishizaki, H., Lee, M. J., and Yang, C. S., 1989, Formation of hydroxy derivatives, aldehydes, and nitrite from *N*-nitrosomethyl-*n*-amylamine by rat liver microsomes and by purified cytochrome P-450 IIB1, *Cancer Res.*, 49:5299-5304.
13. Yang, C. S., Tu, Y. Y., Koop, D. R., and Coon, M. J., 1985, Metabolism of nitrosamines by purified rabbit liver cytochrome P-450 isozymes, *Cancer Res.*, 45:1140-1145.
14. Patten, C. J., Ning, S. M., Lu, A. Y. H., and Yang, C. S., 1986, Acetone-inducible cytochrome P-450: purification, catalytic activity and interaction with cytochrome b_5, *Arch. Biochem. Biophys.*, 251:629-638.
15. Patten, C., Ishizaki, H., Aoyama, T., Lee, M. J., Ning, S. M., Huang, W., Gonzalez, F. J., and Yang, C. S., 1992, Catalytic properties of the human cytochrome P450 2E1 produced by cDNA expression in mammalian cells, *Arch. Biochem. Biophys.*, 299:163-171.
16. Yoo, J.-S. H., and Yang, C. S., 1985, Enzyme specificity in the metabolic activation of *N*-nitrosodimethylamine to a mutagen for Chinese hamster V79 cells, *Cancer Res.*, 45:5569-5574.
17. Yoo, J.-S. H., Ishizaki, H., and Yang, C. S., 1990, Roles of cytochrome P450IIE1 in the dealkylation and denitrosation of *N*-nitrosodimethylamine and *N*-nitrosodiethylamine in rat liver microsomes, *Carcinogenesis (Lond.)*, 11:2239-2243.
18. Yamazaki, H., Inui, Y., Yun, C.-H., Guengerich, F. P., and Shimada, T., 1992, Cytochrome P450 2E1 and 2A6 enzymes as major catalysts for metabolic activation of *N*-nitrosodialkylamines and tobacco-related nitrosamines in human liver microsomes, *Carcinogenesis*, 13:1789-1794.
19. Lee, M., Ishizaki, H., Brady, J. F., and Yang, C. S., 1989, Substrate specificity and alkyl group selectivity in the metabolism of *N*-nitrosodialkylamines, *Cancer Res.*, 49:1470-1474.
20. Wang, M.-H., Wade, D., Chen, L., White, S., and Yang, C. S., 1995, Probing the active sites of rat and human cytochrome P450 2E1 with alcohols and carboxylic acids, *Arch. Biochem. Biophys.*, 317:299-304.
21. Mehta, R., Labuc, G. E., and Archer, M. C., 1984, Tissue and species specificity of the microsomal metabolism of nitrosomethylbenzylamine. In Bartsch, H., and O'Neill, I. K. (eds.) *N-Nitroso compounds: Occurrence and Biological Effects*. IARC Publications, Lyon, France, pp. 473-478.

22. Ludeke, B., Meier, T., and Kleihues, P., 1991, Metabolism of *N*-nitrosomethylbenzylamine (NMBzA) in rat esophageal and liver microsomes. In O'Neill, I. K., Chen, J. C., and Bartsch, H. (eds.) *Relevance to Human Cancer of N-Nitroso Compounds, Tobacco Smoke and Mycotoxins*. IARC Publications, Lyon, France, pp. 286-293.
23. Lu, S. H., Chui, S. X., Yang, W. X., Hu, X. N., Guo, L. P., and Li, F. M., 1991, Relevance of *N*-nitrosamines to esophageal cancer in China. In O'Neill, I. K., Chen, J., and Bartsch, H. (eds.) *Relevance to Human Cancer of N-Nitroso Compounds, Tobacco Smoke and Mycotoxins*. IARC Scientific, Lyon, France, 105, pp. 11-17.
24. Smith, T. J., Guo, Z.-Y., Thomas, P. E., Chung, F.-L., Morse, M. A., Eklind, K., and Yang, C. S., 1990, Metabolism of 4-(methylnitrosamino)-1-(3-pyridyl)-1-butanone in mouse lung microsomes and its inhibition by isothiocyanates, *Cancer Res.*, 50:6817-6822.
25. Guo, Z., Smith, T. J., Thomas, P. E., and Yang, C. S., 1991, Metabolic activation of 4-(methylnitrosamino)-1-(3-pyridyl)-1-butanone as measured by DNA alkylation *in vitro* and its inhibition by isothiocyanates, *Cancer Res.*, 51:4798-4803.
26. Guo, Z., Smith, T. J., Ishizaki, H., and Yang, C. S., 1991, Metabolism of 4-(methylnitrosamino)-1-(3-pyridyl)-1-butanone (NNK) by cytochrome P450IIB1 in a reconstituted system, *Carcinogenesis (Lond.)*, 12:2277-2282.
27. Smith, T. J., Guo, Z., Gonzalez, F. J., Guengerich, F. P., Stoner, G. D., and Yang, C. S., 1992, Metabolism of 4-(methylnitrosamino)-1-(3-pyridyl)-1-butanone (NNK) in human lung and liver microsomes and cytochrome P-450 expressed in hepatoma cells, *Cancer Res.*, 52:1757-1763.
28. Smith, T. J., Guo, Z.-Y., Hong, J.-Y., Ning, S. M., Thomas, P. E., and Yang, C. S., 1992, Kinetics and enzyme involvement in the metabolism of 4-(methylnitrosmino)-1-(3-pyridyl)-1-butanone (NNK) by rat lung and nasal mucosa microsomes, *Carcinogenesis (Lond.)*, 13:1409-1414.
29. Guo, Z., Smith, T. J., Thomas, P. E., and Yang, C. S., 1992, Metabolism of 4-(methylnitrosamino)-1-(3-pyridyl)-1butanone (NNK) by inducible and constitutive cytochrome P450 enzymes in rats, *Arch. Biochem. Biophys.*, 298:279-286.
30. Hong, J.-Y., Smith, T., Lee, M.-J., Li, W., Ma, B.-L., Ning, S.-M., Brady, J. F., Thomas, P. E., and Yang, C. S., 1991, Metabolism of carcinogenic nitrosamines by rat nasal mucosa and the effect of diallyl sulfide, *Cancer Res.*, 51:1509-1514.
31. Hong, J.-Y., Ding, X., Smith, T. J., Coon, M. J., and Yang, C. S., 1992, Metabolism of 4-(methylnitrosamino)-1-(3-pyridyl)-1-butanone (NNK), a tobacco-specific carcinogen, by rabbit nasal microsmes and cytochrome P450s NMa and NMb, *Carcinogenesis (Lond.)*, 13:2141-2144.
32. Penman, B. W., Reece, J., Smith, T., Yang, C. S., Gelboin, H. V., Gonzalez, F. J., and Crespi, C. L., 1992, Characterization of a human cell line expressing high levels of cDNA-derived CYP2D6, *Pharmacogenetics*, 3:28-39.
33. Murphy, S. E., Heiblum, R., and Trushin, N., 1990, Comparative metabolism of *N'*-nitrosonornicotine and 4-(methylnitrosamino)-1-(3-pyridyl)-1-butanone by cultured F344 rat oral tissue and esophagus, *Cancer Res.*, 50:4685-4691.
34. Murphy, S. E., and Heiblum, R., 1990, Effect of nicotine and tobacco-specific nitrosamines on the metabolism of *N'*-nitrosonornicotine and 4-(methylnitrosamino)-1-(3-pyridyl)-1-butanone by rat oral tissue, *Carcinogenesis*, 11:1663-1666.
35. McCoy, G. D., Chen, C.-H. B., and Hecht, S. S., 1981, Influence of mixed-function oxidase inducers on the *in vitro* metabolism of *N'*-nitrosonornicotine by rat and hamster liver microsomes, *Drug Metab. Disp.*, 9:168-169.
36. Chen, C.-H. B., Fung, P. T., and Hecht, S. S., 1979, Assay for microsomal α-hydroxylation of *N'*-nitrosonornicotine and determination of the deuterium isotope effect for α-hydroxylation, *Cancer Res.*, 39:5057-5062.
37. Smith, T. J., Guo, Z., Li, C., Ning, S. M., Thomas, P. E., and Yang, C. S., 1993, Mechanisms of inhibition of 4-(methylnitrosamino)-1-(3-pyridyl)-1-butanone (NNK) bioactivation in mouse by dietary phenethyl isothiocyanate, *Cancer Res.*, 53:3276-3282.
38. Hodgson, R. M., Schweinsberg, F., Wiessler, M., and Kleihues, P., 1982, Mechanism of esophageal tumor induction in rats by *N*-nitrosomethylbenzylamine and its ring-methylated analog *N*-nitrosomethyl(4-methylbenzyl)amine, *Cancer Res.*, 42:2836-2840.
39. Crespi, C. L., Penman, B. W., Gelboin, H. V., and Gonzalez, F. S., 1991, A tobacco smoke-derived nitrosamine, 4-(methylnitrosamino)-1-(3-pyridyl)-1-butanone, is activated by multiple human cytochrome P450s including the polymorphic human cytochrome P450 2D6, *Carcinogenesis*, 12:1197-1201.
40. Peto, R., Gray, R., Brantom, P., and Grasso, P., 1991, Effects on 4080 rat chronic ingestion of *N*-nitrosodiethylamine or *N*-nitrosodimethylamine: A detailed dose-response study, *Cancer Res.*, 51:6415-6451.

ROLE OF MOLECULAR BIOLOGY IN RISK ASSESSMENT

Alvaro Puga, Jana Micka, Ching-yi Chang, Hung-chi Liang, and Daniel W. Nebert

Center for Environmental Genetics and
Department of Environmental Health
University of Cincinnati Medical Center
P.O. Box 670056, Cincinnati, Ohio 45267-0056

INTRODUCTION

Exposure to an ever-increasing number of man-made and natural environmental substances poses a health risk for the exposed individuals. To formulate public policy in order to protect the human population from the adverse effects of these agents, society needs first to gain an understanding of the mechanisms by which toxic agents compromise human health. In environmental health studies, the evaluation of risk results from a complex interplay of factors, including not only scientific components, but also socio-economic, ethical, legal, and geographical. As one of these scientific aspects, molecular biology has become an essential tool for the environmental toxicologist, because the rapidly-expanding advances in our understanding of biological processes at the molecular level have made it possible today to analyze problems that twenty years ago we could not even imagine existed. For example, the technology is now available to answer one of the most challenging questions that toxicologists face, namely: *Are there genes that contribute to increased resistance (or sensitivity) to toxic environmental agents?* Of course, the ultimate goal in this area of risk evaluation is not only to identify these genes, but to develop an understanding of how they function and how they affect human health; this is an eminently feasible goal with our current level of knowledge, given time and adequate resources. As more molecular biologists become attracted to the present challenges of toxicological research, we cannot but expect that many novel advances in molecular biology will be the result of our specific experimental demands, with the consequent opening of unpredictable new frontiers in environmental health research.

The purpose of this article is to delineate some of the problems found in the evaluation of risk, and present the approaches that can be followed to analyze these problems from the molecular biological point of view. Rather than dwell on what has been done in the past, we have chosen to present a more futuristic view, projecting and extrapolating from other fields of research. This is, of course, not just an intellectual exercise, and we will address the status of experiments in our laboratories that have been designed within this framework.

Biological Reactive Intermediates V
Edited by R. Snyder et al., Plenum Press, New York, 1996

THE RESEARCH PLAN

The goals of the Center for Environmental Genetics at the University of Cincinnati are to investigate in various species, but ultimately in humans, the impact of genetic diversity on the response of individuals or populations to toxic environmental agents. The scope of this research plan must of necessity include a multidisciplinary approach, since we are trying to establish links "from the molecule to the human". The fundamental theme of this research is to understand the dissimilarities in response to toxic environmental agents that are due to differences in genetic predisposition. The underlying hypothesis is that genetic predisposition can dramatically influence this response. For this purpose, we take advantage of known, established genetically different subpopulations or of phenotypic differences between *resistant* and *sensitive* groups or individuals. Once these differences are recognized, we are interested in identifying the genes responsible for these differences and their mechanism of action. It is our contention that identification and characterization of these genetic differences in humans are essential for adequate evaluation of the risk of adverse health effects of exposure to environmental substances. This knowledge will provide important keys to understanding fundamental mechanisms of toxicity, which in turn will suggest a rationale for preventive or ameliorative measures and ultimately for molecular intervention and possible clinical use.

THE NATURE OF THE PROBLEM

Individual variation in genetic predisposition often reflects allelic differences in genes encoding proteins involved in critical life functions, such as receptors, drug metabolism, ion channels, multidrug resistance glycoprotein pumps, second messenger pathways, DNA repair, and chelation of metals. This is true of the few dozen cases of human pharmacogenetic differences known to date [1-10]. Therefore, when faced with the task of evaluating, for example, the genetic predisposition of individuals in the Los Angeles population to developing asthma, we could start from an educated guess that the risk might be related to allelic variation in a lung surfactant protein. This approach, although valid, is very risky, because of the inherent limitations of any *ad hoc* hypothesis: if it is wrong, we have to start all over again with a fresh new assumption that has as much chance of being correct as the first one.

An alternative way to approach the problem is to start from the assumption that we know nothing about the biology of the problem, and ask: *Are there traits that can be used to compare and screen normal vs. affected individuals?* This question is about *phenotypes*, and its answer is, of course, yes. We all know that the *dose* of a particular substance is the most important determinant of its toxic effect, and we also know that an increase in the dose of any environmental chemical will generally cause an increase of the toxic response in the individuals exposed to the chemical.

The graph in Fig. 1 illustrates this point, and in our particular case it may represent the severity of asthma as a function of ppm of ozone in the Los Angeles atmosphere. In this graph, each point may represent the average of several observations made on the same individual. The first conclusion that becomes apparent is that some individuals show an exaggerated response to a relatively low dose, whereas others show a slight response to a very high dose. These *outliers* are the key elements to our alternative approach. Clearly, sensitive and resistant outliers can only be detected if we have reliable information about exposure. Whether our measurement is ozone in LA, cigarette packs smoked, or hazardous particles in the work place, if an individual has 10 times the exposure that we thought he

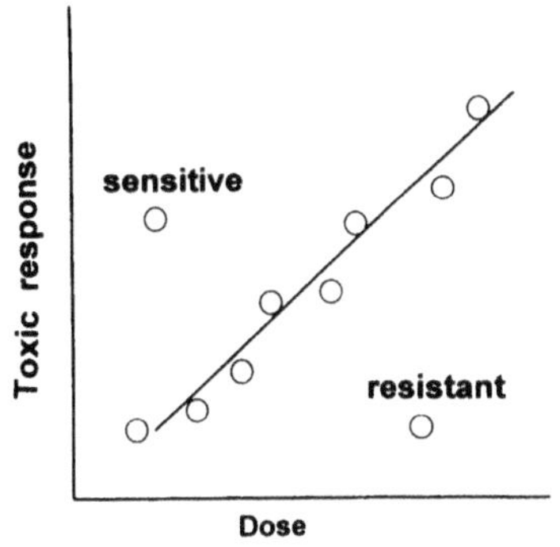

Figure 1. Dose response curve. In general, the toxic response is a linear function of dose; however, some individuals show an elevated response to a low dose and others hardly respond to a very high dose.

did, and thus we diagnose him as the susceptible phenotype when he is not, we will be in big trouble for any future analysis that hinges on erroneous information!

THE REVERSE GENETICS APPROACH

The model that we have applied in the previous section is a non-parametric model. All we did was to analyze one large group of individuals and to break up this group into the three population subgroups of *susceptible*, *normal*, and *resistant*. We determined some characteristic that was different between these groups (for instance, the number or duration of asthma attacks for a given exposure) and we had no preconceived ideas as to the pattern of inheritance, if any, of the trait. Once we have identified the outliers, however, we can apply parametric models to analyze possible inheritance patterns. It cannot be emphasized sufficiently that outliers for resistance/sensitivity are a crucial population resource for finding the genes that control the response to the toxic agent. The outlier becomes the proband for family studies in which we determine vertical inheritance patterns of the trait using the predictions given by Mendelian genetics. The phenotypic question that we posed before becomes now a question about genotypes: *are there genetic/molecular markers that can be used to determine if a certain phenotypic polymorphism is common to all the exposed individuals that are susceptible/resistant to a given environmental agent?* The answer to this question is also, yes, although matters get much more complicated at this point.

The reasons for the complication are illustrated in Table 1. The identification of genes has a major problem of scale. An average gene may be not more than 0.0001% of the human genome, and all that is necessary to cause an observable phenotypic difference is a single base-pair change. How can we get to this critical base pair? The fastest way to narrow down our search is to make use of molecular biological techniques. We can employ for our search a genetic proxy, and measure allelic differences in the number of *short tandem repeats* (STRs) throughout the genome. These STRs consist of head-to-tail repetitions of short sequences, usually 3 or 4 base-pairs long, interspersed throughout the genome without any apparent pattern. Their number may vary from one individual to another, but, unless recombination

Table 1. Gene identification has a major scale problem

Haploid human genome	3 X 10^9 base pairs
Average human chromosome	120 X 10^6 base pairs
Average gene	1 - 200 X 10^3 base pairs
Mutation capable of causing disease	1 base pair

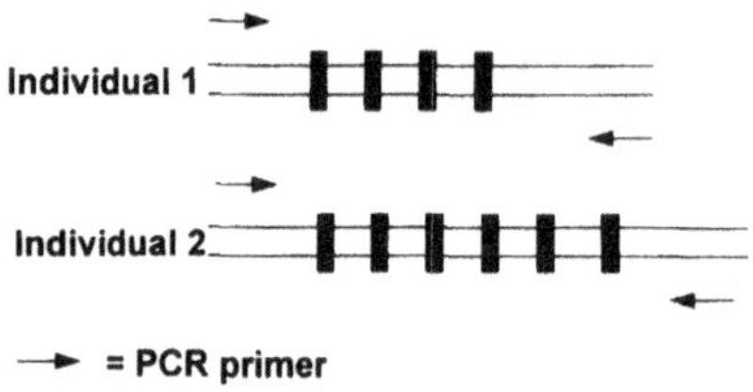

Figure 2. Identification of individuals by the size of their STRs. In the cases shown in the diagram, *individual 1* has four copies of the repeated sequence and *individual 2* has six copies. PCR amplification using primers outside the STR boundaries will results in products of different sizes, that can be used to identify each individual.

events take place within the particular STR region, this number is fixed and it is kept constant within a family throughout generations [11-13]. The usefulness of STRs in our context is that the *size* of each particular STR region in a chromosome can be used as a molecular marker for that particular region of the chromosome. Amplification of the STR region by polymerase chain reaction (PCR) can therefore be used to identify individuals; all we need is a suitable pair of PCR primers. This is represented diagrammatically in Fig. 2. It is clear that the size of the amplified DNA product of individual 1 will be larger than that of individual 2. The sizes can easily be resolved by electrophoresis in polyacrylamide gels.

We can now analyze the family of a particular outlier, for instance, an asthmatic Angeleno affected by high sensitivity to air pollution, for two characters, instead of one. On the one hand, we can determine the distribution of this *affected* phenotype in the family, and on the other, we can measure the size of many STR regions, dispersed through the genome. The question that we ask is, *Is there genetic linkage between the polymorphic marker and the affected phenotype?* A hypothetical diagram of the results that we may find, shown in Fig. 3 appears to indicate that there is, since every individual that has the affected phenotype also has a larger STR region. Of course, things are usually a lot more complicated than this diagram leads us to believe, and rather than absolutely proving that linkage exists, all we can expect to determine is the *likelihood of true linkage*, or linkage odds (or *LOD score*, for short). Our first goal is to establish a linkage map between the affected phenotype and a number of STR markers in a particular chromosome, using a series of genetic linkage analyses algorithms designed to determine LOD scores [14-23]. Once we have the gene localized to a discrete region of a chromosome, our second goal is to clone the gene, using expression libraries, yeast artificial chromosomes, or a number of other techniques beyond the scope of this discussion.

Even if we can identify useful populations of affected individuals with reasonably large families that would allow pedigree analyses of genetic markers, we will need to generate a sufficiently large number of polymorphic STR alleles that would allow us to pinpoint the genes that control complex traits such as environmental toxicity. Since any blind search for linkage would have to cover the totality of the human genome, we would need

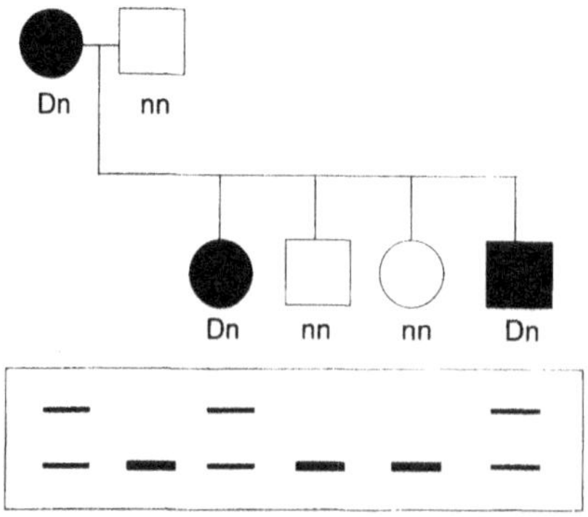

Figure 3. Genetic linkage between polymorphic marker and affected phenotype. In the pedigree shown, all *Dn* individuals have the "affected" (high sensitivity to air pollution) phenotype, as well as an extra STR DNA fragment, as shown by the diagrammatic electrophoresis pattern. We could surmise that the two characters are linked.

approximately 1,500 independent STR markers to cover the genome at the 2-centiMorgan (cM) level. A goal of the Human Genome Project is to saturate the human genome with sets of STRs every 2 cM. Currently, there are some 900 polymorphic probes available, and their number grows daily. Fortunately, a much smaller number is all that might be needed in most cases: less than 300 polymorphic markers, for example, were needed to localize an insulin-dependent diabetes locus to mouse chromosome 4 [24].

We are currently using this approach to study the linkage between the aromatic hydrocarbon receptor (*AhR*) gene, a transcriptional activator that responds to planar halogenated and nonhalogenated aromatic hydrocarbons, and the phenotype of high CYP1A1 enzymatic inducibility, a possible genetic factor in risk of cigarette smoke-induced bronchogenic carcinoma [25-29]. Our results (Micka *et al*, manuscript in preparation) indicate a good probability of linkage with the 7p15-21 region of human chromosome 7, where the Ah receptor gene is localized ([30], and Micka *et al*, manuscript in preparation).

THE DIFFERENTIAL DISPLAY APPROACH

One of the long-term objectives of toxicological research is to elucidate the molecular mechanisms underlying the biological responses to a particular toxic substance. It is only through this elucidation that we would be able to define the biological consequences of exposure and therefore, to properly evaluate risk. We are very interested in understanding the biological effects of dioxin (2,3,7,8-tetrachlorodibenzo-*p*-dioxin; TCDD). TCDD is the prototype congener of all the dioxins, a group of toxic environmental pollutants known to be potent teratogens and tumor promoters. The various biological effects of TCDD appear seemingly unrelated, but they all have a common molecular marker, namely, drastic alterations in the expression of certain genes. Thus, depending on the particular tissue, TCDD may cause hyperexpression of keratin genes (chloracne), inhibition of chondrogenic mesenchyme differentiation (cleft palate), unscheduled cell proliferation (tumor promotion), or programmed death of immature thymocytes (immunosuppression) [25,31-33]. To explain the biological effect of dioxin, we could hypothesize that TCDD modifies gene expression patterns by interfering with control mechanisms that regulate the steady-state levels of different genes in different tissues, including perhaps specific transcription factors. We could generalize this hypothesis, because it is likely to apply to many toxic agents, and say that exposure to a toxic compound might result in changes in the expression of sets of genes and that these changes are responsible for the overall biological effect. Expressed in these terms, we leave room for the likely possibility that, upon exposure, expression of some genes will increase, whereas expression of others will decrease. Our task is, then, to identify the target genes that comprise this primary response—bearing in mind that, since changes in gene expression occur considerably earlier than biological manifestations, any gene that we can identify is potentially an invaluable biomarker of exposure to that particular toxic compound.

The technology is available to identify and isolate genes that are differentially expressed in cells under different types of conditions. We can use a technique known as DDRT-PCR (for **d**ifferential **d**isplay **r**everse **t**ranscriptase PCR) to amplify each and every individual mRNA in a cell [34-37]. The key element here is to use a very special set of oligonucleotide primers. One is a downstream primer anchored at the polyadenylation tail of the mRNA; the other, short and of arbitrary sequence, should be capable of annealing at various possible upstream positions relative to the anchored primer. A combination of four anchored primers and 26 well-defined, arbitrary sequence upstream primers provides a total of 104 primer combinations to: first, make a cDNA copy of the mRNA; and second, amplify the cDNA into PCR products than can be resolved in a polyacrylamide gel. In this manner, and with the use of computer imaging software, some 30,000 mRNA species can be resolved

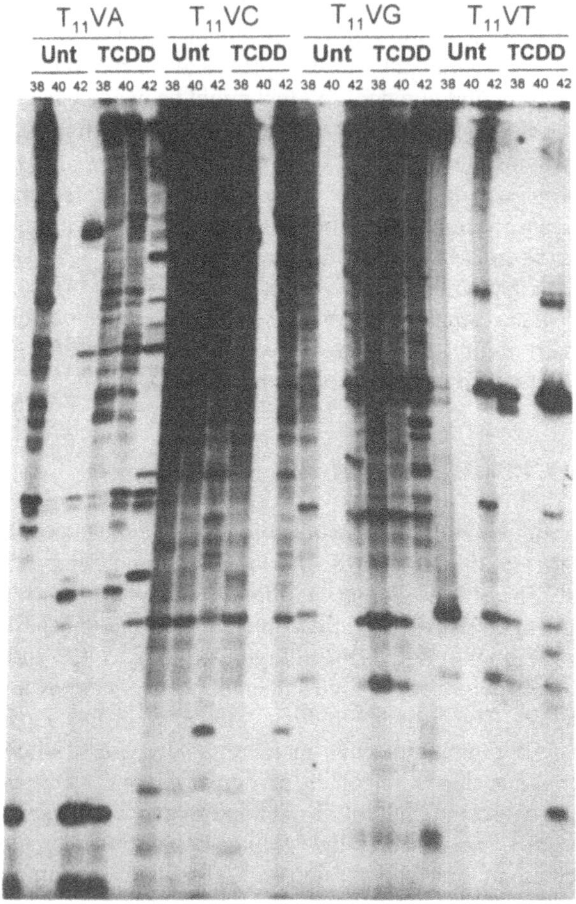

Figure 4. DDRT-PCR electrophoretic pattern of mRNAs from TCDD treated and untreated hepatoma cells. Lanes labelled *U* are from untreated cells and lanes labeled *T* are from cells treated with 10 nM TCDD. In each case, three different temperatures, 38°, 40°, and 42°, were used for the amplification reactions. Four different downstream primers were used, denoted by $T_{11}VA$, $T_{11}VC$, $T_{11}V$, and $T_{11}VT$. In all cases, V=A, C, or G. Several DNA fragments in different lanes show differential amplification; TCDD appears to activate expression of some mRNA species and to silence expression of others.

from a single cell line [34], a number which is close to the estimated total number of genes expressed in tissue culture cells. Fig. 4 shows results of a DDRT-PCR experiment, comparing mouse hepatoma cells treated with 10 nM TCDD with cells that were not. All four anchored primers were used for the reverse transcriptase reaction, and the cDNA was amplified using the corresponding anchored primer and a single arbitrary upstream primer. Several molecular species, possibly corresponding to different mRNA molecules, appear to be up- or down-regulated in the presence of TCDD.

THE TARGETED GENE DISRUPTION APPROACH

Once we have identified genes of interest by either of the two methods outlined above, an efficient assessment of biological risk requires that we find out how these genes function and what is it that they do in the organism. For this purpose, many invaluable studies can be done in tissue-culture cells, but tissue culture studies lack the depth that only the analysis of the complete organism can provide. Multi-enzymatic reactions, resulting from interactions between cells from different tissues, are only one of many pitfalls that befall extrapolation of results in tissue culture to the complete organism. Here, again, novel discoveries in molecular biology provide a unique tool. The gene of interest can be targeted and deleted

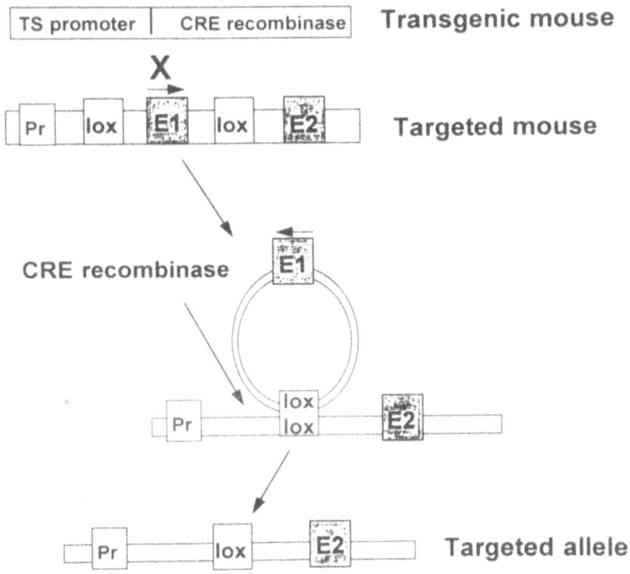

Figure 5. Conditional (tissue-specific) knock-out. A cross between the *transgenic mouse*, carrying the CRE recombinase, and the *targeted mouse*, carrying the *lox* sequences in the gene to be deleted, produces a hybrid mouse in which both transgenes are present. Tissue specific expression of *CRE*, regulated by the *tissue-specific (ts) promoter*, causes a crossover between the two *lox* sequences, followed by recombination and excision of the loop, containing the exon sequence *E1*. Loss of this exon abolishes the function of the targeted gene, but only in the cells where CRE was expressed.

from murine embryonic stem cells and a complete mouse (a "knockout" mouse), carrying a homozygous deletion of the gene, can be developed. To describe the various aspects of gene targeting methodology is well beyond the scope of this article, and can be found in many recent papers [38,39]. We will address in some detail, however, an approach for *conditional, tissue-specific gene targeting*, that may be of great value in toxicological research. This method of knocking out a gene only in certain tissues is shown schematically in Fig. 5.

The actual knockout mouse is a hybrid of two transgenic animals; one is a gain-of-function transgenic that carries a bacteriophage P1 CRE recombinase gene under the control of a tissue-specific promoter. If, for instance, we wanted to knock out a gene specifically in the beta cells of the pancreatic islets of Langerhans, we would use an insulin promoter. The CRE recombinase enzyme promotes homologous recombination between two discrete DNA sequences, termed the *lox* sequences [40,41]. The second partner of the hybrid is a loss-of-function gene-targeted mouse, in which we have inserted within the gene of interest, two *lox* element sequences. In order to preserve expression of this gene in all other tissues, the *lox* elements are inserted within two different introns. Expression of CRE in the targeted tissue of the hybrid mouse would cause the looping-out and excision of the sequences contained between the *lox* elements, effectively deleting one or more exons of the gene in that tissue in which the CRE recombinase is expressed (in our example, the pancreatic beta cells) [42-46]. This conditional gene-targeting approach can be used not only to analyze gene functions expressed in selected tissues, but more importantly, it will be of great value to analyze genes that may have a critical role in development, or that their deletion is lethal. Conditional knock-out in one tissue, or, if an inducible tissue-specific promoter is used, knock-out past a critical age or developmental stage, would allow us to study the function of this gene.

Our laboratories are in the process of developing targeted disruption of several genes involved in the metabolism of foreign chemicals. Among these, we have chosen the cytochrome P450 *Cyp1a2* gene involved in the metabolism of numerous environmentally relevant toxic chemicals. Of particular interest to us are the tobacco smoke-specific 4-(methylnitrosamino)-1-(3-pyridyl)-1-butanone (NNK) and the hepatocarcinogenic mycotoxin aflatoxin B_1 (AFB_1). The long range goal of this project is to understand the role of the CYP1A2 enzyme in environmentally-induced toxicity and carcinogenesis, and eventu-

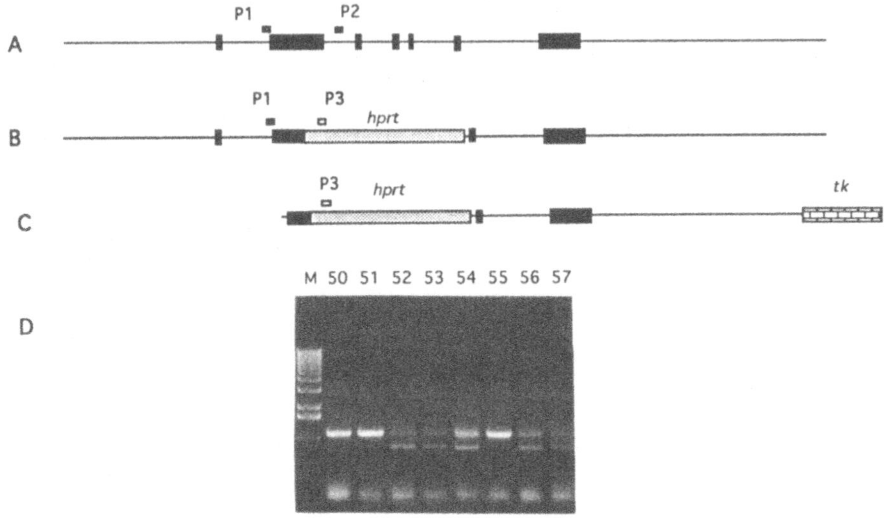

Figure 6. Schematic diagram of the targeted murine *Cyp1a2*. The endogenous *Cyp1a2* locus is shown in **A** and the targeted locus is shown in **B.** Introns are represented by *thin lines* and exons by *filled boxes*. *P1* and *P2* are PCR primers that amplify exclusively the wild type allele: *P1* in combination with *P3* will amplify exclusively the mutant allele; thus, PCR amplification is a very convenient method to genotype the offspring. The replacement vector used for the targeting is shown in **C**. Panel **D** presents the electrophoretic separation of PCR products from a litter of 8 pups derived from the mating of a chimeric animal with a Swiss Black female. The PCR product of the wild type allele is 1.2 kbase pairs; the corresponding product of the mutant allele is 0.9 kbase pairs. The lane labelled *M* contains a size marker. It is clear that some of the pups have one wild type and one mutated allele.

ally to replace the murine *Cyp1a2* gene with its human orthologue. Like other cytochromes P450, the CYP1A2 enzyme not only participates in Phase I detoxification reactions, but in other instances it may employ the same oxygenation reaction to activate procarcinogens and other molecules to carcinogenic, mutagenic, and toxic metabolites. For instance, CYP1A2 appears to provide a detoxification pathway for AFB_1 by catalyzing its 4-hydroxylation; but on the other hand, it appears to activate, via N-hydroxylation, NNK-induced genotoxicity in lung and acetaminophen-induced toxicity in liver. Fig. 6 shows diagrammatically the region of the murine *Cyp1a2* gene used in the targeting, as well as the genotyping by PCR amplification of several founder heterozygous mice (Liang *et al.*, manuscript in preparation).

CONCLUDING REMARKS

Many new developments in molecular biology will revolutionize toxicological research during the next few years. The uses of reverse genetics to find genes of relevance in Toxicology have not even begun to be exploited. All we need in order to uncover these genes are large families having reliable exposure data, with affected phenotypes, and a predictable (Mendelian) mode of inheritance. The rest, polymorphic DNA markers, DNA libraries, chromosomal location, etc. is already accumulating faster than we can imagine. It helps that we do not need to have any knowledge as to the protein product or the function of the gene that we are trying to find. Identification of the critical gene(s) comes later, when we do differential display, tissue culture experiments, or make transgenic animals with our favorite gene. At this point, we can make point mutations to study the effect of functional alterations of the gene, or we can construct tissue-specific knock-outs, as described earlier. We could knock out Phase I and Phase II genes to study toxico-kinetics, mutagenesis, and gene interactions, or we could make transgenes with a reporter gene driven by the control elements of environmentally relevant genes, and use these mice to screen environmentally relevant agents. We might as well start replacing mouse genes for human genes, and effectively "humanize" the mouse to build toxicologic models better adapted to the human condition. The sky is the limit.

ACKNOWLEDGMENTS

Most of the ideas put forth in this article are the result of many informal conversations with our colleagues in the Center for Environmental Genetics. We wish to thank all of them, and most in particular, Drs. Tom Doetschman, Anil Menon, and Steve Potter. This work was supported by NIH Grants ES06273, ES06096, ES06321, and ES06811.

REFERENCES

1. Levy, G. N., Martell, K. J., DeLeon, J. H., and Weber, W. W., 1992, Metabolic, molecular genetic and toxicological aspects of the acetylation polymorphism in inbred mice. *Pharmacogenetics* 2: 197-206.
2. Juberg, D. R., Bond, J.T., and Weber, W. W., 1991, N-acetylation of aromatic amines: Genetic polymorphism in inbred rat strains. *Pharmacogenetics* 1: 50-57.
3. Nebert, D. W., 1991, Polymorphism of human CYP2D genes involved in drug metabolism: possible relationship to individual cancer risk. *Cancer Cells* 3: 93-96.
4. Nebert, D. W., 1991, Role of genetics and drug metabolism in human cancer risk. *Mut. Res.* 247: 267-281.
5. Meyer, U. A., 1994, The molecular basis of genetic polymorphisms of drug metabolism. *J Pharm. Pharmacol.* 46: 409-415.
6. Coutts, R. T., 1994, Polymorphism in the metabolism of drugs, including antidepressant drugs: Comments on phenotyping. *J. Psych. Neuros.* 19: 30-44.
7. Winchester, R., 1994, The molecular basis of susceptibility to rheumatoid arthritis. *Adv. Immunol.* 56: 389-466.
8. Furlong, C. E., Costa, L. G., Hassett, C., Richter, R. J., Sundstrom, J. A., Adler, D. A., Disteche, C. M., Omiecinski, C. J., Chapline, C., and Crabb, J. W., 1993, Human and rabbit paraoxonases: purification, cloning, sequencing, mapping and role of polymorphism in organophosphate detoxification. *Chem. Biol. Interact.* 87: 35-48.
9. Daly, A. K., Cholerton, S., Gregory, W., and Idle, J. R., 1993, Metabolic polymorphisms. *Pharmacol. Ther.* 57: 129-160.
10. Idle, J. R., 1991, Is environmental carcinogenesis modulated by host polymorphism?. *Mut. Res.* 247: 259-266.
11. Pizzuti, A., Friedman, D. L., and Caskey, C. T., 1993, The myotonic dystrophy gene. *Arch. Neurol.* 50: 1173-1179.
12. Tuck-Muller, C. M., Martinez, J. E., Batista, D. A., Kearns, W. G., and Wertelecki, W., 1993, Duplication of the short arm of the X chromosome in mother and daughter. *Hum. Genet.* 91: 395-400.
13. Ross, C. A., McInnis, M. G., Margolis, R. L., and Li, S. H., 1993, Genes with triplet repeats: Candidate mediators of neuropsychiatric disorders. *Trends Neurosc.* 16: 254-260.
14. Curtis, D., 1994 Another procedure for the preliminary ordering of loci based on two point lod scores. *Ann. Hum. Genetics* 58: 65-75.
15. Hildebrandt, F., Pohlmann, A., and Omran, H., 1993, LODVIEW: a computer program for the graphical evaluation of lod score results in exclusion mapping of human disease genes. *Comp. Biomed. Res.* 26: 592-599.
16. Lewis, C. M., and Cannings, C., 1992, The number of loci needed for ELOD calculations. *Ann. Hum. Genetics* 56: 59-69.
17. Collins, A. and Morton, N. E., 1991, Significance of maximal lods. *Ann. Hum. Genetics* 55: 39-41.
18. Risch, N., 1992, Genetic linkage: interpreting lod scores. *Science* 255: 803-804.
19. Schork, N. J., Boehnke, M., Terwilliger, J. D., and Ott, J. 1993, Two-trait-locus linkage analysis: A powerful strategy for mapping complex genetic traits. *Am. J. Hum. Genet.* 53: 1127-1136.

20. Terwilliger, J. D., Speer, M., and Ott, J., 1993, Chromosome-based method for rapid computer simulation in human genetic linkage analysis. *Genet. Epidemiol.* 10: 217-224.
21. Terwilliger, J. D., and Ott, J. 1993, A novel polylocus method for linkage analysis using the lod-score or affected sib-pair method. *Genet. Epidemiol.* 10: 477-482.
22. Ott, J., 1992, The future of multilocus linkage analysis. *Ann. Medicine* 24: 401-403.
23. Keats, B. J., Sherman, S. L., Morton, N. E., Robson, E. B., Buetow, K. H., Cartwright, P. E., Chakravarti, A., Francke, U., Green, P. P., and Ott, J., 1991, Guidelines for human linkage maps. An International System for Human Linkage Maps (ISLM, 1990). *Ann. Hum. Genetics* 55: 1-6.
24. Rodrigues, N. R., Cornall, R. J., Chandler, P., Simpson, E., Wicker, L. S., Peterson, L. B., and Todd, J. A., 1994, Mapping of an insulin-dependent diabetes locus, Idd9, in NOD mice to chromosome 4. *Mammalian Genome* 5: 167-170.
25. Landers, J. P., and Bunce, N. J., 1991, The *Ah* receptor and the mechanism of dioxin toxicity. *Biochem. J.* 276: 273-287.
26. Nebert, D. W., Benedict, W. F., and Kouri, R. E., 1974, Aromatic hydrocarbon-produced tumorigenesis and the genetic differences in aryl hydrocarbon hydroxylase induction. In: *Chemical Carcinogenesis* (Ts'o, P. O. P., and DiPaolo, J. A., Eds.), pp.271-289. Marcel Dekker, New York.
27. Nebert, D. W., Petersen, D. D., and Puga, A., 1991, Human AH locus polymorphism and cancer: inducibility of *CYP1A1* and other genes by combustion products and dioxin. *Pharmacogenetics* 1: 68-78.
28. Swanson, H. I., and Bradfield, C. A., 1993, The AH-receptor: genetics, structure and function. *Pharmacogenetics* 3: 213-230.
29. Dolwick, K. M., Schmidt, J. V., Carver, L. A., Swanson, H. I., and Bradfield, C. A., 1993, Cloning and expression of a human Ah receptor cDNA. *Mol. Pharmacol.* 44: 911-917.
30. Le Beau, M. M., Carver, L. A., Espinosa, R., 3rd, Schmidt, J. V., and Bradfield, C. A., 1994, Chromosomal localization of the human AHR locus encoding the structural gene for the Ah receptor to 7p21—>p15. *Cytogenet. Cell. Genetics* 66: 172-176.
31. Nebert, D. W., 1989, The *Ah* locus: Genetic differences in toxicity, cancer, mutation, and birth defects. *Crit. Rev. Toxicol.* 20: 153-174.
32. Greenlee, W. F., and Neal, R. A., 1985, The *Ah* receptor: a biochemical and biological perspective. In: *The Receptors.* (Conn, P. M., Ed.), pp.89-129. Academic Press, Inc. New York.
33. Whitlock, J. P., Jr., 1991, Genetic and molecular aspects of 2,3,7,8-tetrachlorodibenzo-p-dioxin action. *Ann. Rev. Pharmacol. Toxicol.* 30: 251-277.
34. Bauer, D., Muller, H., Reich, J., Riedel, H., Ahrenkiel, V., Warthoe, P., and Strauss, M., 1993, Identification of differentially expressed mRNA species by an improved display technique (DDRT-PCR). *Nuc. Acids Res.* 21: 4272-4280.
35. Liang, P., Averboukh, L., and Pardee, A. B., 1993, Distribution and cloning of eukaryotic mRNAs by means of differential display: refinements and optimization. *Nuc. Acids Res.* 21: 3269-3275.
36. Liang, P., Averboukh, L., Keyomarsi, K., Sager, R., and Pardee, A. B., 1992, Differential display and cloning of messenger RNAs from human breast cancer versus mammary epithelial cells. *Cancer Res.* 52: 6966-6968.
37. Liang, P., and Pardee, A. B., 1992, Differential display of eukaryotic messenger RNA by means of the polymerase chain reaction. *Science* 257: 967-971.
38. Askew, G. R., Doetschman, T., and Lingrel, J. B., 1993, Site-directed point mutations in embryonic stem cells: a gene-targeting tag-and-exchange strategy. *Mol. Cell. Biol.* 13: 4115-4124.
39. Doetschman, T. C., 1991, Gene targeting in embryonic stem cells. *Biotechnology* 16: 89-101.
40. Adams, D.E., Bliska, J.B., and Cozzarelli, N.R. , 1992, Cre-lox recombination in Escherichia coli cells. Mechanistic differences from the in vitro reaction. *J.Mol.Biol.* 226:661-673.
41. Sternberg, N., 1990, Bacteriophage P1 cloning system for the isolation, amplification, and recovery of DNA fragments as large as 100 kilobase pairs. *Proc. Natl. Acad. Sci. USA* 87: 103-107.
42. Lakso, M., Sauer, B., Mosinger, B., Jr., Lee, E. J., Manning, R. W., Yu, S. H., Mulder, K. L., and Westphal, H., 1992, Targeted oncogene activation by site-specific recombination in transgenic mice. *Proc. Natl. Acad. Sci. USA* 89: 6232-6236.
43. Baubonis, W., and Sauer, B., 1993, Genomic targeting with purified Cre recombinase. *Nucl. Acids Res.* 21: 2025-2029.
44. Sauer, B., 1993, Manipulation of transgenes by site-specific recombination: Use of Cre recombinase. *Methods Enzymol.* 225: 890-900.
45. Orban, P. C., Chui, D., and Marth, J. D., 1992, Tissue- and site-specific DNA recombination in transgenic mice. *Proc. Natl. Acad. Sci. USA* 89: 6861-6865.
46. Sauer, B., and Henderson, N., 1990, Targeted insertion of exogenous DNA into the eukaryotic genome by the Cre recombinase. *New Biologist* 2: 441-449.

HUMAN GSH-TRANSFERASE IN RISK ASSESSMENT

H. M. Bolt

Institut für Arbeitsphysiologie an der Universität Dortmund
Ardeystr. 67, D-44139 Dortmund, Germany

INTRODUCTION

Human cytosolic glutathione-S-transferases (GST) are dimeric enzymes which have been classified, mainly by their isoelectric points, into four classes (Brockmöller et al. 1993):

class α (basic, GSTA);
class μ (near-neutral, new GSTM);
class π (acidic, GSTP);
class θ (slightly basic, GSTT).

In humans, genetic polymorphisms occur among GSTM and GSTT. An increasing number of toxicologically important substrates of these isoenzymes now is identified (Table 1). This has a considerable impact on risk assessment of these substrates.

Class μ Polymorphism (GSTM1)

The GST class μ isozyme is polymorphic in man, as first observed with the classical substrate trans-stilbene oxide (Seidergård et al., 1985). About half the subjects in populations studied so far express the μ isozyme (Bell et al., 1992). The polymorphism is due to delection of the entire GSTM1 gene, resulting in a null allele (Seidergård et al., 1988).

Due to this polymorphism, GST activities in human lymphocytes towards trans-stilbene oxide vary between practically zero and nearly 400 pmol/min/10^6 cells; activities are

Table 1. Substrates of polymorphic glutathione-S-transferases

Isozyme	Test substrates	Toxicological substrates
GSTM1	trans-stilbene oxide ethacrynic acid	styrene oxide, unspecified lung and urinary bladder carcinogens (PAH ?)
GSTT1	epoxy-p-nitro-phenoxy propane	dichloromethane, methyl bromide, ethylene oxide, diepoxybutane

Biological Reactive Intermediates V
Edited by R. Snyder et al., Plenum Press, New York, 1996

low or absent in homozygously deleted subjects, followed by heterozygotes and then by homozygotes for the GSTM1 gene (Bell et al., 1992).

One can distinguish between the allelic variants ψ (GSTM1a) and μ (GSTM1b) which differ by one G-C_{534} exchange, leading to a change of lysine into asparagine in the protein. With test substrates, differences in enzyme activities between these variants are minimal (Seidergård et al., 1988; Brockmöller et al., 1993). There is a discussion as to whether non-expression of the GSTM1 gene is accompanied by a higher risk of lung cancer (Idle et al., 1992; Alexandrie et al., 1994). However, other studies (Brockmöller et al., 1993) are not consistent with this thesis. Inconsistencies between different studies might be due to different extents of tobacco consumption in different populations which seems to be an important factor of the lung cancer risk of the GSTM1 null genotype (Kihara et al., 1994). Moreover, the GSTM1 genetic polymorphism may probably influence the occurrence of colon cancer (Zhong et al., 1994).

The hypothesis has also been put forward that genetic polymorphism of GSTM1 might be linked with the occurrence of human bladder cancer. A study of 229 patients with transitional cell carcinoma of the urinary epithelium and 211 control subjects in the USA led to the result that the GSTM1 O/O genotype conferred a 70% increased risk of bladder cancer, presumably by a slower detoxification of tobacco-smoke related carcinogens. It was hypothesized that 25% of all bladder cancer (in the USA) might be attributed to this particular genotype (Bell et al., 1993).

It has been established that GST class μ catalyses the conjugation of defined epoxides other than trans-stilbene oxide, and especially of styrene oxide, the intermediate epoxide metabolite of styrene (table 1). Warholm et al. (1983) have compared the turnover of styrene oxide in vitro by GST α, μ and π; the class μ enzyme displayed a more than 20-fold higher reaction velocity than class π, and almost no enzymic reaction was seen with class α.

Recent investigations on the excretion of mercapturic acid conjugates of styrene metabolites in styrene-exposed workers have revealed major interindividual variations. This points to large individual differences in the biological inactivation of styrene oxide and is consistent with the long known variations in excretion of mandelic acid, which is the alternative end product of styrene and styrene oxide metabolism (Goergens, 1992).

Class θ Polymorphism (GSTT1)

A hitherto unknown glutathione transferase activity for C_1/C_2 compounds in human erythrocytes has been characterized and later linked with the θ class of GST (Schröder et al., 1992). This enzyme activity displays polymorphism. It is found in approximately three-quarters of the population, called "conjugators", whereas "non-conjugators" lack this specific enzyme activity. No animal model has been found so far which also exhibits this unique human phenomenon in red blood cells. Animals which show class θ related activity in the liver do not express this enzyme in erythrocytes.

Previous investigations in our laboratory have identified a number of substrates for this polymorphic enzyme activity, namely methyl chloride (Peter et al., 1989), methyl bromide, methyl iodide (Hallier et al., 1990), dichlormethane (Thier et al., 1989) and ethylene oxide (Föst et al., 1991). Among these substrates, methyl bromide has the highest affinity for the new glutathione transferase isozyme and can therefore be used as a standard test substrate for the phenotyping of "conjugators" and "non-conjugators" from human blood specimens.

A comparison of the kinetic constants of the initial rate of methyl bromide disappearance from the head-space of incubations with human blood showed a sharp distinction between "conjugators" and "non-conjugators". There are also indications of a further subgroup among the conjugators with very high enzyme activity ("high conjugators"). A

1:2:1 distribution between "non-conjugators", "intermediate conjugators" and conjugators with high enzyme activity ("high-conjugators") has been suggested (Hallier et al., 1993).

Support for the existence of very wide interindividual differences in the metabolism of methyl bromide and methyl chloride may also be inferred from published reports on subjects occupationally exposed to these compounds (review by Bolt and Gansewendt, 1993). A field study on methyl bromide workers was performed in Japan by using the hemoglobin adduct S-methylcysteine as a biological monitor. In a subgroup of seven workers with the highest exposure levels (filling of spray cans and gas cylinders), three individuals had very high adduct levels whereas the other four had only low levels, close to the background of persons not occupationally exposed to methyl bromide.

A field of study (van Doorn et al., 1980) of six workers occupationally exposed to methyl chloride in the Netherlands showed that four of the subjects excreted higher amounts of the metabolite S-methylcysteine in urine whereas two others "hardly excreted any S-methylcysteine after axposure to methyl chloride".

Moreover, six male volunteers had been exposed in a laboratory study to 10 or 50 ppm methyl chloride (Nolan et al., 1985). Two of these subjects, called "slow metabolizers" by the authors, showed 2-3 times higher levels of methyl chloride in both blood and alveolar air during exposure, which decreased more slowly after termination of exposure compared to the four other subjects, called "rapid metabolizers".

Such interindividual differences have also been reported for ethylene oxide in vivo. In a clinical study, hemodialysis patients showing immediate-type allergic reactions during treatment, called "reactors" had significantly higher levels of ethylene oxide adducts with serum albumin than those without allergic reactions ("non-reactors") (Grammer et al., 1984). In investigations of DNA damage using the "alkaline elution" method in subjects occupationally exposed to ethylene oxide, indications were found for the existence of two subgroups of differential susceptibility to ethylene oxide within the investigated group of non-smokers (Fuchs et al., 1994).

Incubation of whole blood samples with any of the three chemicals methyl bromide, ethylene oxide and dichloromethane led to a marked induction of sister chromatid exchanges in the lymphocytes of "non-conjugators." An effect of the chemicals on the frequency of SCE was not or only hardly observable in the lymphocytes of the "conjugators" (Hallier et al., 1993). This was thought to be due to the reduction of the target concentrations of the chemicals in the blood samples of the conjugators through metabolic elimination by the GST in erythrocytes. The highly reactive substance methyl bromide led to practically no cytogenetic effect on the lymphocytes of conjugators due to its high affinity for the metabolizing enzyme in erythrocytes. From previous investigations in vitro it was known that the distribution of radiolabelled ethylene oxide (Föst et al., 1991) and dichloromethane (Thier et al., 1989) in the cellular and plasma compartments of human blood, upon incubation, differed strikingly between blood from "conjugators" and from "non-conjugators".

In this context, it is of great interest that similar interindividual differences in cytosolic GST in human/livers were also found, using the substrate dichloromethane (Bogaards et al., 1993).

The human polymorphic GST catalysing conjugation of dichloromethane was later characterized as a θ enzyme (GSTT1) by means of molecular biology. A cDNA was isolated which enclosed the human class θ GSTT1 and which had 82% identity to rat GSTsu5. Using PCR and Southern blot analyses it was shown that the GSTT1 gene is absent from approximately one-third of the population, It was also shown that the "conjugator" and "non-conjugator" phenotypes (v.s.) are coincident with the presence and absence of the GSTT1 gene, respectively (Pemble et al., 1994).

The introduction of a plasmid encoding for the rat class θ GST 5-5 into S. typhimurium TA 1535 made it possible to demonstrate the effectiveness of GST class θ to transform dichloromethane to mutagenic metabolites (Thier et al., 1993).

The influence of GSTT1 enzyme polymorphism on the cytogenetic toxicity of methyl bromide, ethylene oxide and dichloromethane in vitro suggests that this may be an important factor in individual susceptibility to the toxic effects of these substances. It is now clear that about 25-30% of the European population does not possess the GSTT1 gene. The gene product in individuals possessing this gene is expressed at least in liver and in erythrocytes. Human genotyping and phenotyping is easily possible on the basis of blood sampling.

A relevant active intermediate of dichloromethane metabolism formed by the enzyme in erythrocytes is formaldehyde (Hallier et al., 1994).

CONCLUSION

Recent reviews are focussed on the influence of distinct human enzyme polymorphisms on metabolism of chemical carcinogens and human cancer susceptibility (Caporaso et al., 1991; Idle et al., 1992; Bolt, 1994). Beyond this important field of current research, polymorphisms of enzymes involved in metabolism of typical industrial compounds are now becoming increasingly important in industrial toxicology. Accordingly, human enzyme polymorphisms largely influence the extrapolation of risk of a number of chemicals from animal experiments to humans (Hallier et al., 1994).

REFERENCES

Alexandrie, A.K., Sundberg, M.J., Seidegård, J., Tornling, G., Rannug, A., 1994, Genetic susceptibility to lung cancer with special emphasis on CYP1A1 and GSTM1: A study on host factors in relation to age at onset, gender and histological cancer types, Carcinogenesis. 15:1785-1790

Bell, D.A., Taylor, J.A., Paulson, D.F. et al., 1993, Genetic risk and carcinogen exposure: a common inherited defect of the carcinogen-metabolism gene glutathione-S-transferase M 1 (GSTM1) that increases susceptibility to bladder cancer, Natl Cancer Inst USA. 85: 1159-1164.

Bell, D.A., Thompson, C.L., Taylor, J. et al., 1992, Genetic monitoring of human polymorphic cancer susceptibility genes by polymerase chain reaction: Application to glutathione transferase μ, Environ Health Perspect 98: 113-117.

Bogaards, J.J.P., van Ommen, B., van Bladeren, P.J., 1993, Interindividual differences in the in vitro conjugation of methylene chloride with glutathione S-transferase in 22 human liver samples, Biochem Pharmacol 45: 2166-2169.

Bolt, H.M., Gansewendt, B., 1993, Mechanisms of carcinogenicity of methyl halides, CRC Crit Rev Toxicol 23: 237-253.

Bolt, H.M., 1994, Genetic and individual differences in the process of biotransformation and their relevance for occupational medicine, Med. Lavoro 85: 37-48.

Brockmöller, J., Kerb, R., Drakoulis, N. et al., 1993, Genotype and phenotype of glutathione S-transferase class μ isoenzymes μ and ψ in lung cancer patients and controls, Cancer Res, 53: 1004-1011.

Caporaso, N., Landi, M.T., Vineis, P., 1991, Relevance of metabolic polymorphisms to human carcinogenesis: Evaluation of epidemiologic evidence, Pharmacogenetics, 1: 4-19.

van Doorn, R., Borm, P.J.A., Leijdekkers, C.M. et al., 1980, Detection and identification of S-methylcysteine in urine of workers exposed to methyl chloride, Int Arch Occup Environ Health, 46: 99-109.

Föst, U., Hallier, E., Ottenwälder, H. et al., 1991, Distribution of ethylene oxide in human blood and its implications for biomonitoring, Hum Exp Toxicol 10: 25-31.

Fuchs, J., Wullenweber, U., Hengstler, J.G., Bienfait, H.G., Hiltl, G., Oesch, F., 1994, Genotoxic risk for humans due to workplace exposure to ethylene oxide: remarkable individual differences in susceptibility, Arch Toxicol 68: 343-348.

Goergens, H.W., 1992, Stereochemische Aspekte bei der biologischen Überwachung beruflicher Styrolexposition, Thesis, Universität des Saarlandes, Homburg/Saar, Germany.

Grammer, L.C., Roberts, M., Nicholls, A.J. et al., 1984, IgE against ethylene oxide-altered human serum albumin in patients who have had acute dialysis reactions, J Allergy Clin Immunol 1984, 74: 544-549.

Hallier, E., Schröder, K., Asmuth, K., Dommermuth, A., Aust, B., Goergens, H.W., 1994, Metabolism of dichloromethane (methylene chloride) to formaldehyde in human erythrocytes: Influence of polymorphism of glutathione transferase theta (GSTT1-1). Arch Toxicol 68: 423-427.

Hallier, E., Deutschmann, S., Reichel, C., et al., 1990, A comparative investigation of the metabolism of methyl bromide and methyl iodide in human erythrocytes, Int Arch Occup Environ Health, 62: 221-225.

Hallier, E., Langhof, R., Dannappel, D. et al., 1993, Polymorphism of glutathione conjugation of methyl bromide, ethylene oxide and dichloromethane in human blood: influence on the induction of sister chromatid exchanges (SCE) in lymphocytes, Arch Toxicol 67: 173-178.

Idle, J.R., Armstrong, M., Boddy, A.V., et al., 1992, The pharmacogenetics of chemical carcinogenesis, Pharmacogenetics 2: 246-258.

Kihara, M., Noda, K., 1994, Lung cancer risk of GSTM1 null genotype is dependent on the extent of tobacco smoke exposure, Carcinogenesis 15: 415-418.

Nolan, R.J., Rick, D.L., Landry, T.D., et al.,1985, Pharmacokinetics of inhaled methyl chloride (CH_3Cl) in male volunteers, Fundam Appl Toxicol 5: 61-369.

Pemble, S., Schroeder, K.R., Spencer, S.R., Meyer, D.J., Hallier, E., Bolt, H.M., Ketterer, B., Taylor, J.B., 1994, Human glutathione S-transferase theta (GSTT1): cDNA cloning and the characterization of a genetic polymorphism, Biochem J 300: 271-276.

Peter, H., Deutschmann, S., Reichel, C., Hallier, E., 1989, Metabolism of methyl chloride in human erythrocytes, Arch Toxicol 63: 351-355.

Schröder, K., Hallier, E., Peter, H., Bolt, H.M., 1992, Dissociation of a new glutathione S-transferase activity in human erythrocytes, Biochem Pharmacol 43; 1671-1674.

Seidergård, J., De Pierre, J.W., Pero, R.W., 1985, Hereditary interindividual differences in the glutathione transferase activity towards trans-stilbene oxide in resting human mononuclear leucocytes are due to a particular isozyme(s), Carcinogenesis 6: 12-1216.

Seidergård, J., Vorachek, W.R., Pero, R.W., Pearson, W.R., 1988, Hereditary differences in the expression of the human glutathione S-transferase activity on trans-stilbene oxide are due to a gene deletion, Proc Natl Acad Sci USA 85: 7293-7297.

Thier, R., Föst, U., Deutschmann, S., et al., 1989, Distribution of methylene chloride in humane blood, Arch Toxicol, Suppl. 14: 254-258.

Thier, R., Taylor, J.B., Pemble, S.E. et al., 1993, Expression of the rat theta class glutathione S-transferase 5-5 in S. typhimurium TA 1535 leads to base-pair mutation upon exposure to dihalomethanes, Biol Chem Hoppe-Seyler 374: 796 (abstract F 105).

Warholm, M.,Guthenberg, C., Mannervik, B., 1983, Molecular and catalytic properties of glutathione transferase μ from human liver an enzyme conjugation epoxides, Biochem 22: 3610-3617.

Zhong, S., Wyllie, A.H., Barnes, D., Wolf, C.R., Spurr, N.K., 1994, Relationship between the GSTM1 genetic polymorphism and susceptibility to bladder, breast and colon cancer. Carcinogenesis 14: 1821-1824.

COMPARATIVE ESTIMATION OF THE NEUROTOXIC RISKS OF N-HEXANE AND N-HEPTANE IN RATS AND HUMANS BASED ON THE FORMATION OF THE METABOLITES 2,5-HEXANEDIONE AND 2,5-HEPTANEDIONE

J. G. Filser,* Gy. A. Csanády,† W. Dietz, W. Kessler, P. E. Kreuzer, M. Richter, and A. Störmer

GSF-Institute of Toxicology
Neuherberg, D-85758 Oberschleissheim, Germany

ABSTRACT

In rats and humans, inhalation kinetics of n-hexane (HEX) and n-heptane (HEP) were compared with urinary excretion of 2,5-hexanedione (HDO) and 2,5-heptanedione (HPDO), respectively. Furthermore, the reactivities of HDO and HPDO with Nα-acetyl-L-lysine towards the formation of pyrrolyc adducts was studied. By means of the data gained, the potency of HEP for inducing peripheral neuropathy is compared with the well known one of HEX. In rats, kinetic analysis revealed two different metabolic processes for HEX and HEP, one process characterized by high affinity and low capacity (maximal rate of metabolism Vmax1: HEX 84, HEP 112 μmol/h/kg) and one by low affinity and high capacity (Vmax2: HEX 456 μmol/h/kg). For HEP, Vmax2 cannot be given, since the deviation from linearity of the curve representing the rate of metabolism versus the exposure concentration was too small within the concentration range studied of up to 10000 ppm. Urinary excretion of HDO resulting from exposure to HEX correlated with the first process, whereas the corresponding excretion of HPDO as a metabolite of HEP correlated with the second process. In humans, rates of metabolism of HEX and HEP increased linearly with the exposure concentrations up to the tested values of 300 ppm (HEX) and 500 ppm (HEP), the pulmonary retention at steady state being 23% (HEX) and 35% (HEP) at rest. Of totally metabolized

* Correspondence to: J. G. Filser

† On leave from Central Research Institute of Chemistry, Hungarian Academy of Sciences, Budapest, Hungary

HEX during and after HEX exposure to 300 ppm, about 0.5% was excreted as HDO in urine. Of totally metabolized HEP during and after HEP exposure up to 500 ppm, only about 0.01% was excreted as HPDO in urine. Background excretion of HPDO was found in urine of rats and of both γ-diketones in urine of humans; the sources are still unknown. In rats, urinary excretion of HPDO resulting from exposure to 500 ppm HEP was about 7 times less and in humans about 4 times less than that of HDO resulting from exposure to 50 ppm HEX over the same time span. In vitro, the rate of pyrrole formation from the reaction of HPDO with Nα-acetyl-L-lysine was about half that obtained with HDO. This indicates a lower neurotoxic potency of HPDO. From our findings it becomes intelligible that HEP was not neurotoxic in rats in contrast to HEX. Furthermore, for humans we also conclude the neurotoxic potency of HEP to be significantly lower than that of HEX.

INTRODUCTION

Both saturated alkanes n-hexane (HEX) and n-heptane (HEP) are lipophilic solvents with high vapor pressure. Consequently, occupational exposure to vapors of HEX and HEP may occur. Chronic exposure to HEX can lead to peripheral neuropathy in laboratory animals and in humans which is attributed to the appearance of 2,5-hexanedione (HDO), a metabolic intermediate (Krasavage et al. 1980; reviews in Spencer et al. 1980, Couri and Milks 1982; Graham et al. 1995). Reaction of this γ-diketone with primary amines of neurofilamental protein resulting in pyrrolyc adducts is regarded to be the first step leading to this type of peripheral neuropathy (DeCaprio et al. 1982; Graham et al. 1982; DeCaprio et al. 1983; for review see Graham et al. 1995). It is characterized by distal sensorimotor axonopathy with axonal accumulation of disorganized neurofilaments followed by degeneration of the distal axon (reviewed in Spencer et al. 1980; Couri and Milks 1982; Graham et al. 1995). Biotransformation of HEP by rats and humans also yields a γ-diketone, 2,5-heptanedione (HPDO; Bahima et al. 1984, Perbellini et al. 1986, Störmer et al. 1994). This compound produced peripheral neuropathy, also, following chronic administration to rats (O'Donoghue and Krasavage 1979). However, no neurotoxicity was observed in several studies with rats exposed to the metabolic precursor HEP (reviewed in DGMK 1986). For humans, no data on toxic effects due to exposure to pure HEP are published.

The aim of our study was to explain the difference in neurotoxic efficiency between HEX and HEP in rats and to compare quantitatively the neurotoxic potencies of both alkanes in rats and humans. Therefore, we studied the toxicokinetics of HEX and HEP in both species. Considering the urinary excretion of HDO and HPDO to be a measure of their body burden, we investigated both γ-diketones in urine based on dependence of the exposure to vapors of HEX and HEP. Since the rate of pyrrole formation resulting from the reaction of γ-diketones with primary amines has been demonstrated to correlate with the neurotoxic potency of such compounds (DeCaprio and Weber 1980; DeCaprio et al. 1982; Anthony et al. 1983a, 1983b; Sayre et al. 1986; Genter et al. 1987; DeCaprio et al. 1988; Genter St. Clair et al. 1988; for a review see Graham et al. 1995), we determined pyrrole formation in vitro using Nα-acetyl-L-lysine as reactant for HDO and HPDO.

MATERIAL AND METHODS

Test Compounds

n-Hexane (95%) was purchased from Riedel-de Haën (Seelze, FRG), n-heptane (99.9%) and 5-methyl-3-heptanone (97%) from Aldrich (Steinheim, FRG), 2,5-hexanedione

(>99%) and cyclohexanone (>99.5%) from Fluka (Neu-Ulm, FRG). 2,5-Heptanedione was synthesized as described in Stetter and Kuhlmann (1976) by a catalyzed addition of propanal (97%) to 1-butene-3-on (99%), both obtained from Aldrich. 5-(2-Hydroxyethyl)-4-methyl-1,3-thiazol (98%) and benzyl chloride (99%) used for the synthesis of the catalytic agent according to Stetter and Kuhlmann (1975) were also from Aldrich. The synthesized 2,5-heptanedione (yield 18.4%) was identified by gas chromatography/mass spectrometry (GC/MS). It was also compared with a sample of 2,5-heptanedione kindly provided as a gift by Prof. Dr. L. Perbellini (Institute of Occupational Medicine, University of Padua-Verona, Italy). The product was identified and analyzed for purity by GC/MS (95%) and by nuclear magnetic resonance (93%). For detailed descriptions see Störmer (1995). Nα-Acetyl-L-lysine (>99%) and boron trifluoride 14% in methanol were purchased from Sigma (Deisenhofen, FRG), p-dimethylaminobenzaldehyde was obtained from Merck (Darmstadt, FRG).

Animals

Male Wistar rats (190-320 g) were obtained from GSF-Versuchstierzucht (Neuherberg, FRG). Before and after exposure, animals had free access to standard chow Nr. 1320 from Altromin (Lage, FRG) and tap water.

Exposure of Rats to Constant Vapor Concentrations of N-Heptane and Collection of Urine

Using closed all-glass metabolic inhalation chambers of 4.6 l (Hallier at al. 1981) animals were exposed for 8 h to "constant" vapor concentrations of HEP (one animal per chamber and experiment) using the same procedure as described previously for HEX (Kessler et al., 1990). Atmospheric vapor concentrations of HEP (1000, 2730, 4520 and 8750 ppm) were maintained within a standard deviation of up to 17% by repeated injections of calculated amounts of liquid HEP into the inhalation chambers. During and after exposure, urine was collected in ice-cooled fractions up to 95 h following start of the experiments. Prior to exposure, control urine had been obtained fractionated over time periods of up to 96 h from the identical animals.

Exposure of Humans to N-Hexane and N-Heptane and Collection of Urine

Healthy volunteers (all non-smokers, age 21-39 y) were exposed individually via breathing masks (Jaeger, Würzburg, FRG) to constant concentrations of HEX (135 and 137 ppm for 2 h, 157 ppm for 2.5 h, 181 ppm for 3 h, 297 and 303 ppm for 3.33 h) or HEP (101, 101, 247, 248, 249, 499 and 503 ppm for 4 h each). The open exposure system consisted of a pump producing an air flow of 70 l/min and an all-glass evaporation chamber to which liquid HEX or HEP was constantly added using an infusion pump. In each experiment one volunteer was linked by the breathing mask parallel to the vapor stream. The mask was equipped with two valves allowing the separation of inhaled and exhaled air. Every 1.5-2 min at the beginning of exposure and every 8-10 min later on, 2-4 breaths of mixed exhaled air were collected into a gas bag of 2.5 l, made of polyethylene-coated metal foil (Linde, München, FRG). Vapor concentrations within the vapor stream (inhaled concentration) and in the samples of exhaled air (average exhaled concentration from 2-4 breaths) were measured by gas chromatography. Gas samples were injected via a sample loop of 1.5 ml into a GC-8A gas chromatograph (Shimadzu, Duisburg, FRG) equipped with a flame ionization detector kept at 210°C. Separation was done on a stainless steel column 2.5 x 1/8" packed with Tenax-GC 60-80 mesh (Scima, Egling, FRG). Oven temperature was kept at 180°C for HEX and 200°C for HEP. Gas flow-rates were: nitrogen as carrier gas 50 ml/min, hydrogen 60 ml/ml and air 600 ml/min. Under these conditions retention time was 1 min for HEX and 1.1 min for HEP. Peaks were recorded and integrated using a 4290 integrator (Varian, Darmstadt, FRG). Quantification was done by means of calibration curves obtained separately by injecting defined amounts of pure HEX or HEP into desiccators of known volume.

Urine of volunteers was collected from up to 61 h before until up to 75 h after start of exposure. Samples were kept frozen at -20°C until analysis.

Determination of HDO and HPDO in Urine

Excretion of HDO in urine of rats following exposure to HEX was taken from Figure 4 in Kessler et al. (1990).

In humans, excretion of HDO was determined in urine of the two volunteers exposed to 300 ppm HEX. Excretion of HPDO in urine was determined in rats and in all HEP-exposed volunteers. Analysis of the γ-diketones was performed by GC/MS as described in detail for HDO by Dietz (1994) and for HPDO by Störmer (1995). After addition of NaCl (0.3 g/ml) to untreated urine samples (determination of HDO: man: 750 μl; of HPDO: rat 250 μl, man 5 ml), these were extracted with methylene chloride (determination of HDO: 750 μl; HPDO: 500 μl) containing cyclohexanone (as internal standard for the determination of HDO) or 5-methyl-3-heptanone (as internal standard for the determination of HPDO). Of the extracted organic phase, 1 μl was injected at 40°C into a GC HP 5890II (Hewlett Packard, Böblingen, FRG) using a temperature programmable injector (KAS 2, Gerstel, Mühlheim, FRG). After 2 sec (HDO) or 10 sec (HPDO), the injector was heated with a rate of 10°C/sec to 180°C which was held for 120 sec. The purge valve was closed until 40 sec after injection. Separation was done on a BP 10-capillary column (length 12 m, ID 0.22 mm, film 0.25 μm; SGE, Weiterstadt, FRG) connected to a deactivated precolumn (length 2.4 m, ID 0.53 mm; Amchro, Sulzbach, FRG). After injection, the column temperature was kept at 40°C for 0.5 min and then heated with a rate of 15°C/min to 130°C (HDO) and 135°C (HPDO), respectively. Subsequently, the final temperature of 180°C was reached with a heating rate of 40°C/min (HDO) or 20°C/min (HPDO). HDO, cyclohexanone, HPDO, and 5-methyl-3-heptanone (retention times 4.0, 3.2, 5.1, and 3.5 min) were detected using a mass selective detector HP 5970B (Hewlett Packard, Böblingen, FRG) in the single ion mode. Peak integration was done using ion masses m/z 99 and 114 for HDO, m/z 98 for cyclohexanone, m/z 71 and 99 for HPDO and for 5-methyl-3-heptanone. Calibration curves were constructed from ratios of the peak area of HDO or HPDO to that of the corresponding internal standard. The detection limit given as three times the background noise obtained in standard solutions was 0.078 (rat) and 0.004 (man) μmol/l urine. That of HDO (man) was 0.035 μmol/l.

The excreted amounts of HDO and HPDO were calculated by multiplying the concentrations with the urine volumes. The background excretion rate was determined as quotient of the sum of excreted γ-diketone in controls and the sampling time.

Determination of the Pyrrole Formation from the Reaction of HDO and HPDO with Nα-Acetyl-L-Lysine

Incubations were performed in gas-tight vials at 37°C under nitrogen atmosphere to avoid autoxidation of pyrroles. Nα-acetyl-L-lysine at three concentrations (3, 10, and 30 mmol/l) was incubated with HDO or HPDO at four concentrations (100, 300, 600, and 1000 mmol/l) in phosphate buffer (0.15 mol/l, pH 7.5). Pyrrole concentration in the incubate was measured up to 180 min by a spectrophotometric assay according to Mattocks and White (1970) which is based on the coloured reaction products of pyrroles with p-dimethylaminobenzaldehyde. Samples of 250 µl were added to 250 µl of Ehrlich's reagent (0.75 g p-dimethylaminobenzaldehyde dissolved in 10 ml methanolic 14% boron trifluoride and diluted to 25 ml with ethanol) and heated for 1 min at 95°C. After 9 min of cooling to room temperature, absorbance was measured at the wave length of absorption maxima (528 nm). The values of the molecular extinction coefficients (ε) for the condensation products of the derivatives of HDO and HPDO were determined to be 0.0856 ± 0.010 l/µmol/cm (n = 3) and 0.0908 ± 0.0024 l/µmol/cm (n = 3), respectively. Second order rate constants (k) for the pyrrole formation were calculated from the given concentrations of γ-diketone [γD] and Nα-acetyl-L-lysin [AcLys] in the incubation baths and the measured changes of absorbance per time (ΔA/Δt) of the pyrrole-p-dimethylaminobenzaldehyde condensation products in cuvettes having a thickness (d) of 1 cm: $k = \Delta A/\Delta t/([\gamma D]\cdot[AcLys]\cdot\varepsilon\cdot d)$. More details are given in Richter (1995).

Computer Hardware, Software and Statistics

All writings, drawings and calculations were carried out on Macintosh Computers (Apple Computer GmbH, München, FRG) using the programs Word 5.1, Excel 3.0 (both from Microsoft, München, FRG), SigmaPlot 4.11 (Jandel, Erkrath, USA), StatView SE+Graphics (Abacus concepts, Berkeley, USA).

Mean values and standard deviations were determined using StatView SE+Graphics. Results are presented as arithmetic means ± standard deviations (n = number of determinations). Regressions through the origin were calculated with SigmaPlot using the equations for slope (a) and standard deviation of the slope (s_a) given in Doerffel (1990) and the equation for the regression coefficient (r) given in Sachs (1974):

$$a = \frac{\Sigma x_i \cdot y_i}{\Sigma x_i \cdot x_i} \quad \text{and} \quad s_a = \sqrt{\frac{s_{y.x}^2}{\Sigma x_i^2}} \tag{1}$$

$$r = \sqrt{\frac{s_y^2 - s_{y.x}^2}{s_y^2}} \tag{2}$$

$$\text{with: } s_y^2 = \frac{\Sigma(y_i - \Sigma y_i/n)^2}{n-1} \quad \text{and} \quad s_{y.x}^2 = \frac{\Sigma(y_i - a\cdot x_i)^2}{n-1} \tag{3}$$

Toxicokinetic Analysis

Using the "closed chamber technique" (CCT) which is presented in Filser (1992) toxicokinetic parameters of inhalation kinetics of HEX and HEP in male Wistar rats were determined by means of a two-compartment model reanalyzing concentration-time courses generated earlier for HEX (Kessler et al. 1989) and HEP (Csanády et al. 1992) in the atmospheres of closed all-glass exposure chambers occupied by rats. Toxicokinetic analysis was carried out as described in detail in Filser et al. (1993) by means of the program "SOLVEKIN" (Csanády and Filser 1990) which was installed on a Macintosh computer. In CCT, compartment 1 represents the atmosphere with the compound concentration y1 and the volume V1 and compartment 2 the animal(s) with the volume V2 in which the chemical concentration is given as a single average concentration y2. Its actual value depends on the rates of three processes: the inhalative uptake of a gaseous compound from the atmosphere into the organism and the eliminations out of it via exhalation and via metabolism. Rates of uptake and of exhalation are assumed to be proportional to the compound's concentration in the atmosphere and to its average concentration in the organism, respectively. The proportionality factors (standardized for one (n'=1) rat of 250 ml) are designated "clearance of uptake" (k'12·V'1) and "clearance of exhalation" (k'21·V'2). Metabolic elimination of HEX and of HEP, respectively, is described by the sum of two different saturable kinetic processes according to Michaelis-Menten. Rates of metabolism depend on the actual average concentration y2, the maximum metabolic rates (V'max1 and V'max2) and the apparent Michaelis-Menten constants (Kmapp1 and Kmapp2), both of which are related to y2.

RESULTS

Rats

Inhalation Kinetics of HEX and HEP. Concentration-time courses following the generation of various initial concentrations of HEX and HEP in the atmospheres of closed chambers occupied by two rats in each experiment are shown as semilogarithmic plots in figure 1 together with the corresponding fits obtained by the two-compartment model.

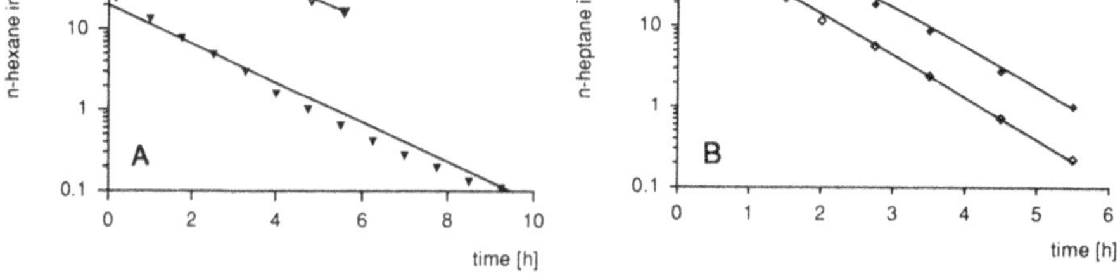

Figure 1. Concentration-time courses of inhaled n-hexane (A) and n-heptane (B) vapors following administration of various initial concentrations in the atmosphere of closed exposure chambers (6.4 l) containing 135 g soda lime and two male Wistar rats in each experiment [individual animal weights: HEX 180-220 g, 187 ± 9.3 g (n = 30) ; HEP: 230-260 g, 240 ± 13 g (n = 8)]. Symbols: measured values, each set of symbols represents 3 and 1 gas uptake experiment using HEX and HEP, respectively; data from Kessler et al. 1989 (HEX) and Csanády et al. 1992 (HEP). Solid lines: best fits obtained by analysis of the measured values using the program SOLVEKIN for the two compartment toxicokinetic model presented in Filser (1992).

Table 1. Toxicokinetic parameters of n-hexane (HEX) and n-heptane (HEP) in one male Wistar-rat ($n' = 1$, $V'2 = 250$ ml) according to the two compartment model of Filser (1992)

Parameter [a]	Formula a)	Value [b] HEX $\bar{X} \pm$ S.D.	HEP $\bar{X} \pm$ S.D.	Dimension
Clearance of uptake (related to atmospheric conc.)	$k'12 \cdot V'1$	2.1±0.14	4.9±0.9	l/h
Clearance of exhalation (related to the average conc. in the organism)	$k'21 \cdot V'2$	0.23±0.1	0.16±0.06	l/h
Maximum rate of metabolism of low capacity and high affinity process	V'max1	21±4.5	28±3.4	µmol/h
Apparent Michaelis-Menten constant of low capacity and high affinity process (related to the average conc. in the organism)	Kmapp1	0.008±0.003	0.035±0.12	µmol/ml
Maximum rate of metabolism of high capacity and low affinity process	V'max2	114±52	n. d. [c]	µmol/h
Apparent Michaelis-Menten constant of high capacity and low affinity process (related to the average conc. in the organism)	Kmapp2	1.3±0.27	n. d. [c]	µmol/ml
First order rate constant of the metabolic process of high capacity and low affinity	$\frac{V'max2}{V'2 \cdot Kmapp2}$	0.35±0.17	0.08±0.01	1/h

[a] for designations and symbols see Material and Methods
[b] Standard deviations are calculated by Gauß's error propagation
[c] n. d.: not determined

For both compounds, two different metabolic processes could be distinguished, one with high affinity (Kmapp1) and low capacity (Vmax1) and one with low affinity (Kmapp2) and high capacity (Vmax2). In contrast to HEX, only the ratio of Vmax2 to Kmapp2 could be quantified for HEP since the deviation from linear kinetics was too small for identifying Vmax2 of HEP. By means of these parameters rates of metabolism were plotted for both pathways based on dependence of the atmospheric exposure concentrations of HEX and HEP at steady state (figure 2).

Excretion of HDO and HPDO in Urine. The excretion of HDO resulting from 8 h exposures to constant concentrations of HEX indicated saturation kinetics (Kessler et al. 1990; figure 2). A linear relation between HEX vapor concentration and HDO excretion was observed only for HEX concentrations of below 250 ppm. At steady state exposure to 50 ppm HEX over 8 h (MAK-value: German Maximum Workplace Concentration, Deutsche Forschungsgemeinschaft 1994), 30 µmol/animal of 250 g are metabolized as calculated using the toxicokinetic parameters given in table 1. From this amount 0.32% (0.10 µmol; calculated from Kessler et al. 1990) are excreted as HDO.

Cumulative HPDO excretion resulting from exposures of four animals (185, 255, 323 and 296 g) to constant concentrations of 1000, 2730, 4520 and 8750 ppm HEP over 8 h could be described by e-functions of the formula: $m_{(t)} = m_{(\infty)} \cdot (1 - e^{-k \cdot (t - tlag)})$. In this equation $m_{(t)}$ represents the amount excreted at time point t, $m_{(\infty)}$ the total HPDO amount excreted due to HEP exposure, k the rate constant (i.e., ln2 divided by the half-life of HPDO excretion), and t_{lag} the lag-time between start of the exposure to HEP and first appearance of HPDO in urine. The HPDO amounts ($m_{(\infty)}$) excreted after HEP exposure to the concentrations mentioned above were 0.02, 0.11, 0.10 and 0.33 µmol (background HPDO excretion

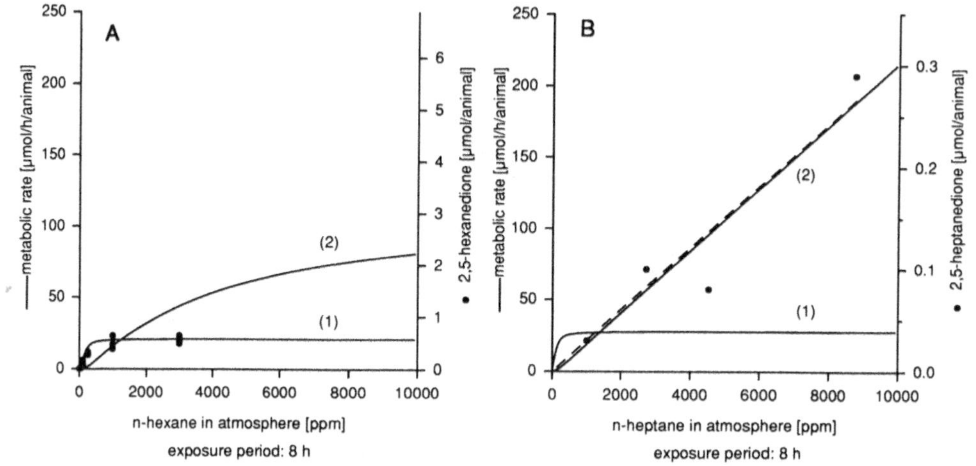

Figure 2. Rates of metabolism of n-hexane (A) and n-heptane (B) at steady state in a single male Wistar rat of 250 g versus atmospheric exposure concentrations calculated from the kinetic parameters in table 1 and excretion of 2,5-hexanedione (A) and 2,5-heptanedione (B) caused by 8 h exposures of male Wistar rats to various constant concentrations of n-hexane and n-heptane, respectively (symbols; every data point one determination in one animal after normalization to a body weight of 250 g). For calculation of rates of metabolism at steady state as functions of the atmospheric exposure concentrations see Filser (1992), Filser et al. (1993), and Laib et al. (1992). Solid lines: calculated rates of metabolism at steady state (1) biotransformation kinetics with high affinity and low capacity; (2) biotransformation kinetics with low affinity and high capacity. Dashed line: Regression through the origin representing the relation between the amounts of 2,5-heptanedione and the atmospheric exposure concentration of n-heptane.

subtracted, see below). After normalizing these data to a body weight of 250 g using surface factors of (250/individual body weight)$^{2/3}$ a linear regression with the exposure-time product of HEP was constructed through the origin ($y = a \cdot x$; $r = 0.9586$) with $a \pm s_a$ being $3.8 \cdot 10^{-6} \pm 0.4 \cdot 10^{-6}$ µmol/(ppm·h). The x-value (ppm·h) represents the product of the atmospheric HEP concentration with the exposure time, and y (µmol) the resulting amount of urinary HPDO excreted per rat. The lag-time for HPDO in urine (t_{lag}) was 5 ± 1.2 h (n = 4). Mean half-life of HPDO in urine was 4.1 ± 1.7 h (n = 4) independent of the exposure concentration. For a rat of 250 g exposed over 8 h to 500 ppm HEP (MAK-value, Deutsche Forschungsgemeinschaft 1994), the share of HPDO excreted in urine to the amount of HEP metabolized is calculated to be 0.006% (0.015 µmol).

Not only HEP exposed but also control animals excreted HPDO in urine. The mean concentration in 20 samples of control urine was 0.23 ± 0.11 µmol/l (n = 20) and the resulting mean HPDO excretion rate per animal was 0.11 ± 0.05 nmol/h (n = 20).

Together with the steady state kinetics of HEX and HEP, figure 2 shows also the urinary γ-diketone excretion resulting from exposures of rats to constant concentrations of HEX and HEP for time periods of 8 h. Excretion of both diketones reflect the biotransformation kinetics which are characterized by high affinity and low capacity for HEX and low affinity and high capacity for HEP. The difference in the amounts of γ-diketones excreted as metabolites of HEX and HEP is considerable.

Humans

Kinetic Parameters of HEX and HEP. For studying uptake and metabolism of HEX and HEP the open exposure system described above was used. Subjects were exposed

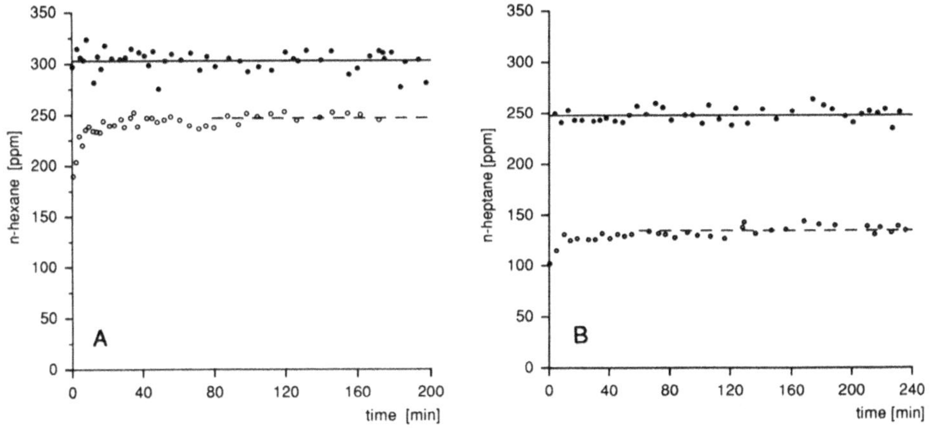

Figure 3. n-Hexane (A) and n-heptane (B) concentrations in inspired (●) and mixed expired (○) air during exposure to these solvents. A) Exposure of one male volunteer (72 kg) to n-hexane (3.33 h); Solid line: mean inspired concentration: 303 ppm; Dashed line: mean mixed exhaled concentration after 80 min: 246 ppm. B) Exposure of one male volunteer (57 kg) to n-heptane (4 h); Solid line: mean inspired concentration: 248 ppm; Dashed line: mean mixed exhaled concentration after 60 min: 135 ppm.

individually to constant concentrations in the inspired air of between 135 and 303 ppm (HEX) and between 100 and 500 ppm (HEP). Typical concentration-time profiles in the ex- and inspired air are given in figure 3.

After 60-90 min of exposure to constant concentrations of both alkanes, pseudo-steady state conditions were reached as becomes evident from the plateaus of the exhalation curves. Pulmonary ventilation determined in exposure experiments with HEX was 7.5 ± 1.2 l/min (n = 5) in 1 female (53 kg) and in 4 male subjects (72, 72, 73 and 75 kg). In experiments with HEP the 3 female (47, 56 and 56 kg) and 4 male subjects (57, 65, 72 and 75 kg) had a pulmonary ventilation of 6.2 ± 2.0 l/min (n = 7). Under these conditions pulmonary retention defined as the difference between inhaled and mixed exhaled concentration divided by the inhaled concentration at steady state was 23% ± 3% (n = 5) for HEX and 35% ± 7% (n = 7) for HEP. Rate of metabolism at steady state is given by the product of pulmonary retention, pulmonary ventilation and the actual atmospheric concentration of the compound. Within the concentration ranges studied, this rate was directly proportional to the exposure concentrations of both compounds.

HDO and HPDO Excretion in Urine. Background concentration of HDO in urine was found to be 88 ± 44 nmol/l (n = 6). The corresponding urinary excretion rate was studied in two male subjects (73 and 72 kg) to be 6.0 and 3.5 nmol/h, respectively. Exposure to 300 ppm HEX for 3.33 h (external exposure dose: 1000 ppm·h) of the identical two volunteers resulted in an urinary excretion of 12.0 and 17.4 μmol HDO (background subtracted). The estimated ratio of mean urinary HDO excretion to external exposure dose was 0.015 μmol/(ppm·h) (figure 4 A).

Half-lives of HDO were 13.3 and 8.3 h. Of totally metabolized HEX during and after HEX exposure, about 0.5% were excreted in urine as HDO.

For HPDO, too, a background excretion in control urine was found. Its mean value ± standard deviation determined from the mean background values of the 7 subjects described above was 43 ± 20 nmol/l (n = 7). The corresponding excretion rate of HPDO was 4.5 ± 4.8 nmol/h (n = 7). Exposures to HEP concentrations of 100, 250 and 500 ppm for 4.00

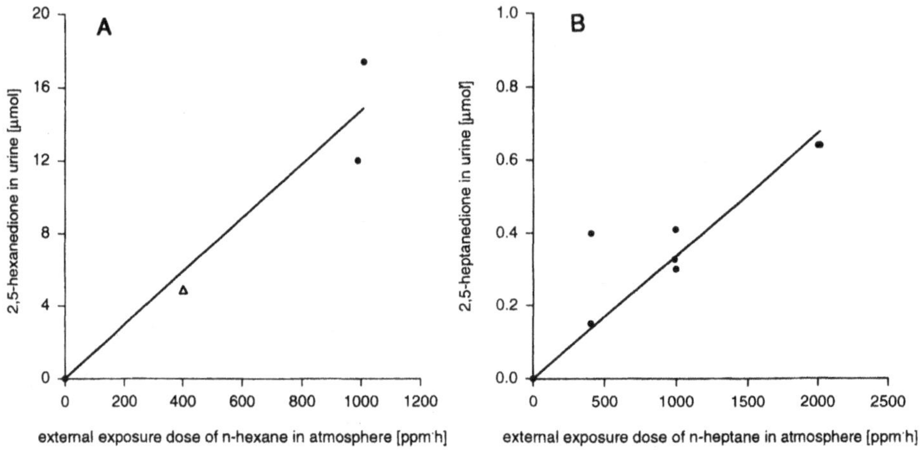

Figure 4. Excreted amounts of 2,5-hexanedione (A) and 2,5-heptanedione (B) in untreated urine of humans exposed to n-hexane and n-heptane versus the external exposure dose (product of exposure concentration and time). Symbols: (●) experimentally obtained values in volunteers caused by inhalative exposure to constant concentrations of n-hexane (300 ppm, 3.33 h) and n-heptane (100 - 500 ppm, 4 h), respectively (background excretion subtracted); (Δ) mean amount of 2,5-hexanedione excreted in urine of workers exposed to an atmospheric concentration of 50 ppm n-hexane (8 h/d, 5 d/w), estimated from data published by Cardona et al. (1993) and Perbellini et al. (1993); for details see discussion. Lines: regressions through the origin.

h each (external exposure doses: 400, 1000 and 2000 ppm·h) led to urinary excretions of 0.15 and 0.40 (2 males), 0.30, 0.33 and 0.41 (2 females, 1 male), and twice to 0.64 (1 female, 1 male) µmol HPDO (individual background values have been subtracted). Using these excretion data (y) a linear regression with the exposure dose (x) of HEP was constructed through the origin (y = a·x; r= 0.874) having a slope of $3.4 \cdot 10^{-4} \pm 0.3 \cdot 10^{-4}$ µmol/(ppm·h) (figure 4 B). Mean half-life of HPDO was 3.4 ± 1.5 h (n = 7). Of totally metabolized HEP during and after HEP exposure, only 0.009% ± 0.004% (n = 7) were excreted as HPDO in urine.

Comparative Reactivity of HDO and HPDO with the ε-Amino Group of Nα-acetyl-L-lysine. The formation of pyrrolyl adducts of both γ-diketones tested with Nα-acetyl-L-lysine followed second order kinetics. The rate constants determined as described above were $0.524 \cdot 10^{-2} \pm 0.075 \cdot 10^{-2}$ l/h/mol (n = 6) for HDO and $0.268 \cdot 10^{-2} \pm 0.033 \cdot 10^{-2}$ l/h/mol (n = 6) for HPDO. This signifies the reactivity of HDO to be almost twice that of HPDO at the physiological pH-value.

DISCUSSION

Rats: Toxicokinetics of HEX and HEP and Urinary Excretion of HDO and HPDO

The main advantage of using SOLVEKIN over a visual fitting of modelled curves through the measured data lies in a mathematically understandable procedure which gives the values of the calculated parameters together with their intervals of confidence. The mean values of the toxicokinetic parameters of HEX in male Wistar rats calculated using SOLVEKIN are somewhat lower than those obtained earlier (Kessler et al. 1989) by visual fitting of model constructed curves through the measured data. However, the previously estimated values are within the ranges given by the standard deviations. Detailed discussions of HEX and HEP toxicokinetics are published in Kessler et al. (1989) and in Csanády et al. (1992). For both compounds two metabolic processes could be distinguished kinetically from the analysis of the gas uptake curves measured in rats. In the same species, two such processes had been observed also for the homologous alkane n-pentane (Filser et al. 1983). The findings are in accordance with in-vitro results of Frommer et al. (1970, 1972 and 1974) and Toftgård et al. (1986): Using rat liver microsomes several NADPH dependent microsomal isozymes could be distinguished which hydroxylate these three alkanes. With the exception of 1-pentanol, all isomeric alcohols were found. Hydroxylation at the C2 position was preferred. For HEX, kinetics in liver microsomes were studied in detail by Toftgård et al. (1986). These authors found two hydroxylation processes characterized by high affinity (low Km1) and low capacity (Vmax1) leading to some 1- and predominantly to 2-hexanol and three processes of low affinity (high Km2) and high capacity (Vmax2) yielding all three hexanols. From HEX hydroxylated by the low affinity processes, more 1- together with 3-hexanol were produced than 2-hexanol at concentrations below 300 μmol/l in the microsomal incubates. From the sum of the in-vitro values of both Vmax1 (SVmax1 = 1.09 nmol/mg/min) and the three Vmax2 (SVmax2 = 6.3 nmol/mg/min) corresponding values for V'max1 and V'max2 in vivo of 20 and 113 μmol/h per rat of 250 g can be calculated using the conversion factors of Kreuzer et al. (1991) for microsomal incubations to in vivo conditions. Although it is not possible to equate Km-values of hydrophobic compounds determined in aqueous microsomal incubations with Kmapp-values representing - according to the underlying toxicokinetic model - average body concentrations in vivo, a comparison between the ratios of Km2 to Km1 in vitro and Kmapp2 to Kmapp1 in vivo seems reasonable. The former are 750 and 183 for 1- and 2-hexanol, respectively. The latter is 163. Obviously, the kinetic parameters gained in vitro match those obtained in vivo from gas uptake studies. Urinary excretion of HDO correlated with the high affinity/low capacity biotransformation of HEX, that of HPDO with the low affinity/high capacity metabolism of HEP (figure 2). These kinetic processes represent the initial hydroxylation steps of HEX and HEP. Consequently, it might be interpreted the rate limiting steps leading to both γ-diketones to be the initial hydroxylation of both alkanes. A prerequisite for the formation of HPDO is the hydroxylation of HEP at the positions ω-1 and ω-2 of the carbon chain whereas oxidative attack at ω-1 positions of HEX is sufficient for HDO. Together with the obtained correlations it can therefore be deduced that ω-1 hydroxylation is represented within the observed concentration range mainly by the high affinity/low capacity process and ω-2 hydroxylation by the low affinity/high capacity one which is less favoured, at least in the lower concentration range. Since hydroxylation at ω-1 is preferred to that at other positions, one should expect more 2,6-heptanedione than HPDO to be formed from HEP. This is the case indeed as has been already demonstrated by Bahima et al. (1984) who exposed female Wistar rats to 2000 ppm HEP for 12 weeks and determined a series of metabolites in urine. HEP was hydroxylated to all isomeric alcohols but mainly at the ω-1 and to a lesser extent at the ω-2 carbon atom. Of all metabolites investigated in urine, HPDO occurred in least amounts.

A comparison of our findings with those of Perbellini et al. (1986) and of Bahima et al. (1984) concerning urinary HPDO excretion following exposure of rats to HEP gives similar results: The former authors exposed male Sprague-Dawley rats for 6 h to 1860 ppm HEP; immediately thereafter, rats were placed in metabolism cages, urine was collected for 24 h, acidified and analyzed for HPDO which was obtained to be 4.4 ± 0.8 μg per rat. The latter authors exposed female Wistar rats over 12 weeks (6 h/d, 5 d/w) to 2000 ppm HEP and collected urine on day 5 of each week for 18 h following end of exposure. HPDO was determined after acidic treatment of urine. The excretion of HPDO, averaged over the 12 weekly determination days, was 2.4 ± 2.0 μg per rat. In both studies, no urine was collected during the exposure periods. Respecting the above obtained half-life of urinary excreted HPDO and the lag-time of 5 h, it can be estimated 16% of totally excreted HPDO to be excreted during the exposure periods of 6 h. Subtracting 16% from the amounts HPDO excreted which can be obtained from the above given relation to the corresponding exposure-time products of HEP, 4.6 and 4.9 μg HPDO should be expected for the exposure conditions of Perbellini et al. (1986) and of Bahima et al. (1984), respectively. These calculated values are similar to those published by the two working groups. Interestingly, the presented calculation is done on free HPDO, determined in untreated, not acidified urine. Acidic treatment of rat urine seems not to increase the yield of HPDO in contrast to the observations made with the HDO excretion resulting from HEX exposure: The urinary HDO content increased about 10 times after urine was treated with acid. Fedtke and Bolt (1987) proved this increase to originate from 4,5-dihydroxy-2-hexanone, another HEX metabolite, which hydrolyses to HDO under acidic conditions.

Humans: Toxicokinetics of HEX and HEP and Urinary Excretion of HDO and HPDO

At steady state, pulmonary retention or the equivalent alveolar retention which is a 3/2 of the former one (Fiserova-Bergerova 1983) is a measure for the rate of metabolism of inhaled compounds (see results). Pulmonary retention of HEX at rest determined in this study at concentrations between 135 and 303 ppm were 23%. Using an open exposure system, too, Veulemans et al. (1982) found a pulmonary retention at rest of 24% and 22% at exposure concentrations of 100 and 200 ppm, respectively. Using a closed exposure system, we determined at HEX concentrations below 2 ppm a "clearance of metabolism related to the atmospheric concentration at steady state" of 2.2 l/min for a human of 79 kg (Filser et al. 1987). This clearance, divided by the alveolar ventilation gives the alveolar retention (Filser et al. 1993). According to Schmidt and Thews (1977), the alveolar ventilation at rest for a reference man of 70 kg is 5 l/min. By means of this value and using for body weight scaling a factor of $(\text{body weight})^{2/3}$ an alveolar retention of HEX of 40.5% equating an pulmonary retention of 27% is obtained for the reference man. These findings do not point to saturation kinetics for the initial biotransformation steps of HEX up to a concentration of 300 ppm since pulmonary retention does not significantly decrease with increasing concentration as would be the case if saturation kinetics became evident. Furthermore, from the low blood:air partition coefficient of 0.8 (Perbellini et al. 1985) it can be concluded that in this concentration range blood flow through the lungs represents the rate limiting step for HEX metabolism (Filser et al. 1994). Therefore, as has been proven by Veulemans et al. (1982), increasing pulmonary ventilation resulting from workload cannot result in a drastic increase of the metabolism of HEX. Further details on the toxicokinetics of HEX are given in Kessler et al. (1989).

Pulmonary retention of HEP determined within an exposure concentration range between 100 and 500 ppm at steady state was 35% ± 7% (n = 7), independently of the atmospheric HEP concentration. From a gas uptake study using a closed exposure system, a "clearance of metabolism related to the atmospheric concentration at steady state" of 3.35 ± 0.82 l/min (n = 3) was obtained for a human of 69 kg, validated at exposure concentrations between 0.15 and 0.4 ppm (Csanády et al. 1992). Using this value and an alveolar ventilation of 5 l/min, a pulmonary retention of 45% ± 11% (n = 3) is calculated. With respect to the standard deviations of both pulmonary retentions, we conclude rate of metabolism of HEP to be proportional to the exposure concentration of up to 500 ppm at least. Within the tested concentration ranges, rates of the initial biotransformation of both compounds HEX and HEP, are limited not by enzymic capacities but by transport to the metabolizing enzymes.

Therefore, the higher pulmonary retention of HEP, compared to that of HEX, can be attributed to its higher blood:air partition coefficient (Filser et al. 1994) which has been determined by Perbellini et al. (1985) to be 1.9. A physiologic toxicokinetic model for uptake of HEP vapors by humans is presented in Csanády et al. (1992).

Background excretion of HDO in acidified urine of non exposed humans was first described by Fedtke and Bolt (1986). The excretion of HDO in untreated urine of human controls was determined from Kezic and Monster (1991) to be 56 ± 23 nmol/l. These values are in the same range as those determined in the present work. At least in part, background excretion of HDO might result from HEX evaporating from fuel tanks during filling. Background excretion of HPDO was about half that of HDO. Since only 0.01% of HEP metabolized following inhalation are excreted as HPDO compared to 0.5% HDO deriving from metabolized HEX, external exposure to environmental HEP vapors alone cannot explain the relatively high HPDO background concentration. An internal source seems probable.

The relation of external HEX burden and urinary excretion of free HDO obtained in this study (figure 4) can be compared with data gained from several cohorts of shoe workers in Spain and in Italy which are published as a result of a collaboration of two working groups: After acid treatment of urine from 189 shoe factory employees (46% males) Cardona et al. (1993) determined a "mean urinary 2,5-HDO concentration of 5.8 mg/l" (i.e., 50.9 μmol/l) resulting from daily inhalation exposure (8 h/d; 5d/w) to 50 ppm HEX. According to Perbellini et al. (1993) the corresponding mean concentration of free HDO determined in urine samples of 92 shoe workers is 8.1% of total HDO. This is equivalent to a mean urinary concentration of free HDO of 4.1 μmol/l or a daily excretion of 4.9 μmol free HDO, taking into account urine production of 1200 ml/d, which is the mean value of adults (males 1400, females 1000 ml/d; International Commission on Radiological Protection 1974). The value computed fits excellently to the regression line in figure 4 derived using our data only.

For the urinary excretion of free HPDO no other values than those presented here were found in the literature.

Pyrrole Formation Rates of HDO and HPDO in Vitro

The molar extinction coefficients (ε) of the condensation products of p-dimethylaminobenzaldehyde with the pyrroles formed as reaction products of HDO and HPDO with Nα-acetyl-L-lysine did not differ very much and were similar to ε of 0.0991 ± 0.0030 l/μmol/cm (n = 6) obtained using dimethylpyrrole as reactant for p-dimethylaminobenzaldehyde. The wave lengths at the absorbance maxima of the three condensation products being 528 nm (derivatives of pyrrolic adducts of Nα-acetyl-L-lysine with HDO and with HPDO, respectively) and 523 nm (dimethylpyrrole derivative), all measured in phosphate buffer (0.15 mol/l, pH 7.5), were also very near to each other (Richter, 1995). The side chains bound to the pyrrolyl amine groups of the condensation products with p-dimethylaminobenzaldehyde hardly seem to influence ε. This conclusion is supported also by findings of Anthony et al (1983b): In dimethylsulfoxide as solvent ε-values of 0.0446 and 0.0445 l/μmol/cm were determined for the p-dimethylaminobenzaldehyde products with 1-hydroxyethyl-2,5-dimethylpyrrole and 1-benzyl-2,5-dimethylpyrrole, respectively. The corresponding absorption maxima were at 530 and 527 nm.

In incubations containing Nα-acetyl-L-lysine together with HDO second order reaction rate constant of pyrrole formation was about twice as high as in those incubations with HPDO as reactant. The difference can be explained by the reactivities of the δ-carbonyl groups of both γ-diketones: The positive induction effect of the methyl residue of HDO is weaker than that of the ethyl residue of HPDO. Therefore, the δ-carbonyl group of HPDO is comparatively less electrophilic and reacts more slowly with nucleophilic amines. Our results agree with data published by DeCaprio et al. (1982) who investigated differences in the reactivities of several diketones with ε-aminocaproic acid by measuring the loss of primary amino groups over time. After identical reaction times, HDO led to a loss, being twice as high as that resulting from HPDO.

Respecting the relation between neurotoxicity of γ-diketones and their pyrrole formation rates, we conclude the neurotoxic potential of HPDO being significantly less than that of HDO.

Comparison between the Urinary Excretion of HDO and HPDO and Between the Risk of Peripheral Neuropathy Resulting from HEX and HEP Exposure of Rat and Man

In rats and in humans, by far more HDO is excreted in urine due to exposure to HEX than HPDO resulting from HEP exposure. Even at exposure conditions to both alkanes according to their MAK values, which are 50 ppm for HEX and 500 ppm for HEP, there are still 6.6 times (rats) and 4.4 times (humans) more HDO excreted than HPDO. The body burden of HDO resulting from exposure of humans to 50 ppm HEX is considered to be too low for any peripherial neurotoxic effect (Deutsche Forschungsgemeinschaft 1982). This conclusion is based particularly on a careful investigation on humans carried out by Sanagi et al. (1980) from which the authors concluded polyneuropathy at HEX exposure concentrations below 100 ppm to be improbable. In rats, exposure to 126 ppm HEX (21 h/d, 7 d/w) for up to 34 weeks (API 1978) or to 106 ppm (12 h/d, 7 d/w) for 24 weeks (Takeuchi et al. 1983) did not lead to neurotoxic effects. However, 200 ppm HEX (12 h/d, 24 w) were high enough to produce peripheral neuropathy (Ono et al. 1982). From these data together with the results presented here on the excretion of HDO, rats and humans seem to be comparable with respect to their sensitivity to HDO.

Although HEP is metabolized to HPDO, which has been proven to produce peripheral neuropathy in rats (O'Donoghue and Krasavage 1979), several studies during which rats were exposed to HEP vapor concentrations up to 3000 ppm did not show any clinical or histopathological changes compared to controls (Bio/dynamics Inc. 1980: 2970 ppm, 6 h/d, 5 d/w, 26 w; Takeuchi et al. 1980, 1981: 3000 ppm, 12h/d, 7d/w, 16 w; Frontali et al. 1981: 1500 ppm, 9 h/d, 5 d/w, 30 w; Bahima et al. 1984: 2000 ppm, 6 h/d, 5 d/w, 12 w). Obviously, the body burden of HPDO was too low to result in neuropathy. This is in accordance with our toxicokinetic findings; only 0.07 μmol/day HPDO are calculated from the above given relationship (compare figure 2) to be excreted by rats exposed over 6 h to 3000 ppm HEP. Of HDO, the compound with the higher neurotoxic potency, a body burden leading to a urinary excretion of 0.6 μmol/day is calculated to have been necessary for the neurotoxic effects observed in the study of Ono et al. (1982) in which rats were exposed to 200 ppm HEX (12 h/d). A body burden of HPDO yielding an urinary HPDO excretion of 0.6 μmol/day would require a daily 12 h exposure of rats to 13000 ppm HEP, a concentration within the range of those at which acute toxicity leading to anaesthesia and death were observed in mice (Fühner 1921, Lazarew 1929). Furthermore, in urine of rats exposed to HEX and HEP, respectively, pyrroles were determined which might be considered as internal biomarkers of the neurotoxic potencies of the metabolic precursors HEX and HEP: Exposure to 2750 ppm HEP over 8 h resulted in an urinary excretion of half as much pyrroles as were found from exposure to 50 ppm HEX over the same time span (Störmer et al. 1993). Therefore, we conclude that it would hardly be possible to ever demonstrate peripheral neurotoxic effects in rats induced by pure HEP.

For humans, no data on peripheral neurotoxic effects of pure HEP are published. In the future, also, we do not expect peripheral neurotoxicity from exposure to the pure compound at the current MAK value of 500 ppm HEP, since there is a safety factor of more than 4 compared to the neurotoxic efficiency of HEX at its MAK value of 50 ppm.

ACKNOWLEDGMENTS

The authors are grateful to Prof. Dr. L. Perbellini, University of Padua-Verona, for kindly supplying us with a sample of 2,5-heptanedione, to Dr. H. Braun, Technische Universität München, for the literature research concerning synthesis of 2,5-heptanedione and to Mr. C. Pütz for his excellent technical assistance. This work was financially supported by the Berufsgenossenschaft der Chemischen Industrie (Heidelberg, Projekt Nr. 134, n-Heptan).

REFERENCES

Anthony D. C., Boekelheide K., and Graham D. G., 1983a, The effect of 3,4-dimethyl substitution on the neurotoxicity of 2,5-hexanedione. I. Accelerated clinical neuropathy is accompanied by more proximal axonal swellings, *Toxicol. Appl. Pharmacol.* 71: 362-371.

Anthony D. C., Boekelheide K., Anderson C. W., and Graham D. G., 1983b, The effect of 3,4-dimethyl substitution on the neurotoxicity of 2,5-hexanedione. II. Dimethyl substitution accelerates pyrrole formation and protein crosslinking, *Toxicol. Appl. Pharmacol.* 71: 372-382.

API, 1978, 26 week inhalation toxicity study of n-hexane in the rat, Washington DC, American Petroleum Institute, *API Medical Research Report* No 28-30077.

Bahima J., Cert A., and Menéndez-Gallego M., 1984, Identification of volatile metabolites of inhaled n-heptane in rat urine, *Toxicol. Appl. Pharmacol.* 76: 473-482.

Bio/dynamics Inc., 1980, A 26-week inhalation toxicity study of heptane in the rat, *Project No 78-7233*, Bio/dynamics Inc., Division of Biology and Safety Evaluation, East Millstone, NJ, USA.

Cardona A., Marhuenda D., Martí J., Brugnone F., Roel J., and Perbellini L., 1993, Biological monitoring of occupational exposure to n-hexane by measurement of urinary 2,5-hexanedione, *Int. Arch. Occup. Environ. Health* 65: 71-74.

Couri D., and Milks M., 1982, Toxicity and metabolism of the neurotoxic hexacarbons n-hexane, 2-hexanone, and 2,5-hexanedione, *Ann. Rev. Pharmacol. Toxicol.* 22: 145-166.

Csanády G. A., and Filser J. G., 1990, SOLVEKIN: a new program for solving pharmaco- and toxicokinetic problems, Presented on the International Workshop on Pharmacokinetic Modelling in Occupational Health, March 4-8, Leysin, Switzerland.

Csanády Gy., Kessler W., and Filser J. G., 1992, Inhalationskinetik von n-Heptan bei Ratte und Mensch, *Verhandlungen der Deutschen Gesellschaft für Arbeitsmedizin und Umweltmedizin e. V.*, 32. Jahrestagung, Genter Verlag, Stuttgart, 613-616.

DeCaprio A. P., and Weber P., 1980, In vitro studies on the amino group reactivity of a neurotoxic hexacarbon solvent, *Pharmacologist* 22: 222.

DeCaprio A. P., Olajas E. J., and Weber P., 1982, Covalent binding of a neurotoxic n-hexane metabolite: Conversion of primary amines to substituted pyrrole adducts by 2,5-hexanedione, *Toxicol. Appl. Pharmacol.* 65: 440-450.

DeCaprio A. P., Strominger N. L., and Weber P., 1983, Neurotoxicity and protein binding of 2,5-hexanedione in the hen, *Toxicol. Appl. Pharmacol.* 68: 297-307.

DeCaprio A. P., Briggs R. G., Jackowski S. J., and Kim J. C. S., 1988, Comparative neurotoxicity and pyrrole-forming potential of 2,5-hexanedione and perdeuterio-2,5-hexanedione in the rat, *Toxicol. Appl. Pharmacol.* 92: 75-85.

Deutsche Forschungsgemeinschaft, 1982, n-Hexan, in: Gesundheitsschädliche Arbeitsstoffe. Toxikologisch-Arbeitsmedizinische Begründungen von MAK-Werten. 11. Lieferung (ed.: D. Henschler), VCH Verlagsgesellschaft Weinheim.

Deutsche Forschungsgemeinschaft, 1994, List of MAK and BAT values 1994: Maximum concentrations and biological tolerance values at the workplace, VCH Verlagsgesellschaft Weinheim.

DGMK (Deutsche Gesellschaft für Mineralölwissenschaft und Kohlechemie e. V.), 1986, Effects of n-heptane on man and animals, DGMK-Project 174-3, Hamburg.

Dietz W. G., 1994, 2,5-Hexandion und Pyrrole im Urin des Menschen: Analytische Bestimmung und Untersuchung der Ausscheidung nach Exposition gegen n-Hexan, GSF-Bericht 7/94, (ed.: GSF - Forschungszentrum für Umwelt und Gesundheit, GmbH), Neuherberg, FRG.

Doerffel K., 1990, Statistik in der analytischen Chemie, 5th ed., Dt. Verlag f. Grundstoffind., Leipzig.

Fedtke N., and Bolt H. M., 1986, Detection of 2,5-hexanedione in the urine of persons not exposed to n-hexane, *Int. Arch. Occup. Environ. Health* 57: 143-148.

Fedtke N., and Bolt H. M., 1987, The relevance of 4,5-dihydroxy-2-hexanone in the excretion kinetics of n-hexane metabolites in rat and man, *Arch. Toxicol.* 61: 131-137.

Filser J. G., Bolt H. M., Muliawan H., and Kappus H., 1983, Quantitative evaluation of ethane and n-pentane as indicators of lipid peroxidation in vivo, *Arch. Toxicol.* 52: 135-147.

Filser J. G., Peter H., Bolt H.M., and Fedtke N., 1987, Pharmacokinetics of the neurotoxin n-hexane in rat and man, *Arch. Toxicol.* 60: 77-80.

Filser J. G., 1992, The closed chamber technique - uptake, endogenous production, excretion, steady-state kinetics and rates of metabolism of gases and vapors, *Arch. Toxicol.* 66: 1-10.

Filser J. G., Schwegler U., Csanády Gy. A., Greim H., Kreuzer P. E., and Kessler W., 1993, Species-specific pharmacokinetics of styrene in rat and man, *Arch. Toxicol.* 67: 517-530.

Filser J. G., Csanády Gy. A., Kessler W., Kreuzer P. E., Störmer A., and Greim H., 1994, Interspecies extrapolation from rodents to humans demonstrated on selected industrial chemicals. *ISSX Proceedings* 6: 38.

Fiserova-Bergerova V., 1983, Modeling of metabolism and excretion in vivo. In: Modeling of inhalation exposure to vapors: Uptake, distribution, and elimination. Vol. I (ed. V. Fiserova-Bergerova), CRC Press, Boca Raton, Florida.

Frommer U., Ullrich V., and Staudinger H., 1970, Hydroxylation of aliphatic compounds by liver microsomes I. The distribution pattern of isomeric alcohols. *Hoppe-Seyler's Z. Physiol. Chem.* 351: 903-912.

Frommer U., Ullrich V., Staudinger H., and Orrenius S., 1972, The monooxygenation of n-heptane by rat liver microsomes, *Biochim. Biophys. Acta* 280: 487-494.

Frommer U., Ullrich V., and Orrenius S., 1974, Influence of inducers and inhibitors on the hydroxylation pattern of n-hexane in rat liver microsomes, *FEBS Letters* 41: 14-16.

Frontali N., Amantini M. C., Spagnolo A., Guarcini A. M., Saltari M. C., Brugnone F., and Perbellini L., 1981, Experimental neurotoxicity and urinary metabolites of the C5-C7 aliphatic hydrocarbons used as glue solvents in shoe manufacture, *Clin. Toxicol.* 18: 1357-1367.

Führer H., 1921, Die narkotische Wirkung des Benzins und seiner Bestandteile (Pentan, Hexan, Heptan, Oktan), *Biochem. Z.* 115: 235-261.

Genter M. B., Szakal-Quin G., Anderson C. W., Anthony D. C., and Graham D. G., 1987, Evidence that pyrrole formation is a pathogenic step in gamma-diketone neuropathy, *Toxicol. Appl. Pharmacol.* 87, 351-362.

Genter St.Clair M. B., Amarnath V., Moody M. A., Anthony D. C., Anderson C. W., and Graham D. G., 1988, Pyrrole oxidation and protein cross-linking as necessary steps in the development of γ-diketone neuropathy, *Chem. Res. Toxicol.* 1: 179-185.

Graham D. G., Anthony D. C., Boekelheide K., Maschmann N. A., Richards R. G., Wolfram J. W., and Shaw B. R., 1982, Studies of the molecular pathogenesis of hexane neuropathy. II. Evidence that pyrrole derivatization of lysyl residues leads to protein crosslinking, *Toxicol. Appl. Pharmacol.* 64: 415-422.

Graham D. G., Amarnath V., Valentine W. M., Pyle S. J., and Anthony D. C., 1995, Pathogenetic studies of hexane and carbon disulfide neurotoxicity, *Crit. Rev. Toxicol.* 25: 91-112.

Hallier E., Filser J. G., and Bolt H. M., 1981, Inhalation pharmacokinetics based on gas uptake studies. II. Pharmacokinetics of acetone in rats, *Arch. Toxicol.* 47: 293-304.

International Commission on Radiological Protection, 1974, Report of the task group on reference man, Pergamon Press, Oxford.

Kessler W., Denk B., and Filser J. G., 1989, Species-specific inhalation pharmacokinetics of 2-nitropropane, methyl ethyl ketone, and n-hexane, In: Biologically based methods for cancer risk assesment (ed. C. C. Travis), NATO ASI Series, Plenum Press, New York, London, 123-139.

Kessler W., Heilmaier H., Kreuzer P., Shen J. H., Filser M., and Filser J. G., 1990, Spectrophotometric determination of pyrrole-like substances in urine of rat and man: An assay for the evaluation of 2,5-hexanedione formed from n-hexane, *Arch. Toxicol.* 64: 242-246.

Kezic S., and Monster A. C., 1991, Determination of 2,5-hexanedione in urine and serum by gas chromatography after derivatization with O-(pentafluorobenzyl)-hydroxylamine and solid-phase extraction, *J. Chromatogr.* 563: 199-204.

Krasavage W. J., O'Donoghue J. L., DiVincenzo G. D., and Terhaar C. J., 1980, The relative neurotoxicity of methyl n-butyl ketone, n-hexane and their metabolites, *Toxicol. Appl. Pharmacol.* 52: 433-441.

Kreuzer P. E., Kessler W., Welter H. F., Baur C., and Filser J. G., 1991, Enzyme specific kinetics of 1,2-epoxybutene-3 in microsomes and cytosol from livers of mouse, rat, and man, *Arch. Toxicol.* 65: 59-67.

Laib R. J., Tucholski M., Filser J. G., and Csanády G. A., 1992, Pharmacokinetic interaction between 1,3-butadiene and styrene in Sprague-Dawley rats, *Arch. Toxicol.* 66: 310-314.

Lazarew N. W., 1929, Über die Giftigkeit verschiedener Kohlenwasserstoffdämpfe, *Arch. Exp. Pathol. Pharmacol.* 143: 223-233.

Mattocks A. R., and White I. N. H., 1970, Estimation of metabolites of pyrrolizidine alkaloids in animal tissues, *Anal. Biochem.* 38: 529-535.

O'Donoghue J. L., and Krasavage W. J., 1979, The structure-activity relationship of aliphatic diketones and their potential neurotoxicity, *Toxicol. Appl. Pharmacol.* 48: A55.

Ono Y., Takeuchi Y., Hisanaga N., Iwata M., Kitoh J., and Sugiura Y., 1982, Neurotoxicity of petroleum benzine compared with n-hexane, *Int. Arch. Occup. Environ. Health* 50: 219-229.

Perbellini L., Brugnone F., Caretta D., and Maranelli G., 1985, Partition coefficients of some industrial aliphatic hydrocarbons (C5-C7) in blood and human tissues, *Br. J. Ind. Med.* 42: 162-167.

Perbellini L., Brugnone F., Cocheo V., De Rosa E., and Bartolucci G. B., 1986, Identification of the n-heptane metabolites in rat and human urine, *Arch. Toxicol.* 58: 229-234.

Perbellini L., Pezzoli G., Brugnone F., and Canesi M., 1993, Biochemical and physiological aspects of 2,5-hexanedione: endogenous or exogenous product? *Int. Arch. Occup. Environ. Health* 65: 49-52.

Richter M., 1995, In-vitro-Untersuchungen zur Pyrrolbildungsgeschwindigkeit von 2,5-Hexandion und 2,5-Heptandion mit Nα-Acetyl-L-lysin als Voraussetzung für eine vergleichende Abschätzung der neurotoxischen Potentiale beider γ-Diketone, Dissertation an der Fakultät für Medizin der Technischen Universität München.

Sachs L., 1974, Angewandte Statistik. 4th ed., Springer Verlag, Berlin.

Sanagi S., Seki Y., Sugimoto K., and Hirata M., 1980, Peripheral nervous system functions of workers exposed to n-hexane at a low level, *Int. Arch. Occup. Environ. Health* 47: 69-79.

Sayre L. M., Shearson C. M., Wongmongkolrit T., Medori R., and Gambetti P., 1986, Structural basis of γ-diketone neurotoxicity: Non-neurotoxicity of 3,3-dimethyl-2,5-hexanedione, a γ-diketone incapable of pyrrole formation, *Toxicol. Appl. Pharmacol.* 84: 36-44.

Schmidt R. F., and Thews G., 1977, Physiologie des Menschen. 19th ed., Springer Verlag, Berlin.

Spencer P. S., Schaumburg H. H., Sabri M. I., and Veronesi B. V., 1980, The enlarging view of hexacarbon neurotoxicity, *Crit. Rev. Toxicol.* 7: 279-356.

Stetter H., and Kuhlmann H., 1975, Addition von Aldehyden an aktivierte Doppelbindungen: VII; Eine neue einfache Synthese von cis-Jasmon und Dihydrojasmon, *Synthesis* 379-380.

Stetter H., and Kuhlmann H., 1976, Addition von Aldehyden an aktivierte Doppelbindungen, XI; Addition aliphatischer, heterocyclischer und aromatischer Aldehyde an Butenon, *Chem. Ber.* 109: 3426-3431.

Störmer A., Kreuzer P. E., Kessler W., Csanády Gy. A., and Filser J. G., 1993, Bestimmung von Pyrrolen im Urin von Ratten nach Exposition gegen n-Heptan und n-Hexan - ein Mittel zum Vergleich des neurotoxischen Potentials beider Kohlenwasserstoffe, *Verhandlungen der Deutschen Gesellschaft für Arbeitsmedizin und Umweltmedizin e. V.*, 33. Jahrestagung, Genter Verlag, Stuttgart, 481-486.

Störmer A., Kessler W., and Filser J. G., 1994, 2,5-Heptandion im Urin des Menschen nach Exposition gegen n-Heptan - Vergleich der neurotoxischen Potentiale von n-Heptan und n-Hexan, *Verhandlungen der Deutschen Gesellschaft für Arbeitsmedizin und Umweltmedizin e. V.*, 34. Jahrestagung, Genter Verlag, Stuttgart, 363-365.

Störmer A., 1995, Pyrrole und 2,5-Heptandion im Urin der Ratte und 2,5-Heptandion im Urin des Menschen: Analytische Bestimmung der Ausscheidung nach Exposition gegen n-Heptan, Dissertation an der Fakultät für Chemie, Biologie und Geowissenschaften der Technischen Universität München.

Takeuchi Y., Ono Y., Hisanga N., Kitoh J., and Sugiura Y., 1980, A comparative study on the neurotoxicity of n-pentane, n-hexane, and n-heptane in the rat, *Br. J. Ind. Med.* 37: 241-247.

Takeuchi Y., Ono Y., Hisanga N., Kitoh J., and Sugiura Y., 1981, A comparative study on the toxicity of n-pentane, n-hexane, and n-heptane to the peripheral nerve of the rat, *Clin. Toxicol.* 18: 1395-1402.

Takeuchi Y., Ono Y., Hisanga N., Iwata M., Aoyama M., Kitoh J., and Sugiura Y., 1983, An experimental study of the combined effects of n-hexane and methyl ethyl ketone, *Br. J. Ind. Med.* 40: 199-203.

Toftgård R., Haaparanta T., Eng L., and Halpert J., 1986, Rat lung and liver microsomal cytochrome P-450 isozymes involved in the hydroxylation of n-hexane, *Biochem. Pharmacol.* 35: 3733-3738.

Veulemans H., can Vlem E., Janssens H., Masschelein R., and Leplat A., 1982, Experimental human exposure to n-hexane, *Int. Arch. Occup. Environ. Health* 49: 251-263.

THE USE OF DATA ON BIOLOGICALLY REACTIVE INTERMEDIATES IN RISK ASSESSMENT

J. Robert Buchanan and Christopher J. Portier

Laboratory of Quantitative and Computational Biology
National Institute of Environmental Health Sciences
Research Triangle Park, North Carolina 27709

INTRODUCTION

As defined by the US National Academy of Sciences (1994), risk assessment is "the evaluation of scientific information on the hazardous properties of environmental agents and on the extent of human exposure to those agents". The goal of risk assessment is a scientifically defensible regulatory exposure level (Pease *et al.*, 1991). The eventual product of any risk assessment should be the probability that populations exposed to the agent will be harmed and to what degree. In achieving this goal, three rather disjoint elements must cooperate and attempt to understand each other; science, politics and policy. The scientific element of risk assessment is usually divided into the three basic categories of hazard identification, dose-response modeling and risk characterization. Hazard identification is the process of determining if there exists a health risk from exposure to an agent and determining that there is exposure to the same agent. Dose-response modeling then uses the available data on exposure and hazard to quantify the risk for varying levels of exposure. Risk characterization concludes the scientific component of the risk assessment by summarizing the degree of risk in populations exposed to the agent.

In order for a risk assessment to be useful and valid, the risk and exposure estimates must be understood by the political community so that an effective policy can be formulated and implemented. The scientific characterization of risk is subject to interpretation in the political arena. Issues such as risk perception, economic factors and the current political climate can have a strong impact on the eventual management of that risk. The political and social issues involved in this component of the risk assessment process are beyond the purview of this discussion and will not be mentioned further.

Any scientifically valid risk assessment can be characterized as a set of mathematical models. This does not pertain solely to dose-response modeling, in which there is an obvious use of mathematical models, but also in the areas of hazard identification and risk characterization where the mathematical models which describe areas of statistics and probability figure prominently. Thus, to understand the scientific basis of risk assessment and to

Biological Reactive Intermediates V
Edited by R. Snyder et al., Plenum Press, New York, 1996

understand how mechanistic data can be used in risk assessment, one needs at least a basic understanding of the development and use of mathematical models in addition to an understanding of the scientific data being modeled.

Mathematical models are a collection of assumptions and formulae used as aids in describing and understanding biological observations. For the purposes of risk assessment, these models can be broken into four basic classes; models of exposure, models of distribution and metabolism, models of effect and statistical models. Models of exposure cover the transport and fate of environmental agents to the point of delivery to the human body. Models of distribution and metabolism concern the transport of the agent into, within and eventually out of the human body, the biochemical reactions that agent undergoes and the distribution and excretion of the reaction products. Models of effect deal directly with the toxicity of the agent covering such concerns as cellular viability, changes in tissue function and structure, failure or modification of organ systems and eventually morbidity and/or mortality to the exposed human. Statistical models are critical in relating the scientific data (observed environmental exposures, laboratory studies, epidemiological data, etc.) to the parameters which make up the other three model classes. All of the statistical methods routinely used in analyzing data and estimating model parameters are based upon a series of mathematical models and assumptions which make them optimal for certain types of data.

Data on biologically reactive intermediates are of increasing importance in the scientific component of risk assessment. The models which directly relate to these data fall into the class of models of distribution and metabolism of the agent. Reactive metabolites are generally used as dose-surrogates in models relating to cellular effects.

Statistical methods ranging from the estimations of simple means and variances to more complicated methods like nonlinear least-squares estimation and likelihood-maximization are used to estimate rate constants from data on metabolites. The rate of delivery to humans, derived from the exposure models, can have a strong effect on which metabolites are formed and under what conditions. Finally, relating the production of reactive intermediates to expressions of toxicity is critical to the use of these data in risk assessment since one must quantify the effects due to metabolites in establishing risk estimates.

Current attempts to improve the scientific validity of risk assessments are focused on a number of themes including the identification of default options, the identification of the data which is most relevant for risk assessment, quantification of interindividual variability, understanding the interactions which occur from exposure to multiple agents, validation of the methods and models used in risk assessment and characterization of the scientific uncertainty in a risk assessment. In what follows, these themes will be discussed in the context of the utility of toxicokinetic data (data on reactive intermediates are part of this class of data) in risk assessment. This discussion will focus on how some of these themes relate to the models and the data pertaining to toxicokinetics. Some of the current research being done to address these issues is detailed.

TOXICOKINETIC DATA IN RISK ASSESSMENT

The biologically effective dose (BED) refers to an internal measure of exposure which is directly related to the exposure and the risk being studied. Examples of BED's routinely used in risk assessment include the tissue concentration of an agent, the concentration of some final metabolic product of that agent or the concentration of some biologically reactive intermediate. It is generally assumed that the risk of a toxic effect is related to the area under the time-concentration curve (AUC) for the BED for rodents and for humans. This is an assumption which is used routinely, but seldom noted in risk assessments in which a BED is calculated as part of the process. Alternatives to the AUC include the maximum concentration or average concentration of parent chemical or metabolite in a target organ. A basic outline for performing risk assessments using BED's follows.

The initial step in any risk assessment is to study the underlying toxic/carcinogenic process (if known) for the agent being evaluated. Data in the literature relating to the mechanism of action for the agent must be collected, analyzed and assessed concerning its overall quality and the suitability for use in quantifying risks. In many cases, data are pertinent to aiding in describing the mechanism of action of a toxin/carcinogen but do not contribute to quantification of this mechanism. These qualitative theories generally must be discarded in favor of theories or parts of the theories which can be quantified. A good example of a detailed theory which cannot be used for quantitative risk assessment is that of 2,3,7,8-tetrachlorodibenzo-p-dioxin (dioxin) as outlined in Lucier et al., 1993. This theory details the series of events which occur as dioxin enters the liver, binds to a number of liver proteins and modifies expression of many proteins in the liver. Events including binding to the Ah-receptor, loss of heat shock proteins, acquisition of transport proteins, binding to DNA, formation of M-RNA and expression of proteins have been characterized, but few have been quantified and theoretical models relating concentration of dioxin to these events cannot be constructed. In a case like this, use of one or two of the key components of this cascade are the only practical solution for estimating risks.

Once a mechanism (presumed or verified) of action has been identified, the next step is to model the relationship between exposure dose and the BED in the test species. Toxicokinetic modeling plays an important role in utilizing this relationship for risk assessment by describing the absorption, distribution, metabolism and elimination of the environmental agents. One of the most common models used in this area is the physiologically-based pharmacokinetic (PBPK) model (Andersen *et al.*, 1987). PBPK models provide a uniform description of the dynamics of chemicals and their metabolites in blood, specific tissues of interest and in urine/feces. They are becoming widely used in risk assessment due to their ability to provide a common framework for modeling chemical dynamics across species.

A continuing problem with these models is a lack of biological realism in the biochemistry of the cell (Kohn and Portier, 1994). This is especially important when considering reactive intermediates in the model. Common practice in using PBPK models for risk assessment is to include a generalized "metabolism" component to the model without regard for quantitative evidence on reactive intermediates and final metabolic products.

Greater realism in PBPK modeling will require incorporation of saturation of biochemical processes, cofactor depletion, hormonal effects, endocrine pathways and the inclusion of effects due to aging. The data needed to support this expansion will be diverse and varied, but if care is taken to obtain levels of constitutive expression of key enzymes and hormones in control animals/populations, the task of considering biochemical changes in the normal animal/human will be simplified. Finally, care should be taken to consider the variability in predictions from these models due to uncertainty in the form of the model (misspecification) and the values of model parameters.

The third step in a risk assessment using a BED is to characterize the relationship between the BED and the risk of a significant health outcome such as cancer. In general, this step is done following characterization of the underlying toxic/carcinogenic process of the chemical being studied. This is not required and, in fact, is a reduction in the "statistical information" available from the data. Although difficult, it would be better to characterize the PBPK model and the toxic effects model simultaneously.

The fourth step is to relate levels of the BED with expressed toxic/carcinogenic effects. The no observed adverse effects level (NOAEL) is an important level of BED. Through the use of the original PBPK model (if it is for human data) or an extrapolated model (from the animal PBPK model to a human model), the exposure dose which achieves this target level of BED in humans can be estimated. In a similar manner the exposure dose

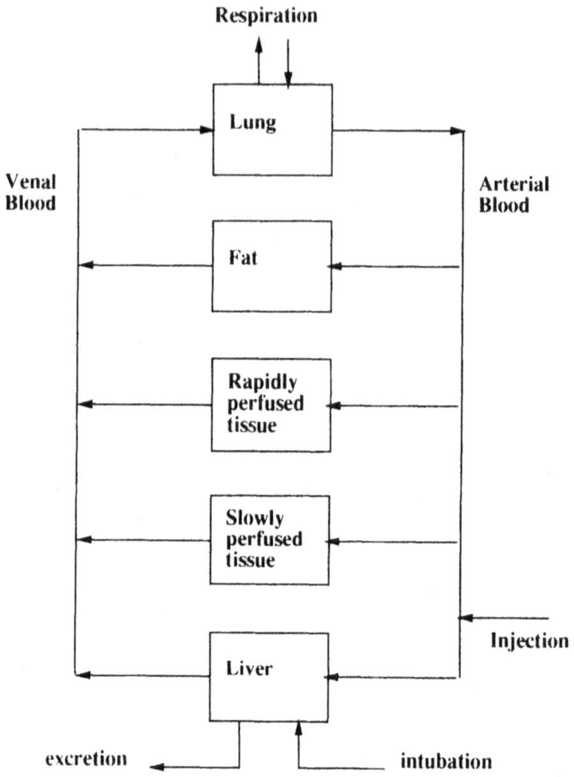

Figure 1. A Simple PBPK model.

which produces a specified increased individual health risk above background levels can be estimated.

MODELING TOXICOKINETIC DATA USING PBPK MODELS

A PBPK model divides the body into physiologically similar compartments. Each compartment represents an organ, tissue or group of tissues. These compartments are linked by the blood using both arterial and venal components. Figure 1 illustrates a simple PBPK model.

PBPK models are characterized using physiological parameters (e.g., blood flow rates, body and organ weights) and biochemical parameters (e.g., partition coefficients, metabolic rates). The compartments and parameters form a mathematical description of the interaction of animal and chemical which consists of a system of differential equations with constraints forcing mass balance. The basic organ compartment of a PBPK model is illustrated by Figure 2.

A tissue is characterized by its size, U_i, and perfusion rate, Q_i. A compound enters a tissue via the arterial blood, $A(t)$, while the compound and possibly metabolites are removed via the venal blood, $V_i(t)$. Chemicals can also enter tissues through means such as absorption and leave tissues via excretion mechanisms. These processes are denoted $Y_i(t)$ and $Z_i(t)$, respectively, in Figure 2. The mass balance for the typical organ in a PBPK model is given by the differential equation

Figure 2. A general organ compartment in a PBPK model where $C_i(t)$ is the concentration of compound in the tissue described by the compartment, U_i is the volume of the tissue, $A(t)$ is the concentration of the compound in the arterial blood, $V_i(t)$ is the concentration of the compound in the venal blood leaving the tissue, Q_i is the perfusion rate of the tissue, $Y_i(t)$ is the amount of compound entering the tissue other than via the arterial blood, and $Z_i(t)$ is the amount of compound leaving the tissue other than via the venal blood.

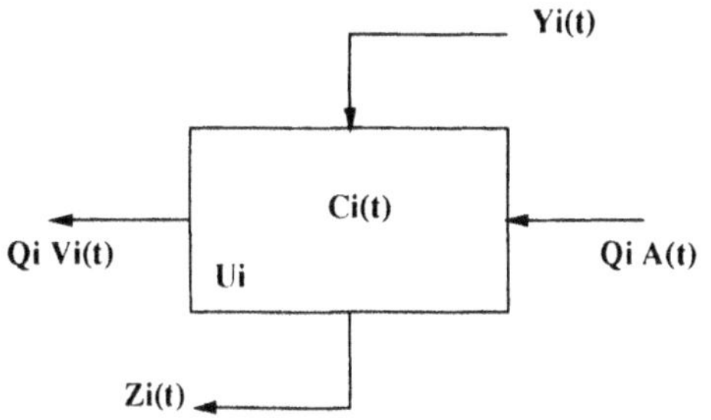

$$U_i \frac{dC_i}{dt} = Q_i(A(t) - V_i(t)) + \frac{dY_i}{dt} - \frac{dZ_i}{dt}.$$

The choice of tissues to include in a model depends on the chemical under study. Administration of radiolabeled chemical and analytical identification of parent and metabolites in some tissues and excreta can highlight organs or tissues where significant sequestration, elimination, or metabolism take place. The body can also be divided into tissue classes based on the time required to replace the volume of blood in the tissue, U_i/Q_i. When this ratio is small as for example in the liver and kidneys, the tissue belongs to the class of rapidly perfused tissues. Tissues such as fat and skin belong to the slowly perfused class. Since the liver is a primary site of metabolism, it is usually modeled as a separate organ. For volatile compounds or the inhalation route of exposure the lung compartment is modeled separately. The organ and tissue volumes, their fractions of body weight, and perfusion rates of the lung, liver, fat, rapidly and slowly perfused tissues form a set of common values present in most PBPK models.

Biotransformation and elimination processes are chemical and tissue dependent. Metabolism rates are usually measured using *in vitro* techniques. Elimination processes are frequently expressed as first order linear or saturable nonlinear kinetics.

$$\frac{dZ_i}{dt} = k_i C_i(t)$$

$$\frac{dZ_i}{dt} = \frac{V_{max} C_i(t)}{k_m + C_i(t)}$$

In a first order linear process, a chemical is eliminated at a rate proportional to its concentration. In a saturable nonlinear process the elimination rate, as expressed in the Michaelis-Menten function above, is approximately proportional to concentration at low concentration, but approaches a maximum of V_{max} at high concentrations. Entry of a compound into a tissue can be described by first order linear absorption, saturable nonlinear absorption, constant rate infusion, Dirac delta function for rapid IV injection, or time-dependent feeding/drinking activity. For some tissues, one or both of Y_i and Z_i may be zero. In general the concentration of compound in the venal blood leaving a tissue is not identical to the concentration in the tissue. The ratio of these concentrations is assumed constant. This ratio is the partition coefficient and describes the relative affinities of a compound for the blood and the tissue. Thus in equation(1), $V_i(t)$ can be replaced by $C_i(t)/P_i$ where P_i is the tissue:blood partition coefficient. Estimation or measurement of rates associated with chemical administration, metabolism, partitioning, and elimination are important to the development of a PBPK model.

If a PBPK model is intended to simulate the concentration of parent chemical and metabolic products in the tissues, then linked submodels are used for each chemical or metabolite to be modeled. Once the physiological and biochemical parameters of the model are specified the model can be used to predict time dependent parent chemical and metabolite concentrations in each tissue included in the model. The reliability and validity of a PBPK model is often determined by comparing model predictions against data sets from animal experiments utilizing a range of dose levels and administration routes (Corley *et al.*, 1990). The effects of uncertainty in parameter values are sometimes investigated numerically by selecting values from a set with normal distribution and specified mean and standard deviation. The mean and standard deviation can be derived from physical/chemical measurements of model parameters or, in the absence of a dataset of repeated measurements, the accepted parameter value with a standard deviation of 10% is often used. Pharmacokinetic modeling with randomly sampled kinetic parameters can reveal differences in response to chemical exposure due to interindividual variability (see for example Portier and Kaplan, 1989; Hattis and Burmaster, 1994). Deliberate and planned alteration of model parameters can provide insight into differences in exposure response in subpopulations for whom the average population's parameter values are not characteristic. For example, decreasing the kidneys' elimination rate for a compound and observing the concomitant effect this change has on tissue concentrations of the compound could provide insight into exposure responses of aged individuals or individuals suffering from kidney disease. Exposure effects on sensitive subpopulations could be investigated by modification of model parameters. For example, since the conjugating capacity of newborns and infants is lower than in adults, then lowering the baseline glutathione level in a model may provide insight into age-related differences to toxic effect due to chemical exposure, particularly if the suspected toxic agent is a reactive metabolite rather then the parent chemical.

USING PBPK MODELS IN RISK ASSESSMENT

A validated PBPK model for humans or an extrapolated model from animals can be used in setting guidelines for human chemical exposure levels. Generally dose surrogates are evaluated for their correlation with the dose-response of tissue toxicity. In cases in which there is the formation of reactive metabolites, toxicity may correlate with the maximum value of metabolite concentration, average value, or cumulative metabolite formation in the target organ. For example, Anderson et al (1987) used the cumulative formation of unnamed metabolite formed via the mixed function oxidase pathway in the liver as the effective dose surrogate for liver tumor formation following exposure to dichloromethane. Reactive metabolites stable enough to leave the tissue in which they form may cause toxicity at remote sites. For example, the naphthalene oxide metabolite of naphthalene is sufficiently stable to diffuse across intact hepatic membranes and travel via the venal blood to the lungs where it causes selective toxicity of the Clara cells in mice (Buckpitt and Warren, 1983; Kanekal *et al.*, 1991; Cho *et al.*, 1994). Another dose surrogate could be binding of metabolites to cellular proteins. For instance formation of reactive chloroform metabolites leads to binding of cellular macromolecules and ultimately cell death (Corley *et al.*, 1990). Route-to-route extrapolation using PBPK models and monitoring of dose surrogates may give insight into the relationship between level of toxicity and route of exposure.

Choice of the proper measure to use as a dose-surrogate is generally hampered by a lack of data aimed at evaluating that question. For example, to determine if the maximal level of reactive intermediate is more important than the cumulative formation, one would need a toxicity study with numerous times of observation and a variety of dosing schemes. This is seldom available and the choice of dose-surrogate is based upon assumptions about the underlying biology. In most cases, the cumulative formation of reactive intermediate is used as it is assumed that each additional amount of reactive intermediate formed increases the risk proportionately. However, in cases such as reactive intermediate induced cytotoxicity, the relationship between effect and metabolite may not be linear in cumulative formation but will be more closely related to peak values. This must be evaluated on a case-by-case basis (Hattis, 1990).

There are several strategies for setting acceptable exposure levels based on animal data and mathematical models. In using PBPK models and animal data with any of these strategies, one must assume that there exists some internal level of exposure in rodents and in humans which, when modeled against the toxic effect, yields the same model. With this assumption, the species differences in response are then a function of the species differences in the formation of that internal level of exposure. Often, this internal level of exposure will simply be the cumulative formation of some reactive intermediate.

One of the simplest strategies for setting acceptable exposure levels involves dividing the lowest dose observed to yield no effect (NOAEL) by uncertainty factors (UFs) to account for variations in intraspecies sensitivity, extrapolation from animals to humans (if the data or model are not derived for humans), and extrapolation from subchronic to chronic exposure scenarios (if chronic exposures were not tested). The number of UFs and their magnitudes will depend on the investigators' confidence in the adequacy of the animal study for predicting human response, the design of the animal study, and the quality of the data. Often it is assumed the NOAEL must be the administered chemical, but this may be inappropriate if the putative agent is a metabolite. PBPK models incorporating the metabolic pathway of the chemical can estimate the level of metabolites and correlate their levels with toxicity/carcinogenicity. In this way a NOAEL for the metabolite rather than the parent chemical can be established. One limitation of this approach is the lack of an estimate of the degree of risk posed by the regulatory level instead focusing on the idea that, by use of the UF's, hazard can be avoided all together. In addition, this method is highly dependent upon the experimental design and fails to utilize the trend seen in the overall dose-response data.

The benchmark dose (BD) is another exposure level sometimes calculated as part of a risk assessment and has been proposed by several groups (e.g. IRLG, 1979; Crump, 1984). The BD is a lower confidence limit on a dose inducing a specified (e.g., 10%) increase in risk of toxicity or carcinogenicity. An important aspect of the BD calculation is the mathematical modeling of the dose-response curve. As stated for the NOAEL/UF approach, a dose surrogate derived from a PBPK model rather than the administered dose can be used when modeling the dose-response curve. Most of the mathematical curves used for modeling dose-response share the property of linear or sub-linear response in the low-dose region. The Weibull model is frequently used (Van Ryzin, 1980).

$$P(dose) = P_o + (1 - P_o)(1 - \exp(-(dose / d)^{\beta}))$$

where P is the probability of toxicity/carcinogenicity in an exposed group, P_0 is the background incidence of the health effect, β is the Weibull exponent (sometimes constrained to be ≥1 to ensure linearity or sub-linearity at low doses), and d is the dose or dose surrogate giving an extra risk of 63% over background. The Weibull exponent can be fitted to the dose-response data using nonlinear least squares or maximum likelihood estimation techniques. Alternative models include the logit model (Wilson and Worchester, 1943) and the probit model (Finney, 1971). Like the NOAEL approach, the BD approach is not generally used to achieve a direct estimate of risk for a specified dose; instead, uncertainty factors are still applied to this level to account for animal-to-human extrapolation and variation in sensitivity among humans.

Direct approaches are also used in the estimation of risks from exposure to environmental agents. Models such as the Weibull, logit and probit mentioned above have been used to extrapolate down to lower doses in an attempt to directly estimate the risk of toxicity from a given exposure. Mechanistically-based models have also been used in this context; most notably the multistage model of carcinogenesis (Armitage and Doll, 1954), a version of the multistage modelwhich provides an upper bound on risk through constraining the first-order term in the model known as the linearized multistage model (Crump 1981) and the two-stage model of carcinogenesis (due to several authors, see for example Moolgavkar and Venzon, 1979). As for the other methods described above, models describing the formation ofreactive intermediates can be used to replace dose in these models.

CONCLUSION

The validity and defensibility of regulatory exposure levels calculated during a risk assessment can be improved when based on a model of the mechanism of action of a chemical. PBPK models provide a framework for simulating the interaction and biochemistry of animal and chemical. In cases for which the putative agent is known or believed to be a reactive metabolite, PBPK models can simulate the production, distribution, and elimination of the metabolite in individual tissues and organs. Many physiological and biochemical parameters such as organ sizes and tissue perfusion rates are common to many PBPK models. Modeling of the mechanisms of action for a chemical are important to successful risk assessment. Incorporation of saturable processes, cofactor depletion, hormonal effects, endocrine pathways, and aging effects will increase the biochemical realism of a PBPK model.

A validated PBPK model provides a way of establishing the correlation between the toxic/carcinogenic endpoint and exposure to administered chemical or a metabolite. Human risk can then be estimated using this correlation.

REFERENCES

Andersen, M. E., Clewell III, H. J., Gargas, M. L., Smith, F. A., and Reitz, R. H. (1987). Physiologically-based pharmacokinetics and the risk assessment process for methylene chloride. *Toxicology and Applied Pharmacology* , 87, 185-205.

Buckpitt, A. R. and Warren, D. L. (1983). Evidence for hepatic formation, export and covalent binding of reactive naphthalene metabolites in extraheaptic tissues in vivo. *Journal of Pharmacology and Experimental Therapeutics* , 225(1), 8-16.

Cho, M. et al. (1994). Covalent interactions of reactive naphthalene metabolites with proteins. *The Journal of Pharmacology and Experimental Therapeutics*, 269(2), 881-889.

Corley, R. A., Mendrala, A. L., Smith, F. A., Staats, D. A., Gargas, M. L., Conolly, R. B., Andersen, M. E., and Reitz, R. H. (1990). Development of a physiologically based pharmacokinetic model for chloroform. *Toxicology and Applied Pharmacology* , 103, 512-527.

Crump, K. S. (1981). An improved procedure for low-dose carcinogenic risk assessment from animal data. *Journal of Environmental Pathology and Toxicology*, 52, 675-684.

Crump, K. S. (1984). A new method for determining allowable daily intakes. *Fundamental and Applied Toxicology* , 4, 854-871.

Finney, D. J. (1971). *Probit Analysis* . Cambridge University Press, London, 3rd edition.

Hattis, D. (1990). Pharmacokinetic Principles for dose rate extrapolation of carcinogenic risk from genetically active agents. *Risk Analysis*, 10, 303-316.

Hattis, D., and Burmaster, D. E. (1994). Assessment of variability and uncertainty distributions for practical risk analyses. *Risk Analysis*, 14(5), 713-730.

Kanekal, S. et al. (1991). Metabolism and cytotoxicity of naphthalene oxide in the isolated perfused mouse lung. *Journal of Pharamcology and Experimental Therapeutics* , 256(1), 391-401.

Kohn, M. C. and Portier, C. J. (1994). A model of effects of TCDD on expression of rat liver proteins. *Progress in Clinical and Biological Research* , 387, P211-P222.

Lucier, G. W., Portier, C. J., and Gallo, M. A. (1993). Receptor mechanisms and dose-response models for the effects of dioxins. *Environmental Health Perspectives,* 101(1), 36-44.

Moolgavkar, S. and Venzon, D. (1979). Two-event models for carcinogenesis: Incidence curves for childhood and adult tumors. *Mathematical Biosciences*, 47, 55-77.

National Research Council, (1994). Science and judgement in risk assessment.

National Academy Press, 2101 Constitution Ave., N.W. Washington, DC 20418

Pease, W., Vendenberg, J., and Hooper, K. (1991). Comparing alternative approaches to establishing regulatory levels for reproductive toxicants: DBCP as a case study. *Environmental Health Perspectives* , 91, 141-155.

Portier, C. J. and Kaplan, N. L. (1989). Variability of safe dose estimates when using complicated models of the carcinogenic process. Fundamental and Applied Toxicology, 13, 533-544.

Van Ryzin, J. (1980). Quantitative risk assessment. *Journal of Occupational Medicine* , 22(5), 321-326.

Wilson, E. B. and Worchester, J. (1943). The determination of LD 50 and its sampling error in bioassay. *Proceedings of the National Academy of Sciences of the United States of America* , 29, 78-85.

METABOLISM OF AFLATOXIN B_1 BY HUMAN HEPATOCYTES IN PRIMARY CULTURE

S. Langouet,[1] B. Coles,[2] F. Morel,[1] K. Maheo,[1] B. Ketterer,[2] and A. Guillouzo[1]

[1] INSERM U49 Hopital Pontchaillou
35033 RENNES Cedex France
[2] Department of Biochemistry and Molecular Biology
UCL
London W1P 6DB, United Kingdom

INTRODUCTION

Aflatoxin B_1 (AFB_1) is a potent liver carcinogen in the rat (Garner *et al* 1972) and, with hepatitis B infections, a co-carcinogen in human hepatocellular carcinoma (Qian *et al* 1994). It is metabolized by cytochrome P450 isoenzymes to give a number of products including AFM_1, AFP_1 and AFQ_1 and the *exo*- and *endo*-8,9 AFB_1 oxides (AFBOs), of which *exo*-AFBO is the ultimate carcinogenic metabolite (Busby & Wogan 1984; Raney *et al* 1992a).

The present study has investigated the effects of glutathione S-transferases (GSTs) inducers on AFB_1 metabolism in primary human hepatocyte cultures, an *in vitro* model known to retain phase I and phase II enzymes of xenobiotic metabolism and their responsiveness to inducers for at least one week (Guillouzo *et al* 1995; Morel *et al* 1990).

The AFBOs are substrates for GSTs (Coles *et al* 1985; Raney *et al* 1992b) and it has been proposed that induction of GSTs may be an important mechanism of protection against the neoplastic effects of AFB_1. Indeed administration of the inducers, ethoxyquin (Kensler *et al* 1986) and oltipraz (Kensler *et al* 1987), has been shown to prevent the development of hepatocellular carcinoma in the rat (*for review see* Hayes *et al* 1991). The activity of GSTs towards AFBO in man is low relative to the rat (Raney *et al* 1992b). Furthermore the most effective human isoenzyme is GSTM1 (*for nomenclature see* Mannervik *et al* 1992), which exhibits genetic polymorphism and is present in only about 50% of the population (Board 1981; Ketterer & Christodoulides 1994). It is important to know to what extent inducers of GSTs might be effective in reducing the risk of AFB_1-associated hepatocellular carcinoma in man and the importance for this of the GSTM1 polymorphism.

MATERIALS AND METHODS

Cell Isolation and Culture

Human hepatic tissue biopsies were obtained from patients undergoing liver resection for primary and secondary hepatomas, in agreement with the National Ethics Committee. Hepatocytes were isolated by a two-step collagenase perfusion procedure (Guguen-Guillouzo *et al* 1982).

After 36-48 h of culture, in a serum-free medium supplemented with dexamethasone, hepatocytes were exposed to 3-methylcholanthrene (MC) and the dithiolethione, oltipraz (OPZ), at final concentrations of 5 and 50 μM respectively either for 24 h or, with medium renewal, for 48 h.

AFB_1 Metabolite Analysis

Metabolites were analysed by C-18 reverse phase h.p.l.c. eluting with a linear gradient of 10% to 75% methanol : water, the eluant being 10mM with respect to ammonium acetate, pH 6.5 throughout. The two isomers *endo-* and *exo*-AFBO glutathione conjugates (AFBSG) were separated essentially as described by Raney *et al* (1992a). The eluate was monitored by fluorescence, enabling the detection of AFB_1, AFM_1, AFP_1 and *endo-* and *exo*-AFBSG.

GST Subunit Analysis

Supernatants from cell homogenates were passed through a glutathione agarose affinity column to obtain a GST fraction, the subunit content of which was analysed by quantitative reverse phase hplc according to Kispert *et al* 1989.

RESULTS AND DISCUSSION

AFB_1 Metabolism by Untreated Hepatocytes

Hepatocytes derived from a GSTM1 positive (HL-1) and a GSTM1 negative (HL-2) liver were put into primary culture and exposed to a medium made 5 μM with respect to AFB_1 for 8 h. AFB_1 metabolism was extensive 77% and 66% being metabolized by the GSTM1 positive and GSTM1 negative cells, respectively (Table I). The hplc system used, detected three major metabolites in the GSTM1 positive cells, namely AFM_1 and the GSH conjugates of *endo-* and *exo*-AFBO; these two isomers were in similar proportion in the untreated hepatocytes. AFP_1 was a minor component and AFQ_1 may have been present, but was not detected by the system used. Only one major metabolite was observed in the GSTM1 negative cells, namely AFM_1.

Effect of MC and OPZ on AFB_1 Metabolism

Pre-treatment of GSTM1 positive cells with MC increased overall metabolism to 97% with two fold increases in both AFM_1 and AFBSG, the latter being exclusively *endo*-AFBSG. In contrast, OPZ decreased metabolism, producing negligible amounts of AFM_1 and low levels of AFBSG, the latter being exclusively *exo*-AFBSG (Table I).

In GSTM1 negative cells the effects of both inducers on AFM_1 were similar to those observed in the GSTM1 positive cells but AFBSG levels were negligible, suggesting that in

Table 1. Metabolism of AFB_1 by cultured human hepatocytes both untreated and after exposure to MC or OPZ (each value is a mean of duplicate determinations [maximum variation ± 0.03])

Samples	AFB₁ and metabolite (μM)				Metabolism (%)
	AFB_1	AFBSG	AFM_1	AFP_1	
HL-1 (GST M1 +)					
control	1.17	0.43	0.36	<0.01	76.6
MC-treated	0.16	1.02	0.74	<0.01	96.8
OPZ-treated	1.91	0.07	<0.01	<0.01	61.8
HL-2 (GST M1 -)					
control	1.7	<0.01	0.036	nd	65.90
MC-treated	1.45	<0.01	0.049	nd	70.98
OPZ-treated	2.49	<0.01	<0.01	nd	50.18

GSTM1 null individuals, detoxication of AFBO by GSH conjugation may not be effective (Table I).

Effect of MC and OPZ on GST Subunit Levels

MC did not induce GST protein, in agreement with a previous study, where small effects on GST mRNA were observed (Morel *et al* 1993). OPZ, on the other hand, increased all GSTs. In the GSTM1 positive cells GSTs A1, A2 and M1 were increased 4.6-, 2.2- and 2.0-fold, respectively, while in the GSTM1 negative cells GSTs A1 and A2 were increased 2.3- and 1.7-fold, respectively (Table II).

Despite its ability to induce GST subunits, OPZ caused a marked reduction of AFBSG levels. This, together with a similar marked decrease in production of AFM_1, suggests that OPZ also acts as an inhibitor of cytochrome P450. Studies are being carried out in order to determine if CYP1A2 and/or CYP3A4, the major enzymes of oxidation of AFB_1 in humans, are indeed affected (Guengerich *et al* 1995). The apparent selective effect of inducers on the production of AFBO isomers is of particular interest.

On the basis of these results OPZ might be expected to be a very effective chemo-protective agent for AFB_1-associated human hepatocellular carcinoma.

Table 2. Analysis of MC and OPZ induction of GST M1, A1 and A2 by reverse phase h.p.l.c. Each value was expressed as μg GST per mg total cytosolic protein

Samples	GST subunits			
	GST M1a	GST M1b	GST A1	GST A2
HL–1(GST M1+)				
control	0.38	—	1.8	0.52
MC	0.4	—	1.6	0.5
OPZ	0.8	—	3.9	2.4
HL–2 (GST M1 –)				
control	—	—	2.75	0.95
MC	—	—	3.8	0.74
OPZ	—	—	6.24	1.68

ACKNOWLEDGMENTS

This work was supported by the Association pour la Recherche sur le Cancer (ARC) and a grant from the Cancer Research Campaign. S. Langouët was a recipient of a fellowship from the Ministère de la Recherche et de l'Espace (MRE) and was granted a short term visiting fellowship from the European Science Foundation (ESF) to carry out part of the work with the London group.

REFERENCES

Board P.G. Biochemical genetics of glutathione S-transferase in man. *Am. Hum. Gen.* 33, 36-43, 1981.

Busby W.F.Jr & Wogan G.N. Aflatoxin. In *Chemical Carcinogenesis*, Vol 2, Second Edition, ed. Searle C.E., ACS Monographs, Washington DC, pp 945-1093, 1984.

Garner R C., Miller E.C., and Miller J.A. Liver microsomal metabolism of aflatoxin B_1 to a reactive derivative toxic to Salmonella typhimurium TA 1530 *Cancer Res.* 32 2058-2062, 1972.

Coles B., Meyer D.J., Ketterer B., Stanton C.A & Garner R.C., Studies in the detoxication of microsomally-activated aflatoxin B_1 *Carcinogenesis*, 6, 693-697, 1985.

Guengerich F.P., Ueng Y.-F., Kim B.-R., Langouët S., Coles B., Iyer R., Thier R., Harris T.M., Shimada T., Yamazaki H., Ketterer B. and Guillouzo A. Activation of toxic chemicals by cytochrome P450 enzymes. Regio- and stereoselective oxidation of aflatoxin B_1. 1995 *see this volume.*

Guguen-Guillouzo C., Campion J.P., Brissot P., Glaise D., Bourel M. and Guillouzo A. High yield preparation of isolated human adult hepatocytes by enzymatic perfusion of the liver. *Cell Biol. Int. Rep.* 6, 625-628, 1982.

Guillouzo A., Langouët S., Morel F., Fardel O., Abdel-Razzak Z. and Corcos L. The isolated human cell as a tool to predict *in vivo* metabolism of drugs. In *Advances in Drug Metabolism in Man* ed. Pacifici G.M. 1995 *in press.*

Hayes J.D., Judah D.J., McLellan L.I. and Neal G.E. Contribution of the glutathione S-transferases to the mechanisms of resistance to aflatoxin B_1. *Pharmac. Ther.* 50, 443-472, 1991.

Kensler T.W., Egner P.A., Davidson N.E., Roebuck B.D., Pikul A. and Groopman J.D. Modulation of aflatoxin metabolism, aflatoxin-N_7-guanine formation and hepatic tumorigenesis in rats fed ethoxyquin: Role of induction of glutathione S-transferases. *Cancer Res.* 3924-3931, 1986.

Kensler T.W., Egnert P.A., Dolan P.M., Groopman J.D. and Roebuck B.D. Mechanism of protection against aflatoxin tumorigenicity in rats fed 5-(2-pyrazinyl)-4-methyl-1, 2-dithiol-3-thione (oltipraz) and related 1,2-dithiol-3-thiones and 1,2-dithiol-3-ones. *Cancer Res.*, 47, 4271-4277, 1987.

Ketterer B. and Christodoulides L. Enzymology of cytosolic glutathione S-transferases. *Advs. Pharmacol.* 27, 37-69, 1994.

Kispert A., Meyer D.J., Lalor E., Coles B., Ketterer B. Purification and characterization of a labile rat glutathione transferase of the mu class. *Biochem. J.* 260, 789-793, 1989.

Mannervik B., Awasthi Y.C., Board P.G., Hayes J.D., Di Ilio C., Ketterer B., Listowsky I., Morgenstern R., Muramatsu, M Pearson W.R., Pickett C.B., Sato K., Widersten M. and Wolf C.R. Nomenclature for human glutathione transferases. *Biochem. J.* 282, 305-308, 1992.

Morel F., Beaune P., Ratanasavanh D., Flinois J.P., Yang C.S., Guengerich F P. and Guillouzo A. Expression of cytochrome P-450 enzymes in cultured human hepatocytes *Eur. J. Biochem* 191, 437-444, 1990.

Morel F., Fardel O., Meyer D.J., Langouët S., Gilmore K S., Tu C.P.D., Kensler T.W., Ketterer B. and Guillouzo A. Preferential increase of glutathione S-transferase class α transcripts in cultured human hepatocytes by phenobarbital, 3-methylcholanthrene and dithiolethiones. *Cancer Res.* 53, 231-234, 1993.

Qian G.-S., Ross R.K., Yu M.C., Yaun J.-M., Gao Y.-T., Henderson B.E. Wogan G.N. and Groopman, J.D. A follow-up study of urinary markers of aflatoxin exposure and liver cancer risk in Shanghai, People's Republic of China. *Cancer Epidemol. Biomarkers* and Prevention 3, 3-10, 1994.

Raney K.D., Coles B., Guengerich F.P. and Harris T.H. The *endo*-8,9-epoxide of aflatoxin B_1 *Chem. Res. Toxicol.* 5, 333-335, 1992a.

Raney K.D., Meyer D.J., Ketterer B., Harris T.M. and Guengerich F.P. Glutathione conjugation of aflatoxin B_1 *exo-* and *endo-* epoxide by rat and human glutathione S-transferases *Chem. Res. Toxicol.* 5, 470-478, 1992b.

INDUCTION OF CYTOCHROMES P450 BY DIOXINS IN LIVER AND LUNG OF MARMOSET MONKEYS (*Callithrix jacchus*)

Thomas G. Schulz,*[1] Diether Neubert,[2] Donald S. Davies,[1] and Robert J. Edwards[1]

[1] Department of Clinical Pharmacology
Royal Postgraduate Medical School
Du Cane Road, London W12 0NN, United Kingdom
[2] Institute of Toxicology and Embryopharmacology
Free University Berlin
Garystrasse 5, 14195 Berlin, Germany

The induction of specific P450 enzymes is difficult to study in man but non-human primates may provide a useful model for this purpose. At present little is known about the inducibility of P450 in the liver or extrahepatic tissues of primates. Previous investigations have shown that several P450s are inducible in the marmoset monkey (*Callithrix jacchus*) by compounds such as phenobarbital or 2,3,7,8-tetrachlorodibenzo-*p*-dioxin (**TCDD**) (Schulz-Schalge & Webb, 1989; Schulz-Schalge et al., 1991a).

Here, we report on the results of studies in which the effects of TCDD and 2,3,7,8-tetrabromodibenzo-*p*-dioxin (**TBDD**) on the levels and activities of CYP1A1 and CYP1A2 were determined in the liver and lungs of marmoset monkeys.

MATERIALS AND METHODS

Adult marmosets weighing 300 to 400 g were treated with a single subcutaneous injection of 1.6 nmol TCDD or TBDD/kg body wt dissolved in DMSO+toluene, 2+1. Control animals received vehicle only. Groups of 2 animals per substance were killed either one or four weeks after treatment.

Microsomal fractions were prepared from liver and lung tissue and ethoxyresorufin O-deethylase (**EROD**), methoxyresorufin O-demethylase (**MROD**), pentoxyresorufin O-dealkylase (**PROD**), benzoyloxyresorufin O-dealkylase (**BROD**), ethoxycoumarin O-deethylase (**ECOD**) and high affinity phenacetin O-deethylase (**POD**) activities were

*Thomas Schulz is a European Science Foundation Research Fellow.

Table 1. Enzyme activities in hepatic microsomes

Activity	Control	Dioxin
EROD	110 ± 59	1300 ± 390
MROD	130 ± 67	1200 ± 380
ECOD	1400 ± 290	1200 ± 200
POD	93 ± 27	300 ± 88

All values are expressed in pmole product / (mg protein x min). The results are shown as mean and SD. The control group consists of 6 animals. The dioxin group consists of 4 TCDD- and 4 TBDD-treated animals.

measured as described previously (Schulz-Schalge et al., 1991b; 1991c; Murray & Boobis, 1986).

Immunoblotting was performed using an anti-peptide antibody with proven specificity for human CYP1A2 (Edwards et al., 1993; 1994) and an antibody directed towards the C-terminus of human CYP1A1.

RESULTS

Liver

Induction of EROD activity in marmoset liver was very similar one and four weeks after treatment. Furthermore, the inductive potency of TCDD and TBDD was comparable for all enzyme activities measured. Therefore, for the comparison of other enzyme activities the results from samples taken at one and four weeks were combined and both treatment groups were also combined (Table 1).

Treatment with dioxins (TCDD or TBDD) caused a 12-fold increase in EROD activity, a 9-fold increase in MROD activity, and a 3-fold increase in high affinity POD activity, but no difference in ECOD activity compared with control animals (Table 1).

There was an excellent correlation between EROD and MROD activities (r=0.99), when values from marmosets investigated at 1, 4, 12, 20 or 28 weeks after treatment were compared. There was also a good correlation between EROD and high affinity POD activity (r=0.82). Immunoblotting with an antibody which recognizes CYP1A2 in many species showed that CYP1A2 was present in control marmoset liver and that the level was increased following dioxin treatment (Figure 1a). Immunoblotting using an antibody targeted towards human CYP1A1 gave positive reactions with hepatic microsomal samples from dioxin-treated marmosets but not control marmosets (Figure 1b).

Lung

Dioxin treatment caused induction of EROD and MROD activities (15- and 23-fold, respectively), but ECOD and high affinity POD activities were undetectable in treated and untreated animals (Table 2). Interestingly, PROD and BROD were detectable in lung microsomal fractions of treated animals but not controls (Table 2). There was an excellent correlation (r=0.99) between EROD and MROD activities, but the ratio of the activities (EROD:MROD) was different from that observed in liver, i.e. 2 in lung compared with 1 in liver.

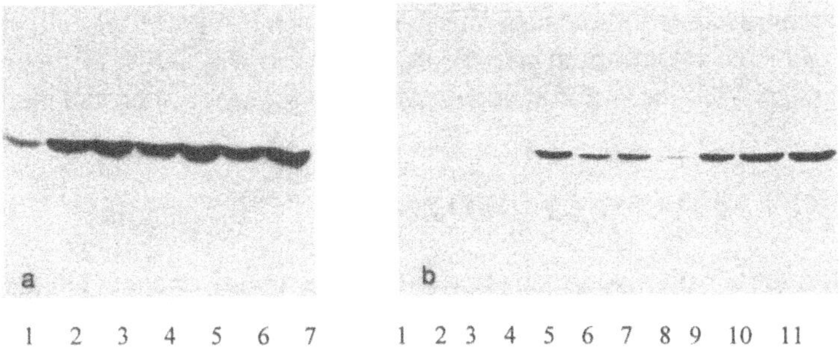

Figure 1. Immunoblots of hepatic microsomes. Microsomal fractions were resolved by SDS-PAGE (9% acrylamide), transfered to nitrocellulose filters and developed with antibodies against CYP1A2 (a). Lane 1: 25 μg liver microsomes from control marmosets; lane 2-7: 25 μg liver microsomes from TCDD-treated marmosets and CYP1A1 (b). Lane 1-4: 25 μg liver microsomes from control marmosets; lane 5-11: 25 μg liver microsomes from TCDD-treated marmosets.

Table 2. Enzyme activities in lung microsomes

Activity	Control	Dioxin
EROD	4.0 ± 1.0	74 ± 22
MROD	1.3 ± 1.0	38 ± 12
BROD	n.d.	1.0 ± 0.1
PROD	n.d.	0.95 ± 0.10
ECOD	n.d.	n.d.
POD	n.d.	n.d.

All values are expressed as pmole product / (mg protein x min). Results are shown as mean and SD.
The control group consists of 4 animals. The dioxin group consists of 8 animals (TCDD and TBDD-treated animals). n.d. = not detectable.

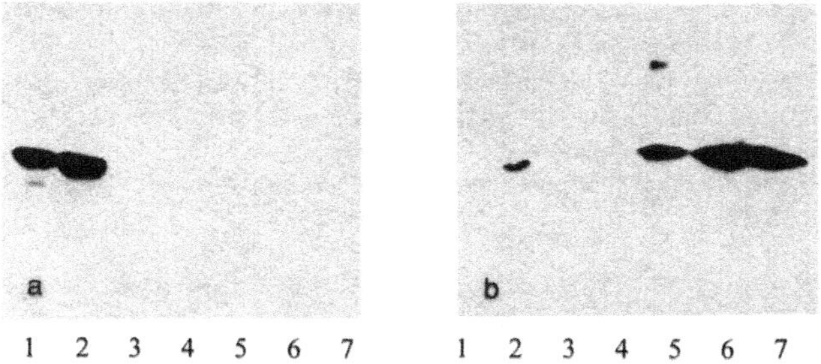

Figure 2. Immunoblots of lung microsomal fractions. Microsomal fractions were resolved by SDS-PAGE (9% acrylamide), transfered to nitrocellulose filters and developed with antibodies against CYP1A2 (a) and CYP1A1 (b). Lane 1: 25 μg liver microsomes from TCDD-treated rats; lane 2: 25 μg liver microsomes from TCDD-treated marmosets; lane 3 and 4: 200 μg lung microsomes from control marmosets; lane 5-7: 200 μg lung microsomes from TCDD-treated marmosets.

Immunoblotting experiments showed that CYP1A2 was absent from both the lung of untreated and treated marmosets (Figure 2a). However, CYP1A1 was clearly detected in the lung of treated but not untreated marmosets (Figure 2b). The anti-CYP1A1 antibody does not bind to rat CYP1A1 due to differences between the sequence of rat and human CYP1A1.

DISCUSSION AND CONCLUSIONS

CYP1A2 is constitutively expressed in the marmoset liver but not in the lung. TCDD and TBDD induce both CYP1A1 and CYP1A2 in the liver of marmoset monkeys. However, the level of induction is much lower than in the rat (Schulz-Schalge et al., 1991a). In the marmoset lung both dioxins induce CYP1A1 apoprotein and this is accompanied by an increase in EROD and MROD activities, but neither high affinity POD activity nor ECOD activity could be detected. Therefore, CYP1A1 in the lung is capable of catalysing EROD as well as MROD and, to a minor extent, BROD and PROD activities.

The results also suggest that in the marmoset liver CYP1A1 and CYP1A2 catalyse both EROD and MROD activities, but neither P450 catalyses ECOD activity, unlike in other species. MROD activity in the marmoset does not specifically reflect CYP1A2 activity as it does in rat liver microsomes. However, marmoset CYP1A2 catalyses high affinity POD activity as does human CYP1A2. Thus the marmoset monkey seems to be a suitable model for investigations of the metabolism of TCDD and possibly other xenobiotics.

REFERENCES

Edwards RJ, Murray BP, Murray S, Schulz T, Neubert D, Gant TW, Thorgeirsson SS, Boobis AR, Davies DS (1994) Contribution of CYP1A1 and CYP1A2 to the activation of heterocyclic amines in monkeys and human. *Carcinogenesis* **15**: 829-836

Edwards RJ, Murray BP, Singleton AM, Murray S, Davies DS, Boobis AR (1993) Identification of the epitope of an anti-peptide antibody which binds to CYP1A2 in many species including man. *Biochem. Pharmacol.* **46**: 213-220

Murray S, Boobis AR (1986) An assay for paracetamol, produced by the O-deethylation of phenacetin in vitro, using gas chromatography/electron capture negative ion chemical ionization mass spectrometry. *Biomed. Environm. Mass Spectr.* **13**: 91-93

Schulz-Schalge T, Koch E, Golor G, Wiesmüller T, Hagenmaier HP, Neubert D (1991a) Comparison of the induction of cytochrome P450 and ethoxyresorufin O-deethylase by a single subcutaneous administration of TCDD in liver microsomes of marmoset monkeys (*Callithrix jacchus*) and rats. *Chemosphere* **23**: 1933-1939

Schulz-Schalge T, Koch E, Schwind KH, Hutzinger O, Neubert D (1991b) Inductive potency of TCDD, TBDD and three 2,3,7,8-mixed-halogenated dioxins in liver microsomes of male rats. Enzyme kinetic considerations. *Chemosphere* **23**: 1925-1931

Schulz-Schalge T, Heger W, Webb J, Kastner M, Neubert D (1991c) Ontogeny of some monooxygenase activities in the marmoset monkey (*Callithrix jacchus*). *J. Med. Primatol.* **20**: 325-333

Schulz-Schalge T, Webb J (1989) Metabolism of some phenoxazone ethers in liver microsomes of untreated and phenobarbital-treated marmoset monkeys (*Callithrix jacchus*). *Naunyn Schmiedeberg's Arch. Pharmacol.* **339**: R 9

ARYL HYDROCARBON RECEPTOR mRNA LEVELS IN DIFFERENT TISSUES OF 2,3,7,8-TETRACHLORODIBENZO-P-DIOXIN-RESPONSIVE AND NONRESPONSIVE MICE

Olaf Döhr, Wei Li, Susanne Donat, Christoph Vogel, and Josef Abel

Medical Institute of Environmental Hygiene at the Heinrich Heine University
Department of Toxikology, Auf'm Hennekamp 50
40225 Düsseldorf, Germany

SUMMARY

The AhR mRNA contents were measured in different tissues of 2,3,7,8-tetrachlorodibenzo-p-dioxin (TCDD)-responsive C57BL/6J and nonresponsive DBA/2J mice. Out of all examined tissues the highest AhR mRNA levels were found in lung, with concentrations of $31.7 \pm 11.0 \times 10^3$ molecules / 100 ng RNA and $20.3 \pm 8.9 \times 10^3$ molecules / 100 ng RNA in C57BL/6J and DBA/2J mice, respectively. The AhR mRNA contents were 5-10 fold lower in heart, liver, thymus, brain and placenta. Low levels were found in spleen, kidney and muscle. Since no significant differences in AhR mRNA expression between the two strains were observed, factors other than regulation of AhR gene expression seem to be responsible for the observed different susceptibility toward TCDD.

INTRODUCTION

2,3,7,8-Tetrachlorodibenzo-p-dioxin (TCDD) is the prototype for a class of polycyclic aryl hydrocarbons which in part elicit their biological responses, e.g., induction of xenobiotic-metabolizing enzymes like cytochrome P450 1A1, through a cellular protein, the aryl hydrocarbon receptor (AhR) (rev. in 1). In the murine system four alleles for the genetic locus of AhR have been identified, causing different susceptibility of strains toward TCDD (rev. in 2). Although cDNA sequence of responsive Ahr^{b-1} allele has successfully been determined (3) the molecular basis for these interstrain differences is under current investigation.

Reverse transcription-polymerase chain reaction (RT-PCR) is the method of choice to analyse the mRNA expression of low copy genes. For quantitation of mRNA levels the quantitative competitive RT-PCR (QC-PCR) has been introduced, in which a competitor fragment is coamplified in PCR reaction. The content of specific transcripts can be calculated for different probes from a calibration curve of a set of different concentrations of template with a constant amount of competitor (4). Here, we describe the use of QC-PCR to determine the AhR mRNA contents in several tissues of TCDD-responsive C57BL/6J (Ahr^{b-1} allele) and nonresponsive DBA/2J mice (Ahr^{d} allele) to investigate whether differences in AhR mRNA expression exist that possibly explain the different sensitivity of these mouse strains to TCDD.

METHODS

Animals

Female C57BL/6J and DBA/2J mice (6 - 8 weeks old) were maintained under standard conditions.

Oligonucleotides

AhR primers were selected from the reported cDNA sequence (3) and sequences of ß-actin primers were as published (5).

QC-PCR

An AhR competitor PCR-fragment of 536 bp length was constructed from the 710 bp PCR-fragment, and QC-PCR was performed as described (6). Briefly, total RNA was prepared, reverse transcribed and each cDNA was spiked with a constant concentration of competitor fragment and coamplified for 33 cycles. The contents of AhR mRNA in different samples were calculated from a calibration curve. Expression of the housekeeping gene ß-actin was analysed to control reverse transcription.

RESULTS AND DISCUSSION

In this study we used a QC-PCR to determine the AhR mRNA levels in several tissues of TCDD-responsive C57BL/6J and nonresponsive DBA/2J mice. A typical result obtained from PCR analysis is shown in figure 1. The AhR mRNA was detectable in all tissues investigated, and mean values of AhR mRNA levels are given in table 1. The results show that out of all examined tissues the highest AhR mRNA levels were found in lung, with

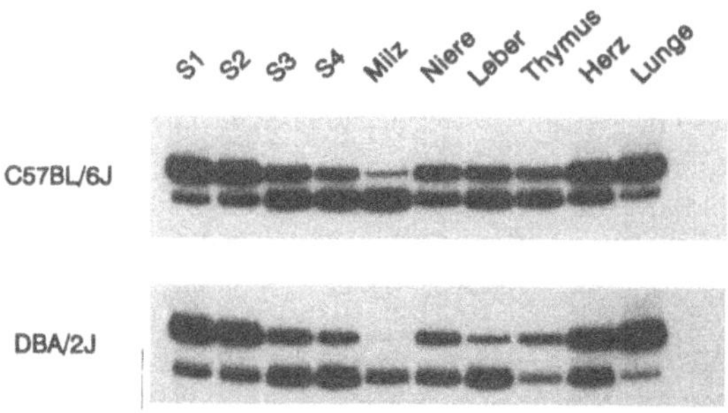

Figure 1. Expression of AhR mRNA in different tissues of C57BL/6J and DBA/2J mice. Each cDNA sample was spiked with 44,000 molecules of competitor fragment and coamplified. To obtain a calibration curve a set of four different standards (S1 - S4) was amplified simultaneously. The standards contained 45,000 (S1), 22,500 (S2), 11,200 (S3), and 5600 (S4) molecules AhR template and 45,000 molecules competitor fragment (upper band: 710 bp AhR PCR-fragment; lower band: 536 bp competitor fragment).

Table 1. AhR mRNA content in different tissues of C57BL/6J and DBA/2J mice

Tissue	C57BL/6J (n = 3)	DBA/2J (n = 5)	P[a]
Heart	1.6 ± 0.2[b]	6.2 ± 4.8	0.50
Liver	3.9 ± 2.8	1.4 ± 0.5	0.47
Spleen	0.6 ± 0.2	1.5 ± 0.6	0.29
Lung	31.7 ± 11.0	20.3 ± 8.9	0.46
Kidney	0.5 ± 0.2	1.3 ± 0.4	0.26
Brain	1.7 ± 0.2	2.6 ± 0.7	0.36
Thymus	3.0 ± 0.6	2.4 ± 0.4	0.45
Muscle	0.4 ± 0.2	0.4 ± 0.2	0.97
Placenta[c]	2.1 ± 0.5	not determined	—

[a] Statistical evaluation using Student's t test.
[b] Values are expressed as molecules x 10^3/ 100 ng RNA (mean ± SD).
[c] On day 15 of gestation.

concentrations of 31.7 ± 11.0 x 10^3 molecules / 100 ng RNA and 20.3 ± 8.9 x 10^3 molecules / 100 ng RNA in C57BL/6J and DBA/2J mice, respectively. In heart, liver, thymus, brain and placenta the AhR mRNA contents were about 5 - 10 fold lower than in lung. Low levels were found in spleen, kidney and muscle. Similar distribution patterns were found by Northern blot analysis in rat and human tissues (3, 7) with high levels of AhR mRNAs in lungs of all three species, indicating that this organ may be a sensitive target for the toxic action of TCDD or other AhR agonists.

Statistical analysis of our data indicate no significant differences in AhR mRNA expression between the two mouse strains implicating that other factors may be responsible for different susceptibility to TCDD. Recently, cDNA sequencing of the four murine AhR alleles revealed that the lower ligand binding affinity of Ahr^d receptor is due to differences in the nucleotide sequence compared to the other alleles (8, 9). These mutations may lead to a lower affinity of Ahr^d receptor to TCDD.

Taken together, these and our results indicate that differences in AhR mRNA expression are not responsible for conspicuous different susceptibility of mouse strains toward AhR binding ligands. Nevertheless the present method will be a useful tool in studying changes of AhR mRNA expression influenced by several endogenous and xenobiotic compounds.

REFERENCES

1. Landers JP, Bunce NJ (1991). The Ah receptor and mechanism of dioxin toxicity. Biochem. J. 276:273-287
2. Swanson HI, Bradfield CA (1993). The Ah-receptor: genetics, structure and function. Pharmacogenetics 3:213-230
3. Ema M, Sogawa K, Watanabe N, Chujoh Y, Matsushita N, Gotho O, Funae Y, Fujii-Kuriyama Y (1992). cDNA cloning and structure of mouse putative Ah receptor mRNA. Biochem. Biophys. Res. Commun. 184:246-253.
4. Zachar V, Thomas RA, Goustin AS (1993). Absolute quantification of target DNA: A simple competitive PCR for efficient analysis of multiple samples. Nucleic Acids Res. 21:2017-2018
5. Vaughan TJ, Pascall RA, Brown KD (1992). Nucleotide sequence and tissue distribution of mouse transforming growth factor-Â. Biochim. Biophys. Acta 1132:322-324.
6. Li W, Donat S, Döhr O, Unfried K, Abel J (1994). Ah receptor in different tissues of C57BL/6J and DBA/2J mice: Use of competitive polymerase chain reaction to measure Ah-receptor mRNA expression. Arch. Biochem. Biophys. 315:279-284.

7. Dolwick KM, Swanson HI, Bradfield CA (1993). Cloning and expression of a human Ah receptor cDNA. Proc. Natl. Acad. Sci. USA 90:8566-8570.
8. Chang C, Smith DR, Prasrad VS, Sidman CL, Nebert DW, Puga A (1993). Ten nucleotide differences, five of which cause amino acid changes, are associated with the Ah receptor locus polymorphism of C57BL/6 and DBA/2 mice. Pharmacogenetics 3:312-321.
9. Poland A, Palen D, Glover ED (1994). Analysis of the four alleles of the murine aryl hydrocarbon receptor. Mol. Pharmacol. 46:915-921.

BIOMONITORING WORKERS EXPOSED TO ARYLAMINES

Application to Hazard Assessment

O. Sepai[1] and G. Sabbioni[2]

[1] Department of Environmental and Occupational Medicine
The Medical School, University of Newcastle upon Tyne
Newcastle NE2 4HH, United Kingdom
[2] Institut für Toxikologie und Pharmakologie
Universität Würzburg, Versbacher Str. 9
D-97078 Würzburg, Germany

Arylamines are important intermediates in the production of plastics and polyurethanes. Employees in many factories are subjected to low level chronic exposure to a wide variety, often a mixture, of arylamines. We have developed methods to measure hemoglobin (Hb) adducts of over 30 arylamines. Using these methods we can biomonitor exposure to low levels of these chemicals. Since, in most cases the air monitoring values were below the detection limit, we have used the protein adduct levels to estimate the daily-dose, which was then compared to the daily dose which yields tumors in 50 % of rodents (TD_{50}).

METHODOLOGY

Biological samples were collected from groups of workers. With each study we endeavored to collect control samples from clerical or medical staff. Blood samples were worked up using methods reported in Schütze et al., (1995), Sepai et al., (1995), and Sabbioni and Beyerbach, (1995). The Hb, plasma and urine samples were stored at -20 °C. Acid or base hydrolyzed samples were extracted at basic pH into organic solvents, derivatized with perfluorinated acid anhydride and analyzed by gas chromatography - mass spectrometry, with negative chemical ionization. For each compound a calibration line was established at five concentration levels covering the expected levels of adducts or metabolites in the samples.

RESULTS AND DISCUSSION

A high incidence of bladder cancer was reported in employees from a factory producing plastic ware. Blood samples were available from 73 workers of this factory. The

Biological Reactive Intermediates V
Edited by R. Snyder et al., Plenum Press, New York, 1996

Hb-adducts of 2-methylaniline (2MA), aniline (A) and 4-aminobiphenyl (4ABP) were investigated. Extremely high amounts (0.1 - 200 ng/g Hb) of 2MA and (0.1-35 ng/g Hb) of A were found (unpublished results). The levels of 4ABP were equivalent to the controls.

The use of environmental air monitoring as a means of dose estimation has a number of drawbacks: (I) it is often not sensitive enough, (II) does not give an indication of the effective dose but only the total dose inhaled, and (III) gives no indication of the metabolism of the compound in question. Hb adduct levels are dosimeters for the bioavailability of reactive xenobiotics and possibly also a dosimeter for the DNA-adducts at the site of tumor formation. Furthermore, with the knowledge from animal data it is possible to estimate the daily-dose from the measured Hb-adduct levels. However, it is necessary to make the following assumptions:

1. The adduct level result from steady-state exposures (A_c),

$$A_c = 0.5 \text{ x } A \text{ x } T_{er}$$

(Where A is the average daily increment per total Hb, and T_{er} is the lifetime of an erythrocyte.) Thus, to calculate the single dose the adduct level has to be divided by 60 (Tannenbaum et al., 1986).

2. Modified Hb has the same life-span as unmodified Hb and the adducts are stable to repair mechanisms.
3. The pharmacokinetics of 2MA are comparable in rats and humans, so that the dose associated with the Hb can be divided by 0.00059 which is the proportion of an administered dose found associated with the Hb in rats (Sabbioni, 1992).

Thus in a 70 kg individual the human exposure dose is 68 μg 2MA/kg/day.

In the absence of epidemiological data and as it is not possible to produce dose-response relationships in human carcinogenesis models it is necessary to characterize human health risk from animal data. The daily dose given to rodents that causes a fifty percent greater likelihood of the development of a tumor is termed the TD_{50}. The human exposure dose expressed as a percentage of the rodent potency dose (TD_{50}) has been termed the HERP index by Ames et al., (1987). The value of this type of comparison has been discussed extensively by many workers (Ames et. al., 1987, Goodman and Wilson, 1991, Crabtree et al., 1991, Gold et al., 1992 Talaska et al., 1994). The possibility of such comparisons is very useful, as long as the many assumptions that are required are not ignored. Using the HERP index it was possible to rank the risk or hazard potential for the workers exposed to the levels of amines for which we have established an internal-exposure dose from the adduct levels.

The daily dose of 2MA for the workers with the highest adduct levels in this study was 1/300 of the TD_{50} value determined for rats (Gold, et al., 1993), that is, a HERP of 0.3%.

Other groups of workers investigated were exposed to 4,4'-methylenedianiline (MDA) or 4,4'-methylenediphenyl diisocyanate (MDI) (Schütze et al., 1995). The air levels of MDA and MDI were below detection limits; however, adduct and metabolite levels were detected in a high percentage of the samples. Hb-adducts of MDA were found in 97% and N-acetyl-MDA (AcMDA) in 65% of the MDA-workers (Fig. 1). Hb-adducts of MDA were found in 38% of the MDI-workers (Fig. 1).

The presence of the acetylated compound was below detection limits in the Hb of the MDI exposed workers. Only the adducts of the primary amine from the hydrolysis of MDI were analyzed in this study. The presence of MDI adducts or other isocyanate metabolites were not investigated. Urine, collected at the same time as the blood samples from these MDA and MDI workers was extracted at alkaline pH with and without preceding acid treatment. MDA and AcMDA were found in the urine of 84% of the MDA-workers and 78% of the MDI-workers. In order to release MDA and AcMDA from possible conjugates

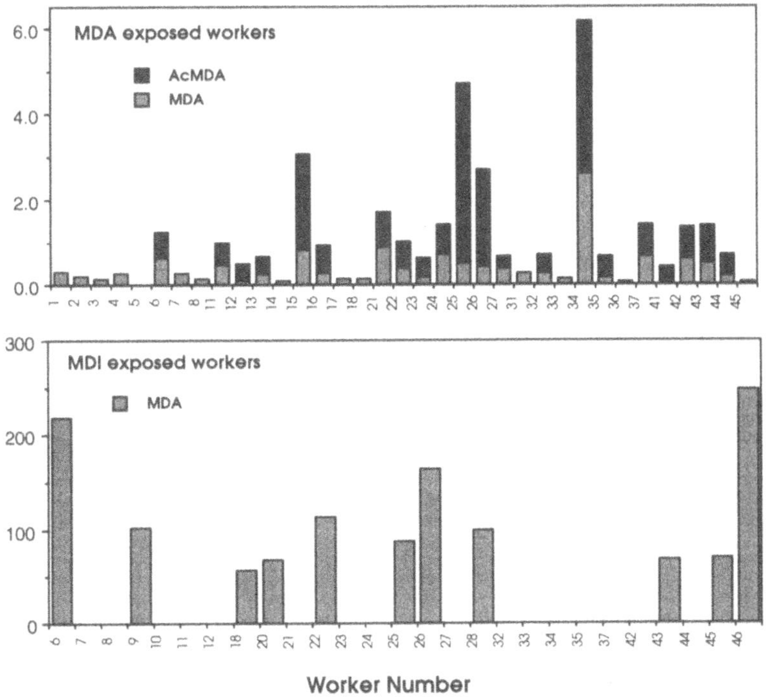

Figure 1. MDA and AcMDA analysis from Hb of workers exposed to MDA or MDI. Blood was collected from a group of 33 workers exposed to MDA plus two members of medical stuff (number 8 and 18) as factory controls. A second group of 27 workers were exposed to MDI: from this factory no controls were available.

urine was treated under strong acidic conditions. Following this procedure higher levels of MDA were found than the sum of MDA and AcMDA after base extraction alone. Urinary metabolites are indicators of recent exposure: it is not advisable to estimate an average daily dose from these values as there is a likelihood of over or under estimation (Cocker et al., 1994). In rats 0.044% of MDA-dose was found as Hb adducts (Schütze, 1993). Assuming that the pharmacokinetics of MDA are similar in rats and humans, the daily MDA-dose of these workers is at least 10 000 times below the TD_{50} (Gold et al., 1993) found for rats (i.e., a HERP of 0.01%).

Another topic of concern is the non-occupational exposure to aromatic amines released from medical devices made of polyurethane (PU). We demonstrated the presence of degradation products of PU —namely monomeric toluendiamines: 2,4-toluenediamine (24TDA), a suspected human carcinogen and 2,6-toluenediamine (26TDA)— in the blood and urine up to two years post operation from patients with PU-covered breast implants (Sepai et al., 1995). From our results we can estimate a potential risk from these implants. Following a lag period of approximately 20 days, where no TDAs above background levels were detected in the plasma, the levels of both 24TDA and 26TDA rose, reaching a maximum of 4.4 (2.1) ng/ml plasma for 24TDA (26TDA) (Fig. 2).

A period of 82 days after one patient received two PU-coated prostheses the plasma contained 4.4 ng 24TDA/ml) (Fig. 2). This value corresponds to a daily dose of 70400 ng for a 60 kg person. This was calculated by comparison to the adduct level found associated with plasma in rats dosed with radioactive TDA (Grantham et al., 1978) and a steady state to single dose conversion factor for albumin adducts of ca. 29 (Sabbioni et al., 1987). This

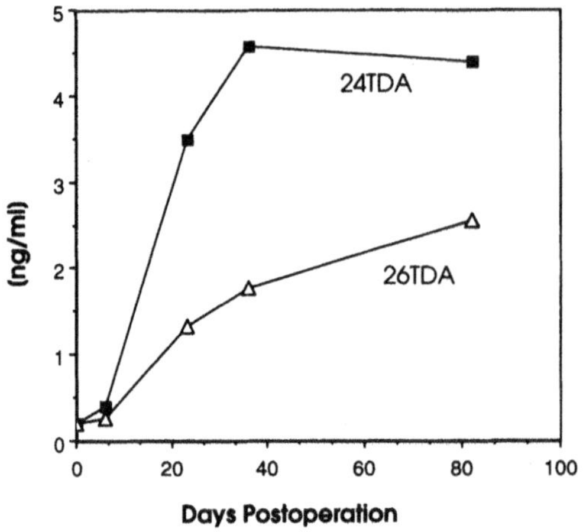

Figure 2. 24TDA and 26TDA analysis from the plasma of a patient with two PU-coated breast implants showing the initial lag period followed by an increase of the TDAs reaching a plateau after approximately 40 days.

value is ca. 300 times larger than the daily-dose estimation from two implants reported by an Expert Panel which reviewed data and risk assessments from the Food and Drug Administration (FDA) in the United States (Expert Panel of the Canadian Medical Association, 1991). The Expert Panel ascribes five additional breast cancers in 10 million patients with two implants. Our risk estimation is therefore, ca 1500 additional breast cancer cases in 10 million women. This additional risk is, of course, minuscule when 1 in 10 woman are likely to be inflicted with breast cancer in the normal western population. However, the risk of 24TDA was only related to breast cancer. We should keep in mind that the primary site of action of this suspected carcinogen is the liver or the kidney. This implies that the risk of liver cancer may be of more concern. Our calculated dose level is ca 1200 times lower than the TD_{50} in rats (Gold et al., 1993), that is a HERP of 0.08%.

CONCLUSIONS

In several cases ambient air monitoring is a poor exposure measure for humans. Determining the internal dose takes into account different modes of exposure. We found hemoglobin adducts in different groups of workers for which in most cases the air levels were below detection limits. From the measured Hb-adduct levels we estimated the daily dose which was then compared to rodent carcinogenic potency data. The HERP indexes for the arylamines we have studied were 0.3% for 2MA, 0.01% for MDA and 0.08% for 24TDA. The genotoxic risk resulting from these chemicals without taking into account synergistic effects might be comparable to the hazard from the average content of aflatoxin in a (U.S.) peanut butter sandwich (HERP index of 0.03%), from symphytine—a pyrrolizidine alkaloid—in a daily cup of comfrey tea (HERP index of 0.033%), and from the amount of formaldehyde in conventional home air (HERP index of 0.6%) (Ames et al., 1987).

ACKNOWLEDGMENTS

We are grateful to the Körber-Stiftung (Hamburg, Germany) for their financial support of these studies.

REFERENCES

Ames, B.N., Magaw, R. and Gold, L.S. (1987) Ranking possible carcinogenic hazards. Science, 236, 271-280.

Cocker, J., Nutley, B.P. and Wilson, H.K. (1994) A biological monitoring assessment of exposure to methylene dianiline in manufacturers and users, Occup. Environ. Med., 51, 519-522.

Crabtree, H.C., Hart, D., Thomas, M.C., Witham, B.H., McKensie, I.G. and Smith, C.P. (1991) Carcinogenic ranking of aromatic amines and nitro compounds, Mutat. Res., 264, 155-162.

Expert panel of the Canadian Medical Association. (1991) Safety of polyurethane-covered, breast implants. Can. Med. Assoc. J. 145, 1125-1128.

Gold, L.S., Slone, T.H., Stern, B. R., Manley, N.B. and Ames, B.N. (1992) Rodent carcinogens: setting priorities, Science, 258, 261-265.

Gold, L.S., Manley, N.B., Slone, T.H., Garfinkel, G.B., Rohrbach, L. and Ames, B.N. (1993) The fifth plot of the carcinogenic potency database: results of animal bioassays published in the general literature through 1988 and by the National Toxicology Program through 1989, Environ. Health Perspect., 100, 65-135.

Goodman, G. and Wilson, R. (1991) Quantitative prediction of human cancer risk from rodent carcinogenic potencies: a closer look at the epidemiological evidence for some chemicals not definitively carcinogenic in humans, Regul. Toxicol. Pharmacol., 14, 118-146.

Grantham, P.H., Mohan, L., Benjamin, T., Roller, P.P., Miller, J.R., and Weisburger E.K. (1978) Comparison of the metabolism of 2,4-toluenediamine in rats and mice. J. Environ. Pathol. Toxicol. 3, 149-166.

Sabbioni, G. (1992) Quantitative structure activity relationship of the N-oxidation of aromatic amines. Chem.-Biol. Interact., 81, 91-117.

Sabbioni, G. (1994) Hemoglobin binding of nitroarenes and quantitative structure-activity relationships. Chem. Res. Toxicol., 7, 267-274

Sabbioni, G. and Beyerbach, A. (1995) Determination of hemoglobin adducts of arylamines in humans. J. Chromatog. B. 667, 75-83.

Sabbioni, G., Skipper, P.L., Büchi, G., and Tannenbaum, S.T. (1987) Isolation and characterization of the major serum albumin adducts formed by aflatoxin B1 in vivo in rats. Carcinogenesis 8, 819-824.

Schützer, D. (1993) Expositions nachweis und Risikoermittlung bei bis(4-aminophenyl)methanen und davon abgeleiteten Industrie chemikalien, Ph.D. Thesis. Universität Würzburg, Germany.

Schütze, D., Sepai, O., Lewalter, J., Miksche, L., Henschler, D. and Sabbioni, G. (1995) Biomonitoring of workers exposed to 4,4'-methylenedianiline and 4,4'-methylenediphenyl diisocyanate, Carcinogenesis, 16, 573-582.

Sepai, O., Czech, S., Eckert, P., Henschler, D. and Sabbioni, G. (1995) Exposure to toluenediamins in patients with polyurethane covered breast implants, Toxicol. Lett., 77, 371-378.

Talaska, G., Schamer, M., Casetta, G., Tizzani, A. and Vineis, P. (1994) Carcinogen-DNA adducts in bladder biopsies and urothelial cells: a risk asssessment exercise, Cancer Lett., 84, 93-97.

Tannenbaum, S. R. Bryant, M.S., Skipper, P.L. and M. Maclure. (1986) Hemoglobin adducts of tobacco-related aromatic amines: application to molecular epidemiology, Banbury Rep., 26, 63-75.

AUTHOR INDEX

SUBJECT INDEX

GPSR Compliance

The European Union's (EU) General Product Safety Regulation (GPSR) is a set of rules that requires consumer products to be safe and our obligations to ensure this.

If you have any concerns about our products, you can contact us on ProductSafety@springernature.com

In case Publisher is established outside the EU, the EU authorized representative is:

Springer Nature Customer Service Center GmbH
Europaplatz 3
69115 Heidelberg, Germany

Batch number: 10761599

Printed by Printforce, the Netherlands